Springer Series in Statistics

Springer Series in Statistics

Anderson: Continuous-Time Markov Chains: An Applications-Oriented Approach.
Andrews/Herzberg: Data: A Collection of Problems from Many Fields for the Student and Research Worker.
Anscombe: Computing in Statistical Science through APL.
Berger: Statistical Decision Theory and Bayesian Analysis, 2nd edition.
Brémaud: Point Processes and Queues: Martingale Dynamics.
Brockwell/Davis: Time Series: Theory and Methods, 2nd edition.
Daley/Vere-Jones: An Introduction to the Theory of Point Processes.
Dzhaparidze: Parameter Estimation and Hypothesis Testing in Spectral Analysis of Stationary Time Series.
Farrell: Multivariate Calculation.
Fienberg/Hoaglin/Kruskal/Tanur (Eds.): A Statistical Model: Frederick Mosteller's Contributions to Statistics, Science, and Public Policy.
Goodman/Kruskal: Measures of Association for Cross Classifications.
Grandell: Aspects of Risk Theory.
Hall: The Bootstrap and Edgeworth Expansion.
Härdle: Smoothing Techniques: With Implementation in S.
Hartigan: Bayes Theory.
Heyer: Theory of Statistical Experiments.
Jolliffe: Principal Component Analysis.
Kres: Statistical Tables for Multivariate Analysis.
Leadbetter/Lindgren/Rootzén: Extremes and Related Properties of Random Sequences and Processes.
Le Cam: Asymptotic Methods in Statistical Decision Theory.
Le Cam/Yang: Asymptotics in Statistics: Some Basic Concepts.
Manoukian: Modern Concepts and Theorems of Mathematical Statistics.
Miller, Jr.: Simultaneous Statistical Inference, 2nd edition.
Mosteller/Wallace: Applied Bayesian and Classical Inference: The Case of *The Federalist Papers*.
Pollard: Convergence of Stochastic Processes.
Pratt/Gibbons: Concepts of Nonparametric Theory.
Read/Cressie: Goodness-of-Fit Statistics for Discrete Multivariate Data.
Reiss: Approximate Distributions of Order Statistics: With Applications to Nonparametric Statistics.
Ross: Nonlinear Estimation.
Sachs: Applied Statistics: A Handbook of Techniques, 2nd edition.
Särndal/Swensson/Wretman: Model Assisted Survey Sampling.
Seneta: Non-Negative Matrices and Markov Chains.
Siegmund: Sequential Analysis: Tests and Confidence Intervals.
Tong: The Multivariate Normal Distribution.
Vapnik: Estimation of Dependences Based on Empirical Data.
West/Harrison: Bayesian Forecasting and Dynamic Models.
Wolter: Introduction to Variance Estimation.

(continued after index)

Carl-Erik Särndal Bengt Swensson
Jan Wretman

Model Assisted Survey Sampling

Springer-Verlag
New York Berlin Heidelberg London Paris
Tokyo Hong Kong Barcelona Budapest

Carl-Erik Särndal
Département de mathématiques et de statistique
Université de Montréal
Montréal, Québec H3C 3J7
Canada

Bengt Swensson
Department of Data Analysis
University of Örebro
701 30 Örebro
Sweden

Jan Wretman
Department of Statistics
Stockholm University
106 91 Stockholm
Sweden

The work on this book was supported in part by Statistics Sweden.

With 8 illustrations.

Mathematics Subject Classification: 62D05

Library of Congress Cataloging-in-Publication Data
Särndal, Carl-Erik, 1937–
Model assisted survey sampling / Carl-Erik Särndal, Bengt Swensson, Jan Wretman.
p. cm.—(Springer series in statistics)
Includes bibliographical references and indexes.
ISBN 0-387-97528-4 (alk. paper)
1. Sampling (Statistics) I. Swensson, Bengt. II. Wretman, Jan Håkan, 1939– . III. Title. IV. Series.
QA276.6.S37 1991
001.4'222—dc20 91-7851

Printed on acid-free paper.

Production managed by Bill Imbornoni; manufacturing supervised by Bob Paella.
Typeset by Asco Trade Typesetting Ltd., Hong Kong.
Printed and bound by R.R. Donnelley & Sons, Harrisonburg, VA.
Printed in the United States of America.

9 8 7 6 5 4 3 2 (corrected printing 1993)

ISBN 0-387-97528-4 Springer-Verlag New York Berlin Heidelberg
ISBN 3-540-97528-4 Springer-Verlag Berlin Heidelberg New York

Preface

This text on survey sampling contains both basic and advanced material. The main theme is estimation in surveys. Other books of this kind exist, but most were written before the recent rapid advances. This book has four important objectives:

1. To develop the central ideas in survey sampling from the unified perspective of unequal probability sampling. In a majority of surveys, the sampling units have different probabilities of selection, and these probabilities play a crucial role in estimation.
2. To write a basic sampling text that, unlike its predecessors, is guided by statistical modeling in the derivation of estimators. The model assisted approach in this book clarifies and systematizes the use of auxiliary variables, which is an important feature of survey design.
3. To cover significant recent developments in special areas such as analysis of survey data, domain estimation, variance estimation, methods for nonresponse, and measurement error models.
4. To provide opportunity for students to practice sampling techniques on real data. We provide numerous exercises concerning estimation for real (albeit small) populations described in the appendices.

This book grew in part out of our work as leaders of survey methodology development projects at Statistics Sweden. In supervising younger colleagues, we repeatedly found it more fruitful to stress a few important general principles than to consider every selection scheme and every estimator formula as a separate estimation problem. We emphasize a general approach.

This book will be useful in teaching basic, as well as more advanced, university courses in survey sampling. Our suggestions for structuring such courses are given below. The material has been tested in our own courses in Montréal,

Örebro, and Stockholm. Also, this book will provide a good source of information for persons engaged in practical survey work or in survey methodology research.

The theory and methods discussed in this book have their primary field of application in the large surveys typically carried out by government statistical agencies and major survey institutes. Such organizations have the resources to collect data for the large probability samples drawn by the complex survey designs that are often required. However, the issues and the methods discussed in the book are also relevant to smaller surveys.

Statistical modeling has strongly influenced survey sampling theory in recent years. In this book, sampling theory is assisted by modeling. It becomes simple to explain how the auxiliary information available in a given survey will lead to a particular estimation technique. The teaching of sampling and the style of presentation in journal articles on sampling have changed a great deal by this new emphasis. Readers of this book will become familiar with this new style.

We use the randomization theory, or design-based, point of view. This is the traditional mode of inference in surveys, ever since the sampling theory breakthroughs in the 1930s and 1940s. The reasoning is familiar to survey statisticians in government and elsewhere.

A body of basic and more advanced knowledge is defined that is useful for the survey sampling methodologist, and a broad range of topics is covered, without going into great detail on any one subject. Some topics to which this book devotes a single chapter have been the subject of specialized treatises.

The material should be accessible to a wide audience. The presentation is rich in statistical concepts, but the mathematical level is not advanced. We assume that readers of this book have been exposed to at least one course in statistical inference, covering principles of point estimation and confidence intervals. Some familiarity with statistical linear models, in particular regression analysis, is also desirable. A previous exposure to finite population sampling theory is not required for otherwise well-prepared readers. Some prior knowledge of sampling techniques is, of course, an advantage.

A collection of exercises is placed at the end of each chapter. The understanding of sampling theory is facilitated by analyzing data. Some of the exercises involve sampling and analysis of data from three populations of Swedish municipalities, named MU284, MU281, and MU200, and one population of countries named CO124. These populations are necessarily small in comparison to populations in real-world surveys, but the issues invoked by the exercises are real. Appendices B, C, and D present the populations. Other exercises are theoretical; some of them ask the reader to derive an expression or verify an assertion in the text.

There are various ways in which the book can be used for teaching courses in survey sampling. A first (one-semester or one-quarter) course can be based on the following chapters: 1, 2, 3, parts of 4, 5, 6, 7, and 8, and, at the instructor's discretion, a selection of material from later chapters. To mention at least a few of the issues in chapters 10, 14, and 15 seems particularly important in a first course. A second course (one-semester or one-quarter) may use topics from chapters 4 to 7 not covered in the first course, followed by a selection of material from chapters 8 to 17.

Certain sections usually placed toward the end of a chapter provide further detail on ideas or derivations presented earlier. Examples are sections 3.8, 5.12, and 6.8. Such sections are not essential for the main flow of the argument, and they may be omitted in a first as well as in a second course.

The Monte Carlo simulations and other computation for this book were carried out on an Apple Macintosh II, using MicroAPLs APL.68000-Macintosh II 68020+68881 version 7.20A.

Statistics Sweden generously supported this project. We are truly grateful for their cooperation as well as for support from University of Örebro, Stockholm University, Université de Montréal, Svenska Handelsbanken, and the Natural Sciences and Engineering Research Council of Canada.

Many individuals helped and encouraged us during the work on this book. In particular, we are indebted to Christer Arvas and Lars Lyberg for their supportive view of the project, to Jean-Claude Deville, Eva Elvers, Michael Hidiroglou, Klaus Krickeberg, and Ingrid Lyberg for critical appraisal of some of the chapters, to Kerstin Averbäck, Bibi Thunell and the late Sivan Rosén for typing, and to Patricia Dean and Nathalie Gaudet for editorial assistance. We have benefitted from discussions with survey statisticians at Statistics Canada and Statistics Sweden.

Montréal — Carl-Erik Särndal
Örebro — Bengt Swensson
Stockholm — Jan Wretman

Contents

PART I

Principles of Estimation for Finite Populations and Important Sampling Designs

CHAPTER 1

Survey Sampling in Theory and Practice

1.1. Surveys in Society

The need for statistical information seems endless in modern society. In particular, data are regularly collected to satisfy the need for information about specified sets of elements, called finite populations. For example, our objective might be to obtain information about the households in a city and their spending patterns, the business enterprises in an industry and their profits, the individuals in a country and their participation in the work force, or the farms in a region and their production of cereal.

One of the most important modes of data collection for satisfying such needs is a sample survey, that is, a partial investigation of the finite population. A sample survey costs less than a complete enumeration, is usually less time consuming, and may even be more accurate than the complete enumeration. This book is an up to date account of statistical theory and methods for sample surveys. The emphasis is on new developments.

Over the last few decades, survey sampling has evolved into an extensive body of theory, methods, and operations used daily all over the world. As Rossi, Wright and Anderson (1983) point out, it is appropriate today to speak of a worldwide survey industry with different sectors: a government sector, an academic sector, a private and mass media sector, and a residual sector consisting of ad hoc and in-house surveys.

In many countries, a central statistical office is mandated by law to provide statistical information about the state of the nation, and surveys are an important part of this activity. For example, in Canada, the 1971 Statistics Act mandates Statistics Canada to "collect, compile, analyze, abstract, and publish statistical information relating to the commercial, industrial, financial, social, economic, and general activities and condition of the people."

Thus, national statistical offices regularly produce statistics on important national characteristics and activities, including demography (age and sex distribution, fertility and mortality), agriculture (crop distribution), labor force (employment), health and living conditions, and industry and trade. We owe much of the essential theory of survey sampling to individuals who are or were associated with government statistical offices.

In the academic sector, survey sampling is extensively used, especially in sociology and public opinion research, but also in economics, political science, psychology, and social psychology. Many academically affiliated survey institutes are heavily engaged in survey sampling activity. In the private and mass media sectors, we find television audience surveys, readership surveys, polls, and marketing surveys. The contents of ad hoc and in-house surveys vary greatly. Examples include payroll surveys and surveys for auditing purposes.

Survey sampling has thus grown into a universally accepted approach for information gathering. Extensive resources are devoted every year to surveys. We do not have accurate figures to illustrate the scope of the industry. However, as an example from the United States, the National Research Council in 1981 estimated that the survey industry in the country conducted roughly 100 million interviews in one year. If we assume a cost of \$20 to \$40 per interview, interviewing alone (which is only one component of the total survey operation) represents 2 to 4 billion dollars per year.

News media provide the public with the results of new or recurring surveys. It is widely accepted that a sample of fairly modest size is sufficient to give an accurate picture of a much larger universe; for example, a well-selected sample of a few thousand individuals can portray with great accuracy a total population of millions. However data gathering is costly. Therefore, it makes a great difference if a major national survey uses 20,000 observations, when 15,000 or even 10,000 observations might suffice. For reasons of cost effectiveness, it is imperative to use the best methods available for sampling design and estimation, to profit from auxiliary information, and so on.

Here statistical knowledge and insight become highly important. The expert survey statistician must have a good grasp of statistical concepts in general, as well as the particular reasoning used in survey sampling. A good measure of practical experience is also necessary. In this book, we present a basic body of knowledge concerning survey sampling theory and methods.

1.2. Skeleton Outline of a Survey

To start, we need a skeleton outline of a survey and some basic terminology. The terms "survey" and "sample survey" are used to denote statistical investigations with the following methodologic features (key words are italicized):

i. A survey concerns a finite set of *elements* called a *finite population.* An enumeration rule exists that unequivocally defines the elements belonging to the population. The goal of a survey is to provide information about the finite population in question or about subpopulations of special interest, for example, "men" and "women" as two subpopulations of "all persons." Such subpopulations are called *domains of study* or just *domains*.

ii. A value of one or more *variables of study* is associated with each population element. The goal of a survey is to get information about unknown *population characteristics* or *parameters*. Parameters are functions of the study variable values. They are unknown, quantitative measures of interest to the investigator, for example, total revenue, mean revenue, total yield, number of unemployed, for the entire population or for specified domains.

iii. In most surveys, access to and observation of individual population elements is established through a *sampling frame*, a device that associates the elements of the population with the *sampling units* in the frame.

iv. From the population, a *sample* (that is, a subset) of elements is selected. This can be done by selecting sampling units in the frame. A sample is a *probability sample* if realized by a chance mechanism, respecting the basic rules discussed in Section 1.3.

v. The sample elements are *observed*, that is, for each element in the sample, the variables of study are *measured* and the values *recorded*. The measurement conforms to a well-defined *measurement plan*, specified in terms of measurement instruments, one or more measurement operations, the order between these, and the conditions under which they are carried out.

vi. The recorded variable values are used to calculate (*point*) *estimates* of the finite population parameters of interest (totals, means, medians, ratios, regression coefficients, etc.). Estimates of the precision of the estimates are also calculated. The estimates are finally published.

In a sample survey, observation is limited to a subset of the population. The special type of survey where the whole population is observed is called a *census* or a *complete enumeration*.

EXAMPLE 1.2.1. Labor force surveys are conducted in many countries. Such a survey aims at answering questions of the following type: How many persons are currently in the labor force in the country as a whole and in various regions of the country? What proportion of these are unemployed? In this case, some of the key concepts may be as follows. *Population*: All persons in the country with certain exceptions (such as infants, people in institutions). *Domains of interest*: age/sex groups of the population, occupational groups in the population, and regions of the country. *Variables*: Each person can be described at the time of the survey as (a) belonging to the labor force or not, and (b) employed or not. Correspondingly, there is a variable of interest that takes the value "one" for a person in the labor force, "zero" for a person not in the labor force. To measure unemployment, a second variable of interest is defined as taking the value "one" if a person is unemployed, "zero" otherwise. Precise definitions are essential. If the purpose is to estimate unemployment in a given month, and if an interviewed person states that he worked one week during that month, but that he is unemployed the day of the interview, there must be a clear rule stating whether he is to be recorded as unemployed or not. *Population characteristics of interest*: Number of persons in the labor force. Number of unemployed persons in the labor force. Proportion of unemployed persons in the labor force. *Sample*: A sample of persons is selected from the population in an efficient manner given existing devices for observational access to the persons in the country. *Observations*: Each person included in the sample is visited by a trained interviewer who asks questions following a standardized questionnaire and records the answers. *Data processing and estimation*: The recorded data are edited, that is, prepared for the estimation phase; rules for handling of nonresponse are observed; estimates of the population characteristics are calculated. Indicators of the uncertainty of the estimates (variance estimates) are calculated. The results are published.

EXAMPLE 1.2.2 Consider a household survey whose aim is to obtain information about planned household expenditures in the coming year for specified durable goods. Here, some of the basic concepts may be as follows. *Population*: All households in the country. *Variables*: Planned expenditure amounts for specified goods, such as automobiles, refrigerators, etc. *Population characteristics of interest*: Total of planned household expenditures for the specified durable goods. *Sample*: A sample of households is obtained by initially selecting a sample of geographic areas, then by subsampling of households in selected areas. *Observations*: Each household in the sample receives a self-administered questionnaire. For a majority of households, the questions are answered and the questionnaire returned. Households not returning the questionnaire are followed up by telephone or visited by a trained interviewer to obtain the desired information. *Data processing and estimation*: Data are edited. The calculation of point estimates and precision takes into account the two-stage design of the survey.

The methodologic features (i) to (vi) identified above prompt a few comments.

1. The complexity of a survey can vary greatly, depending on the size of the population and the means of accessing that population. To survey the members of a professional society, the hospitals in a region, or the residents in a small municipality may be a relatively simple matter. At the other extreme are complex nationwide surveys, with a population of many millions spread over a large territory; such surveys are typically carried out by government statistical agencies and require extensive administrative and financial resources.
2. Although a survey involves observations on individual population elements, the purpose of a survey is not to use such data for decision-making

about individual elements, but to obtain summary statistics for the population or for specific subgroups.

3. In the same survey, there are often many variables of study and many domains of interest. The number of characteristics to estimate may be large, in the hundreds or even in the thousands.
4. Finite population parameters are quantitative measures of various aspects of the population. Prior to a survey they are unknown. In this book, we examine the estimation of different types of parameters: the total of a variable of study, the mean of the variable, the median of the variable, the correlation coefficient between two variables, and so on. The exact value of a finite population parameter can be obtained in a special case, namely, if we observe the complete population (i.e., the survey is a census), and there are no measurement errors and no nonresponse. A census does not automatically mean "estimation without error."
5. Most people are aware of the term "census" in a particular sense, namely, as a fact finding operation about a country's population, in particular about such sociodemographic characteristics as the age distribution, education levels, special skills, mother tongue, housing conditions, household structure, migration patterns. In these situations there is often a "census proper," done through a "short form" (a questionnaire with few questions) going out to all individuals, while a "long form" may be administered to a 20% sample with a request for more extensive information.
6. A sample is any subset of the population. It may or may not be drawn by a probability mechanism. A simple example of a probability sampling scheme is one that gives every sample of a fixed size the same probability of selection. This is *simple random sampling* without replacement. In practice, selection schemes are usually more complex. Probability sampling has over the years proved to be a highly accurate instrument and is the focus of this book. The reasons for probability sampling are discussed later in this chapter and in Chapter 14. An example of a nonprobability sample is one that is designated by an expert as representative of the population. Only in fortunate circumstances will nonprobabilistic selection yield accurate estimates.
7. To correctly measure and record the desired information for all sampled elements may be difficult or impossible. False responses may be obtained. For some elements designated for the survey, measurements may be missing because of, for example, impossibility to contact or refusal to respond. These *nonsampling errors* may be considerable.
8. Advances in computer technology have made it possible to produce a great deal of official statistics (for example, in the economic sector) from administrative data files. Several files may be used. For example, elements are matched in two complete population registers, and the information combined. The matched files give a more extensive base for the production of statistics. (For populations of individuals, matching may conflict with privacy considerations.) Information from a sample survey may also be com-

bined with the information from one or more complete administrative registers. The administrative data may then serve as auxiliary information to strengthen the survey estimates.

1.3. Probability Sampling

Probability sampling is an approach to sample selection that satisfies certain conditions, which, for the case of selecting elements directly from the population, are described as follows:

1. We can define the set of samples, $\mathscr{S} = \{s_1, s_2, \ldots, s_M\}$, that are possible to obtain with the sampling procedure.
2. A known probability of selection $p(s)$ is associated with each possible sample s.
3. The procedure gives every element in the population a nonzero probability of selection.
4. We select one sample by a random mechanism under which each possible s receives exactly the probability $p(s)$.

A sample realized under these four requirements is called a *probability sample.*

If the survey functions without disturbance, we can go out and measure each element in the realized sample and obtain true observed values for the study variables. We assume a formula exists for computing an estimate of each parameter of interest. The sample data are inserted in the formula, yielding, for every possible sample, a unique estimate.

The function $p(\cdot)$ defines a probability distribution on $\mathscr{S} = \{s_1, s_2, \ldots, s_M\}$. It is called a *sampling design*, or just *design*; a more rigorous definition is given in Section 2.3.

The probability referred to in point 3 is called the *inclusion probability* of the element. Under a probability sampling design, every population element has a strictly positive inclusion probability. This is a strong requirement, but one that plays an important role in the probability sampling approach. In practice there are sometimes compelling reasons for not adhering strictly to the requirement. Cut-off sampling (see Section 14.4) is an occasionally used technique in which certain elements are deliberately excluded from selection. In that case, valid conclusions are limited to the part of the population that can be sampled.

The randomization referred to in point 4 is usually carried out by an easily implemented algorithm. A common type of algorithm is one in which a randomized experiment is performed for each element listed in the frame, leading either to inclusion or noninclusion. Different algorithms are discussed in Chapters 2 and 3.

Sampling is often carried out in two or more stages. Clusters of elements are selected in an initial stage. This may be followed by one or more subsampling stages; the elements themselves are sampled at the ultimate stage. To have a probability sampling design in that case, conditions 1 to 4 above must apply to each stage. The procedure as a whole must give every population element a strictly positive inclusion probability.

Probability sampling has developed into an important scientific approach. Two important reasons for randomized selection are (1) the elimination of selection biases, and (2) randomly selected samples are "objective" and acceptable to the public. Some milestones in the development of the approach are noted in Chapter 14. Most of the arguments and methods in this book are based on the probability sampling philosophy.

1.4. Sampling Frame

The *frame* or the *sampling frame* is any material or device used to obtain observational access to the finite population of interest. It must be possible with the aid of the frame to (1) identify and select a sample in a way that respects a given probability sampling design and (2) establish contact with selected elements (by telephone, visit at home, mailed questionnaire, etc.). The following more detailed definition is from Lessler (1982):

Frame. The materials or devices which delimit, identify, and allow access to the elements of the target population. In a sample survey, the units of the frame are the units to which the probability sampling scheme is applied. The frame also includes any auxiliary information (measures of size, demographic information) that is used for (1) special sampling techniques, such as stratification and probability proportional-to-size sample selections, or for (2) special estimation techniques, such as ratio or regression estimation.

As in the preceding definition, we call *elements* the entities that make up the population, and *units* or *sampling units* the entities of the frame. The latter term stresses that the sample is actually selected from the frame.

EXAMPLE 1.4.1. The Swedish Register of the Total Population is a large sampling frame that lists some eight million individuals. This frame gives, for each individual, information on variables such as date of birth, sex, marital status, address, and taxable income. It is a reasonably correct frame. There are few omissions, and few are listed in the frame who do not rightfully belong in it. An attractive feature of this frame is that it gives direct access to the country's entire population. Stratified sampling from this frame is often used for Swedish surveys. Sampled elements (individuals) can be contacted with relative ease.

Example 1.4.2. The Central Frame Data Base is a sampling frame compiled by Statistics Canada for use in business surveys. It is a fairly complex frame, consists of several parts, and is based on information from several sources. The construction of this frame is based on two types of Canadian tax returns: corporate and individual, the latter for the self-employed. The frame has two main components: an "integrated" component (containing all of the large business establishments) and a "nonintegrated" component, which is further divided into two separate parts using information from Revenue Canada Taxation. Business firms reporting small total operating revenue are considered "out-of-scope" for business surveys. Continuous updating is required to register "births" (starting of new business activity), "deaths" (termination of business activity), and changes in classification based on geography, industry, or size.

We use the term *direct element sampling* to denote sample selection from a frame that directly identifies the individual elements of the population of interest. That is, the units in the frame are objects of the same kind as those that we wish to measure and observe. A selection of elements can take place directly from the frame. Ideally, the set of elements identified in the frame is equal to the set of elements in the population of interest.

For example, if the population of interest is Swedish individuals, we can carry out direct element sampling from the frame "Register of the Total Population" mentioned in Example 1.4.1. Here, unit equals element which equals individual. (The two sets are actually not exactly equal, but differences are minor so the frame is nearly perfect.) The frame in Example 1.4.2 can be used for direct element sampling with the objective of studying the population of business establishments in Canada; in that case, unit equals element which equals business establishment.

The following is a list of properties that a frame for direct element sampling should have. Minimum requirements are that:

1. The units in the frame are identified, for example, through an identifier k running from 1 to N_F where N_F is the number of sampling units.
2. All units can be found, if selected in a sample. That is, the address, telephone number, location on map, or other device for making contact is specified in the frame or can be made available.

It simplifies many sample selection procedures if the following feature is present:

3. The frame is organized in a systematic fashion, for example, the units are ordered by geography or by size.

Other information is sometimes available in the frame and will often improve estimates. The following is desirable:

4. The frame contains a vector of additional information for each unit; such

information may be used for efficiency improvement such as stratification or to construct estimators that involve auxiliary variables.

5. When estimation is required for domains (subpopulations), the frame specifies the domain to which each unit belongs.

Other desirable properties involve the relationship between the units in the frame and the population elements:

6. Every element in the population of interest is present only once in the frame.
7. No element not in the population of interest is present in the frame.

The preceding two features will simplify many selection and estimation procedures.

8. Every element in the population of interest is present in the frame.

The last mentioned property is particularly important, because in its absence, the frame does not give access to the whole population of interest. Then not even observation of all elements in the frame will make it possible to calculate the true value of a finite population parameter of interest.

In practice, a frame is often in the form of a computer data file. At a minimum, it is a file with an element identifier k running from 1 to N_F. It may contain other information, as mentioned in points 4 and 5. We can state all that is available in the frame about the kth element as a vector $\mathbf{x}_k = (x_{1k}, x_{2k}, \ldots, x_{jk}, \ldots, x_{qk})$. Here, x_{jk} is the value of the jth x-variable for the kth element. The value x_{jk} may be quantitative (for example, date of birth or salary for individual k) or qualitative (for example, address for individual k). The frame can be seen as a matrix arrangement with N_F rows (records), and with each row is associated $q + 1$ data entries (fields); one entry for the identifier and q entries for the components of the row vector $\mathbf{x}_k$, as follows:

Identifier	Known Vector
1	$\mathbf{x}_1$
2	$\mathbf{x}_2$
$\vdots$	$\vdots$
k	$\mathbf{x}_k$
$\vdots$	$\vdots$
N_F	$\mathbf{x}_{N_F}$

It is important to construct a good sampling frame. As Jessen (1978) puts it: "In many practical situations the frame is a matter of choice to the survey planner, and sometimes a critical one, since frames not only should provide us with a clear specification of what collection of objects or phenomena we are undertaking to study, but they should also help us carry out our study efficiently and with determinable reliability. Some very worthwhile investigations are not undertaken at all because of the lack of an apparent frame;

others, because of faulty frames, have ended in a disaster or in a cloud of doubt." Guidelines for frame construction and maintenance are discussed in Section 14.7.

1.5. Area Frames and Similar Devices

The following distinction is important:

i. A frame as a direct listing or identification of elements.
ii. A frame as a listing or identification of (smaller or larger) sets of elements.

In case (i), direct element sampling can be carried out. In case (ii), access to the elements is more roundabout, namely, by selecting sets of elements and by observing all or some of the elements in the selected sets. In many situations, case (ii) is the only option, since it may not be possible to find or construct (without prohibitive cost) a direct list of elements. The total number of population elements is often unknown in case (ii). For example, let us think of the population of households in a large metropolitan area. In many cities, there is nothing that comes close to a complete register of the households. Other sampling units than individual households must be considered. One way is to define sampling units as city blocks on a map and select a sample of such units. With relative ease, we may then gain access to the households in (a modest number of) selected blocks. A variation of the same idea occurs when segments are identified on a forest map, and a sample of segments is drawn with the objective of observing individual trees in selected segments. The concept of area frame is defined as follows:

Area Frame. A geographic frame consisting of area units; every population element belongs to an area unit, and it can be identified after inspection of the area unit. The area units may vary in size and in the number of elements that they contain.

Area sampling entails sampling from an area frame, such as a city map, a forest map, or an aerial photograph. The sets of elements drawn with the aid of an area frame are often called *clusters*. In a secondary selection step, the selected clusters may be subsampled. A sample of still smaller areas could be defined and sampled, and so on, until the elements themselves are finally sampled in the ultimate step.

Maps are, of course, not always used when sets (clusters) of elements are sampled; a succession of lists may be used instead. A frame for studying a population of high school students may in the first step consist of a list of school districts, then a list of all schools in each selected district, then a list of all classes in each selected school, then in the fourth and final step one would gain access to students. "Frame" here refers to a device with four successive layers. School districts are the first stage sampling units, schools the second stage sampling units, classes the third stage units. The individual elements (the students) are the sampling units in the fourth and final stage of sampling. In a selection consisting of several stages, each stage has its own type of sampling unit.

A finite population is made up of elements. They are sometimes also called *units of analysis*, which underscores that they are entities that are measured and for which values are recorded. For example, if one is interested in estimating the population total of the variable "household income," the element (or unit of analysis) is naturally the household. The frame is an instrument for gaining access (more or less directly) to these elements. One way is to first select city blocks and then observe households in selected blocks.

Our examples so far have perhaps left the impression that in practice "element" is always something "smaller than or at most equal to" a "sampling unit." This is not necessarily so, as the following example shows.

EXAMPLE 1.5.1. The Swedish household survey HINK is a national survey of household income. In the absence of a good complete list of households, it has proved convenient to use the Swedish Register of the Total Population described in Example 1.4.1 as a sampling frame. This register is a list of individuals. A probability sample (ordinarily, a stratified random sample) of individuals is selected. The households to which the selected individuals belong are identified, and income-related variables are measured for these households. Here, the sampling unit is the individual, and the element (unit of analysis) is the household. That is, an element comprises one or more sampling units. This is a particular case of "network sampling" (see Sirken 1972, Levy 1977, and Rosén 1987).

1.6. Target Population and Frame Population

It becomes necessary at this point to distinguish target population from frame population. The *target population* is the set of elements about which information is wanted and parameter estimates are required. The *frame population* is the set of all elements that are either listed directly as units in the frame or can be identified through a more complex frame concept, such as a frame for selection in several stages.

Frame quality can be studied through the relations that exist between the target population, denoted U, and the frame population denoted U_F. If each element in the frame population corresponds to one and only one element in the target population, then $U = U_F$, and the frame is perfect for the target population. In all other cases, there is *frame imperfection*. Frame imperfections are discussed in more detail in Section 14.7. At this stage it suffices to note that points 6, 7, and 8 in Section 1.4 point out three common frame imperfections, namely, undercoverage, overcoverage, and duplicate listings.

The frame has *undercoverage* when some target population elements are not in the frame population. The frame has *overcoverage* when elements not in the target population are in the frame population. A *duplicate listing* occurs when a target population element is listed in the frame more than once. For example, if unit equals element which equals business enterprise, it may be that a unit in the frame is a firm that recently went out of business. Since no longer involved in business activity, this firm is not in the target population, but still exists as a unit in the frame. Another frame unit may be a firm that exists, but in a different industry than the one defined by the target population. Both firms are part of the overcoverage. Newly established firms not yet listed in the frame are part of the undercoverage.

If a probability sample is selected from the frame population, valid statistical inference can be made about the frame population. If the frame population is different from the target population, valid inference about the target population may be impossible, so the goal of the survey may be missed. The problem is particularly serious if the frame gives access only to parts of the target population.

To come up with a perfect sampling frame is not always possible in practice. Minor imperfections are often tolerated, since a perfect frame may not be obtained without excessive cost. However, it is imperative that frame imperfections be minor.

To construct a high-quality frame for the target population is an important aspect of survey planning, and adequate resources must be set aside for this activity. Viewed somewhat differently, when the target population is defined, a realistic goal should be set. There is no sense in fixing a target population for which a good frame cannot realistically be obtained within budget restrictions. Grossly invalid conclusions may result from samples drawn from faulty frames. Inexpensive, easy-to-come-by frames should be avoided if they give only fragmentary access to the target population.

Frame construction and maintenance are discussed in Chapter 14. Sample selection is sometimes carried out with the aid of several partially overlapping frames; such multiple frame sampling is also considered in Chapter 14.

1.7. Survey Operations and Associated Sources of Error

A survey consists of a number of survey operations. Especially in a large survey, the operations may extend over a considerable period of time, from the planning stage to the ultimate publication of results. The operations affect the quality of survey estimates. We distinguish five phases of survey operations, as follows. With each phase we can associate sources of errors in the estimates.

i. *Sample Selection*

This phase consists of the execution of a preconceived sampling design. The sample size necessary to obtain the desired precision is determined. A sample of elements is drawn with the given sampling design using a suitable sampling frame, which may be an already existing frame or one that is constructed specifically for the survey. Errors in estimates associated with this phase are (1) frame errors, of which undercoverage is particularly serious, and (2) sampling error, which arises because a sample, not the whole population, is observed.

ii. *Data Collection*

There is a preconceived measurement plan with a specified mode of data collection (personal interview, telephone interview, mail questionnaire, or other). The field work is organized; interviewers are selected and interviewer assignments are determined. Data are collected according to the measurement plan, for the elements in the sample. The data are recorded and transmitted to the central statistical office. Errors in estimates resulting from this phase include (1) measurement errors, for instance, the respondent gives (intentionally or unintentionally) incorrect answers, the interviewer understands or records incorrectly, the interviewer influences the responses, the questionnaire is misinterpreted, (2) error due to nonresponse (missing observations).

iii. *Data Processing*

This phase serves to prepare collected data for estimation and analysis and includes the following elements:

Coding and data entry (transcription of information from questionnaire to a medium suited for estimation and data analysis).

Editing (consistency checks to see if observed values conform to a set of logical rules, handling of values that do not pass the edit check, outlier detection and treatment).

Renewed contact with respondents to get clarification if necessary.

Imputation (substitution of good artificial values for missing values).

Errors in estimates associated with this phase include transcription errors (keying errors), coding errors, errors in imputed values, errors introduced by or not corrected by edit.

iv. *Estimation and Analysis*

This phase entails the calculation of survey estimates according to the specified point estimator formula, with appropriate use of auxiliary information and adjustment for nonresponse, as well as a calculation of measures of precision in the estimates (variance estimate, coefficient of variation of the estimate, confidence interval). Statistical analyses may be carried out, such as comparison of subgroups of the population, correlation and regression analyses, etc. All errors from the phases (i) to (iii) above will affect the point estimates and they should ideally be accounted for in the calculation of the measures of precision.

v. *Dissemination of Results and Postsurvey Evaluation*

This phase includes the publication of the survey results, including a general declaration of the conditions surrounding the survey. This declaration often follows a set of specified guidelines for quality declaration.

Errors in survey estimates are traditionally divided into two major categories: sampling error and nonsampling error. The *sampling error* is, as mentioned, the error caused by observing a sample instead of the whole population. The sampling error is subject to sample-to-sample variation. The *nonsampling errors* include all other errors. The two principal categories of nonsampling errors are:

a. *Errors due to nonobservation.* Failure to obtain data from parts of the target population.
b. *Errors in observations.* This kind of error occurs when an element is selected and observed, but the finally recorded value for that element (the value that goes into the estimation and analysis phase) differs from the true value. Two major types are (b1) measurement error (error arising in the data collection phase) and (b2) processing error (error arising in the data processing phase).

There are two principal types of *nonobservation*, namely, (1) undercoverage, that is, failure of the frame to give access to all elements that belong to the target population (such elements will obviously not be selected, much less observed, and they have zero inclusion probability), and (2) nonresponse, that is, some elements actually selected for the sample turn out to be nonobservations because of refusal or incapacity to answer, not being at home, and so on. Nonobservation generally results in biased estimates.

Measurement error can be traced to four principal sources:

1. The interviewer.
2. The respondent.
3. The questionnaire.
4. The mode of the interview, that is, whether telephone, personal interview, self-administered questionnaire, or other medium is used.

Processing error comprises the errors arising from coding, transcription, imputation, editing, outlier treatment, and other types of preestimation data handling. In modern computer assisted data collection methods (CATI and CAPI; see Section 14.9), the data collection and data processing phases tend to be merged, and, as a consequence, processing errors may be reduced.

The basic estimation theory, taking into account the sampling error, is presented in Part I of this book (Chapters 2 to 5). Various extensions are given in Parts II and III (Chapters 6 to 13). Estimation in the presence of nonsampling errors is treated in Part IV (Chapters 14 to 17). In particular, Chapter 14 contains a more detailed discussion of nonsampling errors and their impact on the survey estimates.

1.8. Planning a Survey and the Need for Total Survey Design

A survey usually has its background in some practical problem. Someone—a member of parliament, a researcher, an administrator, a decisionmaker—formulates a question in the course of his work, a problem to which no answer is readily available. A study is needed. It can be an experiment, a survey, or some other form of fact finding. In either case, it is imperative that the problem be clearly stated. If a survey is the instrument chosen for the study, the survey statistician needs to work with a clearly stated objective. What exactly is the problem? Exactly what information is wanted?

For example, suppose a survey of housing conditions for the elderly is proposed. This description is vague and too general. The concepts involved must be given clear definitions. Precisely what is the population of aged individuals of interest? What age groups are to be included? Should we look separately at single-person households, two-person households, households where the elderly cohabit with younger persons, etc.? What is the more complete specification of housing conditions? Do we mean age of dwelling or some other quality measure of the dwelling? What time period is to be studied? Should one distinguish between an urban elderly population and a rural elderly population? As answers are obtained to these questions, the survey statistician begins to work toward a reformulation of the original question into one stated in precise terms that can be answered by a survey. The statistician's formulation must be unequivocal on the following:

i. The finite population and the subpopulations for which information is required.
ii. The kinds of information needed for this population, that is, the variables to be measured and the parameters to be estimated.

Once the operational definitions are clearly stated, the survey statistician can work toward the specification of a suitable survey design, including sampling frame, data collection method, staff required, sample selection, estimation method, and determination of the sample size required to obtain the desired precision in the survey results. Before going ahead with the survey, the statistician will make sure that his "translation" corresponds fully to what the problem originator had in mind: Will the survey give at least an approximate solution to the right problem? In the words of W. E. Deming (1950, p. 3), "The requirement of a plain statement of what is wanted (the *specification* of the survey) is perhaps one of the greatest contributions of modern theoretical statistics".

Some important aspects of *survey planning* are as follows:

Specifying the objective of the survey.

Translation of the subject-matter problem into a survey problem.

Specification of target population, known variables (auxiliary variables), study variables, population parameters to be estimated.

Construction of sampling frame, if none is available;

Inventory of resources available in terms of budget, staff, data processing, and other equipment.

Specifications of requirements to be met, for instance, time schedule and accuracy of estimates.

Specification of data collection method, including questionnaire construction.

Specification of sampling design, sample selection mechanism , and sample size determination.

Specification of data processing methods, including edit and imputation.

Specification of formulas for point estimator and measures of precision (variance estimator).

Training of personnel, organization of field work.

Allocation of resources to different survey operations.

Allocation of resources to control and evaluation.

The survey planning should ideally lead to a decision for each of the survey operations. Statistical theory can guide us to important conclusions about *some* of these decisions, in particular with regard to sample selection, choice of estimator, different sources of error and their associated components of variance, methods for assessment of the accuracy of the estimates, and statistical analysis of survey data.

The planning process should try to foresee difficulties that may arise. Resources should be set aside and back-up procedures should be identified to deal with perceived difficulties. For example, some nonresponse can almost certainly be expected, and the survey planning should take this into account; the nonresponse should be kept at low levels. Procedures for follow up and renewed contact with nonrespondents should be identified and included in budget considerations. Estimator formulas that attempt to adjust for nonresponse should be identified.

Ideally, survey planning should lead to an optimal specification for the survey as a whole. The goal is to obtain the best possible accuracy under a fixed budget. In a major survey, the decision problem is, however, so complex that an optimum, in the sense of a mathematical solution to a closed-form problem, is inconceivable. There are too many interrelated decisions and too many variables to take into account. The concept of total survey design, discussed in the next section, can be seen as a tool in the search for an overall optimization of a survey.

1.9. Total Survey Design

The term *total survey design* has come to be used for planning processes that aim at overall optimization in a survey. The concept arose out of a desire for an overall control of all sources of errors in a survey. Instrumental in this regard were efforts at the United States Bureau of the Census. Key references are Hansen et al. (1951), Hansen, Hurwitz, and Bershad (1962), Hansen, Hurwitz, and Pritzker (1964). A detailed discussion of total survey design is found in Dalenius (1974).

Total survey design is concerned with obtaining the best possible precision in the survey estimates while striking an overall economic balance between sampling and nonsampling errors. For a view of total survey design, it is helpful to consider a survey from three perspectives:

1. The requirements.
2. The survey specifications.
3. The survey operations.

By the requirements are meant the needs for information about the finite population, usually originating in some subject-matter problem. Corresponding to these requirements is a conceptual survey which will achieve the *ideal goal*, if carried out under the best possible circumstances.

The survey specifications are a set of rules and operations, which together constitute a *defined goal* of the survey. Because of actual conditions, the defined goal may be somewhat different from the ideal goal. The defined goal specifies key elements of the survey, such as population, sampling design, measurement procedures, estimators, and auxiliary variables.

Several survey designs usually exist for realizing the defined goal. The survey statistician selects from a set of operationally feasible survey designs one that comes as close as possible to realizing the defined goal. The selected design gives rise to a series of *survey operations*. The essential ones are those that we identified in Section 1.7. The survey is finally carried out executing these operations as carefully as possible; this constitutes the *survey operations proper* (see Figure 1.1).

Following Dalenius (1974) we can summarize the total survey design process in a diagram as shown in Figure 1.1.

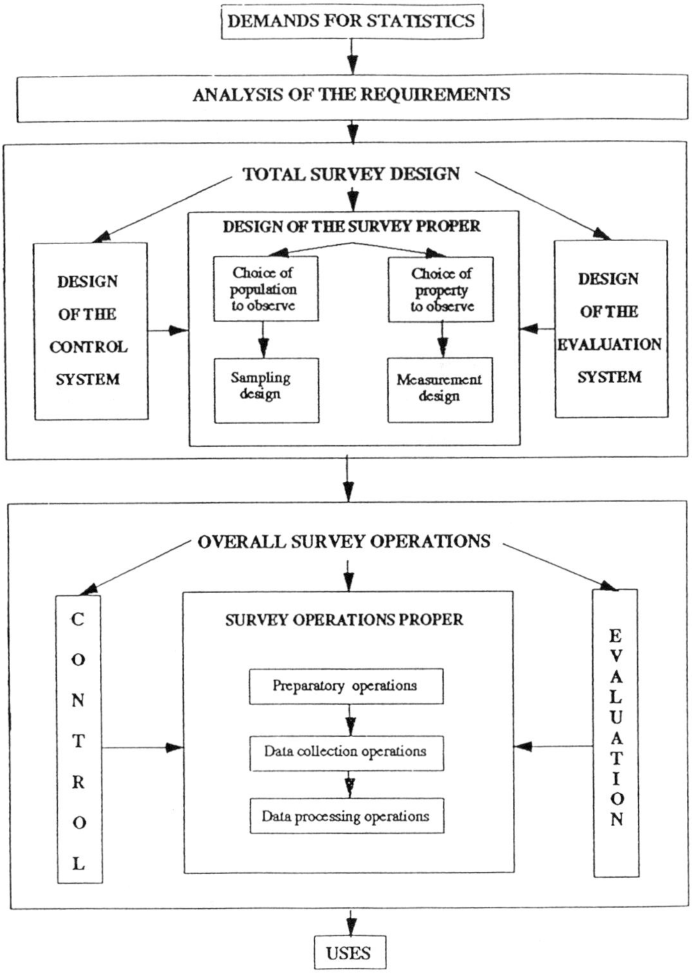

Figure 1.1. The Total Survey Design Process.

1.10. The Role of Statistical Theory in Survey Sampling

There is no unified theory covering all survey operations simultaneously. The current state of the art offers partial solutions, obtained sometimes under restrictive assumptions and idealized conditions. This book covers a crucial aspect of the total survey picture, namely, the part where statistical theory, especially estimation theory, is used to obtain answers.

Estimation theory is the organized study of variability in estimates. In surveys, the variability is tied to various errors identified in Section 1.7.

Suppose that the probability distributions of the various errors can be given, if not a complete specification, at least some general features. A stochastic structure is thereby defined, and we can work toward obtaining probability statements about the total error in an estimate. This is the traditional goal of statistical inference.

To study the variability of survey estimates, we need to specify as accurately as possible the stochastic features of the various errors. Let us see how this can be done.

i. *Stochastic Structure Relative to Sampling Error*

How was the sample selected? In the probability sampling approach, we know the answer. The sample is selected according to the given probability sampling design. We know or can determine the probabilities given to the different possible samples. This is the key to describing the sample-to-sample variation of a proposed estimator. The statistical properties of the estimator can be worked out: its expected value, its bias (if any), its variance, and so on. We can also obtain an estimate of the variance and a confidence interval. These notions are explained in detail in Chapter 2.

We mentioned that two reasons for randomized sample selection are protection against selection bias and the fact that randomly selected samples are viewed as objective. In this book, the probability distribution associated with the randomized sample selection has another very important function. It provides the basis for probabilistic conclusions (inferences) about the target population. For instance, a 95% confidence interval will have the property of covering the unknown population quantity in 95% of all samples that can be drawn with the given probability sampling design. The term used in the literature for this kind of conclusion is *design-based inference* or *randomization inference*. Design-based inference is objective; nobody can challenge that the sample was really selected according to the given sampling design. The probability distribution associated with the design is "real," not modeled or assumed.

ii. *Stochastic Structure Relative to Nonsampling Errors*

How do measurement errors and data processing errors arise? In most situations, we do not know. Any answer will involve hypothetical assumptions, called *model assumptions*, about these errors. Models for errors in observations are discussed in Chapter 16.

How does nonresponse arise? What is the mechanism that generates response from elements in the sample? Again, in most situations we do not know, and any answer will have to be in the form of a model. Models of this kind are called *response mechanism models*; see Chapter 15.

iii. *Stochastic Structure Relative to the Origin of Finite Population Variable Values*

How were the population variable values generated? An attempt answering this question is to say that the N population values $y_1, \ldots, y_N$ of the study variable y are generated from a superpopulation. If the idea of a superpopulation is accepted, we must concede that the properties of this population are unknown or at least partially unknown. *Superpopulation modeling* is the activity whereby one specifies or assumes certain features of the mechanism thought to have generated $y_1, \ldots, y_N$. Superpopulation modeling leads to model-based inference, which is discussed in Section 14.5.

A number of important distinctions have been made in this chapter. They help in getting a perspective on how the material is organized in the chapters to come. Chapters 2 to 12 are based on the idea of probability sampling from a high quality sampling frame in the absence of nonsampling errors. Chapters 2 and 3 present elementary estimation theory for finite populations. Methods for estimation of totals and means are presented for important sampling designs. In Chapter 5, we look at the estimation of other parameters of interest, such as medians, variances, correlation, and regression coefficients. Estimation with auxiliary information is an important theme in Chapters 6 to 8. Chapters 4 and 8 deal with sampling in two stages. Chapters 9 to 13 are devoted to significant special topics, such as estimation in two phases, estimation for domains, variance estimation techniques, questions of optimal sampling design, and analysis of survey data. In Chapter 14, we assess the limitations and extensions of the probability sampling approach, and the nature of nonsampling errors is examined. Complete chapters are devoted to two important types of nonsampling error, namely, nonresponse (Chapter 15) and measurement error (Chapter 16). Finally, the importance of quality declaration of survey data is stressed (Chapter 17).

Exercises

1.1. Consider an agriculture yield survey aimed at obtaining information about the yield of different crops (wheat, rye, etc.) in a given country and in a given year. Specify definitions that might be appropriate in such a survey for key concepts such as population, variable(s) of study, parameter(s) of interest. Try to be as specific as possible in your definitions.

1.2. For a survey of the type indicated in Exercise 1.1, discuss what device(s) reasonably may be used to constitute a sampling frame. Discuss possible methods for accurate data collection, with consideration given to avoidance of measurement error.

1.3. Consider a retail sales survey in a given country, that is, a survey aimed at finding out about the amount of sales in retail stores. Specify definitions that might be realistic in this case for key concepts such as population, variable(s) of study, and parameter(s) of interest. Try to be as specific as possible in your definitions.

1.4. For a survey of the type indicated in Exercise 1.3, discuss what device(s) reasonably may be used to constitute a sampling frame. Discuss possible alternatives for data collection, with consideration given to avoidance of measurement error.

1.5. Contrast a "simple survey" (say, a survey of approximately 500 members of a professional society) with a "complex survey" (say, a survey of approximately 10 million inhabitants of a country). Think through the list of survey operations and specify where you foresee that great differences in complexity may arise.

CHAPTER 2

Basic Ideas in Estimation from Probability Samples

2.1. Introduction

This chapter introduces some basic notions in survey sampling, such as sampling design and sampling scheme, estimation by π-expanded sums, design variance, design effect, and design-based confidence intervals. Mastery of these concepts is essential for an understanding of the subsequent chapters. Basic designs, such as simple random sampling without replacement and Bernoulli sampling, are discussed.

An estimate's error is the deviation of the estimate from the unknown value of the parameter that we wish to estimate. In this chapter, we focus on the sampling error and ignore any errors caused by faulty measurement, nonresponse, or other nonsampling sources. The sampling error is caused by calculating the estimate from data for a subset of the population only. There is no sampling error when the estimate is based on data for the entire population, as in a census.

2.2. Population, Sample, and Sample Selection

Let us consider a population consisting of N elements labeled $k = 1, \ldots, N$,

$$\{u_1, \ldots, u_k, \ldots, u_N\}$$

For simplicity, we let the kth element be represented by its label k. Thus, we denote the finite population as

$$U = \{1, \ldots, k, \ldots, N\}$$

In this chapter, the population size N is treated as known at the outset. But for many populations encountered in practice, N is unknown, and methods of estimation for this case are developed in later chapters.

Let y denote a variable, and let y_k be the value of y for the kth population element. For instance, if U is a population of households and y is the variable "disposable income," then y_k quantifies the disposable income of the kth household. We assume that the values y_k, $k \in U$, are unknown at the outset. Suppose that an estimate is needed for the population total of y,

$$t = \sum_U y_k$$

or for the population mean of y,

$$\bar{y}_U = t/N = \sum_U y_k/N$$

In these expressions, $\sum_U y_k$ is abbreviated notation for $\sum_{k \in U} y_k$. More generally, if A is any set of population elements such that $A \subseteq U$, we write $\sum_A$ for $\sum_{k \in A}$.

In our example, $t = \sum_U y_k$ is the total disposable income of the households in the population U and $\bar{y}_U$ is the mean disposable income of the households in U. The unknown quantities t and $\bar{y}_U$ are simple examples of finite population parameters. The variable y is called the *study variable.* To estimate the parameters, we observe y_k for a subset of U, rather than all of U, which would usually prove too expensive or otherwise impractical. A subset of the population, called a *sample,* is first selected, and the value y_k is observed for each element k in that subset. Finally, this limited set of y_k values is used to calculate estimates of t and $\bar{y}_U$.

Our general notation for a sample is s. A sample can be any subset of the population U. However, most of the time we consider samples that are realized by a probabilistic (randomized) selection scheme. We assume in Chapters 2 and 3 that a frame is available that lists the population elements from 1 to N. The sample selection can then be carried out as direct element sampling, as discussed in Section 1.4. In practice, a wide variety of selection schemes are used. For example, in stratified sampling (see Section 3.7 for details), the population is first divided into nonoverlapping subsets (strata), a sample selection method is chosen for each stratum, and finally a sample is selected in each stratum according to the given method.

Ordinarily, sample selection is carried out by a series of randomized experiments. Collectively, the series of experiments will result in a selected sample s. In practice, computers perform the series of experiments with great speed. There are two basic types of sample selection scheme: draw sequential scheme and list sequential scheme.

A *draw-sequential scheme* is carried out by a series of randomized experiments, called draws, from the entire population or from a given subset of the population. Each draw results in an element selected for the sample.

EXAMPLE 2.2.1. One of the several ways to carry out *simple random sampling without replacement* is by means of the following draw sequential scheme. "Without replacement" means that each draw is carried out among elements that have not already been chosen.

1. Select with equal probability, $1/N$, a first element from the N elements.
2. Select with equal probability, $1/(N-1)$, a second element from the remaining $N-1$ elements.

⋮

n. Select with equal probability, $1/(N-n+1)$, an nth element from the $N-n+1$ elements that remain after the first $n-1$ selections.

A *list-sequential scheme* consists of a series of randomized experiments executed in the following way. Proceed down the frame list of elements, although not necessarily to the end of the list, and carry out one experiment for each element, which will result either in the selection or in the nonselection of the element in question.

EXAMPLE 2.2.2 (Bernoulli sampling). The elements appear in the frame in a given order, $k = 1, \ldots, N$. Let π be a fixed constant such that $0 < \pi < 1$. Let $\varepsilon_1, \varepsilon_2, \ldots, \varepsilon_N$ be N independent realizations of a Unif (0, 1) random variable, where Unif (0, 1) denotes the random variable with a uniform distribution on the (0, 1) interval. The selection, or non-selection, of the kth element is decided by the following rule: If $\varepsilon_k < \pi$, the element is selected, otherwise not, $k = 1, \ldots, N$. Clearly, the probability of selection is

$$\Pr(\varepsilon_k < \pi) = \pi$$

for each of the N elements, and, for $k \neq l$, the events "k is selected" and "l is selected" are independent. The number of selected elements is a binomially distributed random variable with parameters N and π.

Simple random sampling without replacement, although conceived originally as a series of draws, is nowadays usually carried out list sequentially with the aid of a computer, for example, by the algorithm given in Section 3.3.1.

EXAMPLE 2.2.3. An experiment was carried out as follows. Five independent Bernoulli samples were drawn, with $\pi = 0.3$, from a population of $N = 10$ elements. Once drawn, each sample was replaced into the population. The results are shown in the following table, where a "1" denotes that the element was selected, and a "0" denotes that the element was not selected.

	Sample				
Element	1	2	3	4	5
1	1	1	0	1	1
2	0	0	0	0	0
3	0	1	1	1	1
4	0	0	0	0	0
5	0	1	1	0	0
6	0	0	0	0	0
7	0	0	1	1	0
8	0	1	0	0	0
9	0	1	1	1	0
10	0	0	0	0	0
Number of selected elements	1	5	4	4	2

In this experiment, the realized sample sizes vary from 1 to 5. Each sample size is a realization of a Bin(10, 0.3) random variable, that is, a binomially distributed random variable with parameters 10 and 0.3, so the expected size of each sample is $10 \cdot 0.3 = 3$. The number of times that a given element is picked varies in this example between 0 and 4. This number is a realization of a Bin(5, 0.3) random variable. The expected number of times each element is selected over the five repeated samples is thus $5 \cdot 0.3 = 1.5$.

2.3. Sampling Design

Given a sample selection scheme, it is possible (although not always simple) to state the probability of selecting a specified sample s. It is convenient to have special notation for this probability. We shall use the notation $p(s)$.

In other words, we assume that there is a function $p(\cdot)$ such that $p(s)$ gives the probability of selecting s under the scheme in use. The function $p(\cdot)$ will be called the *sampling design*. It plays a central role because it determines the essential statistical properties (sampling distribution, expected value, variance) of random quantities calculated from a sample, such as the sample mean, the sample median, and the sample variance.

EXAMPLE 2.3.1. Consider the simple random sampling without replacement scheme in Example 2.2.1. Any given ordered sequence of n selected elements has the probability $(N - n)!/N!$. A specified set s with n elements can be obtained as the result of $n!$ different (but equiprobable) sequences. The sampling design is thus

$$p(s) = \begin{cases} 1\Big/\dbinom{N}{n} & \text{if } s \text{ has } n \text{ elements} \\ 0 & \text{otherwise} \end{cases} \tag{2.3.1}$$

We denote this design as *SI*. It is a discrete uniform probability distribution on the set of $\binom{N}{n}$ samples s of the fixed size n.

For a given sampling design $p(\cdot)$, we can regard any sample s as the outcome of a set-valued random variable S, whose probability distribution is specified by the function $p(\cdot)$. Let $\mathscr{S}$ be the set of all samples, s. (Thus $\mathscr{S}$ is the set of 2^N subsets of U, if we include the empty set, as well as U itself.) Then we have

$$\Pr(S = s) = p(s)$$

for any $s \in \mathscr{S}$. Because $p(s)$ is a probability distribution on $\mathscr{S}$, we have

$$\text{i. } p(s) \geq 0, \quad \text{all} \quad s \in \mathscr{S}$$

and

$$\text{ii. } \sum_{s \in \mathscr{S}} p(s) = 1$$

Note that many of the 2^N sets s contained in $\mathscr{S}$ may in fact have zero probability. The subset of $\mathscr{S}$ composed of those s for which $p(s)$ is strictly greater than zero constitutes the set of possible samples. They are the only ones that can be drawn given the specified design.

The *sample size*, to be denoted n_s, is the number of elements in s, that is, the cardinal of the set s. However, n_s is not necessarily the same for all possible samples. For the design *SI*, all possible samples do have the same size, namely, n, but this is not the case for Bernoulli sampling.

Example 2.3.2. In Bernoulli sampling (for the definition, see Example 2.2.2), the sampling design is obtained as follows. Let n_s be the size of s. The probability of selecting any element is π, and the experiments in the list sequential procedure are independent. Thus,

$$p(s) = \underbrace{\pi \cdots \pi}_{n_s \text{ times}} \underbrace{(1 - \pi) \cdots (1 - \pi)}_{N - n_s \text{ times}} = \pi^{n_s}(1 - \pi)^{N - n_s} \tag{2.3.2}$$

This design will be referred to as *BE*. The probability that the random sample size is exactly n_s is given by

$$\binom{N}{n_s} \pi^{n_s}(1 - \pi)^{N - n_s}$$

for $n_s = 0, 1, \ldots, N$. Thus, the sample size is binomially distributed, with the expected value $N\pi$ and the variance $N\pi(1 - \pi)$.

The sampling design $p(\cdot)$ determines the statistical properties of quantities calculated from the sample. However, $p(\cdot)$ is mainly a mathematical tool and not very practical in itself for selecting the sample.

For a given design $p(\cdot)$, sample selection could in theory (but only in theory unless N is very small) proceed according to the following algorithm, which focuses directly on the fact that $p(\cdot)$ is a specified probability distribution on $\mathscr{S}$.

1. Enumerate the members s of the set $\mathscr{S}_0$ of possible samples, that is, the set of samples s with $p(s) > 0$.
2. Attach the appropriate probability $p(s)$ to each sample $s \in \mathscr{S}_0$.
3. Select one sample $s \in \mathscr{S}_0$ by a random mechanism obeying the probability distribution $p(s)$, using a random number table or a computer-assisted randomized selection procedure.

This algorithm is almost never a practical possibility for sample selection. The simple reason is that in virtually all situations met in practice, the number of samples is too large. It is often "astronomically large." If, for example, $N =$ 1,000, and $\mathscr{S}_0$ is the set of all samples of the fixed size $n = 40$, then the number of samples is

$$\binom{1000}{40} \doteq 5.6 \times 10^{71}$$

If $N = 5{,}000$ and $n = 200$ (thus a sampling fraction, as in the previous case, of $200/5{,}000 = 4\%$), the number of samples increases to

$$\binom{5{,}000}{200} \doteq 1.4 \times 10^{363}$$

Clearly the enumeration of all these samples is a hopeless task, even for a computer.

Remark 2.3.1. It has become standard in modern literature to call the probability distribution $p(\cdot)$ the *sampling design.* Some statisticians use the similar term *sample design* with a different meaning. For example, Hansen, Hurwitz and Madow (1953, vol. II, page 7) say: "The sample design will consist of the sampling plan and the method of estimation." "Sampling plan" corresponds roughly to our concept sampling design. Used in this way, sample design is thus a broad concept that includes the choice of sampling design, the actual selection of a sample, as well as the estimation.

By way of terminology, the *design stage* of a survey is sometimes used to designate the period during which the sample selection procedure (and thereby the sampling design) is decided and the sample selected. By contrast, the *estimation stage* refers to the period when the data are already collected and the required estimates are calculated.

It is important to note that several different sample selection schemes may conform to one and the same sampling design $p(\cdot)$. If the sampling design $p(\cdot)$ is taken as a starting point, one has to specify a suitable sample selection scheme that implements the design. The scheme should be efficient in terms of computer execution, cost, and other practical aspects.

EXAMPLE 2.3.3. The design *SI* given in Example 2.3.1 can be implemented by the following draw-sequential scheme, an alternative to the scheme in Example 2.2.1. Select with equal probability $1/N$, a first element from the N elements in the population. Replace the element obtained. Repeat this step until n distinct elements are obtained. If v denotes the number of draws required, we have $v \geq n$ with probability one, since already drawn elements may be selected again. Here, v is a random variable with a fairly complex probability distribution. It can be shown that the selection scheme just stated conforms to the conditions of the design *SI*. Other possible executions of *SI* exist. An often used list sequential scheme is stated in Section 3.3.1.

The objective of a survey is ordinarily to estimate one or more population parameters. Two of the many important choices that must be made in a survey are as follows:

1. The choice of a sampling design and a sample selection scheme to implement the design.
2. The choice of a formula (an estimator) by which to calculate an estimate of a given parameter of interest.

The two choices are not independent of each other. For example, the choice of estimator will usually depend on the choice of sampling design.

A *strategy* is the combination of a sampling design and an estimator. For a given parameter, the general aim is to find the best possible strategy, that is, one that estimates the parameter as accurately as possible.

2.4. Inclusion Probabilities

An interesting feature of a finite population of N labeled elements is that the elements can be given different probabilities of inclusion in the sample. The sampling statistician often takes advantage of the identifiability of the elements by deliberately attaching different inclusion probabilities to the various elements. This is one way to obtain more accurate estimates.

Suppose that a certain sampling design has been fixed. That is, $p(s)$, the probability of selecting s, has a given mathematical form. The inclusion of a given element k in a sample is a random event indicated by the random variable I_k, defined as

$$I_k = \begin{cases} 1 & \text{if } k \in S \\ 0 & \text{if not} \end{cases} \tag{2.4.1}$$

Note that $I_k = I_k(S)$ is a function of the random variable S. We call I_k the *sample membership indicator* of element k.

The probability that element k will be included in a sample, denoted π_k, is obtained from the given design $p(\cdot)$ as follows:

$$\pi_k = \Pr(k \in S) = \Pr(I_k = 1) = \sum_{s \ni k} p(s) \tag{2.4.2}$$

Here, $s \ni k$ denotes that the sum is over those samples s that contain the given k. The probability that both of the elements k and l will be included is denoted π_{kl} and is obtained from the given $p(\cdot)$ as follows:

$$\pi_{kl} = \Pr(k\&l \subset S) - \Pr(I_k I_l = 1) = \sum_{s \ni k\&l} p(s) \tag{2.4.3}$$

We have $\pi_{kl} = \pi_{lk}$ for all k, l.

Note that (2.4.3) applies also when $k = l$, for in that case

$$\pi_{kk} = \Pr(I_k^2 = 1) = \Pr(I_k = 1) = \pi_k$$

That is, in the following, π_{kk} should be interpreted as identical to π_k, for $k = 1, \ldots, N$.

Remark 2.4.1. The writing "$k \in S$" in, for example, equation (2.4.2) should be interpreted as the random event "$S \ni k$," which is the event "a sample containing k is realized".

With a given design $p(\cdot)$ are associated the N quantities

$$\pi_1, \ldots, \pi_k, \ldots, \pi_N$$

They constitute the set of first-order inclusion probabilities. Moreover, with $p(\cdot)$ are associated the $N(N-1)/2$ quantities

$$\pi_{12}, \pi_{13}, \ldots, \pi_{kl}, \ldots, \pi_{N-1,N}$$

which are called the set of second-order inclusion probabilities. Inclusion probabilities of higher order can be defined and calculated for a given design $p(\cdot)$. However, they play a less important role and will not be discussed in this book.

Example 2.4.1. Consider the *SI* design defined in Example 2.3.1. Let us calculate the inclusion probabilities of first and second orders. There are exactly $\binom{N-1}{n-1}$ samples s that include the element k, and exactly $\binom{N-2}{n-2}$ samples s that include the elements k and l ($k \neq l$). Since all samples of size n have the same probability, $1\Big/\binom{N}{n}$, we obtain from (2.4.2) and (2.4.3)

$$\pi_k = \sum_{s \ni k} p(s) = \binom{N-1}{n-1}\Big/\binom{N}{n} = n/N; \qquad k = 1, \ldots, N \tag{2.4.4}$$

and

$$\pi_{kl} = \sum_{s \ni k\&l} p(s) = \binom{N-2}{n-2} \bigg/ \binom{N}{n} = \frac{n(n-1)}{N(N-1)}; \quad k \neq l = 1, \ldots, N \quad (2.4.5)$$

EXAMPLE 2.4.2. The *BE* sampling design (see Example 2.3.2) can be characterized as a design in which the indicators I_k are independently and identically distributed, each I_k obeying the Bernoulli distribution with parameter π. All N elements have the same first-order inclusion probability,

$$\pi_k = \pi; \qquad k = 1, \ldots, N$$

moreover, since the I_k are independent,

$$\pi_{kl} = E(I_k I_l) = E(I_k)E(I_l) = \pi^2$$

for any $k \neq l$.

A sampling design is often chosen to yield certain desired first- and second-order inclusion probabilities. Although $p(\cdot)$ may be complicated, we can attain one of the primary goals, namely, to determine expected value and variance of certain quantities calculated from the sample, from a knowledge of the π_k and the π_{kl} alone.

Unless otherwise stated, we assume that the sampling design is such that all first-order inclusion probabilities π_k are strictly positive, that is,

$$\pi_k > 0, \quad \text{all} \quad k \in U \quad (2.4.6)$$

This requirement ensures that every element has a chance to appear in the sample. In order that a sampling design be called a *probability sampling design*, the π_k must satisfy (2.4.6) (cf. condition 3 in Section 1.3). A sample s realized by such a design is called a *probability sample*.

Remark 2.4.2. In practice, one nevertheless sometimes uses designs where the requirement $\pi_k > 0$ for all $k \in U$ is not met. One example is cut-off sampling. For instance, in a population of business enterprises, the smallest firms are sometimes cut off, that is, given a zero inclusion probability, because their contribution to the whole is deemed trivial, and the cost of constructing and maintaining a frame that lists the numerous small enterprises may be too high. Cut-off sampling leads to some bias in the estimates and must be used only with great caution. For further discussion of cut-off sampling and other nonprobability sampling designs, the reader is referred to Section 14.4.

Remark 2.4.3. In direct element sampling, all π_k, $k = 1, \ldots, N$, are ordinarily known prior to sampling. However, in more complex design procedures (notably multistage and multiphase sampling, see Chapters 4 and 7), the sampling is often carried out in such a way that π_k cannot be calculated at the outset for all k. In multistage sampling, for example, the inclusion probabilities are known a priori for the sampling units in each stage, which does not imply that they are known a priori for all elements $k \in U$.

Another important property of a design occurs when the condition

$$\pi_{kl} > 0 \quad \text{for all} \quad k \neq l \in U \tag{2.4.7}$$

holds. A sampling design is said to be *measurable* if (2.4.6) and (2.4.7) are satisfied. A measurable design allows the calculations of valid variance estimates and valid confidence intervals based on the observed survey data. The notion of measurability is further discussed in Section 14.3.

The N indicators can be summarized in vector form as

$$\mathbf{I} = (I_1, \ldots, I_k, \ldots, I_N)'$$

The event $S = s$ is clearly equivalent to the event $\mathbf{I} = \mathbf{i}_s$, where

$$\mathbf{i}_s = (i_{1s}, \ldots, i_{Ns})'$$

with $i_{ks} = 1$ if $k \in s$ and $i_{ks} = 0$ if not. Then the probability distribution $p(\cdot)$ introduced in Section 2.3 can be written in terms of the random vector $\mathbf{I}$ as

$$\Pr(\mathbf{I} = \mathbf{i}_s) = \Pr(S = s) = p(s) \quad \text{for} \quad s \in \mathscr{S}$$

Remark 2.4.4. Unless otherwise stated, the designs considered in this book are such that the probability

$$\Pr(S = s) = \Pr(\mathbf{I} = \mathbf{i}_s)$$

does not depend explicitly on the values of the study variables, y, z, and so on. Such designs are called noninformative. The probability $\Pr(S = s)$ may depend on auxiliary variable values, that is, on other variable values known beforehand for the population elements.

An example of a design that is informative is the following. Two elements are drawn sequentially, the first by giving equal selection probability to every element, the second by assigning to the various elements selection probabilities that depend on the value y_k of the element, k_1, obtained in the first draw. In practical survey work, the noninformative designs dominate.

2.5. The Notion of a Statistic

The general theory of statistics uses the term "statistic" to refer to a real-valued function whose value may vary with the different outcomes of a certain experiment. Moreover, an essential requirement for a statistic is that it must be computable for any given outcome of the experiment. The same general idea of a statistic will serve the purposes of this book. We want to examine how a statistic varies from one realization s to another of the random set S. In other words, it is sample-to-sample variation that is of interest.

If $Q(S)$ is a real-valued function of the random set S, we call any such function a statistic, provided that the value $Q(s)$ can be calculated once s, an outcome of S, has been specified and the data for the elements of s have been collected.

A simple but important statistic is $I_k(S)$, the random variable defined by (2.4.1). It indicates membership or nonmembership in the sample s of the kth element.

The sample size (that is, the cardinal of S), defined as

$$n_S = \sum_U I_k(S)$$

is another simple example of a statistic.

Other examples of statistics are $\sum_S y_k$, the sample total of the variable y, and $\sum_S y_k / \sum_S z_k$, the ratio of the sample totals of two variables y and z. By contrast, $\sum_S y_k / \sum_U z_k$ is not a statistic, unless the population total of z happens to be known from other sources.

When a sample is drawn in practice, exactly one realization s of the random set S occurs. Once s has been realized, we assume that it is possible to observe and measure certain variables of interest, for example, y and z, for each element k in s. Thus, in the case of the statistic $Q(S) = \sum_S y_k / \sum_S z_k$, for example, we can, after measurement, calculate the realized value of the statistic, namely, $Q(s) = \sum_s y_k / \sum_s z_k$.

Note in this example that y and z are variables in the sense of taking possibly different values y_k and z_k for the various elements k. However, y and z are not treated as *random* variables. The random nature of a statistic $Q(S)$ stems solely from the fact that the set S is random.

Remark 2.5.1. If we have two variables of study, y and z, the example $Q(S) = \sum_S y_k / \sum_S z_k$ shows that a more telling (but too cumbersome) notation for a statistic would be

$$Q(S) = Q[(k, y_k, z_k): k \in S]$$

Expressed in words, $Q(S)$ is a function of S, $\mathbf{y} = (y_1, \ldots, y_N)$ and $\mathbf{z} = (z_1, \ldots, z_N)$ that depends on $\mathbf{y}$ and $\mathbf{z}$ only through those values y_k, z_k for which $k \in S$. The realized value of $Q(S)$ is computed from the set of pairs (y_k, z_k) associated with the elements k in the realized sample s. For simplicity, we write simply $Q(S)$ for the statistic and $Q(s)$ for the realized value.

Because a statistic $Q(S)$ is a random variable in the sense just described, it has various statistical properties, such as an expected value and a variance. These concepts are detailed in the following definition.

Definition 2.5.1. The expectation and the variance of a statistic $Q = Q(S)$ are defined, respectively, by

$$E(Q) = \sum_{s \in \mathscr{S}} p(s)Q(s)$$

$$V(Q) = E\{[Q - E(Q)]^2\}$$

$$= \sum_{s \in \mathscr{S}} p(s)[Q(s) - E(Q)]^2$$

The covariance between two statistics $Q_1 = Q_1(S)$ and $Q_2 = Q_2(S)$ is defined by

$$C(Q_1, Q_2) = E\{[Q_1 - E(Q_1)][Q_2 - E(Q_2)]\}$$
$$= \sum_{s \in \mathscr{S}} p(s)[Q_1(s) - E(Q_1)][Q_2(s) - E(Q_2)]$$

Note that these definitions refer to the variation over all possible samples that can be obtained under the given sampling design, $p(s)$. To emphasize this, the terms *design expectation, design variance*, and *design covariance* are often used in the literature. Here, we generally suppress the word "design" in these and similar terms, as there is no risk of misinterpretation. The design expectation $E(Q)$, for example, is the weighted average of the values $Q(s)$, the weight of $Q(s)$ being the probability $p(s)$ with which s is chosen.

When estimators are examined and compared, it is often of interest to determine the value of an expectation $E(Q)$, a variance $V(Q)$, or a covariance $C(Q_1, Q_2)$.

For simple statistics $Q(S)$, the expected value and the variance can often be evaluated easily as closed form analytic expressions. This is true in particular for the *linear statistics* that we examine in detail later. Whether $Q(S)$ is simple or not, there exists a "long run frequency interpretation" of the expected value $E(Q)$ and the variance $V(Q)$. Suppose we let a computer draw 10,000 independent samples, each of size n, with the *SI* design, from a population of size N. Once drawn, a sample is replaced into the population, the next sample is drawn, and so on. For each sample, the value of $Q(S)$ is computed.

At the end of the run, we can let the computer calculate the average and the variance of the 10,000 obtained values of $Q(S)$. The values thus obtained will closely approximate the expectation $E[Q(S)]$ and the variance $V[Q(S)]$, respectively. This method of approximating the value of quantities that may be hard to calculate by analytic means is known as a Monte Carlo simulation.

In survey sampling, we have the following *frequency interpretation* of the expected value and the variance of a statistic: In a long run of repeated samples drawn from the finite population with a given sampling design $p(\cdot)$, the average of the values $Q(s)$ and the variance of the values $Q(s)$ will closely approximate their theoretical counterparts.

The number of realized samples is the crucial factor in determining the accuracy of these approximations. For instance, if samples of size $n = 6$ are drawn from a population of size $N = 20$, the number of possible samples is $\binom{20}{6} = 38{,}760$, and the 10,000 samples realized in the simulation represent at most $10{,}000/38{,}760 = 26\%$ of the possible samples. By contrast, if samples of size $n = 200$ are drawn from a population of $N = 5{,}000$, the number of possible samples is of the order of 10^{363}. Close approximations to the true expected value and the true variance would, however, generally be obtained in this case also with 10,000 repeated samples, although they now represent a vanishingly small fraction of all the possible samples.

The massive calculations required in a Monte Carlo simulation are not part of an actual survey. In a survey, there is ordinarily one and only one sample from which conclusions are drawn about the finite population. Monte Carlo simulation is, however, a highly useful tool in the evaluation of the statistical properties of complicated estimators. More detail on this subject is given in Section 7.9.1.

2.6. The Sample Membership Indicators

The estimators that we are interested in examining can be expressed as functions of the sample membership indicators defined by (2.4.1). It is therefore important to describe the basic properties of the statistics $I_k = I_k(S)$, for $k = 1, \ldots, N$, as in the following result.

Result 2.6.1. *For an arbitrary sampling design $p(s)$, and for $k, l = 1, \ldots, N$,*

$$E(I_k) = \pi_k$$

$$V(I_k) = \pi_k(1 - \pi_k)$$

$$C(I_k, I_l) = \pi_{kl} - \pi_k \pi_l$$

PROOF. Note that $I_k = I_k(S)$ is a Bernoulli random variable. Thus, $E(I_k) = \Pr(I_k = 1) = \pi_k$, using (2.4.2). Because $E(I_k^2) = E(I_k) = \pi_k$, it follows that $V(I_k) = E(I_k^2) - \pi_k^2 = \pi_k(1 - \pi_k)$. Moreover, $I_k I_l = 1$, if and only if both k and l are members of s. Thus $E(I_k I_l) = \Pr(I_k I_l = 1) = \pi_{kl}$ by equation (2.4.3), so that

$$C(I_k, I_l) = E(I_k I_l) - E(I_k)E(I_l) = \pi_{kl} - \pi_k \pi_l \qquad \square$$

Note that for $k = l$, the covariance equals the variance, that is,

$$C(I_k, I_k) = V(I_k)$$

Depending on the design, the covariances $C(I_k, I_l)$ can be zero, positive, or negative.

Remark 2.6.1. It will save space to have a special symbol for the variances and covariances of I_k. We use the symbol Δ. That is, for any $k, l \in U$, we set

$$C(I_k, I_l) = \pi_{kl} - \pi_k \pi_l = \Delta_{kl} \tag{2.6.1}$$

When $k = l$, this implies (because $\pi_{kk} = \pi_k$) that

$$C(I_k, I_k) = V(I_k) = \pi_k(1 - \pi_k) = \Delta_{kk} \tag{2.6.2}$$

Double sums over a set of elements will now be needed. Let a_{kl} be any quantity associated with k and l, where $k, l \in U$. Let A be any set of elements,

$A \subseteq U$. For example, A may be the whole population U, or A may be a sample s. The double sum notation

$$\sum\sum_A a_{kl}$$

will be our shorthand for

$$\sum_{k \in A} \sum_{l \in A} a_{kl} = \sum_{k \in A} a_{kk} + \sum_{k \in A} \sum_{\substack{l \in A \\ k \neq l}} a_{kl}$$

The double sum on the right-hand side of this expression we denote more concisely as

$$\sum\sum_{\substack{A \\ k \neq l}} a_{kl}$$

Thus we have

$$\sum\sum_A a_{kl} = \sum_A a_{kk} + \sum\sum_{\substack{A \\ k \neq l}} a_{kl}$$

Another simple statistic is the sample size n_S. It can be expressed in terms of the indicators I_k as

$$n_S = \sum_U I_k$$

The first two moments of the statistic n_S follow easily from Result 2.6.1. We have

$$E(n_S) = \sum_U \pi_k \tag{2.6.3}$$

and

$$\begin{aligned} V(n_S) &= \sum_U \pi_k(1 - \pi_k) + \sum\sum_{\substack{U \\ k \neq l}} (\pi_{kl} - \pi_k \pi_l) \\ &= \sum_U \pi_k - \left(\sum_U \pi_k\right)^2 + \sum\sum_{\substack{U \\ k \neq l}} \pi_{kl} \end{aligned} \tag{2.6.4}$$

EXAMPLE 2.6.1. We return to the design BE considered in Examples 2.2.2, 2.3.2, and 2.4.2. The sample size n_S is a binomially distributed random variable with parameters N and π. It follows that

$$E_{BE}(n_S) = N\pi$$

and

$$V_{BE}(n_S) = N\pi(1 - \pi)$$

The results can be obtained alternatively from equations (2.6.3) and (2.6.4). For example, from (2.6.4), using $\pi_k = \pi$ for all k and $\pi_{kl} = \pi^2$ for all $k \neq l$,

$$V_{BE}(n_S) = N\pi - (N\pi)^2 + N(N - 1)\pi^2 = N\pi(1 - \pi)$$

Another example of random sample size occurs in single-stage cluster sampling, as described in Chapter 4. A cluster is a set of elements, for example, the households in a city block or the students in a class. If clusters are selected and if the sample consists of all elements in the selected clusters, then the sample size, that is, the number of observed elements, will be variable if the clusters are of unequal sizes.

Practitioners avoid designs in which the sample size varies extensively. One reason is that variable sample size will cause an increase in variance for certain types of estimators. More importantly, survey statisticians dislike being in a situation where the number of observations is highly unpredictable when the survey is planned.

A fixed (sample) size design is such that whenever $p(s) > 0$, the sample s will contain a fixed number of elements say n. That is, a sample is realizable under a fixed size design only if its size is exactly n. But all samples of size n need not be realizable to have a fixed size design.

In the case of a fixed size design the inclusion probabilities obey some simple relations, which are stated in the following result.

Result 2.6.2. *If the design $p(s)$ has the fixed size n, then*

$$\sum_U \pi_k = n$$

$$\sum\sum_{\substack{U \\ k \neq l}} \pi_{kl} = n(n-1)$$

$$\sum_{\substack{l \in U \\ l \neq k}} \pi_{kl} = (n-1)\pi_k$$

Proof. If $p(s)$ is of the fixed size n, then $n_S = n$ with probability one. Thus $E(n_S) = n$ and $V(n_S) = 0$. Using (2.6.3) and (2.6.4), we obtain the first two results of the theorem. The third result follows from the derivation

$$\sum_{\substack{l \in U \\ l \neq k}} \pi_{kl} = \sum_{\substack{l \in U \\ l \neq k}} E(I_k I_l) = E[I_k(\textstyle\sum_U I_l - I_k)]$$

$$= nE(I_k) - E(I_k^2) = (n-1)\pi_k \qquad \square$$

The *SI* design is an example of a fixed size design. That the three parts of Result 2.6.2 hold for this design in particular is easily checked by use of the inclusion probabilities π_k and π_{kl} calculated in Example 2.4.1.

2.7. Estimators and Their Basic Statistical Properties

Most of the statistics that we examine in this book are different kinds of estimators. An estimator is a statistic thought to produce values that, for most samples, lie near the unknown population quantity that one wishes to estimate. Such quantities are called parameters. The general notation for a parameter will be θ.

If there is only one study variable, y, we can think of θ as a function of

$y_1, \ldots, y_N$, the N values of y. Thus,

$$\theta = \theta(y_1, \ldots, y_N)$$

Examples include the population total,

$$\theta = t = \textstyle\sum_U y_k$$

the population mean,

$$\theta = \bar{y}_U = \textstyle\sum_U y_k/N$$

and the population variance,

$$\begin{aligned}\theta = S_{yU}^2 &= \textstyle\sum_U (y_k - \bar{y}_U)^2/(N-1) \\ &= \textstyle\sum_U y_k^2/(N-1) - (\sum_U y_k)^2/N(N-1)\end{aligned}$$

A parameter can be a function of the values of two or more variables of study, as in the case of the ratio of the population totals of y and z,

$$\theta = \frac{\sum_U y_k}{\sum_U z_k} \tag{2.7.1}$$

We denote an estimator of θ by

$$\hat{\theta} = \hat{\theta}(S)$$

If s is a realization of the random set S, we assume it is possible to calculate $\hat{\theta}$ from the study variable values $y_k, z_k, \ldots$ associated with the elements $k \in s$. For example, under the *SI* design,

$$\hat{\theta} = N\frac{\sum_s y_k}{n}$$

is an often-used estimator of the parameter $\theta = \sum_U y_k$ and

$$\hat{\theta} = \frac{\sum_s y_k}{\sum_s z_k}$$

is an often-used estimator of the parameter θ given by (2.7.1).

It is of considerable interest to describe the sample-to-sample variations of a proposed estimator $\hat{\theta}$. An estimator that varies little around the unknown value θ is on intuitive grounds "better" than one that varies a great deal.

By the sampling distribution of a estimator $\hat{\theta}$ we mean a specification of all the possible values of $\hat{\theta}$, together with the probability that $\hat{\theta}$ attains each of these values under the sampling design in use $p(s)$. That is, the exact statement of the sampling distribution of $\hat{\theta}$ requires, for each possible value c of $\hat{\theta}$, a specification of the probability

$$\Pr(\hat{\theta} = c) = \sum_{s \in \mathscr{S}_c} p(s) \tag{2.7.2}$$

where $\mathscr{S}_c$ is the set of samples s for which $\hat{\theta} = c$.

For a given population of y-values, $y_1, \ldots, y_N$, a given design, and a given estimator, it is possible in theory to produce the exact sampling distribution of a statistic $\hat{\theta}$. But, computationally, it would be a formidable task for most finite populations of practical interest. Ordinarily, both the number of possible samples and the number of possible values of $\hat{\theta}$ are extremely large. However, summary measures that describe important aspects of the sampling distribution of an estimator $\hat{\theta}$ are needed, for example, when comparisons are made with competing estimators. These summary measures are usually unknown, theoretical quantities. The following summary measures are derived from Definition 2.5.1. The *expectation* of $\hat{\theta}$ is given by

$$E(\hat{\theta}) = \sum_{s \in \mathscr{S}} p(s)\hat{\theta}(s)$$

It is a weighted average of the possible values $\hat{\theta}(s)$ of $\hat{\theta}$, with the probabilities $p(s)$ as weights. The *variance* of $\hat{\theta}$ is given by

$$V(\hat{\theta}) = \sum_{s \in \mathscr{S}} p(s)\{\hat{\theta}(s) - E(\hat{\theta})\}^2$$

Two important measures of the quality of an estimator $\hat{\theta}$ are the bias and the mean square error. The *bias* of $\hat{\theta}$ is defined as

$$B(\hat{\theta}) = E(\hat{\theta}) - \theta$$

An estimator $\hat{\theta}$ is said to be *unbiased* for θ if

$$B(\hat{\theta}) = 0 \quad \text{for all} \quad \mathbf{y} = (y_1, \ldots, y_N)' \in \mathscr{R}^N$$

The *mean square error* of $\hat{\theta}$ is defined as

$$\text{MSE}(\hat{\theta}) = E[\hat{\theta} - \theta]^2 = \sum_{s \in \mathscr{S}} p(s)[\hat{\theta}(s) - \theta]^2$$

An easily verified result is that

$$\text{MSE}(\hat{\theta}) = V(\hat{\theta}) + [B(\hat{\theta})]^2 \tag{2.7.3}$$

If $\hat{\theta}$ is unbiased for θ, it follows from equation (2.7.3) that $\text{MSE}(\hat{\theta}) = V(\hat{\theta})$.

Remark 2.7.1. Note the distinction between an estimate and an estimator. By the *estimate* produced by the *estimator* $\hat{\theta} = \hat{\theta}(S)$ is meant the number $\hat{\theta}(s)$ that can be calculated after a specific outcome s of the random set S has been observed and the study variable values y_k, z_k, ... have been recorded for the elements $k \in s$. For example, for an *SI* sample of n elements, the random variable

$$\hat{\theta}(S) = N\frac{\sum_S y_k}{n} = N\frac{\sum_U I_k y_k}{n}$$

is an estimator of $\theta = \sum_U y_k$; the estimate obtained for a particular outcome s is the number

$$\hat{\theta}(s) = N\frac{\sum_s y_k}{n}$$

In the following, we ignore the typographic distinction between S, the random set, and s, a realization of S. For simplicity, we use the lower case character to designate both the random set and its realization. There is little risk of misunderstanding.

An estimator is unbiased if its weighted average (over all possible samples using the probabilities $p(s)$ as weights) is equal to the unknown parameter value. The most important estimators in survey sampling are unbiased or approximately unbiased. It is characteristic of an approximately unbiased estimator that the bias is unimportant in large samples. For most of the approximately unbiased estimators that we consider, the bias is actually very small, even for modest sample sizes.

Remark 2.7.2. The statement that an estimator $\hat{\theta}$ is unbiased is a statement of average performance, namely, over all possible samples. The probability-weighted average of the deviations $\hat{\theta} - \theta$ is nil. However, to say that an estimate is biased is strictly speaking incorrect. An *estimate* is a constant value obtained for a particular sample realization. This value can be off the mark, in the sense of deviating from the unknown parameter value θ. Because an estimate is a number, it has no variation and no bias. The term *biased estimate* is nevertheless used occasionally. The only way that the term makes sense is if it is interpreted as "an estimate calculated from an estimator that is biased."

Although unbiasedness or approximate unbiasedness are desirable properties, it is clear that these properties say nothing about another important aspect of the sampling distribution, namely, how widely dispersed the various values of the estimator are. The variance is a measure of this dispersion.

When choosing between several possible estimators $\hat{\theta}_1, \hat{\theta}_2, \ldots$ for one and the same parameter θ, the statistician will normally want to single out one for which the sampling distribution is narrowly concentrated around the unknown value. θ. This suggests using the criterion of "small mean square error" to select an estimator, because, if $\text{MSE}(\hat{\theta}) = V(\hat{\theta}) + [B(\hat{\theta})]^2$ is small, there is strong reason to believe that the sample actually drawn in a survey will have produced an estimate near the true value. However, even if the sampling distribution is tightly concentrated around θ there is always a small possibility that our particular sample was "bad," so that the estimate falls in one of the tails of the distribution, rather far removed from θ. The statistician must live with this possibility.

The survey statistician should avoid estimators that are considerably biased, because valid confidence intervals cannot be obtained if the bias is substantial (for a further explanation, see Section 5.2). Therefore, typically the statistician will seek among estimators that are at least approximately unbiased and choose one that has a small variance.

Remark 2.7.3. The square root of the variance $[V(\hat{\theta})]^{1/2}$ is called the *standard error* of the estimator $\hat{\theta}$. The ratio of the standard error of the estimator to the expected value, $CV(\hat{\theta}) = [V(\hat{\theta})]^{1/2}/E(\hat{\theta})$, is called the *relative standard error* or the *coefficient of variation* of the estimator.

As part of the estimation phase, an estimator of the variance is usually computed from the survey data. Denote this estimate $\hat{V}(\hat{\theta})$. If the variable of interest is measured in dollars, $\hat{V}(\hat{\theta})$ has the dimension of (dollars)^2. In many applications, it is hard to relate to the often very large number $\hat{V}(\hat{\theta})$. The more telling dimensionless quantity

$$cve(\hat{\theta}) = [\hat{V}(\hat{\theta})]^{1/2}/\hat{\theta} \tag{2.7.4}$$

is therefore often computed and expressed as a percentage. It is a convenient and widely used indicator of the precision attained in the survey, when the unbiased or nearly unbiased estimator $\hat{\theta}$ is used for the parameter θ. In fact, survey statisticians often call the quantity *cve* the coefficient of variation, but more accurately one would describe it is an estimate (although not unbiased) of the "theoretical" coefficient of variation $[V(\hat{\theta})]^{1/2}/E(\hat{\theta})$, which is a complex finite population characteristic. Thus, a statistician may express the opinion that a computed *cve* value of 2% is good, considering the constraints of the survey, whereas a *cve* of 9% may have been considered unacceptable. Many numerical examples and exercises throughout the book require a calculation of the *cve* according to the definition (2.7.4).

2.8. The π Estimator and Its Properties

Let us consider the estimation of the population total $t = \sum_U y_k$, and let us examine the estimator

$$\hat{t}_\pi = \sum_s \frac{y_k}{\pi_k} \tag{2.8.1}$$

This estimator can be expressed as a linear function of the indicators I_k,

$$\hat{t}_\pi = \sum_U I_k \frac{y_k}{\pi_k} \tag{2.8.2}$$

Because $E(I_k) = \pi_k$ and $\pi_k > 0$ for all $k \in U$, it follows immediately that $\hat{t}_\pi$ is unbiased for $t = \sum_U y_k$. The quantity y_k/π_k may appropriately be called the "π-expanded y-value for the kth element." The estimator (2.8.1) is simply the sample sum of the π-expanded y-values.

The expansion of a variable value through division by the inclusion probability is an operation that recurs so often in this book that a special symbol is needed. We chose the symbol ˇ. That is, $\check{y}_k = y_k/\pi_k$, $\check{x}_k = x_k/\pi_k$, and so on.

The estimator shown in (2.8.1) or (2.8.2) then takes the form

$$\hat{t}_\pi = \sum_s \check{y}_k = \sum_U \check{y}_k I_k \tag{2.8.3}$$

Here, the random element is expressed by s or, equivalently, by the indicators $I_1, \ldots, I_k, \ldots, I_N$. The π-expanded values $\check{y}_k$ are fixed constants.

The estimator given in (2.8.1) and the equivalent forms (2.8.2) and (2.8.3) will be referred to as the *π estimator* of the total $t = \sum_U y_k$.

The π expansion has the effect of increasing the importance of the elements in the sample. Because the sample contains fewer elements than the population, an expansion is required to reach the level of the whole population. The kth element, when present in the sample, will, as it were, represent $1/\pi_k$ population elements. The formula (2.8.1) and the equivalent forms (2.8.2) and (2.8.3) embody an extremely important principle, namely, the use of π-expanded sample values to obtain an unbiased estimator of a population total when sampling is done with arbitrary positive inclusion probabilities. Many uses of this principle are found in this book.

Horvitz and Thompson (1952) used the principle of π expansion to estimate the total $t = \sum_U y_k$, and formula (2.8.1) is often called the Horvitz-Thompson estimator. A similar principle had been used earlier, in particular for unequal probability sampling with replacement, by Hansen and Hurwitz (1943) (see Section 2.9). We use the simpler term, π estimator, for an expression of the form shown in equation (2.8.1).

The π estimator is linear in I_k, which simplifies the derivation of the variance, as shown by the proof of the important Result 2.8.1 below. In Result 2.8.1, and throughout, recall that

$$\sum\sum_U a_{kl}$$

is to be understood as

$$\sum_{k \in U} \sum_{l \in U} a_{kl} = \sum_{k \in U} a_{kk} + \sum_{k \in U} \sum_{\substack{l \in U \\ k \neq l}} a_{kl} = \sum_U a_{kk} + \sum\sum_{\substack{U \\ k \neq l}} a_{kl}$$

Also, for all $k, l \in U$, define the expanded Δ value

$$\check{\Delta}_{kl} = \Delta_{kl}/\pi_{kl}$$

where Δ_{kl}, by the definition given in (2.6.1), is the covariance between I_k and I_l. It follows that

$$\check{\Delta}_{kl} = 1 - (\pi_k \pi_l / \pi_{kl}) \quad \text{for } k \neq l; \qquad \check{\Delta}_{kk} = 1 - \pi_k \tag{2.8.4}$$

Result 2.8.1. *The π estimator*

$$\hat{t}_\pi = \sum_s \check{y}_k$$

is unbiased for $t = \sum_U y_k$ with the variance

$$V(\hat{t}_\pi) = \sum\sum_U \Delta_{kl} \check{y}_k \check{y}_l \tag{2.8.5}$$

where Δ_{kl} is defined by (2.6.1). Provided that $\pi_{kl} > 0$ for all $k, l \in U$, an unbiased estimator of $V(\hat{t}_\pi)$ is given by

$$\hat{V}(\hat{t}_\pi) = \sum\sum_s \check{\Delta}_{kl} \check{y}_k \check{y}_l \tag{2.8.6}$$

where $\check{\Delta}_{kl} = \Delta_{kl}/\pi_{kl}$.

PROOF. The unbiasedness has already been established. To find the variance, use

$$\hat{t}_\pi = \sum_U \check{y}_k I_k$$

where the $\check{y}_k$ are constants and the I_k are the indicator random variables defined by (2.4.1). The standard formula for the variance of a linear combination of random variables gives

$$V(\hat{t}_\pi) = \sum_U V(I_k)\check{y}_k^2 + \sum\sum_{\substack{U\\k \neq l}} C(I_k, I_l)\check{y}_k \check{y}_l$$

Now, by Remark 2.6.1, $C(I_k, I_l) = \Delta_{kl}$ and $V(I_k) = \Delta_{kk}$, so

$$\begin{aligned} V(\hat{t}_\pi) &= \sum_U \Delta_{kk}\check{y}_k^2 + \sum\sum_{\substack{U\\k \neq l}} \Delta_{kl}\check{y}_k \check{y}_l \\ &= \sum\sum_U \Delta_{kl}\check{y}_k \check{y}_l \qquad (2.8.7) \end{aligned}$$

It remains to verify the unbiasedness of the variance estimator. Using the indicators I_k, we can write equation (2.8.6) as

$$\hat{V}(\hat{t}_\pi) = \sum\sum_U I_k I_l \check{\Delta}_{kl}\check{y}_k \check{y}_l$$

provided that $\pi_{kl} > 0$ for all k and $l \in U$ (and not only for k and l in some subset of U). Now, the desired result

$$E\{\hat{V}(\hat{t}_\pi)\} = V(\hat{t}_\pi)$$

follows easily from the fact that

$$E(I_k I_l \check{\Delta}_{kl}) = \pi_{kl}\check{\Delta}_{kl} = \Delta_{kl}$$ □

The variance (2.8.5) can alternatively be expressed directly in terms of the original (non-expanded) values y_k as

$$\begin{aligned} V(\hat{t}_\pi) &= \sum\sum_U \left(\frac{\pi_{kl}}{\pi_k \pi_l} - 1\right) y_k y_l \\ &= \sum\sum_U \frac{\pi_{kl}}{\pi_k \pi_l} y_k y_l - \left(\sum_U y_k\right)^2 \qquad (2.8.8) \end{aligned}$$

The variance estimator (2.8.6) can be written in terms of the original y-values as

$$\hat{V}(\hat{t}_\pi) = \sum\sum_s \frac{1}{\pi_{kl}}\left(\frac{\pi_{kl}}{\pi_k \pi_l} - 1\right) y_k y_l \qquad (2.8.9)$$

Remark 2.8.1. Estimators that involve π-expanded sample sums will be used frequently in this book. In discussing their variances and variance estimators, we repeatedly obtain expressions like (2.8.5) and (2.8.6). The reader should get used to the structure of these double-sum expressions and note that they incorporate one part due to the variances of the random variables I_k and one part due to the covariances between all pairs of variables I_k and I_l. The term due to the covariances $C(I_k, I_l) = \Delta_{kl}$ ($k \neq l$) plays an important role in the variance of $\hat{t}_\pi$. It is usually far from negligible. In general, the I_k are neither independent nor identically distributed. If all y_k values are positive, then negative covariances Δ_{kl} work in the direction of a smaller variance. This is the case for the *SI* design with the sampling fraction $f = n/N$, and we have $\Delta_{kk} = f(1-f)$ for all k and $\Delta_{kl} = -f(1-f)/(N-1)$ for all $k \neq l$. For the *BE* design the I_k are indeed independent and identically distributed, so $\Delta_{kk} = \pi(1-\pi)$ for all k and $\Delta_{kl} = 0$ for all $k \neq l$.

Remark 2.8.2. In this book, a number of results are presented in the manner of Result 2.8.1. That is, an estimator, its variance, and a variance estimator are presented. The variance is an ordinarily unknown quantity, because it depends on the complete set of population values $y_1, \ldots, y_N$. The variance can serve as a tool in theoretical comparisons of several strategies. By contrast, the estimated variance is computable from sample data, and serves as an indicator of the quality of a survey estimate.

For a sampling design of fixed size, one can give an alternative expression for the variance $V(\hat{t}_\pi)$ that leads directly to an alternative estimator of the variance, as specified in the following result.

Result 2.8.2. *If $p(s)$ is a fixed size sampling design, then the variance of the π estimator can be written alternatively as*

$$V(\hat{t}_\pi) = -\tfrac{1}{2}\sum\sum_U \Delta_{kl}(\check{y}_k - \check{y}_l)^2 \tag{2.8.10}$$

Provided that $\pi_{kl} > 0$ for all $k \neq l \in U$, an unbiased estimator of $V(\hat{t}_\pi)$ is given by

$$\hat{V}(\hat{t}_\pi) = -\tfrac{1}{2}\sum\sum_s \check{\Delta}_{kl}(\check{y}_k - \check{y}_l)^2 \tag{2.8.11}$$

Remark 2.8.3. Because $\check{y}_k - \check{y}_l = 0$ if $k = l$, the terms in which $k = l$ contribute nothing to the double sums of Result 2.8.2. One could thus equivalently write $\underset{k \neq l}{\sum\sum_U}$ in the formula for $V(\hat{t}_\pi)$, and $\underset{k \neq l}{\sum\sum_s}$ in the formula for $\hat{V}(\hat{t}_\pi)$. The variance estimator (2.8.11) is due to Yates and Grundy (1953) and Sen (1953). The variance estimator (2.8.6) is attributed to Horvitz and Thompson (1952).

PROOF. Let us first verify the equivalence of equations (2.8.5) and (2.8.10) when the size of the design is fixed at n. Developing the square in (2.8.10) and summing, we have

$$V(\hat{t}_\pi) = \sum\sum_U \Delta_{kl}\check{y}_k\check{y}_l - \sum\sum_U \Delta_{kl}\check{y}_k^2 \tag{2.8.12}$$

Now

$$\sum\sum_U \Delta_{kl}\check{y}_k^2 = \sum_{k \in U} \check{y}_k^2 \sum_{l \in U} \Delta_{kl}$$

Using Result 2.6.2, we have, for every fixed k,

$$\sum_{l \in U} \Delta_{kl} = \sum_{l \in U} \pi_{kl} - \sum_{l \in U} \pi_k \pi_l = n\pi_k - n\pi_k = 0$$

and the equivalence of equations (2.8.5) and (2.8.10) follows. To prove the unbiasedness of the variance estimator (2.8.11), it suffices to note (a) that

$$\hat{V}(\hat{t}_\pi) = -\tfrac{1}{2}\sum\sum_U I_k I_l \check{\Delta}_{kl}(\check{y}_k - \check{y}_l)^2$$

provided that $\pi_{kl} > 0$ for all $k, l \in U$, and (b) that

$$E(I_k I_l \check{\Delta}_{kl}) = \Delta_{kl}$$

□

Note that the two variances given by (2.8.5) and (2.8.10) are identical whenever the design is of fixed size. However, for such a design, the two variance estimators (2.8.6) and (2.8.11) are not necessarily identical, but both are unbiased.

EXAMPLE 2.8.1. Let us examine the π estimator for the SI design. The inclusion probabilities are given by equations (2.4.4) and (2.4.5). The π estimator takes the form

$$\hat{t}_\pi = N\bar{y}_s$$

where $\bar{y}_s = \sum_s y_k/n$ is the sample mean of y. The design has the fixed size n, and we can use (2.8.10) and (2.8.11) to find the variance and the variance estimator.

Let $f = n/N$ be the sampling fraction, and let

$$S_{yU}^2 = \frac{1}{N-1}\sum_U (y_k - \bar{y}_U)^2$$

be the population variance. Using that

$$\Delta_{kl} = -\frac{f(1-f)}{N-1} \quad \text{for any} \quad k \neq l$$

we get from (2.8.10)

$$\begin{aligned} V_{SI}(\hat{t}_\pi) &= -\tfrac{1}{2}\sum\sum_U \Delta_{kl}(\check{y}_k - \check{y}_l)^2 \\ &= -\tfrac{1}{2}\left[-\frac{f(1-f)}{N-1}\right]\frac{1}{f^2}\sum\sum_U [(y_k - \bar{y}_U) - (y_l - \bar{y}_U)]^2 \\ &= \frac{1-f}{2f(N-1)} \cdot 2N(N-1)S_{yU}^2 = N^2\frac{1-f}{n}S_{yU}^2 \end{aligned}$$

A similar derivation, using (2.8.11), gives the variance estimator

$$\hat{V}_{SI}(\hat{t}_\pi) = N^2\frac{1-f}{n}S_{ys}^2$$

where

$$S_{ys}^2 = \frac{1}{n-1}\sum_s (y_k - \bar{y}_s)^2$$

is the sample variance of y.

Remark 2.8.4. There exist designs for which the variance estimators given by (2.8.6) and (2.8.11) are identical. Two important examples are the *SI* design and the stratified simple random sampling (*STSI*) design, see Section 3.7.

It is an easy exercise to show that for the *SI* design both (2.8.6) and (2.8.11) give the result

$$\hat{V}_{SI}(\hat{t}_\pi) = N^2\left(\frac{1}{n} - \frac{1}{N}\right)S_{ys}^2 \qquad (2.8.13)$$

where S_{ys}^2 is the variance of y in s.

Remark 2.8.5. In practice, a variance estimate is seldom calculated directly from (2.8.6) or (2.8.11). The double-sum feature of these formulas is computationally inconvenient. For example, if a sample s contains $n = 1{,}000$ observations, over $5 \cdot 10^5$ terms would need to be calculated to obtain the variance estimate. Instead, for each particular design of interest, the formula is developed, with the aid of the appropriate $\check{\Delta}_{kl}$, into an expression that is suitable for rapid computation. For example, for the *SI* design with the sampling fraction $f = n/N$, $\check{\Delta}_{kk} = \Delta_{kk}/\pi_{kk} = 1 - f$ for all k and $\check{\Delta}_{kl} = \Delta_{kl}/\pi_{kl} = -(1-f)/(n-1)$ for all $k \neq l$, and both (2.8.6) and (2.8.11) reduce to the expression given in (2.8.13), which is easy to compute. Simplifications of this kind occur when the $\check{\Delta}_{kl}$ are equal within subsets of the set of all pairs (k, l). Stratified simple random sampling is another example; see Section 3.7.

Remark 2.8.6. Both of the variance estimators (2.8.6) and (2.8.11) require that

$$\pi_{kl} > 0 \quad \text{for all} \quad k \neq l \in U \qquad (2.8.14)$$

For any s that has been selected, $p(s)$ is necessarily positive and thus $\pi_{kl} > 0$ for all $k \neq l \in s$, whereas the stronger requirement stated in (2.8.14) may not be satisfied. The variance estimators (2.8.6) and (2.8.11) can be computed for any s. However, unless the condition (2.8.14) is verified, these variance estimates should not be used. They can be totally misleading. An example occurs for systematic sampling, see Section 3.4.

Remark 2.8.7. For many of the standard sampling designs, (2.8.6) and (2.8.11) give variance estimates that are always nonnegative. But there exist sampling designs and configurations $y_1, \ldots, y_N$ for which (2.8.6) or (2.8.11) could lead to a negative variance estimate for some samples. Obviously such variance estimates are unacceptable. It is easily seen that the Yates-Grundy estimator (2.8.11) is always nonnegative if the design is such that $\Delta_{kl} = \pi_{kl} - \pi_k\pi_l < 0$

for all $k \neq l \in U$. It is an advantage if a fixed size sampling design has the property $\Delta_{kl} \leq 0$ for all $k \neq l$, because this guarantees the nonnegativity of the Yates-Grundy formula. For instance, we note that for the *SI* design, all Δ_{kl} are negative for $k \neq l$ and equal to $-f(1-f)/(N-1)$. Several authors have investigated the conditions under which (2.8.6) or (2.8.11) are always nonnegative. For a discussion, see Lanke (1975), Chaudhuri (1981), and Rao (1982). Even if $\Delta_{kl} < 0$ for all $k \neq l \in U$ does not hold, the possibility that (2.8.11) yields a negative variance estimate is remote in medium to large samples. In practice, one need not worry much about the occurrence of negative variance estimates. For further discussion, see Section 3.6.

We conclude this section by stating another useful result concerning sums of π-expanded values.

Result 2.8.3. (a) *Let* $a_1, \ldots, a_k, \ldots, a_N$ *be fixed numbers and* $\check{a}_k = a_k/\pi_k$ *(with* $\pi_k > 0$*) for* $k = 1, \ldots, N$. *Then* $\sum_s \check{a}_k$ *is unbiased for* $\sum_U a_k$. (b) *Let* $a_{11}, a_{12}, \ldots, a_{kl}, \ldots, a_{NN}$ *be fixed numbers and* $\check{a}_{kl} = a_{kl}/\pi_{kl}$ *(with* $\pi_{kl} > 0$*) for* $k, l \in U$. *Then* $\sum\sum_s \check{a}_{kl}$ *is unbiased for* $\sum\sum_U a_{kl}$.

The proofs are simple and based on principles already used in proving Results 2.8.1 and 2.8.2.

EXAMPLE 2.8.2. If $a_k = y_k$, part (a) of Result 2.8.3 states that $\sum_s \check{y}_k$ is unbiased for $\sum_U y_k$ (see Result 2.8.1). For example, letting $a_k = y_k^2$, the conclusion is that $\sum_s (y_k^2/\pi_k)$ is unbiased for $\sum_U y_k^2$. As an example illustrating part (b) of Result 2.8.3, take $a_{kl} = \Delta_{kl}(\check{y}_k - \check{y}_l)^2$, which leads to the conclusion seen in Result 2.8.2 that $-(1/2)\sum\sum_s \check{\Delta}_{kl}(\check{y}_k - \check{y}_l)^2$ is unbiased for $-(1/2)\sum\sum_U \Delta_{kl}(\check{y}_k - \check{y}_l)^2$.

2.9. With-Replacement Sampling

We noted in Section 2.2 that two basic classes of schemes exist for sample selection, called draw sequential schemes and list sequential schemes. A draw sequential scheme consists of a number of randomized selections (draws) from the population; one population element is chosen in each draw. Here it is important to distinguish with-replacement schemes from without-replacement schemes. Already-drawn elements can be reselected in a with-replacement scheme. In a without-replacement scheme, an already drawn element cannot be selected again.

With-replacement schemes are not so important in sampling practice. Nevertheless, there are reasons for examining some procedures of this kind. One is that certain estimators for the with-replacement case have extremely simple statistical properties, and results concerning these estimators will prove useful later in the book. Another reason is that the fundamental distinction with-replacement versus without-replacement needs to be addressed in the context of survey sampling.

EXAMPLE 2.9.1. Consider the scheme consisting of a given number, say m, of independent draws such that, in each draw, every one of the N population elements has the same selection probability, that is, $1/N$. Once drawn, an element is replaced into the population so that all N elements participate in each draw. Clearly, the same element may be drawn more than once. This is a draw-sequential implementation of *simple random sampling with replacement*, a design which we denote SIR. The probability that any one element is drawn exactly r times in the m trials is

$$\binom{m}{r}\left(\frac{1}{N}\right)^r\left(1-\frac{1}{N}\right)^{m-r}$$

In particular, the probability that any given element is not drawn at all is given by

$$\left(1-\frac{1}{N}\right)^m$$

Consequently, the probability that the kth element is drawn at least once, which is the inclusion probability of the element under the procedure, is

$$\pi_k = 1 - \left(1-\frac{1}{N}\right)^m \tag{2.9.1}$$

for $k = 1, \ldots, N$.

In the case of with-replacement sampling schemes, we must be more careful with the interpretation of the word "sample." Let k_i denote the element selected in the ith draw, $i = 1, \ldots, m$, where m is the number of draws. The vector of elements drawn,

$$os = (k_1, k_2, \ldots, k_m) \tag{2.9.2}$$

constitutes what we call an *ordered sample*. It contains information about the drawing order, as well as about the multiplicities, that is, the number of times that each element is drawn. One and the same element may occur more than once in the sample given by (2.9.2). The probability distribution of the ordered samples is called the *ordered (sampling) design* induced by the scheme.

We can also consider the usual set-theoretical notion of a sample used so far in the chapter. Seen in this way, the sample realized by the with-replacement scheme is the set of distinct elements contained in the ordered sample (2.9.2). In other words, given the ordered sample (2.9.2), we obtain the corresponding set-sample, s, as

$$s = \{k\colon k = k_i \text{ for some } i;\ i = 1, \ldots, m\} \tag{2.9.3}$$

The size, n_s, of s is a random variable, and $n_s \leq m$ with probability one. Now, the set s contains no information about either the drawing order or the multiplicities. However, for estimation purposes, this need not entail a loss of information (see Section 3.8). The ordered sampling design induces a certain probability distribution $p(\cdot)$ for the set-samples s. In those few instances in this book where with-replacement sampling is considered, it is important to make a distinction between ordered sample and set-sample.

Certain estimators with extremely simple statistical properties can be obtained from the ordered sample, whereas estimators based on the corresponding set-sample are more complicated to analyze.

EXAMPLE 2.9.2. Consider again simple random sampling with replacement (with m draws), as defined in Example 2.9.1. There are N^m different (but equiprobable) ordered samples of the fixed length m. The probability of any one ordered sample of size m is therefore $1/N^m$. Any other ordered sample has probability zero. The ordered design associated with simple random sampling with replacement, denoted SIR, is a uniform distribution on the set of all ordered samples of fixed size m. By contrast, the set-sample distribution $p(s)$ induced by SIR is more complex. We do not need the exact form of this distribution, but it is not hard to show that the inclusion probabilities of first and second order are, respectively,

$$\pi_k = 1 - \left(1 - \frac{1}{N}\right)^m; \qquad k = 1, \ldots, N,$$

and

$$\pi_{kl} = 1 - 2\left(1 - \frac{1}{N}\right)^m + \left(1 - \frac{2}{N}\right)^m; \qquad k \neq l = 1, \ldots, N. \qquad (2.9.4)$$

It is a simple step to generalize simple random sampling with replacement to a procedure that allows unequal probabilities for drawing the different elements, while still retaining the independence of the draws. The general form of sampling with replacement has the following features. Suppose that $p_1, \ldots, p_k, \ldots, p_N$ are given positive numbers satisfying

$$\textstyle\sum_U p_k = 1$$

The selection is carried out by drawing a first element in such a way that

$$\Pr(\text{selecting element } k) = p_k; \qquad k = 1, \ldots, N$$

The selected element, k_1, is replaced. The same set of probabilities is used to select a second element k_2, ... and an mth element k_m. The m draws are independent. The probability of obtaining a specified ordered sample $(k_1, k_2, \ldots, k_m)$ is thus

$$\Pr[(k_1, k_2, \ldots, k_m)] = p_{k_1} p_{k_2} \cdots p_{k_m}$$

This describes the ordered sampling design.

The corresponding set-sample distribution is complicated. However, it is not hard to see that the probability that the kth element is drawn at least once is given by

$$\pi_k = 1 - (1 - p_k)^m \tag{2.9.5}$$

This is the inclusion probability of the kth element. If $m = 1$, then $\pi_k = p_k$. If $m > 1$ and if p_k is very small, then $\pi_k \doteq m p_k$.

Let us construct a suitable estimator of the population total, $t = \sum_U y_k$, under sampling with replacement. Define first the p-expanded value of the kth element as

$$\frac{y_k}{p_k}$$

(Note that for $m \geq 2$, y_k/p_k will differ from the π-expanded value, $\check{y}_k = y_k/\pi_k$.) Averaging the m p-expanded values leads to

$$\hat{t}_{\text{pwr}} = \frac{1}{m} \sum_{i=1}^{m} \frac{y_{k_i}}{p_{k_i}}$$

This estimator, due to Hansen and Hurwitz (1943), will be called the pwr estimator, where pwr refers to "p-expanded with replacement." It is a function of the ordered sample, so it is the ordered design that determines the properties of the pwr estimator. Alternatively, we can write

$$\hat{t}_{\text{pwr}} = \frac{1}{m} \sum_{i=1}^{m} Z_i$$

where Z_i is the random variable such that

$$Z_i = y_k/p_k \qquad \text{if } k_i = k \tag{2.9.6}$$

that is, if the element k is selected in the ith draw. The distribution of Z_i is, for $i = 1, \ldots, m$, such that

$$\Pr\left(Z_i = \frac{y_k}{p_k}\right) = p_k; \qquad k = 1, \ldots, N$$

For simplicity all N values y_k/p_k are assumed to be different.

The random variables $Z_1, \ldots, Z_m$ are distributed independently and identically, because independent draws are carried out with the same set of probabilities $p_1, \ldots, p_N$ in each draw. This is the key to the simple statistical properties of the pwr estimator shown in the following result.

Result 2.9.1. *The pwr estimator*

$$\hat{t}_{\text{pwr}} = \frac{1}{m} \sum_{i=1}^{m} \frac{y_{k_i}}{p_{k_i}} \tag{2.9.7}$$

is unbiased for $t = \sum_U y_k$. *Its variance is given by*

$$V(\hat{t}_{\text{pwr}}) = \frac{V_1}{m}$$

where

$$V_1 = \sum_U \left(\frac{y_k}{p_k} - t\right)^2 p_k \tag{2.9.8}$$

An unbiased estimator of the variance is given by

$$\hat{V}(\hat{t}_{\text{pwr}}) = \frac{\hat{V}_1}{m}$$

where

$$\hat{V}_1 = \frac{1}{m-1} \sum_{i=1}^{m} \left(\frac{y_{k_i}}{p_{k_i}} - \hat{t}_{\text{pwr}}\right)^2 \tag{2.9.9}$$

Remark 2.9.1. Note the appealing form of the variance estimator, $\hat{V}(\hat{t}_{\text{pwr}}) = \hat{V}_1/m$, which is simply $1/m$ times the variance of the m p-expanded values.

Proof. Using the independent and identically distributed random variables Z_i defined by (2.9.6), we have

$$\hat{t}_{\text{pwr}} = \frac{1}{m} \sum_{i=1}^{m} Z_i = \bar{Z}$$

Now, for $i = 1, \ldots, m$, the mean and variance of Z_i are, respectively,

$$E(Z_i) = \sum_U \frac{y_k}{p_k} p_k = t$$

$$V(Z_i) = E(Z_i - t)^2$$

$$= \sum_U \left(\frac{y_k}{p_k} - t\right)^2 p_k = V_1$$

Since $\hat{t}_{\text{pwr}}$ is the arithmetic mean of m independent and identically distributed random variables, it follows that

$$E(\hat{t}_{\text{pwr}}) = t$$

and

$$V(\hat{t}_{\text{pwr}}) = V_1/m$$

as claimed by the result. To show the unbiasedness of the variance estimator, note simply that if $Z_i, \ldots, Z_m$ are independent and identically distributed random variables, each with variance V_1, then

$$\frac{1}{m-1} \sum_{i=1}^{m} (Z_i - \bar{Z})^2 = \hat{V}_1$$

is an unbiased estimator of V_1. □

Remark 2.9.2. It is easily seen that $V(\hat{t}_{\text{pwr}}) = 0$ if the y_k values satisfy

$$y_k = cp_k, \qquad k = 1, \ldots, N$$

for some constant c. In other words, the variance of $\hat{t}_{\text{pwr}}$ is zero if y_k is exactly proportional to p_k. In practice it is not possible to choose p_k proportional to y_k, because $y_1, \ldots, y_N$ are ordinarily not known up to a proportionality constant. However, if the p_k can be chosen so that y_k is nearly proportional to p_k, the variance would still be quite small. This is why the pwr estimator has advantages when a known set of positive "size measures" $x_1, \ldots, x_N$ exists, such that the ratios y_k/x_k are roughly constant throughout the population. We can then let the p_k be determined by

$$p_k = x_k/\textstyle\sum_U x_k; \qquad k = 1, \ldots, N$$

and the procedure should yield a small variance. This particular application of with-replacement sampling is called *probability proportional-to-size sampling*, or pps sampling.

Remark 2.9.3. In with-replacement sampling, an alternative to the pwr estimator is the π estimator,

$$\hat{t}_\pi = \textstyle\sum_s \check{y}_k \tag{2.9.10}$$

where s is the set of distinct elements in the ordered sample $(k_1, k_2, \ldots, k_m)$, and $\check{y}_k = y_k/\pi_k$ with π_k given by (2.9.5). The size of s is random. For $m \geq 2$, the π estimator is not identical to the pwr estimator. Which is the better estimator? There is no simple answer. Both are unbiased. As far as the variance is concerned, no general conclusion is possible. Which of the two estimators has the smaller variance will depend on the configuration of the y-values $y_1, \ldots, y_N$. For further discussion of these matters, see Section 3.8.

2.10. The Design Effect

Under the *SI* design, with n elements drawn from N, the π estimator of the population total t was found in Section 2.8 to be given by the expanded mean,

$$\hat{t}_\pi = N\bar{y}_s$$

where $\bar{y}_s = \sum_s y_k/n$ is the sample mean of y. The variance is

$$V_{SI}(\hat{t}_\pi) = N^2\left(\frac{1}{n} - \frac{1}{N}\right)S_{yU}^2$$

The strategy consisting of the *SI* design and the expanded mean, $N\bar{y}_s$, is often taken as a point of reference when alternative ways of estimating t are considered. Let $p(s)$ denote some other design than *SI* (with the same expected sample size, $\sum_U \pi_k = n$, to ensure a fair comparison). If $\hat{t}_\pi$ is the π estimator for that design, let

$$\text{deff}(p, \hat{t}_\pi) = \frac{V_p(\hat{t}_\pi)}{V_{SI}(N\bar{y}_s)} = \frac{\sum\sum_U \Delta_{kl}\check{y}_k\check{y}_l}{N^2\left(\frac{1}{n} - \frac{1}{N}\right)S_{yU}^2} \tag{2.10.1}$$

This variance ratio is called the *design effect*. The design effect, $\text{deff}(p, \hat{t}_\pi)$, expresses how well the strategy composed of the design p and the estimator $\hat{t}_\pi$ fares in comparison to the reference strategy consisting of SI sampling and the estimator $N\bar{y}_s$. When $\text{deff}(p, \hat{t}_\pi)$ exceeds unity, precision is lost by not using the SI design; if $\text{deff}(p, \hat{t}_\pi)$ is less than unity, precision is gained compared to the SI design. The π estimator is considered under both designs; it is the sampling design and the sampling distribution of the π estimator that differs from one strategy to the other.

EXAMPLE 2.10.1. Consider the BE design previously considered in Examples 2.2.2, 2.3.2, 2.4.2, and 2.6.1. We have $\pi_k = \pi$ for all k, $\Delta_{kk} = \pi(1 - \pi)$ for all k, and $\Delta_{kl} = 0$ for all $k \neq l$. The π estimator is

$$\hat{t}_\pi = \frac{1}{\pi}\sum_s y_k$$

Using equation (2.8.5), we obtain the variance

$$V_{BE}(\hat{t}_\pi) = \frac{1 - \pi}{\pi}\sum_U y_k^2$$

Now,

$$\sum_U y_k^2 = (N - 1)S_{yU}^2 + N(\bar{y}_U)^2 = \left[1 - \frac{1}{N} + \frac{1}{(cv_{yU})^2}\right]NS_{yU}^2$$

where

$$cv_{yU} = S_{yU}/\bar{y}_U$$

is the coefficient of variation of y in the population. This important descriptive measure, which expresses the standard deviation as a fraction of the mean, is useful for populations of nonnegative values $y_1, \ldots, y_N$. Consequently, we can write

$$V_{BE}(\hat{t}_\pi) = N\frac{1 - \pi}{\pi}S_{yU}^2\left[1 - \frac{1}{N} + \frac{1}{(cv_{yU})^2}\right]$$

To obtain a fair comparison with the SI design (n elements chosen from N), we fix the expected sample size for the BE design, which is $N\pi$, in such a way that $N\pi = n$. Then

$$\text{deff}(BE, \hat{t}_\pi) = \frac{V_{BE}(\hat{t}_\pi)}{V_{SI}(\hat{t}_\pi)} = 1 - \frac{1}{N} + \frac{1}{(cv_{yU})^2} \doteq 1 + \frac{1}{(cv_{yU})^2} \tag{2.10.2}$$

For many variables and many populations the coefficient of variation satisfies $0.5 \le cv_{yU} \le 1.0$. With $cv_{yU} = 1.0$, the design effect of *BE* is deff(*BE*) = 2.0. If $cv_{yU} = 0.5$, deff(*BE*) = 5.0. This illustrates that *BE* sampling is often considerably less precise than *SI* sampling when the π estimator is used. The loss of precision is explained by the variability of the size of the *BE* sample.

2.11. Confidence Intervals

A basic course in statistics introduces the idea of a confidence interval as a random interval having a stated probability (usually near unity) containing the unknown value of a parameter of interest. The same idea underlies confidence intervals for finite population parameters. Here we discuss confidence intervals as they relate to estimation of the finite population total $t = \sum_U y_k$; however, the ideas apply more generally to estimating any finite population parameter θ.

A confidence interval is a random interval $CI(s) = [t_L(s), t_U(s)]$, where the lower endpoint $t_L(s)$ and the upper endpoint $t_U(s)$ are two given statistics such that $t_L(s) \le t_U(s)$ for every s. The probability that the unknown total t is contained in the interval $CI(s)$ is

$$\Pr[CI(s) \ni t]$$

This probability is called the *confidence level* or the *coverage probability* of the interval. Normally, one wants the confidence level to be near unity. The statistician accepts some small risk of noncoverage in exchange for a relatively narrow interval. Note that the random entity in the interval is s, the randomly selected sample. A confidence interval in the context of survey sampling is interpreted with reference to the sampling design $p(s)$, as follows.

Suppose for a moment that we have access to all N y_k-values. Thus $t = \sum_U y_k$ is known. Suppose also that we can calculate the interval $CI(s) = [t_L(s), t_U(s)]$ for every possible sample s, that is, every sample with $p(s) > 0$. Denote this set of samples by $\mathscr{S}_0$. For each $s \in \mathscr{S}_0$, we observe whether the calculated interval covers the parameter t or not. Let $\mathscr{S}_{oc} \subset \mathscr{S}_0$ be the set consisting of those samples s for which the computed interval includes the value t, and let $\mathscr{S}_{oc}^* = \mathscr{S}_0 - \mathscr{S}_{oc}$ be the complement set. The confidence level is then

$$\Pr[\mathrm{CI}(s) \ni t] = 1 - \alpha \tag{2.11.1}$$

where α is the accumulated probability of those samples s for which the interval fails to include t, that is,

$$\alpha = \sum_{s \in \mathscr{S}_{oc}^*} p(s)$$

In practice, the total t is, of course, unknown, and we need a practical method for constructing the upper and the lower endpoints, $t_U(s)$ and $t_L(s)$, so that a desired confidence level $1 - \alpha$ (say, 95%) is attained. For the estimators typically used in survey sampling, it is difficult to give a method that yields an exact $1 - \alpha$ confidence level. We usually have to work with approximate procedures.

Let $\hat{t}$ be the point estimator for the unknown t. A confidence interval for t, at the approximate level $1 - \alpha$, is often computed as

$$\hat{t} \pm z_{1-\alpha/2}[\hat{V}(\hat{t})]^{1/2} \tag{2.11.2}$$

where $z_{1-\alpha/2}$ is the constant exceeded with probability $\alpha/2$ by the $N(0, 1)$ random variable. Usually, one chooses a small value for α, such as 5% or 1%. That is, in the interval (2.11.2), we use $z_{0.975} = 1.96$ for $1 - \alpha = 95\%$ and $z_{0.995} = 2.576$ for $1 - \alpha = 99\%$. In the following, we often calculate intervals with the aid of formula (2.11.2).

The interval (2.11.2) will contain the unknown total t for an approximate proportion of $1 - \alpha$ of repeated samples s drawn with the given design, if the following two conditions are verified.

1. The sampling distribution of $\hat{t}$ is approximately a normal distribution with mean t and variance $V(\hat{t})$.
2. There exists a consistent variance estimator $\hat{V}(\hat{t})$ for $V(\hat{t})$.

The first condition is essentially equivalent to saying that the Central Limit Theorem applies for the random variable $\hat{t}$. The Central Limit Theorem would state, in this case, that the limiting distribution of $\hat{t}/N$, when the sample size increases, is a normal distribution. The second condition states, essentially, that with a probability tending to one as the sample size increases, the ratio $\hat{V}(\hat{t})/V(\hat{t})$ is within the limits $1 \pm \varepsilon$, for some arbitrary positive number ε. Asymptotic concepts, including consistency, are further discussed in in Section 5.3.

The role that the two conditions play for the confidence interval (2.11.2) become clear if we note that

$$\frac{\hat{t} - t}{\sqrt{\hat{V}(\hat{t})}} = \frac{\hat{t} - t}{\sqrt{V(\hat{t})}}\left(\frac{V(\hat{t})}{\hat{V}(\hat{t})}\right)^{1/2} \tag{2.11.3}$$

Under the first condition, $(\hat{t} - t)/[V(\hat{t})]^{1/2}$ has an approximate $N(0, 1)$ distribution. Under the second condition, $[\hat{V}(\hat{t})/V(\hat{t})]^{1/2}$ is near one with high probability when the sample is sufficiently large. We can then treat the variable in equation (2.11.3) as an approximate $N(0, 1)$ random variable, which justifies the use of the normal deviate $z_{1-\alpha/2}$ in (2.11.2).

Remark 2.11.1. If $V(\hat{t})$ is known, a confidence interval can also be calculated as

$$\hat{t} \pm z_{1-\alpha/2}[V(\hat{t})]^{1/2} \tag{2.11.4}$$

For a limited sample size, we can expect the interval (2.11.4) to come closer to the desired $1 - \alpha$ level than the interval (2.11.2), which requires an estimate of the variance. Put differently, the approach to normality may be slower when $\hat{V}(\hat{t})$ rather than $V(\hat{t})$ is used. However, in practice, there is usually little choice but to use interval (2.11.2), because the variance $V(\hat{t})$ is ordinarily unknown.

Remark 2.11.2. The exactness of the normal approximation used in (2.11.2) depends significantly on the shape of the finite population. If a histogram of the N values $y_1, \ldots, y_N$ is highly skewed and there exist outlying values or other abnormal features, we can expect the approach to normality of the variable (2.11.3) to be slower. That is, highly nonnormal finite populations may require larger sample sizes before the normal approximation can be used. In simple cases, one can make explicit statements about the sample size required for the normal approximation to apply. Several authors have considered the effect of population skewness on the confidence interval. Cochran (1977, p. 58) and Dalén (1986) consider the case where an unknown finite population proportion is being estimated under the *SI* design. This case is simple to analyze because the study variable is dichotomous, that is, the value of y_k is either 0 or 1.

Remark 2.11.3. How can we check if the confidence interval procedure defined by (2.11.2) is at least approximately valid, that is, if the desired confidence level $1 - \alpha$ is roughly attained for a given sample size, sampling design, and population shape? Two methods are (a) theoretical validation and (b) empirical validation.

A form of theoretical validation is to verify whether the estimator $\hat{t}$ obeys the first condition under the given design, that is, whether the Central Limit Theorem holds. Some results of this kind are available; for example, see Hájek (1960, 1964) and Rosén (1972). Although such results can establish asymptotic normality under certain assumptions, they do not answer the practitioner's question of what sample size is required to make the normal approximation acceptable. Empirical validation can be carried out by Monte Carlo simulation. A long series of K samples (say, $K = 10{,}000$ samples) of a given size are drawn from a completely known finite population, according to a given design $p(s)$. For each sample, the estimator $\hat{t}$, the variance estimator $\hat{V}(\hat{t})$, and the confidence interval (2.11.2) are computed. Since the entire finite population is known, the parameter value $t = \sum_U y_k$ is known. Consequently, for each of the K intervals computed at a desired $1 - \alpha$ confidence level, one can observe whether t is included in the interval or not. If R of the K intervals are found to contain t, the empirical confidence level of the experiment is defined as the proportion R/K. This proportion should lie near the desired confidence interval level $1 - \alpha$. Experiments of this kind have been carried out for a variety of estimators, sampling designs, population shapes, and sample sizes. The results of such an experiment are reported in Section 7.9.1.

Exercises

2.1. In planning an office network study, the following draw sequential sampling scheme was proposed for selecting a random sample of two nonadjacent office hour intervals from the eight intervals 9–10, 10–11, ..., 16–17 (labeled 1, 2, ..., 8):

1. Draw the first hour interval for the sample with equal probability from the eight intervals.
2. Draw, without replacement, the second hour interval with equal probability from the intervals not adjacent in time to the one selected in the first draw.

(a) Determine the first-order inclusion probabilities. (b) Determine the second-order inclusion probabilities. Is the design induced by the proposed sampling scheme measurable? (c) Determine the covariances of the sample membership indicators. (d) Verify that Result 2.6.2 holds in the present application.

2.2. Consider the design in Example 1.5.2 (the HINK survey.) A sample s of n individuals is drawn by the design *SI* from a frame that lists N individuals. The households corresponding to selected individuals are identified. Compute the inclusion probability of a household composed of M individuals, where $M < n$. Obtain approximate expressions for the inclusion probability for $M = 1, 2$, or 3, supposing that both N and n are large with $n/N = f > 0$.

2.3. Consider a population U composed of three disjoint subpopulations U_1, U_2, and U_3 of sizes $N_1 = 600$, $N_2 = 300$, and $N_3 = 100$, respectively. Thus U is of size $N = 1{,}000$. For each element $k \in U$, the inclusion or noninclusion in the sample, s, is determined by a Bernoulli experiment that gives the element k the probability π_k of being selected. The experiments are independent. (a) Let $\pi_k = 0.1$ for all $k \in U_1$, $\pi_k = 0.2$ for all $k \in U_2$, and $\pi_k = 0.8$ for all $k \in U_3$. Find the expected value and the variance of the sample size, n_s, under this design. (b) Suppose that π_k is constant for all $k \in U$. Determine this constant so that the expected value of the sample size agrees with the expected sample size obtained in case (a). Obtain the variance of the sample size; compare with the variance in case (a).

2.4. A population of 1,600 individuals is divided into 800 clusters (households) such that there are N_a clusters of size a ($a = 1, 2, 3, 4$) according to the following table:

a	N_a
1	250
2	350
3	150
4	50

A sample of individuals is selected as follows: 300 clusters are drawn from the 800 by the *SI* design, and all individuals in selected clusters are to be interviewed. If n_s denotes the total number of individuals to be interviewed, calculate $E(n_s)$ and $V(n_s)$.

2.5. Consider a population of size $N = 3$, $U = \{1, 2, 3\}$. Let $s_1 = \{1, 2\}$, $s_2 = \{1, 3\}$, $s_3 = \{2, 3\}$, $s_4 = \{1, 2, 3\}$, $p(s_1) = 0.4$, $p(s_2) = 0.3$, $p(s_3) = 0.2$, and $p(s_4) = 0.1$. (a) Calculate all the π_k and all the π_{kl}. (b) Find the value of $E(n_s)$ in two ways: (i) by a direct calculation, using the definition and (ii) by use of the formula that expresses $E(n_s)$ as a function of the π_k.

2.6. Consider the population and the design in Exercise 2.5 above. Let the values of the study variable y be $y_1 = 16$, $y_2 = 21$, and $y_3 = 18$. Then we have $t = 55$. (a) Compute the expected value and the variance of the π estimator from the definitions in Section 2.7. (b) Compute the variance of the π estimator using Result 2.8.1. (c) Compute the coefficient of variation of the π estimator. (d) Compute a variance estimate $\hat{V}(\hat{t}_\pi)$ using the variance estimator (2.8.6) for each of the four possible samples. Determine the expected value of the variance estimator in the present situation using the definition of expected value in Section 2.7.

2.7. Let s be a sample realized by the BE design with $\pi_k = \pi$ for all $k \in U$. Let n_s denote the random size of s. Show that the conditional probability of obtaining s given n_s is the same as the probability of a SI sample of the fixed size n_s from N.

2.8. Consider the design SI, with n elements drawn from N. Then

$$E_{SI}\{S_{ys}^2\} = S_{yU}^2$$

Prove this using (i) that

$$\sum\sum_s (y_k - y_l)^2 = 2n(n-1)S_{ys}^2$$

with an analogous expression for S_{yU}^2, and (ii) that

$$\sum\sum_s (y_k - y_l)^2 = \sum\sum_U I_k I_l (y_k - y_l)^2$$

This exercise illustrates the importance of the second-order inclusion probabilities.

2.9. Let s be a sample drawn by the BE design with $\pi_k = \pi$ for all k. Let

$$S_{ys}^2 = \frac{\sum_s (y_k - \bar{y}_s)^2}{n_s - 1} \qquad \text{if } n_s \geq 2$$

and $S_{ys}^2 = 0$ if $n_s \leq 1$. Show that the relative bias of S_{ys}^2 is given by

$$\frac{E_{BE}\{S_{ys}^2\} - S_{yU}^2}{S_{yU}^2} = -\Pr(n_s \leq 1) = -(1-\pi)^{N-1}[1 + (N-1)\pi]$$

2.10. Let s ($s \subset U$) be an SI sample, and let s_1 ($s_1 \subset U - s$) be an SI sample from the remaining portion of the population. Let

$$\hat{\bar{y}}_U = \sum_s y_k/n \qquad \text{and} \qquad \hat{\bar{y}}_{U1} = \sum_{s_1} y_k/n_1$$

where n and n_1 are the respective sizes (fixed) of s and s_1. Find the covariance between $\hat{\bar{y}}_U$ and $\hat{\bar{y}}_{U1}$.

2.11. Show that the variance V_1 given by (2.9.8) and the variance estimator $\hat{V}_1$ given by (2.9.9) can be written

$$V_1 = \sum_U \frac{y_k^2}{p_k} - t^2$$

and

$$\hat{V}_1 = \frac{1}{m-1}\left[\sum_{i=1}^{m}\left(\frac{y_{k_i}}{p_{k_i}}\right)^2 - m\hat{t}_{pwr}^2\right]$$

respectively.

2.12. To estimate the average income per household ($\sum_U y_k/N$) for a population of $N = 200$ households, a listing of the 600 persons that belong to the 200 households was used as follows.

An *SIR* sample of size $m = 10$ persons was drawn. The households of the selected persons were identified, and information on average household income, y_k/x_k, was collected, where y_k is the total household income in dollars, and x_k is the number of persons in the household. The results are as follows:

Draw i	Average household income (y_{k_i}/x_{k_i})
1	7,000
2	8,000
3	6,000
4	5,000
5	9,000
6	4,000
7	7,000
8	8,000
9	4,000
10	2,000

Compute an estimate of the average income per household based on the pwr estimator as well as the corresponding *cve*.

2.13. Show that the second-order inclusion probabilities for general with-replacement sampling are, for $k \neq l$,

$$\pi_{kl} = 1 - (1 - p_k)^m - (1 - p_l)^m + (1 - p_k - p_l)^m$$

where p_k is the probability of selecting k in each of the m independent draws, and $\sum_U p_k = 1$.

CHAPTER 3

Unbiased Estimation for Element Sampling Designs

3.1. Introduction

In this chapter, we discuss unbiased estimation for direct element sampling. There are two characteristics of direct element sampling: (1) there exists a sampling frame identifying every population element and (2) in the selection of the sample, the population elements are the sampling units. The following direct element sampling designs will be considered in this chapter. The abbreviations that we use are given in parentheses.

i. Bernoulli sampling (*BE*).
ii. Simple random sampling—without replacement (*SI*) and with replacement (*SIR*).
iii. Systematic sampling (*SY*).
vi. Poisson sampling (*PO*).
v. Probability proportional-to-size sampling—without replacement (πps) and with replacement (*pps*).
vi. Stratified sampling (*ST*).

Most of these designs are used extensively in survey practice. The *BE* and *PO* designs lead to random sample size. However, these designs are well suited to illustrate certain basic ideas in survey sampling, and they are useful as well as models for response mechanisms in a later discussion of estimation in the presence of nonresponse (see Chapter 15).

The parameter of principal interest in this chapter is the population total of the study variable y, that is, $t = \sum_U y_k$. Our subject in this chapter is *unbiased* estimation of t, with special concentration on the π estimator,

$$\hat{t}_\pi = \sum_s \check{y}_k = \sum_s y_k/\pi_k \tag{3.1.1}$$

and (in with-replacement sampling) on the pwr estimator,

$$\hat{t}_{\text{pwr}} = \frac{1}{m} \sum_{i=1}^{m} y_{k_i}/p_{k_i} \tag{3.1.2}$$

For each of the designs, we present the estimator, its variance, and the appropriate variance estimator(s). In principle, the results for $\hat{t}_\pi$ can always be obtained from the general Result 2.8.1 by applying the π_k and the π_{kl} that are induced by the specific design.

The actual selection of a sample s from the population U requires a sampling frame that identifies the sampling units, which are the population elements. The frame often consists of a magnetic tape or a disk on which the information is stored sequentially, element by element. In such cases, computerized list-sequential selection is convenient. Such selection is executed by means of a series of independent realizations

$$\varepsilon_1, \varepsilon_2, \ldots$$

of a Unif(0, 1) random variable, that is, a variable distributed uniformly on the (0, 1) interval.

Results are also given in this chapter for certain linear parameters of the form

$$\theta = \sum_U c_k y_k \tag{3.1.3}$$

where the c_k are constants. If the c_k are known at least for $k \in s$, the π-estimation method can immediately be applied; the π estimator is

$$\hat{\theta}_\pi = \sum_s \frac{c_k y_k}{\pi_k} = \sum_s c_k \check{y}_k$$

The variance and the estimated variance of $\hat{\theta}_\pi$ follow easily from Result 2.8.1 by substituting $c_k y_k$ for y_k and $c_k \check{y}_k$ for $\check{y}_k$.

3.2. Bernoulli Sampling

An extremely simple design is Bernoulli sampling (*BE*). Let us summarize what Examples 2.2.2, 2.3.2, and 2.4.2 stated about this design: In *BE* sampling, the sample membership indicators $I_1, \ldots, I_N$ are independent and identically distributed random variables. If π (such that $0 < \pi < 1$) is a given constant, each I_k follows the same Bernoulli distribution,

$$\Pr(I_k = 1) = \pi; \qquad \Pr(I_k = 0) = 1 - \pi$$

If n_s denotes the (random) sample size, the design *BE* is expressed by

$$p(s) = \pi^{n_s}(1 - \pi)^{N - n_s} \tag{3.2.1}$$

The inclusion probabilities are $\pi_k = \pi$ for all k and $\pi_{kl} = \pi^2$ for all $k \neq l$.

To select a Bernoulli sample, we can use the simple list-sequential scheme described in Example 2.2.2. In Bernoulli sampling, the sample size, n_s, is a random variable, distributed binomially, with mean and variance given by

$$E_{BE}(n_s) = N\pi; \qquad V_{BE}(n_s) = N\pi(1 - \pi)$$

The range of probable values of n_s can be assessed by an interval. Using the normal distribution to approximate the binomial, we have n_s contained within the limits

$$N\pi \pm z_{1-(\alpha/2)}[N\pi(1 - \pi)]^{1/2}$$

with a probability of roughly $1 - \alpha$, where the constant $z_{1-(\alpha/2)}$ is exceeded with probability $\alpha/2$ by the unit normal random variable. For example, if $N = 10{,}000$ and $\pi = 0.2$, the 99% interval is

$$2{,}000 \pm 2.58(1{,}600)^{1/2} \doteq 2{,}000 \pm 103$$

so the probable variation in this case is within roughly 5% of the expected number.

In some surveys it may be a drawback not to be able to tell at the outset exactly how large the selected sample will be. In addition, the variation in the sample size has a tendency to increase the variance of the π estimator. The basic properties of the π estimator under BE sampling are given in Result 3.2.1. They follow easily from Result 2.8.1.

Result 3.2.1. *Under the BE design, the π estimator of the population total $t = \sum_U y_k$ takes the form*

$$\hat{t}_\pi = \frac{1}{\pi} \sum_s y_k \tag{3.2.2}$$

The variance is given by

$$V_{BE}(\hat{t}_\pi) = \left(\frac{1}{\pi} - 1\right) \sum_U y_k^2 \tag{3.2.3}$$

An unbiased variance estimator is

$$\hat{V}_{BE}(\hat{t}_\pi) = \frac{1}{\pi}\left(\frac{1}{\pi} - 1\right) \sum_s y_k^2 \tag{3.2.4}$$

The π estimator is often inefficient under BE sampling; an improved estimator of t is given by equation (3.2.6) below.

If $n = N\pi$ denotes the expected sample size, the variance (3.2.3) can be written as

$$V_{BE}(\hat{t}_\pi) = N^2\left(\frac{1}{n} - \frac{1}{N}\right) S_{yU}^2 \left[1 - \frac{1}{N} + (cv_{yU})^{-2}\right] \tag{3.2.5}$$

where

$$cv_{yU} = S_{yU}/\bar{y}_U$$

is the coefficient of variation of y in the population U.

EXAMPLE 3.2.1. A university professor who is correcting 600 written examinations decides to get a preliminary idea of the passing rate on the test. He decides to use a simple randomized scheme to single out a smaller number of exam copies for first-hand correction. In passing through the pile of exams, he tosses an ordinary six-sided die, once for each exam copy. If the die shows a 6, he corrects the corresponding exam, otherwise not. Suppose the sample selected in this way consists of 90 students and that 60 out of these are found to have passed; let us compute a 95% confidence interval, based on the normal approximation, for the number of passing students (among the 600). We note that the sampling design is BE with $\pi = 1/6$. Let the study variable y be defined by $y_k = 1$ if student k passes, $y_k = 0$ if student k fails. Then $\sum_s y_k = \sum_s y_k^2 = 60$. Since $N = 600$ and $1/\pi = 6$, Result 3.2.1 gives an approximately 95% confidence interval with the endpoints

$$\frac{1}{\pi}\sum_s y_k \pm 1.96\left[\frac{1}{\pi}\left(\frac{1}{\pi} - 1\right)\sum_s y_k^2\right]^{1/2}$$

$$= 6 \cdot 60 \pm 1.96\{6 \cdot 5 \cdot 60\}^{1/2} \doteq 360 \pm 83$$

This result may seem a bit surprising if the following reasonable argument was made instead. The passing rate in the sample is $60/90 = 2/3$. Thus a reasonable point estimate of the number of passing students in the population is $600 \cdot 2/3 = 400$. By comparison, the π estimate was 360, which is as much as 10% lower.

Which estimate, 360 or 400, is closer to the truth? We cannot tell, since the actual number of passing students in the population remains unknown until all 600 exams have been corrected. However, one can say that the point estimate 400 is better in the sense that it stems from an estimator with less sampling variability than that of the π estimator.

For the design BE, a better estimator than $\hat{t}_\pi$ is given by

$$\hat{t}_{\text{alt}} = N\left(\textstyle\sum_s y_k/n_s\right) = N\bar{y}_s \tag{3.2.6}$$

The estimate 400 in Example 3.2.1 was obtained by this formula. The effect of placing the random size n_s in the denominator is to reduce the part of the variability of $\hat{t}_\pi$ that stems from sample size variation. Note that

$$\hat{t}_{\text{alt}} = (n/n_s)\hat{t}_\pi$$

where $n = N\pi = E_{BE}(n_s)$ is the expected sample size. An approximate expression for the variance of $\hat{t}_{\text{alt}}$, derived later in Example 7.4.1, is

$$V_{BE}(\hat{t}_{\text{alt}}) \doteq N\left(\frac{1}{\pi} - 1\right) S_{yU}^2 = N^2\left(\frac{1}{n} - \frac{1}{N}\right) S_{yU}^2 \tag{3.2.7}$$

Using (3.2.5), we note that

$$V_{BE}(\hat{t}_{\pi})/V_{BE}(\hat{t}_{\text{alt}}) \doteq 1 + (cv_{yU})^{-2} \tag{3.2.8}$$

That is, the efficiency gain realized by $\hat{t}_{\text{alt}}$ over $\hat{t}_{\pi}$ is particularly pronounced when the coefficient of variation cv_{yU} is small. The variance $V_{BE}(\hat{t}_{\text{alt}})$ is very nearly the same as when a fixed number n of elements are selected with the *SI* design (see Section 3.3).

Our discussion illustrates the following important fact: Although the π estimator may be excellent for fixed size designs, it is subject to a variance penalty for sampling designs with variable sample size. The calculated confidence intervals tend to be wide. Nevertheless, this in itself is not a reason to avoid random sample size designs. If the estimator is chosen appropriately, there is no variance penalty from the variability of the sample size.

EXAMPLE 3.2.2. If we assume, in Example 3.2.1, that the true number of passing students in the population is 390, we get

$$cv_{yU} = S_{yU}/\bar{y}_U = \left(\frac{600 \cdot 7}{599 \cdot 13}\right)^{1/2} = 0.734$$

Thus the variance ratio (3.2.8) becomes

$$\frac{V_{BE}(\hat{t}_{\pi})}{V_{BE}(\hat{t}_{\text{alt}})} \doteq 1 + \frac{1}{(0.734)^2} = 2.85$$

The alternative estimator $\hat{t}_{\text{alt}}$ is considerably more precise.

Although one of the drawbacks with variable sample size may be corrected by a better choice of estimator, another drawback remains, namely, some loss of control over the field work. For example, the budget may be considerably surpassed if the realized sample size greatly exceeds the expected.

Although it is good practice to avoid designs with highly variable sample size, it is a fact that the ideal of a fixed sample size is often spoiled for other reasons. In most surveys, there is a degree of nonresponse and the number of actually responding elements cannot be exactly predicted. Also, if specific subgroups (domains) of the finite population require separate estimates, the selection in such subgroups can usually not be fully controlled. The number of observations in a domain is often random. Estimation for domains is discussed in Chapter 10. In practice, many of the estimations required in a survey are thus carried out with an a priori unpredictable number of observations. In surveys with nonresponse, the response behavior in different population subgroups is often modeled as a Bernoulli sample selection. This technique is discussed in Chapter 15.

3.3. Simple Random Sampling

The *BE* design belongs to the category of designs that may be termed equal probability sampling designs. The common feature of such designs is that the first-order inclusion probabilities are all equal, that is, $\pi_k = \text{constant}$ ($k = 1, \ldots, N$). We now consider two additional equal probability sampling designs, namely, simple random sampling without replacement (*SI*) and with replacement (*SIR*), which were discussed briefly in Chapter 2. The *SI* design is often taken as a point of reference when discussing alternative designs.

3.3.1. Simple Random Sampling without Replacement

Under the *SI* design (see Examples 2.2.1, 2.3.1, and 2.4.1) every sample s of the fixed size n receives the same probability of being selected. That is,

$$p(s) = \begin{cases} 1\Big/\binom{N}{n} & \text{if } s \text{ is of size } n \\ 0 & \text{otherwise} \end{cases} \tag{3.3.1}$$

The inclusion probabilities are

$$\pi_k = \frac{n}{N} = f; \qquad k = 1, \ldots, N$$

$$\pi_{kl} = \frac{n(n-1)}{N(N-1)}; \qquad k \neq l = 1, \ldots, N$$

We call $f = n/N$ the *sampling fraction*. The *SI* design can be carried out by the draw-sequential scheme in Example 2.2.1. Several other schemes can be used to implement the design *SI*. For a large population of elements stored sequentially on a magnetic tape, list-sequential selection is usually more convenient than draw-sequential selection.

The following list-sequential scheme gives an *SI* sample of size n.

Scheme A

Let $\varepsilon_1, \varepsilon_2, \ldots,$ be independent random numbers drawn from the Unif(0, 1) distribution. If $\varepsilon_1 < n/N$, the element $k = 1$ is selected, otherwise not. For subsequent elements, $k = 2, 3, \ldots,$ let n_k be the number of elements selected among the first $k - 1$ elements in the population list. If

$$\varepsilon_k < \frac{n - n_k}{N - k + 1}$$

the element k is selected, otherwise not. The procedure terminates when $n_k = n$.

This scheme, given by Fan, Muller and Rezucha (1962), can be shown to conform to the definition of the *SI* design. If N is unknown, a preliminary pass through the file is necessary to determine N. McLeod and Bellhouse (1983) proposed a scheme that implements an *SI* selection of n elements from a list whose length N is unknown. That is, in their method, no preliminary pass through the file is required to determine N.

Yet another implementation of the *SI* design is by the following scheme, which has the advantage that it permits a simultaneous selection of several nonoverlapping simple random samples. Nonoverlapping samples are desirable when several different surveys of the same population need to be carried out within a short space of time. Little or no overlap reduces respondent burden. For a population consisting of households or business firms as elements, response rates may be increased if an element is not approached too often. Samples drawn without overlap are called *negatively coordinated.*

The scheme supposes that N independent Unif(0, 1) random numbers $\varepsilon_1, \ldots, \varepsilon_k, \ldots, \varepsilon_N$ are first drawn, where ε_k is tied to the kth element. These numbers are then ordered according to size

$$\varepsilon_{(k_1)} < \varepsilon_{(k_2)} < \cdots < \varepsilon_{(k_N)}$$

The notation implies that the ith smallest of the N ε-values is tied to the element k_i; $i = 1, \ldots, N$. Now, the n smallest ε-values correspond to a set of elements, $\{k_1, \ldots, k_n\}$, which one can show to be an *SI* selection of n from N. The set of n elements corresponding to the next n ε-values, $\{k_{n+1}, \ldots, k_{2n}\}$, is a second *SI* sample, without overlap with the first, and so on. Any n preassigned positions in the ordered sequence identifies an *SI* sample. This can be accepted by intuition; we do not give a formal proof here. For further discussion of the method of negatively coordinated samples, the reader is referred to Atmer, Thulin and Bäcklund (1975) and Sunter (1977a,b).

For the design *SI*, we now look at π estimation of some important parameters. Case 1 concerns the population total, $t = \sum_U y_k$. Case 2 deals with the population mean, $\bar{y}_U = t/N$, and case 3 discusses parameters of interest for a subpopulation (domain).

Case 1. Estimation of the Population Total

We summarize the important conclusions; they follow from Result 2.8.1 and were discussed in Example 2.8.1.

Result 3.3.1. *Under the SI design, the π estimator of the population total $t = \sum_U y_k$ can be written*

$$\hat{t}_\pi = N\bar{y}_s = \frac{1}{f}\sum_s y_k \tag{3.3.2}$$

where $f = n/N$ is the sampling fraction. The variance is given by

$$V_{SI}(\hat{t}_\pi) = N^2\left(\frac{1}{n} - \frac{1}{N}\right)S_{yU}^2 = N^2\frac{1-f}{n}S_{yU}^2 \tag{3.3.3}$$

An unbiased variance estimator is

$$\hat{V}_{SI}(\hat{t}_\pi) = N^2\frac{1-f}{n}S_{ys}^2 \tag{3.3.4}$$

Example 3.3.1 (This example uses the population of Swedish municipalities given in Appendix B.) To estimate the total of the variable CS82 ($= y =$ number of conservative seats in municipal councils) for the MU200 population (the 200 smallest Swedish municipalities according to the 1975 population), an SI sample of size 50 was selected. The distribution of CS82 in the sample was:

CS82	3	4	5	6	7	8	9	10	11	12	13	14	16
Frequency	4	7	3	6	7	10	4	2	2	1	1	1	2

Let us compute an approximately 95% confidence interval for the CS82 total. We note that $\sum_s y_k = 369$ and $S_{ys}^2 = 9.914$. The endpoints of the interval are $4 \cdot 369 \pm 1.96[200^2(1 - 0.25) \cdot 9.914/50]^{1/2}$, which gives $1{,}476 \pm 151$.

Case 2. Estimation of the Population Mean, $\bar{y}_U = \sum_U y_k/N$

An unbiased estimator of $\bar{y}_U$ is obtained directly by dividing the π estimator (3.3.2) by N,

$$\hat{\bar{y}}_{U\pi} = \frac{\hat{t}_\pi}{N} = \frac{\sum_s y_k}{n} = \bar{y}_s$$

For the variance and the variance estimator, we have only to divide the corresponding expressions in Result 3.3.1 by N^2.

Result 3.3.2. *Under the SI design, the sample mean* $\bar{y}_s$ *is an unbiased estimator of the population mean* $\bar{y}_U$. *The variance is given by*

$$V_{SI}(\bar{y}_s) = \frac{1-f}{n}S_{yU}^2 \tag{3.3.5}$$

An unbiased variance estimator is

$$\hat{V}_{SI}(\bar{y}_s) = \frac{1-f}{n}S_{ys}^2 \tag{3.3.6}$$

Remark 3.3.1. As a general rule in this chapter, the results concerning estimation of the population mean $\bar{y}_U = \sum_U y_k/N = t/N$ are obtained directly from the corresponding results concerning π estimation of t. The π estimator of t is divided by N; the variance and the variance estimator are divided by N^2.

Case 3. Estimation for Domains (Subpopulations)

In most surveys, estimates are wanted not only for the whole population U, but also for specific subpopulations, called domains. Estimation for domains is an important subject considered in depth in Chapter 10. Here we give only some simple results illustrating the use of the π estimator in the SI design, for estimating (i) the absolute and relative size of a domain and (ii) a domain total and a domain mean.

Let the specific subpopulation of interest (the domain) be denoted U_d, where $U_d \subset U$; let N_d be the size of U_d. That is, N_d is the number of elements in U_d. Let $P_d = N_d/N$ be the relative size of U_d, that is, the proportion of elements in U which belong to U_d. We assume that N is known and N_d unknown, which is often the case in practice.

i. *Estimation of the Absolute Size and the Relative Size of a Domain*

The estimation of the two new parameters of interest, N_d and P_d, can be handled as special cases of the estimation of a population total and a population mean. To see this, we introduce a domain indicator variable, z_d, whose value for the kth element is defined as

$$z_{dk} = \begin{cases} 1 & \text{if } k \in U_d \\ 0 & \text{otherwise} \end{cases} \quad (k = 1, \dots, N) \tag{3.3.7}$$

Now,

$$\textstyle\sum_U z_{dk} = N_d \tag{3.3.8}$$

and

$$\bar{z}_{dU} = \textstyle\sum_U z_{dk}/N = N_d/N = P_d \tag{3.3.9}$$

This identifies N_d as the population total of the new variable z_d, and P_d as the population mean of z_d. Consequently, the results of cases 1 and 2 above can be applied directly. Nevertheless, it is instructive to go one step further and rewrite the results in terms of new notation. Let $Q_d = 1 - P_d$, let $n_d = \sum_s z_{dk}$, the number of elements in the sample that belong to U_d, let $p_d = n_d/n$, the proportion of elements in the sample that belong to U_d, and let $q_d = 1 - p_d$.

Given the definition of z_{dk} given in (3.3.7), some simple algebra will show that

$$S_{z_dU}^2 = \frac{N}{N-1} P_d Q_d \tag{3.3.10}$$

and that

$$S_{z_ds}^2 = \frac{n}{n-1} p_d q_d \tag{3.3.11}$$

The following Result 3.3.3 is now a simple consequence of Result 3.3.1 and equations (3.3.10) and (3.3.11).

Result 3.3.3. *Under the SI design, the π estimator of the absolute domain size N_d takes the form*

$$\hat{N}_d = Np_d \tag{3.3.12}$$

The variance is given by

$$V_{SI}(\hat{N}_d) = N^2 \frac{N-n}{N-1} \frac{P_d Q_d}{n} \tag{3.3.13}$$

An unbiased variance estimator is

$$\hat{V}_{SI}(\hat{N}_d) = N^2(1-f)\frac{p_d q_d}{n-1} \tag{3.3.14}$$

We can use Result 3.3.3 to obtain easily corresponding results for π estimation of the relative domain size, $P_d = N_d/N$. The implied estimator is simply $\hat{P}_d = p_d = n_d/n$. The variance and the variance estimator are N^2 times smaller than the corresponding expressions in Result 3.3.3. These conclusions are important in that they give the method for estimating a population proportion, under the *SI* design. For example, suppose that an estimator is required for the proportion $P_d = N_d/N$ of persons aged 65 and over in a population of N individuals. "Aged 65 and over" defines the domain, and the estimation of P_d can be seen as a simple exercise in domain estimation.

ii. *Estimation of a Domain Total and a Domain Mean When the Domain Size Is Unknown*

In a survey, an estimate is often wanted for the total of a specified domain, $t_d = \sum_{U_d} y_k$, and for the corresponding domain mean, $\bar{y}_{U_d} = \sum_{U_d} y_k/N_d$. For instance, in a survey of households where the study variable y is household disposable income, one may want to estimate the total disposable income of households with three or more children or the average disposable income of households with three or more children.

We can use the results on the π estimator to estimate t_d and $\bar{y}_{U_d}$. To this end, introduce the new variable y_d such that

$$y_{dk} = \begin{cases} y_k & \text{if } k \in U_d \\ 0 & \text{otherwise} \end{cases}$$

Then t_d is the total (over the entire population U) of the new variable y_d, that is,

$$t_d = \sum_{U_d} y_k = \sum_U y_{dk}$$

Result 3.3.1 can now be applied directly with the new variable y_d in place of

y. The π estimator of t_d becomes

$$\hat{t}_{d\pi} = \sum_s y_{dk}/\pi_k = \frac{N}{n}\sum_s y_{dk} = \frac{N}{n}\sum_{s_d} y_k \tag{3.3.15}$$

where $s_d = U_d \cap s$. That is, s_d is the subset of the sample s that falls in the domain U_d. The variance and the estimated variance follow easily from Result 3.3.1. Their derivation is left as an exercise.

Remark 3.3.2. In many applications the domain size N_d is unknown. If, however, N_d happens to be known, one normally prefers an alternative to the π estimator (3.3.15), namely,

$$\hat{t}_{d,\text{alt}} = N_d\left(\sum_{s_d} y_k/n_d\right) = N_d\bar{y}_{s_d} \tag{3.3.16}$$

where n_d is the size of s_d. Note that n_d is a random variable. To derive the variance of $\hat{t}_{d,\text{alt}}$, which is often much smaller than that of $\hat{t}_{d\pi}$, requires methods introduced later in Section 5.8.

Example 3.3.2 (This example makes use of the population of 124 countries, CO124, given in Appendix D.) An *SI* sample of 50 countries was drawn from the CO124 population. The variable P83 (= y = 1983 population) was observed for the sample countries. Let s_d be the set of sampled countries in the domain "Europe" (corresponding to continent 5). The following summary data were calculated:

$$\sum_{s_d} y_k = 205.2 \quad \text{and} \quad \sum_{s_d} y_k^2 = 6{,}232.84$$

The π estimate of the total of P83 for Europe (t_d) calculated from (3.3.15) is

$$\hat{t}_{d\pi} = \frac{124}{50}205.2 = 508.9$$

A variance estimate is

$$\hat{V}_{SI}(\hat{t}_{d\pi}) = 124^2\frac{1 - 50/124}{50}\frac{6{,}232.84 - 205.2^2/50}{49}$$

$$= 20{,}189.8$$

The value of *cve* is 28%, indicating a rather imprecise estimate.

To estimate the domain mean

$$\bar{y}_{U_d} = t_d/N_d \tag{3.3.17}$$

we could try the π estimator of t_d, divided by N_d, that is,

$$\hat{\bar{y}}_{U_d\pi} = \frac{N}{nN_d}\sum_{s_d} y_k \tag{3.3.18}$$

This estimator cannot be used if N_d is unknown. Moreover, a better estimator of $\bar{y}_{U_d}$, whether N_d is known or not, is obtained by dividing (3.3.16) by N_d. This leads to

$$\hat{\bar{y}}_{U_d \text{alt}} = \sum_{s_d} y_k/n_d = \bar{y}_{s_d} \tag{3.3.19}$$

This intuitively sound estimator is simply the mean of the y-values observed in the domain. The variance and the variance estimator are derived later with the aid of Result 5.8.1.

Remark 3.3.3. We note that several of the parameters encountered in cases 2 and 3 can be expressed as

$$\theta = \sum_U c_k y_k$$

where $c_1, \ldots, c_N$ are constants. For the population mean dealt with in case 2, c_k has the known value $1/N$ for all k. In case 3, estimation of the domain total $\sum_{U_d} y_k$, we have $c_k = 1$ for $k \in U_d$ and $c_k = 0$ otherwise. Normally, the domain membership indicator c_k is not known beforehand for all $k \in U$. But for elements k in the sample s, it may be possible to determine c_k. This is sufficient for the π-estimation method to function. By contrast, for estimation of the domain mean, $\bar{y}_{U_d}$, in case 3, we have $c_k = 1/N_d$ for $k \in U_d$ and $c_k = 0$ otherwise. If N_d is unknown, the c_k remain unknown, even for elements k appearing in the sample. This is why the π-estimation method fails when N_d is unknown.

3.3.2. Simple Random Sampling with Replacement

Simple random sampling with replacement (SIR), considered briefly in Examples 2.9.1 and 2.9.2, is the ordered design that gives the same selection probability $1/N^m$ to every ordered sample

$$os = (k_1, \ldots, k_i, \ldots, k_m)$$

where k_i is the element obtained in the ith draw. An element may be drawn more than once. The ordered sample contains at most m distinct elements.

The SIR design can be implemented by the draw-sequential scheme given in Example 2.9.1 with m independent with-replacement draws, such that each draw gives any one element a chance of $1/N$ to be drawn. In m draws, any given element will appear in the ordered sample a certain number of times, say r, such that r is a binomially distributed random variable with mean m/N and variance

$$\frac{m}{N}\left(1 - \frac{1}{N}\right) \doteq \frac{m}{N}$$

When N is moderate to large, the Poisson distribution with mean m/N will be an excellent approximation to the distribution of r. Under the SIR design, the pwr estimator (see Section 2.9) for the population total $t = \sum_U y_k$ takes the form

$$\hat{t}_{\text{pwr}} = N\bar{y}_{os} \tag{3.3.20}$$

where $\bar{y}_{os}$ is the ordered sample mean, repeated elements included,

$$\bar{y}_{os} = \frac{1}{m} \sum_{i=1}^{m} y_{k_i} \tag{3.3.21}$$

Defining the ordered sample variance as

$$S_{os}^2 = \frac{1}{m-1} \sum_{i=1}^{m} (y_{k_i} - \bar{y}_{os})^2 \tag{3.3.22}$$

we have the following result, which follows from Result 2.9.1.

Result 3.3.4. *Under the SIR design, the pwr estimator of the population total* $t = \sum_U y_k$ *takes the form of equation* (3.3.20). *The variance is given by*

$$V_{SIR}(\hat{t}_{\text{pwr}}) = N(N-1)S_{yU}^2/m \tag{3.3.23}$$

An unbiased variance estimator is

$$\hat{V}_{SIR}(\hat{t}_{\text{pwr}}) = N^2 S_{os}^2/m \tag{3.3.24}$$

Let us compare sampling with and without replacement. If n, the sample size in the without-replacement case, equals m, the ordered sample size in the with-replacement case, then

$$\frac{V_{SIR}(N\bar{y}_{os})}{V_{SI}(N\bar{y}_s)} = \frac{1-N^{-1}}{1-f} \doteq \frac{1}{1-f}$$

This shows that the two estimation strategies are roughly of equal efficiency when the sampling fraction $f = n/N$ is very small. On the other hand, if f is substantial, considerable efficiency is lost in with-replacement sampling. For example, if $f = 50\%$, the with-replacement variance is double the without-replacement variance.

Remark 3.3.4. For the *SIR* design, there exist other unbiased estimators than $N\bar{y}_{os}$. These are discussed in Section 3.8.

3.4. Systematic Sampling

3.4.1. Definitions and Main Result

Systematic sampling refers to a set of procedures that offer several practical advantages, particularly its simplicity of execution. We concentrate on systematic sampling in its basic form. A first element is drawn at random, and with equal probability, among the first a elements in the population list. The positive integer a is fixed in advance and is called the *sampling interval.* No further random draw is needed. The rest of the sample is determined by systematically taking every ath element thereafter, until the end of the list. Thus there are only a possible samples, each having the same probability $1/a$ of being selected. The simplicity of only one random draw is a great advantage. It is easy, for example, for an interviewer to select a systematic sample while out in the field.

For a more formal definition of this type of systematic sampling, let a be the fixed sampling interval and let n be the integer part of N/a, where N is the population size. Then

$$N = na + c$$

where the integer c satisfies $0 \leq c < a$. If $c = 0$, a sample of size n will be drawn by the procedure that we now present. If $c > 0$, the sample size is going to be either n or $n + 1$. The selection, which can be seen as list sequential, is as follows:

i. Select with equal probability $1/a$ a random integer, say r, between 1 and a (inclusively).
ii. The selected sample is composed as

$$s = \{k\colon k = r + (j - 1)a \leq N;\, j = 1, 2, \ldots, n_s\} = s_r \tag{3.4.1}$$

say, where the sample size n_s is either $n + 1$ (when $r \leq c$), or n (when $c < r \leq a$).

The integer r is called the *random start*.

Example 3.4.1. Let us return to the university professor in Example 3.2.1. Instead of selecting a sample of students by Bernoulli sampling, which requires 600 throws of the die, he could have used systematic sampling as follows: To select the first student, throw the die once, suppose that a "2" is realized. The systematic sample is then completely determined; it consists of the 2nd, the 8th, the 14th, ..., and the 596th student. The sample size is 100; and $a = 6$, $c = 0$.

The set of possible samples, denoted $\mathscr{S}_{SY}$, consists of the a different (non-overlapping) sets that can be obtained in the manner of equation (3.4.1), namely,

$$\mathscr{S}_{SY} = \{s_1, \ldots, s_r, \ldots, s_a\}$$

Ordinarily, this represents an extremely small number of possible samples, compared to, for example, *SI* sampling.

The sampling design, which we denote *SY*, is given by

$$p(s) = \begin{cases} 1/a & \text{if } s \in \mathscr{S}_{SY} \\ 0 & \text{for any other sample } s \end{cases}$$

To determine the π estimator and its variance, we need the first- and second-order inclusion probabilities. Because each element k belongs to one and only one of the a equally probable systematic samples, we have, for every $k \in U$,

$$\pi_k = 1/a$$

whereas, for every $k \neq l \in U$,

$$\pi_{kl} = \begin{cases} 1/a & \text{if } k \text{ and } l \text{ belong to the same sample } s \\ 0 & \text{otherwise} \end{cases} \tag{3.4.2}$$

The desirable property that all $\pi_{kl} > 0$ (see Section 2.4) is thus not fulfilled.

Remark 3.4.1. There is no overlap between any two samples and the a samples together make up the whole population U. That is,

$$s_1, \ldots, s_r, \ldots, s_a$$

represents a partition of U into a subpopulations. That is,

$$U = \bigcup_{r=1}^{a} s_r$$

and we can write $t = \sum_U y_k$ as

$$t = \sum_{r=1}^{a} t_{s_r} \quad \text{with} \quad t_{s_r} = \sum_{s_r} y_k$$

The SY design can thus be described as a random, equal probability selection of one out of a different subpopulations. The selected one is completely surveyed. The situation is illustrated by the following table for the case where $c = 0$.

	Sample, s				
	s_1	$\cdots$	s_r	$\cdots$	s_a
y values	y_1	$\cdots$	y_r	$\cdots$	y_a
	y_{a+1}	$\cdots$	y_{a+r}	$\cdots$	y_{2a}
	$\vdots$		$\vdots$		$\vdots$
	$y_{(n-1)a+1}$	$\cdots$	$y_{(n-1)a+r}$	$\cdots$	y_N
Sample total	t_{s_1}	$\cdots$	t_{s_r}	$\cdots$	t_{s_a}

Then, as a consequence of the general Result 2.8.1, we have the following result.

Result 3.4.1. *Under the SY design, with the sampling interval a, the π estimator of the population total $t = \sum_U y_k$ takes the form*

$$\hat{t}_\pi = a t_s \tag{3.4.3}$$

where $t_s = \sum_s y_k$ *is the sample total of* y, *and* s *is a member of the set of possible samples* $\{s_1, \ldots, s_r, \ldots, s_a\}$, s_r *being defined by* (3.4.1). *The variance is given by*

$$V_{SY}(\hat{t}_\pi) = a \sum_{r=1}^{a} (t_{s_r} - \bar{t})^2 \tag{3.4.4}$$

where $\bar{t} = \sum_{r=1}^{a} t_{s_r}/a = t/a$ *is the average of the sample totals* $t_{s_r} = \sum_{s_r} y_k$.

The variance can also be written

$$V_{SY}(\hat{t}_\pi) = a(a-1)S_t^2$$

where

$$S_t^2 = \frac{1}{a-1} \sum_{r=1}^{a} (t_{s_r} - \bar{t})^2$$

is the variance of the sample totals. The variance is small if the sample totals are nearly equal.

PROOF. Since $\pi_k = 1/a$ $(k = 1, \ldots, N)$ we have

$$\hat{t}_\pi = \sum_s \check{y}_k = \sum_s y_k/\pi_k = at_s$$

To determine the design variance, we use (2.8.8) and (3.4.2) to obtain

$$\begin{aligned} V_{SY}(\hat{t}_\pi) &= \sum\sum_U \frac{\pi_{kl}}{\pi_k \pi_l} y_k y_l - t^2 \\ &= a \sum_{r=1}^{a} \{\sum\sum_{s_r} y_k y_l\} - t^2 \\ &= a \sum_{r=1}^{a} t_{s_r}^2 - t^2 = a \sum_{r=1}^{a} (t_{s_r} - \bar{t})^2 \end{aligned}$$ □

Remark 3.4.2. Since the condition $\pi_{kl} > 0$ for all $k \neq l$ is not verified, one should not use the variance estimator of Result 2.8.1. The formula gives, in this case, a nonsensical result. Variance estimation in systematic sampling is discussed in Section 3.4.4.

3.4.2. Controlling the Sample Size

From the definition of the SY design it follows that if $c = 0$, then $N = na$, and all a possible samples have the same size n. If, on the other hand, $c > 0$, the sample size will be either $n + 1$ (if $r \leq c$) or n (if $r > c$). It is also clear that requiring a to be a positive integer will in extreme situations lead to sample sizes that may differ substantially from what is desired.

For example, suppose that $N = 149$, and that the desired sample size is 60. Then the choice $a = 2$ gives a sample size of either 74 or 75, whereas $a = 3$ gives a sample size of 49 or 50. Sizes between 50 and 74 are impossible. Different methods exist to handle this problem; two are presented below. Obviously, if N is very large compared to n, the problem is minor.

i. *Fractional Interval Method*

This method permits a fractional value of a. Let $a = N/n$, where n is the desired sample size. Draw a random number ξ from the uniform distribution on the interval $(0, a)$. The selected sample will then consist of those elements k for which

$$k - 1 < \xi + (j - 1)a \leq k; \qquad j = 1, \ldots, n$$

The method is equivalent to drawing, with equal probability $1/N$, a random integer r between 1 and N (inclusively), and selecting the elements k for which

$$(k - 1)n < r + (j - 1)N \leq kn; \qquad j = 1, \ldots, n$$

For example, the element $k = 1$ is selected if r satisfies $0 < r \leq n$, which occurs with probability n/N.

Similarly, any element k will have the chance $\pi_k = 1/a = n/N$ of being chosen, and every possible sample will be exactly size n.

ii. *Circular Systematic Sampling Method*

In this method, the frame is laid out circularly, that is, the last element ($k = N$) is followed by the first ($k = 1$), and so on. A random number r between 1 and N (inclusively) is drawn with equal probability. Let a be the integer nearest to N/n. Then the sample consists of the elements k such that, for $j = 1, \ldots, n$,

$$k = r + (j - 1)a \quad \text{if} \quad r + (j - 1)a \leq N$$

or

$$k = r + (j - 1)a - N \quad \text{if} \quad r + (j - 1)a > N$$

As in the fractional interval method, every sample will have the size n, and $\pi_k = n/N$ for every k.

Remark 3.4.3. To find the variance of the π estimator for the fractional interval and circular methods, the π_{kl} must first be derived. This requires some special attention, since the possible samples are not necessarily pairwise disjoint in these two methods. In the case $c = 0$, all three systematic sampling methods presented in this section are equivalent. When N is large compared to n, the difference between the three methods is slight.

3.4.3. The Efficiency of Systematic Sampling

The variance (3.4.4) is near zero if the sample totals t_{s_r} are approximately equal. When does this happen? Recall that the samples s_r were formed by taking systematically every ath element. If the ordering of the N population elements is such that the resulting systematic samples have roughly the same y-total, the variance will be small. In other words, the efficiency of systematic sampling depends greatly on the particular ordering of the N elements on which the systematic selection is applied.

We now study the efficiency of systematic sampling as a function of population ordering. For simplicity, let us consider the case where $N = an$, where a is an integer. By Result 3.4.1, the π estimator for the sample $s = s_r$ is given by

$$\hat{t}_\pi = N \sum_{s_r} y_k/n = N\bar{y}_{s_r} \tag{3.4.5}$$

and the variance is

$$V_{SY}(\hat{t}_\pi) = N^2 \frac{1}{a} \sum_{r=1}^{a} (\bar{y}_{s_r} - \bar{y}_U)^2 \tag{3.4.6}$$

with $\bar{y}_U = \sum_U y_k/N$, which is the population mean.

The variance shown in equation (3.4.6) is near zero if all sample means $\bar{y}_{s_r}$ are nearly equal. To further analyze the variance (3.4.6), note that

$$\sum_U (y_k - \bar{y}_U)^2 = \sum_{r=1}^{a} \sum_{s_r} (y_k - \bar{y}_{s_r})^2 + \sum_{r=1}^{a} n(\bar{y}_{s_r} - \bar{y}_U)^2 \tag{3.4.7}$$

That is, the total variation in the population can be decomposed into the variation within systematic samples and the variation between systematic samples, as in a standard one-way analysis of variance. This may be written

$$SST = SSW + SSB \tag{3.4.8}$$

where SS represents sum of squares; T, total; W, within; and B, between. For a given population, $SST = (N - 1)S_{yU}^2$ is fixed. Hence, an increase in the within variation SSW, is accompanied by a corresponding decrease in the between variation SSB.

Apart from a fixed multiplicative constant, SSB determines the variance (3.4.6), which we can now write as

$$V_{SY}(\hat{t}_\pi) = N \cdot SSB$$

In other words, the more homogeneous the elements within systematic samples are, the less efficient the systematic sampling is. Homogeneous is used here to connote the tendency to have equal y-values. Thus, to achieve a favorable population ordering for systematic sampling, we should strive for an ordering that entails a low degree of homogeneity among elements within the same systematic sample.

How do we measure homogeneity? We discuss two related measures. The first is

$$\rho = 1 - \frac{n}{n-1} \cdot \frac{SSW}{SST} \tag{3.4.9}$$

which is called the *intraclass correlation coefficient*, since it can be written alternatively as

$$\rho = \frac{2 \sum_{r=1}^{a} \left[\sum\sum_{k<l} {}_{s_r} (y_k - \bar{y}_U)(y_l - \bar{y}_U) \right]}{(n-1)(N-1)S_{yU}^2} \tag{3.4.10}$$

We can interpret ρ as a measure of the correlation between pairs of elements within the same systematic sample. A positive value of ρ is obtained when elements in the same sample tend to have similar y-values. At the one extreme, $\rho = 1$ if $SSW = 0$, that is, there is complete homogeneity (no variation) within systematic samples. At the other extreme, $\rho = -1/(n-1)$ if $SSB = 0$, that is, complete heterogeneity within samples.

The homogeneity measure that we prefer is closely related to ρ, namely,

$$\delta = 1 - \frac{N-1}{N-a} \cdot \frac{SSW}{SST} \tag{3.4.11}$$

Introducing the *intra-sample variance*,

$$S_{yW}^2 = \frac{SSW}{N-a} \tag{3.4.12}$$

and noting that the overall population variance is

$$S_{yU}^2 = \frac{SST}{N-1} \tag{3.4.13}$$

we thus have the simple equation

$$S_{yW}^2 = (1-\delta)S_{yU}^2$$

or

$$\frac{\text{intrasample variance}}{\text{overall variance}} = 1 - \delta \tag{3.4.14}$$

An advantage with δ (rather than ρ) as a measure of homogeneity is that the representation of ρ in equation (3.4.10), which is reminiscent of a correlation coefficient, holds only for classes (samples) of equal size. By contrast, the easy-to-grasp representation of δ in equation (3.4.14) applies whether the s_r are of the same size or not.

The extreme values of δ are

$$\delta_{\text{min}} = -\frac{a-1}{N-a}$$

which occurs if $SSB = 0$ ($\bar{y}_s$ = constant for all s), and

$$\delta_{\text{max}} = 1$$

which occurs if $SSW = 0$, that is, there is complete homogeneity.

The following result expresses $V_{SY}(\hat{t}_\pi)$ as a simple function of δ and is useful for the comparison with *SI* sampling.

Result 3.4.2. *Under the SY design (with $N = an$, where a is an integer), the variance of the π estimator of the population total can be written*

$$V_{SY}(\hat{t}_\pi) = \frac{N^2 S_{yU}^2}{n}[(1 - f) + (n - 1)\delta] \tag{3.4.15}$$

where $f = n/N = 1/a$ is the sampling fraction.

The proof, which follows from equation (3.4.6) and the definition of δ, is left as an exercise.

From (3.4.15) we see, once again, that the more homogeneous, or the less heterogeneous, the elements are within systematic samples, the less efficient systematic sampling is.

Let us compare the designs SY and SI. To shorten the notation, let $V_{SY} = V_{SY}(\hat{t}_\pi)$ and $V_{SI} = V_{SI}(\hat{t}_\pi)$ be the respective variances of the π estimator of $t = \sum_U y_k$. In both cases, the estimator is $\hat{t}_\pi = N\bar{y}_s$. Nevertheless, the sampling distribution is different in the two cases. Since

$$V_{SI} = N^2 \frac{1 - f}{n} S_{yU}^2$$

the design effect of SY (see Section 2.10 for the definition) is obtained, with the aid of (3.4.15), as

$$\text{deff}(SY, \hat{t}_\pi) = \frac{V_{SY}}{V_{SI}} = 1 + \frac{n - 1}{1 - f}\delta$$

Hence SY is more efficient than SI if $\delta < 0$, that is, if $S_{yW}^2 > S_{yU}^2$. To create a situation where this condition holds, we must (if possible) arrange the population so that the y_k within each systematic sample exhibit considerable heterogeneity. This will often be the case if the arrangement is such that neighboring elements have y_k-values close to each other. Conversely, SI is more efficient than SY if $\delta > 0$, that is, if $S_{yW}^2 < S_{yU}^2$. However, in practice the statistician often lacks the information needed to create an ordering with given properties, for instance, an ordering favorable to SY sampling. The ordering is often given once and for all.

EXAMPLE 3.4.2. This example shows the effect of different population orderings. Suppose that $N = 100$, that the variable y takes the values $1, 2, \ldots, 100$, and that $n = 10$. Then the number of samples is $a = N/n = 10$. We have

$$S_{yU}^2 = \frac{N(N + 1)}{12} = \frac{100 \cdot 101}{12}$$

Regardless of the ordering of the population, the variance under *SI* sampling of $n = 10$ from $N = 100$ is

$$V_{SI} = N^2 \frac{1-f}{n} S_{yU}^2 = 7.575 \cdot 10^5$$

Let us examine a few orderings.

(a) Suppose the ordering is such that $y_k = k$ ($k = 1, \ldots, 100$), that is, a perfect linear trend in the values y_k. The following table shows the ten possible systematic samples, and t_{s_r}, which is the total of the rth sample.

	r									
	1	2	3	4	5	6	7	8	9	10
y_k	1	2	3	4	5	6	7	8	9	10
	11	12	13	14	15	16	17	18	19	20
	21	22	23	24	25	26	27	28	29	30
	31	32	33	34	35	36	37	38	39	40
	41	42	43	44	45	46	47	48	49	50
	51	52	53	54	55	56	57	58	59	60
	61	62	63	64	65	66	67	68	69	70
	71	72	73	74	75	76	77	78	79	80
	81	82	83	84	85	86	87	88	89	90
	91	92	93	94	95	96	97	98	99	100
t_{s_r}	460	470	480	490	500	510	520	530	540	550

In this case, $V_{SY} = 8.25 \cdot 10^4$, which means that V_{SI} is more than nine times V_{SY}. The measure δ of homogeneity is $\delta = -0.089$, which is not far from $\delta_{\text{min}} = -0.1$.

(b) An optimal (minimal variance) ordering for systematic sampling is given in the following table.

	r									
	1	2	3	4	5	6	7	8	9	10
y_k	1	2	3	4	5	6	7	8	9	10
	20	19	18	17	16	15	14	13	12	11
	21	22	23	24	25	26	27	28	29	30
	40	39	38	37	36	35	34	33	32	31
	41	42	43	44	45	46	47	48	49	50
	60	59	58	57	56	55	54	53	52	51
	61	62	63	64	65	66	67	68	69	70
	80	79	78	77	76	75	74	73	72	71
	81	82	83	84	85	86	87	88	89	90
	100	99	98	97	96	95	94	93	92	91
t_{s_r}	505	505	505	505	505	505	505	505	505	505

Since all t_{s_r} are equal, $V_{SY} = 0$, and $\delta = \delta_{\text{min}} = -0.1$.

(c) A large positive δ value is associated with the ordering in the following table.

	r									
	1	2	3	4	5	6	7	8	9	10
y_k	1	11	21	31	41	51	61	71	81	91
	2	12	22	32	42	52	62	72	82	92
	3	13	23	33	43	53	63	73	83	93
	4	14	24	34	44	54	64	74	84	94
	5	15	25	35	45	55	65	75	85	95
	6	16	26	36	46	56	66	76	86	96
	7	17	27	37	47	57	67	77	87	97
	8	18	28	38	48	58	68	78	88	98
	9	19	29	39	49	59	69	79	89	99
	10	20	30	40	50	60	70	80	90	100
t_{s_r}	55	155	255	355	455	555	655	755	855	955

We get $V_{SY} = 8.25 \cdot 10^6$, which is almost 11 times V_{SI}. Here $\delta = 0.989$, which is close to the maximum $\delta_{\max} = 1$.
(d) A haphazard ordering is given in the final table.

	r									
	1	2	3	4	5	6	7	8	9	10
y_k	48	14	71	13	40	59	18	45	6	53
	38	23	11	58	70	22	24	88	77	84
	10	51	98	65	93	68	25	32	99	9
	17	26	8	78	34	87	96	39	20	54
	56	79	31	86	43	66	2	62	57	5
	73	7	80	27	60	89	76	81	85	83
	3	28	33	90	55	1	21	69	61	92
	74	37	44	94	12	72	100	30	63	97
	75	41	16	82	35	95	67	50	64	29
	49	42	15	19	46	36	47	91	52	4
t_{s_r}	443	348	407	612	488	595	476	587	584	510

Here, $V_{SY} = 7.1766 \cdot 10^5$, which is close to V_{SI}, and $\delta = -0.005$. This arrangement was created by a random permutation of the integers 1 to 100. The population can be said to be in random order, thus, δ is expected to be near 0.

The example shows that care must be taken when using systematic sampling. The price for the simplicity of systematic sampling can be a high loss of efficiency. On the other hand, if the population is ordered in a way favorable to systematic sampling, large efficiency gains can be realized.

3.4.4. Estimating the Variance

There is no unbiased estimator of the design variance $V_{SY}(\hat{t}_\pi)$. This is another trade off for the simplicity of systematic sampling. We cannot assess the sampling variability of the point estimate. There exist some less-than-perfect approaches for handling this problem. One is to use a biased variance estimator; another is to modify the systematic selection to permit unbiased variance estimation.

Several biased variance estimators have been proposed in the literature; let us mention one. Suppose there is strong reason to believe that, in a specific application, SY is at least as good as SI when small variance is the criterion. If s_r is the selected systematic sample, the sample variance is

$$S^2_{ys_r} = \frac{1}{n-1} \sum_{s_r} (y_k - \bar{y}_{s_r})^2$$

As an estimator of $V_{SY}(\hat{t}_\pi)$, we might try

$$\hat{V} = \frac{N^2(1-f)}{n} S^2_{ys_r} \tag{3.4.16}$$

which is the variance estimator appropriate for the SI design.

Suppose we are in a situation where SY sampling is more efficient than SI sampling, that is,

$$V_{SY}(\hat{t}_\pi) < V_{SI}(\hat{t}_\pi)$$

which holds if and only if $\delta < 0$. One can then show that $\hat{V}$ given by equation (3.4.16) will *overestimate* $V_{SY}(\hat{t}_\pi)$. A variance estimator $\hat{V}$ is said to overestimate if its expected value exceeds the variance for which $\hat{V}$ is used as an estimator. That is, for the variance estimator $\hat{V}$ given by Equation (3.4.16) we have

$$E_{SY}(\hat{V}) > V_{SY}(\hat{t}_\pi)$$

if $\delta < 0$. When SY sampling is used and $\delta < 0$, a confidence interval for t calculated by means of equation (3.4.16) will be what is known as *conservative*. A confidence interval is called conservative if a 95% level say is aimed for in the calculation of the interval—the constant $z_{0.975} = 1.96$ would then be used, assuming approximate normality—and the real confidence level is greater than 95%. In repeated samples, such an interval contains the unknown parameter value at a rate of more than 95% of all samples.

EXAMPLE 3.4.3. For the population in Example 3.4.2, we have (approximately) the following table:

Ordering	$E_{SY}(\hat{V})$	$V_{SY}(\hat{t}_\pi)$	
(a)	$8.3 \cdot 10^6$	$8.3 \cdot 10^4$	overestimation
(b)	$8.3 \cdot 10^5$	0	overestimation
(c)	$8.3 \cdot 10^3$	$8.3 \cdot 10^6$	underestimation
(d)	$7.6 \cdot 10^5$	$7.2 \cdot 10^5$	roughly unbiased

In case (d), where SY and SI are approximately equally efficient designs (and δ very near zero), there is good agreement between $E_{SY}(\hat{V})$ and $V_{SY}(\hat{t}_\pi)$. In the other cases, $\hat{V}$ either overestimates or underestimates $V_{SY}(\hat{t}_\pi)$, depending on whether $\delta < 0$ or $\delta > 0$. This is confirmed by the figures in the table. For instance, in case (a), the value of $E_{SY}(\hat{V})$ greatly exceeds that of $V_{SY}(\hat{t}_\pi)$, whereas the relationship is inversed for case (c). Alternatives to $\hat{V}$ have been proposed in the literature. A review is given in Wolter (1985), who devotes a chapter to variance estimation for systematic sampling.

Another solution to the variance estimation problem is to modify the design SY. For instance, instead of using only one random start and the sampling interval a, we can use $m > 1$ random starts and the sampling interval ma. This gives a sample consisting of m systematic "slices," each of size n/m. Let us for simplicity assume that n/m and $a = N/n$ are integers. An SI sample of m integers is selected from the integers 1 to ma, inclusively. Say that the selected numbers are $r_1, r_2, \ldots, r_m$. The sample is then given by

$$s = \{k: k = r_i + (j-1)ma; i = 1, \ldots, m, j = 1, \ldots, n/m\}$$

In this case $\pi_k = m/ma = n/N$ for every k, whereas

$$\pi_{kl} = \begin{cases} \dfrac{n}{N} & \text{if } k\&l \text{ belong to the same sample } s \\[2ex] \dfrac{n}{N} \cdot \dfrac{m-1}{ma-1} & \text{if } k\&l \text{ belong to different samples} \end{cases}$$

In the terminology of cluster sampling introduced in the next chapter, systematic sampling with m random starts is equivalent to grouping the population into ma clusters, each of size n/m, and taking an SI sample of m clusters. All elements in the selected clusters are surveyed.

The single random start variety of systematic sampling, that is, the design that we denote SY, is also a special case of cluster sampling, namely, with one single cluster drawn at random. A drawback with several random starts is that it often leads to a higher variance than a single random start.

As shown in the next chapter, in some surveys, the problem of variance estimation is not as serious as it may appear. Often systematic sampling is used for the selection of the last stage units in a multistage sampling design. In such situations, it is often possible to get good variance estimators for the π estimator despite the use of systematic sampling in the final stage.

Systematic sampling in two dimensions is appropriate in certain applications. A field or a forest may be divided into smaller squares or rectangles of equal size. A random starting point (consisting of a random horizontal coordinate and a random vertical coordinate) is determined for the first square; the selected point in the first square will then systematically identify a sampled point in each of the other squares. Bellhouse (1977, 1981) has developed designs for spatial sampling; a review of systematic sampling in one or more dimensions is found in Bellhouse (1988).

3.5. Poisson Sampling

Bernoulli sampling, simple random sampling, and systematic sampling are equal probability designs, that is, all π_k are equal in these designs. Equal inclusion probabilities lead to simple estimators but are not typical of survey sampling. Most designs used in practice are unequal probability designs, because they are usually more efficient.

Our first example of an unequal probability sampling design is Poisson sampling (*PO*). This random size design is a generalization of Bernoulli sampling. Let π_k be a predetermined positive inclusion probability for the kth element, where $k = 1, \ldots, N$. Poisson sampling is defined as follows: the sample membership indicators I_k are independent, with I_k distributed as

$$\Pr(I_k = 1) = \pi_k, \qquad \Pr(I_k = 0) = 1 - \pi_k$$

$k = 1, \ldots, N$. The sampling design *PO* is such that the sample s has the probability

$$p(s) = \prod_{k \in s} \pi_k \prod_{k \in U-s} (1 - \pi_k) \tag{3.5.1}$$

where $s \in \mathscr{S}$, the set of all 2^N subsets of U. Because of the independence, $\pi_{kl} = \pi_k \pi_l$ for any $k \neq l$. Because the π_k can be specified in a variety of ways, Poisson sampling corresponds to a whole class of designs.

Given a set of inclusion probabilities $\pi_1, \ldots, \pi_N$, the design *PO* has a simple list-sequential implementation. Let $\varepsilon_1, \ldots, \varepsilon_N$ be independent random numbers drawn from the Unif(0, 1) distribution. If $\varepsilon_k < \pi_k$, the element k is selected, otherwise not; $k = 1, \ldots, N$.

In Poisson sampling the sample size n_s is random, with mean

$$E_{PO}(n_s) = \textstyle\sum_U \pi_k \tag{3.5.2}$$

and variance

$$V_{PO}(n_s) = \sum_U \pi_k(1 - \pi_k) \tag{3.5.3}$$

The following result concerning the π estimator under Poisson sampling follows easily from the general Result 2.8.1 by observing that $\pi_{kl} = \pi_k\pi_l$ for every $k \neq l$.

Result 3.5.1. *Under the PO design, the π estimator of the population total $t = \sum_U y_k$ is given by*

$$\hat{t}_\pi = \sum_s \check{y}_k = \sum_s y_k/\pi_k \tag{3.5.4}$$

The variance is given by

$$V_{PO}(\hat{t}_\pi) = \sum_U \pi_k(1 - \pi_k)\check{y}_k^2 = \sum_U \left(\frac{1}{\pi_k} - 1\right) y_k^2 \tag{3.5.5}$$

An unbiased variance estimator is

$$\hat{V}_{PO}(\hat{t}_\pi) = \sum_s (1 - \pi_k)\check{y}_k^2 \tag{3.5.6}$$

Here it is appropriate to warn that $V_{PO}(\hat{t}_\pi)$ may be unduly large because of the variability in the sample size. A better, but slightly biased, estimator for Poisson sampling is given by (3.5.9) below. In Bernoulli sampling, the design is completely specified as soon as we have fixed the expected sample size (assuming N to be known). By contrast, in Poisson sampling there is a host of choices for the π_k, for a fixed expected sample size. What then is the best choice? An answer is obtained by minimizing the variance (3.5.5) for a fixed expected sample size, $n = \sum_U \pi_k$. This is equivalent to minimizing the product

$$(\textstyle\sum_U y_k^2/\pi_k)(\sum_U \pi_k)$$

But, by the Cauchy-Schwartz inequality,

$$(\textstyle\sum_U y_k^2/\pi_k)(\sum_U \pi_k) \geq (\sum_U y_k)^2$$

with equality if and only if $y_k/\pi_k = \lambda$, a constant. Assuming that $y_k > 0$ for all k, we have $\pi_k = y_k/\lambda$. Finally, since $n = \sum_U \pi_k$, we get

$$\pi_k = ny_k/\textstyle\sum_U y_k \tag{3.5.7}$$

$k = 1, \ldots, N$, assuming also that $y_k \leq \sum_U y_k/n$ for all k.

Now, the y_k are unknown, so the solution given by (3.5.7) is of purely academic interest. However, in some surveys, we have access to one or more auxiliary variables, that is, variables whose values are known for the entire population. Suppose $x_1, \ldots, x_N$ are known positive values of an auxiliary variable x. There may be reason also to assume that y is approximately proportional to x. In such a case, we can let π_k be proportional to the known x_k. That is, for $k = 1, \ldots, N$,

$$\pi_k = nx_k/\textstyle\sum_U x_k \tag{3.5.8}$$

assuming $x_k \leq \sum_U x_k/n$ for all k. (If $x_k > \sum_U x_k/n$, we should take $\pi_k = 1$.) If y_k/x_k is nearly constant, the resulting π estimator will have a small variance. Inclusion probabilities determined by (3.5.8) are said to be *proportional-to-size*. The value x_k is considered a size measure of the kth element. Typical size measures are total assets or number of employees for a population of business firms, total acreage for a population of farms, and so on.

Although this reasoning is correct, there is one catch, namely, that Poisson sampling has the same drawback as Bernoulli sampling, that is, a random sample size. To illustrate this drawback, suppose it were possible to choose the π_k optimally according to (3.5.7), with $n = \sum_U \pi_k$, the expected sample size. In this extreme case, the π estimator becomes

$$\hat{t}_\pi = \sum_s y_k/\pi_k = (n_s/n) \sum_U y_k = (n_s/n)t$$

Hence, the sample-to-sample variation of $\hat{t}_\pi$ will merely consist of the variation in the sample size n_s.

This argument leads us to expect the π estimator $\hat{t}_\pi = \sum_s \check{y}_k$ to perform extremely well if it were possible to construct a fixed size design with inclusion probabilities π_k that are, at least very nearly, proportional to y_k. If the π_k were exactly proportional to y_k in a fixed size design, the π estimator would have zero variance.

Remark 3.5.1. Under Poisson sampling, an alternative estimator of $t = \sum_U y_k$ is

$$\hat{t}_{\text{alt}} = N \frac{\sum_s \check{y}_k}{\hat{N}} \tag{3.5.9}$$

where

$$\hat{N} = \sum_s (1/\pi_k)$$

The approximate variance (see Result 7.4.1) is

$$V(\hat{t}_{\text{alt}}) \doteq \sum_U \frac{(y_k - \bar{y}_U)^2}{\pi_k} - NS_{yU}^2$$

which is ordinarily smaller than the variance of $\hat{t}_\pi$ given by (3.5.5). Thus, $\hat{t}_{\text{alt}}$ is generally preferred to $\hat{t}_\pi$.

3.6. Probability Proportional-to-Size Sampling

3.6.1. Introduction

Our discussion of Poisson sampling showed that if the study variable y is approximately proportional to a positive and known auxiliary variable x, there is some merit in selecting the elements with probability proportional to

x. For example, x may be an inexpensive rough measure or an eye estimate of the study variable. Probability proportional-to-size sampling is advantageous not only for Poisson sampling, but for other sampling designs as well. To realize more clearly these advantages, consider probability proportional-to-size sampling in two contexts: (i) fixed-size without-replacement designs coupled with the π estimator, and (ii) with-replacement designs coupled with the pwr estimator.

i. *Fixed-Size Without-Replacement Designs Coupled with the π Estimator*

Consider the π estimator

$$\hat{t}_\pi = \sum_s y_k/\pi_k \tag{3.6.1}$$

Suppose it is possible to construct a fixed-size without-replacement design, and a sampling scheme to implement that design, such that

$$y_k/\pi_k = c, \qquad k = 1, \ldots, N \tag{3.6.2}$$

where c is a constant. Then, for any sample s, we would have

$$\hat{t}_\pi = nc$$

where n is the fixed size of s. Since $\hat{t}_\pi$ has no variation, the variance would be zero. Obviously, a design (and a corresponding sampling scheme) satisfying (3.6.2) cannot be found, because it requires advance knowledge of all the y_k. However, suppose that the auxiliary variable x is known to be approximately proportional to y. Then choosing π_k proportional to the known value x_k will lead to approximately constant ratios y_k/π_k. As a result, the variance of the π estimator will be small. A problem that we return to in Section 3.6.2 is how to find an easily implemented without-replacement scheme having π_k proportional to preassigned positive numbers x_k.

ii. *With-Replacement Designs Coupled with the pwr Estimator*

We can make a similar case for with-replacement sampling. Consider the pwr estimator (see Section 2.9),

$$\hat{t}_{\text{pwr}} = \frac{1}{m} \sum_{i=1}^{m} y_{k_i}/p_{k_i} \tag{3.6.3}$$

where m is the fixed number of with replacement draws and p_{k_i} is the probability with which the element k_i is drawn. Now, if it is possible to find a with-replacement sampling design and a corresponding sampling scheme such that

$$y_k/p_k = c, \qquad k = 1, \ldots, N \tag{3.6.4}$$

where c is a constant, then, for any ordered sample $os = (k_1, \ldots, k_m)$, we would have

$$\hat{t}_{\text{pwr}} = c$$

That is, $\hat{t}_{\text{pwr}}$ would have no sample-to-sample variation and consequently a zero variance. Again, the proportionality expressed in (3.6.4) is an ideal that cannot be attained, for the same reason as in (i) above. And again a solution that approaches the ideal is to take $p_k \propto x_k$, where x_k is an auxiliary variable value roughly proportional to y_k.

Remark 3.6.1. That the π estimator (3.6.1) has zero variance under the proportionality expressed by (3.6.2) can alternatively be seen from Result 2.8.2, which gives the variance of $\hat{t}_\pi$ in the case of a fixed-size design as

$$V(\hat{t}_\pi) = -\frac{1}{2}\sum\sum_U \Delta_{kl}\left(\frac{y_k}{\pi_k} - \frac{y_l}{\pi_l}\right)^2$$

This variance is zero if (3.6.2) holds true. As for the with-replacement case, Result 2.9.1 states that the variance of the pwr estimator is

$$V(\hat{t}_{\text{pwr}}) = \frac{1}{m}\sum_U p_k\left(\frac{y_k}{p_k} - t\right)^2$$

which can also be written as

$$V(\hat{t}_{\text{pwr}}) = \frac{1}{2m}\sum\sum_U p_k p_l\left(\frac{y_k}{p_k} - \frac{y_l}{p_l}\right)^2$$

From this last expression, it is evident that the variance is zero under the proportionality stated in (3.6.4).

Our discussion has shown that there is considerable incentive to study sampling designs such that the probabilities π_k, or p_k, are proportional to a positive size measure x_k. Such designs are called probability proportional-to-size designs, and the sampling schemes used to implement them are called probability proportional-to-size schemes. To distinguish the without-replacement case from the with-replacement case, we use the terms *πps sampling* (that is, $\pi_k \propto x_k$) and *pps sampling* (that is, $p_k \propto x_k$).

Remark 3.6.2. For populations with large variation in x, it may not be possible to select a sample in strict agreement with the πps rule $\pi_k \propto x_k$. The required proportionality factor is $n/\sum_U x_k$ (because $\sum_U \pi_k = n =$ the fixed sample size). In other words,

$$\pi_k = \frac{n x_k}{\sum_U x_k} \tag{3.6.5}$$

Now, obviously $\pi_k \leq 1$ must be satisfied. When $n = 1$, this is the case for all k. If $n > 1$, and some x_k values are very large, it may be true for some elements that

$$\frac{nx_k}{\sum_U x_k} > 1$$

contradicting the requirement $\pi_k \leq 1$.

There are methods to deal with the conflict. We could set $\pi_k = 1$ for all k such that $nx_k > \sum_U x_k$, and let π_k be proportional to x for the remaining elements k. That is, for a desired fixed sample size of n, take

$$\pi_k = (n - n_A)\frac{x_k}{\sum_{U-A} x_k}$$

for $k \in U - A$, where A is the set of n_A elements such that $nx_k > \sum_U x_k$. If necessary, the procedure is repeated until all $\pi_k \leq 1$. Another method would be to partition the population U into subpopulations, and to draw a πps sample in each subpopulation, making sure that the conflict does not arise for any of the subpopulations.

In the rest of this chapter we simply assume that $nx_k/\sum_U x_k \leq 1$ for every $k \in U$.

3.6.2. πps Sampling

In πps sampling, the inclusion probabilities satisfy $\pi_k \propto x_k$, where $x_1, \ldots, x_N$ are known, positive numbers. It is not a simple matter to devise a fixed-size πps sampling scheme having all of the following desirable properties:

a. The actual selection of the sample is relatively simple.
b. The first-order inclusion probabilities π_k $(k = 1, \ldots, N)$ are strictly proportional to x_k.
c. The second-order inclusion probabilities satisfy $\pi_{kl} > 0$ for all $k \neq l$ (needed to make possible unbiased variance estimation).
d. The π_{kl} generated by the scheme can be computed exactly without very heavy calculations.
e. $\Delta_{kl} = \pi_{kl} - \pi_k \pi_l < 0$ for all $k \neq l$ to guarantee that the Sen-Yates-Grundy variance estimator given by (2.8.11) takes always a nonnegative value.

In the following we discuss three cases: (i) $n = 1$, (ii) $n = 2$, (iii) $n > 2$, where n is the fixed sample size.

Remark 3.6.3. At first glance, it may seem unrealistic to consider sample sizes as small as $n = 1$ or $n = 2$. This is not quite true. Stratified sampling (see Section 3.7) with a very large number of strata is used in some surveys. For example, in multistage sampling (Chapter 4), the primary sampling units are often stratified (partitioned) into a large number of strata. It then makes sense to select a sample of only 1 or 2 primary sampling units from each stratum. Even with only one unit selected from each stratum, it is sometimes possible, in the presence of strong auxiliary information, to construct a variance estimator with a small bias using the collapsed stratum technique described in Remark 3.7.2.

i. *Sample Size $n = 1$*

The *cumulative total method* can be used to select a πps sample of size $n = 1$. This method requires cumulation of the size measures x_k as follows:

1. Set $T_0 = 0$, and compute $T_k = T_{k-1} + x_k$, $k = 1, \ldots, N$.
2. Draw a Unif(0, 1) random number, ε. If $T_{k-1} < \varepsilon T_N \le T_k$, element k is selected.

Since

$$\pi_k = \Pr(T_{k-1} < \varepsilon T_N \le T_k) = \frac{T_k - T_{k-1}}{T_N} = \frac{x_k}{\sum_U x_k}$$

the design is strictly πps. Note that $\pi_{kl} = 0$ for all $k \neq l$, implying that no unbiased variance estimate can be obtained.

EXAMPLE 3.6.1. Suppose we have a population of $N = 5$ elements, with $x_1 = 10$, $x_2 = 25$, $x_3 = 5$, $x_4 = 40$, and $x_5 = 20$. The following table gives k, x_k, and T_k.

k	x_k	T_k	Selection interval
1	10	10	01–10
2	25	35	11–35
3	5	40	36–40
4	40	80	41–80
5	20	100	81–100

All x_k are integers in this simple example; therefore, we can alternatively draw the sample of one by means of a single two-digit random number. The element $k = 1$ is associated with the two-digit numbers 01–10, ..., the element $k = 5$ with 81–99 and 00. Select at random a two-digit random number r, and let us suppose the result is $r = 27$. Then element $k = 2$ is selected, since $10 < r \le 35$.

Remark 3.6.4. If N is large, and if the selection must be done without the aid of a computer, the cumulation of the size measures will be time-consuming. There exist sampling schemes which do not require cumulation. One such scheme, suggested by Lahiri (1951), is considered in an end-of-chapter exercise.

The above scheme for selecting a πps sample is very simple, and it leads to strict proportionality. However, with $n = 1$, we have $\pi_{kl} = 0$, for all $k \neq l$, which means that unbiased variance estimation is impossible. A necessary (but not sufficient) condition for $\pi_{kl} > 0$ $(1 \leq k \neq l \leq N)$ is that the sample size be at least two. We now turn to the case $n = 2$.

ii. *Sample Size* $n = 2$

A number of schemes have been proposed for fixed size πps sampling with $n = 2$, some quite simple, some rather complicated. Here, we present only one, proposed by Brewer (1963a, 1975). For a comprehensive overview of different schemes, and an evaluation of their advantages and disadvantages, see Brewer and Hanif (1983). In Brewer's scheme, which is draw-sequential, the drawing probabilities in each of the two draws are suitably adjusted so that, jointly over the two draws, the desired inclusion probabilities $\pi_k = 2x_k/\sum_U x_k$, $k = 1, \ldots, N$, are obtained. We assume for simplicity that $x_k < \sum_U x_k/2$ for all k. For the first draw, we require the quantities

$$c_k = \frac{x_k(T_N - x_k)}{T_N(T_N - 2x_k)} \tag{3.6.6}$$

where $T_N = \sum_U x_k$. The scheme is specified by the following.

1. In the first draw, give the element k the probability

$$p_k = c_k/\textstyle\sum_U c_k$$

of being selected, where c_k is given by Equation (3.6.6).

2. Without replacing the first drawn element (say k_1), give the element l the probability

$$p_{l|k_1} = x_l/(T_N - x_{k_1})$$

of being selected in the second draw.

For this scheme, one can show that, for $k = 1, \ldots, N$

$$\pi_k = 2x_k/T_N \tag{3.6.7}$$

as required under πps sampling with $n = 2$, and, for all $k \neq l$,

$$\pi_{kl} = \frac{2x_k x_l}{T_N(\sum_U c_k)} \frac{T_N - x_k - x_l}{(T_N - 2x_k)(T_N - 2x_l)} \tag{3.6.8}$$

The scheme satisfies $\Delta_{kl} < 0$ for all $k \neq l$ (see Rao, 1965). Consequently, the Sen-Yates-Grundy variance estimator given in (2.8.11) is always nonnegative. Thus, the scheme has all of the desirable properties a–e listed above.

EXAMPLE 3.6.2. For the population in Example 3.6.1, we have $\sum_U c_k = 289/144$. Suppose we have drawn the elements $k = 3$ and $k = 4$ using Brewer's scheme, and that $y_3 = 4$ and $y_4 = 18$. Let us compute the π estimate of $t = \sum_U y_k$, and an unbiased variance estimate. We have

$$\hat{t}_\pi = \sum_s \check{y}_k = \frac{4}{0.1} + \frac{18}{0.8} = 40 + 22.5 = 62.5$$

and

$$\hat{V}(\hat{t}_\pi) = -\frac{1}{2} \sum\sum_s \check{\Delta}_{kl}(\check{y}_k - \check{y}_l)^2 = -\check{\Delta}_{34}(40 - 22.5)^2$$

Using Equation (3.6.8) first to obtain $\pi_{34} = 1{,}584/26{,}010$, we then have

$$\check{\Delta}_{34} = 1 - \frac{\pi_3 \pi_4}{\pi_{34}} = 1 - \frac{0.1 \cdot 0.8}{\frac{1{,}584}{26{,}010}}$$

Hence

$$\hat{V}(\hat{t}_\pi) = \left(\frac{2{,}601}{1{,}584} \cdot 0.8 - 1\right)(40 - 22.5)^2 = 96.05$$

Note that a considerable amount of calculation is needed to carry out the scheme and the estimation, even though n is as small as 2.

iii. *Sample Size* $n > 2$

Most existing schemes for fixed-size πps sampling with $n > 2$ (requiring π_k to be strictly proportional to x_k) are complicated. The majority of these are draw-sequential, and the calculations of the second-order inclusion probabilities rapidly become cumbersome as the sample size increases. We do not present any such draw-sequential scheme here. The interested reader is referred to Brewer and Hanif (1983).

However, if we are willing to relax the requirement of strict proportionality between the first-order inclusion probabilities and x, there is one quite simple scheme at our disposal. The scheme, proposed by Sunter (1977a, b), can be seen as a generalization of the list-sequential Scheme A for the *SI* design (see beginning of Section 3.3.1). For the most important part of the population, Sunter's scheme gives the inclusion probability π_k strictly proportional to x_k. For a small portion of the population, corresponding to the elements with the smallest x_k-values, equal, rather than x_k-proportional, inclusion probability is applied to gain simplicity. The scheme has been used with good results, and we believe that in most cases no important drawback (loss of efficiency, etc.) will follow from the simplification. This scheme, which is list-sequential, is now presented.

a. Order the N elements by the size of x, in descending order, and let $k = 1, \ldots, N$ index the elements ordered in this way. That is, $k = 1$ corresponds to the element with the largest x-value and so on.

b. For the element $k = 1$, generate a Unif(0, 1) random number ε_1 and compute

$$\pi_1 = \frac{nx_1}{T_N} \tag{3.6.9}$$

where $T_N = \sum_U x_k$. If $\varepsilon_1 < \pi_1$, the element $k = 1$ is selected, otherwise not.

c. For each $k = 2, 3, \ldots$, define the "reduced population" $U_k = \{k, k+1, \ldots, N\}$, generate an independent Unif(0, 1) random number ε_k, and compute

$$\pi'_k = \frac{(n - n_k)x_k}{t_k} \tag{3.6.10}$$

where

$$t_k = x_k + x_{k+1} + \cdots + x_N = \sum_{j \in U_k} x_j$$

and n_k is the number of elements selected among the first $k - 1$ elements in the population list obtained in step a. If $\varepsilon_k < \pi'_k$, the element k is selected, otherwise not.

d. The process described in steps b and c ends when $n_k = n$ or $k = k^*$, whichever occurs first, where $k^* = \min\{k_0, N - n + 1\}$, with k_0 equal to the smallest k for which $nx_k/t_k \geq 1$.

e. If $n_{k^*} < n$, the process in steps b and c has not produced the full size n sample. The final $n - n_{k^*}$ elements are selected from the remaining $N - k^* + 1$ elements by the *SI* design according to Scheme A. This means that, for each element $k = k^*, k^* + 1, \ldots$, we generate an independent Unif(0, 1) random number ε_k and compute

$$\pi_k^0 = \frac{n - n_k}{N - k + 1} \tag{3.6.11}$$

If $\varepsilon_k < \pi_k^0$, the element k is selected, otherwise not. The process ends when $n_k = n$.

One can show that the first order inclusion probabilities are

$$\pi_k = \begin{cases} nx_k/T_N, & k = 1, \ldots, k^* - 1 \\ n\bar{x}_{k^*}/T_N, & k = k^*, \ldots, N \end{cases}$$

where

$$\bar{x}_{k^*} = t_{k^*}/(N - k^* + 1)$$

Thus the scheme does not lead to strict πps sampling, unless the $N - k^* + 1$ smallest elements happen to have the same x_k-value. Smoothing out the π_k for the last $N - k^* + 1$ elements is only a minor price to pay for a scheme that is workable in practice.

EXAMPLE 3.6.3. For the population in Example 3.6.1, let us compute the first-order inclusion probabilities when sampling $n = 2$ elements according to Sunter's sampling scheme. After reordering the population according to step a, we have

k	x_k	t_k	nx_k/t_k
1	40	100	80/100
2	25	60	50/60
3	20	35	40/35 > 1
4	10	15	20/15 > 1
5	5	5	10/5 > 1

Thus $k_0 = \min\{k\colon nx_k/t_k \geq 1\} = 3$, and $k^* = \min\{k_0, N - n + 1\} = \min\{3, 4\} = 3$; hence $\bar{x}_{k^*} = 35/3$. The first-order inclusion probabilities are $\pi_1 = 0.8$, $\pi_2 = 0.5$, $\pi_3 = \pi_4 = \pi_5 = 7/30 = 0.2333$.

For the computation of the π_{kl}, set $g_1 = 1/t_2$ and, for $k = 2, 3, \ldots, k^* - 1$,

$$g_k = \left(1 - \frac{x_1}{t_2}\right)\left(1 - \frac{x_2}{t_3}\right)\cdots\left(1 - \frac{x_{k-1}}{t_k}\right)\Big/t_{k+1} = g_{k-1}\frac{t_k - x_{k-1}}{t_{k+1}} \tag{3.6.12}$$

It can be shown that

$$\pi_{kl} = \begin{cases} \dfrac{n(n-1)}{T_N} g_k x_k x_l & \text{for } 1 \leq k < l < k^* \\[2ex] \dfrac{n(n-1)}{T_N} g_k x_k \bar{x}_{k^*} & \text{for } 1 \leq k < k^* \leq l \leq N \\[2ex] \dfrac{n(n-1)}{T_N} g_{k^*-1} \dfrac{t_{k^*} - x_{k^*-1}}{t_{k^*} - \bar{x}_{k^*}} (\bar{x}_{k^*})^2 & \text{for } k^* \leq k < l \leq N \end{cases} \tag{3.6.13}$$

Thus the g_k can be computed and stored simultaneously with the drawing of the sample. Sunter (1977b) also shows that, for all $k \neq l$, $\pi_{kl} > 0$ and $\Delta_{kl} = \pi_{kl} - \pi_k \pi_l < 0$ for all $k \neq l$, which guarantees that the Sen-Yates-Grundy variance estimator given by (2.8.11) is unbiased and always nonnegative.

EXAMPLE 3.6.4. Let us compute π_{kl} $(1 \leq k < l \leq N)$ for the design and the population involved in Example 3.6.2. We have $k^* = 3$ and $\bar{x}_{k^*} = 35/3$. Furthermore,

$$g_1 = 1/60 \quad \text{and} \quad g_2 = \frac{1}{60} \cdot \frac{60 - 40}{35} = 1/105$$

We get from (3.6.13)

$$\pi_{12} = \frac{2}{100} \cdot \frac{1}{60} \cdot 40 \cdot 25 = 30/90$$

$$\pi_{13} = \pi_{14} = \pi_{15} = \frac{2}{100} \cdot \frac{1}{60} \cdot 40 \cdot \frac{35}{3} = 14/90$$

$$\pi_{23} = \pi_{24} = \pi_{25} = \frac{2}{100} \cdot \frac{1}{105} \cdot 25 \cdot \frac{35}{3} = 5/90$$

$$\pi_{34} = \pi_{35} = \pi_{45} = \frac{2}{100} \cdot \frac{1}{105} \cdot \frac{35 - 25}{35 - 35/3} \cdot \left(\frac{35}{3}\right)^2 = 1/90$$

EXAMPLE 3.6.5. For the population in Example 3.6.2, suppose that we have drawn the elements $k = 3$ and $k = 4$ using Sunter's scheme and that $y_3 = 4$ and $y_4 = 18$. The indices refer to the ordering in Example 3.6.2. Let us compute the π estimate of t and an unbiased variance estimate. We obtain

$$\hat{t}_\pi = 30 \cdot 4/7 + 18/0.8 = 39.64$$

$$\hat{V}(\hat{t}_\pi) = 0.2(120/7 - 22.5)^2 = 5.740$$

Compare the results obtained by Brewer's method, Example 3.6.2.

Although Sunter's method described here (compared to some other fixed-size πps schemes for $n > 2$) is manageable in practice, the fact remains that a large amount of calculation, including the buildup and the storing of the π_{kl}, is required to estimate the variance in πps sampling. This is a considerable drawback of the πps sampling method.

Sunter (1986) presents further list-sequential schemes similar to the one described above. One of these schemes achieves, with considerable calculation effort, a strict proportionality between π_k and x_k.

We close this section with a description of another πps scheme, popular because of its simplicity. A generalization of the equal probability systematic selection scheme of Section 3.4, this new scheme may be described as systematic πps selection.

Let $T_0 = 0$ and $T_k = T_{k-1} + x_k$, $k = 1, \ldots, N$, as in the cumulative total method (see πps sampling with $n = 1$). Fix a sampling interval a, where a is a positive integer. Let n be the integer part of T_N/a, where $T_N = \sum_U x_k$. Then

$$T_N = \sum_U x_k = na + c$$

where c satisfies $0 \le c < a$. If $c = 0$, a sample of size n will be drawn. If $c > 0$, the sample size is going to be either n or $n + 1$. Assume that $nx_k \le T_N - c = na$ for all k, which corresponds to the assumption $nx_k \le \sum_U x_k$ made in Section 3.6.1. For simplicity we also assume that every x_k is an integer, or, with little practical consequence, rounded off to the nearest integer.

Systematic πps selection is then defined as follows:

a. Select with equal probability, $1/a$, an integer, say r, between 1 and a (inclusively).

b. The selected sample is then

$$s = \{k: T_{k-1} < r + (j-1)a \leq T_k \text{ for some } j = 1, 2, \ldots, n_s\} = s_r$$

say, where the sample size n_s is either $n + 1$ (when $r \leq c$) or n (when $c < r \leq a$).

We can describe the scheme by saying that the distances x_k are laid out, one after another, on a horizontal axis, starting at the origin and ending at $T_N = \sum_U x_k$. When $c = 0$, the total distance T_N is divided into n intervals of equal length a. A random start is chosen in the first interval, and we proceed systematically thereafter, taking the elements k identified by points at the constant distance a apart.

In auditing applications, the method has been found useful to select a sample of accounts for audit. Then x_k is the size, in dollars, of the kth account, known from a previous occasion, say. In this context, the method and its variations are known as Dollar Unit Sampling.

Systematic πps selection implies a sampling design with essentially fixed size because n_s is either n or $n + 1$, and with the inclusion probabilities

$$\pi_k = \frac{n x_k}{T_N - c}$$

That is, we have a probability proportional-to-size design.

Systematic πps selection raises the same issues as those encountered in equal probability systematic selection (see Section 3.4), namely, controlling the sample size, the question of a good ordering of the population, and the problem of estimating the design variance of the π estimator. The variance estimation problem is discussed at some length in Wolter (1985).

3.6.3. *pps* Sampling

In *pps* sampling, the sampling is with-replacement (see Section 2.9) and $p_k \propto x_k$, for known positive numbers $x_1, \ldots, x_N$. That is,

$$p_k = \frac{x_k}{\sum_U x_k}; \qquad k = 1, \ldots, N$$

If we want to select only one element, the cumulative total method given in Section 3.6.2 can be used, since *pps* sampling and πps sampling are then equivalent. If the cumulative total method is repeated independently m times, we obtain a general draw sequential *pps* sampling scheme.

The results for the pwr estimator using *pps* selection are obtained directly from Result 2.9.1. Let the ordered sample resulting from the *pps* selection be

$$os = (k_1, k_2, \ldots, k_m)$$

The pwr estimator of the total $t = \sum_U y_k$ is then

$$\hat{t}_{\text{pwr}} = \frac{1}{m}\sum_{i=1}^{m}\frac{y_{k_i}}{p_{k_i}} = \left(\sum\nolimits_U x_k\right)\frac{1}{m}\sum_{i=1}^{m}\frac{y_{k_i}}{x_{k_i}}$$

that is, the known total $\sum_U x_k$ times the ordered sample mean of the ratios y_{k_i}/x_{k_i}. The variance and the variance estimator are calculated from the expressions in Result 2.9.1, with $p_k = x_k/\sum_U x_k$.

EXAMPLE 3.6.6. To estimate the total of the variable IMP ($= y =$ imports) for the CO124 population, a *pps* sample of $m = 10$ countries was drawn with probability proportional to the value of GNP ($= x =$ gross national product). The results were

i	Label	GNP	IMP
1	106	75,709	152,899
2	104	62,731	103,734
3	111	15,428	60,743
4	52	305,690	269,878
5	74	6,970	11,539
6	48	19,635	8,201
7	124	156,300	80,410
8	52	305,690	269,878
9	124	156,300	80,410
10	119	53,673	105,477

The population total of GNP is 1,269,864. A point estimate of the total of IMP is

$$\hat{t}_{\text{pwr}} = \left(\sum\nolimits_U x_k\right)\frac{1}{m}\sum_{i=1}^{m}\frac{y_{k_i}}{x_{k_i}} = 1{,}269{,}864 \cdot 0.1 \cdot 14.443376 \doteq 1.834 \cdot 10^6$$

whereas a variance estimate is

$$\hat{V}(\hat{t}_{\text{pwr}}) = \left(\sum\nolimits_U x_k\right)^2 \frac{1}{m(m-1)}\left[\sum_{i=1}^{m}\left(\frac{y_{k_i}}{x_{k_i}}\right)^2 - \frac{1}{m}\left(\sum_{i=1}^{m}\frac{y_{k_i}}{x_{k_i}}\right)^2\right] \doteq 1.845 \cdot 10^{11}$$

which gives $cve = 23\%$. The cve is fairly high, reflecting the small sample size, and the use of with-replacement sampling (two countries were selected twice). The strategy is nevertheless considerably more precise than SI sampling of 10 countries and the estimator $N\bar{y}_s$.

Remark 3.6.5. No heavy calculations are needed to compute a variance estimate for the *pwr* estimator. This is in sharp contrast to the cumbersome calculation of a variance estimate for the π estimator, n which a large number of cross-products $\Delta_{kl}\check{y}_k\check{y}_l/\pi_{kl}$ must be computed. This simplicity speaks in favor of the *pwr* estimator and *pps* sampling. On the other hand, the *pwr* estimator is usually less efficient than the π estimator. Sometimes one tries to make the best of two worlds in the following way:

a. A fixed size (m) πps sampling design is used such that $\pi_k = mp_k = mx_k/\sum_U x_k$
b. The π estimator is used to estimate the population total t.
c. The variance of the π estimator is estimated by the *pps* sampling formula

$$v = \frac{1}{m(m-1)} \sum_s \left(\frac{y_k}{p_k} - \frac{1}{m} \sum_s \frac{y_k}{p_k} \right)^2 \qquad (3.6.14)$$

The variance estimator v, borrowed from the theory of *pps* sampling, is not without bias. Durbin (1953) showed that the bias of v as an estimator of $V(\hat{t}_\pi)$ is given by

$$\mathrm{B}(v) = E(v) - V(\hat{t}_\pi) = \frac{m}{m-1}[V(\hat{t}_{\mathrm{pwr}}) - V(\hat{t}_\pi)] \qquad (3.6.15)$$

If the population is such that the π estimator is more efficient than the *pwr* estimator [that is, that $V(\hat{t}_{\mathrm{pwr}}) > V(\hat{t}_\pi)$], then v overestimates $V(\hat{t}_\pi)$. In other words v has a positive bias as an estimator of $V(\hat{t}_\pi)$. A conservative confidence interval (see Section 3.4) can thus be constructed with the aid of the easily computed expression given in (3.6.14).

3.6.4. Selection from Randomly Formed Groups

We now describe a selection mechanism that can be seen as an attempt to circumvent the difficulty of constructing a πps without-replacement selection with fixed size. A sample of n elements is to be selected. The population U is partitioned at random into n pairwise disjoint subgroups, U_i, $i = 1, \ldots, n$. That is, U_1 is an *SI* selection from U, U_2 is an *SI* selection from $U - U_1$, and so on. Let N_i be the size of group U_i. The sample is selected as follows. A single element is drawn from each group with probability proportional to the size of x, which is easy to do using the cumulative total method: The element k in the group U_i is given the probability $x_k/N_i\bar{x}_{U_i}$, where $\bar{x}_{U_i} = \sum_{U_i} x_k/N_i$. Let k_i be the selected element from U_i. The unbiased estimator of $t = \sum_U y_k$ is given by

$$\hat{t}_{\mathrm{gr}} = \sum_{i=1}^{n} N_i \bar{x}_{U_i} \frac{y_{k_i}}{x_{k_i}}$$

It is best to let the group sizes be either m or $m + 1$, where m is the integer part of N/n. This combination of design and estimator was suggested by Dalenius (1953) for $n = 2$ and elaborated more generally for any n by Rao,

Hartley and Cochran (1962). Ordinarily the procedure entails some loss of efficiency compared to πps selection with π_k strictly proportional to x_k and the estimator $\hat{t}_\pi = \sum_s y_k/\pi_k$, but as we have seen, the strict proportionality selection is hard to implement. On the other hand, the procedure is usually more efficient than the strategy composed of *pps* with-replacement selection and the pwr estimator.

3.7. Stratified Sampling

3.7.1. Introduction

In stratified sampling, the population is divided into nonoverlapping subpopulations called strata. A probability sample is selected in each stratum. The selections in the different strata are independent. Stratified sampling is a powerful and flexible method that is widely used in practice. Let us examine a few of the reasons why stratified sampling is so popular.

i. Suppose that estimates of specified precision are wanted for specified subpopulations (domains of study). Each domain of study can be treated as a separate stratum if domain membership is specified in the frame. An adequate probability sample can then be selected from each stratum.
ii. In a survey, practical aspects related to response, measurement, and auxiliary information may differ considerably from one subpopulation to another. Nonresponse rates and measurement problems may be more pronounced in some subpopulations than in others. The extent of the auxiliary information may differ greatly. These factors suggest that the choice of sampling design and estimator perhaps should be made differently in different subpopulations to increase the efficiency of the estimation. One may thus want to treat each subpopulation as a separate stratum.
iii. For administrative reasons, the survey organization may have divided its total territory into several geographic districts with a field office in each district. Here, it is natural to let each district be a stratum.

An additional reason in favor of stratified sampling is that most of the potential gain in efficiency of πps sampling can be captured through stratified selection with *SI* sampling within well-constructed strata. Stratified sampling is in several respects simpler than and consequently preferred to πps sampling. For example, the stratified design does not require the heavy computation for variance estimation that we found in the case of πps sampling.

To fully capitalize on the potential of the stratified sampling technique, the sampler must first resolve a number of technical questions. The objective is to select an efficient, yet practical, stratified sample. Let us list some of these questions.

i. *The Construction of Strata*

1. If there is a choice, which stratification variable should be used? By stratification variable is meant the characteristic or the characteristics used for subdividing the population into strata. For example, would an age and sex stratification be preferred to a stratification by occupational groups?
2. How should strata be demarcated? If the stratification uses age groups, what age intervals should be used to set up the strata?
3. How many strata should there be? How many age groups should there be, if age is a stratification variable?

ii. *Choice of Sampling and Estimation Methods within Strata*

1. A sampling design and a sample size must be specified in each stratum. Often the same type of sampling design is applied in all of the strata.
2. An estimator must be specified for each stratum. Often this choice is also made uniformly for all strata.

The solutions to these problems are usually interdependent. When an overall solution is worked out in practice, a number of considerations are usually taken into account. Some of these include requirements on the precision of estimates for different domains of study and for different variables of study, cost considerations, and the administrative restrictions.

In a large survey, the variety of demands that arise often makes it impossible to arrive at a global optimum for the stratified sampling design. In this chapter and in Chapter 12, we examine some of the issues and their solutions.

3.7.2. Notation, Definitions, and Estimation

Let us first introduce some notation and definitions. By a *stratification* of a finite population $U = \{1, \ldots, k, \ldots, N\}$ we mean a partitioning of U into H subpopulations, called strata and denoted $U_1, \ldots, U_h, \ldots, U_H$, where $U_h = \{k : k \text{ belongs to stratum } h\}$. By *stratified sampling* we mean that a probability sample s_h is selected from U_h according to a design $p_h(\cdot)$ ($h = 1, \ldots, H$) and that the selection in one stratum is independent of the selections in all other strata.

The resulting total sample, denoted s as usual, will thus be composed as

$$s = s_1 \cup s_2 \cup \cdots \cup s_H$$

and, because of the independence feature,

$$p(s) = p_1(s_1)p_2(s_2) \cdots p_H(s_H)$$

The number of elements in stratum h, called the size of stratum h, is denoted N_h, which is assumed to be known. Since the strata form a partition of U we have

$$N = \sum_{h=1}^{H} N_h$$

Furthermore, the population total can be decomposed as

$$t = \sum_U y_k = \sum_{h=1}^{H} t_h = \sum_{h=1}^{H} N_h \bar{y}_{U_h} \tag{3.7.1}$$

where $t_h = \sum_{U_h} y_k$ is the stratum total, and $\bar{y}_{U_h}$ the stratum mean. Finally, let $W_h = N_h/N$ denote the relative size of the stratum U_h. Then the population mean has the decomposition

$$\bar{y}_U = \sum_{h=1}^{H} W_h \bar{y}_{U_h} \tag{3.7.2}$$

Result 3.7.1. *In stratified sampling, the π estimator of the population total $t = \sum_U y_k$ can be written*

$$\hat{t}_\pi = \sum_{h=1}^{H} \hat{t}_{h\pi} \tag{3.7.3}$$

where $\hat{t}_{h\pi}$ is the π estimator of $t_h = \sum_{U_h} y_k$. The variance can be written

$$V_{ST}(\hat{t}_\pi) = \sum_{h=1}^{H} V_h(\hat{t}_{h\pi}) \tag{3.7.4}$$

where $V_h(\hat{t}_{h\pi})$ is the variance of $\hat{t}_{h\pi}$. An unbiased variance estimator is given by

$$\hat{V}_{ST}(\hat{t}_\pi) = \sum_{h=1}^{H} \hat{V}_h(\hat{t}_{h\pi}) \tag{3.7.5}$$

supposing there exists an unbiased variance estimator $\hat{V}_h(\hat{t}_{h\pi})$ for every h.

Proof. From the definition of stratified sampling it is clear that, for $k \in U_h$,

$$\pi_k = \Pr(k \in s) = \Pr(k \in s_h)$$

Thus the π estimator of t is

$$\hat{t}_\pi = \sum_s \check{y}_k = \sum_{h=1}^{H} \sum_{s_h} \check{y}_k = \sum_{h=1}^{H} \hat{t}_{h\pi}$$

where $\hat{t}_{h\pi}$ is the π estimator of t_h. Because the selections in different strata are independent, the random variables $\hat{t}_{h\pi}$ are independent, and the results (3.7.4) and (3.7.5) follow easily. □

These results also follow from the general Result 2.8.1 by observing that, for k and l belonging to different strata, we have $\pi_{kl} = \pi_k \pi_l$ and thus $\Delta_{kl} = 0$, cancelling all terms in $\sum\sum_U$ for which k and l belong to different strata. Result 3.7.1 is widely applicable in the sense that it leaves the sampling designs unspecified in the different strata.

Often the same design is used in all strata. For example, we may apply the *SI* design in all strata. This selection method, perhaps the most important example of stratified sampling, will be denoted *STSI*. As another example, we may have the *SY* design given in Section 3.4.1 in each stratum; thereby we obtain the stratified design denoted *STSY*. Analogously, *STBE* is the stratified design using Bernoulli sampling in each stratum, and so on. The stratified estimator (3.7.3), the variance (3.7.4) and the variance estimator (3.7.5) can be worked out in each particular case, given what we know from earlier sections about the *SI*, *SY*, *BE*, and other designs.

Since the *STSI* design is so widely used, let us summarize the principal results in this particular case. Let n_h be the fixed size of the *SI* sample in stratum h. As a consequence of Results 3.7.1 and 3.3.1, we then have the following result.

Result 3.7.2. *Under the STSI design, the* π *estimator of the population total* $t = \sum_U y_k$ *is*

$$\hat{t}_\pi = \sum_{h=1}^{H} N_h \bar{y}_{s_h} \tag{3.7.6}$$

where $\bar{y}_{s_h} = \sum_{s_h} y_k / n_h$. *The variance is*

$$V_{STSI}(\hat{t}_\pi) = \sum_{h=1}^{H} N_h^2 \frac{1 - f_h}{n_h} S_{yU_h}^2 \tag{3.7.7}$$

$$= \sum_{h=1}^{H} N_h^2 S_{yU_h}^2 / n_h - \sum_{h=1}^{H} N_h S_{yU_h}^2 \tag{3.7.8}$$

where $f_h = n_h / N_h$ *is the sampling fraction in stratum* h *and*

$$S_{yU_h}^2 = \frac{1}{N_h - 1} \sum_{U_h} (y_k - \bar{y}_{U_h})^2$$

is the stratum variance; $\bar{y}_{U_h} = \sum_{U_h} y_k / N_h$. *An unbiased variance estimator is*

$$\hat{V}_{STSI}(\hat{t}_\pi) = \sum_{h=1}^{H} N_h^2 \frac{1 - f_h}{n_h} S_{ys_h}^2 \tag{3.7.9}$$

where

$$S_{ys_h}^2 = \frac{1}{n_h - 1} \sum_{s_h} (y_k - \bar{y}_{s_h})^2$$

is the sample variance in stratum h.

Example 3.7.1. *STSI* sampling with a total sample size of $n = 30$ was used to estimate the total of the variable P83 for the CO124 population. The population was stratified into four strata according to the value of the variable P80 (see the table below). From strata 1 to 3, *SI* samples of 6, 6, and 11 countries, respectively, were drawn, while every country in stratum 4 was selected for the survey (take-all stratum). The following results were recorded:

Stratum h	P80	N_h	n_h	$\sum_{s_h} y_k$	$\sum_{s_h} y_k^2$
1	0– 9.99	70	6	31.1	234.95
2	10–29.99	29	6	119.6	2,613.26
3	30–99.99	18	11	621.0	40,122.36
4	100–···	7	7	2,671.4	1,769,052.90

Let us compute an approximately 95% confidence interval for the population total of P83. Using (3.7.6) we get the point estimate 4,628, and equation (3.7.9) gives the variance estimate 21,911. Hence the confidence limits are 4,628 ± 290. The distribution of P83 is strongly skewed, and the total sample size is modest. Nevertheless the *cve* (the estimated relative standard error; see Section 2.7) is only 3%. The stratification variable P80 is strongly correlated with the study variable P83. As a result, the variance is greatly reduced by stratification and is much smaller than the variance that would be obtained with *SI* sampling with $n = 30$ and the simple expansion estimator $N\bar{y}_s$.

3.7.3. Optimum Sample Allocation

Consider a population with fixed strata. Suppose that the designs used within strata have been decided. The plan is to use the π estimator of the population total. Prior to sample selection, we must then determine the fixed (expected) sample sizes n_h ($h = 1, \ldots, H$). We examine this allocation problem in the case where the stratified sampling design is such that

$$V_{ST}(\hat{t}_\pi) = \sum_{h=1}^{H} A_h/n_h + B = V \tag{3.7.10}$$

say, where A_h and B do not depend on the n_h. The *STSI* and *STBE* designs are among those that give rise to a variance of the form given in (3.7.10).

Finally, suppose that the total cost of the survey can be expressed as

$$C = c_0 + \sum_{h=1}^{H} n_h c_h \tag{3.7.11}$$

where c_0 is a fixed overhead cost and $c_h > 0$ is the cost of surveying one element in stratum h.

The problem of *optimum sample allocation* is formulated as the determination of the n_h that minimize the variance V subject to a fixed cost C, or, conversely, the determination of the n_h that minimize the cost C subject to a fixed precision V.

Result 3.7.3. *Under a stratified sampling design for which* $V_{ST}(\hat{t}_\pi)$ *is of the form* (3.7.10), *the optimum sample allocation under the linear cost function* (3.7.11) *is achieved by choosing* n_h *proportional to* $(A_h/c_h)^{1/2}$.

PROOF. Let $V^* = V - B$ and $C^* = C - c_0$. Our optimization problem is then equivalent to minimizing the product

$$V^*C^* = \left(\sum_{h=1}^{H} A_h/n_h\right)\left(\sum_{h=1}^{H} n_h c_h\right) \qquad (3.7.12)$$

Now, the Cauchy inequality states that

$$\left(\sum a_h^2\right)\left(\sum b_h^2\right) \geq \left(\sum a_h b_h\right)^2$$

Taking $a_h = (A_h/n_h)^{1/2}$ and $b_h = (n_h c_h)^{1/2}$, we have

$$V^*C^* \geq \left[\sum_{h=1}^{H} (A_h c_h)^{1/2}\right]^2$$

where the quantity on the right-hand side of the inequality is independent of the n_h. Equality holds if and only if b_h/a_h is constant for every h, that is,

$$\left(\frac{n_h c_h}{A_h/n_h}\right)^{1/2} = \text{constant}$$

or

$$n_h \propto (A_h/c_h)^{1/2} \quad \square \qquad (3.7.13)$$

We can now apply Result 3.7.3 separately for each of the two versions of the optimum sample allocation problem. The conclusions are summarized as follows. The proof is left as an exercise.

Result 3.7.4. *Minimizing the variance V given by equation* (3.7.10) *for a fixed cost C given by equation* (3.7.11) *gives the solution*

$$n_h = (C - c_0)(A_h/c_h)^{1/2} \Big/ \sum_{h=1}^{H} (A_h c_h)^{1/2}; \qquad h = 1, \ldots, H \qquad (3.7.14)$$

and the minimum variance is

$$V_{opt} = \frac{1}{C - c_0}\left[\sum_{h=1}^{H} (A_h c_h)^{1/2}\right]^2 + B \qquad (3.7.15)$$

Minimizing the cost C for fixed variance V gives the solution

$$n_h = (A_h/c_h)^{1/2}\left[\sum_{h=1}^{H} (A_h c_h)^{1/2}\right] \Big/ (V - B); \qquad h = 1, \ldots, H \qquad (3.7.16)$$

and the minimum cost is

$$C_{opt} = c_0 + \frac{1}{V - B}\left[\sum_{h=1}^{H} (A_h c_h)^{1/2}\right]^2 \tag{3.7.17}$$

In Results 3.7.3 and 3.7.4, it has been tacitly assumed that the solution satisfies $n_h \leq N_h$ for every h. Cases in which this does not hold are discussed in Section 12.7.

Let us consider in particular the *STSI* design. In this case the variance formula (3.7.7) can be written in the form of equation (3.7.10) with

$$A_h = N_h^2 S_{yU_h}^2, \qquad B = -\sum_{h=1}^{H} N_h S_{yU_h}^2$$

Minimizing the variance V for a fixed cost C gives

$$n_h = (C - c_0)\frac{N_h S_{yU_h}/(c_h)^{1/2}}{\sum_{h=1}^{H} N_h S_{yU_h}(c_h)^{1/2}} \tag{3.7.18}$$

That is, the larger the variation within a stratum, the larger n_h. Also, the lower the stratum cost, the larger n_h.

3.7.4. Alternative Allocations under *STSI* Sampling

Let us examine in some further detail the *STSI* design, and alternative sample allocations that may be advantageous in this case. We shall assume that all stratum costs c_h are equal. Let n be the total sample size, so that

$$n = \sum_{h=1}^{H} n_h$$

i. *Optimum Allocation*

The *optimum allocation*, obtained from equation (3.7.18) with c_h a constant, is given by

$$n_h = n \cdot \frac{N_h S_{yU_h}}{\sum_{h=1}^{H} N_h S_{yU_h}} \tag{3.7.19}$$

The result is also called the *Neyman allocation*, following the important contribution by Neyman (1934). The calculation of the optimal n_h from (3.7.19) will require that the stratum standard deviations, S_{yU_h}, are known. The optimum, that is, the minimum variance, can be attained only if the exact S_{yU_h} are available. This will never be the case in practice. But in repeated surveys, one may be able to use past experience to state close approximations to the true standard deviations. Based on such values, it is thus possible to calculate an allocation that may be close to the optimal.

Let us consider some alternative, but ordinarily less than optimal, allocations.

ii. *x-Optimal Allocation*

An alternative with practical advantages over the allocation (3.7.19) is what we call x-optimal allocation. Suppose that x is an auxiliary variable, highly correlated with y, and the stratum standard deviations of x, S_{xU_h}, are known. Using the structure of formula (3.7.19), the x-optimal allocation is obtained as

$$n_h = n \frac{N_h S_{xU_h}}{\sum_{h=1}^{H} N_h S_{xU_h}} \tag{3.7.20}$$

If the correlation between x and y is perfect ($y_k = a + bx_k$, $k = 1, \ldots, N$), then the allocation given in equation (3.7.20) is in fact optimal. If the correlation is strong but less than perfect, (3.7.20) may still lead to a near-optimal allocation; the method is often used in practice with good results.

iii. *Proportional Allocation*

The proportional allocation is defined by

$$n_h = n \cdot N_h / N \tag{3.7.21}$$

Since the stratum sizes N_h are assumed to be known, this simple allocation can always be calculated. We note from (3.7.19) that if the stratum standard deviations S_{yU_h} are all equal, then proportional allocation is optimal. Otherwise proportional allocation leads to a higher variance than the Neyman allocation given by (3.7.19), especially if the S_{yU_h} differ considerably.

iv. *Allocation Proportional to the y-Total*

Allocation proportional to the y-total is defined by

$$n_h = n \frac{\sum_{U_h} y_k}{\sum_U y_k} \tag{3.7.22}$$

Here the y variable is assumed to be always positive. This allocation is optimum, that is, it agrees with the allocation (3.7.19), if the coefficient of variation

$$cv_{yU_h} = \frac{S_{yU_h}}{\bar{y}_{U_h}}$$

is constant in all strata. Now, the stratum y-totals are ordinarily unknown, so again the allocation (3.7.22) cannot be used in practice.

v. *Allocation Proportional to the x-Total*

Allocation proportional to the x-total is analogous to (3.7.22) and defined as

$$n_h = n \frac{\sum_{U_h} x_k}{\sum_U x_k} \tag{3.7.23}$$

The assumption here is that x is an always positive auxiliary variable with known stratum totals $\sum_{U_h} x_k$. The easily calculated allocation (3.7.23) has been found useful in practice. Its justification is that if x and y are highly correlated and if the coefficient of variation is about the same in all strata, then (3.7.23) should be not far from optimal.

With the aid of Result 3.7.2, one shows easily that the following variance expressions hold for the stratified estimator

$$\hat{t}_\pi = \sum_{h=1}^{H} N_h \bar{y}_{s_h} = N \sum_{h=1}^{H} W_h \bar{y}_{s_h}$$

1. Under the optimal allocation given by (3.7.19),

$$V_{STSI,o}(\hat{t}_\pi) = \frac{N^2}{n}\left(\sum_{h=1}^{H} W_h S_{yU_h}\right)^2 - N \sum_{h=1}^{H} W_h S_{yU_h}^2 \tag{3.7.24}$$

2. Under the proportional allocation (3.7.21),

$$V_{STSI,p}(\hat{t}_\pi) = N^2\left(\frac{1}{n} - \frac{1}{N}\right) \sum_{h=1}^{H} W_h S_{yU_h}^2 \tag{3.7.25}$$

Since the allocation in (1) is optimal, the variance (3.7.24) cannot be greater than the variance (3.7.25). Equality is obtained if all stratum variances are equal, but often (3.7.24) will be considerably smaller than (3.7.25), especially if there are wide discrepancies in the stratum variances.

When the sample allocation is carried out sensibly, for example, proportionally, the *STSI* design will ordinarily be much more efficient than the basic *SI* design, for a common sample size n. For the *SI* design, the π estimator is $N\bar{y}_s = N\sum_s y_k/n$. One can show that

$$V_{SI}(N\bar{y}_s) - V_{STSI,p}\left(\sum_{h=1}^{H} N_h \bar{y}_{s_h}\right)$$

$$= \frac{N^3}{N-1}\left(\frac{1}{n} - \frac{1}{N}\right)\left[\sum_{h=1}^{H} W_h(\bar{y}_{U_h} - \bar{y}_U)^2 - \frac{1}{N}\sum_{h=1}^{H}(1 - W_h) S_{yU_h}^2\right] \tag{3.7.26}$$

where $\bar{y}_{U_h}$ is the mean in stratum U_h and $\bar{y}_U$ is the overall population mean. The expression (3.7.26) follows from the analysis of variance decomposition

$$(N-1)S_{yU}^2 = \sum_U (y_k - \bar{y}_U)^2$$
$$= \sum_{h=1}^{H} N_h(\bar{y}_{U_h} - \bar{y}_U)^2 + \sum_{h=1}^{H} (N_h - 1)S_{yU_h}^2 \tag{3.7.27}$$

or

$$SST = SSB + SSW$$

if we use the standard analysis of variance notation.

Now, equation (3.7.26) shows that it is theoretically possible for the proportionally allocated stratified sample to lead to a slightly larger variance than the *SI* sample, namely, if all stratum means $\bar{y}_{U_h}$ are equal or very nearly equal. However, this equality is a remote possibility, and in many instances $\sum W_h(\bar{y}_{U_h} - \bar{y}_U)^2$ will considerably exceed $N^{-1}\sum(1 - W_h)S_{yU_h}^2$.

If *SSB* accounts for a significant proportion of *SST*, then differences among the strata means are a major reason for the variability of *y*. In such a case, *STSI* sampling with proportional allocation yields a substantially smaller variance than *SI* sampling.

Remark 3.7.1. In surveys where there are many effective criteria for stratification, the statistician may work with a large number of strata. Even the extreme case, with only one observation per stratum, is sometimes used. However, in that case, the variance estimator given by equation (3.7.5) cannot be used. Instead, a *collapsed strata technique* has been proposed, in which pairs of strata are formed. Using the two observations in a pair, a measure of variance can be obtained. The technique is as follows. Let the population total $t = \sum_{h=1}^{H} t_h$ be estimated by $\hat{t}_\pi = \sum_{h=1}^{H} \hat{t}_h$, where $\hat{t}_h = y_k/\pi_k$ is unbiased for the stratum total t_h, assuming a single element k is selected with probability π_k from the stratum. The π_k add to unity in the stratum. In particular, $\pi_k = 1/N_h$ for all k if simple random selection is used. After pairing and relabeling, let $j1$ and $j2$ refer to the two strata in the jth pair, $j = 1, \ldots, M = H/2$, if we assume H to be even. Consider the variance estimator

$$\hat{V}_{\text{coll}} = \sum_{j=1}^{M} (\hat{t}_{j1} - \hat{t}_{j2})^2$$

where the subscript "coll" indicates collapsed. It is easy to show that

$$E(\hat{V}_{\text{coll}}) = V(\hat{t}_\pi) + \sum_{j=1}^{M} (t_{j1} - t_{j2})^2$$

Thus $\hat{V}_{\text{coll}}$ will ordinarily overestimate the variance $V(\hat{t}_\pi)$ and lead to conservative confidence intervals. If the strata can be paired so that $t_{j1} = t_{j2}$ for every pair j, then $\hat{V}_{\text{coll}}$ is unbiased, and it is only in this case that the collapsed strata variance estimator is fully satisfactory, that is, free of bias. To eliminate the bias of $\hat{V}_{\text{coll}}$ is hard in most cases, since it would require knowledge of the strata totals.

3.8. Sampling without Replacement versus Sampling with Replacement

We found in Section 2.9 that a with-replacement sampling scheme with m draws yields an ordered sample where the same element may occur more than once,

$$os = (k_1, k_2, \ldots, k_m)$$

The ordered sample can be "reduced" to a set-sample, namely,

$$s = \{k: k = k_i \text{ for some } i = 1, \ldots, m\}$$

How should an estimator of $t = \sum_U y_k$ be formed in this case? One approach is to take into account all of the m draws in the ordered sample os. The y_k-value of an element that happens to be drawn twice, say, will then appear twice in the estimator formula. The distributional properties of the estimator are determined by the ordered sampling design, that is, the distribution of the ordered samples that can be obtained under the with-replacement scheme. Or should one consider only the set s of distinct elements when the estimator is formed? The y_k-value of an element will appear in the estimator in the same way regardless of how many times it was drawn, given that it was drawn at least once. In this case, the sampling distribution of the estimator is determined by the set-sample distribution induced by the scheme. We use the case of the SIR design to illustrate the different options.

3.8.1. Alternative Estimators for Simple Random Sampling with Replacement

For simple random sampling with replacement (the ordered design denoted SIR), we can form at least three unbiased estimators of t. The first is based on the ordered sample, the other two on the set-sample.

i. *Unbiased pwr Estimator*

The unbiased pwr estimator uses all m draws, so that

$$\hat{t}_{\text{pwr}} = N\bar{y}_{os} = \frac{N}{m} \sum_{i=1}^{m} y_{k_i} \tag{3.8.1}$$

The distributional properties of $\hat{t}_{\text{pwr}}$ are determined by the SIR design, which gives the same probability, N^{-m}, to every ordered sample of the given length m. The variance and the variance estimator are stated in Result 3.3.4.

ii. *Unbiased π Estimator*

If we base the estimator on the set sample s, one option is the unbiased π estimator, $\hat{t}_\pi$. Its distributional properties are determined by the set sample distribution $p(s)$ induced by the *SIR* scheme. The first- and second-order inclusion probabilities (see Example 2.9.2) are

$$\pi_k = 1 - \left(1 - \frac{1}{N}\right)^m; \qquad k = 1, \dots, N \qquad (3.8.2)$$

$$\pi_{kl} = 1 - 2\left(1 - \frac{1}{N}\right)^m + \left(1 - \frac{2}{N}\right)^m; \qquad k \neq l = 1, \dots, N \qquad (3.8.3)$$

It follows that the π estimator is given by

$$\hat{t}_\pi = \textstyle\sum_s y_k \Big/ \left\{1 - \left(1 - \frac{1}{N}\right)^m\right\} \qquad (3.8.4)$$

The variance and the variance estimator for $\hat{t}_\pi$ are obtained from the general Result 2.8.1 using π_k and π_{kl} from equations (3.8.2) and (3.8.3).

For $m \geq 2$, the π estimator is not identical to the pwr estimator. Which estimator is better? The question is not so simple. Both are unbiased. As far as the variance is concerned, an examination will show that there is no generally valid conclusion. Which of the two estimators has the smaller variance will depend on the configuration of the y-values $y_1, \dots, y_N$, as the comparisons in Section 3.8.2 show.

iii. *Alternative Unbiased Estimator Based on the Set-Sample*

An alternative estimator based on the set-sample is

$$\hat{t}_{\text{alt}} = N\bar{y}_s = \frac{N}{n_s}\textstyle\sum_s y_k \qquad (3.8.5)$$

that is, N times the mean of y for the $n_s \leq m$ distinct elements in $os = (k_1, \dots, k_m)$. This estimator, too, is unbiased. Moreover, one can show that regardless of the configuration of the population values $y_1, \dots, y_N$, $\hat{t}_{\text{alt}}$ has a variance that never exceeds that of $\hat{t}_{\text{pwr}}$. It carries too far here to prove this result, which is due to Basu (1958) and Raj and Khamis (1958); see Cassel, Särndal, and Wretman (1977) for a proof.

For most populations $y_1, \dots, y_N$, $\hat{t}_{\text{alt}}$ will actually have a distinctly smaller variance than both $\hat{t}_{\text{pwr}}$ and $\hat{t}_\pi$. If $n = N[1 - (1 - 1/N)^m]$ denotes the expected size of the set sample, we have that $\hat{t}_{\text{alt}}$ and $\hat{t}_\pi$ are related by

$$\hat{t}_{\text{alt}} = \frac{n}{n_s}\hat{t}_\pi \qquad (3.8.6)$$

We have already contrasted $\hat{t}_\pi$ with $\hat{t}_{\text{alt}}$ in connection with the BE design in Section 3.2. It was noted there that dividing $\hat{t}_\pi$ by n_s reduces the variability.

3.8.2. The Design Effect of Simple Random Sampling with Replacement

A different perspective of SIR sampling is obtained if we compare it with SI sampling, for a comparable number of observed elements. The expected size of the set-sample generated by m SIR draws is $N\phi$, where

$$\phi = 1 - \left(1 - \frac{1}{N}\right)^m$$

is the common value of all the π_k. An equitable comparison is obtained if we assume SI sampling with a fixed sample size n equal to the expected size under SIR, that is, $N\phi$. The required number of SIR draws is then

$$m = \frac{\ln(1 - n/N)}{\ln(1 - N^{-1})} \tag{3.8.7}$$

In reality, m is the integer nearest to $\ln(1 - f)/\ln(1 - N^{-1})$. In the following analysis, we assume for simplicity that m is given exactly by equation (3.8.7).

Using Results 3.3.1 and 3.3.4, we obtain the following design effect when the pwr estimator is used,

$$\text{deff}(SIR, \hat{t}_{\text{pwr}}) = \frac{V_{SIR}(N\bar{y}_{os})}{V_{SI}(N\bar{y}_s)} = \frac{N(N-1)S_{yU}^2/m}{N^2(1 - n/N)S_{yU}^2/n} = \frac{n(1 - N^{-1})}{m(1 - n/N)}$$

where m is given by (3.8.7). Note that $N\bar{y}_s$ is the unbiased estimator for the SI design.

A numerical calculation for $N = 1{,}000$ and an expected sampling fraction of $\phi = n/N = 20\%$ gives $m = 223.0$ and

$$\text{deff}(SIR, \hat{t}_{\text{pwr}}) = 1.120 \tag{3.8.8}$$

This deff value tells us that SIR combined with $\hat{t}_{\text{pwr}}$ leads to a nonnegligible loss of efficiency compared to SI sampling combined with $N\bar{y}_s$.

The second option we identified was to use the π estimator (3.8.4). It follows after some calculation that

$$\text{deff}(SIR, \hat{t}_\pi) = \frac{V_{SIR}(\hat{t}_\pi)}{V_{SI}(N\bar{y}_s)} = \left(1 - \frac{1}{N}\right)\left(1 - \rho\right) + \frac{1 + (N-1)\rho}{(cv_{yU})^2}$$

where $cv_{yU} = S_{yU}/\bar{y}_U$ is the coefficient of variation in the finite population, and $\rho = \phi^{-1}[(1 - \phi)^{a-1} - 1 + \phi]$ with the constant a determined by

$$a = \frac{\ln\left(1 - \dfrac{2}{N}\right)}{\ln\left(1 - \dfrac{1}{N}\right)}$$

The expression for ρ, which is the correlation between any pair of indicator variables, I_k and I_l, $k \neq l$, can be obtained with the aid of (3.8.3).

Calculating again for $N = 1{,}000$ and an expected sampling fraction of $\phi = n/N = 20\%$, we obtain

$$\text{deff}(SIR, \hat{t}_\pi) \doteq 1 + \frac{0.107}{(cv_{yU})^2}$$

The magnitude of the coefficient of variation cv_{yU} is the key factor in this expression. The design effect exceeds unity, so in this case too there is a loss of precision compared to SI sampling combined with $N\bar{y}_s$. When $cv_{yU} > 0.95$, $\text{deff}(SIR, \hat{t}_\pi)$ is smaller than the deff for the pwr estimator given by (3.8.8).

The third option for the SIR design was the estimator (3.8.5) defined as N times the mean of y for the distinct elements. One can show that

$$\text{deff}(SIR, \hat{t}_{\text{alt}}) = \frac{V_{SIR}(\hat{t}_{\text{alt}})}{V_{SI}(N\bar{y}_s)} = \frac{E\left(\dfrac{1}{n_s}\right) - \dfrac{1}{N}}{\dfrac{1}{n} - \dfrac{1}{N}}$$

where n_s is the random size of the set-sample s.

Since

$$E\left(\frac{1}{n_s}\right) \geq \frac{1}{E(n_s)} = \frac{1}{n}$$

the deff exceeds unity, but it is very near unity if n is fairly large. The distribution of n_s, although not very simple, is known; for details see Lanke (1975). Pathak (1961) shows that

$$E\left(\frac{1}{n_s}\right) = \frac{1}{N} \sum_{j=1}^{N} \left(\frac{j}{N}\right)^{m-1}$$

which can be used to calculate the design effect.

For the case $N = 1{,}000$ and $\phi = n/N = 20\%$, we get

$$\text{deff}(SIR, \hat{t}_{\text{alt}}) = 1.01$$

That is, SIR with the estimator $\hat{t}_{\text{alt}}$ is practically equal in efficiency to SI with $N\bar{y}_s$, assuming a common expected sample size for the two strategies. In this example, $\hat{t}_{\text{alt}}$ is distinctly better than the other two estimators that we identified for SIR sampling, namely, $\hat{t}_{\text{pwr}}$ and $\hat{t}_\pi$.

Remark 3.8.1. So far we have defined an ordered design only in the context of sampling with replacement. But more generally an ordered design is any design that yields an ordered sample, $os = (k_1, \ldots, k_i, \ldots, k_m)$, where k_i is the element obtained in the ith of m experiments (draws), such that repetitions of the same element may be allowed (as in with-replacement sampling) or disallowed (as in without-replacement sampling).

For each element $k = 1, \ldots, N$, let δ_k be the number of times that k appears in the ordered sample $(k_1, \ldots, k_m)$. The expected value of the count variable δ_k is denoted $\alpha_k = E(\delta_k)$. Define the estimator

$$\hat{t}_\alpha = \sum_{i=1}^{m} \frac{y_{k_i}}{\alpha_{k_i}}$$

It is not difficult to verify that $\hat{t}_\alpha$ is unbiased for t. Two interesting special cases are the following.

Special Case 1. The m draws are without replacement. Any unit can occur at most once in the ordered sample. Then $\alpha_k = \pi_k$, the inclusion probability of the kth element. Thus

$$\hat{t}_\alpha = \sum_{i=1}^{m} \frac{y_{k_i}}{\alpha_{k_i}} = \sum\nolimits_s \frac{y_k}{\pi_k} = \hat{t}_\pi$$

which is the π estimator. Here, the set sample s consists of m distinct elements $k_1, \ldots, k_m$.

Special Case 2. The m independent draws are with replacement, and each draw uses the same drawing probabilities, $p_1, \ldots, p_k, \ldots, p_N$. Then $\alpha_k = mp_k$, and we obtain

$$\hat{t}_\alpha = \sum_{i=1}^{m} \frac{y_{k_i}}{\alpha_{k_i}} = \frac{1}{m} \sum_{i=1}^{m} \frac{y_{k_i}}{p_{k_i}} = \hat{t}_{\text{pwr}}$$

which is the pwr estimator.

Exercises

3.1. Suppose you want to draw a Bernoulli sample from the CO124 population in order to estimate the total of the variable $IMP(= y)$ with a relative standard error of 10% by (a) the π estimator and (b) the alternative estimator shown in equation (3.2.6). Using $\sum_U y_k \doteq 1.81 \cdot 10^6$ and $\sum_U y_k^2 \doteq 1.69 \cdot 10^{11}$, compute the necessary expected sample sizes. Comment on your result. The relative standard error is defined as $[V(\hat{t})]^{1/2}/t$.

3.2. Using a π-value of 0.1, a Bernoulli sample was drawn from the MU284 population in order to estimate the totals of the two variables P85 and RMT85. The following results were obtained.

Variable	$\sum_s y_k$	$\sum_s y_k^2$
P85	564	15,790
RMT85	4,178	878,452

Compute unbiased estimates of the two totals, as well as the corresponding unbiased variance estimates, and the corresponding *cve*'s.

3.3. A Bernoulli sample of size $n_s = 71$, drawn with a π-value of 0.5 from the CO124 population, contained 21 countries from Asia (continent 1). Compute an unbiased point estimate and an approximately 95% confidence interval for the percentage of countries belonging to Asia.

3.4. For the MU284 population, the coefficient of variation of the variable CS82 equals 0.543. Determine the minimum (expected) sample size needed to estimate the total of CS82 so that the relative standard error of the π estimator, $[V(\hat{t}_\pi)]^{1/2}/t$, is no greater than 10% under (a) simple random sampling (*SI*) and (b) Bernoulli sampling (*BE*). (c) What is the expected sample size required under Bernoulli sampling, if the alternative estimator (3.2.6) is used instead of the π estimator?

3.5. An *SI* sample of size 30 was drawn in order to estimate the total of the variable S82 ($= y$) for the MU284 population. The sample totals $\sum_s y_k = 1{,}424$ and $\sum_s y_k^2 = 70{,}758$ were calculated. Compute an approximately 95% confidence interval for the total of S82, the total number of seats in the municipal councils.

3.6. For the MU200 population, compute an approximately 95% confidence interval for the mean of the variable CS82, that is, the average number of conservative seats per municipality, using the data in Example 3.3.1.

3.7. Using the information in Example 3.3.1, compute 99% confidence intervals for (a) the number and (b) the proportion of municipalities where the conservatives have less than 8 seats.

3.8. Consider Example 3.3.2. There are 24 countries in Europe. Use this information and the data in Example 3.3.2 to estimate the mean of the variable P83 for Europe. Also, compute a variance estimate as well as the corresponding *cve*.

3.9. Under the *SI* design, the customary $100(1-\alpha)\%$ confidence interval for the population mean $\bar{y}_U$ can be written as $\bar{y}_s(1 \pm A)$ with

$$A = z_{1-\alpha/2} cv_{ys} \sqrt{\frac{1-f}{n}}$$

where $cv_{ys} = S_{ys}/\bar{y}_s$ is the coefficient of variation of y in the sample, and $f = n/N$. Let us assume that $cv_{ys} \doteq cv_{yU} = S_{yU}/\bar{y}_U$ and that f is negligible. Let $\alpha = 5\%$ ($z_{0.975} = 1.96$). What sample size n is required, approximately, to attain the precision $A \leq 3\%$ if (a) $cv_{yU} = 0.5$; (b) $\text{cv}_{yU} = 1.0$; (c) $cv_{yU} = 1.5$?

3.10. Consider the population MU280, that is, the population of Swedish municipalities after removal of the four largest municipalities. The *SY* design is used with the sampling interval 10 for estimating the total of the variable RMT85 under three different population orderings.

0_1. The ordering given in the Appendix.
0_2. A size ordering of the population from the smallest to the largest value of the variable S82.
0_3. A size ordering of the population from the smallest to the largest value of the variable ME84.

For each population ordering there are 10 possible systematic samples. The following table gives the total of RMT85 for each of the possible samples for each of the three orderings.

	Ordering		
Sample no.	0_1	0_2	0_3
1	4,826	4,507	4,633
2	4,938	5,006	4,770
3	4,531	5,503	5,036
4	7,821	5,736	4,873
5	7,051	6,149	5,182
6	5,235	5,673	5,504
7	5,592	5,047	5,222
8	4,432	4,538	5,443
9	3,037	4,499	5,649
10	4,411	5,216	5,562

(a) Compute the standard error of the π estimator for each ordering. (b) The population variance of RMT85 is 35,932. Compute the standard error of the π estimator under the *SI* design with $n = 28$. (c) Compute the homogeneity measure δ for each of the three orderings, and compute $\delta_{\min}$ as defined in Section 3.4.

3.11. To estimate the total of the variable P83 ($= y$) for the CO124 population, systematic sampling with two random starts was carried out as follows. The population was size-ordered from the smallest to the largest value of P80. An *SI* sample of two integers was then drawn from the integers 1, 2, 3, and 4. The selected integers were 2 and 3. The size-ordered elements with serial numbers 2, 6, 10, ..., 122 form the first slice of the systematic sample, the size-ordered elements with serial numbers 3, 7, 11, ..., 123 form the second slice of the systematic sample. These slices are denoted s_1 and s_2. The totals

$$\textstyle\sum_{s_1} y_k = 825.7, \qquad \sum_{s_2} y_k = 1{,}320.4$$

were calculated. Compute an unbiased estimate of the total of P83 as well as an unbiased variance estimate. Also, compute the *cve*.

3.12. Two strategies, (BE, $\hat{t}_\pi$) and (PO, $\hat{t}_\pi$), are considered for estimating the total of the variable IMP ($= y$) for the CO124 population. The first strategy calls for *BE* sampling with expected size $n = 20$. The second strategy calls for *PO* sampling with expected size $n = 20$ as follows: the seven largest countries are selected with $\pi_k = 1$, whereas all other countries are selected with π_k proportional to the value of the variable MEX $(= x)$, that is,

$$\pi_k = \begin{cases} 1 & \text{if } k \in U_{\text{large}} \\ 13x_k / \sum_{U_{\text{small}}} x_k & \text{if } k \in U_{\text{small}} \end{cases}$$

where U_{large} consists of the seven largest countries, $U_{\text{small}} = U - U_{\text{large}}$, and $\sum_{U_{\text{small}}} x_k = 132{,}725$. Compute the ratio $V_{BE}(\hat{t}_\pi)/V_{PO}(\hat{t}_\pi)$ using the following information $\sum_{U_{\text{large}}} y_k^2 \doteq 1.266 \cdot 10^{11}$, $\sum_{U_{\text{small}}} y_k^2 \doteq 4.26 \cdot 10^{10}$, and $\sum_{U_{\text{small}}} y_k^2/\pi_k \doteq 1.349 \cdot 10^{11}$.

3.13. A Poisson sample of expected size 10 was drawn from the MU284 population to estimate the total of the variable REV84 $(= y)$. The inclusion probabilities were proportional to the value of the variable P75. This led to the selection of 12 municipalities, for which the following observations were recorded.

P75	REV84
54	5,246
671	59,877
28	2,208
27	2,546
29	2,903
62	6,850
42	3,773
48	4,055
33	4,014
446	38,945
12	1,162
46	4,852

(a) Compute an unbiased estimate of the total of REV84. (The total of P75 is 8,182.) (b) Compute an unbiased estimate of the variance of the estimator used in (a), and use this to calculate the associated *cve*. (c) Show that

$$[\widehat{V_{SI}(\hat{t}_\pi)}]_{PO} = \frac{N}{N-1} \frac{1 - n/N}{2n} \sum\sum_s \frac{(y_k - y_l)^2}{\pi_k \pi_l}$$

is an unbiased estimator of the variance of $\hat{t}_\pi$ that would be obtained for an *SI* sample of size n. (d) Use the data from the Poisson sample above to calculate a variance estimate of the π estimator under *SI* with $n = 12$.

3.14. As measured by the 1975 population, the four largest municipalities in the MU284 population are those with the labels 16, 137, 114, and 29. To estimate the total of the variable P85 for this population of size four, a πps sample of size $n = 2$ was drawn using Brewer's scheme, with P75 as the known size variable. The result was as follows:

k	y_k	x_k
16	653	671
114	229	247

(a) Compute an unbiased estimate of the total of P85. (b) Compute an unbiased estimate of the variance. Also, compute the associated *cve*.

3.15. To estimate the total of the variable SS82 ($= y$), the MU284 population was stratified into four strata, using S82 ($= x$) as stratification variable. Some information about these strata is given in the following table.

Number of seats	N_h	$\sum_{U_h} x_k$	$\sum_{U_h} x_k^2$
31–40	44	1,518	52,764
41–50	168	7,524	339,344
51–70	56	3,198	184,168
71–···	16	1,260	100,016

(a) The *STSI* design and the π estimator $\sum N_h \bar{y}_{s_h}$ are used. Allocate a total sample size of $n = 40$ according to (1) proportional allocation, (2) x-optimal allocation, and (3) allocation proportional to the x-total. (b) Compare the three strategies in (a) with the strategy defined by *SI* sampling with the size $n = 40$ and the π estimator $N\bar{y}_s$. Use the information in the following table to calculate the variances of the four strategies.

Number of seats	N_h	$\sum_{U_h} y_k$	$\sum_{U_h} y_k^2$
31–40	44	756	13,784
41–50	168	3,383	72,223
51–70	56	1,545	44,529
71–···	16	617	24,137

(c) x-Optimal allocation was finally decided on, and an *STSI* sample of size $n = 40$ was drawn. The variable SS82 ($= y$) was observed, and the following results were calculated:

Number of seats	$\sum_{s_h} y_k$	$\sum_{s_h} y_k^2$
31–40	89	1,647
41–50	441	9,735
51–70	280	8,294
71–···	152	5,794

Compute an unbiased estimate of the total of SS82. Also, compute an unbiased variance estimate, and the corresponding *cve*.

3.16. (a) Compute the variance of the π estimator for the total of the variable P83 ($= y$) for the CO124 population for the following *STSI* design with two strata: Stratum 1 consists of the 105 smallest countries as measured by the value of P80, and stratum 2 consists of the 19 largest countries. From stratum 1, an *SI* sample of size 11 is to be drawn, whereas every country in stratum 2 is to be surveyed (take-all stratum). Use the following data:

Stratum h	$\sum_{U_h} y_k$	$\sum_{U_h} y_k^2$
1	1,098.9	21,855.05
2	3,445.9	1,822,736.83

(b) If *SI* sampling and the π estimator $N\bar{y}_s$ were used instead, what sample size would be required to achieve the same variance as the one obtained in (a)?

3.17. Consider a survey that uses the *SI* design. Suppose that it is required to estimate the population mean $\bar{y}_U$ with a precision such that

$$\left|\frac{\bar{y}_s - \bar{y}_U}{\bar{y}_U}\right| \leq c$$

holds with an (approximate) probability of at least $1 - \alpha$, where c is a specified constant. Show that the requirement implies a sample size n satisfying

$$n \geq \frac{\left(\dfrac{z \cdot cv_{yU}}{c}\right)^2}{\left[1 + \dfrac{1}{N}\left(\dfrac{z \cdot cv_{yU}}{c}\right)^2\right]}$$

where $z = z_{1-\alpha/2}$ and cv_{yU} is the population coefficient of variation.

3.18. Let the domain indicator variables z_{dk} be defined by (3.3.7). Verify the variance expressions shown in equations (3.3.10) and (3.3.11).

3.19. Show that the variance of the π estimator (3.3.15) for t_d under the *SI* design (n from N) is

$$V_{SI}(\hat{t}_{d\pi}) = N^2 \frac{1-f}{n} \frac{1}{N-1}\left[\sum_{U_d} y_k^2 - \frac{(\sum_{U_d} y_k)^2}{N}\right]$$

Also, show that an unbiased variance estimator is given by

$$\hat{V}_{SI}(\hat{t}_{d\pi}) = N^2 \frac{1-f}{n} \frac{1}{n-1}\left[\sum_{s_d} y_k^2 - \frac{(\sum_{s_d} y_k)^2}{n}\right]$$

3.20. Show that under the *SY* design with the sampling interval a, the bias of

$$\hat{V} = \sum\sum_s \check{\Delta}_{kl} \check{y}_k \check{y}_l$$

as an estimator for $V_{SY}(\hat{t}_\pi)$ is given by

$$B(\hat{V}) = E_{SY}(\hat{V}) - V_{SY}(\hat{t}_\pi) = t^2 - \sum_{r=1}^{a} t_{s_r}^2 = \sum\sum_{r \neq r'} t_{s_r} t_{s_{r'}}$$

3.21. If $N = n \cdot a$, where a is an integer, show that the homogeneity measures δ and ρ given by equations (3.4.9) and (3.4.11), respectively, are linearly related as

$$\delta = [1 + (N - 1)\rho]/N$$

and that consequently $\delta \doteq \rho$ when N is large.

3.22. Verify that the intraclass correlation coefficient ρ defined by equation (3.4.9) alternatively can be written as equation (3.4.10).

3.23. Prove Result 3.4.2.

3.24. If $N = n \cdot a$, where a is an integer, show that the variance $V_{SY}(\hat{t}_\pi)$ given by (3.4.6) can be written as

$$V_{SY}(\hat{t}_\pi) = \frac{1}{2} \sum_{r=1}^{a} \sum_{r'=1}^{a} (t_{s_r} - t_{s_{r'}})^2 \quad \text{where} \quad t_{s_r} = \textstyle\sum_{s_r} y_k.$$

3.25. For the SY design, consider the variance estimator $\hat{V}$ given by (3.4.16). If $N = n \cdot a$, where a is an integer, show that

$$E_{SY}(\hat{V}) \gtreqless V_{SY}(\hat{t}_\pi)$$

according as

$$V_{SI}(\hat{t}_\pi) \gtreqless V_{SY}(\hat{t}_\pi)$$

where $V_{SI}(\hat{t}_\pi)$ and $V_{SY}(\hat{t}_\pi)$ are given by (3.3.3) and (3.4.6), respectively. What is the consequence of using $\hat{V}$ as an estimator of $V_{SY}(\hat{t}_\pi)$ when $V_{SI}(\hat{t}_\pi) < V_{SY}(\hat{t}_\pi)$?

3.26. Specify the entire set of samples that are possible with the circular systematic sampling method described in Section 3.4.2 if $N = 10$ and $n = 4$. Specify all π_{kl} for this method.

3.27. For Brewer's πps scheme for $n = 2$, show that the first- and second-order inclusion probabilities π_k and π_{kl} are given by (3.6.7) and (3.6.8), respectively.

3.28. The cumulative total method for drawing one element with probability proportional-to-size (x) can be time-consuming if N is large (see Remark 3.6.4). The following scheme, due to Lahiri (1951), does not require cumulation of the size measures: Start by choosing a number M known to be $\geq \max(x_1, \ldots, x_k, \ldots, x_N)$. Then proceed as follows:

1. Draw with equal probability one of the N population elements. Let η be the selected element.
2. Draw a Unif(0, 1) random number ε.
3. If $\varepsilon M \leq x_\eta$, element η is selected, and we are through.
4. If $\varepsilon M > x_\eta$, element η is rejected, and the procedure is repeated, starting at (1).

Prove that this scheme, for $k = 1, 2, \ldots, N$, gives $\pi_k = x_k / \sum_U x_k$.

3.29. Specify the inclusion probabilities π_k and π_{kl} for the *STSI* design. Then use the general Result 2.8.1 to obtain the variance and the variance estimator given in Result 3.7.2.

3.30. A certain survey uses the *STSI* design. Suppose that it is required to estimate the population mean $\bar{y}_U$ with a precision such that

$$\left| \sum_{h=1}^{H} W_h \bar{y}_{s_h} - \bar{y}_U \right| \leq a$$

with probability at least $1 - \alpha$ ($W_h = N_h/N$), where a is a given constant. Show that the sample size required to achieve this is

$$n \geq \frac{\left(\frac{z}{a}\right)^2 \sum_{h=1}^{H} \frac{W_h^2 S_{yU_h}^2}{w_h}}{1 + \frac{1}{N}\left(\frac{z}{a}\right)^2 \sum_{h=1}^{H} W_h S_{yU_h}^2}$$

where $w_h = n_h/n$ and $z = z_{1-\alpha/2}$.

3.31. (This exercise is a continuation of 3.30.) Let $N = 1{,}000$, $H = 2$, $W_1 = 1 - W_2 = 0.8$, $S_{yU_1}^2 = 4$, $S_{yU_2}^2 = 16$, $z = 1.96$, $a = 0.5$. Study the required sample size n as a function of $w_1 = 1 - w_2$, as w_1 varies from 0 to 1.

3.32. Prove Result 3.7.4.

3.33. Use Result 3.7.2 concerning the *STSI* design to verify the variance expressions (3.7.24) and (3.7.25).

3.34. Compare proportional allocation *STSI* sampling with *SI* sampling. Verify equation (3.7.26) for the difference between the variances obtained under the two designs.

3.35. Compare the two variances

$$V_p = V_{STSI,p}\left(\sum_{h=1}^{H} N_h \bar{y}_{s_h}\right) = N^2\left(\frac{1}{n} - \frac{1}{N}\right) \sum_{h=1}^{H} W_h S_{yU_h}^2$$

and

$$V = V_{SI}(N\bar{y}_s) = N^2\left(\frac{1}{n} - \frac{1}{N}\right) S_{yU}^2$$

Suppose that all stratum variances are equal, that is $S_{yU_h}^2 = S_0^2$, say for $h = 1, \ldots, H$. (a) Show that

$$\frac{V_p}{V} \leq 1 + \frac{H-1}{N-H}$$

(b) Under what condition is equality attained in (a)? (c) Under what condition on $A = S_0^2/S_{yU}^2$ do we have $V_p/V < 1$?

3.36. Show that, under the *SIR* design with m independent draws, the estimator $\hat{t}_{\text{alt}}$ shown in equation (3.8.5) is unbiased for the population total t and that its variance is given by

$$V_{SIR}(\hat{t}_{\text{alt}}) = N^2\left[E\left(\frac{1}{n_s}\right) - \frac{1}{N}\right]S_{yU}^2$$

where n_s is the random size of the set sample s. Finally, show that

$$\text{deff}(SIR, \hat{t}_{\text{alt}}) = \frac{E\left(\frac{1}{n_s}\right) - \frac{1}{N}}{\frac{1}{n} - \frac{1}{N}}$$

3.37. Under the conditions of Exercise 3.36 above, show that the following are unbiased variance estimators for $\hat{t}_{\text{alt}}$:

(a) $\hat{V}_1 = N^2\left\{\frac{1}{n_s} - \frac{1}{N}\right\}S_{ys}^2$

(b) $\hat{V}_2 = \frac{n_s(n_s - 1)}{N - 1}\left[NE\left(\frac{1}{n_s}\right) - 1\right]\frac{S_{ys}^2}{\bar{\pi}}$

(c) $\hat{V}_3 = \left[\frac{N^2}{n_s} - N^2 + \frac{n_s(n_s - 1)}{\bar{\pi}}\right]S_{ys}^2$

where

$$\bar{\pi} = 1 - 2\left(1 - \frac{1}{N}\right)^m + \left(1 - \frac{2}{N}\right)^m$$

and

$$S_{ys}^2 = \frac{1}{n_s - 1}\sum\nolimits_s (y_k - \bar{y}_s)^2$$

The estimators in (b) and (c) were proposed by Rao (1979a, 1982).

3.38. For the SIR design with m independent draws, two unbiased estimators of $t = \sum_U y_k$ are

$$\hat{t}_1 = \frac{N\sum_{i=1}^{m} y_{k_i}}{m} \quad \text{and} \quad \hat{t}_2 = \frac{\sum_s y_k}{1 - \left(1 - \frac{1}{N}\right)^m}$$

One cannot conclude that one of these estimators is uniformly better than the other. That is, it does not hold that the variance of one is at most equal to the variance of the other for all $\mathbf{y} = (y_1, \ldots, y_N) \in \mathbb{R}^N$. Specify a vector $\mathbf{y} \in \mathbb{R}^N$ such that $V_{SIR}(\hat{t}_1) < V_{SIR}(\hat{t}_2)$; specify another vector $\mathbf{y} \in \mathbb{R}^N$ such that the opposite is true, that is, such that $V_{SIR}(\hat{t}_1) > V_{SIR}(\hat{t}_2)$.

3.39. Consider general with-replacement sampling, with given probabilities $p_1, \ldots, p_k, \ldots, p_N$ applied in each of m independent draws. For $k = 1, \ldots, N$ define the random variable J_k as the number of times that the element k is selected in the m draws, and note that

$$\hat{t}_{\text{pwr}} = \frac{1}{m}\sum\nolimits_U J_k y_k / p_k$$

Use the N dimensional joint distribution of $J_1, \ldots, J_N$ to obtain an alternative proof of Result 2.9.1. That is, prove that $\hat{t}_{\text{pwr}}$ is unbiased for t, with the variance stated in Result 2.9.1.

3.40. Consider general with-replacement sampling with m independent draws, as in Exercise 3.39 above. Show that an alternative unbiased variance estimator to (2.9.9) is given by

$$\hat{V} = \frac{1}{m} \sum\sum_{k<l}{}_s \frac{p_k p_l}{\pi_{kl}} \left(\frac{y_k}{p_k} - \frac{y_l}{p_l} \right)^2$$

where, for all $k \neq l$,

$$\pi_{kl} = 1 - (1 - p_k)^m - (1 - p_l)^m + (1 - p_k - p_l)^m$$

and s is the set of distinct elements in the ordered sample. The result comes from Rao (1979a).

3.41. To estimate the total $t = \sum_U y_k$ of the population $U = \{1, \ldots, k, \ldots, N\}$, consider the following technique, supposing that $x_1, \ldots, x_k, \ldots, x_N$ are known, strictly positive values. Poisson sampling is used as follows: For $k = 1, \ldots, N$, a Bernoulli experiment is carried out with the probability $\pi_k = Cx_k$ of selecting k, where C is a positive constant. The experiments are independent. (a) Determine C so that $E(n_s) = n$, where n is a fixed integer and n_s is the size of s. Assume that $nx_k < \sum_U x_k$ for all k. (b) Determine the constant A so that the estimator $\hat{t} = A \sum_s (y_k/x_k)$ is unbiased for $t = \sum_U y_k$. (c) Having found A, calculate the variance of $\hat{t}$. (d) Suppose that $y_k = Bx_k$, $k = 1, \ldots, N$, where B is a constant. Then show that the proposed estimation method gives a smaller variance than the strategy consisting of SI sampling (with n elements chosen from N) and the estimator $N\bar{y}_s = (N/n)\sum_s y_k$, provided $cv_{yU} = S_{yU}/\bar{y}_U > \{(1 - f)/(1 - N^{-1}f)\}^{1/2} \doteq (1 - f)^{1/2}$. (In practice, exact proportionality between y and the known variable x is unlikely, but if y is roughly proportional to x, then the proposed method may improve over SI sampling combined with $N\bar{y}_s$, provided cv_{yU} is sufficiently large.)

CHAPTER 4

Unbiased Estimation for Cluster Sampling and Sampling in Two or More Stages

4.1. Introduction

All of the designs discussed in Chapter 3 assume that direct element sampling is possible. That is, the population elements can be used as sampling units in a single stage of sampling. However, in many medium- to large-scale sample surveys, direct element sampling is not used for one or both of the following reasons:

i. There exists no sampling frame that identifies each and every population element, and the cost of producing such a frame is prohibitive.
ii. The population elements are scattered over a wide area, in which case direct element sampling will result in a widely scattered sample. Hence the cost of the field work would be prohibitive because of very high travel expenses if personal interviews are required. Also, efficient supervision of the field work might be difficult, which could result in high nonresponse rates and severe measurement errors.

A variety of sampling designs are available for surveys in which direct element sampling is impossible or impractical. These range from cluster sampling to highly complex multistage sampling designs using unequal probability sampling at the various stages of selection. Let us introduce some important terms. In *cluster sampling*, the finite population is grouped into subpopulations called clusters. A probability sample of clusters is selected, and every population element in the selected clusters is surveyed.

EXAMPLE 4.1.1. The Swedish Board of Education sponsors annual surveys in Sweden to measure drug use among ninth-grade students. In this survey, data on drug use is collected through anonymous questionnaires from every student in a sample of ninth-grade classes. The sampling frame consists of a list of all ninth-grade classes. This is an example of cluster sampling with classes as clusters and with students as elements.

Cluster sampling is also called *single-stage cluster sampling*. By contrast, in *two-stage sampling*, the sample of elements is obtained as a result of two stages of sampling:

i. The population elements are first grouped into disjoint subpopulations, called *primary sampling units* (PSUs). A probability sample of PSUs is drawn (*first-stage sampling*).
ii. For each PSU in the first-stage sample, the type of sampling unit to be used in the second-stage sampling is decided upon. These *second-stage sampling units* (SSUs) may be elements or clusters of elements. A probability sample of SSUs is drawn (*second-stage sampling*) from each PSU in the first-stage sample. When the SSUs are clusters, every element in the selected SSUs is surveyed.

Remark 4.1.1. When every SSU is an element, we use the term *two-stage element sampling*; when every SSU is a cluster of elements we have *two-stage cluster sampling*.

EXAMPLE 4.1.2. Consider the Swedish Board of Education surveys mentioned in Example 4.1.1. Suppose that a sample of students is drawn from every selected class and that data on drug use is collected for every selected student. This is an example of two-stage element sampling with classes as PSUs and with students as elements.

Alternatively, suppose that ninth graders for this type of survey are selected as follows: (i) from a frame containing every school in the country, a sample of schools is drawn; (ii) from every selected school, a sample of ninth-grade classes is drawn, and all students in the selected classes are surveyed. This is an example of two-stage cluster sampling, with schools as PSUs and with classes as SSUs.

Multistage sampling consists of three or more stages of sampling. There is a hierarchy of sampling units: primary sampling units, secondary sampling units within the PSUs, tertiary sampling units within the SSUs, and so on. The sampling units in the last-stage sampling are called *ultimate sampling units* and those in the next to the last stage are called *penultimate sampling units.*

Remark 4.1.2. The term multistage element sampling is applicable when the ultimate sampling units are the elements. In multistage cluster sampling, the ultimate sampling units are clusters of elements.

EXAMPLE 4.1.3. To continue Example 4.1.2, suppose that a sample of ninth graders is drawn from every selected class, and that data on drug use is collected from all selected students. This is an example of three-stage element sampling with schools as PSUs, with classes as SSUs. The students are tertiary sampling units or ultimate sampling units.

In this chapter we concentrate on estimation of the population total $t = \sum_U y_k$, using for the most part the π estimator $\hat{t}_\pi$. Section 4.5 is somewhat different, in that sampling with replacement is considered at stage one, and the estimator used in that case is reminiscent of the pwr estimator.

Note that in this chapter, the estimation of the population mean $\bar{y}_U = t/N$ cannot ordinarily be obtained as a simple by-product through dividing $\hat{t}_\pi$ by N and $V(\hat{t}_\pi)$ and $\hat{V}(\hat{t}_\pi)$ by N^2. This complication arises because the population size N is usually unknown in surveys that require selection of clusters. With N unknown, the parameter $\bar{y}_U = t/N$ is a ratio of two unknowns, for which separate estimators are required. The estimation of $\bar{y}_U$ for cluster and two-stage sampling designs is deferred until Chapters 5 and 8.

4.2. Single-Stage Cluster Sampling

4.2.1. Introduction

In single-stage cluster sampling, the finite population $U = \{1, \ldots, k, \ldots, N\}$ is partitioned into N_{I} subpopulations, called clusters, and denoted $U_1, \ldots, U_i, \ldots, U_{N_{\text{I}}}$. The set of clusters is symbolically represented as

$$U_{\text{I}} = \{1, \ldots, i, \ldots, N_{\text{I}}\}$$

This represents a population of clusters from which a sample of clusters is selected. The index I will be used to identify entities associated with the clusters. The reason for I instead of the perhaps more natural C is that the former index facilitates the transition to two-stage and multistage sampling, where I will naturally refer to the first stage, II to the second stage, and so on. The number of population elements in the ith cluster U_i is denoted N_i. The partitioning of U is expressed by the equations

$$U = \bigcup_{i \in U_{\text{I}}} U_i \qquad \text{and} \qquad N = \sum_{i \in U_{\text{I}}} N_i$$

Just as $\sum_A$ denotes $\sum_{k \in A}$ when A is a set of elements, we will write simply $\sum_B$ for $\sum_{i \in B}$ when B is a set of clusters, for example,

$$N = \textstyle\sum_{U_{\text{I}}} N_i = \displaystyle\sum_{i \in U_{\text{I}}} N_i.$$

Single-stage cluster sampling (or simply *cluster sampling*) is now defined in the following way:

i. A probability sample s_{I} of clusters is drawn from U_{I} according to the design $p_{\text{I}}(\cdot)$. The size of s_{I} is denoted by n_{I}, for a fixed size design, or by $n_{s_{\text{I}}}$ for a variable size design.
ii. Every population element in the selected clusters is observed.

Here, $p_{\text{I}}(\cdot)$ may be any of the conventional designs, that is, simple random sampling without replacement, systematic sampling, stratified sampling and so on. We continue to use s as a symbol for the set of elements that are observed. That is,

$$s = \bigcup_{i \in s_{\text{I}}} U_i$$

The size of s is

$$n_s = \sum_{s_{\text{I}}} N_i$$

Note that even if $p_{\text{I}}(\cdot)$ is a fixed size design, the number of observed elements n_s in general will not be fixed, because the cluster sizes N_i may vary.

The first- and second-order cluster inclusion probabilities induced by the design $p_{\text{I}}(\cdot)$ are

$$\pi_{\text{I}i} = \sum_{s_{\text{I}} \ni i} p_{\text{I}}(s_{\text{I}})$$

and for two clusters i and j,

$$\pi_{\text{I}ij} = \sum_{s_{\text{I}} \ni i \,\&\, j} p_{\text{I}}(s_{\text{I}})$$

We have $\pi_{\text{I}ii} = \pi_{\text{I}i}$. Let us turn to the element inclusion probabilities. Since the sample s contains every element in the selected clusters, we have, for every k in U_i,

$$\pi_k = \Pr(k \in s) = \Pr(i \in s_{\text{I}}) = \pi_{\text{I}i} \tag{4.2.1}$$

The second-order inclusion probabilities are given by

$$\pi_{kl} = \Pr(k \,\&\, l \in s) = \Pr(i \in s_{\text{I}}) = \pi_{\text{I}i} \tag{4.2.2}$$

if both k and l belong to the same cluster U_i, and

$$\pi_{kl} = \Pr(k \,\&\, l \in s) = \Pr(i \,\&\, j \in s_{\text{I}}) = \pi_{\text{I}ij} \tag{4.2.3}$$

if k and l belong to different clusters U_i and U_j. Note that $\pi_{kk} = \pi_k$. It is convenient to introduce the simplified notation

$$t_i = \sum_{U_i} y_k$$

for the ith cluster total. The population total to be estimated can then be expressed as

$$t = \sum_U y_k = \sum_{U_{\text{I}}} t_i$$

Let $\check{\Delta}_{\text{I}ij} = \Delta_{\text{I}ij}/\pi_{\text{I}ij}$, with $\Delta_{\text{I}ij} = \pi_{\text{I}ij} - \pi_{\text{I}i}\pi_{\text{I}j}$, and define $\check{t}_i = t_i/\pi_{\text{I}i}$. The notation $\sum\sum_B a_{ij}$ is shorthand for

$$\sum_{i \in B} \sum_{j \in B} a_{ij}$$

where B is a set of clusters. We now have the following result.

Result 4.2.1. *In cluster sampling, the π estimator of the population total $t = \sum_U y_k$ can be written*

$$\hat{t}_\pi = \sum_{s_\text{I}} \check{t}_i = \sum_{s_\text{I}} t_i/\pi_{\text{I}i} \tag{4.2.4}$$

The variance is given by

$$V(\hat{t}_\pi) = \sum\sum_{U_\text{I}} \Delta_{\text{I}ij}\check{t}_i\check{t}_j \tag{4.2.5}$$

An unbiased variance estimator is

$$\hat{V}(\hat{t}_\pi) = \sum\sum_{s_\text{I}} \check{\Delta}_{\text{I}ij}\check{t}_i\check{t}_j \tag{4.2.6}$$

PROOF. To determine the π estimator,

$$\hat{t}_\pi = \sum_s \check{y}_k = \sum_{s_\text{I}} \sum_{U_i} y_k/\pi_k$$

we use Equation (4.2.1) and obtain

$$\hat{t}_\pi = \sum_{s_\text{I}} \left(\sum_{U_i} y_k\right)/\pi_{\text{I}i} = \sum_{s_\text{I}} t_i/\pi_{\text{I}i} = \sum_{s_\text{I}} \check{t}_i$$

The rest of the result follows immediately from Result 2.8.1 by observing that $\sum_{s_\text{I}} \check{t}_i$ is the π estimator of $\sum_{U_\text{I}} t_i$. The appropriate πs are the cluster inclusion probabilities. □

If $p_\text{I}(\cdot)$ is a fixed-size design, the variance $V(\hat{t}_\pi)$ in Result 4.2.1. can also be expressed as

$$V(\hat{t}_\pi) = -\tfrac{1}{2}\sum\sum_{U_\text{I}} \Delta_{\text{I}ij}(\check{t}_i - \check{t}_j)^2 \tag{4.2.7}$$

with the unbiased variance estimator

$$\hat{V}(\hat{t}_\pi) = -\tfrac{1}{2}\sum\sum_{s_\text{I}} \check{\Delta}_{\text{I}ij}(\check{t}_i - \check{t}_j)^2 \tag{4.2.8}$$

Result 4.2.1 leads to some interesting conclusions about the efficiency of cluster sampling. We assume that the π estimator is used and that $p_\text{I}(\cdot)$ is a fixed-size design, so that equations (4.2.7) and (4.2.8) apply.

a. From (4.2.7), we see that if all $\check{t}_i = t_i/\pi_{\text{I}i}$ are equal, then $V(\hat{t}_\pi) = 0$. Thus, if we can choose $\pi_{\text{I}i}$ approximately proportional to the cluster totals t_i, then cluster sampling will be highly efficient.
b. If the cluster sizes N_i are known at the planning stage, one can choose a design with $\pi_{\text{I}i} \propto N_i$. Since $t_i = N_i\bar{y}_{U_i} = \sum_{U_i} y_k$, this is a good choice if there is little variation among the cluster means $\bar{y}_{U_i}$. If all $\bar{y}_{U_i}$ are equal, we would in effect have $V(\hat{t}_\pi) = 0$.

c. An equal probability cluster sampling design (that is, one where all $\pi_{\text{I}i}$ are equal) is often a poor choice when the clusters are of different sizes. For such a design to be highly efficient, one must have $\bar{y}_{U_i}$ roughly proportional to N_i^{-1}. This, however, is seldom the case in practice.

4.2.2. Simple Random Cluster Sampling

This section further illustrates point (c) above, that equal probability cluster sampling may often be inefficient. We consider simple random (without-replacement) cluster sampling. This design is abbreviated *SIC*. That is, an *SI* sample s_{I} of fixed size n_{I} is drawn from the N_{I} clusters in U_{I}, and all elements in selected clusters are observed. It follows from Result 4.2.1 and the results on the *SI* design in Section 3.3 that the π estimator of the population total is given by

$$\hat{t}_\pi = N_{\text{I}}\bar{t}_{s_{\text{I}}} \tag{4.2.9}$$

where $\bar{t}_{s_{\text{I}}} = \sum_{s_{\text{I}}} t_i/n_{\text{I}}$ is the mean of the cluster totals t_i in s_{I}. The variance can be written

$$V_{SIC}(\hat{t}_\pi) = N_{\text{I}}^2 \frac{1 - f_{\text{I}}}{n_{\text{I}}} S_{tU_{\text{I}}}^2 \tag{4.2.10}$$

where $f_{\text{I}} = n_{\text{I}}/N_{\text{I}}$ is the cluster sampling fraction and

$$S_{tU_{\text{I}}}^2 = \frac{1}{N_{\text{I}} - 1} \sum_{U_{\text{I}}} (t_i - \bar{t}_{U_{\text{I}}})^2 \tag{4.2.11}$$

with $\bar{t}_{U_{\text{I}}} = \sum_{U_{\text{I}}} t_i/N_{\text{I}}$. The unbiased variance estimator is

$$\hat{V}_{SIC}(\hat{t}_\pi) = N_{\text{I}}^2 \frac{1 - f_{\text{I}}}{n_{\text{I}}} S_{ts_{\text{I}}}^2 \tag{4.2.12}$$

where

$$S_{ts_{\text{I}}}^2 = \frac{1}{n_{\text{I}} - 1} \sum_{s_{\text{I}}} (t_i - \bar{t}_{s_{\text{I}}})^2$$

EXAMPLE 4.2.1. The following procedure was used to estimate the total number of conservative seats in the Swedish municipality councils, that is, the total of the variable CS82 ($=y$) for the MU284 population. An *SIC* sample of size $n_{\text{I}} = 16$ was drawn from the $N_{\text{I}} = 50$ clusters defined in Appendix C, with the following result:

i:	46	49	42	23	38	43	30	17	12	22	27	32	33	36	19	18
$\sum_{U_i} y_k$:	19	13	18	57	46	38	42	107	74	70	39	32	29	48	56	63

With $t_i = \sum_{U_i} y_k$ we have $\sum_{s_I} t_i = 751$ and $\sum_{s_I} t_i^2 = 44{,}047$. Hence the π estimate of the population total, using equation (4.2.9), is

$$\hat{t}_\pi = \frac{50}{16} 751 = 2{,}347$$

whereas, using equation (4.2.12), its unbiased variance estimate is

$$\hat{V}_{SIC}(\hat{t}_\pi) = \frac{50 \cdot 34}{16 \cdot 15} \left\{ 44{,}047 - \frac{751^2}{16} \right\} = 62{,}312$$

The *cve* is approximately 11%. Factors contributing to the rather high *cve* value are the small sample size and the fact that the *SIC* design is rather inefficient for this variable.

Remark 4.2.1. As noted in Section 3.4, systematic sampling with one random start corresponds formally to *SIC* with $n_I = 1$, where the N_I clusters correspond to the a possible systematic samples. Furthermore, systematic sampling with m random starts (see Section 3.4.4) can be looked upon as *SIC* with $n_I = m$ and $N_I = ma$. In this case, equation (4.2.12) gives an unbiased variance estimator for the π estimator of the population total.

We now take a closer look at the design *SIC* and compare it to a direct simple random selection of elements. It is useful to work with the homogeneity coefficient δ, defined in the present context as

$$\delta = 1 - \frac{S_{yW}^2}{S_{yU}^2} \tag{4.2.13}$$

where

$$S_{yW}^2 = \frac{1}{N - N_I} \sum_{U_I} \sum_{U_i} (y_k - \bar{y}_{U_i})^2 \tag{4.2.14}$$

is the pooled within-cluster variance. Here, $\bar{y}_{U_i} = \sum_{U_i} y_k / N_i$ is the ith cluster mean. Note that if

$$S_{yU_i}^2 = \frac{1}{N_i - 1} \sum_{U_i} (y_k - \bar{y}_{U_i})^2$$

denotes the variance of y within the cluster U_i, we can write equation (4.2.14) as

$$S_{yW}^2 = \frac{\sum_{U_I} (N_i - 1) S_{yU_i}^2}{\sum_{U_I} (N_i - 1)} \tag{4.2.15}$$

This gives a simple interpretation of S_{yW}^2. It is the weighted average of the N_I cluster variances $S_{yU_i}^2$, with the weights $N_i - 1$. Consequently, δ is positive or negative as the average within-cluster variance is smaller than or greater than the overall variance S_{yU}^2, respectively.

Remark 4.2.2. Readers familiar with regression analysis will identify δ as the coefficient of determination adjusted for degrees of freedom, often denoted R^2_{adj}, when fitting the linear regression of y on N_{I} dummy variables (indicating cluster membership) to the entire population of N data points.

The homogeneity coefficient δ satisfies

$$-\frac{N_{\mathrm{I}} - 1}{N - N_{\mathrm{I}}} \leq \delta \leq 1 \tag{4.2.16}$$

The upper bound on δ follows from equation (4.2.13) and the fact that $S^2_{yW} \geq 0$. To verify the lower bound on δ, we can utilize the standard ANOVA decomposition $SST = SSW + SSB$, which in this particular case is

$$(N - 1)S^2_{yU} = (N - N_{\mathrm{I}})S^2_{yW} + SSB \tag{4.2.17}$$

where

$$SSB = \textstyle\sum_{U_{\mathrm{I}}} N_i(\bar{y}_{U_i} - \bar{y}_U)^2$$

Since $SSB \geq 0$, it follows from equation (4.2.17) that

$$\frac{S^2_{yW}}{S^2_{yU}} \leq \frac{N - 1}{N - N_{\mathrm{I}}}$$

Hence,

$$\delta = 1 - \frac{S^2_{yW}}{S^2_{yU}} \geq 1 - \frac{N - 1}{N - N_{\mathrm{I}}} = -\frac{N_{\mathrm{I}} - 1}{N - N_{\mathrm{I}}}$$

Remark 4.2.3. A small δ-value means that elements in the same cluster are dissimilar with respect to the study variable, that is, have a low degree of homogeneity. A large δ-value means that elements in the same cluster are similar, that is, have a high degree of homogeneity. At the one extreme, $\delta = 1$ means that the variation is zero within every cluster; at the other extreme, $\delta = -(N_{\mathrm{I}} - 1)/(N - N_{\mathrm{I}})$ implies that all cluster means are equal. The lower extreme $-(N_{\mathrm{I}} - 1)/(N - N_{\mathrm{I}})$ is usually close to zero especially if N is large compared to N_{I}. The value $\delta = 0$ is obtained when the average within-cluster variance equals the variance in the entire population U.

Let $\bar{N} = N/N_{\mathrm{I}}$ denote the average number of elements per cluster, let $K_{\mathrm{I}} = N_{\mathrm{I}}^2(1 - f_{\mathrm{I}})/n_{\mathrm{I}}$, and let

$$\mathrm{Cov} = \frac{1}{N_{\mathrm{I}} - 1}\textstyle\sum_{U_{\mathrm{I}}}(N_i - \bar{N})N_i\bar{y}^2_{U_i} \tag{4.2.18}$$

be the covariance between N_i and $N_i\bar{y}^2_{U_i}$. It is easily verified that

$$S^2_{tU_{\mathrm{I}}} = \bar{N}S^2_{yU}\left(1 + \frac{N - N_{\mathrm{I}}}{N_{\mathrm{I}} - 1}\delta\right) + \mathrm{Cov} \tag{4.2.19}$$

Insertion of this expression into (4.2.10), which we call V_{SIC}, gives

$$V_{SIC} = \left(1 + \frac{N - N_{\mathrm{I}}}{N_{\mathrm{I}} - 1}\delta\right)\bar{N}K_{\mathrm{I}}S_{yU}^2 + K_{\mathrm{I}}\mathrm{Cov} \tag{4.2.20}$$

The expected number of observed elements under SIC, with n_{I} clusters drawn from N_{I}, is

$$E_{SIC}(n_s) = n_{\mathrm{I}}\bar{N} = n$$

To obtain a fair comparison, consider direct SI sampling, with the (fixed) sample size $n = n_{\mathrm{I}}\bar{N}$. The π estimator of t is then $N\bar{y}_s$, and the variance is

$$V_{SI} = V_{SI}(N\bar{y}_s) = \bar{N}K_{\mathrm{I}}S_{yU}^2$$

Hence, a third expression for V_{SIC} is

$$V_{SIC} = \left(1 + \frac{N - N_{\mathrm{I}}}{N_{\mathrm{I}} - 1}\delta\right)V_{SI} + K_{\mathrm{I}}\mathrm{Cov} \tag{4.2.21}$$

whereby we obtain the design effect of SIC:

$$\mathrm{deff}(SIC, \hat{t}_\pi) = \frac{V_{SIC}}{V_{SI}} = 1 + \frac{N - N_{\mathrm{I}}}{N_{\mathrm{I}} - 1}\delta + \frac{\mathrm{Cov}}{\bar{N}S_{yU}^2} \tag{4.2.22}$$

These expressions lead to some interesting conclusions about the efficiency of SIC sampling.

Case 1

Suppose that all cluster sizes N_i are equal, that is, $N_i = \bar{N}$ for every i. In this case, Cov = 0, and we obtain

$$\mathrm{deff}(SIC, \hat{t}_\pi) = \frac{V_{SIC}}{V_{SI}} = 1 + \frac{N - N_{\mathrm{I}}}{N_{\mathrm{I}} - 1}\delta \doteq 1 + (\bar{N} - 1)\delta \tag{4.2.23}$$

This shows that $V_{SIC} < V_{SI}$ if and only if $\delta < 0$, that is, if and only if there is a sufficiently large within-cluster variation. However, many clusters encountered in practice are formed by "nearby" elements, and, because such elements tend to resemble each other more or less, it is likely that $\delta > 0$. Consequently, V_{SIC} is greater than V_{SI}, often considerably greater. For instance, even if δ is weakly positive, say $\delta = 0.08$, we get, with an average cluster size of $\bar{N} = 300$,

$$\mathrm{deff}(SIC, \hat{t}_\pi) \doteq 25$$

This shows a large loss of efficiency due to cluster sampling, because the average cluster size is rather large in this case.

Case 2

Suppose that the clusters vary in size. If the correlation between N_i and $A_i = N_i\bar{y}_{U_i}^2$ is positive, as is often the case, the variance increase due to selection of clusters may be worse than in Case 1, since the second term in formula (4.2.21) can be large. To highlight the effect of variation in cluster size, consider the extreme case of minimal homogeneity, that is, $\delta = \delta_{\min} = -(N_{\mathrm{I}} - 1)/(N - N_{\mathrm{I}})$. In this case, all $\bar{y}_{U_i}$ are equal to $\bar{y}_U$ and (4.2.21) can be written

$$V_{SIC} = \bar{y}_U^2 K_{\mathrm{I}} S_{NU_{\mathrm{I}}}^2 \tag{4.2.24}$$

which will be large if the cluster size variance

$$S_{NU_{\mathrm{I}}}^2 = \frac{1}{N_{\mathrm{I}} - 1} \sum_{U_{\mathrm{I}}} (N_i - \bar{N})^2$$

is large. In this case,

$$\mathrm{deff}(SIC, \hat{t}_\pi) = \frac{V_{SIC}}{V_{SI}} = \bar{N} \left(\frac{cv_N}{cv_y} \right)^2$$

where the two coefficients of variation are $cv_N = S_{NU_{\mathrm{I}}}/\bar{N}$ and $cv_y = S_{yU}/\bar{y}_U$. The ratio V_{SIC}/V_{SI} can be much greater than unity, especially if $\bar{N}$ is large.

Our discussion shows that the strategy $(SIC, \hat{t}_\pi)$ is likely to be inefficient in many situations, especially if the clusters are homogeneous and/or of unequal sizes. However, from a cost efficiency point of view, the strategy $(SIC, \hat{t}_\pi)$ may have advantages, since it is often much cheaper to survey clusters of elements than to survey the geographically scattered sample that may arise from a simple random selection of elements.

However, the efficiency of cluster sampling can be improved when auxiliary information is available. The choice of strategy then depends on the information available. A simple case is when an approximate measure of size u_i is available for each cluster $i = 1, \ldots, N_{\mathrm{I}}$. If u_i is roughly proportional to t_i, we can reduce the variance of the π estimator by using πps cluster sampling with inclusion probabilities $\pi_{\mathrm{I}i} \propto u_i$. An alternative is to use stratified cluster sampling with strata of clusters formed so that the variation of u_i is small in each stratum.

Another option is to retain the SIC design but change the estimator to reflect the known values u_i. An approximately unbiased estimator based on this idea is

$$\hat{t}_{\mathrm{alt}} = \sum_{U_{\mathrm{I}}} u_i \frac{\sum_{s_{\mathrm{I}}} t_i/\pi_{\mathrm{I}i}}{\sum_{s_{\mathrm{I}}} u_i/\pi_{\mathrm{I}i}}$$

which will be further discussed in Section 8.5.

4.3. Two-Stage Sampling

4.3.1. Introduction

In the preceding section, we noted that, in general, $V_{SIC} > V_{SI}$. This was explained by the tendency for elements in the same cluster to resemble each other, which implies that the homogeneity measure δ is positive, and by the variation in the cluster sizes. The variance of the π estimator under SIC can always be reduced by selecting more clusters. However, the increased cost of taking a bigger sample may be unacceptable under the available budget.

To control cost and at the same time increase the number of selected clusters, we may subsample within the selected clusters, instead of surveying all elements in the selected clusters. Then we must estimate the cluster totals t_i from the subsamples. If the variation within the clusters is small, the estimates $\hat{t}_i$ have a small variance, even for rather modest subsample sizes. If often pays to use two-stage sampling instead of cluster sampling.

Notation and estimation in two-stage sampling are slightly more complex than in cluster sampling. There are two sources of variation. The first-stage sampling variation arises from the selection of primary sampling units (PSUs). The second-stage sampling variation arises from the subsampling of secondary sampling units (SSUs) within selected PSUs. Nevertheless, the principles that underlie the corresponding new estimators and their variances are straightforward, because of the hierarchic structure of the sampling.

The population of elements $U = \{1, \ldots, k, \ldots, N\}$ is partitioned into N_{I} PSUs, denoted $U_1, \ldots, U_i, \ldots, U_{N_{\mathrm{I}}}$. The whole set of PSUs (the population of PSUs) is denoted $U_{\mathrm{I}} = \{1, \ldots, i, \ldots, N_{\mathrm{I}}\}$. The size of U_i, that is, the number of elements in U_i, is denoted N_i. We have

$$N = \sum_{U_{\mathrm{I}}} N_i$$

We start with a very general class of two-stage sampling designs.

First stage. A sample s_{I} of PSUs is drawn from U_{I} ($s_{\mathrm{I}} \subset U_{\mathrm{I}}$) according to the design $p_{\mathrm{I}}(\cdot)$.

Second stage. For every $i \in s_{\mathrm{I}}$, a sample s_i of elements is drawn from U_i ($s_i \subset U_i$) according to the design $p_i(\cdot \mid s_{\mathrm{I}})$.

The resulting sample of elements, denoted s, is composed as

$$s = \bigcup_{i \in s_{\mathrm{I}}} s_i$$

Our formulation allows any first-stage design and any subsampling design for every $i \in s_{\mathrm{I}}$. The subsampling design may depend on the outcome s_{I} of the first-stage sampling. It may vary from one first-stage sample to another. Moreover, subsampling in U_i is not necessarily independent of subsampling in U_j ($i \neq j$).

This generality is not necessary in the present chapter. We narrow down the class of second-stage designs by requiring invariance and independence. These concepts are defined as follows. *Invariance* of the second-stage design means that for every i, and for every $s_{\mathrm{I}} \ni i$, we have $p_i(\cdot \mid s_{\mathrm{I}}) = p_i(\cdot)$. In words, every time the ith PSU is included in a first stage sample, identically the same subsampling design $p_i(\cdot)$ must be used. For example, the requirement may be that SI sampling with a predetermined size n_i must be used whenever the ith PSU is selected, irrespective of what other PSUs were obtained. *Independence* of the second-stage design means that, for every first-stage sample s_I,

$$\Pr\left(\bigcup_{i \in s_{\mathrm{I}}} s_i \,|\, s_{\mathrm{I}}\right) = \prod_{i \in s_{\mathrm{I}}} \Pr(s_i \,|\, s_{\mathrm{I}})$$

In words, subsampling in a given PSU is carried out independently of subsampling in any other PSU.

In the rest of this chapter, we tacitly assume that the requirements of invariance and independence are satisfied. When the requirements are not met, we can use the theory of two-phase sampling presented in Chapter 9. Two-phase sampling is a general name for sampling followed by subsampling, of which two-stage sampling with invariance and independence is an important special case.

In the rest of this section, we restrict ourselves to two-stage element sampling. That is, the SSUs are the elements. Some results for two-stage cluster sampling are given in Section 4.4.

Sample sizes are denoted as follows. The number of PSUs in s_{I} is denoted $n_{s_{\mathrm{I}}}$, or simply n_{I}, if the design $p_{\mathrm{I}}(\cdot)$ is of fixed size. The number of elements in s_i is denoted n_{s_i}, or simply n_i, if $p_i(\cdot)$ is of fixed size. The total number of elements in s, denoted n_s, is

$$n_s = \sum_{i \in s_{\mathrm{I}}} n_{s_i}$$

We need the inclusion probabilities associated with two-stage sampling. For the first-stage design $p_{\mathrm{I}}(\cdot)$, let us denote the inclusion probabilities as $\pi_{\mathrm{I}i}$, $\pi_{\mathrm{I}ij}$. Let

$$\Delta_{\mathrm{I}ij} = \pi_{\mathrm{I}ij} - \pi_{\mathrm{I}i}\pi_{\mathrm{I}j}$$

with $\Delta_{\mathrm{I}ii} = \pi_{\mathrm{I}i}(1 - \pi_{\mathrm{I}i})$, and

$$\check{\Delta}_{\mathrm{I}ij} = \Delta_{\mathrm{I}ij}/\pi_{\mathrm{I}ij}$$

Correspondingly, for the second-stage design $p_i(\cdot)$, we use the notation $\pi_{k|i}$ and $\pi_{kl|i}$. The Δ-quantities are

$$\Delta_{kl|i} = \pi_{kl|i} - \pi_{k|i}\pi_{l|i}$$

with $\Delta_{kk|i} = \pi_{k|i}(1 - \pi_{k|i})$, and, finally,

$$\check{\Delta}_{kl|i} = \Delta_{kl|i}/\pi_{kl|i}$$

4.3.2. Two-Stage Element Sampling

To obtain the π estimator, its variance, and a corresponding variance estimator we can apply the general Result 2.8.1, with the particular π_k and π_{kl} that hold under two-stage element sampling. It follows from the invariance and independence properties (see Section 4.3.1) that the element inclusion probabilities are

$$\pi_k = \pi_{Ii}\pi_{k|i} \quad \text{if } k \in U_i \tag{4.3.1}$$

and

$$\pi_{kl} = \begin{cases} \pi_{Ii}\pi_{k|i} & \text{if } k = l \in U_i \\ \pi_{Ii}\pi_{kl|i} & \text{if } k \;\&\; l \in U_i,\ k \neq l \\ \pi_{Iij}\pi_{k|i}\pi_{l|j} & \text{if } k \in U_i \text{ and } l \in U_j\ (i \neq j) \end{cases} \tag{4.3.2}$$

We can now apply Result 2.8.1. However, some algebraic manipulation is required to obtain expressions that clearly show the variation contributed by each of the two stages. A quicker derivation of the variance of the π estimator is obtained with the aid of the following well-known results from general statistical theory; see, for example, Kendall and Stuart, (1976, p. 196).

i. The expected value of a random variable can be expressed as the expected value of conditional expectations.
ii. The variance of a random variable can be expressed as the sum of the variance of conditional expectations and the expected value of conditional variances.

If X is a random variable and if we condition on the event A, then (i) and (ii) can be expressed more formally as

$$E(X) = E_A[E(X|A)]$$

$$V(X) = V_A[E(X|A)] + E_A[V(X|A)]$$

where E_A and V_A indicate mean and variance with respect to the probability distribution of the event A.

In two-stage sampling, we condition on the event that the sample s_I is realized in stage one. Let $\check{y}_{k|i} = y_k/\pi_{k|i}$ and let

$$\hat{t}_{i\pi} = \sum_{s_i} \check{y}_{k|i} \tag{4.3.3}$$

be the π estimator with respect to stage two of the PSU total $t_i = \sum_{U_i} y_k$. In repeated subsampling of U_i, according to the design $p_i(\cdot)$, $\hat{t}_{i\pi}$ is unbiased for t_i. The variance with respect to stage two is

$$V_i = \sum\sum_{U_i} \Delta_{kl|i}\check{y}_{k|i}\check{y}_{l|i} \tag{4.3.4}$$

which is unbiasedly estimated by

$$\hat{V}_i = \sum\sum_{s_i} \check{\Delta}_{kl|i}\check{y}_{k|i}\check{y}_{l|i} \tag{4.3.5}$$

As usual, alternative formulas can be given for fixed-size designs. If the design $p_i(\cdot)$ is of fixed size, V_i can also be written

$$V_i = -\tfrac{1}{2}\sum\sum_{U_i} \Delta_{kl|i}(\check{y}_{k|i} - \check{y}_{l|i})^2 \tag{4.3.6}$$

which is unbiasedly estimated by

$$\hat{V}_i = -\tfrac{1}{2}\sum\sum_{s_i} \check{\Delta}_{kl|i}(\check{y}_{k|i} - \check{y}_{l|i})^2 \tag{4.3.7}$$

The variance of the π estimator is written in the following result as the sum of two components, V_{PSU} and V_{SSU}, which reflect the two sources of variation. The index 2st is used to indicate two-stage. Note the difference between $\check{t}_i = t_i/\pi_{\mathrm{I}i}$ and $\hat{t}_{i\pi}$ given by Equation (4.3.3).

Result 4.3.1. *In two-stage element sampling, the π estimator of the population total $t = \sum_U y_k$ can be written as*

$$\hat{t}_\pi = \textstyle\sum_{s_\mathrm{I}} \hat{t}_{i\pi}/\pi_{\mathrm{I}i} \tag{4.3.8}$$

where $\hat{t}_{i\pi}$ is the π estimator of t_i with respect to stage two. The variance of $\hat{t}_\pi$ can be written as the sum of two components,

$$V_{2st}(\hat{t}_\pi) = V_{\mathrm{PSU}} + V_{\mathrm{SSU}} \tag{4.3.9}$$

with

$$V_{\mathrm{PSU}} = \textstyle\sum\sum_{U_\mathrm{I}} \Delta_{\mathrm{I}ij}\check{t}_i\check{t}_j \tag{4.3.10}$$

where $\check{t}_i = t_i/\pi_{\mathrm{I}i}$, and

$$V_{\mathrm{SSU}} = \textstyle\sum_{U_\mathrm{I}} V_i/\pi_{\mathrm{I}i} \tag{4.3.11}$$

where V_i is given by equation (4.3.4). The first component V_{PSU} is unbiasedly estimated by

$$\hat{V}_{\mathrm{PSU}} = \sum\sum_{s_\mathrm{I}} \check{\Delta}_{\mathrm{I}ij} \frac{\hat{t}_{i\pi}}{\pi_{\mathrm{I}i}} \frac{\hat{t}_{j\pi}}{\pi_{\mathrm{I}j}} - \sum_{s_\mathrm{I}} \frac{1}{\pi_{\mathrm{I}i}}\left(\frac{1}{\pi_{\mathrm{I}i}} - 1\right)\hat{V}_i \tag{4.3.12}$$

where $\hat{V}_i$ is given by equation (4.3.5), and the second component V_{SSU} is unbiasedly estimated by

$$\hat{V}_{\mathrm{SSU}} = \textstyle\sum_{s_\mathrm{I}} \hat{V}_i/\pi_{\mathrm{I}i}^2 \tag{4.3.13}$$

An unbiased estimator of $V_{2\mathrm{st}}(\hat{t}_\pi)$ is

$$\hat{V}_{2\mathrm{st}}(\hat{t}_\pi) = \hat{V}_{\mathrm{PSU}} + \hat{V}_{\mathrm{SSU}} = \sum\sum_{s_\mathrm{I}} \check{\Delta}_{\mathrm{I}ij} \frac{\hat{t}_{i\pi}}{\pi_{\mathrm{I}i}} \frac{\hat{t}_{j\pi}}{\pi_{\mathrm{I}j}} + \sum_{s_\mathrm{I}} \frac{\hat{V}_i}{\pi_{\mathrm{I}i}} \tag{4.3.14}$$

Some remarks are in order before the proof.

Remark 4.3.1. In many applications, as in a pilot study preceding a new major survey, or in the current planning of an established survey, it is of interest to form an idea of the variance contributed by each of the two stages of sampling. That is, estimates are required of V_{PSU} and V_{SSU} separately. The estimators given in (4.3.12) and (4.3.13) are useful for this purpose. Note that (4.3.12) does not always give a positive estimate.

Remark 4.3.2. It is illustrative to note the conditions under which each of the variance components are zero: (a) if $s_\mathrm{I} = U_\mathrm{I}$ with probability 1, then $\pi_{\mathrm{I}i} =$

$\pi_{\mathrm{I}ij} = 1$ for all i and j. Then, $V_{\mathrm{PSU}} = 0$ and $V_{\mathrm{SSU}} = \sum_{U_{\mathrm{I}}} V_i$. This is simply the variance of the π estimator in stratified sampling, with N_{I} PSUs as a set of strata; (b) if $s_i = U_i$ with probability 1 for all i, then $V_{\mathrm{SSU}} = 0$. In this case, V_{PSU} is the variance of the π estimator in single stage cluster sampling given by equation (4.2.5).

Remark 4.3.3. The proof of Result 4.3.1 is facilitated if we use the notational conventions

$$E_{\mathrm{I}}E_{\mathrm{II}}(\hat{t}_\pi) = E_{p_{\mathrm{I}}}[E(\hat{t}_\pi \mid s_{\mathrm{I}})]$$

$$V_{\mathrm{I}}E_{\mathrm{II}}(\hat{t}_\pi) = V_{p_{\mathrm{I}}}[E(\hat{t}_\pi \mid s_{\mathrm{I}})]$$

and

$$E_{\mathrm{I}}V_{\mathrm{II}}(\hat{t}_\pi) = E_{p_{\mathrm{I}}}[V(\hat{t}_\pi \mid s_{\mathrm{I}})]$$

That is, the subscript I indicates expected value or variance with respect to the design $p_{\mathrm{I}}(\cdot)$ used in stage one, and II indicates conditional expected value or conditional variance with respect to the set of designs $p_i(\cdot)$, $i \in s_{\mathrm{I}}$, used in stage two, given s_{I}. In the following derivations, invariance and independence are used as indicated

$$E_{\mathrm{II}}(\hat{t}_\pi) = E(\hat{t}_\pi \mid s_{\mathrm{I}}) = \sum_{s_{\mathrm{I}}} E_{p_i}(\hat{t}_{i\pi}/\pi_{\mathrm{I}i} \mid s_{\mathrm{I}}) \underset{\substack{\uparrow \\ \text{invariance}}}{=} \sum_{s_{\mathrm{I}}} E_{p_i}(\hat{t}_{i\pi}/\pi_{\mathrm{I}i}) = \sum_{s_{\mathrm{I}}} t_i/\pi_{\mathrm{I}i} = \sum_{s_{\mathrm{I}}} \check{t}_i \tag{4.3.15}$$

$$V_{\mathrm{II}}(\hat{t}_\pi) = V(\hat{t}_\pi \mid s_{\mathrm{I}}) \underset{\substack{\uparrow \\ \text{independence}}}{=} \sum_{s_{\mathrm{I}}} V(\hat{t}_{i\pi}/\pi_{\mathrm{I}i} \mid s_{\mathrm{I}}) \underset{\substack{\uparrow \\ \text{invariance}}}{=} \sum_{s_{\mathrm{I}}} V_{p_i}(\hat{t}_{i\pi}/\pi_{\mathrm{I}i}) = \sum_{s_{\mathrm{I}}} V_i/\pi_{\mathrm{I}i}^2 \tag{4.3.16}$$

PROOF. Using equation (4.3.1), we find the π estimator

$$\hat{t}_\pi = \sum_s \check{y}_k = \sum_s y_k/\pi_k = \sum_{s_{\mathrm{I}}} \sum_{s_i} y_k/(\pi_{\mathrm{I}i}\pi_{k|i})$$
$$= \sum_{s_{\mathrm{I}}} \left(\sum_{s_i} \check{y}_{k|i}\right)/\pi_{\mathrm{I}i} = \sum_{s_{\mathrm{I}}} \hat{t}_{i\pi}/\pi_{\mathrm{I}i}$$

For the variance, we use the conditional moments (4.3.15) and (4.3.16) to obtain

$$V_{2\mathrm{st}}(\hat{t}_\pi) = V_{\mathrm{I}}E_{\mathrm{II}}(\hat{t}_\pi) + E_{\mathrm{I}}V_{\mathrm{II}}(\hat{t}_\pi)$$
$$= V_{\mathrm{I}}\left(\sum_{s_{\mathrm{I}}} \check{t}_i\right) + E_{\mathrm{I}}\left(\sum_{s_{\mathrm{I}}} V_i/\pi_{\mathrm{I}i}^2\right)$$
$$= \sum\sum_{U_{\mathrm{I}}} \Delta_{\mathrm{I}ij}\check{t}_i\check{t}_j + \sum_{U_{\mathrm{I}}} V_i/\pi_{\mathrm{I}i}$$

which proves the variance expression (4.3.9). Let us turn to the variance component estimators. Because of the independence property,

$$E_{\mathrm{II}}(\hat{t}_{i\pi}\hat{t}_{j\pi}) = \begin{cases} t_i^2 + V_i & \text{for } i = j \\ t_i t_j & \text{for } i \neq j \end{cases}$$

We now have (a)

$$E\left(\sum\sum_{s_\mathrm{I}} \check{\Delta}_{\mathrm{I}ij} \frac{\hat{t}_{i\pi}}{\pi_{\mathrm{I}i}} \frac{\hat{t}_{j\pi}}{\pi_{\mathrm{I}j}}\right) = E_\mathrm{I}\left[\sum\sum_{s_\mathrm{I}} \check{\Delta}_{\mathrm{I}ij} \frac{E_{\mathrm{II}}(\hat{t}_{i\pi}\hat{t}_{j\pi})}{\pi_{\mathrm{I}i}\pi_{\mathrm{I}j}}\right]$$

$$= E_\mathrm{I}\left(\sum\sum_{s_\mathrm{I}} \check{\Delta}_{\mathrm{I}ij} \frac{t_i}{\pi_{\mathrm{I}i}} \frac{t_j}{\pi_{\mathrm{I}j}}\right) + E_\mathrm{I}\left(\sum_{s_\mathrm{I}} \check{\Delta}_{\mathrm{I}ii} \frac{V_i}{\pi_{\mathrm{I}i}^2}\right)$$

$$= \sum\sum_{U_\mathrm{I}} \Delta_{\mathrm{I}ij} \check{t}_i \check{t}_j + \sum_{U_\mathrm{I}} \left(\frac{1}{\pi_{\mathrm{I}i}} - 1\right) V_i$$

$$= V_{\mathrm{PSU}} + \sum_{U_\mathrm{I}} \left(\frac{1}{\pi_{\mathrm{I}i}} - 1\right) V_i$$

and (b)

$$E\left[-\sum_{s_\mathrm{I}} \frac{1}{\pi_{\mathrm{I}i}}\left(\frac{1}{\pi_{\mathrm{I}i}} - 1\right)\hat{V}_i\right] = -E_\mathrm{I}\left[\sum_{s_\mathrm{I}} \frac{1}{\pi_{\mathrm{I}i}}\left(\frac{1}{\pi_{\mathrm{I}i}} - 1\right)E_{\mathrm{II}}(\hat{V}_i)\right]$$

$$= -E_\mathrm{I}\left[\sum_{s_\mathrm{I}} \frac{1}{\pi_{\mathrm{I}i}}\left(\frac{1}{\pi_{\mathrm{I}i}} - 1\right)V_i\right] = -\sum_{U_\mathrm{I}}\left(\frac{1}{\pi_{\mathrm{I}i}} - 1\right)V_i$$

Together, (a) and (b) imply that $E(\hat{V}_{\mathrm{PSU}}) = V_{\mathrm{PSU}}$. Next, we prove that $\hat{V}_{\mathrm{SSU}}$ is unbiased for V_{SSU}:

$$E(\hat{V}_{\mathrm{SSU}}) = E_\mathrm{I}\left[\sum_{s_\mathrm{I}} \frac{E_{\mathrm{II}}(\hat{V}_i)}{\pi_{\mathrm{I}i}^2}\right]$$

$$= E_\mathrm{I}\left(\sum_{s_\mathrm{I}} \frac{V_i}{\pi_{\mathrm{I}i}^2}\right) = \sum_{U_\mathrm{I}} \frac{V_i}{\pi_{\mathrm{I}i}} = V_{\mathrm{SSU}}$$

Finally, we see that

$$E\{\hat{V}_{2\mathrm{st}}(\hat{t}_\pi)\} = E(\hat{V}_{\mathrm{PSU}} + \hat{V}_{\mathrm{SSU}}) = V_{\mathrm{PSU}} + V_{\mathrm{SSU}} = V_{2\mathrm{st}}(\hat{t}_\pi)$$

Thus, the total variance $V_{2\mathrm{st}}(\hat{t}_\pi)$ is unbiasedly estimated by $\hat{V}_{2\mathrm{st}}(\hat{t}_\pi)$, and the proof is completed. □

Remark 4.3.4. The variance estimator (4.3.14) can also be obtained, after some algebra, from equation (2.8.6) in the general Result 2.8.1, by noting that the π_k and π_{kl} are expressed by (4.3.1) and (4.3.2) when two-stage element sampling is involved.

The computation of a variance estimate from formula (4.3.14) may become cumbersome, especially since a variance estimate $\hat{V}_i$ must be calculated for every $i \in s_\mathrm{I}$. Thus there is a need for a computationally simpler variance estimator. Let us try the first term of (4.3.14), that is,

$$\hat{V}^* = \sum\sum_{s_\mathrm{I}} \check{\Delta}_{\mathrm{I}ij} \frac{\hat{t}_{i\pi}}{\pi_{\mathrm{I}i}} \frac{\hat{t}_{j\pi}}{\pi_{\mathrm{I}j}} \tag{4.3.17}$$

The quantities $\check{\Delta}_{Iij}$ are determined by the first-stage design, and the only other information required for this simplified variance estimator is the estimated PSU totals. We have, from (a) in the proof of Result 4.3.1,

$$E(\hat{V}^*) = V_{2st}(\hat{t}_\pi) - \sum_{U_I} V_i$$

Thus the bias of $\hat{V}^*$ is given by

$$B(\hat{V}^*) = -\sum_{U_I} V_i \tag{4.3.18}$$

which means that $\hat{V}^*$ underestimates the unknown variance of $\hat{t}_\pi$. This undesirable property may lead to undue optimism in judging the precision of a computed estimate. However, a look at the relative bias of the variance estimator

$$\frac{B(\hat{V}^*)}{V_{2st}(\hat{t}_\pi)} = -\frac{\sum_{U_I} V_i}{\sum\sum_{U_I} \Delta_{Iij}\check{t}_i\check{t}_j + \sum_{U_I} V_i/\pi_{Ii}} \tag{4.3.19}$$

reveals that the underestimation obtained from $\hat{V}^*$ may in many cases be unimportant. The numerator in the expression (4.3.19) will often be small compared to the denominator if the π_{Ii} are small, with little or even negligible underestimation as a consequence. For instance, suppose the clusters are drawn with *SI* sampling at the rate $n_I/N_I = 10\% = \pi_{Ii}$. If $V_{PSU}/V_{SSU} = 5$, then the relative bias obtained from equation (4.3.19) is only $-1/60 = -0.017$.

We see from equations (4.3.12) and (4.3.13) that $\hat{V}^*$ overestimates the first component V_{PSU} but does not "cover" the sum of the two components.

Naturally, the ideal would be to find a variance estimator that is exactly unbiased, and as simple as $\hat{V}^*$, that is, dependent only on the estimated PSU totals and on the first-stage quantities Δ_{Iij}. We must then introduce additional variation into the estimated PSU totals to compensate for the underestimation.

Srinath and Hidiroglou (1980) devised such a method for the case where the first-stage design is general (but of fixed size), and the second stage involves *SI* sampling in the selected PSUs with n_i elements drawn from N_i in the ith PSU. The additional variation is introduced by further subsampling, selecting, again by *SI*, n_i' from the n_i sampled elements in the ith PSU. If $\bar{y}_i'$ denotes the mean of the n_i' subsampled elements, then

$$\hat{\hat{t}}_i = N_i\bar{y}_i'$$

is unbiased for t_i. One can show that

$$\hat{V}' = -\frac{1}{2}\sum\sum_{s_I} \check{\Delta}_{Iij}\left(\frac{\hat{\hat{t}}_i}{\pi_{Ii}} - \frac{\hat{\hat{t}}_j}{\pi_{Ij}}\right)^2$$

is unbiased for V_{2st}, provided the subsampled number is taken as

$$n_i' = \frac{n_i(1 - \pi_{Ii})}{1 - \pi_{Ii}f_i}$$

where $f_i = n_i/N_i$. Equivalently, one should exclude (by simple random exclusion)

$$n_i - n_i' = n_i \frac{\pi_{Ii}(1 - f_i)}{1 - \pi_{Ii} f_i}$$

of the sampled elements in the ith cluster and calculate the mean $\bar{y}_i'$ of the remaining n_i' elements. If π_{Ii} is small, little exclusion is needed.

Remark 4.3.5. If $\hat{V}^*$ given by equation (4.3.17) is deemed satisfactory (despite the underestimation) as an estimator of $V_{2\text{st}}(\hat{t}_\pi)$, some computational simplicity has been gained. Another advantage of $\hat{V}^*$ is that we may use a second-stage design for which no unbiased variance estimator formula exists, for example, systematic sampling with a single random start, and still be able to assess the precision of $\hat{t}_\pi$. (In that case, it is not possible, however, to obtain an unbiased estimate of the second component V_{SSU}.) The variance estimator $\hat{V}^*$ may still be computationally rather heavy because of the mixed terms $i \neq j$. Thus there is a need for a still simpler variance estimator. One such estimator is proposed in Section 4.5.

Remark 4.3.6. Many two-stage or multistage surveys use what is called a *self-weighting design.* A self-weighting two-stage selection is carried out as follows. Suppose that u_i is a known (rough) measure of size of the ith PSU. We may then use a first-stage design with $\pi_{Ii} = cu_i$ (for a suitable constant c), and, in the second stage, *SI* subsampling of n_i from N_i elements so that

$$\frac{n_i}{N_i} = \frac{1}{u_i} \tag{4.3.20}$$

If a fixed size n_I is used in the first stage, $c = n_I/\sum_{U_I} u_i$. We then have

$$\pi_k = \pi_{Ii}\pi_{k|i} = cu_i \cdot \frac{n_i}{N_i} = c$$

The π estimator of t becomes

$$\hat{t}_\pi = \textstyle\sum_s y_k/\pi_k = \frac{1}{c}\sum_{s_I}\sum_{s_i} y_k$$

that is, the sample values y_k are uniformly weighted. One reason to use this approach is to gain control of the field work. If we can assume that

$$\frac{N_i}{u_i} \doteq \text{constant}$$

for all i, it follows from equation (4.3.20) that

$$n_i \doteq \text{constant}$$

That is, the number of interviews to be carried out is roughly the same in all

PSUs selected. If there is one interviewer per PSU, the work loads are thus roughly equal.

Example 4.3.1. We denote by SI, SI the design consisting of simple random sampling without replacement in both stages. In stage one, take an SI sample s_{I} of n_{I} from the N_{I} clusters, then for $i \in s_{\mathrm{I}}$, take an SI sample of n_i from the N_i elements in the ith cluster. Applying Result 4.3.1, we find that the π estimator of the population total t can be written

$$\hat{t}_\pi = \frac{N_{\mathrm{I}}}{n_{\mathrm{I}}} \sum_{s_{\mathrm{I}}} N_i \bar{y}_{s_i} = \frac{N_{\mathrm{I}}}{n_{\mathrm{I}}} \sum_{s_{\mathrm{I}}} \hat{t}_{i\pi} \tag{4.3.21}$$

with

$$\hat{t}_{i\pi} = N_i \bar{y}_{s_i} = N_i \left(\sum_{s_i} y_k\right)/n_i$$

The variance is

$$V_{2\mathrm{st}}(\hat{t}_\pi) = N_{\mathrm{I}}^2 \frac{1 - f_{\mathrm{I}}}{n_{\mathrm{I}}} S_{tU_{\mathrm{I}}}^2 + \frac{N_{\mathrm{I}}}{n_{\mathrm{I}}} \sum_{U_{\mathrm{I}}} N_i^2 \frac{1 - f_i}{n_i} S_{yU_i}^2 \tag{4.3.22}$$

where $f_{\mathrm{I}} = n_{\mathrm{I}}/N_{\mathrm{I}}$; $f_i = n_i/N_i$,

$$S_{tU_{\mathrm{I}}}^2 = \frac{1}{N_{\mathrm{I}} - 1} \sum_{U_{\mathrm{I}}} (t_i - \bar{t}_{U_{\mathrm{I}}})^2$$

is the variance in U_{I} of the cluster totals t_i, with $\bar{t}_{U_{\mathrm{I}}} = \sum_{U_{\mathrm{I}}} t_i / N_{\mathrm{I}}$, and

$$S_{yU_i}^2 = \frac{1}{N_i - 1} \sum_{U_i} (y_k - \bar{y}_{U_i})^2$$

with $\bar{y}_{U_i} = \sum_{U_i} y_k / N_i$.

The unbiased variance estimator is

$$\hat{V}_{2\mathrm{st}}(\hat{t}_\pi) = N_{\mathrm{I}}^2 \frac{1 - f_{\mathrm{I}}}{n_{\mathrm{I}}} S_{\hat{t}s_{\mathrm{I}}}^2 + \frac{N_{\mathrm{I}}}{n_{\mathrm{I}}} \sum_{s_{\mathrm{I}}} N_i^2 \frac{1 - f_i}{n_i} S_{ys_i}^2 \tag{4.3.23}$$

where

$$S_{\hat{t}s_{\mathrm{I}}}^2 = \frac{1}{n_{\mathrm{I}} - 1} \sum_{s_{\mathrm{I}}} \left[\hat{t}_{i\pi} - \left(\sum_{s_{\mathrm{I}}} \hat{t}_{i\pi}/n_{\mathrm{I}}\right)\right]^2$$

is the variance in s_{I} of the estimated cluster totals $\hat{t}_{i\pi} = N_i \bar{y}_{s_i}$, and

$$S_{ys_i}^2 = \frac{1}{n_i - 1} \sum_{s_i} (y_k - \bar{y}_{s_i})^2$$

Note that the second term of the variance estimator (4.3.23) is not unbiased for the second component of the variance (4.3.22).

Example 4.3.2. A small two-stage SI, SI sample was drawn to estimate the total of the variable S82 ($= y$) for the MU284 population of municipalities.

In stage one, an *SI* sample s_I of size $n_\mathrm{I} = 5$ was drawn from the $N_\mathrm{I} = 50$ PSUs (= clusters) defined in Appendix C. From every cluster in s_I, an *SI* sample of size $n_i = 3$ was drawn from the N_i elements in the ith PSU. The following data were recorded:

i	N_i	y_k
19	5	41, 49, 49
45	8	49, 49, 45
47	5	31, 31, 35
50	9	39, 41, 61
31	7	49, 51, 33

To calculate the π estimate and the associated variance estimate, we first compute $\hat{t}_{i\pi}$ and $S^2_{ys_i}$ for every $i \in s_\mathrm{I}$:

i	$\hat{t}_{i\pi}$	$S^2_{ys_i}$
19	695/3	64/3
45	1144/3	16/3
47	485/3	16/3
50	423	148
31	931/3	292/3

Using equation (4.3.21) we get

$$\hat{t}_\pi = 10 \cdot \left(\frac{695}{3} + \frac{1144}{3} + \frac{485}{3} + 423 + \frac{931}{3} \right) = 15{,}080$$

and

$$S^2_{\hat{t}s_\mathrm{I}} = \frac{1}{n_\mathrm{I} - 1} \sum_{s_\mathrm{I}} [\hat{t}_{i\pi} - (\sum_{s_\mathrm{I}} \hat{t}_{i\pi}/n_\mathrm{I})]^2 = 11{,}410.9$$

Finally, formula (4.3.23) gives

$$\hat{V}_{2\mathrm{st}}(\hat{t}_\pi) = \frac{50 \cdot 45}{5} 11{,}410.9 + 10 \cdot \left(\frac{5 \cdot 2}{3} \frac{64}{3} + \frac{8 \cdot 5}{3} \frac{16}{3} + \frac{5 \cdot 2}{3} \frac{16}{3} + \frac{9 \cdot 6}{3} 148 + \frac{7 \cdot 4}{3} \frac{292}{3} \right) = 5{,}172{,}234$$

The *cve* for this small two-stage sample is 15%.

What can be said about the efficiency of the design *SI*, *SI*? We note that the first component of the variance (4.3.22) is equal to the cluster sampling variance given in equation (4.2.10). The analysis in Section 4.2, in particular formula (4.2.22), showed that the cluster sampling variance is inflated in part by a considerable within-cluster homogeneity ($\delta > 0$), in part by high variation in the cluster sizes N_i.

The efficiency is often improved by the following procedure. The clusters are first stratified by some measure of size, so that clusters of comparable size are grouped together. Then, in each stratum of clusters, use *SI* sampling, in each of the two stages. Let $s_{\mathrm{I}h}$ of size $n_{\mathrm{I}h}$ denote an *SI* sample selected among the $N_{\mathrm{I}h}$ clusters in the hth stratum of clusters ($h = 1, \ldots, H$). The π estimator is then

$$\hat{t}_\pi = \sum_{h=1}^{H} \frac{N_{\mathrm{I}h}}{n_{\mathrm{I}h}} \sum\nolimits_{s_{\mathrm{I}h}} N_i \bar{y}_{s_i}$$

Its variance and unbiased variance estimator are easily found with the aid of Result 4.3.1.

Alternatively, cluster size measures can be used to build an estimator of the ratio type, in the following way. Let u_i denote a measure of size, known for all N_{I} PSUs. With *SI* sampling in both stages, an improvement on the π estimator (4.3.21) is

$$\hat{t}_{\mathrm{alt}} = \sum\nolimits_{U_{\mathrm{I}}} u_i \frac{\sum_{s_{\mathrm{I}}} N_i \bar{y}_{s_i}}{\sum_{s_{\mathrm{I}}} u_i}$$

This estimator is further discussed in Section 8.5.

4.4. Multistage Sampling

4.4.1. Introduction and a General Result

Despite their complexity, sampling designs in three or more stages are often used in large-scale surveys. For example, the Canadian Labor Force Survey earlier used four stages in rural areas. In the current design, the number of stages for rural areas has been reduced to three, namely, a first stage with stratified, geographically defined PSUs, a second stage with Census Enumeration Areas as SSUs, and a third stage with dwellings as tertiary units. This requires good-quality lists of dwellings for entire rural Census Enumeration Areas. Three-stage sampling is presented in Section 4.4.2. The objective in this section is to obtain a general result for sampling in r stages, where $r \geq 2$.

As in two-stage sampling, the population is partitioned into primary sampling units (PSUs), $U_1, \ldots, U_i, \ldots, U_{N_{\mathrm{I}}}$. Their sizes are often unknown before sampling begins. The set of PSUs will be represented symbolically as

$$U_{\mathrm{I}} = \{1, \ldots, i, \ldots, N_{\mathrm{I}}\}$$

Let s_{I}, $p_{\mathrm{I}}(\cdot)$, $\pi_{\mathrm{I}i}$, $\pi_{\mathrm{I}ij}$, $\Delta_{\mathrm{I}ij}$, $\check{\Delta}_{\mathrm{I}ij}$, t_i, and $\check{t}_i$ have the same meanings as in Section 4.3.

For Result 4.4.1 below, we do not need detailed notation for the subsequent $r - 1$ stages of sampling. We do assume, however, that we can construct an estimator $\hat{t}_i$ of the PSU total t_i such that $\hat{t}_i$ is unbiased with respect to the final $r - 1$ stages of selection, that is,

$$E(\hat{t}_i | s_\mathrm{I}) = t_i$$

for all i. In Result 4.4.1, $\hat{t}_i$ is not necessarily the π estimator; any estimator that is unbiased, given s_I, is allowed. The ultimate-stage sampling units are not necessarily elements; they can also be clusters of elements.

Finally, let $V_i = V(\hat{t}_i | s_\mathrm{I})$ be the variance of $\hat{t}_i$ due to the last $r - 1$ stages of selection, and let $\hat{V}_i$ be an unbiased estimator of V_i given s_I, that is, $E(\hat{V}_i | s_\mathrm{I}) = V_i$.

We assume invariance and independence of sampling stages subsequent to the first stage. Whenever a certain PSU is selected, subsampling of that PSU follows an invariant rule, and subsampling of one PSU is independent of subsampling of all other PSUs. It is easy to show that an unbiased estimator of the population total t is given by

$$\hat{t} = \sum_{s_\mathrm{I}} \hat{t}_i / \pi_{\mathrm{I}i} \tag{4.4.1}$$

Its properties are spelled out in the following result. The index rst indicates r-stage.

Result 4.4.1. *In r-stage sampling* ($r \geq 2$), *an unbiased estimator of the population total t is given by*

$$\hat{t} = \sum_{s_\mathrm{I}} \hat{t}_i / \pi_{\mathrm{I}i}$$

where $E(\hat{t}_i | s_\mathrm{I}) = t_i$. *With* $\check{t}_i = t_i / \pi_{\mathrm{I}i}$, *the variance can be written*

$$V_{r\mathrm{st}}(\hat{t}) = \sum\sum_{U_\mathrm{I}} \Delta_{\mathrm{I}ij} \check{t}_i \check{t}_j + \sum_{U_\mathrm{I}} V_i / \pi_{\mathrm{I}i} \tag{4.4.2}$$

where the first term represents the variance contributed by the first-stage sampling, and the second term combines variance contributed by all subsequent stages of sampling. An unbiased variance estimator is given by

$$\hat{V}_{r\mathrm{st}}(\hat{t}) = \sum\sum_{s_\mathrm{I}} \check{\Delta}_{\mathrm{I}ij} \frac{\hat{t}_i}{\pi_{\mathrm{I}i}} \frac{\hat{t}_j}{\pi_{\mathrm{I}j}} + \sum_{s_\mathrm{I}} \hat{V}_i / \pi_{\mathrm{I}i} \tag{4.4.3}$$

where $\hat{V}_i$ *satisfies* $E(\hat{V}_i | s_\mathrm{I}) = V_i$ *for all i.*

Proof. As the proof closely resembles that of Result 4.3.1, many details will be left out.

We have

$$\begin{aligned} V_{r\mathrm{st}}(\hat{t}) &= V_\mathrm{I}\{E(\hat{t} | s_\mathrm{I})\} + E_\mathrm{I}\{V(\hat{t} | s_\mathrm{I})\} \\ &= V_\mathrm{I}(\textstyle\sum_{s_\mathrm{I}} t_i / \pi_{\mathrm{I}i}) + E_\mathrm{I}(\textstyle\sum_{s_\mathrm{I}} V_i / \pi_{\mathrm{I}i}^2) \\ &= \textstyle\sum\sum_{U_\mathrm{I}} \Delta_{\mathrm{I}ij} \check{t}_i \check{t}_j + \textstyle\sum_{U_\mathrm{I}} V_i / \pi_{\mathrm{I}i} \end{aligned}$$

and

$$
\begin{aligned}
E\{\hat{V}_{\text{rst}}(\hat{t})\} &= E_{\text{I}}\{E(\hat{V}_{\text{rst}}(\hat{t})|s_{\text{I}})\} \\
&= E_{\text{I}}\{\textstyle\sum\sum_{s_{\text{I}}} \check{\Delta}_{\text{I}ij}\check{t}_i\check{t}_j + \sum_{s_{\text{I}}}(1-\pi_{\text{I}i})V_i/\pi_{\text{I}i}^2 + \sum_{s_{\text{I}}} V_i/\pi_{\text{I}i}\} \\
&= \textstyle\sum\sum_{U_{\text{I}}} \Delta_{\text{I}ij}\check{t}_i\check{t}_j + \sum_{U_{\text{I}}} V_i/\pi_{\text{I}i} = V_{\text{rst}}(\hat{t})
\end{aligned}
$$

which completes the proof. □

Although it says little about the exact nature of the last $r - 1$ stages, Result 4.4.1 is of practical interest for the following reason. The computational effort associated with multistage sampling is usually considerable. In particular, this holds for the variance estimation. To calculate the $\hat{V}_i$ necessary for using (4.4.3) may in fact be a formidable task. For example, if $\hat{t}_i$ is the π estimator relative to the final $r - 1$ stages, the calculation of $\hat{V}_i$ requires the second-order inclusion probabilities of all $r - 1$ designs.

A simplified variance estimator that does not involve the $\hat{V}_i$ would bring about a great simplification. To this end, we can try the variance estimator consisting of the first term of (4.4.3) only, that is,

$$
\hat{V}^* = \sum\sum_{s_{\text{I}}} \check{\Delta}_{\text{I}ij} \frac{\hat{t}_i}{\pi_{\text{I}i}} \frac{\hat{t}_j}{\pi_{\text{I}j}} \tag{4.4.4}
$$

which requires only the PSU total estimates and the Δ for stage one. One can show that the relative bias of $\hat{V}^*$ is, under the present conditions too, given by (4.3.19). Often $\hat{V}^*$ will lead to an unimportant underestimation.

Generalizing the procedure given for two stages in Section 4.3, Srinath and Hidiroglou (1980) proposed a procedure to remove the bias of $\hat{V}^*$ through subsampling of the samples obtained in the final $r - 1$ stages.

4.4.2. Three-Stage Element Sampling

Going beyond two stages of sampling does not introduce any significant difficulties for estimation of the population total, as long as we restrict ourselves to r-stage sampling such that the designs in the second, third, fourth, and so on stage each have the invariance and independence properties. However, the notation for a general simple r-stage sampling design becomes unwieldy, and we give detailed results only for three-stage element sampling, where "element" means that the ultimate, third-stage sampling units are the population elements.

The following examination of three-stage sampling shows how the π estimator, its variance, and the variance estimator are constructed according to a stage-by-stage hierarchy. The three types of sampling unit are described by (i) to (iii) below. The typical PSU, SSU, and tertiary unit is indexed as i, q, and k, respectively.

i. The N elements of U are partitioned into the PSUs $U_1, \ldots, U_i, \ldots, U_{N_{\text{I}}}$. The set of PSUs is symbolically represented by

$$U_{\mathrm{I}} = \{1, \ldots, i, \ldots, N_{\mathrm{I}}\}$$

Let N_i be the size of U_i; we then have

$$N = \sum_{U_{\mathrm{I}}} N_i$$

ii. The N_i elements in U_i ($i = 1, \ldots, N_{\mathrm{I}}$) are partitioned into $N_{\mathrm{II}i}$ secondary sampling units (SSUs),

$$U_{i1}, \ldots, U_{iq}, \ldots, U_{iN_{\mathrm{II}i}}$$

The set of SSUs formed by the partitioning of U_i is symbolically represented by

$$U_{\mathrm{II}i} = \{1, \ldots, q, \ldots, N_{\mathrm{II}i}\}$$

Letting N_{iq} be the size of U_{iq}, we have

$$N_i = \sum_{U_{\mathrm{II}i}} N_{iq}$$

iii. The tertiary sampling units are the population elements.

The sampling proceeds as follows: (*First-stage*) A sample s_{I} of PSUs is drawn from U_{I} ($s_{\mathrm{I}} \subset U_{\mathrm{I}}$) according to the design $p_{\mathrm{I}}(\cdot)$. (*Second stage*) For $i \in s_{\mathrm{I}}$, a sample $s_{\mathrm{II}i}$ of SSUs is drawn from $U_{\mathrm{II}i}$ ($s_{\mathrm{II}i} \subset U_{\mathrm{II}i}$) according to the design $p_{\mathrm{II}i}(\cdot)$ (where the subscripts indicate design at stage II for the ith PSU). (*Third-stage*) For $q \in s_{\mathrm{II}i}$, a sample s_{iq} of elements is drawn from U_{iq} ($s_{iq} \subset U_{iq}$) according to the design $p_{iq}(\cdot)$ (an even more detailed notation would have been $p_{\mathrm{III}iq}(\cdot)$, for design at stage III in the iqth SSU).

The elements ultimately observed are the elements $k \in s$, where

$$s = \bigcup_{i \in s_{\mathrm{I}}} \bigcup_{q \in s_{\mathrm{II}i}} s_{iq}$$

To obtain the π estimator of the population total t and its variance, we need the inclusion probabilities for each of the three stages. The following table shows the notation, and the corresponding Δ quantities. Here, i and j denote distinct PSUs, q and r denote distinct SSUs, and k and l denote distinct tertiary (ultimate) units (TSUs).

		Inclusion probabilities		
Stage	Design	First order	Second order	Δ Quantities
I	$p_{\mathrm{I}}(\cdot)$	$\pi_{\mathrm{I}i}$	$\pi_{\mathrm{I}ij}$	$\Delta_{\mathrm{I}ij} = \pi_{\mathrm{I}ij} - \pi_{\mathrm{I}i}\pi_{\mathrm{I}j}$
II	$p_{\mathrm{II}i}(\cdot)$	$\pi_{\mathrm{II}q\|i}$	$\pi_{\mathrm{II}qr\|i}$	$\Delta_{\mathrm{II}qr\|i} = \pi_{\mathrm{II}qr\|i} - \pi_{\mathrm{II}q\|i}\pi_{\mathrm{II}r\|i}$
III	$p_{iq}(\cdot)$	$\pi_{k\|iq}$	$\pi_{kl\|iq}$	$\Delta_{kl\|iq} = \pi_{kl\|iq} - \pi_{k\|iq}\pi_{l\|iq}$

We have $\pi_{\mathrm{I}ii} = \pi_{\mathrm{I}i}$; $\pi_{\mathrm{II}qq|i} = \pi_{\mathrm{II}q|i}$; $\pi_{kk|iq} = \pi_{k|iq}$.

As before, $\check{\Delta}$ is defined as Δ divided by the appropriate second-order inclusion probability, that is,

$$\check{\Delta}_{\mathrm{I}ij} = \Delta_{\mathrm{I}ij}/\pi_{\mathrm{I}ij}; \qquad \check{\Delta}_{\mathrm{II}qr|i} = \Delta_{\mathrm{II}qr|i}/\pi_{\mathrm{II}qr|i}; \qquad \check{\Delta}_{kl|iq} = \Delta_{kl|iq}/\pi_{kl|iq}$$

Finally, introduce the totals

$$t_{iq} = \sum_{U_{iq}} y_k; \qquad t_i = \sum_{U_{\mathrm{II}i}} t_{iq}; \qquad t = \sum_{U_\mathrm{I}} t_i$$

The π estimator of $t = \sum_U y_k$ is built up in stages, proceeding backward from the last to the first stage. The π estimator of t_{iq} with respect to stage III is

$$\hat{t}_{iq\pi} = \sum_{s_{iq}} y_k/\pi_{k|iq} \tag{4.4.5}$$

from which the π estimator of t_i with respect to stages II and III is constructed as

$$\hat{t}_{i\pi} = \sum_{s_{\mathrm{II}i}} \hat{t}_{iq\pi}/\pi_{\mathrm{II}q|i} \tag{4.4.6}$$

Finally, the π estimator of t with respect to all three stages is given by

$$\hat{t}_\pi = \sum_{s_\mathrm{I}} \hat{t}_{i\pi}/\pi_{\mathrm{I}i} \tag{4.4.7}$$

Let us find the variance of $\hat{t}_\pi$ and an unbiased estimator thereof. Here, Result 4.4.1 with $r = 3$ applies, but it does not give explicit expressions for V_i and $\hat{V}_i$. However, since we have now specified that π estimation is involved in each of the three stages, we can easily find V_i and $\hat{V}_i$, using Result 4.3.1 on two-stage element sampling. Let

$$V_{iq} = \sum\sum_{U_{iq}} \Delta_{kl|iq} \frac{y_k}{\pi_{k|iq}} \frac{y_l}{\pi_{l|iq}} \tag{4.4.8}$$

be the variance of $\hat{t}_{iq\pi}$ in repeated subsampling from U_{iq}, and let

$$V_{\mathrm{II}i} = \sum\sum_{U_{\mathrm{II}i}} \Delta_{\mathrm{II}qr|i} \frac{t_{iq}}{\pi_{\mathrm{II}q|i}} \frac{t_{ir}}{\pi_{\mathrm{II}r|i}} \tag{4.4.9}$$

be the variance of

$$\sum_{s_{\mathrm{II}i}} t_{iq}/\pi_{\mathrm{II}q|i} \tag{4.4.10}$$

in repeated subsampling from $U_{\mathrm{II}i}$. Since $\hat{t}_{i\pi}$ is the π estimator with respect to the last two stages, we have

$$V_i = V_{\mathrm{II}i} + \sum_{U_{\mathrm{II}i}} V_{iq}/\pi_{\mathrm{II}q|i}$$

If this expression is inserted into formula (4.4.2) with $r = 3$, we obtain the three variance components stated in the following result. The total variance is indexed 3st for three-stage.

Result 4.4.2. *In three-stage element sampling, the variance of the π estimator* (4.4.7) *of the population total t can be written*

$$V_{3\mathrm{st}}(\hat{t}_\pi) = V_{\mathrm{PSU}} + V_{\mathrm{SSU}} + V_{\mathrm{TSU}} \tag{4.4.11}$$

where

$$V_{PSU} = \sum\sum_{U_I} \Delta_{Iij} \check{t}_i \check{t}_j \tag{4.4.12}$$

with $\check{t}_i = t_i/\pi_{Ii}$ *gives the variance contribution due to the first-stage sampling,*

$$V_{SSU} = \sum_{U_I} V_{IIi}/\pi_{Ii} \tag{4.4.13}$$

gives the variance contribution due to the second-stage sampling, and

$$V_{TSU} = \sum_{U_I} (\sum_{U_{IIi}} V_{iq}/\pi_{IIq|i})/\pi_{Ii} \tag{4.4.14}$$

gives the variance contribution due to the third-stage sampling.

Turning to variance estimation, we use equation (4.4.3) with $r = 3$. Thus, an unbiased variance estimator is given by

$$\hat{V}_{3st}(\hat{t}_\pi) = \sum\sum_{s_I} \check{\Delta}_{Iij} \frac{\hat{t}_{i\pi}}{\pi_{Ii}} \frac{\hat{t}_{j\pi}}{\pi_{Ij}} + \sum_{s_I} \frac{\hat{V}_i}{\pi_{Ii}} \tag{4.4.15}$$

where the appropriate $\hat{V}_i$, structured in the manner of the two-stage formula (4.3.14), is

$$\hat{V}_i = \sum\sum_{s_{IIi}} \check{\Delta}_{IIqr|i} \frac{\hat{t}_{iq\pi}}{\pi_{IIq|i}} \frac{\hat{t}_{ir\pi}}{\pi_{IIr|i}} + \sum_{s_{IIi}} \frac{\hat{V}_{iq}}{\pi_{IIq|i}} \tag{4.4.16}$$

with

$$\hat{V}_{iq} = \sum\sum_{s_{iq}} \check{\Delta}_{kl|iq} \frac{y_k}{\pi_{k|iq}} \frac{y_l}{\pi_{l|iq}} \tag{4.4.17}$$

For the planning of future surveys, it will be of interest to get separate estimates of the three variance components V_{PSU}, V_{SSU} and V_{TSU}. It is left as an exercise to show that unbiased estimators of the three components are as stated in the following result.

Result 4.4.3. *Unbiased estimators of the variance components in three-stage element sampling (see Result* 4.4.2) *are*

$$\hat{V}_{TSU} = \sum_{s_I} \{\sum_{s_{IIi}} \hat{V}_{iq}/\pi^2_{IIq|i}\}/\pi^2_{Ii} \tag{4.4.18}$$

$$\hat{V}_{SSU} = \sum_{s_I} \hat{V}_i/\pi^2_{Ii} - \hat{V}_{TSU} \tag{4.4.19}$$

and

$$\hat{V}_{PSU} = \hat{V}_{3st}(\hat{t}_\pi) - \hat{V}_{SSU} - \hat{V}_{TSU} \tag{4.4.20}$$

where $\hat{V}_{3st}(\hat{t}_\pi)$, $\hat{V}_i$ *and* $\hat{V}_{iq}$ *are given, respectively, by equations* (4.4.15), (4.4.16), *and* (4.4.17).

Variance and variance estimation in multistage surveys is discussed in Rao (1975, 1979a) and Gray (1975). Gray and Platek (1976) analyze design effects in multistage surveys. A computer package for calculation of the variance estimates was developed by Bellhouse (1980); it has been adapted for IBM microcomputers under the name TREES, see Rylett and Bellhouse (1988). Principles for computing variance estimates are discussed in Bellhouse (1985).

Remark 4.4.1. Illustrating the discussion following Result 4.4.1, we can see from equations (4.4.18) to (4.4.20) that the computation of an unbiased variance estimator in three-stage element sampling will be cumbersome, even on a large computer. It requires the calculation of many components, including the $\hat{V}_{iq}$ for every $i \in s_{\mathrm{I}}$ and every $q \in s_{\mathrm{II}i}$, and the $\hat{V}_i$ for every $i \in s_{\mathrm{I}}$. Thus, there is a great need for a simpler variance estimator. We have already discussed (Section 4.4.1) the merits of the somewhat negatively biased formula $\hat{V}^*$ given by equation (4.4.4), which drastically reduces the computational effort. Other simplified approaches are given in Section 4.6.

Remark 4.4.2. As a byproduct of Results 4.4.2 and 4.4.3, we easily get the results for two-stage cluster sampling, which differs from simple three-stage sampling in that the SSUs U_{iq}, which are clusters of elements, are surveyed completely. Hence the exact total t_{iq} can be determined for every U_{iq} in the sample. The π estimator of the population total t can still be written as

$$\hat{t}_\pi = \sum_{s_{\mathrm{I}}} \hat{t}_{i\pi}/\pi_{\mathrm{I}i} \tag{4.4.21}$$

but with

$$\hat{t}_{i\pi} = \sum_{s_{\mathrm{II}i}} t_{iq}/\pi_{\mathrm{II}q|i} \tag{4.4.22}$$

Because the variance contribution from third-stage sampling is nil, the variance of the π estimator is

$$V(\hat{t}_\pi) = \sum\sum_{U_{\mathrm{I}}} \Delta_{\mathrm{I}ij}\check{t}_i\check{t}_j + \sum_{U_{\mathrm{I}}} V_{\mathrm{II}i}/\pi_{\mathrm{I}i} \tag{4.4.23}$$

where $\check{t}_i = t_i/\pi_{\mathrm{I}i}$ and $V_{\mathrm{II}i}$ is given by (4.4.9). An unbiased variance estimator is given by

$$\hat{V}(\hat{t}_\pi) = \sum\sum_{s_{\mathrm{I}}} \check{\Delta}_{\mathrm{I}ij}\frac{\hat{t}_{i\pi}}{\pi_{\mathrm{I}i}}\frac{\hat{t}_{j\pi}}{\pi_{\mathrm{I}j}} + \sum_{s_{\mathrm{I}}} \frac{\hat{V}_{\mathrm{II}i}}{\pi_{\mathrm{I}i}} \tag{4.4.24}$$

where

$$\hat{V}_{\mathrm{II}i} = \sum\sum_{s_{\mathrm{II}i}} \check{\Delta}_{\mathrm{II}qr|i}\frac{t_{iq}}{\pi_{\mathrm{II}q|i}}\frac{t_{ir}}{\pi_{\mathrm{II}r|i}} \tag{4.4.25}$$

4.5. With-Replacement Sampling of PSUs

It is good practice in the reporting of survey results to supply at least the most important point estimates with their estimated standard errors, that is, the square root of the estimated variances. This enables the user to evaluate the quality of the results. Numerous point estimates are usually computed in a medium-to-large scale survey. In multistage sampling, the computation of standard errors can be time-consuming and costly, especially when some of the stages involve sampling without replacement.

One way to save time and cost is to use simplified, but somewhat biased, variance estimators. Examples of such estimators were discussed in Section 4.4. Another way to save resources is to use sampling with replacement at one or more stages. There is usually some loss of efficiency when sampling is done with replacement. However, this is compensated for by a reduction in the computation required for the variance estimates.

We consider the following type of sampling in r stages, where $r \geq 2$.

i. In the first sampling stage, an ordered sample, $os_{\mathrm{I}} = (i_1, \ldots, i_\nu, \ldots, i_{m_{\mathrm{I}}})$, of PSUs is drawn according to a with-replacement sampling scheme such that, at every draw, p_i $(i = 1, \ldots, N_{\mathrm{I}})$ is the probability of selecting the ith PSU $(\sum_{U_{\mathrm{I}}} p_i = 1)$.
ii. The sampling in stages following the first has the properties of invariance and independence.
iii. If a PSU is selected more than once in the ordered first-stage sample, it is independently subsampled as many times as it is drawn.

Let $\hat{t}_{i_\nu}$ be an unbiased estimator of t_{i_ν}, given os_{I}. Let $V_{i_\nu} = V(\hat{t}_{i_\nu} | os_{\mathrm{I}})$ be the variance of $\hat{t}_{i_\nu}$ contributed by the last $r - 1$ stages of sampling, given os_{I}.

Result 4.5.1. *In r-stage sampling satisfying the specifications* (i), (ii), *and* (iii) *above, an unbiased estimator of the population total* $t = \sum_U y_k$ *is given by*

$$\hat{t} = \frac{1}{m_{\mathrm{I}}} \sum_{\nu=1}^{m_{\mathrm{I}}} \hat{t}_{i_\nu} / p_{i_\nu} \tag{4.5.1}$$

The variance of $\hat{t}$ *is given by*

$$V(\hat{t}) = \frac{1}{m_{\mathrm{I}}} \sum_{i=1}^{N_{\mathrm{I}}} p_i \left(\frac{t_i}{p_i} - t \right)^2 + \frac{1}{m_{\mathrm{I}}} \sum_{i=1}^{N_{\mathrm{I}}} \frac{V_i}{p_i} \tag{4.5.2}$$

and an unbiased variance estimator is given by

$$\hat{V}(\hat{t}) = \frac{1}{m_{\mathrm{I}}(m_{\mathrm{I}} - 1)} \sum_{\nu=1}^{m_{\mathrm{I}}} \left(\frac{\hat{t}_{i_\nu}}{p_{i_\nu}} - \hat{t} \right)^2 \tag{4.5.3}$$

PROOF. The proof follows the same lines of argument as that of Result 2.9.1. Let

$$\hat{Z}_\nu = \hat{t}_{i_\nu} / p_{i_\nu} \quad \text{and} \quad Z_\nu = t_{i_\nu} / p_{i_\nu}$$

Now, $\hat{Z}_\nu$ $(\nu = 1, \ldots, m_{\mathrm{I}})$ are independently and identically distributed random variables, with

$$E(\hat{Z}_\nu) = E\{E(\hat{Z}_\nu | os_{\mathrm{I}})\} = E(Z_\nu) = t$$

and

$$V(\hat{Z}_v) = V\{E(\hat{Z}_v|os_{\text{I}})\} + E\{V(\hat{Z}_v|os_{\text{I}})\}$$
$$= V(Z_v) + E(V_{i_v}/p_{i_v}^2)$$
$$= \sum_{i=1}^{N_{\text{I}}} p_i\left(\frac{t_i}{p_i} - t\right)^2 + \sum_{i=1}^{N_{\text{I}}} V_i/p_i$$

Since $\hat{t}$ is the mean of the m_{I} independently and identically distributed random variables $\hat{Z}_v$, the rest of the proof follows from arguments already used in proving Result 2.9.1. □

EXAMPLE 4.5.1. To estimate the total of the variable RMT85 ($=y$) for the MU284 population, a small two-stage sample was selected as follows. In the first stage a pps sample of $m_{\text{I}} = 5$ PSUs was drawn. The PSUs are the clusters defined in Appendix C, and the drawing probabilities were proportional to the value of the variable TME84. From each PSU, i_v ($v = 1, \ldots, 5$), two municipalities were selected using pps sampling with drawing probabilities proportional to ME84 ($=x$). The following data were recorded:

v	i_v	t_{xi_v}	k	x_k	y_k
1	24	48,977	137	47,074	6,720
			137	47,074	6,720
2	24	48,977	137	47,074	6,720
			135	418	65
3	40	5,253	225	2,771	418
			224	1,039	153
4	48	11,781	268	5,292	764
			270	4,777	592
5	4	59,505	16	45,324	6,263
			18	3,994	532

The total of TME84 is 505,226.

To calculate an unbiased estimate of the total of RMT we first observe that for $v = 1$

$$\hat{t}_{i_v}/p_{i_v} = \frac{505{,}226}{48{,}997}\,\frac{48{,}997}{2}\left(\frac{6{,}720}{47{,}074} + \frac{6{,}720}{47{,}074}\right) = 72{,}123$$

whereas the values of $\hat{t}_{i_v}/p_{i_v}$ for $v = 2, 3, 4$, and 5 are

ν:	2	3	4	5
$\hat{t}_{i_\nu}/p_{i_\nu}$:	75,343	75,305	67,775	68,555

Using equations (4.5.1) and (4.5.3), we get

$$\hat{t} = 71{,}820 \quad \text{and} \quad \hat{V}(\hat{t}) = 2{,}583{,}541.$$

The corresponding *cve* is approximately 2%.

4.6. Comparing Simplified Variance Estimators in Multistage Sampling

Let us return to multistage sampling with PSUs selected without replacement, which is normally preferred to sampling with replacement because of higher efficiency. As already mentioned, the calculation of the unbiased variance estimator (4.4.3) can be extremely heavy. Simplified procedures are needed. Although Result 4.5.1 deals with sampling with replacement, it suggests a simplified variance estimator that can be used for selection without replacement.

Suppose that the first-stage sampling calls for the selection of a fixed number n_{I} of PSUs with inclusion probabilities $\pi_{\mathrm{I}i}$.

For example, we could have a πps sampling design such that $\pi_{\mathrm{I}i}$ is taken proportional to u_i, an approximate size measure for the ith PSU. This implies that $\pi_{\mathrm{I}i} = n_{\mathrm{I}} u_i / \sum_{U_{\mathrm{I}}} u_i$. We assume that the π estimator

$$\hat{t}_\pi = \sum_{s_{\mathrm{I}}} \hat{t}_{i\pi}/\pi_{\mathrm{I}i}$$

is used to estimate the population total t, where $\hat{t}_{i\pi}$ is the π estimator with respect to the subsequent $r - 1$ stages. Options for estimating the variance of $\hat{t}_\pi$, $V_{\mathrm{rst}}(\hat{t}_\pi)$ given by equation (4.4.2), include the following:

a. Use the unbiased variance estimator (4.4.3). The calculations can be heavy, because $\hat{V}_i$, the estimated variance of $\hat{t}_{i\pi}$, must first be calculated for every selected PSU. The required estimates $\hat{V}_i$ may be numerous and may depend on several subsequent stages.
b. Use the abridged formula obtained from (4.4.3) by simply dropping the term that involves the $\hat{V}_i$. Because the size of the first-stage design is fixed, we can consider

$$\hat{V}^* = -\tfrac{1}{2} \sum\sum_{s_{\mathrm{I}}} \check{\Delta}_{\mathrm{I}ij} \left(\frac{\hat{t}_{i\pi}}{\pi_{\mathrm{I}i}} - \frac{\hat{t}_{j\pi}}{\pi_{\mathrm{I}j}} \right)^2 \tag{4.6.1}$$

Now, $\hat{V}^*$ underestimates $V(\hat{t}_\pi)$. Nevertheless, in many instances the underestimation is unimportant.

c. Use a formula that is computationally even simpler than (4.6.1), namely,

$$\hat{V}^{**} = \frac{1}{n_{\mathrm{I}}(n_{\mathrm{I}} - 1)} \sum_{s_{\mathrm{I}}} \left(\frac{\hat{t}_{i\pi}}{p_i} - \hat{t}_{\pi} \right)^2 \tag{4.6.2}$$

where p_i is determined by $\pi_{\mathrm{I}i} = n_{\mathrm{I}} p_i$. Here, $\hat{V}^{**}$ is obtained by mimicking the expression (4.5.3) that applies when PSUs are sampled with replacement.

The use of $\hat{V}^{**}$ may lead to over- or underestimation of $V(\hat{t}_{\pi})$, depending on the sampling design at stage one. One can show that $\hat{V}^{**}$ overestimates $V(\hat{t}_{\pi})$ if and only if selection of n_{I} clusters without replacement is more efficient for estimating the total t than drawing $m_{\mathrm{I}} = n_{\mathrm{I}}$ clusters with replacement. An analogous conclusion for single-stage sampling is expressed by (3.6.15).

Under *SI* sampling of PSUs, with n_{I} PSUs chosen from N_{I}, one can show that the variance estimators given by equations (4.6.2) and (4.6.1) can be written, respectively, as

$$\hat{V}^{**} = N_{\mathrm{I}}^2 \frac{1}{n_{\mathrm{I}}(n_{\mathrm{I}} - 1)} \sum_{s_{\mathrm{I}}} [\hat{t}_{i\pi} - (\textstyle\sum_{s_{\mathrm{I}}} \hat{t}_{i\pi}/n_{\mathrm{I}})]^2$$

and

$$\hat{V}^{*} = \left(1 - \frac{n_{\mathrm{I}}}{N_{\mathrm{I}}}\right) \hat{V}^{**}$$

Here, $\hat{V}^{**}$ overestimates $V(\hat{t}_{\pi})$. If the sampling fraction $n_{\mathrm{I}}/N_{\mathrm{I}}$ is small, the numerical difference between the two estimators is unimportant.

Another design for which $\hat{V}^{**}$ overestimates $V(\hat{t}_{\pi})$ is systematic πps sampling when the PSUs are randomly ordered before the first-stage sample is drawn. Yet other designs for which $\hat{V}^{**}$ leads to overestimates are described in Sukhatme and Sukhatme (1970, pp. 64–67); see also Jönrup (1974).

Exercises

4.1. To estimate the total of the variable SS82 ($= y$) for the MU284 population, an *SIC* sample of size $n_{\mathrm{I}} = 20$ was drawn from the $N_{\mathrm{I}} = 50$ clusters defined in Appendix C. The following results were recorded:

$$\sum_{s_{\mathrm{I}}} t_i = 2{,}450 \quad \text{and} \quad \sum_{s_{\mathrm{I}}} t_i^2 = 327{,}296$$

Compute an unbiased estimate of the total of the variable SS82. Also, compute an unbiased variance estimate as well as the corresponding *cve*.

4.2. Consider the partitioning of the MU284 population defined by the 50 clusters given in Appendix C. Compute the coefficient of homogeneity (δ) and the design effect deff$(SIC, \hat{t}_{\pi})$ for estimation of the totals of the variables RMT85 and CS82.

Use the following data:

	RMT85	CS82
S^2_{yU}	355,612	24.369
S^2_{yW}	337,288	13.993
Cov	−261,547	93.630

4.3. The total Swedish population in 1985 equals the total of the variable P85 for the MU284 population. To estimate this total, systematic cluster sampling with three random starts was used as follows. The 50 clusters defined in Appendix C were ordered according to the values of the variable TP75. From the integers 1, 2, ..., 10, three were selected using *SI* sampling. The selected integers were 2, 8, and 5. Each of these integers determines one "slice" of the whole systematic sample. The first slice is composed of the 2nd, the 12th, the 22nd, the 32nd, and the 42nd size-ordered clusters, which correspond to the clusters labeled 29, 13, 1, 14, and 50. The second slice consists of the clusters labeled 39, 18, 10, 6, and 20, and the third slice consists of the clusters labeled 42, 26, 11, 38, and 7. (a) Compute an unbiased estimate of the 1985 Swedish population (using data in the Appendix). (b) Compute an unbiased estimate of the variance of the estimator used in (a) above, as well as the corresponding *cve*.

4.4. Consider the MU284 population and the variable REV84. We want to estimate the population total using cluster sampling, with the clusters defined in Appendix C. If *SIC* sampling is used with $n_I = 25$, the variance of the π estimator is $8.5 \cdot 10^9$. Compare this variance with the variance of the π estimator under *STSIC*, supposing that the 50 clusters are stratified into four strata according to the values of the variable TP75. Use the information in the following table, where S_{th} denotes the standard deviation of the cluster totals t_i in stratum h:

TP75	N_{Ih}	S_{th}	n_{Ih}
0–75	16	2,268	4
75–175	13	4,729	8
175–275	17	4,848	9
275–···	4	22,420	4
	50		25

4.5. Cluster sampling was used as follows to estimate the MU284 population total for the variable ME84. The 50 clusters defined in Appendix C were stratified into two strata according to the values of the variable TP75. The four largest clusters were assigned to stratum 1, namely, the clusters labeled 4, 9, 20, and 24. The remaining clusters were assigned to stratum 2. From stratum 1, a πps sample of $n_I = 2$ clusters was drawn by Brewer's scheme with TP75 as the size measure. This led to the following results:

Label	TP75	TME84
20	382	32,647
4	886	59,505

From stratum 2, a pps sample of $m_{\text{I}} = 5$ clusters was drawn with TP75 as the size measure. The results were as follows:

Label	TP75	TME84
43	225	12,116
23	80	4,434
22	217	11,745
2	193	14,286
38	204	11,376

(a) Using the data given above compute an unbiased estimate of the ME84 population total. Use the π estimator in stratum 1 and the pwr estimator in stratum 2. The stratum totals of the variable TP75 are 2,100 and 6,082. (b) Compute an unbiased estimate of the variance of the estimator in (a), as well as the corresponding *cve*.

4.6. A population of 1,010 sausages was partitioned into two primary units, the first of size 1,000, the second of size 10. The following sampling design was used to estimate the mean number of sausage ends for this population. With equal probability (1/2), one primary unit is selected. From the selected primary unit, an *SI* sample of 2 sausages is selected. Mr. A carries out the design. The first primary unit is selected. He observes that both sausages in the second-stage sample have two ends. Mr. A calculates the sample mean number of ends and obtains the result 2. He claims that this value is an unbiased estimate of the population mean. Mr. B, using the same sample information as Mr. A, is of the opinion that an unbiased estimate of the mean number of ends is

$$2 \cdot 1{,}000 \cdot 2/1{,}010 = 3.96$$

Discuss the two methods of estimation. Identify the logic behind each of them.

4.7. Use the plant size data in Exercise 8.1 to calculate the two variance components and the total variance in estimation according to the unbiased two-stage strategies 1 and 3 defined in Exercise 8.1.

4.8. The objective is to estimate the total of the variable CS82 ($= y$) for the MU284 population. Two-stage sampling is used, with *SI* sampling of $n_{\text{I}} = 5$ from $N_{\text{I}} = 50$ PSUs. The PSUs are the clusters defined in Appendix C. In every selected PSU, an *SI* sample of size $n_i = 2$ is drawn. The following results were recorded:

i	N_i	k	y_k
11	6	61	8
		58	10
5	5	24	13
		23	12
20	5	113	11
		112	8
48	5	266	2
		270	7
14	7	77	16
		81	10

(a) Compute an unbiased estimate of the total of the variable CS82. (b) Compute unbiased estimates of V_{PSU}, V_{SSU}, and $V = V_{PSU} + V_{SSU}$. Also, compute the corresponding *cve*. (c) Compute simplified variance estimates according to equations (4.6.1) and (4.6.2), as well as the corresponding *cve*s. Comment on your conclusions.

4.9. Let us define the employment rate in the Swedish labor force as the proportion of individuals of age 20 to 64 years who work at least 20 hours per week. To estimate this employment rate, a sample survey was carried out with the following two-stage design: (*Stage* 1) The 284 Swedish municipalities were size-ordered according to the number of persons in the 20 to 64 age group. (This variable is not in the Appendix.) The total of that variable is 4,804,685. The municipalities were then divided into four strata of equal sizes, so that stratum 1 contained the 71 first (smallest) municipalities, stratum 2 the next 71 municipalities, and so on. From each stratum, an *SI* sample of the three municipalities was drawn. (*Stage* 2) From each of the selected municipalities, an *SI* sample of 100 individuals from the age group 20 to 64 years was drawn. For every sampled individual, information on employment status was obtained. The results were as follows:

Stratum	No. individuals in selected municipality	No. employed in the sample from the municipality
1	2,927	72
	5,436	72
	5,010	74
2	8,731	75
	5,749	68
	6,703	73

Stratum	No. individuals in selected municipality	No. employed in the sample from the municipality
3	10,823	75
	14,763	70
	9,446	72
4	18,511	74
	26,687	74
	67,144	75

(a) Estimate the employment rate (in %) in the country using π estimation. (b) Calculate an unbiased variance estimate, as well as the corresponding *cve*. (c) The design is not very efficient. Why? Propose a better design for π estimation, using the same number of municipalities for the first stage sampling.

4.10. To estimate the percentage of Swedish income earners (according to the 1984 taxation), a two-stage sampling design was used as follows: (*Stage* 1) A Poisson sample of municipalities was drawn from the 284 Swedish municipalities. The inclusion probability of the ith municipality was proportional to the population size (the population total) of the municipality, and the expected sample size was 10. (*Stage* 2) From each selected municipality, an *SI* sample of 100 individuals was selected, and the number of income earners was observed. The results were as follows for a total population of 8,327,484 in 284 municipalities:

Selected municipality	Population size	No. of income earners
10	59,665	72
19	26,052	83
86	29,885	76
89	15,151	74
101	48,839	76
114	230,381	81
129	27,784	74
133	12,639	75
157	30,355	75
160	13,408	80
236	87,621	79
242	18,600	75
266	8,664	78
283	9,969	75

(a) Compute an unbiased estimate of the percentage of income earners in the total population. (b) Compute an unbiased estimate of the variance of the estimator used in (a). Note: An alternative estimate from the same data is derived in Exercise 5.15.

4.11. To estimate the total of the variable P85 ($= y$) for the MU284 population, the following two-stage sampling design was used. (*Stage* 1) The clusters in Appendix C were used as PSUs. A pps sample of five PSUs was drawn with selection probabilities proportional to the value of the variable TP75. (*Stage* 2) From each selected PSU, a πps sample of two municipalities was drawn according to Brewer's scheme with inclusion probabilities proportional to the value of the variable P75 ($= x$). The following data were collected for the selected municipalities:

i_v	k	y_k	x_k
33	186	10	10
	187	13	14
21	117	105	102
	119	26	26
43	236	88	85
	239	28	28
12	68	11	12
	69	66	62
48	268	84	74
	270	74	72

(a) Compute an unbiased estimate of the total of the variable P85. (b) Compute an unbiased estimate of the variance of the estimator used in (a), as well as the corresponding *cve*.

4.12. A graduate student has 401 books on his bookshelf. To obtain an estimate of the total number of words in his collection of books, he uses the following three-stage sampling design. (*Stage* 1) He selects five out of the 401 books by *SI* sampling. (*Stage* 2) From each selected book he takes an *SI* sample of two pages. The number of pages in each selected book is given in the table below. (*Stage* 3) From each selected page he selects lines by *SY* sampling with two random starts as follows: He selects an *SI* sample of two out of the first 10 integers, say r and r'. The systematic sample consists of the lines $r, r + 10, r + 20, \ldots$ and $r', r' + 10, r' + 20, \ldots$. He finally counts the number of words on each selected line. The results were:

Selected book	Number of pages	Selected pages	Number of words in systematic sample: 1	2
81	178	36	13	11
		82	18	21
186	128	94	29	28
		117	29	21

Selected book	Number of pages	Selected pages	Number of words in systematic sample	
			1	2
195	278	61	22	24
		212	25	27
288	243	99	19	15
		111	20	24
387	191	71	34	29
		178	34	35

(a) Compute an unbiased estimate of the total number of words in the book collection. (b) Compute an unbiased estimate of the variance of the estimator in (a) above, as well as the corresponding *cve*. (c) Compute the two simplified variance estimators discussed in Section 4.6. (d) Compute unbiased estimates of V_{PSU}, V_{SSU}, and V_{TSU}.

4.13. Starting from Result 2.8.1, derive expressions for the cluster-sampling variance given by (4.2.5) to (4.2.8) by using the second-order inclusion probabilities (4.2.2) and (4.2.3).

4.14. Prove that (4.2.19) holds, that is, that the variance of the cluster totals can be written

$$S_{tU_I}^2 = \bar{N} S_{yU}^2 \left\{ 1 + \frac{N - N_I}{N_I - 1} \delta \right\} + \text{Cov}$$

where Cov is given by equation (4.2.18).

4.15. Show that the expected number of elements under the *SIC* design with n_I clusters drawn from N_I is $E_{SIC}(n_s) = n_I N / N_I$.

4.16. Verify the expressions (4.3.1) and (4.3.2) for the element inclusion probabilities in two-stage element sampling.

4.17. Derive the variance estimator (4.3.14) from Result 2.8.1 by using the inclusion probabilities given by (4.3.1) and (4.3.2).

4.18. For the two-stage element sampling design *SI*, *SI*, give unbiased estimators for each of the two components of the variance (4.3.22).

4.19. Derive the variance of the unbiased estimator

$$\hat{t}_\pi = \sum_{h=1}^{H} \frac{N_{Ih}}{n_{Ih}} \sum\nolimits_{s_{Ih}} N_i \bar{y}_{s_i}$$

given in Section 4.3.2 for stratified two-stage element sampling. Also derive an unbiased estimator of this variance.

4.20. Prove that the variance estimators given by equations (4.4.18) to (4.4.20) in Result 4.4.3 are unbiased for V_{TSU}, V_{SSU}, and V_{PSU}, respectively.

4.21. Consider multistage sampling with a fixed-size design at the first stage. Show that the simplified variance estimator (4.6.1), that is,

$$\hat{V}^* = -\frac{1}{2}\sum\sum_{s_\text{I}} \check{\Delta}_{\text{I}ij}\left(\frac{\hat{t}_{i\pi}}{\pi_{\text{I}i}} - \frac{\hat{t}_{j\pi}}{\pi_{\text{I}j}}\right)^2$$

underestimates the variance of the π estimator.

4.22. Show that in multistage sampling where the first-stage design is SI (of n_I PSUs from N_I), the simplified variance estimator

$$\hat{V}^{**} = N_\text{I}^2 \frac{1}{n_\text{I}(n_\text{I} - 1)} \sum_{s_\text{I}} [\hat{t}_{i\pi} - (\textstyle\sum_{s_\text{I}} \hat{t}_{i\pi}/n_\text{I})]^2$$

overestimates the variance of the π estimator.

4.23. A population U of size 1,000 is divided into 250 clusters, each cluster containing four elements. The cluster means $\bar{y}_{U_i}$ are specified as follows: 50 cluster means equal $M - 2$, 50 equal $M - 1$, 50 equal M, 50 equal $M + 1$, and the last 50 cluster means equal $M + 2$, where M is a constant. The within cluster variance $S^2_{yU_i}$ is the same for all 250 clusters and equal to V_0. Compare two strategies for estimating the total $t = \sum_U y_k$: (i) SI sampling of 25 out of the 250 clusters, followed by observation of the $25 \times 4 = 100$ elements in the selected clusters. (ii) A direct SI sampling of 100 out of the 1,000 elements followed by observation of the 100 selected elements. Which of the two strategies gives the smaller variance for the π estimator of t supposing that (a) $V_0 = 5$; (b) $V_0 = 10$?

CHAPTER 5

Introduction to More Complex Estimation Problems

5.1. Introduction

The dominating question in Chapters 2 through 4 was the estimation of one particular parameter, the population total $t = \sum_U y_k$. To this end, we examined the properties of the π estimator $\hat{t}_\pi = \sum_s \check{y}_k$ under a variety of sampling designs. We now turn our attention to a number of other parameters of common interest, namely, ratios of totals, domain means, population variances, regression coefficients, medians, and other population quantiles. These parameters are not as simply structured as the population total.

The estimators in this chapter are constructed with a number of π estimators as building blocks. For this reason, the results for π estimation discussed in earlier chapters provide important background material.

In Chapters 3 and 4, we emphasized different designs; in this chapter, we discuss different parameters, keeping the design general. It is up to the reader to place the results in this chapter into the context of a particular design, for example, two-stage sampling with *STSI* in stage one and *SI* in stage two. Sections 5.2 to 5.5 present preparatory material and provide a general background to the specific estimation problems treated in Sections 5.6 to 5.11.

As a simple example of the type of parameter that we want to estimate in this chapter, consider the ratio between two unknown population totals,

$$R = \frac{\sum_U y_k}{\sum_U z_k} = \frac{t_y}{t_z}$$

where y and z are two variables of study. For example, in a population of individuals, y_k may represent the savings and z_k the income of the kth individual. Then R is the savings portion of an income dollar in the finite population. An obvious way to compose an estimator of R is to estimate the totals t_y and

t_z by their respective π estimators $\hat{t}_{y\pi} = \sum_s \check{y}_k$ and $\hat{t}_{z\pi} = \sum_s \check{z}_k$, where $\check{y}_k = y_k/\pi_k$ and $\check{z}_k = z_k/\pi_k$. The resulting estimator of R is

$$\hat{R} = \frac{\hat{t}_{y\pi}}{\hat{t}_{z\pi}} = \frac{\sum_s \check{y}_k}{\sum_s \check{z}_k}$$

Although composed of two unbiased components, $\hat{t}_{y\pi}$ and $\hat{t}_{z\pi}$, the estimator $\hat{R}$ is not unbiased for R. As one might guess, however, $\hat{R}$, is *approximately* unbiased for R, under certain conditions, as we show in Section 5.6.

A simple formula for the variance of $\hat{R}$ cannot be given, although the variances of $\hat{t}_{y\pi}$ and $\hat{t}_{z\pi}$ are well-known expressions. However, without much difficulty, we can obtain an approximate variance (see Section 5.6), which also leads to a procedure for variance estimation. Confidence intervals that conform approximately to a desired confidence level can thereby be obtained.

The procedure that led from the parameter R to the estimator $\hat{R}$ is a simple example of an important general principle for estimation of a population parameter θ that can be expressed as a function of several population totals, $t_1, t_2, \ldots, t_q$,

$$\theta = f(t_1, \ldots, t_j, \ldots, t_q)$$

where

$$t_j = \sum_U y_{jk}$$

and $y_{1k}, \ldots, y_{qk}$ are values for the kth element of the study variables $y_1, \ldots, y_q$. The principle is simply the following. In the function $f(\cdot, \ldots, \cdot)$, replace each unknown total t_j by the corresponding π estimator,

$$\hat{t}_{j\pi} = \sum_s \check{y}_{jk} = \sum_s y_{jk}/\pi_k$$

The resulting estimator of θ is

$$\hat{\theta} = f(\hat{t}_{1\pi}, \ldots, \hat{t}_{j\pi}, \ldots, \hat{t}_{q\pi})$$

Naturally, we are interested in finding the statistical properties (bias, variance, etc.) of $\hat{\theta}$. This is easy if f happens to be a linear function; no new tools are required. But parameters often encountered in practice correspond to nonlinear functions f. By first-order Taylor approximation, $\hat{\theta}$ can then be approximated by a more easily handled linear function. Approximate bias and approximate variance expressions can thus be obtained. The general procedure is given in Section 5.5.

5.2. The Effect of Bias on Confidence Statements

Unbiasedness is a characteristic of the π estimator. Although the property is desirable, the importance of exact unbiasedness should not be exaggerated. There are two important reasons why it is not reasonable, in many cases, to strive for an exactly unbiased estimator.

1. Many parameters have a structure that makes it difficult to find an unbiased estimator.
2. An estimator with some bias can often have a smaller variance and mean square error (MSE) than an unbiased estimator.

Many estimators useful in practice are in fact only *approximately* unbiased, as will be evident from this and subsequent chapters.

On the other hand, it is generally agreed that greatly biased estimators should be avoided. Hájek (1971) expresses this rule resolutely:

> ... greatly biased estimates are poor no matter what other properties they have.

Then, how much bias should one accept? An ideal estimator is generally thought of as one whose sampling distribution is tightly concentrated around the unknown parameter value. This guarantees a high probability of a close estimate. Let $\hat{\theta}$ be an estimator of θ with variance $V(\hat{\theta})$ and bias $B(\hat{\theta}) = E(\hat{\theta}) - \theta$. A customary measure of the accuracy of $\hat{\theta}$ is the mean square error (MSE)

$$\text{MSE}(\hat{\theta}) = E[(\hat{\theta} - \theta)^2] = V(\hat{\theta}) + [B(\hat{\theta})]^2$$

which depends on both the variance and the bias. If the MSE were the sole concern, we would consider how the bias and the variance cooperate in producing a small MSE. A nonzero, and perhaps considerable, bias could be compensated for by a small variance.

But the MSE does not tell the full story. In addition to a small MSE, we also require that the bias be small relative to the standard error. This is important for confidence intervals to be valid, as the following argument shows.

Let us define the bias ratio by

$$\text{BR}(\hat{\theta}) = \frac{B(\hat{\theta})}{[V(\hat{\theta})]^{1/2}} \tag{5.2.1}$$

This quantity is often of interest for the following reason. As long as $\text{BR}(\hat{\theta})$ is small, a calculated confidence interval will not be greatly in error, despite a nonzero bias. To see this, we make the simplifying assumption that

$$Z = \frac{\hat{\theta} - E(\hat{\theta})}{[V(\hat{\theta})]^{1/2}} \tag{5.2.2}$$

follows the $N(0, 1)$ distribution. As an approximation, this will generally be true. The probability that the unknown θ is contained in the interval

$$(\hat{\theta} - z_{1-\alpha/2}[V(\hat{\theta})]^{1/2}, \hat{\theta} + z_{1-\alpha/2}[V(\hat{\theta})]^{1/2}) \tag{5.2.3}$$

is often called the coverage probability and is given by

$$\begin{aligned} P_0 &= \Pr\{\hat{\theta} - z_{1-\alpha/2}[V(\hat{\theta})]^{1/2} < \theta < \hat{\theta} + z_{1-\alpha/2}[V(\hat{\theta})]^{1/2}\} \\ &= \Pr\{-z_{1-\alpha/2} - \text{BR}(\hat{\theta}) < Z < z_{1-\alpha/2} - \text{BR}(\hat{\theta})\} \end{aligned}$$

where Z is the $N(0, 1)$ random variable. Note that (5.2.3) is truly a confidence interval only if the variance $V(\hat{\theta})$ is known. Normally in a confidence interval procedure, $V(\hat{\theta})$ must be replaced by an estimated variance $\hat{V}(\hat{\theta})$.

It follows that the coverage probability P_0 equals the nominal, desired confidence level, $1 - \alpha$, only if $\text{BR}(\hat{\theta})$ is zero. A nonzero bias ratio distorts the coverage probability somewhat, but the effect is minor when the bias ratio is near zero. The following table shows the coverage probability P_0 for selected values of $\text{BR}(\hat{\theta})$, when $1 - \alpha = 95\%$.

Table 5.1. The Coverage Probability P_0 as a Function of the Bias Ratio $\text{BR}(\hat{\theta})$.

$\lvert\text{BR}(\hat{\theta})\rvert$	P_0
0.00	0.9500
0.05	0.9497
0.10	0.9489
0.30	0.9396
0.50	0.9210
1.00	0.8300

As the table shows, the effect of the bias ratio on the coverage probability P_0 may be essentially ignored if $\lvert\text{BR}(\hat{\theta})\rvert \leq 0.1$, because, in that case, the coverage probability lies between 0.9489 and 0.95. In effect, even with a bias ratio as great as 0.5, the distorting effect is not extremely pronounced, because then $P_0 = 0.9210$.

Table 5.1 is meant only as a rough guide to the effect that bias can have on confidence statements. In practice, the bias ratio is unknown, making it impossible to tell the true value of the coverage probability P_0. The main message from Table 5.1 is the importance of a small bias ratio for obtaining valid confidence statements. For the estimators that we examine, sufficiently large samples will usually lead to bias ratios that are small. (For an illustration, see Result 5.6.1).

When the sample size increases (in the case of a fixed-size design) or when the expected sample size increases (in the case of a random size design), the variance $V(\hat{\theta})$ will typically tend to zero, as will the bias $B(\hat{\theta})$, and in such a way that the bias ratio $B(\hat{\theta})/[V(\hat{\theta})]^{1/2}$ approaches zero.

Remark 5.2.1. When a somewhat biased estimator is used, one may want to consider other types of interval estimation procedures than

$$\hat{\theta} \pm z_{1-\alpha/2}[\hat{V}(\hat{\theta})]^{1/2}$$

For example, if $\hat{\theta}$ is biased for θ, and $\widehat{\text{MSE}}(\hat{\theta})$ is a good estimator of

$$\text{MSE}(\hat{\theta}) = V(\hat{\theta}) + [B(\hat{\theta})]^2$$

then an intuitive interval for θ is

$$\hat{\theta} \pm z_{1-\alpha/2}[\widehat{\text{MSE}}(\hat{\theta})]^{1/2} \tag{5.2.4}$$

Naturally, one would like to know the coverage properties of the interval given by (5.2.4) as a function of the bias.

To simplify the question, suppose $\text{MSE}(\hat{\theta})$ is known. We are then interested in knowing the probability that the interval

$$\hat{\theta} \pm z_{1-\alpha/2}[\text{MSE}(\hat{\theta})]^{1/2}$$

contains the true value θ. The value of the probability can again be expressed in terms of the bias ratio $\text{BR}(\hat{\theta})$. For example, with $1 - \alpha = 0.95$ ($z_{0.975} = 1.96$), studies assuming that $\hat{\theta}$ is normally distributed have shown that the coverage probability lies between 0.939 and 0.95 if $|\text{BR}(\hat{\theta})| \leq 1.0$.

5.3. Consistency and Asymptotic Unbiasedness

We first recall the concepts of consistency and asymptotic unbiasedness from the general theory of statistical inference. Let a parameter τ be estimated by an estimator $\hat{\tau}_n$, that is, a function of n independent and identically distributed random variables $\xi_1, \xi_2, \ldots, \xi_n$. The estimator $\hat{\tau}_n$ is then said to be *asymptotically unbiased* for τ if

$$\lim_{n\to\infty} E(\hat{\tau}_n) = \tau$$

and $\hat{\tau}_n$ is said to be *consistent* for τ if, for any fixed $\varepsilon > 0$,

$$\lim_{n\to\infty} \Pr(|\hat{\tau}_n - \tau| > \varepsilon) = 0$$

(Although statisticians often express themselves in these terms, strictly speaking it is not the single estimator $\hat{t}_n$ but the sequence $\hat{\tau}_1, \hat{\tau}_2, \ldots$ that is asymptotically unbiased or consistent.)

In practice, n is always finite, although perhaps large. The practical importance of asymptotic unbiasedness and consistency is the following. If the estimator is known to be asymptotically unbiased, then it can be considered approximately unbiased when n is large enough. And if consistency holds, then the sampling distribution of the estimator can be considered tightly concentrated around τ, when n is large enough.

Now, let us return to survey sampling. It turns out that the above definitions of asymptotic unbiasedness and consistency cannot be immediately carried over to samples from a finite population. For if $\hat{\theta}_n$ is an estimator of θ based on a sample of n from a given finite population of size N, then we cannot let $n \to \infty$ without further ado, since $n \leq N$, and N is fixed and finite.

In this context, asymptotic results require a more complex machinery with a sequence of increasing populations so that both n and N tend to infinity. A complete treatment of these matters is not presented here, but we do give an idea of the conceptual framework for asymptotic reasoning in survey sampling.

We start with the idea of an infinite sequence of elements, labeled $k = 1, 2, 3, \ldots$, and an associated infinite sequence of y values, denoted $y_1, y_2, y_3, \ldots$, where y_k is the value tied to the kth element.

Consider a sequence of populations $U_1, U_2, U_3, \ldots$, where U_ν consists of the N_ν first elements from the infinite sequence of elements just mentioned, that is, $U_\nu = \{1, 2, \ldots, N_\nu\}$. We assume that $U_1 \subset U_2 \subset U_3 \subset \cdots$ and hence $N_1 < N_2 < N_3 < \cdots$. Let θ_ν be the value of a certain parameter in population U_ν, that is, θ_ν is a function of the values $y_1, y_2, \ldots, y_{N_\nu}$.

For each population U_ν, consider a probability sampling design $p_\nu(\cdot)$ that assigns a certain probability $p_\nu(s_\nu)$ to each possible sample s_ν of elements from U_ν. Let $\pi_{\nu k}$ and $\pi_{\nu kl}(k, l = 1, 2, \ldots, N_\nu)$ be the inclusion probabilities determined by the design $p_\nu(\cdot)$. For simplicity, we assume that the sample size is fixed and denoted by n_ν. Also assume that $n_1 < n_2 < n_3 < \cdots$. Clearly, $\nu \to \infty$ means that both $n_\nu \to \infty$ and $N_\nu \to \infty$. Let $\hat{\theta}_\nu$ be an estimator of θ_ν, based on the observed y_k values, that is, those for which $k \in s_\nu$.

For example, the parameter θ_ν may be the population mean

$$\theta_\nu = \bar{y}_{U_\nu} = \textstyle\sum_{U_\nu} y_k / N_\nu$$

and $\hat{\theta}_\nu$ may be the π estimator

$$\hat{\theta}_\nu = \textstyle\sum_s y_k / N_\nu \pi_{\nu k} \tag{5.3.1}$$

With reference to the sequence of populations and sampling designs described above, we can now state the following definitions:

a. An estimator $\hat{\theta}_\nu$ is *asymptotically unbiased* for θ_ν, if

$$\lim_{\nu \to \infty} [E_{p_\nu}(\hat{\theta}_\nu) - \theta_\nu] = 0$$

b. An estimator $\hat{\theta}_\nu$ is *consistent* for θ_ν, if, for any fixed $\varepsilon > 0$,

$$\lim_{\nu \to \infty} \Pr(|\hat{\theta}_\nu - \theta_\nu| > \varepsilon) = 0$$

These definitions are still somewhat nebulous, because the limit process has not been fully specified. Whether an estimator is consistent or asymptotically unbiased depends on how we specify the sequence $\{y_k\}$ of y values and the sequence $\{p_\nu\}$ of sampling designs. Conditions are usually required on the limiting behavior of the moments of the finite population and of the inclusion probabilities. Isaki and Fuller (1982) and Robinson and Särndal (1983) have stated conditions for the consistency of the π estimator (5.3.1) (and also of the regression estimator to be discussed in the following chapters). Robinson and

Särndal (1983) also give conditions for asymptotic unbiasedness. Another reference of interest in this context is Brewer (1979).

In this book, we sometimes state without proving formally that an estimator is asymptotically unbiased (approximately unbiased) or consistent. Mathematically rigorous proofs of such statements are beyond the scope of this book.

Remark 5.3.1. Consistency will be important later in this book for the following reason. If consistent estimators $\hat{\theta}_1, \ldots, \hat{\theta}_r$ are available for the parameters $\theta_1, \ldots, \theta_r$, then, for many functions f,

$$f(\hat{\theta}_1, \ldots, \hat{\theta}_r)$$

is consistent for $f(\theta_1, \ldots, \theta_r)$. In other words, a function of consistent estimators is itself consistent.

For example, if $\hat{\bar{y}}_U$ and $\hat{\bar{z}}_U$ are consistent estimators of the population means $\bar{y}_U$ and $\bar{z}_U$, respectively, then $\dfrac{\hat{\bar{y}}_U}{\hat{\bar{z}}_U}$ will be consistent for $\dfrac{\bar{y}_U}{\bar{z}_U}$.

It is not hard to show (with the help of Chebyshev's inequality) that if an estimator $\hat{\theta}_\nu$ is asymptotically unbiased for θ_ν and its variance tends to zero as ν tends to infinity, then $\hat{\theta}_\nu$ is consistent (see Exercise 5.18).

A different type of consistency, also applicable in the context of survey sampling, is *finite population consistency*. To define this type of consistency we need not consider a sequence of increasing populations but only a fixed finite population, for which we let the sample size increase so that it ultimately equals the population size. The definition is as follows.

c. An estimator $\hat{\theta}$ of θ is consistent for a finite population under a given class of designs if $s = U$ implies that $\hat{\theta} = \theta$.

EXAMPLE 5.3.1. Consider simple random sampling without replacement (*SI*). More specifically, we consider the class of all *SI* designs corresponding to values of the fixed sample size n such that $1 \leq n \leq N$. In this class, the event $s = U$ can occur only under one particular design, namely, the one with $n = N$, meaning that $\pi_k = 1$ for $k = 1, \ldots, N$. When this is the case, the π estimator takes the value

$$\hat{t}_\pi = \sum\nolimits_s \frac{y_k}{\pi_k} = \sum\nolimits_U \frac{y_k}{1} = t$$

which shows that $\hat{t}_\pi$ is consistent in the sense of definition (c) for the class of *SI* designs.

This property actually holds for any class of fixed-size designs such that $1 \leq n \leq N$. In such a class, the event $s = U$ can occur only under one particular design, namely, the one with $\pi_k = 1$ for $k = 1, \ldots, N$, and it follows that $\hat{t}_\pi$ is consistent in the sense of definition (c).

In a class of designs in which the event $s = U$ can occur, with nonzero probability, even when all π_k are not equal to one, the π estimator $\hat{t}_\pi$ is no longer consistent for t in the sense of definition (c).

EXAMPLE 5.3.2. Consider the BE design with a given value π such that $0 < \pi < 1$. Recall that π is the inclusion probability shared by all $k \in U$. The event $s = U$ can occur, with probability $\pi^N > 0$. If $s = U$ occurs,

$$\hat{t}_\pi = \sum_s \frac{y_k}{\pi_k} = \sum_U \frac{y_k}{\pi} = \frac{t}{\pi} > t$$

hence, $\hat{t}_\pi$ is not consistent in the sense of definition (c) for a BE design as long as $\pi < 1$.

Both criteria of consistency should be applied with some caution in practice. An estimator that is consistent may still be unsatisfactory when the sample size is small.

EXAMPLE 5.3.3. Consider once more the BE design, with $0 < \pi < 1$. The unweighted estimator $\hat{t} = \sum_s y_k$ is consistent for t in the sense of definition (c), because $\hat{t} = t$ whenever $s = U$. Since π is the expected sampling fraction, in practice, a π-value would ordinarily be rather small, say, $\pi = 10\%$. Clearly, in that case the bias of $\hat{t}$,

$$E_{BE}(\hat{t}) - t = -(1 - \pi)t$$

will be considerable; $\hat{t} = \sum_s y_k$ will systematically underestimate t, despite consistency in the sense of definition (c).

Remark 5.3.2. The infinite sequence of increasing populations that served above as a conceptual framework for defining consistency can also provide the setting for a mathematically rigorous definition of the asymptotic normality of estimators, (see, for example, Fuller, 1975; Hidiroglou, 1974; and Isaki and Fuller, 1982).

5.4. π Estimators for Several Variables of Study

Most surveys involve not one but several study variables. In this section we examine the case in which the total of each study variable is estimated by its corresponding π estimator. Suppose that there are q study variables, denoted $y_1, \ldots, y_j, \ldots, y_q$. Let the N population values of the jth variable be $y_{j1}, \ldots, y_{jk}, \ldots, y_{jN}; j = 1, \ldots, q$. The objective is to estimate the q components of the vector of unknown totals

$$\mathbf{t} = (t_1, \ldots, t_j, \ldots, t_q)'$$

where

$$t_j = \sum_U y_{jk}$$

As usual, a probability sample s is drawn from the population U according to the design $p(s)$ with the inclusion probabilities π_k and π_{kl}. For each $k \in s$, we observe the value of the vector

$$\mathbf{y}_k = (y_{1k}, \ldots, y_{jk}, \ldots, y_{qk})'$$

Supposing that each total is estimated by the corresponding π estimator, the vector of estimators is

$$\hat{\mathbf{t}}_\pi = (\hat{t}_{1\pi}, \ldots, \hat{t}_{j\pi}, \ldots, \hat{t}_{q\pi})'$$

where

$$\hat{t}_{j\pi} = \sum_s \check{y}_{jk}$$

with $\check{y}_{jk} = y_{jk}/\pi_k$. Clearly,

$$E(\hat{\mathbf{t}}_\pi) = \mathbf{t}$$

That is, $\hat{\mathbf{t}}_\pi$ is an unbiased vector of estimators. The following result deals with the variance–covariance matrix associated with $\hat{\mathbf{t}}_\pi$ and with the unbiased estimation of that matrix.

Result 5.4.1. *Let*

$$\hat{\mathbf{t}}_\pi = (\hat{t}_{1\pi}, \ldots, \hat{t}_{j\pi}, \ldots, \hat{t}_{q\pi})'$$

be the vector of π estimators corresponding to the q variables $y_1, \ldots, y_j, \ldots, y_q$, where

$$\hat{t}_{j\pi} = \sum_s \check{y}_{jk}$$

with $\check{y}_{jk} = y_{jk}/\pi_k$. Then the variance–covariance matrix

$$V(\hat{\mathbf{t}}_\pi) = E\{(\hat{\mathbf{t}}_\pi - \mathbf{t})(\hat{\mathbf{t}}_\pi - \mathbf{t})'\}$$

is a symmetric matrix such that the jth diagonal element is given by the variance of $\hat{t}_{j\pi}$,

$$V(\hat{t}_{j\pi}) = \sum\sum_U \Delta_{kl}\check{y}_{jk}\check{y}_{jl}$$

and the off-diagonal element jj' is given by the covariance of $\hat{t}_{j\pi}$ and $\hat{t}_{j'\pi}$,

$$C(\hat{t}_{j\pi}, \hat{t}_{j'\pi}) = \sum\sum_U \Delta_{kl}\check{y}_{jk}\check{y}_{j'l}$$

The matrix $V(\hat{\mathbf{t}}_\pi)$ is estimated unbiasedly by the matrix $\hat{V}(\hat{\mathbf{t}}_\pi)$ such that the jth diagonal element is

$$\hat{V}(\hat{t}_{j\pi}) = \sum\sum_s \check{\Delta}_{kl}\check{y}_{jk}\check{y}_{jl}$$

and the off-diagonal element jj' is

$$\hat{C}(\hat{t}_{j\pi}, \hat{t}_{j'\pi}) = \sum\sum_s \check{\Delta}_{kl}\check{y}_{jk}\check{y}_{j'l}$$

where $\check{\Delta}_{kl} = \Delta_{kl}/\pi_{kl}$.

PROOF. The results concerning the diagonal elements of $V(\hat{\mathbf{t}}_\pi)$ and $\hat{V}(\hat{\mathbf{t}}_\pi)$ follow immediately from Result 2.8.1. If $j \neq j'$, the element jj' of the matrix $V(\hat{\mathbf{t}}_\pi)$ is

$$\begin{aligned} C(\hat{t}_{j\pi}, \hat{t}_{j'\pi}) &= C(\textstyle\sum_U I_k \check{y}_{jk}, \sum_U I_k \check{y}_{j'k}) \\ &= \textstyle\sum\sum_U C(I_k, I_l)\check{y}_{jk}\check{y}_{j'l} \\ &= \textstyle\sum\sum_U \Delta_{kl}\check{y}_{jk}\check{y}_{j'l} \end{aligned}$$

An unbiased estimator of this covariance is obtained, analogously to Result 2.8.1, by replacing U by s and Δ_{kl} by $\check{\Delta}_{kl}$. □

In the special case of a fixed-size design, formulas analogous to those in Result 2.8.2 are derived in Exercise 5.21.

EXAMPLE 5.4.1. One application of Result 5.4.1 occurs when estimates are required for several population totals based on the same study variable, y. Suppose, for example, that we want to estimate the three totals

$$t_1 = \textstyle\sum_U y_k, \qquad t_2 = \sum_U y_k^2, \qquad t_3 = \sum_U y_k^3$$

We can regard this problem as an application of Result 5.4.1, with $q = 3$ variables, y_1, y_2, and y_3, such that $y_{1k} = y_k$, $y_{2k} = y_k^2$, and $y_{3k} = y_k^3$.

EXAMPLE 5.4.2. Another important application of Result 5.4.1 occurs in estimation for domains. Let $U_1, \ldots, U_d, \ldots, U_D$ represent a partitioning of U into D domains, say, a grouping of N households of a population U into D different types of households. Let y_k be the amount of money spent by the kth household for a certain type of service. The objective is to estimate, for $d = 1, \ldots, D$, the total amount spent by households in the dth group,

$$t_d = \textstyle\sum_{U_d} y_k = \sum_U z_{dk} y_k$$

where

$$z_{dk} = \begin{cases} 1 & \text{if } k \in U_d \\ 0 & \text{if not} \end{cases}$$

In this setting, we can identify D variables, $y_1, \ldots, y_d, \ldots, y_D$, where, for $d = 1, \ldots, D$; $k = 1, \ldots, N$,

$$y_{dk} = z_{dk} y_k$$

Having drawn the sample s, we observe y_k as well as $z_{1k}, \ldots, z_{dk}, \ldots, z_{Dk}$ for each $k \in s$. That is, we observe the amount spent by the kth household, y_k, as well as the group to which the household belongs. Result 5.4.1 now leads to the following simple conclusions. The π estimator of the dth domain total,

$$\hat{t}_{d\pi} = \textstyle\sum_s \check{y}_{dk} = \sum_{s_d} \check{y}_k$$

has the variance

$$V(\hat{t}_{d\pi}) = \textstyle\sum\sum_U \Delta_{kl}\check{y}_{dk}\check{y}_{dl} = \sum\sum_{U_d} \Delta_{kl}\check{y}_k\check{y}_l$$

The covariance between two domain total estimators is obtained from Result 5.4.1 as

$$C(\hat{t}_{d\pi}, \hat{t}_{d'\pi}) = \sum\sum_U \Delta_{kl}\check{y}_{dk}\check{y}_{d'l} = \sum_{k \in U_d} \sum_{l \in U_{d'}} \Delta_{kl}\check{y}_k\check{y}_l$$

where $\check{y}_{dk} = y_{dk}/\pi_k$, $\check{y}_k = y_k/\pi_k$, and s_d is the subset of the sample s that comes from the dth domain. In general, there is a nonzero correlation between two estimated domain totals. It is left as an exercise to work out the expression for the correlation between $\hat{t}_{d\pi}$ and $\hat{t}_{d'\pi}$ under the *SI* design.

5.5. The Taylor Linearization Technique for Variance Estimation

We now examine the often-occurring problem of estimating a population parameter θ that can be expressed as a function of q population totals, $t_1, \ldots, t_q$

$$\theta = f(t_1, \ldots, t_q)$$

where $t_j = \sum_U y_{jk}$ $(j = 1, \ldots, q)$. We assume that the vector $(y_{1k}, \ldots, y_{jk}, \ldots, y_{qk})'$ can be observed for $k \in s$, enabling us to form the π estimators

$$\hat{t}_{j\pi} = \sum_s \check{y}_{jk} = \sum_s y_{jk}/\pi_k; \qquad j = 1, \ldots, q$$

By the principle already mentioned in Section 5.1, our estimator of θ is then

$$\hat{\theta} = f(\hat{t}_{1\pi}, \ldots, \hat{t}_{q\pi}) \tag{5.5.1}$$

Analyzing the properties of $\hat{\theta}$ is easy when f is a linear function, that is, when

$$\theta = a_0 + \sum_{j=1}^{q} a_j t_j$$

The estimator (5.5.1) is then

$$\hat{\theta} = a_0 + \sum_{j=1}^{q} a_j \hat{t}_{j\pi} \tag{5.5.2}$$

which is unbiased for θ, with variance

$$V(\hat{\theta}) = V\left(\sum_{j=1}^{q} a_j \hat{t}_{j\pi}\right) = \sum_{j=1}^{q} \sum_{j'=1}^{q} a_j a_{j'} C(\hat{t}_{j\pi}, \hat{t}_{j'\pi}) \tag{5.5.3}$$

where the covariances $C(\hat{t}_{j\pi}, \hat{t}_{j'\pi})$ are those given in Result 5.4.1. When $j = j'$, $C(\hat{t}_{j\pi}, \hat{t}_{j\pi}) = V(\hat{t}_{j\pi})$, that is, the variance of $\hat{t}_{j\pi}$. The variance (5.5.3) is estimated by using the estimated covariances $\hat{C}(\hat{t}_{j\pi}, \hat{t}_{j'\pi})$ also given in Result 5.4.1

$$\hat{V}(\hat{\theta}) = \sum_{j=1}^{q} \sum_{j'=1}^{q} a_j a_{j'} \hat{C}(\hat{t}_{j\pi}, \hat{t}_{j'\pi}) \tag{5.5.4}$$

When $j = j'$, $\hat{C}(\hat{t}_{j\pi}, \hat{t}_{j\pi})$ is the estimated variance $\hat{V}(\hat{t}_{j\pi})$.

Remark 5.5.1. If f is a linear function, the variance given by (5.5.3), as well as its estimator, have alternative expressions that yield computational advantages. Note that the estimator (5.5.2) can be written as

$$\hat{\theta} = a_0 + \sum_s \check{u}_k$$

where $\check{u}_k = u_k/\pi_k$, with

$$u_k = \sum_{j=1}^{q} a_j y_{jk}$$

Then the variance (5.5.3) can be expressed as

$$V(\hat{\theta}) = V(\sum_s \check{u}_k) = \sum\sum_U \Delta_{kl}\check{u}_k\check{u}_l \tag{5.5.5}$$

which is estimated by

$$\hat{V}(\hat{\theta}) = \sum\sum_s \check{\Delta}_{kl}\check{u}_k\check{u}_l \tag{5.5.6}$$

A calculation of $\hat{V}(\hat{\theta})$ according to equation (5.5.6) requires prior computation of $\check{u}_k$ for $k \in s$, a fairly simple operation. Once the $\check{u}_k$ are available, a single double sum operation gives the desired estimate (5.5.6). Clearly, this is computationally simpler than using equation (5.5.4), which calls for computation of $q(q+1)/2$ estimated variances and covariances $\hat{C}(\hat{t}_{j\pi}, \hat{t}_{j'\pi})$, each of which is a laborious double sum.

Our main concern in this section is the case where $\theta = f(t_1, \ldots, t_q)$ is a nonlinear function of the q totals. Then it is often impossible to obtain exact results on the bias and variance of the estimator $\hat{\theta} = f(\hat{t}_{1\pi}, \ldots, \hat{t}_{q\pi})$. To circumvent much difficulty, we use the *Taylor linearization technique*, which yields an approximate expression for the variance of $\hat{\theta}$ and also an approximate estimator of this variance. This technique also makes the calculation of approximate confidence intervals for θ possible. The Taylor linearization technique has been used in various fields of statistics for a long time, and it dates back to Gauss at least. Examples of its general applicability in survey sampling are given by Tepping (1968) and Woodruff (1971).

The technique approximates the nonlinear estimator $\hat{\theta}$ by a pseudo-estimator, denoted $\hat{\theta}_0$, which is a linear function of $\hat{t}_1, \ldots, \hat{t}_q$, thus easy to handle. ($\hat{\theta}_0$ will ordinarily depend on certain unknowns; hence it is not a true estimator.) If the approximation is good, $\hat{\theta}_0$ will perform approximately as $\hat{\theta}$ and we can use the easily found variance $V(\hat{\theta}_0)$ as an approximation of $V(\hat{\theta})$. A variance estimator $\hat{V}(\hat{\theta})$ is also easily found.

The technique for finding $\hat{\theta}_0$ consists of the first-order Taylor approximation of the function f, expanding around the point $(t_1, \ldots, t_q)$, and neglecting the remainder term. We obtain

$$\hat{\theta} \doteq \hat{\theta}_0 = \theta + \sum_{j=1}^{q} a_j(\hat{t}_{j\pi} - t_j) \tag{5.5.7}$$

where

$$a_j = \frac{\partial f}{\partial \hat{t}_{j\pi}}\bigg|_{(\hat{t}_{1\pi}, \ldots, \hat{t}_{q\pi}) = (t_1, \ldots, t_q)} \tag{5.5.8}$$

Thus, for large samples (when $\hat{t}_{1\pi}, \ldots, \hat{t}_{q\pi}$ with high probability take values near $t_1, \ldots, t_q$), the estimator $\hat{\theta}$ will perform approximately as the linear random variable $\hat{\theta}_0$. The numeric accuracy of the approximation (5.5.7) will vary from one outcome s to another. We shall assume that the bias and the variance of $\hat{\theta}$ can be approximated by the corresponding easily derived quantities for the linear statistic $\hat{\theta}_0$.

In the following, we often use the variance of a linearized statistic as an approximation to that of a more complex (nonlinear) estimator. The special symbol

$$AV(\hat{\theta})$$

indicates approximate variance of $\hat{\theta}$, and this approximate variance is equal to the exact variance of the linearized statistic $\hat{\theta}_0$.

For computational convenience, we take the shortcut mentioned in Remark 5.5.1. Set

$$u_k = \sum_{j=1}^{q} a_j y_{jk} \tag{5.5.9}$$

and $\check{u}_k = u_k/\pi_k$. Then, since equations (5.5.3) and (5.5.5) are equivalent, the approximate variance of $\hat{\theta}$ is obtained as

$$\begin{aligned} AV(\hat{\theta}) &= V(\hat{\theta}_0) = V\left(\sum_{j=1}^{q} a_j \hat{t}_{j\pi}\right) \\ &= V(\textstyle\sum_s \check{u}_k) \\ &= \textstyle\sum\sum_U \Delta_{kl} \check{u}_k \check{u}_l \end{aligned} \tag{5.5.10}$$

Remark 5.5.2. The expression for $V(\hat{\theta})$ given in (5.5.3) can also be considered an approximation of the mean square error of $\hat{\theta}$. Since $E(\hat{\theta}_0) = \theta$, it follows that

$$\text{MSE}(\hat{\theta}) \doteq \text{MSE}(\hat{\theta}_0) = V(\hat{\theta}_0)$$

The quantities u_k appearing in (5.5.9) can generally by derived without too much difficulty. However, a problem lies in the estimation of (5.5.10). The u_k depend on $a_1, \ldots, a_q$, quantities which will often depend in their turn on unknown population totals. Hence, the u_k are unknown. The usual remedy is to replace each unknown total on which a_j depends with the corresponding π estimator. Thereby we arrive at an estimator $\hat{a}_j$, say, of a_j, permitting us to form and calculate for all $k \in s$ the variable

$$\hat{u}_k = \sum_{j=1}^{q} \hat{a}_j y_{jk} \tag{5.5.11}$$

The final step, then, is to take the variance estimation formula associated with equation (5.5.10) and replace the unknown u_k by $\hat{u}_k$, which yields

$$\hat{V}(\hat{\theta}) = \sum\sum_s \check{\Delta}_{kl} \frac{\hat{u}_k}{\pi_k} \frac{\hat{u}_l}{\pi_l} \tag{5.5.12}$$

The justification for the procedure is that $\hat{u}_k$, being a (possibly nonlinear) function of π estimators, is consistent for u_k. Now, $\hat{V}(\hat{\theta})$ is a function of the consistent estimators $\hat{u}_k$ and should in large samples behave as if it had been based on the true (unknown) u_k. Thus, $\hat{V}(\hat{\theta})$ can be assumed to be consistent for $V(\hat{\theta})$.

Because the AV expression shown in (5.5.10) was the starting point that led to the estimator (5.5.12), we have, strictly speaking, succeeded in estimating the *approximate* variance. However, because the approximate variance $AV(\hat{\theta})$ and the exact variance $V(\hat{\theta})$ agree well in large samples, (5.5.12) will serve well also as estimator of $V(\hat{\theta})$. This has been demonstrated by simulation studies for different cases. The important technique that we have developed in this section is summarized as follows:

Result 5.5.1. *For the parameter*

$$\theta = f(t_1, \ldots, t_q)$$

where

$$t_1 = \sum_U y_{1k}, \ldots, t_q = \sum_U y_{qk}$$

are population totals, an approximately unbiased estimator is given by

$$\hat{\theta} = f(\hat{t}_{1\pi}, \ldots, \hat{t}_{q\pi})$$

where $\hat{t}_{1\pi}, \ldots, \hat{t}_{q\pi}$ are the corresponding π estimators. Via Taylor linearization as described by equations (5.5.7) *and* (5.5.8), *the approximate variance of $\hat{\theta}$ is obtained as*

$$AV(\hat{\theta}) = \sum\sum_U \Delta_{kl} \frac{u_k}{\pi_k} \frac{u_l}{\pi_l}$$

where $u_k = \sum_{j=1}^{q} a_j y_{jk}$, with coefficients a_j determined by (5.5.8). *A variance estimator is given by*

$$\hat{V}(\hat{\theta}) = \sum\sum_s \check{\Delta}_{kl} \frac{\hat{u}_k}{\pi_k} \frac{\hat{u}_l}{\pi_l} \tag{5.5.13}$$

where

$$\hat{u}_k = \sum_{j=1}^{q} \hat{a}_j y_{jk}$$

with $\hat{a}_j$ obtained from a_j by substituting the appropriate π estimator for each unknown total.

As usual, when the design is of fixed size, an alternative variance estimator is available. In this case, it is given by

$$\hat{V}(\hat{\theta}) = -\frac{1}{2}\sum\sum_s \check{\Delta}_{kl}\left(\frac{\hat{u}_k}{\pi_k} - \frac{\hat{u}_l}{\pi_l}\right)^2 \tag{5.5.14}$$

Applications of Result 5.5.1 are given in this and following chapters. The optional variance estimator (5.5.14) can always be used for a fixed-size design. It agrees with (5.5.13) for the *SI* and *STSI* designs.

Remark 5.5.3. We caution that the Taylor linearization method has a tendency to lead to underestimated variances in not so large samples. (In very large samples, the bias of the variance estimator is nil.) The complexity of the statistic is a factor of importance. For a simple statistic, such as the weighted sample mean $\tilde{y}_s = (\sum_s y_k/\pi_k)/(\sum_s 1/\pi_k)$ (see Section 5.7), the underestimation of the Taylor variance estimator may be without consequence even for modest sample sizes, but for complex statistics such as an estimator of a population variance, covariance, or correlation coefficient, fairly large samples may be required before the bias is negligible.

5.6. Estimation of a Ratio

We return to the problem of estimating a ratio between two unknown population totals,

$$R = \frac{t_y}{t_z} = \frac{\sum_U y_k}{\sum_U z_k} \tag{5.6.1}$$

For example, if U is a population of households, y_k equals the total income of the kth household, and z_k equals the number of persons in the kth household, then R is the per capita income of individuals belonging to the households in the population. Or R may represent the acreage devoted to wheat, $\sum_U y_k$, divided by total farm acreage, $\sum_U z_k$, for a population of N farms.

Now, if the two unknown totals are estimated, respectively, by $\hat{t}_{y\pi} = \sum_s \check{y}_k$ and $\hat{t}_{z\pi} = \sum_s \check{z}_k$, the resulting (nonlinear) estimator of R is

$$\hat{R} = \frac{\hat{t}_{y\pi}}{\hat{t}_{z\pi}} \tag{5.6.2}$$

We now analyze $\hat{R}$ in some depth. Hartley and Ross (1954) established an interesting upper bound on the bias of $\hat{R}$ for the case of *SI* sampling. Here we present this result for the case of an arbitrary design.

Result 5.6.1. *The bias of the statistic $\hat{R}$ given by* (5.6.2) *satisfies*

$$\frac{[E(\hat{R}) - R]^2}{V(\hat{R})} \leq \frac{V(\hat{t}_{z\pi})}{t_z^2} \tag{5.6.3}$$

Proof. Consider the covariance $C(\hat{R}, \hat{t}_{z\pi})$. Since $\hat{R}\hat{t}_{z\pi} = \hat{t}_{y\pi}$, this covariance can be written as

$$\begin{aligned} C(\hat{R}, \hat{t}_{z\pi}) &= E(\hat{R}\hat{t}_{z\pi}) - E(\hat{R})E(\hat{t}_{z\pi}) \\ &= E(\hat{t}_{y\pi}) - E(\hat{R})E(\hat{t}_{z\pi}) \\ &= t_y - E(\hat{R})t_z \\ &= -t_z[E(\hat{R}) - R] \end{aligned}$$

that is,

$$E(\hat{R}) - R = -C(\hat{R}, \hat{t}_{z\pi})/t_z$$

Now, a squared correlation coefficient is bounded upward by unity, so that

$$\begin{aligned} [E(\hat{R} - R)]^2 &= [C(\hat{R}, \hat{t}_{z\pi})]^2/t_z^2 \\ &= [\rho(\hat{R}, \hat{t}_{z\pi})]^2 V(\hat{R}) V(\hat{t}_{z\pi})/t_z^2 \\ &\le V(\hat{R}) V(\hat{t}_{z\pi})/t_z^2 \end{aligned}$$

which proves the result. □

Result 5.6.1 leads to the following interesting conclusion. If

$$\text{BR}(\hat{R}) = \frac{B(\hat{R})}{\{V(\hat{R})\}^{1/2}} = \frac{E(\hat{R}) - R}{\{V(\hat{R})\}^{1/2}}$$

denotes the bias ratio of $\hat{R}$ (see equation (5.2.1) for definition of bias ratio), the result states that

$$[\text{BR}(\hat{R})]^2 \le \frac{V(\hat{t}_{z\pi})}{t_z^2} \tag{5.6.4}$$

That is, if the relative standard error of $\hat{t}_{z\pi}$, $[V(\hat{t}_{z\pi})]^{1/2}/|t_z|$, approaches zero with increasing sample size (as will normally be the case), the bias ratio of $\hat{R}$ will also tend to zero. We recall from Section 5.2 that this is of vital importance for the construction of valid confidence intervals. In other words, here is an example where the bias ratio is easily shown to be small in large samples.

Example 5.6.1. For the *SI* design, the inequality (5.6.4) can be written as

$$[\text{BR}(\hat{R})]^2 \le \left(\frac{1}{n} - \frac{1}{N}\right)(cv_{zU})^2$$

where $cv_{zU} = S_{zU}/\bar{z}_U$ is the coefficient of variation of z. We assume that z is always positive. This shows that the bias ratio of $\hat{R}$ in this case tends to zero as $n^{-1/2}$.

We shall now apply the Taylor linearization technique described in Section 5.5 to find an approximate variance of $\hat{R}$ and to find a variance estimator that can serve for confidence interval calculations. The estimator $\hat{R}$ is a function of the two random variables $\hat{t}_{y\pi}$ and $\hat{t}_{z\pi}$,

$$\hat{R} = \frac{\hat{t}_{y\pi}}{\hat{t}_{z\pi}} = f(\hat{t}_{y\pi}, \hat{t}_{z\pi})$$

The required partial derivatives are

$$\frac{\partial \hat{R}}{\partial \hat{t}_{y\pi}} = \frac{1}{\hat{t}_{z\pi}}; \qquad \frac{\partial \hat{R}}{\partial \hat{t}_{z\pi}} = -\frac{\hat{t}_{y\pi}}{\hat{t}_{z\pi}^2}$$

Evaluating these at the expected value point (t_y, t_z) we get

$$a_1 = \left.\frac{\partial \hat{R}}{\partial \hat{t}_{y\pi}}\right|_{(t_y, t_z)} = \frac{1}{t_z}$$

$$a_2 = \left.\frac{\partial \hat{R}}{\partial \hat{t}_{z\pi}}\right|_{(t_y, t_z)} = -\frac{t_y}{t_z^2} = -\frac{R}{t_z}$$

Now, from (5.5.8) and (5.5.10)

$$u_k = a_1 y_k + a_2 z_k = \frac{1}{t_z}(y_k - R z_k)$$

and

$$\hat{u}_k = \frac{1}{\hat{t}_{z\pi}}(y_k - \hat{R} z_k)$$

and the following result is obtained.

Result 5.6.2. *Using Taylor linearization, the ratio statistic* $\hat{R} = \hat{t}_{y\pi}/\hat{t}_{z\pi}$ *is approximated as follows*

$$\hat{R} \doteq \hat{R}_0 = R + \frac{1}{t_z} \sum_s \frac{y_k - R z_k}{\pi_k} \tag{5.6.5}$$

The estimator $\hat{R}$ *is approximately unbiased for* R*, with the approximate variance*

$$AV(\hat{R}) = V(\hat{R}_0) = \frac{1}{t_z^2} \sum\sum_U \Delta_{kl} \frac{y_k - R z_k}{\pi_k} \frac{y_l - R z_l}{\pi_l} \tag{5.6.6}$$

The variance estimator is

$$\hat{V}(\hat{R}) = \frac{1}{\hat{t}_{z\pi}^2} \sum\sum_s \check{\Delta}_{kl} \frac{y_k - \hat{R} z_k}{\pi_k} \frac{y_l - \hat{R} z_l}{\pi_l} \tag{5.6.7}$$

It is left as an exercise to formulate the alternative expression for the variance and the alternative variance estimator for the case of a fixed size design.

Remark 5.6.1. The following expressions are sometimes useful

$$\hat{R}_0 = R + \frac{1}{t_z}(\hat{t}_{y\pi} - R\hat{t}_{z\pi}) \tag{5.6.8}$$

$$AV(\hat{R}) = V(\hat{R}_0) = \frac{1}{t_z^2}[V(\hat{t}_{y\pi}) + R^2V(\hat{t}_{z\pi}) - 2RC(\hat{t}_{y\pi}, \hat{t}_{z\pi})] \tag{5.6.9}$$

$$\hat{V}(\hat{R}) = \frac{1}{\hat{t}_{z\pi}^2}[\hat{V}(\hat{t}_{y\pi}) + \hat{R}^2\hat{V}(\hat{t}_{z\pi}) - 2\hat{R}\hat{C}(\hat{t}_{y\pi}, \hat{t}_{z\pi})] \tag{5.6.10}$$

Remark 5.6.2. Under the approximation shown in (5.6.5)

$$E(\hat{R}) \doteq E(\hat{R}_0) = R$$

In other words, the bias of $\hat{R}$, although nonzero, is approximated by zero. That the bias of $\hat{R}$ is left undetected indicates a rather rough approximation, at least for small samples. A more refined expression for the bias can be obtained by extending the Taylor expansion to include also second-order terms.

EXAMPLE 5.6.2. Consider the *SI* design, with the sample size $n = fN$. Then $\hat{t}_{y\pi} = N\bar{y}_s$, $\hat{t}_{z\pi} = N\bar{z}_s$, and $\hat{R} = \dfrac{\bar{y}_s}{\bar{z}_s}$. The linear approximation given by (5.6.5) or equivalently by (5.6.8) is

$$\hat{R}_0 = R + \frac{1}{\bar{z}_U}\frac{1}{n}\sum\nolimits_s (y_k - Rz_k) = R + \frac{\bar{y}_s - R\bar{z}_s}{\bar{z}_U}$$

The variance approximation given by (5.6.6) or (5.6.9) is

$$\begin{aligned} AV(\hat{R}) &= \frac{1}{\bar{z}_U^2}\frac{1-f}{n}\frac{1}{N-1}\sum\nolimits_U (y_k - Rz_k)^2 \\ &= \frac{1}{\bar{z}_U^2}\frac{1-f}{n}(S_{yU}^2 + R^2S_{zU}^2 - 2RS_{yzU}) \end{aligned}$$

where S_{yzU} is the population covariance. From (5.6.7) or (5.6.10) we get

$$\begin{aligned} \hat{V}(\hat{R}) &= \frac{1}{\bar{z}_s^2}\frac{1-f}{n}\frac{1}{n-1}\sum\nolimits_s (y_k - \hat{R}z_k)^2 \\ &= \frac{1}{\bar{z}_s^2}\frac{1-f}{n}(S_{ys}^2 + \hat{R}^2S_{zs}^2 - 2\hat{R}S_{yzs}) \end{aligned}$$

where

$$S_{ys}^2 = \frac{1}{n-1}\sum\nolimits_s (y_k - \bar{y}_s)^2$$

$$S_{yzs} = \frac{1}{n-1} \sum_s (y_k - \bar{y}_s)(z_k - \bar{z}_s)$$

and S_{zs}^2 is analogous to S_{ys}^2.

EXAMPLE 5.6.3. The results of this chapter can be applied to sampling in two or more stages. Here let us look at the design *SI,SI* as described in Example 4.3.1. First, an *SI* sample s_{I} of n_{I} PSUs is selected from N_{I}. Within a selected PSU, an *SI* sample s_i of n_i elements is drawn from N_i. The ratio $R = (\sum_U y_k)/(\sum_U z_k) = t_y/t_z$ is then estimated by $\hat{R} = \hat{t}_{y\pi}/\hat{t}_{z\pi}$, where $\hat{t}_{y\pi} = (N_{\text{I}}/n_{\text{I}}) \sum_{s_{\text{I}}} N_i \bar{y}_{s_i}$ and the expression for $\hat{t}_{z\pi}$ is analogous. Thus,

$$\hat{R} = \frac{\sum_{s_{\text{I}}} N_i \bar{y}_{s_i}}{\sum_{s_{\text{I}}} N_i \bar{z}_{s_i}}$$

The variance estimator is obtained by an analogy. The variance of the π estimator, for an arbitrary design, is estimated by $\sum\sum_s \check{\Delta}_{kl} \check{y}_k \check{y}_l$. Compare this with the variance estimator for $\hat{R}$, which is given by (5.6.7). To go from the first formula to the second, we replace y_k by $e_k = y_k - \hat{R} z_k$, and multiply by $1/\hat{t}_{z\pi}^2$. We carry out this operation for equation (4.3.23), which is the expression for $\sum\sum_s \check{\Delta}_{kl} \check{y}_k \check{y}_l$ under the *SI,SI* design. After simplification, the resulting variance estimator for $\hat{R}$ is

$$\hat{V}(\hat{R}) = \frac{1}{(\sum_{s_{\text{I}}} N_i \bar{z}_{s_i}/n_{\text{I}})^2} \left[\left(\frac{1}{n_{\text{I}}} - \frac{1}{N_{\text{I}}}\right) \frac{\sum_{s_{\text{I}}} N_i^2 \bar{e}_{s_i}^2}{n_{\text{I}} - 1} + \frac{1}{n_{\text{I}} N_{\text{I}}} \sum_{s_{\text{I}}} N_i^2 \left(\frac{1}{n_i} - \frac{1}{N_i}\right) \frac{\sum_{s_i} (e_k - \bar{e}_{s_i})^2}{n_i - 1} \right]$$

with $\bar{e}_{s_i} = \bar{y}_{s_i} - \hat{R} \bar{z}_{s_i}$.

The method just presented for estimating the ratio R leads to one of the most widely used estimators for a population total, namely, the ratio estimator. The ratio estimator requires auxiliary information. Estimation with auxiliary information is systematically developed in Chapters 6 through 8, and we only briefly mention the method here. If y is a variable of interest, the population total $t_y = \sum_U y_k$ can be written as

$$t_y = t_z \frac{t_y}{t_z} = t_z R$$

Provided that t_z is a known quantity, R can be estimated by (5.6.2), and therefore t_y can be estimated by the ratio estimator

$$\hat{t}_{yra} = t_z \hat{R} = t_z \frac{\hat{t}_{y\pi}}{\hat{t}_{z\pi}}$$

The total $t_z = \sum_U z_k$ of the auxiliary variable z must be exactly known for the method to work. In addition, y_k and z_k must be available for $k \in s$. We emphasize "exactly," for an inaccurate value of t_z will lead to a non-negligible bias in $\hat{t}_{yra}$. Note, however, that as long as t_z is known, the variable z may be a "cheap" version of y; what is required for a precise estimate with the method is that y_k is roughly proportional to z_k throughout the population (see Remark 5.6.4). For example, z_k may be an "eye estimate" of the study variable or a measurement obtained easily and at low cost for the entire population, whereas y_k is the precise, but perhaps costly measurement of the study variable, obtained for the sample s only.

The basic facts on the ratio estimator follow immediately from Result 5.6.2 and are summarized in the following result.

Result 5.6.3. *If $t_y = \sum_U y_k$ is the unknown total of a study variable y, and if $t_z = \sum_U z_k$ is the known total of an auxiliary variable z, then the ratio estimator*

$$\hat{t}_{yra} = t_z \frac{\hat{t}_{y\pi}}{\hat{t}_{z\pi}} = t_z \hat{R}$$

is approximately unbiased for t_y. The approximate variance is given by

$$AV(\hat{t}_{yra}) = \sum\sum_U \Delta_{kl} \frac{y_k - Rz_k}{\pi_k} \frac{y_l - Rz_l}{\pi_l} \tag{5.6.11}$$

The variance estimator is

$$\hat{V}(\hat{t}_{yra}) = \left(\frac{t_z}{\hat{t}_{z\pi}}\right)^2 \sum\sum_s \check{\Delta}_{kl} \frac{y_k - \hat{R}z_k}{\pi_k} \frac{y_l - \hat{R}z_l}{\pi_l} \tag{5.6.12}$$

Remark 5.6.3. The expression for the AV given by (5.6.11) is zero if the residuals $y_k - Rz_k$ are zero for all $k \in U$. This will not occur in practice, but not infrequently we have a set of residuals that are nonzero but small. Then the AV will also be small. That is, the ratio estimator is very precise when the population points (y_k, z_k) are tightly scattered around a straight line through the origin and with a certain (unknown) slope R. This regression model can be said to generate the ratio estimator (for a further discussion, see Section 7.3).

5.7. Estimation of a Population Mean

A population parameter closely related to the total $t_y = \sum_U y_k$ is the population mean per element

$$\bar{y}_U = \frac{1}{N} \sum_U y_k = \frac{t_y}{N}$$

Provided N is known, a simple unbiased estimator of $\bar{y}_U$ is

$$\hat{\bar{y}}_{U\pi} = \frac{\hat{t}_{y\pi}}{N} = \frac{1}{N} \sum_s \frac{y_k}{\pi_k} \tag{5.7.1}$$

with variance

$$V(\hat{\bar{y}}_{U\pi}) = \frac{1}{N^2} \sum\sum_U \Delta_{kl} \frac{y_k}{\pi_k} \frac{y_l}{\pi_l} \tag{5.7.2}$$

and variance estimator

$$\hat{V}(\hat{\bar{y}}_{U\pi}) = \frac{1}{N^2} \sum\sum_s \check{\Delta}_{kl} \frac{y_k}{\pi_k} \frac{y_l}{\pi_l}$$

An alternative procedure is to estimate t_y as well as N (whether N is known or not). The estimator of $\bar{y}_U$ then becomes

$$\tilde{y}_s = \frac{\hat{t}_{y\pi}}{\hat{N}} = \frac{\sum_s y_k/\pi_k}{\sum_s 1/\pi_k} \tag{5.7.3}$$

where $\hat{N} = \sum_s (1/\pi_k)$ is the π estimator of N. This estimator is called the weighted sample mean. Interestingly, $\tilde{y}_s$ is often better than $\hat{\bar{y}}_{U\pi}$, as we see below. This is true even though $\tilde{y}_s$ foregoes information about N in cases where N is known. In other words, whether N is known or not, (5.7.3) is ordinarily the estimator to use.

Now, $\tilde{y}_s$ is nonlinear, consequently only approximately unbiased, and we cannot give its exact variance expression. By viewing $\bar{y}_U$ as a ratio between two population totals, $\bar{y}_U = t_y/t_z$, where $z_k = 1$ for $k = 1, \ldots, N$, and $\tilde{y}_s$ as the ratio between the corresponding π estimators, we can immediately apply Result 5.6.2 with $\hat{R} = \tilde{y}_s$ to obtain the following result.

Result 5.7.1. *An approximately unbiased estimator of the population mean $\bar{y}_U$ is given by the weighted sample mean*

$$\tilde{y}_s = \frac{\sum_s y_k/\pi_k}{\sum_s 1/\pi_k}$$

The approximate variance is given by

$$AV(\tilde{y}_s) = \frac{1}{N^2} \sum\sum_U \Delta_{kl} \left(\frac{y_k - \bar{y}_U}{\pi_k}\right)\left(\frac{y_l - \bar{y}_U}{\pi_l}\right) \tag{5.7.4}$$

A variance estimator is

$$\hat{V}(\tilde{y}_s) = \frac{1}{\hat{N}^2} \sum\sum_s \check{\Delta}_{kl} \left(\frac{y_k - \tilde{y}_s}{\pi_k}\right)\left(\frac{y_l - \tilde{y}_s}{\pi_l}\right)$$

For some designs, the estimators $\hat{\bar{y}}_{U\pi}$ and $\tilde{y}_s$ are identical, that is, they produce the same value for all samples, as in the *SI* and *STSI* designs. When N is unknown, there is no choice between $\hat{\bar{y}}_{U\pi}$ and $\tilde{y}_s$; only the latter can be used. However, if N is known and the two estimators differ, a choice must be made. As already hinted, $\tilde{y}_s$ is usually the better estimator, despite estimation of an a priori known quantity N. Although it is hard to pinpoint the exact conditions (finite population values $y_1, \ldots, y_N$, etc.) under which $\tilde{y}_s$ is preferred, we give at least a few arguments in favor of $\tilde{y}_s$.

First, we can see from the variances expressed in (5.7.2) and (5.7.4) that $\tilde{y}_s$ is preferred when the values of all $y_k - \bar{y}_U$ tend to be small. An extreme example is the following.

EXAMPLE 5.7.1. Let $y_k = c$ for $k = 1, \ldots, N$. Then

$$\tilde{y}_s = c$$

for any s, whereas

$$\hat{\bar{y}}_{U\pi} = c\frac{\sum_s 1/\pi_k}{N} = c\frac{\hat{N}}{N}$$

and, hence, $\tilde{y}_s$ is a better estimator if $V(\hat{N}) > 0$.

Secondly, $\tilde{y}_s = \hat{t}_{y\pi}/\hat{N}$ performs better than $\hat{\bar{y}}_{U\pi} = \hat{t}_{y\pi}/N$ in cases where the sample size is variable. If the realized sample size n_s happens to be greater than average, both the numerator sum and the denominator sum of $\tilde{y}_s$ will have relatively more terms. Vice versa, if n_s is smallish, both sums have relatively fewer terms. The ratio thereby retains a certain stability. By contrast, $\hat{\bar{y}}_{U\pi}$, whose denominator remains fixed, lacks this stability.

EXAMPLE 5.7.2. Consider again the population $y_k = c$ for $k = 1, \ldots, N$. Consider a variable sample size design with equal inclusion probabilities, that is, $\pi_k = \pi$ for $k = 1, \ldots, N$. (Bernoulli sampling is an example of such a design.) Then

$$\tilde{y}_s = c$$

for any sample s, whereas

$$\hat{\bar{y}}_{U\pi} = c\frac{n_s}{N\pi}$$

Thus, $\hat{\bar{y}}_{U\pi}$ has variability due entirely to the fact that n_s varies; by contrast, $\tilde{y}_s$ is perfectly stable.

A third argument in favor of $\tilde{y}_s$ arises in the case when the π_k are poorly (or negatively) correlated with the y_k values. Suppose that the sample contains an element with a large y_k but a small π_k. The numerator sum of $\tilde{y}_s$ will be very large. However, this will be compensated to some extent by the large value $1/\pi_k$ in the denominator. Thus, $\tilde{y}_s$ enjoys a kind of adaptability, or insensitivity, to "unlucky" samples that is lacking in $\hat{\bar{y}}_{U\pi} = \hat{t}_{y\pi}/N$, because its denominator N remains fixed. The following example shows that $\hat{\bar{y}}_{U\pi}$ may give a totally misleading estimate, whereas $\tilde{y}_s$ performs better.

EXAMPLE 5.7.3. Consider a population of $N = 10$ elements with $y_1 = \cdots = y_9 = c$, and $y_{10} = 2c$. To estimate $\bar{y}_U = 1.1c$, a sample of one single element is to be drawn, and the inclusion probabilities are $\pi_1 = \cdots = \pi_9 = 0.11$ and $\pi_{10} = 0.01$. Thus, the element $k = 10$ has the largest y-value and the smallest inclusion probability. In this rather extreme case,

$$\tilde{y}_s = \begin{cases} c & \text{for } s = \{1\}, \ldots, \{9\} \\ 2c & \text{for } s = \{10\} \end{cases}$$

whereas

$$\hat{\bar{y}}_{U\pi} = \begin{cases} \dfrac{c}{1.1} & \text{for } s = \{1\}, \ldots, \{9\} \\ 20c & \text{for } s = \{10\} \end{cases}$$

Clearly, with $\tilde{y}_s$ we avoid the possibility of the foolish estimate produced by $\hat{\bar{y}}_{U\pi}$ when the sample consists of the element $k = 10$. That $\hat{\bar{y}}_{U\pi}$ is unbiased (whereas $\tilde{y}_s$ has a certain bias) is poor consolation in this case. The variance of $\hat{\bar{y}}_{U\pi}$ is many times that of $\tilde{y}_s$.

An alternative estimator of the population mean that is useful for certain designs with a variable sample size is

$$\hat{\bar{y}}_{U\text{alt}} = \frac{n}{Nn_s} \sum\nolimits_s \frac{y_k}{\pi_k} \tag{5.7.5}$$

where $n = E(n_s) = \sum_U \pi_k$ is the expected sample size. If the sample size is fixed, then $\hat{\bar{y}}_{U\text{alt}}$ is identical to $\hat{\bar{y}}_{U\pi}$ (that is, the two estimators agree for every sample). If all π_k are equal, then $\hat{\bar{y}}_{U\text{alt}}$ is identical to $\tilde{y}_s$. But the approximately unbiased estimator (5.7.5) may have a substantially smaller variance than both $\hat{\bar{y}}_{U\pi}$ and $\tilde{y}_s$ when (i) the sample size is random, (ii) there is considerable variation in the π_k, and (iii) y_k/π_k is roughly constant throughout the population. Note that N must be known in the estimator (5.7.5). For example, $\hat{\bar{y}}_{U\text{alt}}$ is ordinarily preferred to the other two estimators if Poisson sampling is used with $\pi_k \propto x_k$, where $x_1, \ldots, x_N$ are known positive size measures, and the points (y_k, x_k) scatter tightly around a line through the origin. The factor n/Nn_s used in (5.7.5) is then more stable than the corresponding factor $N/\hat{N}$ used in $\tilde{y}_s$. In fact, (5.7.5) with $\pi_k \propto x_k$ is a special case of the ratio estimator (see Section 7.3.2). The ratio estimator is conceived for cases where a linear relationship through the origin exists between the study variable y and a known variable x. This explains in part why the estimator (5.7.5) works well when y_k/π_k is roughly constant.

5.8. Estimation of a Domain Mean

The estimation of a domain total was considered in Sections 3.3 and 5.4. We now also have the necessary machinery for estimating a domain mean. Result 5.6.2 will again be helpful.

Let U be a population with N elements, and let U_d be a domain of U, that is, a subset of U, with $N_d < N$ elements. We assume that the elements belonging to U_d cannot be identified beforehand, and that the domain size N_d is unknown. The population parameter of interest here is the mean per element in the domain U_d, denoted

$$\bar{y}_{U_d} = \frac{1}{N_d} \sum_{U_d} y_k \tag{5.8.1}$$

By introducing the indicator variable z_d defined by

$$z_{dk} = \begin{cases} 1 & \text{for } k \in U_d \\ 0 & \text{for } k \notin U_d \end{cases}$$

we can write $\bar{y}_{U_d}$ as a ratio between two totals over the whole population U

$$\bar{y}_{U_d} = \frac{\sum_U z_{dk} y_k}{\sum_U z_{dk}} = \frac{t_{z_d y}}{t_{z_d}} \tag{5.8.2}$$

As usual, s denotes the sample drawn from the whole of U. Now, s will typically contain some elements from U_d, and some elements from the rest of U. Let s_d be the subset of s consisting of the elements from U_d, that is, $s_d = s \cap U_d$. The number of elements in s_d is random; the probability that s_d is empty is assumed negligible.

For elements $k \in s$, we observe y_k and z_{dk}. (Observing z_{dk} is to observe whether k comes from the domain U_d or not.) The method of Section 5.6 can be used to obtain the following estimator of $\bar{y}_d$

$$\tilde{y}_{s_d} = \frac{\hat{t}_{z_d y, \pi}}{\hat{t}_{z_d \pi}} = \frac{\sum_s z_{dk} y_k / \pi_k}{\sum_s z_{dk} / \pi_k} = \frac{\sum_{s_d} y_k / \pi_k}{\sum_{s_d} 1/\pi_k} \tag{5.8.3}$$

From Result 5.6.2, with y_k, z_k, and R replaced, respectively, by $z_{dk} y_k$, z_{dk}, and $\bar{y}_{U_d}$, we obtain the following result.

Result 5.8.1. *An approximately unbiased estimator of the domain mean $\bar{y}_{U_d}$ is given by the weighted domain sample mean*

$$\tilde{y}_{s_d} = \frac{\sum_{s_d} y_k / \pi_k}{\sum_{s_d} 1/\pi_k}$$

The approximate variance is given by

$$AV(\tilde{y}_{s_d}) = \frac{1}{N_d^2} \sum\sum_{U_d} \Delta_{kl} \left(\frac{y_k - \bar{y}_{U_d}}{\pi_k} \right) \left(\frac{y_l - \bar{y}_{U_d}}{\pi_l} \right) \tag{5.8.4}$$

The variance estimator is

$$\hat{V}(\tilde{y}_{s_d}) = \frac{1}{\hat{N}_d^2} \sum\sum_{s_d} \check{\Delta}_{kl} \left(\frac{y_k - \tilde{y}_{s_d}}{\pi_k} \right) \left(\frac{y_l - \tilde{y}_{s_d}}{\pi_l} \right) \tag{5.8.5}$$

where $\hat{N}_d = \sum_{s_d} 1/\pi_k$.

As always, an alternative variance estimator can be considered when a fixed size design is used. In this case,

$$\hat{V}(\tilde{y}_{s_d}) = -\frac{1}{2\hat{N}_d^2}\sum\sum_{s_d} \check{\Delta}_{kl}\left(\frac{y_k - \tilde{y}_{s_d}}{\pi_k} - \frac{y_l - \tilde{y}_{s_d}}{\pi_l}\right)^2 \tag{5.8.6}$$

5.9. Estimation of Variances and Covariances in a Finite Population

Other important parameters of a finite population are the variance of a study variable and the covariance between two study variables. In this section, we see how such parameters can be estimated, using the general Taylor linearization technique of Section 5.5. We present the approximate variances and the appropriate variance estimators obtained from Result 5.5.1.

For the kth element of the finite population $U = \{1, \ldots, k, \ldots, N\}$, y_k and z_k are the values of the two variables of study y and z. The corresponding finite population variances are defined as

$$\begin{aligned} S_{yy} &= \frac{1}{N-1}\sum_U (y_k - \bar{y}_U)^2 \\ S_{zz} &= \frac{1}{N-1}\sum_U (z_k - \bar{z}_U)^2 \end{aligned} \tag{5.9.1}$$

where $\bar{y}_U = \sum_U y_k/N$ and $\bar{z}_U = \sum_U z_k/N$. In our earlier notation, S_{yy} is equal to S_{yU}^2 and S_{zz} is equal to S_{zU}^2. The finite population covariance between y and z is defined as

$$S_{yzU} = S_{yz} = \frac{1}{N-1}\sum_U (y_k - \bar{y}_U)(z_k - \bar{z}_U) \tag{5.9.2}$$

For simplicity, the shorter notation S_{yz} is used in this section. Let us construct estimators for these new parameters, assuming that (y_k, z_k) is observed for the elements $k \in s$, where s is a probability sample from $U = \{1, \ldots, k, \ldots, N\}$. The estimation of S_{zz} is completely analogous to that of S_{yy}. We can concentrate on the covarance, S_{yz}, and then obtain estimators of the variances, S_{yy} and S_{zz}, as special cases. Now, S_{yz}, S_{yy}, and S_{zz} can be expressed as functions of the population totals

$$\begin{aligned} &t_1 = \sum_U 1 = N, \qquad t_y = \sum_U y_k, \qquad t_z = \sum_U z_k \\ &t_{yy} = \sum_U y_k^2, \qquad t_{zz} = \sum_U z_k^2, \qquad t_{yz} = \sum_U y_k z_k \end{aligned} \tag{5.9.3}$$

namely,

$$S_{yz} = \frac{1}{N-1}t_{yz} - \frac{1}{N(N-1)}t_y t_z \tag{5.9.4}$$

$$S_{yy} = \frac{1}{N-1}t_{yy} - \frac{1}{N(N-1)}t_y^2 \tag{5.9.5}$$

The expression for S_{zz} is analogous. Note that N is seen as a population total as well. We construct estimators of these parameters by substituting π estimators for the unknown totals. These π estimators are

$$\hat{t}_{1\pi} = \sum_s \frac{1}{\pi_k} = \hat{N}, \qquad \hat{t}_{y\pi} = \sum_s \frac{y_k}{\pi_k}, \qquad \hat{t}_{z\pi} = \sum_s \frac{z_k}{\pi_k},$$
$$\hat{t}_{yy\pi} = \sum_s \frac{y_k^2}{\pi_k}, \qquad \hat{t}_{yz\pi} = \sum_s \frac{y_k z_k}{\pi_k} \tag{5.9.6}$$

Estimation of the Finite Population Covariance S_{yz}

To obtain a consistent estimator, insert π estimators into equation (5.9.4). This leads to

$$\tilde{S}_{yz} = \frac{1}{\hat{N} - 1} \hat{t}_{yz\pi} - \frac{1}{\hat{N}(\hat{N} - 1)} \hat{t}_{y\pi} \hat{t}_{z\pi} \tag{5.9.7}$$

or, equivalently,

$$\tilde{S}_{yz} = \frac{1}{\hat{N} - 1} \sum_s \frac{(y_k - \tilde{y}_s)(z_k - \tilde{z}_s)}{\pi_k} \tag{5.9.8}$$

where $\tilde{y}_s = \hat{t}_{y\pi}/\hat{N}$ and $\tilde{z}_s = \hat{t}_{z\pi}/\hat{N}$ are the weighted sample means. Now we employ Result 5.5.1. To obtain the approximate variance of $\tilde{S}_{yz}$, we put

$$u_k = [(y_k - \bar{y}_U)(z_k - \bar{z}_U) - S_{yz}]/(N - 1) \tag{5.9.9}$$

and the corresponding "estimate quantity," required for the variance estimator given by (5.5.13), is

$$\hat{u}_k = [(y_k - \tilde{y}_s)(z_k - \tilde{z}_s) - \tilde{S}_{yz}]/(\hat{N} - 1) \tag{5.9.10}$$

These expressions are easily derived as an exercise, using the technique underlying Result 5.5.1.

EXAMPLE 5.9.1. Consider the *SI* design (n elements drawn from N). Here, equation (5.9.8) is

$$\tilde{S}_{yz} = \frac{N}{N - 1} \frac{n - 1}{n} S_{yzs}$$

with

$$S_{yzs} = \frac{1}{n - 1} \sum_s (y_k - \bar{y}_s)(z_k - \bar{z}_s)$$

Here, $\bar{y}_s$ and $\bar{z}_s$ are straight sample means. The variance estimator is

$$\hat{V}(\tilde{S}_{yz}) = \left(\frac{N}{N - 1}\right)^2 \frac{1 - f}{n} \frac{\sum_s d_k^2}{n - 1}$$

with

$$d_k = (y_k - \bar{y}_s)(z_k - \bar{z}_s) - [(n - 1)/n] S_{yzs}$$

Estimation of the Finite Population Variance S_{yy}

The results that correspond to (5.9.7), (5.9.9) and (5.9.10) are the following. The consistent estimator of the variance S_{yy} is

$$\tilde{S}_{yy} = \frac{1}{\hat{N}-1}\hat{t}_{yy\pi} - \frac{1}{\hat{N}(\hat{N}-1)}\hat{t}^2_{y\pi} = \frac{1}{\hat{N}-1}\sum_s \frac{(y_k - \tilde{y}_s)^2}{\pi_k} \tag{5.9.11}$$

The appropriate quantities for the approximate variance and the variance estimator in Result 5.5.1 are

$$u_k = [(y_k - \bar{y}_U)^2 - S_{yy}]/(N-1) \tag{5.9.12}$$

and

$$\hat{u}_k = [(y_k - \tilde{y}_s)^2 - \tilde{S}_{yy}]/(\hat{N}-1) \tag{5.9.13}$$

In these results, we treat N as an unknown, as is often the case in practice. Even if N is known and $\hat{N} \neq N$, it, nevertheless, is better to replace N by $\hat{N}$, as we have argued earlier.

Note that $\tilde{S}_{yy}$ and $\tilde{S}_{yz}$ are not unbiased, but consistent. It is interesting to observe that, if N is known, unbiased estimation of S_{yz}, S_{yy}, and S_{zz} is possible; the parameters can then be expressed as linear functions of population totals. These unbiased estimators may be of interest when N is known and $\hat{N} = N$, as in the *SI* and *STSI* designs. Let us use S_{yz} as an illustration. Equation (5.9.4) suggests the estimator

$$\hat{S}_{yz} = \frac{1}{N-1}\left\{\hat{t}_{yz} - \frac{1}{N}[\hat{t}_y\hat{t}_z - \hat{C}(\hat{t}_y, \hat{t}_z)]\right\} \tag{5.9.14}$$

where $\hat{t}_{yz}$, $\hat{t}_y$, and $\hat{t}_z$ are any unbiased estimators for the totals t_{yz}, t_y, and t_z, respectively, and $\hat{C}(\hat{t}_y, \hat{t}_z)$ is any unbiased estimator of the covariance between $\hat{t}_y$ and $\hat{t}_z$. As is easily verified, $\hat{S}_{yz}$ is unbiased.

An obvious choice in equation (5.9.14) is to use the π estimators $\hat{t}_y = \hat{t}_{y\pi}$, $\hat{t}_z = \hat{t}_{z\pi}$, and $\hat{t}_{yz} = \hat{t}_{yz\pi}$ defined by (5.9.6) and to let $\hat{C}(\hat{t}_y, \hat{t}_z)$ be the standard covariance estimator for two π estimators, that is,

$$\hat{C}(\hat{t}_y, \hat{t}_z) = \hat{C}(\hat{t}_{y\pi}, \hat{t}_{z\pi}) = \sum\sum_s \check{\Delta}_{kl}\check{y}_k\check{z}_l$$

Substituting these quantities into (5.9.14), we find, after simplification and rearrangement of terms, that

$$\hat{S}_{yz} = \frac{1}{N}\sum_s \frac{y_k z_k}{\pi_k} - \frac{1}{N(N-1)}\sum\sum_{s \atop k \neq l} \frac{y_k z_l}{\pi_{kl}} \tag{5.9.15}$$

For an alternative derivation of this unbiased estimator, note that the parameter can be written as

$$S_{yz} = \frac{1}{N}t_{yz} - \frac{1}{N(N-1)}\sum\sum_{U \atop k \neq l} y_k z_l$$

or, equivalently,

$$S_{yz} = \sum\sum_U c_{kl} y_k z_l$$

if we let $c_{kl} = 1/N$ for $k = l$ and $c_{kl} = -1/N(N-1)$ for $k \neq l$. An unbiased estimator of S_{yz} is then

$$\hat{S}_{yz} = \sum\sum_s c_{kl} y_k z_l / \pi_{kl}$$

which is precisely the estimator (5.9.15).

A slightly different unbiased estimator of S_{yz}, when N is known, follows from the easily verified representation

$$S_{yz} = \frac{1}{2N(N-1)} \sum\sum_U (y_k - y_l)(z_k - z_l) \tag{5.9.16}$$

which leads directly to

$$S_{yz}^* = \frac{1}{2N(N-1)} \sum\sum_s \frac{(y_k - y_l)(z_k - z_l)}{\pi_{kl}} \tag{5.9.17}$$

In cases where all three are applicable, we should not expect any important differences between $\tilde{S}_{yz}$, $\hat{S}_{yz}$, and S_{yz}^*. From a practical point of view, it matters little which estimator is used, especially in large samples. That $\hat{S}_{yz}$ and S_{yz}^* are not identical estimators is shown by the following example.

Example 5.9.2. Let $U = \{1, 2, 3\}$; $s_1 = \{1, 2\}$; $s_2 = \{1, 3\}$; $s_3 = \{2, 3\}$, and $p(s_1) = p(s_2) = 0.4$; $p(s_3) = 0.2$. If the sample s_1 is realized, the unbiased estimators $\hat{S}_{yz}$ and S_{yz}^* are

$$\hat{S}_{yz} = (y_1 z_1 + \tfrac{4}{3} y_2 z_2 - y_1 z_2 - y_2 z_1)/2.4$$

$$S_{yz}^* = (y_1 z_1 + y_2 z_2 - y_1 z_2 - y_2 z_1)/2.4.$$

Because there is at least one sample for which the estimates differ, the two estimators are not identical.

Example 5.9.3. Consider the *STSI* design; we use our usual notation for this design. Let us compare $\tilde{S}_{yz}$ and $\hat{S}_{yz}$ as given by equations (5.9.7) and (5.9.15). After simplification, $\hat{S}_{yz}$ can be written as

$$\hat{S}_{yz} = S_1 + \frac{N}{N-1} S_2.$$

where the two terms contain "between strata" and "within strata" components as follows:

$$S_1 = \sum_{h=1}^{H} W_h S_{yzs_h}$$

$$S_2 = S_{yzb} - \sum_{h=1}^{H} \frac{W_h(1 - W_h)}{n_h} S_{yzs_h}$$

with

$$S_{yzb} = \sum_{h=1}^{H} W_h (\bar{y}_{s_h} - \bar{y}_{st})(\bar{z}_{s_h} - \bar{z}_{st})$$

$$S_{yzs_h} = \frac{1}{n_h - 1} \sum_{s_h} (y_k - \bar{y}_{s_h})(z_k - \bar{z}_{s_h})$$

Here,

$$\bar{y}_{s_h} = \sum_{s_h} y_k / n_h, \qquad \bar{y}_{st} = \sum_{h=1}^{H} W_h \bar{y}_{s_h}, \qquad W_h = N_h / N$$

and $\bar{z}_{s_h}$ and $\bar{z}_{st}$ are analogous. The consistent estimator $\tilde{S}_{yz}$ takes the form

$$\tilde{S}_{yz} = \frac{N}{N-1}(S_1 + S_2')$$

with

$$S_2' = S_{yzb} - \sum_{h=1}^{H} \frac{W_h}{n_h} S_{yzs_h}$$

The difference between $\hat{S}_{yz}$ and $\tilde{S}_{yz}$ is minor and can be ignored for most practical purposes, especially in large samples. If all stratum sample sizes n_h are large, so that the second, negative terms of S_2 and S_2' can be ignored, we have

$$\hat{S}_{yz} \doteq \tilde{S}_{yz} \doteq \sum_{h=1}^{H} W_h S_{yzs_h} + S_{yzb}$$

that is, the sum of a within-strata term and a between-strata term.

Remark 5.9.1. Liu and Thompson (1983) consider a "batch approach" to the estimation of quadratic finite population parameters, such as

$$\sum\sum_U b_{kl}(y_k - y_l)(z_k - z_l) \tag{5.9.18}$$

[Compare equation (5.9.16), which can be written in the form of (5.9.18) if $b_{kl} = 1/2N(N-1)$ for all $k \neq l$.] Here, the parameter (5.9.18) can be described as a batch total over the collection of batches $\{(k, l) : k \neq l\}$. If π_{kl} is the batch inclusion probability, we have immediately that

$$\sum\sum_s b_{kl}(y_k - y_l)(z_k - z_l)/\pi_{kl}$$

is unbiased for (5.9.18). Liu and Thompson (1983) examine the statistical properties, including admissibility, of estimators of this type.

5.10. Estimation of Regression Coefficients

5.10.1. The Parameters of Interest

Suppose that the survey involves $q + 1$ study variables, y and $z_1, \ldots, z_j, \ldots, z_q$. Their respective values for the kth element are denoted $y_k, z_{1k}, \ldots, z_{jk}, \ldots, z_{qk}$. The parameters of interest are the regression coefficients $B_1, \ldots, B_q$ which would be obtained if the hyperplane $y = B_1 z_1 + \cdots + B_q z_q$ were fitted to the N finite population points by ordinary least squares.

As an example, with $q = 2$, suppose that z_k = disposable income and y_k = savings for the kth household in a population of households. We wish to estimate the slope B_2, and the intercept B_1 of the line $y = B_1 + B_2 z$ that gives the best fit (by the least squares criterion) to the whole set of N points (y_k, z_k). That B_1 and B_2 are unknown parameters is obvious; they depend on the whole finite population. Clearly, B_2, for example, is of interest as a measure of the additional savings generated by an extra dollar of disposable income for a household in the population. A sample s is drawn by a given design, and the observed points (y_k, z_k) for $k \in s$ can be used to estimate B_1 and B_2. Some care must be exercised. For example, the sample may be stratified with highly different sampling fractions in the different strata. The differing sampling weights must then be taken into account to prevent bias in the estimators of B_1 and B_2.

For each element $k \in U$, the q-component vector $\mathbf{z}_k$ is defined as

$$\mathbf{z}_k = (z_{1k}, \ldots, z_{qk})'$$

(If the regression is "with intercept," then $z_{1k} = 1$ for all k.) If all N values $\mathbf{z}_k$ and y_k are assembled in matrix form, we obtain a $q \times N$ matrix $\mathbf{Z}$, and an N-vector $\mathbf{y}$ defined as

$$\mathbf{Z} = \begin{bmatrix} z_{11} & \cdots & z_{1N} \\ \vdots & \cdots & \vdots \\ z_{q1} & \cdots & z_{qN} \end{bmatrix} = [\mathbf{z}_1, \ldots, \mathbf{z}_N]; \qquad \mathbf{y} = \begin{bmatrix} y_1 \\ \vdots \\ y_N \end{bmatrix}$$

The population characteristic to be estimated is the q-vector of regression coefficients

$$\mathbf{B} = [B_1, \ldots., B_q]' = (\mathbf{Z}\mathbf{Z}')^{-1}\mathbf{Z}\mathbf{y} \tag{5.10.1}$$

$$= \left(\textstyle\sum_U \mathbf{z}_k \mathbf{z}_k'\right)^{-1} \textstyle\sum_U \mathbf{z}_k y_k \tag{5.10.2}$$

where it is assumed that the matrix $\mathbf{Z}\mathbf{Z}'$ is nonsingular.

By way of interpretation, $\mathbf{B}$ given by equation (5.10.1) is the coefficient vector of the best fitting plane of the type $y = B_1 z_1 + \cdots + B_q z_q$. In other words, among all possible values $\mathbf{B}$, the value given by (5.10.1) is the one that minimizes

$$\textstyle\sum_U (y_k - \mathbf{B}'\mathbf{z}_k)^2$$

as is well-known from least squares theory.

The matrix representation shown in (5.10.1) is familiar from texts on regression analysis; the equivalent expression shown in (5.10.2) is often more convenient in the finite population context. Note that no model assumptions have been introduced about the generation of a value y_k, given $\mathbf{z}_k$. We simply consider $\mathbf{B}$ as a characteristic describing one particular aspect of the actual finite population point scatter $\{(y_k, z_{1k}, \ldots., z_{qk}) : k = 1, \ldots, N\}$. Of course, the choice of $\mathbf{B}$ as a parameter of interest may be motivated by some model thinking, including the decision to consider y the variable to be explained, with $z_1, \ldots, z_q$ as the explanatory variables. But for the moment, we do not need the type of model assumptions usually made in regression analysis. We consider only the design-inference problem of estimating a fixed (but unknown) $\mathbf{B}$, regardless of any underlying regression model. For additional remarks on the interpretation of the parameter $\mathbf{B}$, the reader is referred to Section 13.6.

EXAMPLE 5.10.1 Let $q = 1$ and

$$\mathbf{Z} = [z_1, \ldots, z_N]; \qquad \mathbf{B} = B$$

This corresponds to fitting a straight line through the origin, $y = Bz$, to the population data. Equation (5.10.2) yields the following well-known expression for the slope B

$$B = \frac{\sum_U z_k y_k}{\sum_U z_k^2}$$

EXAMPLE 5.10.2. Let $q = 2$, with the first z variable identically equal to unity. Then

$$\mathbf{Z} = \begin{bmatrix} 1 & 1 & \cdots & 1 \\ z_1 & z_2 & \cdots & z_N \end{bmatrix}; \qquad \mathbf{B} = \begin{bmatrix} B_1 \\ B_2 \end{bmatrix}$$

This corresponds to fitting a straight line with an intercept,

$$y = B_1 + B_2 z$$

to the population data. Equation (5.10.2) now gives the equally well-known expressions

$$B_2 = \frac{N(\sum_U z_k y_k) - (\sum_U z_k)(\sum_U y_k)}{N(\sum_U z_k^2) - (\sum_U z_k)^2} = \frac{\sum_U (z_k - \bar{z}_U)(y_k - \bar{y}_U)}{\sum_U (z_k - \bar{z}_U)^2}$$

for the slope, and

$$B_1 = \bar{y}_U - B_2 \bar{z}_U$$

for the intercept.

5.10.2. Estimation of the Regression Coefficients

We now consider the problem of estimating the population parameter vector $\mathbf{B}$, defined by (5.10.1) or (5.10.2), using data y_k and $\mathbf{z}_k$ for $k \in s$, where s is a probability sample from U. Adopting the technique used earlier in this chapter, we first express $\mathbf{B}$ as a (vector-valued) function of suitably defined population totals. Then $\mathbf{B}$ is estimated by substituting the appropriate π estimators for these totals. The general Result 5.5.1 will apply.

To this end, define, for $j = 1, \ldots, q$, and $j' = 1, \ldots, q$, the population totals

$$t_{jj'} = \sum_U z_{jk} z_{j'k} = t_{j'j}$$

and

$$t_{j0} = \sum_U z_{jk} y_k$$

Furthermore, let $\mathbf{T}$ be the symmetric $q \times q$ matrix with elements $t_{jj'}$ ($j = 1, \ldots, q; j' = 1, \ldots, q$), and let $\mathbf{t}$ be the q-vector with components t_{j0} ($j = 1, \ldots, q$). Since

$$\mathbf{T} = \sum_U \mathbf{z}_k \mathbf{z}_k'$$

and

$$\mathbf{t} = \sum_U \mathbf{z}_k y_k$$

we have, from equation (5.10.2), and assuming that $\mathbf{T}^{-1}$ exists, that

$$\mathbf{B} = \mathbf{T}^{-1}\mathbf{t} \tag{5.10.3}$$

which is a function of $q(q+1)/2$ totals $t_{jj'}$ and q totals t_{j0}.

The totals $t_{jj'}$ and t_{j0} can be unbiasedly estimated by their respective π estimators,

$$\hat{t}_{jj',\pi} = \sum_s \frac{z_{jk} z_{j'k}}{\pi_k}$$

and

$$\hat{t}_{j0,\pi} = \sum_s \frac{z_{jk} y_k}{\pi_k}$$

($j = 1, \ldots, q; j' = 1, \ldots, q$), or, in matrix notation,

$$\hat{\mathbf{T}} = \sum_s \frac{\mathbf{z}_k \mathbf{z}_k'}{\pi_k}$$

and

$$\hat{\mathbf{t}} = \sum_s \frac{\mathbf{z}_k y_k}{\pi_k}$$

(The expression for $\hat{\mathbf{T}}$, for example, means that every element of the $q \times q$ matrix $\mathbf{z}_k \mathbf{z}_k'$ is divided by the scalar value π_k, and then summed over $k \in s$.)

Inserting $\hat{\mathbf{T}}$ and $\hat{\mathbf{t}}$ into (5.10.3) gives the desired estimator,

$$\hat{\mathbf{B}} = \hat{\mathbf{T}}^{-1}\hat{\mathbf{t}} = \left(\sum_s \frac{\mathbf{z}_k \mathbf{z}_k'}{\pi_k}\right)^{-1} \left(\sum_s \frac{\mathbf{z}_k y_k}{\pi_k}\right) \tag{5.10.4}$$

Remark 5.10.1. An alternative way of expressing Equation (5.10.4) is

$$\hat{\mathbf{B}} = (\mathbf{Z}_s \mathbf{\Pi}_s^{-1} \mathbf{Z}_s')^{-1} \mathbf{Z}_s \mathbf{\Pi}_s^{-1} \mathbf{y}_s$$

where $\mathbf{Z}_s$ is the sample part of $\mathbf{Z}$ by which we mean the $q \times n_s$ matrix consisting of those columns $\mathbf{z}_k$ of $\mathbf{Z}$ for which $k \in s$. That is, $\mathbf{Z}_s$ is what remains of $\mathbf{Z}$ after eliminating all columns tied to the $N - n_s$ nonsampled elements. Similarly, the n_s vector $\mathbf{y}_s$ is the sample part of the N vector $\mathbf{y}$. $\mathbf{\Pi}_s$ is an $n_s \times n_s$ diagonal matrix with diagonal elements π_k corresponding to $k \in s$.

The proposed estimator (5.10.4) is not exactly unbiased for $\mathbf{B}$, and we must settle for an approximate variance formula. The Taylor linearization, as described in Section 5.5, can be used to find a linear approximation $\hat{\mathbf{B}}_0$ to $\hat{\mathbf{B}}$, that is, an easily handled random vector assumed to approximate $\hat{\mathbf{B}}$ for large samples. We arrive at the conclusions in Result 5.10.1 below. The proof involves no new principles, but it is rather lengthy and is given in Section 5.12.

Result 5.10.1. *By Taylor linearization, the estimator* $\hat{\mathbf{B}}$, *defined in equation* (5.10.4), *is approximated by*

$$\hat{\mathbf{B}}_0 = \mathbf{B} + \mathbf{T}^{-1}(\hat{\mathbf{t}} - \hat{\mathbf{T}}\mathbf{B})$$

Moreover, $\hat{\mathbf{B}}$ *is an approximately unbiased estimator of* $\mathbf{B}$ *with the approximate variance–covariance matrix*

$$AV(\hat{\mathbf{B}}) = \mathbf{T}^{-1}\mathbf{V}\mathbf{T}^{-1} \tag{5.10.5}$$

where $\mathbf{V}$ *is a symmetric* $q \times q$ *matrix with elements*

$$v_{jj'} = \sum\sum_U \Delta_{kl}\left(\frac{z_{jk}E_k}{\pi_k}\right)\left(\frac{z_{j'l}E_l}{\pi_l}\right) \tag{5.10.6}$$

and E_k *is the population fit residual,*

$$E_k = y_k - \mathbf{z}_k'\mathbf{B}$$

The estimator of the variance–covariance matrix is

$$\hat{V}(\hat{\mathbf{B}}) = \left(\sum_s \frac{\mathbf{z}_k\mathbf{z}_k'}{\pi_k}\right)^{-1} \hat{\mathbf{V}} \left(\sum_s \frac{\mathbf{z}_k\mathbf{z}_k'}{\pi_k}\right)^{-1} \tag{5.10.7}$$

where $\hat{\mathbf{V}}$ *is a symmetric* $q \times q$ *matrix with elements*

$$\hat{v}_{jj'} = \sum\sum_s \check{\Delta}_{kl}\left(\frac{z_{jk}e_k}{\pi_k}\right)\left(\frac{z_{j'l}e_l}{\pi_l}\right) \tag{5.10.8}$$

and e_k *is the sample fit residual,*

$$e_k = y_k - \mathbf{z}_k'\hat{\mathbf{B}}$$

It is left as an exercise to write down the expression that can be used as an alternative to $\hat{v}_{jj'}$ in the case of a fixed-size design.

Example 5.10.3. Result 5.10.1 is given for the case of multiple regression. Let us apply the result to the special case of simple regression mentioned in Example 5.10.1, where a line through the origin was fitted. The slope

$$B = \frac{\sum_U z_k y_k}{\sum_U z_k^2}$$

is then estimated by

$$\hat{B} = \frac{\sum_s z_k y_k/\pi_k}{\sum_s z_k^2/\pi_k}$$

The approximate variance is

$$AV(\hat{B}) = (\textstyle\sum_U z_k^2)^{-2} V(\sum_s z_k E_k/\pi_k)$$
$$= (\textstyle\sum_U z_k^2)^{-2} \sum\sum_U \Delta_{kl}\left(\frac{z_k E_k}{\pi_k}\right)\left(\frac{z_l E_l}{\pi_l}\right)$$

The variance estimator is

$$\hat{V}(\hat{B}) = (\textstyle\sum_s z_k^2/\pi_k)^{-2} \sum\sum_s \check{\Delta}_{kl}\left(\frac{z_k e_k}{\pi_k}\right)\left(\frac{z_l e_l}{\pi_l}\right)$$

where $E_k = y_k - Bz_k$ and $e_k = y_k - \hat{B}z_k$. Note that the estimator $\hat{B}$ can be seen, formally, as a ratio between two estimated totals. That is, we could have used the theory of Section 5.6 to obtain the same results.

EXAMPLE 5.10.4. Let us now consider the case of fitting a line with an intercept, mentioned in Example 5.10.2. The slope

$$B_2 = \frac{\sum_U (z_k - \bar{z}_U)(y_k - \bar{y}_U)}{\sum_U (z_k - \bar{z}_U)^2}$$

and the intercept, $B_1 = \bar{y}_U - B_2\bar{z}_U$, are estimated, respectively, by

$$\hat{B}_2 = \frac{\left(\sum_s \frac{1}{\pi_k}\right)\left(\sum_s \frac{z_k y_k}{\pi_k}\right) - \left(\sum_s \frac{z_k}{\pi_k}\right)\left(\sum_s \frac{y_k}{\pi_k}\right)}{\left(\sum_s \frac{1}{\pi_k}\right)\left(\sum_s \frac{z_k^2}{\pi_k}\right) - \left(\sum_s \frac{z_k}{\pi_k}\right)^2}$$

$$= \frac{\sum_s \frac{(z_k - \tilde{z}_s)(y_k - \tilde{y}_s)}{\pi_k}}{\sum_s \frac{(z_k - \tilde{z}_s)^2}{\pi_k}}$$

and

$$\hat{B}_1 = \tilde{y}_s - \hat{B}_2\tilde{z}_s$$

We present the variance results for $\hat{B}_2$ only; those for $\hat{B}_1$ are left as an exercise. The slope estimator has the approximate variance

$$AV(\hat{B}_2) = \frac{\sum\sum_U \Delta_{kl} \frac{(z_k - \bar{z}_U)E_k}{\pi_k} \frac{(z_l - \bar{z}_U)E_l}{\pi_l}}{[\sum_U (z_k - \bar{z}_U)^2]^2}$$

where

$$E_k = y_k - B_1 - B_2 z_k = y_k - \bar{y}_U - B_2(z_k - \bar{z}_U)$$

is the population fit residual for the kth element. The variance estimator for $\hat{B}_2$ is

$$\hat{V}(\hat{B}_2) = \frac{\sum\sum_s \check{\Delta}_{kl} \frac{(z_k - \tilde{z}_s)e_k}{\pi_k} \frac{(z_l - \tilde{z}_s)e_l}{\pi_l}}{\left[\sum_s \frac{(z_k - \tilde{z}_s)^2}{\pi_k}\right]^2}$$

where

$$e_k = y_k - \hat{B}_1 - \hat{B}_2 z_k = y_k - \tilde{y}_s - \hat{B}_2(z_k - \tilde{z}_s)$$

is the sample fit residual.

Example 5.10.5. Because the design is arbitrary, $\hat{V}(\hat{B}_2)$ in Example 5.10.4 is a complex expression. Simpler formulas are obtained for the *SI* design. Then,

$$\hat{B}_2 = \frac{\sum_s (z_k - \bar{z}_s)(y - \bar{y}_s)}{\sum_s (z_k - \bar{z}_s)^2}$$

which is the familiar "unweighted" regression coefficient, where $\bar{y}_s = \sum_s y_k/n$ is the unweighted mean, and $\bar{z}_s$ is defined analogously. The approximate variance and the variance estimator are, respectively,

$$AV_{SI}(\hat{B}_2) = \frac{N^2 \frac{1-f}{n} \frac{1}{N-1} \sum_U (z_k - \bar{z}_U)^2 E_k^2}{[\sum_U (z_k - \bar{z}_U)^2]^2}$$

where $f = n/N$, and

$$\hat{V}(\hat{B}_2) = \frac{(1-f)\frac{n}{n-1} \sum_s (z_k - \bar{z}_s)^2 e_k^2}{[\sum_s (z_k - \bar{z}_s)^2]^2} \quad (5.10.9)$$

One notes that $\hat{V}(\hat{B}_2)$ does not agree with the customary formula arising from standard regression model analysis, namely,

$$\frac{1}{n-2} \sum_s e_k^2 / \sum_s (z_k - \bar{z}_s)^2 \quad (5.10.10)$$

The difference lies not only in the finite population correction factor $1 - f$, present in equation (5.10.9), but, more fundamentally, the two formulas differ in structure. The reason is that (5.10.9) is *nonparametric*, obtained, in contrast to equation (5.10.10), without invoking the usual regression model, that is, $y_k = \alpha + \beta z_k + \varepsilon_k$, where the ε_k are independent random error terms with $E(\varepsilon_k) = 0$; $E(\varepsilon_k^2) = \sigma^2$.

Folsom (1974) and Shah, Holt and Folsom (1977) were among the first to present results similar to those in this section. Regression analysis in sample surveys is also treated, with some elements of model-based theory, in Hidiroglou (1974) and Fuller (1975). Regression analysis as we have presented it can be performed by a few computer programs, for example, SUPERCARP (Hidiroglou, Fuller, and Hickman, 1980), SURREGR (Holt, 1982), NASSREG (Chu et al., 1985), and REPERR (Van Eck, 1979).

5.11. Estimation of a Population Median

A parameter of practical interest (but not so much discussed in the literature) is the median of a finite population. The population median (with respect to the variable y) is a value M that divides the population in half, so that half of the population elements have values smaller than M, whereas the other half have values larger than M. We devote this section to the problem of estimating a population median, using data from a probability sample.

As usual, let $y_1, y_2, \ldots, y_N$ be the values of the population elements, for the study variable y. For any given number y $(-\infty < y < \infty)$, the population distribution function $F(y)$ is defined as the proportion of elements in the population for which $y_k \leq y$. More formally, the stepwise increasing function $F(y)$ can be written as

$$F(y) = \frac{1}{N}(\#A_y)$$

where A_y is the set of population elements with y_k values not exceeding y, that is, $A_y = \{k\colon k \in U, \text{ and } y_k \leq y\}$ and $\#A_y$ denotes the number of elements in the set A_y.

The population median M is now defined as

$$M = F^{-1}(0.5) \tag{5.11.1}$$

where $F^{-1}(\cdot)$ is the inverse function of $F(\cdot)$.

Figure 5.1 illustrates how M is determined. In case (a), M is uniquely determined through (5.11.1). In case (b), the point 0.5 on the $F(y)$ axis corresponds to a whole interval of y values. So that the median is always uniquely defined, we choose to define M as the value corresponding to the midpoint of the interval in question, as case (b) illustrates.

Our problem now is to estimate the population median M, using data y_k for $k \in s$, where s is a probability sample. The general procedure that we suggest has the following two steps:

i. First, obtain an estimated distribution function, $\hat{F}(y)$.
ii. Then, estimate $M = F^{-1}(0.5)$ by

$$\hat{M} = \hat{F}^{-1}(0.5)$$

where the inverse $\hat{F}^{-1}$ is to be understood in the same way as F^{-1} above.

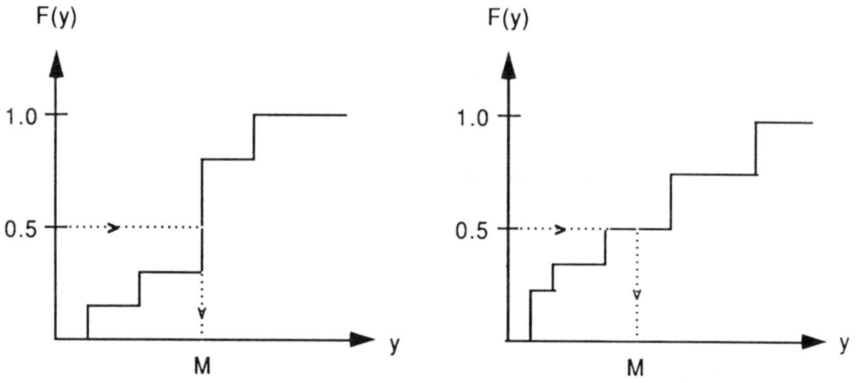

Figure 5.1. Determination of the population median. Case (a) to the left, case (b) to the right.

This method has been in use for a long time; the first published account is probably Woodruff (1952). More recently, Francisco and Fuller (1986), Kuk (1988), and Rao, Kovar and Mantel (1990) considered the estimation of distribution functions and medians for finite populations.

Remark 5.11.1. The technique in (i) and (ii) can be summarized as follows: We first produce an estimated distribution function $\hat{F}$ and then calculate $\hat{M}$ from $\hat{F}$ in the same way as we would have determined M from F, if F had been known. The technique can be generalized to estimation of any parameter θ defined as a function of the population distribution function $F(y)$, for example, any quantile of the population. (The median is the 50% quantile.) For example, to estimate the first quartile, the 25% quantile, we would use (i) and (ii) with 0.25 instead of 0.5. Among other population parameters defined in terms of the distribution function, we might mention, for example, the Gini coefficient, which is of interest in describing income distributions. Brewer (1981), Nygård and Sandström (1985) and Sandström et al. (1988) examine the estimation of the Gini coefficient in a variety of ways.

We now describe in closer detail the mechanics of estimating the median M, including a confidence interval procedure due to Woodruff (1952). A sample s is selected from the population by a sampling design $p(\cdot)$ with inclusion probabilities π_k, π_{kl}. The estimation of $F(y)$, for any given y, can be viewed as the already familiar problem of estimating a population mean. To see this, note that $F(y)$ can be expressed as a population mean,

$$F(y) = \frac{1}{N} \sum_U z_{k,y} = \frac{t_{z_y}}{N} = \bar{z}_{yU} \tag{5.11.2}$$

where $z_{k,y}$ is an indicator variable, defined for $k = 1, \ldots, N$, and for any given real number y, as

$$z_{k,y} = \begin{cases} 1 & \text{if } y_k \leq y \\ 0 & \text{if } y_k > y \end{cases}$$

Consequently let us estimate $F(y) = \bar{z}_{yU}$, for a given y, by an expression corresponding directly to the sample-weighted mean $\tilde{y}_s$ considered in Section 5.7,

$$\hat{F}(y) = \tilde{z}_{y,s} = \frac{\hat{t}_{z_y,\pi}}{\hat{N}}$$

$$= \frac{\sum_s \frac{z_{k,y}}{\pi_k}}{\sum_s \frac{1}{\pi_k}} = \frac{\sum_{s \cap A_y} \frac{1}{\pi_k}}{\sum_s \frac{1}{\pi_k}} \tag{5.11.3}$$

where $s \cap A_y$ is the set of sample elements with values $y_k \leq y$. Now, $\hat{F}(y)$, the estimated distribution function, is a nondecreasing step function climbing from zero to one like $F(y)$. To divide by $\hat{N}$ in equation (5.11.3), rather than by N (if N were known) will often have advantages from a variance point of view, as discussed in Section 5.7. Also, to have $\hat{N}$ in the denominator of (5.11.3) means that $\hat{F}(y)$ reaches the ultimate value of unity as y increases, which is a desirable property. [By contrast, if $F(y)$ were estimated by $N^{-1}\sum_s z_{k,y}/\pi_k$, the ultimate value, $N^{-1}\sum_s 1/\pi_k$, is not necessarily unity.]

The approximate variance of the estimator (5.11.3), as well as an estimator of this variance, can be obtained from the results of Section 5.7. Now, $\hat{F}(y)$ becomes a tool for finding the desired estimator $\hat{M}$ of the median M. As prescribed by step (ii) above, we set

$$\hat{M} = \hat{F}^{-1}(0.5) \tag{5.11.4}$$

where $\hat{F}^{-1}$ is the inverse function of $\hat{F}(y)$ given by (5.11.3). Note that to find $\hat{M}$ this way, we need not compute $\hat{F}(y)$ for the whole range of y-values, but only for the center part of the population.

As with (5.11.1), it may happen that (5.11.4) does not produce a unique value $\hat{M}$. We assume that this problem is resolved in a manner analogous to case (b) of the figure above.

Remark 5.11.2. For computational purposes we can alternatively express the median estimator as follows. Denote the y_k-values of the sampled elements, arranged in increasing order of size, by

$$y_{1:s} \leq y_{2:s} \leq \cdots \leq y_{n_s:s}$$

and the corresponding inclusion probabilities π_k by

$$\pi_{1:s}, \pi_{2:s}, \ldots, \pi_{n_s:s}$$

Set $B_0 = 0$, and define the cumulative sums

$$B_1 = \frac{1}{\pi_{1:s}}$$

$$B_2 = \frac{1}{\pi_{1:s}} + \frac{1}{\pi_{2:s}}$$

and so on; in general, for $l = 1, \ldots, n_s$,

$$B_l = \sum_{j=1}^{l} \frac{1}{\pi_{j:s}}$$

Clearly, $B_{n_s} = \hat{N} = \sum_{j=1}^{n_s} 1/\pi_{j:s} = \sum_s 1/\pi_k$. The median estimator (5.11.4) can now be alternatively written

$$\hat{M} = \begin{cases} y_{l:s} & \text{if } B_{l-1} < 0.5\hat{N} < B_l \\ \frac{1}{2}(y_{l:s} + y_{l+1:s}) & \text{if } B_l = 0.5\hat{N} \end{cases} \tag{5.11.5}$$

In other words, to calculate $\hat{M}$, we first compute $\hat{N}_\pi$, and then examine the cumulative sums B_l until we find *either* that for some l,

$$B_{l-1} < 0.5\hat{N} \quad \text{and} \quad B_l > 0.5\hat{N}$$

in which case $\hat{M} = y_{l:s}$, *or* we find that for some l,

$$B_l = 0.5\hat{N}$$

in which case $\hat{M} = \frac{1}{2}(y_{l:s} + y_{l+1:s})$.

In practice, of course, we need not consider the first values $B_1, B_2, \ldots$. What is essential is to find the point where the cumulative sum rises above the value $0.5\hat{N}$. A small scale example illustrates the principle for calculating the estimate $\hat{M}$.

Example 5.11.1. A sample of size five is drawn from a population of size $N = 50$. The sample obtained is $s = \{2, 17, 28, 29, 33\}$, and the observed data are

k	y_k	π_k
2	8	0.10
17	15	0.05
28	12	0.10
29	4	0.10
33	8	0.20

Rearrangement in increasing order of size of y_k gives the following:

l	$y_{l:s}$	$\pi_{l:s}$	$1/\pi_{l:s}$
1	4	0.10	10
2	8	0.10	10
3	8	0.20	5
4	12	0.10	10
5	15	0.05	20
Σ			55

We calculate first the estimated distribution function $\hat{F}$, then $\hat{M}$ from (5.11.4). Since $\hat{N} = 55$, we have, using equation (5.11.3), the following:

y	$\hat{F}(y)$
$y < 4$	0
$4 \leq y < 8$	10/55
$8 \leq y < 12$	$(10 + 10 + 5)/55 = 25/55$
$12 \leq y < 15$	$(10 + 10 + 5 + 10)/55 = 35/55$
$15 \leq y$	1

or, in graph form,

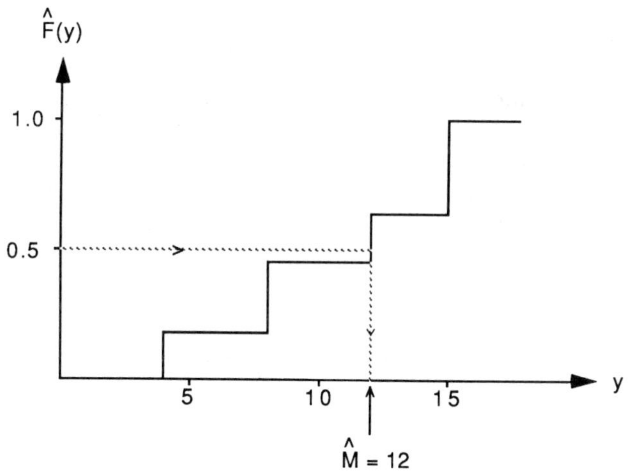

From the table of $\hat{F}$, or from the graph, we find the estimated median to be

$$\hat{M} = \hat{F}^{-1}(0.5) = 12$$

Alternatively, $\hat{M}$ may be calculated from (5.11.5) as follows:

l	$y_{l:s}$	$1/\pi_{l:s}$	B_l
1	4	10	10
2	8	10	$10 + 10 = 20$
3	8	5	$10 + 10 + 5 = 25$
4	12	10	$10 + 10 + 5 + 10 = 35$
5	15	20	$10 + 10 + 5 + 10 + 20 = 55$

Since $B_3 = 25 < 55/2 = \hat{N}/2$, and $B_4 = 35 > 55/2$, we obtain

$$\hat{M} = y_{4:s} = 12$$

It remains to find an approximate confidence interval for M. The reasoning is based on the following relation. For any two constants c_1 and c_2 and for any value of M,

$$\Pr\{c_1 \leq \hat{F}(M) \leq c_2\} \doteq \Pr\{\hat{F}^{-1}(c_1) \leq M \leq \hat{F}^{-1}(c_2)\} \tag{5.11.6}$$

If F and $\hat{F}$ were continuous and strictly increasing, then (5.11.6) would hold exactly; it is the stepwise feature of F and $\hat{F}$ that causes the approximation.

From (5.11.6) it follows that if c_1 and c_2 could be found such that

$$\Pr\{c_1 \leq \hat{F}(M) \leq c_2\} = 0.95$$

then the interval

$$[\hat{F}^{-1}(c_1), \hat{F}^{-1}(c_2)]$$

would be approximately a 95% confidence interval for M. To illustrate, we use the level $1 - \alpha = 95\%$. Usually, however, we cannot determine the constants c_1 and c_2 exactly, so further approximations are required.

We assume that $\hat{F}(M)$ is approximately normally distributed around its expected value, which is $F(M) \doteq 0.5$. This assumption is justified for large samples. We would preferably like to choose

$$c_1 = 0.5 - 1.96\{V[\hat{F}(M)]\}^{1/2} \qquad c_2 = 0.5 + 1.96\{V[\hat{F}(M)]\}^{1/2}$$

where, using the variance approximation (5.7.4),

$$\begin{aligned} V[\hat{F}(M)] &= V\left[\hat{N}^{-1} \sum\nolimits_s \frac{z_{k,M}}{\pi_k}\right] \\ &\doteq N^{-2} \sum\sum\nolimits_U \Delta_{kl} \left(\frac{z_{k,M} - 0.5}{\pi_k}\right)\left(\frac{z_{l,M} - 0.5}{\pi_l}\right) \end{aligned} \tag{5.11.7}$$

To estimate this unknown approximate variance is not so simple, since M is unknown, which makes $z_{k,M}$ an unobservable variable. To circumvent this difficulty, we take the variance estimation formula corresponding to $V[\hat{F}(M)]$, but with $z_{k,M}$ replaced by $z_{k,\hat{M}}$ (for $k \in s$), which gives

$$\hat{V}[\hat{F}(M)] = \hat{N}^{-2} \sum\sum_s \check{\Delta}_{kl} \left(\frac{z_{k,\hat{M}} - 0.5}{\pi_k}\right)\left(\frac{z_{l,\hat{M}} - 0.5}{\pi_l}\right) \tag{5.11.8}$$

or, under a fixed-size design,

$$\hat{V}[\hat{F}(M)] = -\tfrac{1}{2}\hat{N}^{-2} \sum\sum_s \check{\Delta}_{kl} \left(\frac{z_{k,\hat{M}} - 0.5}{\pi_k} - \frac{z_{l,\hat{M}} - 0.5}{\pi_l}\right)^2 \tag{5.11.9}$$

The confidence interval then becomes

$$[\hat{F}^{-1}(0.5 - 1.96\hat{V}^{1/2}), \hat{F}^{-1}(0.5 + 1.96\hat{V}^{1/2})] \tag{5.11.10}$$

where $\hat{V} = \hat{V}[\hat{F}(M)]$ is given by (5.11.8) or (5.11.9). The assumption that justifies this step is that $\hat{M}$ is consistent for M, and that, consequently, $\hat{V}[\hat{F}(M)]$ is consistent for $V[\hat{F}(M)]$. (This procedure has much in common with the linearization technique of Result 5.5.1, in which the unknown constants a_j were replaced by consistent estimates $\hat{a}_j$.) Clearly, as the method rests on several approximations, it should be applied with caution, at least when the sample size is small.

EXAMPLE 5.11.2. Consider the *SI* design, with the sample size $n = fN$. This turns out to be an especially simple case, because there is no need to estimate the variance $V[\hat{F}(M)]$ as it contains no unknowns. In this case, $\hat{F}(M)$ is simply the proportion of elements in s with $y_k \leq M$. Hence, $n\hat{F}(M)$ is a hypergeometrically distributed random variable with

$$E[\hat{F}(M)] = F(M) \doteq 0.5$$

$$V[\hat{F}(M)] = \frac{N-n}{N-1}\frac{1}{n}F(M)\{1 - F(M)\}$$

$$\doteq \frac{1-f}{n}0.25 \tag{5.11.11}$$

If several population elements have $y_k = M$, the value of $F(M)$ need not be exactly equal to 0.5. Alternatively, we could obtain (5.11.11) directly from (5.11.7).

A normal approximation gives us

$$\Pr\{c_1 \leq \hat{F}(M) \leq c_2\} \doteq 0.95$$

for

$$c_1 = 0.5 - 1.96\left(\frac{1-f}{n}0.25\right)^{1/2}$$

$$c_2 = 0.5 + 1.96\left(\frac{1-f}{n}0.25\right)^{1/2}$$

With these values of c_1 and c_2, the approximately 95% confidence interval is

$$[\hat{F}^{-1}(c_1), \hat{F}^{-1}(c_2)]$$

EXAMPLE 5.11.3. Turning now to stratified simple random sampling (the *STSI* design), the technique becomes more complicated than in the preceding example. We can write

$$F(y) = \sum_{h=1}^{H} W_h F_h(y)$$

and

$$\hat{F}(y) = \sum_{h=1}^{H} W_h \hat{F}_h(y) \tag{5.11.12}$$

where $W_h = N_h/N$, F_h is the distribution function for stratum h, and $\hat{F}_h(y)$ is the proportion of elements in the sample from stratum h that have $y_k \leq y$. Then

$$\hat{F}(M) = \sum_{h=1}^{H} W_h \hat{F}_h(M)$$

The random variables $n_h \hat{F}_h(M)$ $(h = 1, \ldots, H)$ are independent and hypergeometrically distributed such that

$$E[\hat{F}_h(M)] = F_h(M)$$

and

$$V[\hat{F}_h(M)] = \frac{N_h - n_h}{N_h - 1} \frac{1}{n_h} F_h(M)[1 - F_h(M)]$$

$$\doteq \frac{1 - f_h}{n_h} F_h(M)[1 - F_h(M)]$$

where $f_h = n_h/N_h$. It follows that

$$E[\hat{F}(M)] = \sum_{h=1}^{H} W_h F_h(M) \doteq 0.5$$

and

$$V[\hat{F}(M)] \doteq \sum_{h=1}^{H} W_h^2 \frac{1 - f_h}{n_h} F_h(M)\{1 - F_h(M)\}$$

The suggested estimator of this variance is

$$\hat{V}[\hat{F}(M)] = \sum_{h=1}^{H} W_h^2 \frac{1 - f_h}{n_h} \hat{F}_h(\hat{M})\{1 - \hat{F}_h(\hat{M})\} \tag{5.11.13}$$

where $\hat{M}$ is defined by $\hat{M} = \hat{F}^{-1}(0.5)$, with $\hat{F}$ in this case determined by equation (5.11.12).

If we set

$$c_1 = 0.5 - 1.96\{\hat{V}[\hat{F}(M)]\}^{1/2}$$
$$c_2 = 0.5 + 1.96\{\hat{V}[\hat{F}(M)]\}^{1/2}$$

then, assuming that the normal approximation is valid,

$$\Pr\{c_1 \leq \hat{F}(M) \leq c_2\} \doteq 0.95$$

With these values of c_1 and c_2,

$$[\hat{F}^{-1}(c_1), \hat{F}^{-1}(c_2)]$$

gives the desired (approximately) 95% confidence interval.

5.12. Demonstration of Result 5.10.1

The following proof of Result 5.10.1 is another application of the general Taylor linearization technique summarized in Result 5.5.1.

The estimator

$$\hat{\mathbf{B}} = \hat{\mathbf{T}}^{-1}\hat{\mathbf{t}}$$

given by (5.10.4), is a function of the $q(q+1)/2$ different π estimators $\hat{t}_{jj',\pi}$ in the (symmetric) $q \times q$ matrix $\hat{\mathbf{T}}$, and of the q π estimators $\hat{t}_{j0,\pi}$ that make up the vector $\hat{\mathbf{t}}$.

Taylor linearization amounts to finding a linear approximation

$$\hat{\mathbf{B}} \doteq \hat{\mathbf{B}}_0 = \mathbf{B} + \sum_{j=1}^{q} \sum_{j' \leq j} \mathbf{a}_{jj'}(\hat{t}_{jj',\pi} - t_{jj'}) + \sum_{j=1}^{q} \mathbf{a}_{j0}(\hat{t}_{j0,\pi} - t_{j0}) \quad (5.12.1)$$

where $\mathbf{a}_{jj'}$ and $\mathbf{a}_{j0}$ are q-vectors defined as

$$\mathbf{a}_{jj'} = \left.\frac{\partial \hat{\mathbf{B}}}{\partial \hat{t}_{jj',\pi}}\right|_{\hat{\mathbf{T}}=\mathbf{T},\hat{\mathbf{t}}=\mathbf{t}}$$

and

$$\mathbf{a}_{j0} = \left.\frac{\partial \hat{\mathbf{B}}}{\partial \hat{t}_{j0,\pi}}\right|_{\hat{\mathbf{T}}=\mathbf{T},\hat{\mathbf{t}}=\mathbf{t}}$$

Using rules of differentiation, given by, for example, Graybill (1983), we obtain the partial derivatives

$$\frac{\partial \hat{\mathbf{B}}}{\partial \hat{t}_{jj',\pi}} = \frac{\partial(\hat{\mathbf{T}}^{-1}\hat{\mathbf{t}})}{\partial \hat{t}_{jj',\pi}} = \left(\frac{\partial \hat{\mathbf{T}}^{-1}}{\partial \hat{t}_{jj',\pi}}\right)\hat{\mathbf{t}}$$
$$= (-\hat{\mathbf{T}}^{-1}\mathbf{\Lambda}_{jj'}\hat{\mathbf{T}}^{-1})\hat{\mathbf{t}}$$
$$= -\hat{\mathbf{T}}^{-1}\mathbf{\Lambda}_{jj'}\hat{\mathbf{B}}$$

where $\mathbf{\Lambda}_{jj'}$ is a $q \times q$ matrix with ones in positions (j, j') and (j', j), and zeros elsewhere; and

$$\frac{\partial \hat{\mathbf{B}}}{\partial \hat{t}_{j0,\pi}} = \frac{\partial(\hat{\mathbf{T}}^{-1}\hat{\mathbf{t}})}{\partial \hat{t}_{j0,\pi}} = \hat{\mathbf{T}}^{-1}\left(\frac{\partial \hat{\mathbf{t}}}{\partial \hat{t}_{j0,\pi}}\right) = \hat{\mathbf{T}}^{-1}\boldsymbol{\lambda}_j$$

where $\boldsymbol{\lambda}_j$ is a q-vector with the jth component equal to one and zeros elsewhere.

Evaluating these derivatives at the point $(\mathbf{T}, \mathbf{t})$ and inserting into (5.12.1) gives

$$\begin{aligned}\hat{\mathbf{B}}_0 &= \mathbf{B} - \sum_{j=1}^{q} \sum_{j' \le j} \mathbf{T}^{-1}\boldsymbol{\Lambda}_{jj'}\mathbf{B}(\hat{t}_{jj',\pi} - t_{jj'}) + \sum_{j=1}^{q} \mathbf{T}^{-1}\boldsymbol{\lambda}_j(\hat{t}_{j0,\pi} - t_{j0}) \\ &= \mathbf{B} - \mathbf{T}^{-1}(\hat{\mathbf{T}} - \mathbf{T})\mathbf{B} + \mathbf{T}^{-1}(\hat{\mathbf{t}} - \mathbf{t}) \\ &= \mathbf{B} + \mathbf{T}^{-1}(\hat{\mathbf{t}} - \hat{\mathbf{T}}\mathbf{B})\end{aligned}$$

which is the linearized form of $\hat{\mathbf{B}}$.

The approximate variance–covariance matrix of $\hat{\mathbf{B}}$ is the exact variance–covariance matrix of $\hat{\mathbf{B}}_0$. We note that

$$\begin{aligned}\hat{\mathbf{t}} - \hat{\mathbf{T}}\mathbf{B} &= \sum\nolimits_s \frac{1}{\pi_k}\mathbf{z}_k y_k - \left(\sum\nolimits_s \frac{1}{\pi_k}\mathbf{z}_k\mathbf{z}_k'\right)\mathbf{B} \\ &= \sum\nolimits_s \frac{1}{\pi_k}\mathbf{z}_k(y_k - \mathbf{z}_k'\mathbf{B}) \\ &= \sum\nolimits_s \frac{1}{\pi_k}\mathbf{z}_k E_k\end{aligned}$$

where $E_k = y_k - \mathbf{z}_k'\mathbf{B}$. Thus $\hat{\mathbf{t}} - \hat{\mathbf{T}}\mathbf{B}$ is a vector composed of the q π estimators

$$\sum\nolimits_s \frac{z_{jk}E_k}{\pi_k}; \qquad j = 1, \ldots, q$$

Using Result 5.4.1 we have

$$\begin{aligned}V(\hat{\mathbf{B}}) \doteq V(\hat{\mathbf{B}}_0) &= V[\mathbf{B} + \mathbf{T}^{-1}(\hat{\mathbf{t}} - \hat{\mathbf{T}}\mathbf{B})] \\ &= \mathbf{T}^{-1}V(\hat{\mathbf{t}} - \hat{\mathbf{T}}\mathbf{B})\mathbf{T}^{-1} \\ &= \mathbf{T}^{-1}\mathbf{V}\mathbf{T}^{-1}\end{aligned}$$

where $\mathbf{V}$ is a $q \times q$ matrix with elements

$$\begin{aligned}v_{jj'} &= C\left(\sum\nolimits_s \frac{z_{jk}E_k}{\pi_k}, \sum\nolimits_s \frac{z_{j'k}E_k}{\pi_k}\right) \\ &= \sum\sum\nolimits_U \Delta_{kl}\left(\frac{z_{jk}E_k}{\pi_k}\right)\left(\frac{z_{j'l}E_l}{\pi_l}\right)\end{aligned}$$

The estimated covariance matrix is

$$\hat{\mathbf{T}}^{-1}\hat{\mathbf{V}}\hat{\mathbf{T}}^{-1}$$

where

$$\hat{\mathbf{T}} = \sum_s \frac{1}{\pi_k} \mathbf{z}_k \mathbf{z}_k'$$

and $\hat{\mathbf{V}}$ is a $q \times q$ matrix with elements

$$\hat{v}_{jj'} = \sum\sum_s \check{\Delta}_{kl} \left(\frac{z_{jk} e_k}{\pi_k}\right)\left(\frac{z_{j'l} e_l}{\pi_l}\right)$$

with $e_k = y_k - \mathbf{z}_k' \hat{\mathbf{B}}$ for $k \in s$.

Exercises

5.1. The objective is to estimate the difference between the totals of the variables SS82 ($= y$) and CS82 ($= z$) for the MU281 population. An *SI* sample of size $n = 40$ was drawn, and y and z were observed for elements k in the sample. The following quantities were calculated:

$$\sum_s y_k = 866, \quad S_{ys}^2 = 42.438, \quad \sum_s z_k = 383, \quad S_{zs}^2 = 29.430, \quad S_{yzs} = 1.309$$

Compute an approximately 95% confidence interval for the difference between the two totals using π estimation.

5.2. A *BE* sample was drawn with the constant inclusion probability $\pi = 0.3$ from the CO124 population. We want to estimate the ratio ($= R$) of the totals of the variables IMP ($= y$) and EXP ($= z$). For the sample of 43 selected countries, the following quantities were calculated:

$$\sum_s y_k = 669{,}281; \quad \sum_s z_k = 608{,}902; \quad \sum_s (y_k - \hat{R} z_k)^2 = 3{,}496{,}228{,}001$$

where

$$\hat{R} = \hat{t}_{y\pi} / \hat{t}_{z\pi}$$

is the ratio between the two π estimates. Compute an estimate of R and the corresponding *cve*.

5.3. For each of the populations MU200, MU281, and MU284, an estimate is required for the per capita revenue from 1985 municipal taxation, that is, the ratio, R, of the RMT85 ($= y$) total to the P85 ($= z$) total. An *SI* sample of size n is to be drawn. For each population, determine the smallest possible n in order to obtain that $|BR(\hat{R})| \leq 0.1$. The following information is given for the three populations:

	$\sum_U z_k$	$\sum_U z_k^2$
MU200	2,680	43,454
MU281	7,033	338,471
MU284	8,339	997,097

5.4. We want to estimate the percentage increase in total population from 1980 to 1983 for the countries in the CO124 population. An *SI* sample of size $n = 40$ is drawn, and the variables P83 (= y) and P80 (= z) are observed for the elements in the sample. The following results are obtained:

$$\sum_s y_k = 1{,}560.4; \quad \sum_s y_k^2 = 609{,}833.24; \quad \sum_s y_k z_k = 558{,}395.94$$

$$\sum_s z_k = 1{,}447.7; \quad \sum_s z_k^2 = 511{,}520.87$$

(a) Compute an estimate of the percentage increase in total population from 1980 to 1983. (b) Calculate an estimate of the variance of the estimator in (a). (c) The total of the variable P80 is 4,308.1. Use this information to calculate an unbiased estimate of the percentage in (a). Also, calculate an unbiased variance estimate of the unbiased estimator. Comment on your results!

5.5. The objective is to estimate the total t_y of the variable REV84 (= y) for the MU281 population. A *BE* sample of expected size $n = 50$ was drawn. Compute an approximately 95% confidence interval for t_y based on each of the following estimators: (a) the π estimator, (b) the ratio estimator with P75 (= x) as the auxiliary variable. The known population total of x is $t_x = 6{,}818$. The following are summary data for the sample:

$$\sum_s y_k = 128{,}080; \quad \sum_s y_k^2 = 640{,}315{,}550;$$

$$\sum_s x_k = 1{,}067; \quad \sum_s x_k^2 = 45{,}041; \quad \sum_s y_k x_k = 4{,}957{,}800$$

5.6. The following scattergram describes the relation for the MU200 population between the variables REV84 (= y) and P75 (= x):

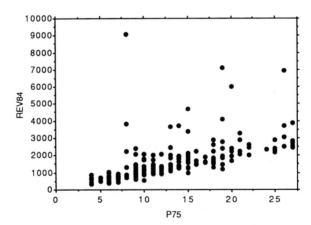

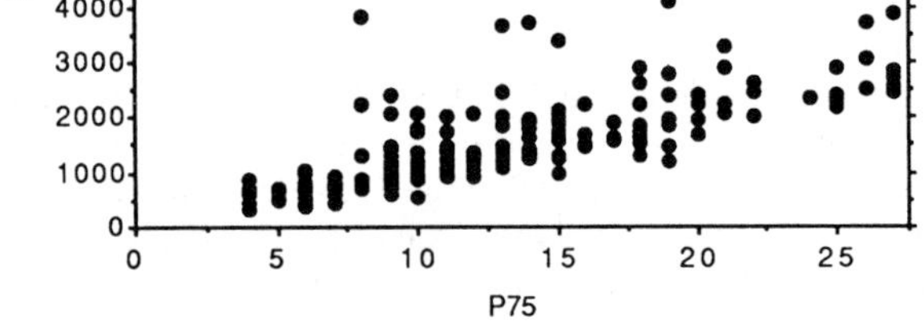

We want to estimate the mean of REV84 from a sample selected by a *PO* design with inclusion probabilities that are proportional to P75. Which of the two estimators $\hat{\bar{y}}_{U\pi}$ and $\tilde{y}_s$ would you expect to be the better one? Explain why.

5.7. A major company needs an estimate of the number of working days lost during the past year due to employee absence from work. Using the sampling fraction 1/25, an *SI* sample s of size 2,000 is drawn from the population of employees. For every employee in the sample, the number of days absent is determined. Let

s_1 denote the male employees in the sample, and let s_2 denote the female employees in the sample. Some summary survey results are:

$$n_{s_1} = 931; \qquad \textstyle\sum_{s_1} y_k = 12{,}604; \qquad \sum_{s_1} y_k^2 = 232{,}226$$
$$n_{s_2} = 1{,}069; \qquad \textstyle\sum_{s_2} y_k = 20{,}412; \qquad \sum_{s_2} y_k^2 = 476{,}208$$

Compute an approximately 95% confidence interval for the male mean number of days absent. The results of Exercises 5.34 and 5.35 may be used.

5.8. Consider the MU281 population and the two variables RMT85 ($= y$) and P85 ($= z$). We want to estimate the slope, $B = (\sum_U z_k y_k)/(\sum_U z_k^2)$, of a regression line through the origin. An *SI* sample of size $n = 50$ is drawn and the following results are obtained:

$$\textstyle\sum_s z_k y_k = 582{,}220; \qquad \sum_s z_k^2 = 72{,}768; \qquad \sum_s z_k^2 y_k^2 = 4.13155 \times 10^{10}$$
$$\textstyle\sum_s z_k^4 = 586{,}102{,}776; \qquad \sum_s z_k^3 y_k = 4{,}911{,}948{,}478$$

Compute an approximately 95% confidence interval for B.

5.9. Consider a survey involving *SI* sampling of $n = 50$ municipalities from the MU281 population. Two study variables are CS82 ($= y_1$) and ME84 ($= y_2$). (a) Compute the coefficients of variation for $\hat{t}_{y_j\pi}$ and the approximate coefficients of variation for $\tilde{S}_{y_j}^2$ ($j = 1, 2$) from the following summary population data, where u_{jk} is given by equation (5.9.12):

j	$\sum_U y_{jk}$	$\sum_U y_{jk}^2$	$\sum_U u_{jk}$	$\sum_U u_{jk}^2$
1	2,508	28,392	−0.07	4.75
2	388,134	1,138,056,730	−7,678	$5.4 \cdot 10^{11}$

(b) Compute the approximate coefficient of variation for $\tilde{S}_{y_1 y_2}$ using the following additional population data, where u_k is given by equation (5.9.9):

$$\textstyle\sum_U y_{1k} y_{2k} = 4{,}699{,}279; \qquad \sum_U u_k = -16; \qquad \sum_U u_k^2 = 420{,}739$$

5.10. A *BE* sample with an expected size of 100 municipalities was drawn from the MU281 population in order to estimate the covariance between the study variables S82 ($= y$) and P85 ($= z$). For the 97 municipalities that were selected, summary sample statistics were calculated as follows:

$$\textstyle\sum_s y_k = 4{,}597; \qquad \sum_s z_k = 2{,}638; \qquad \sum_s y_k z_k = 153{,}126;$$
$$\textstyle\sum_s [(y_k - \bar{y}_s)(z_k - \bar{z}_s) - \tilde{S}_{yz}]^2 = 37{,}994{,}717$$

Compute a point estimate of the population covariance. Also, compute the corresponding *cve*.

5.11. For the countries in the CO124 population, an estimate is required of the median of military expenditures, that is, the median of the variable MEX. An *SI* sample of $n = 40$ was drawn from the $N = 124$ countries. The military expenditures for the sampled countries were, after size ordering:

19	72	110	140	470	1,423	2,184	3,442
22	82	111	164	527	1,426	2,254	4,394
23	84	115	229	675	1,484	2,900	9,461
28	86	115	238	782	1,546	3,175	22,458
35	92	127	290	1,234	2,037	3,424	118,800

Calculate a point estimate and an approximately 95% confidence interval for the population median.

5.12. An estimate is required for the median of the variable P85 ($= y$) for the MU200 population. A Poisson sample of expected size 32 was drawn with probability proportional to the value of the variable P75. Some results of the survey are given below:

$y_{l:s}$	$B_l/\hat{N}$	$y_{l:s}$	$B_l/\hat{N}$
7	0.06575	17	0.66876
8	0.11690	18	0.69433
8	0.17443	18	0.72141
9	0.22557	18	0.74848
10	0.27160	20	0.77150
10	0.31345	21	0.79572
10	0.36459	22	0.81764
13	0.40643	22	0.83956
14	0.43712	23	0.86048
14	0.46780	24	0.87889
15	0.49337	24	0.89660
15	0.52406	25	0.91851
15	0.55474	25	0.94153
16	0.58543	25	0.95994
17	0.61612	26	0.98295
17	0.64319	33	1

Calculate a point estimate and an approximately 95% confidence interval for the population median, using that $\hat{V}[\hat{F}(M)] = 0.012$.

5.13. Continuation of Exercise 4.1. We want to estimate the total of the variable SS82 ($= y$) for the MU284 population using an *SIC* design. The following information on the variable S82 ($= z$) was obtained from a register:

$$\textstyle\sum_{s_I} t_{zi} = 5{,}229;\quad \sum_{s_I} t_{zi}^2 = 1{,}468{,}333;\quad \sum_{s_I} t_{yi}t_{zi} = 687{,}246$$

where t_{zi} and t_{yi} are the totals of z and of y for the ith cluster. (a) Calculate the ratio estimate of the total of SS82, as well as an estimate of the variance of the ratio estimator and the corresponding *cve*. The known population total of z is $\sum_U z_k = 13{,}500$. Compare with the results of Exercise 4.1. (b) Calculate an estimate of the percentage of seats occupied by the Social Democrats, that is, $100 \cdot t_y/t_z$, as well as a variance estimate and the corresponding *cve*.

5.14. Continuation of Exercise 4.9. Estimate the employment rate defined in Exercise 4.9 by replacing the known total 4,804,685 by a sample-based unbiased estimate. Also, compute a variance estimate. Discuss the properties of these new estimates.

5.15. Continuation of Exercise 4.10. Estimate the percentage of income earners by replacing the known population total 8,327,484 by an unbiased estimate. Also, compute a variance estimate. Discuss the properties of these new estimates.

5.16. Verify the values of the coverage probability P_0 given in Table 5.1.

5.17. Calculate a new table, analogous to Table 5.1, giving coverage probabilities of the alternative interval $\hat{\theta} \pm z_{1-\alpha/2}[\text{MSE}(\hat{\theta})]^{1/2}$, assuming that $\hat{\theta}$ is normally distributed.

5.18. Show (for example, using Chebyshev's inequality) that if an estimator $\hat{\theta}_v$ is asymptotically unbiased for θ_v, and its variance tends to zero as v tends to infinity, then $\hat{\theta}_v$ is consistent for θ_v in the sense of definition (b) of Section 5.3.

5.19. Show that the π estimator $\hat{t}_\pi$ is a consistent estimator of the population total t in the sense of definition (c) of Section 5.3 for any class of fixed-size designs containing the complete enumeration design (that is, the design with $\pi_k = 1$ for all $k \in U$) as a member.

5.20. Show that the estimator $\hat{t} = N \sum_s y_k/n_s$ is consistent for the population total t in the sense of definition (c) of Section 5.3 for any Bernoulli design (and any class of Bernoulli designs).

5.21. Show that under a fixed-size design, the covariance of two π estimators $\hat{t}_{j\pi} = \sum_s \check{y}_{jk}$ and $\hat{t}_{j'\pi} = \sum_s \check{y}_{j'k}$ based on the same sample s can be alternatively expressed as

$$C(\hat{t}_{j\pi}, \hat{t}_{j'\pi}) = -\tfrac{1}{2} \sum\sum_U \Delta_{kl}(\check{y}_{jk} - \check{y}_{jl})(\check{y}_{j'k} - \check{y}_{j'l})$$

and that an alternative unbiased estimator of the covariance in this case is

$$\hat{C}(\hat{t}_{j\pi}, \hat{t}_{j'\pi}) = -\tfrac{1}{2} \sum\sum_s \check{\Delta}_{kl}(\check{y}_{jk} - \check{y}_{jl})(\check{y}_{j'k} - \check{y}_{j'l})$$

5.22. Show that under the *SI* design with $f = n/N$, the covariance of the π estimators of the two population totals $t_j = \sum_U y_{jk}$ and $t_{j'} = \sum_U y_{j'k}$ is

$$C(N\bar{y}_{js}, N\bar{y}_{j's}) = N^2 \frac{1-f}{n} S_{y_j y_{j'} U}$$

which is unbiasedly estimated by

$$\hat{C}(N\bar{y}_{js}, N\bar{y}_{j's}) = N^2 \frac{1-f}{n} S_{y_j y_{j'} s}$$

where

$$S_{y_j y_{j'} U} = \sum_U (y_{jk} - \bar{y}_{jU})(y_{j'k} - \bar{y}_{j'U})/(N-1)$$

and

$$S_{y_j y_{j'} s} = \sum_s (y_{jk} - \bar{y}_{js})(y_{j'k} - \bar{y}_{j's})/(n-1)$$

5.23. Show that under the *SI* design with $f = n/N$, the π estimator $\hat{t}_{d\pi}$ (given in Example 5.4.2) of the dth domain total $t_d = \sum_{U_d} y_k$ is

$$\hat{t}_{d\pi} = \frac{N}{n} \sum_{s_d} y_k$$

with variance

$$V(\hat{t}_{d\pi}) = N^2 \frac{1-f}{n} S^2_{y_d U}$$

where

$$S^2_{y_d U} = \sum_U (y_{dk} - \bar{y}_{dU})^2/(N-1); \quad \bar{y}_{dU} = t_d/N; \quad y_{dk} = z_{dk}\, y_k$$

5.24. Show that under the *SI* design with $f = n/N$, an alternative expression for the variance $V(\hat{t}_{d\pi})$ of the preceding exercise is

$$V(\hat{t}_{d\pi}) = N^2 \frac{1-f}{n} \frac{1}{N-1} [(N_d - 1) S^2_{yU_d} + N_d(1 - P_d)\bar{y}^2_{U_d}]$$

$$\doteq N^2 \frac{1-f}{n} P_d [S^2_{yU_d} + (1 - P_d)\bar{y}^2_{U_d}]$$

where

$$P_d = N_d/N, \qquad \bar{y}_{U_d} = t_d/N_d$$

and

$$S^2_{yU_d} = \sum_{U_d} (y_k - \bar{y}_{U_d})^2/(N_d - 1)$$

5.25. Show that under the *SI* design with $f = n/N$, an unbiased estimator of the variance of the π estimator $\hat{t}_{d\pi} = (N/n)\sum_{s_d} y_k$ is

$$\hat{V}(\hat{t}_{d\pi}) = N^2 \frac{1-f}{n} S^2_{y_d s}$$

where

$$S^2_{y_d s} = \sum_s (y_{dk} - \bar{y}_{ds})^2/(n-1)$$

and

$$\bar{y}_{ds} = \sum_s y_{dk}/n = \sum_{s_d} y_k/n$$

Find an expression for $\hat{V}(\hat{t}_{d\pi})$ of the same form as $V(\hat{t}_{d\pi})$ in Exercise 5.24.

5.26. Show that under the *SI* design with $f = n/N$, the covariance of the π estimators $\hat{t}_{d\pi} = (N/n)\sum_{s_d} y_k$ and $\hat{t}_{d'\pi} = (N/n)\sum_{s_{d'}} y_k$ of the two domain totals $t_d = \sum_{U_d} y_k$ and $t_{d'} = \sum_{U_{d'}} y_k$ is

$$C(\hat{t}_{d\pi}, \hat{t}_{d'\pi}) = -N^2 \frac{1-f}{n} \frac{N}{N-1} P_d P_{d'} \bar{y}_{U_d} \bar{y}_{U_{d'}}$$

$$\doteq -N^2 \frac{1-f}{n} P_d P_{d'} \bar{y}_{U_d} \bar{y}_{U_{d'}}$$

which is estimated unbiasedly by

$$\hat{C}(\hat{t}_{d\pi}, \hat{t}_{d'\pi}) = -N^2 \frac{1-f}{n(n-1)} \sum_s (y_{dk} - \bar{y}_{ds})(y_{d'k} - \bar{y}_{d's})$$

$$= -N^2 \frac{1-f}{n} \frac{n}{n-1} \hat{P}_d \hat{P}_{d'} \bar{y}_{s_d} \bar{y}_{s_{d'}}$$

$$\doteq -N^2 \frac{1-f}{n} \hat{P}_d \hat{P}_{d'} \bar{y}_{s_d} \bar{y}_{s_{d'}}$$

where $\hat{P}_d = n_{s_d}/n$ and $\bar{y}_{s_d} = \sum_{s_d} y_k / n_{s_d}$.

5.27. Show that the linear statistic $\hat{\theta}_0$, obtained in Section 5.5 by Taylor linearization of $\hat{\theta}$, has zero bias, that is, $E(\hat{\theta}_0) = \theta$.

5.28. Verify in detail that Result 5.6.2 follows from the Taylor linearization technique as described in Section 5.5.

5.29. Give the alternative expression for $AV(\hat{R})$ for the case of a fixed size design, and also give the alternative variance estimator for that case.

5.30. Show that if second-order terms are also included in the Taylor expansion, then a more refined approximation of the bias of $\hat{R}$ is obtained as

$$B(\hat{R}) \doteq \frac{V(\hat{t}_{z\pi})}{t_z^2} \left[R - \frac{C(\hat{t}_{y\pi}, \hat{t}_{z\pi})}{V(\hat{t}_{z\pi})} \right]$$

Also show that under the *SI* design, this bias expression can be written as

$$B(\hat{R}) \doteq \frac{(1-f)}{n} \frac{S_{zU}^2}{\bar{z}_U^2} \left(R - \frac{S_{yzU}}{S_{zU}^2} \right)$$

Finally, give interpretations of these two bias expressions in terms of least-squares regression.

5.31. Show that under the *BE* design, the weighted sample mean $\tilde{y}_s$, given by (5.7.3), is equal to the sample mean, that is,

$$\tilde{y}_s = \bar{y}_s = \sum_s y_k / n_s$$

whereas the expression for the unbiased estimator $\hat{\bar{y}}_{U\pi}$ given by (5.7.1) is

$$\hat{\bar{y}}_{U\pi} = [n_s / E(n_s)] \bar{y}_s$$

5.32. Show that, under the *BE* design with $\pi_k = \pi$ for $k = 1, \dots, N$, the approximate variance (5.7.4) of the weighted sample mean $\tilde{y}_s = \bar{y}_s$ is

$$AV(\bar{y}_s) = [(1-\pi)/N\pi] S_{yU}^2$$

which is equal to the variance of the sample mean $\bar{y}_s$ under the *SI* design with the same expected sample size.

5.33. Show that under the *BE* design with $\pi_k = \pi$ for all k, Result 5.7.1 gives the following variance estimator for the weighted sample mean $\tilde{y}_s = \bar{y}_s$,

$$\hat{V}(\bar{y}_s) = [(1-\pi)/n_s][(n_s - 1)/n_s]S^2_{ys}$$
$$\doteq [(1-\pi)/n_s]S^2_{ys}$$

5.34. Show that under the *SI* design with $f = n/N$, the estimator (5.8.3) of the domain mean $\bar{y}_{U_d}$ is equal to the sample mean in s_d, that is,

$$\tilde{y}_{s_d} = \bar{y}_{s_d} = \textstyle\sum_{s_d} y_k/n_{s_d}$$

Also show that in this case the approximate variance (5.8.4) is given by

$$AV(\tilde{y}_{s_d}) = \frac{1-f}{nP_d}\frac{N(N_d - 1)}{(N-1)N_d}S^2_{U_d} \doteq \frac{1-f}{nP_d}S^2_{U_d}$$

where

$$S^2_{U_d} = \textstyle\sum_{U_d}(y_k - \bar{y}_{U_d})^2/(N_d - 1)$$

5.35. Show that under the *SI* design with $f = n/N$, the variance estimator (5.8.5) for $\tilde{y}_{s_d}$ is

$$\hat{V}(\tilde{y}_{s_d}) = \frac{1-f}{n_{s_d}}\frac{n(n_{s_d} - 1)}{(n-1)n_{s_d}}S^2_{s_d} \doteq \frac{1-f}{n_{s_d}}S^2_{s_d}$$

where

$$S^2_{s_d} = \textstyle\sum_{s_d}(y_k - \bar{y}_{s_d})^2/(n_{s_d} - 1)$$

5.36. Show that, in the Taylor linearization of the estimator $\tilde{S}_{yz}$ in Section 5.9, the u_k values, defined by equation (5.9.9), are

$$u_k = [(y_k - \bar{y}_U)(z_k - \bar{z}_U) - S_{yz}]/(N-1)$$

where $S_{yz} = S_{yzU}$. Also show that, in the Taylor linearization of $\tilde{S}_{yy}$, the u_k values are

$$u_k = [(y_k - \bar{y}_U)^2 - S_{yy}]/(N-1)$$

where $S_{yy} = S^2_{yU}$.

5.37. Verify the result of Example 5.9.1, namely that, under the *SI* design, Taylor linearization gives the variance estimator

$$\hat{V}(\tilde{S}_{yz}) = [N/(N-1)]^2[(1-f)/n]\textstyle\sum_s d_k^2/(n-1)$$

with

$$d_k = (y_k - \bar{y}_s)(z_k - \bar{z}_s) - [(n-1)/n]S_{yzs}$$

5.38. Show that for any self-weighting design, that is, for any design with equal inclusion probabilities, $\pi_k = \pi$ $(k = 1, \ldots, N)$, the estimator $\tilde{S}_{yz}$, given by equations (5.9.7) and (5.9.8), is practically equal to the sample covariance,

$$\tilde{S}_{yz} = \sum_s (y_k - \bar{y}_s)(z_k - \bar{z}_s)/(n_s - \pi)$$
$$\doteq \sum_s (y_k - \bar{y}_s)(z_k - \bar{z}_s)/(n_s - 1) = S_{yzs}$$

5.39. Let $\hat{t}_y$, $\hat{t}_z$, and $\hat{t}_{yz}$ be any unbiased estimators of the population totals t_y, t_z, and $t_{yz} = \sum_U y_k z_k$, respectively, and let $\hat{C}(\hat{t}_y, \hat{t}_z)$ be any unbiased estimator of the covariance $C(\hat{t}_y, \hat{t}_z)$. Show that $\hat{S}_{yz}$ given by (5.9.14) is an unbiased estimator of S_{yz}.

5.40. Show that the population covariance S_{yz} defined by equation (5.9.2), can be expressed as

$$S_{yz} = \sum\sum_U (y_k - y_l)(z_k - z_l)/2N(N - 1)$$

5.41. Verify that the three expressions (5.10.1) to (5.10.3) for $\mathbf{B}$ are identical, that is,

$$\mathbf{B} = (\mathbf{Z}\mathbf{Z}')^{-1}\mathbf{Z}\mathbf{y} = (\textstyle\sum_U \mathbf{z}_k\mathbf{z}_k')^{-1}(\sum_U \mathbf{z}_k y_k) = \mathbf{T}^{-1}\mathbf{t}$$

5.42. Verify that the following three expressions for the estimator of $\mathbf{B}$ are identical, that is,

$$\hat{\mathbf{B}} = (\mathbf{Z}_s\mathbf{\Pi}_s^{-1}\mathbf{Z}_s')^{-1}\mathbf{Z}_s\mathbf{\Pi}_s^{-1}\mathbf{y}_s = (\textstyle\sum_s(1/\pi_k)\mathbf{z}_k\mathbf{z}_k')^{-1}(\sum_s(1/\pi_k)\mathbf{z}_k y_k) = \hat{\mathbf{T}}^{-1}\hat{\mathbf{t}}$$

5.43. Show that for a given sample s, the estimator $\hat{\mathbf{B}}$, given by equation (5.10.4), satisfies the weighted least-squares criterion. That is, for a given s, the weighted sum of the squared sample fit residuals

$$(\mathbf{y}_s - \mathbf{Z}_s'\mathbf{b})'\mathbf{\Pi}_s^{-1}(\mathbf{y}_s - \mathbf{Z}_s'\mathbf{b}) = \textstyle\sum_s (1/\pi_k)(y_k - \mathbf{z}_k'\mathbf{b})^2$$

is minimized when $\mathbf{b} = \hat{\mathbf{B}}$.

5.44. Verify the results in Example 5.10.3.

5.45. Verify the results in Example 5.10.4, and give the results for $\hat{B}_1$.

5.46. Verify the results in Example 5.10.5.

PART II

Estimation through Linear Modeling, Using Auxiliary Variables

CHAPTER 6

The Regression Estimator

6.1. Introduction

The emphasis laid on the use of auxiliary information for improving the precision of estimates is characteristic of sampling theory. The regression estimator introduced in this chapter is one type of estimator that attempts to make efficient use of auxiliary information about the population. The purpose of this chapter is to introduce the regression estimator and to explore its general properties. Chapters 7 and 8 contain a number of examples of regression estimators illustrating various applications.

6.2. Auxiliary Variables

Generally speaking, an auxiliary variable is any variable about which information is available prior to sampling. Ordinarily, we assume that the a priori information for an auxiliary variable is complete. The value of the variable, say x, is known for each of the N population elements so that the values x_1, $\ldots$, x_N are at our disposal prior to sampling. An auxiliary variable assists in the estimation of the study variable. The goal is to obtain an estimator with increased accuracy.

Some sampling frames are equipped from the outset with one or more auxiliary variables, or with information that can be transformed into auxiliary variables through simple numerical manipulations. That is, the frame provides not only the identification characteristics of the units, but attached to each unit is also the value of one or more auxiliary variables. For example, a register of farms may contain information about the area of each farm. A list of districts may contain information about the number of people living in each district at the time of the latest population census.

EXAMPLE 6.2.1. In Sweden, the Register of the Total Population is an often-used sampling frame in surveys of individuals or households. The register contains information that can easily be turned into a number of auxiliary variables, some quantitative, some categorical. Among the former are age (obtained from the date of birth) and taxable income. Categorical auxiliary variables available in the register are sex, marital status, and residential district.

Auxiliary variable values can be transferred to the sampling frame from administrative or other registers by matching these registers to the sampling frame. There are practical problems associated with matching. For example, the frame and the register may date from different periods in time, elements may be coded differently or erroneously, and so on. In these cases, an element in the frame cannot always be unambiguously identified in the register. These are sometimes difficult problems. For a reference, the reader is referred to Cox and Boruch (1988).

We now assume that one or more auxiliary variables are present. The use of auxiliary information is, however, not new to this chapter. In Chapter 3, we learned that auxiliary information can be used at the design stage of a survey to create a sampling design that increases the precision of the π estimator.

One approach was πps sampling, that is, to make the inclusion probabilities $\pi_1, \ldots, \pi_N$ of the design proportional to known, positive values $x_1, \ldots, x_N$ of an auxiliary variable. The π estimator will then have a small variance if x is more or less proportional to y, the study variable. However, πps sampling was sometimes found to be difficult to carry out.

Another approach in Chapter 3 was to use auxiliary information to construct strata such that the π estimator for a stratified simple random sampling design,

$$\hat{t}_\pi = \sum_{h=1}^{H} N_h \bar{y}_{s_h}$$

obtains a small variance. However, the stratification that is efficient for one study variable may be inefficient for another.

In this chapter, the point of departure is different. The auxiliary information will now be used at the estimation stage. The auxiliary variables will enter explicitly into the estimator formula, not only through the π_k. That is, for a given sampling design, we construct estimators that utilize information from auxiliary variables and bring considerable variance reduction compared to the π estimator.

The basic assumption behind the use of auxiliary variables is that they covary with the study variable and thus carry information about the study variable. Such covariation is used advantageously in the regression estimator. Basic theory for the regression estimator is presented in this chapter. In coming chapters, the technique of regression estimation is applied to a variety of practical situations.

6.3. The Difference Estimator

The main topic of this chapter is the regression estimator. However, we start with the closely related difference estimator, for two reasons, namely,

a. The difference estimator is simpler to handle mathematically.
b. An understanding of the difference estimator simplifies the transition to the regression estimator.

We assume that there are J auxiliary variables, denoted $x_1, \ldots, x_j, \ldots, x_J$. The value of the jth x variable for the kth population element is denoted x_{jk}. For the kth element we define the vector $\mathbf{x}_k = (x_{1k}, \ldots, x_{jk}, \ldots, x_{Jk})'$. As before, the study variable y takes the value y_k for the kth element.

The values $y_1, \ldots, y_N$ are assumed unknown and inaccessible before sampling, whereas $\mathbf{x}_1, \ldots, \mathbf{x}_N$ are known. The population parameter to be estimated is the population total of y,

$$t_y = \textstyle\sum_U y_k$$

A sample s is drawn from U by a sampling design $p(\cdot)$ with inclusion probabilities $\pi_k > 0$ and $\pi_{kl} > 0$. For each $k \in s$, we observe y_k and the attached $\mathbf{x}_k$. The problem is to estimate t_y when we have observed $(y_k, \mathbf{x}_k)$ for the elements $k \in s$, and when we also know $\mathbf{x}_k$ for $k \in U - s$, the nonsample set.

The main idea underlying the difference estimator is to use the auxiliary information to form a set of N proxy y-values, denoted $y_1^0, \ldots, y_N^0$, such that y_k^0 is at least a decent approximation of y_k, and preferably a value that zeroes in on y_k. We start with this modest aspiration; a later task will be to form the proxy values in a more scientific way.

Of particular interest is the formation of the proxy value y_k^0 as a linear combination of the known values $x_{1k}, \ldots, x_{Jk}$

$$y_k^0 = \sum_{j=1}^{J} A_j x_{jk} = \mathbf{A}'\mathbf{x}_k \tag{6.3.1}$$

where we assume for now that $\mathbf{A} = (A_1, \ldots, A_J)'$ is a *known* coefficient vector. Later, the coefficients will be replaced by quantities estimated from the sample. Obviously, y_k^0 can be calculated for all $k \in U$, since the x-values are known for the whole population.

When previous studies or subject matter theory suggest an approximate linear relation for $k = 1, \ldots, N$,

$$y_k \doteq \sum_{j=1}^{J} A_j x_{jk} = \mathbf{A}'\mathbf{x}_k \tag{6.3.2}$$

with a known coefficient vector $\mathbf{A} = (A_1, \ldots, A_J)'$, it is reasonable to choose y_k^0 in accordance with (6.3.1). In other words, we believe that the study variable can be explained to a considerable extent by the auxiliary variables through the linear relation shown in (6.3.2), where the A_j are known.

One of the simplest applications has a single x variable, $x_1 = x$ and $A_1 = 1$. Then (6.3.1) becomes

$$y_k^0 = x_k$$

and the assumption inherent in (6.3.2) is that

$$y_k \doteq y_k^0 = x_k$$

for $k = 1, \ldots, N$. For example, this simple formulation underlies a successful application, in accounting, of the difference estimation method by Godfrey, Roshwalb and Wright (1984). In their application, x_k is the book value and y_k is the audit value for the kth inventory item in a population of $N = 8{,}069$ inventory items from which a sample of items was selected for audit (instead of taking a complete count).

The unknown population total to be estimated can now be written as

$$\begin{aligned} t_y &= \textstyle\sum_U y_k \\ &= \textstyle\sum_U y_k^0 + \sum_U (y_k - y_k^0) \\ &= \textstyle\sum_U y_k^0 + \sum_U D_k \end{aligned} \tag{6.3.3}$$

if we let

$$D_k = y_k - y_k^0$$

for $k = 1, \ldots, N$. The proxy total $\sum_U y_k^0$ in equation (6.3.3) is known (since $y_1^0, \ldots, y_N^0$ are known), whereas the total of the differences, $\Sigma_U D_k$, is unknown (since $y_1, \ldots, y_N$ are unknown).

The immediate thought here is to use π estimation to form an unbiased estimator of the unknown term in (6.3.3). The result is called *the difference estimator*; it is denoted by $\hat{t}_{y,\text{dif}}$ and given by

$$\hat{t}_{y,\text{dif}} = \textstyle\sum_U y_k^0 + \sum_s \check{D}_k$$

where

$$\check{D}_k = D_k/\pi_k = (y_k - y_k^0)/\pi_k$$

The basic results for this new estimator are easily derived.

Result 6.3.1. *The difference estimator,*

$$\hat{t}_{y,\text{dif}} = \textstyle\sum_U y_k^0 + \sum_s \check{D}_k \tag{6.3.4}$$

with

$$\check{D}_k = D_k/\pi_k = (y_k - y_k^0)/\pi_k$$

is unbiased for $t_y = \sum_U y_k$. *The variance is given by*

$$V(\hat{t}_{y,\text{dif}}) = \sum\sum_U \Delta_{kl}\check{D}_k\check{D}_l \tag{6.3.5}$$

which is unbiasedly estimated by

$$\hat{V}(\hat{t}_{y,\text{dif}}) = \sum\sum_s \check{\Delta}_{kl}\check{D}_k\check{D}_l \tag{6.3.6}$$

PROOF. Since $\sum_s \check{D}_k$ is unbiased for $\sum_U D_k$, it follows immediately from (6.3.3) that $\hat{t}_{y,\text{dif}}$ is unbiased for $t_y = \sum_U y_k$. The variance and variance estimation formulas follow directly from π estimation results in Chapter 2, for we notice, from (6.3.4), that

$$V(\hat{t}_{y,\text{dif}}) = V(\textstyle\sum_s \check{D}_k)$$ □

Upon inspection of the variance shown in equation (6.3.5), it is clear that the difference estimator performs well when the differences D_k are zero or very near zero for all k. That is, we want to create the proxy values y_k^0 so that the $D_k = y_k - y_k^0$ are small.

Remark 6.3.1. Under a fixed-size design, the variance (6.3.5) can alternatively be written

$$V(\hat{t}_{y,\text{dif}}) = -\tfrac{1}{2}\sum\sum_U \Delta_{kl}(\check{D}_k - \check{D}_l)^2 \tag{6.3.7}$$

which gives the alternative unbiased variance estimator

$$\hat{V}(\hat{t}_{y,\text{dif}}) = -\tfrac{1}{2}\sum\sum_s \check{\Delta}_{kl}(\check{D}_k - \check{D}_l)^2 \tag{6.3.8}$$

EXAMPLE 6.3.1. If

$$y_k^0 = x_k; \qquad D_k = y_k - x_k$$

it is not difficult to show for the *SI* design with sample size $n = fN$ that

$$V(\hat{t}_{y,\text{dif}}) = N^2\frac{1-f}{n}\{S_{yU}^2 + S_{xU}^2 - 2S_{xyU}\}$$

where S_{yU}^2 and S_{xU}^2 are the population variances of y and x, respectively, and

$$S_{xyU} = \frac{1}{N-1}\sum_U (x_k - \bar{x}_U)(y_k - \bar{y}_U)$$

is the population covariance. If the population correlation

$$r = \frac{S_{xyU}}{S_{xU}S_{yU}}$$

is high, the difference estimator will often yield a large reduction in variance compared to the simple π estimator $\hat{t}_{y\pi}$. We have

$$\frac{V(\hat{t}_{y,\text{dif}})}{V(\hat{t}_{y\pi})} = 1 + \left(\frac{S_{xU}}{S_{yU}}\right)^2 - 2r\frac{S_{xU}}{S_{yU}}$$

For example, if $r = 0.95$ and $S_{xU}/S_{yU} = 1.1$,

$$\frac{V(\hat{t}_{y,\text{dif}})}{V(\hat{t}_{y\pi})} = 0.12$$

In general, under SI sampling and with $y_k^0 = x_k$, $V(\hat{t}_{y,\text{dif}}) < V(\hat{t}_{y\pi})$ if and only if $r > \frac{1}{2}(S_{xU}/S_{yU})$.

An alternative view of the difference estimator is to regard it as a possible improvement over the π estimator

$$\hat{t}_{y\pi} = \sum_s \check{y}_k = \sum_s y_k/\pi_k$$

which makes no explicit use of the auxiliary information. Suppose the proxy values are linear combinations in accordance with (6.3.1). Then (6.3.4) can be written as

$$\hat{t}_{y,\text{dif}} = \hat{t}_{y\pi} + \sum_{j=1}^{J} A_j(t_{x_j} - \hat{t}_{x_j\pi}) \tag{6.3.9}$$

where, for $j = 1, \ldots, J$,

$$\hat{t}_{x_j\pi} = \sum_s \check{x}_{jk} = \sum_s x_{jk}/\pi_k$$

is the π estimator of the x_j total

$$t_{x_j} = \sum_U x_{jk}$$

To be explicit, the difference estimator is equal to the π estimator plus an adjustment term.

Interestingly, in the ideal situation, when (6.3.2) holds exactly, that is, when

$$y_k = y_k^0 = \sum_{j=1}^{J} A_j x_{jk}$$

for $k = 1, \ldots, N$, the adjustment term will exactly cancel the error in the π estimator, so that the difference estimator is then completely error free. This is seen as follows. The error in the π estimator is

$$\hat{t}_{y\pi} - t_y = \sum_s \frac{y_k}{\pi_k} - \sum_U y_k$$

The adjustment term, when (6.3.2) holds exactly, is

$$\sum_{j=1}^{J} A_j(t_{x_j} - \hat{t}_{x_j\pi}) = -\left(\sum_s y_k/\pi_k - \sum_U y_k\right)$$

The error of the difference estimator,

$$\hat{t}_{y,\text{dif}} - t_y$$

is thus zero.

Remark 6.3.2. The difference estimator will perform well when there is an approximate linear relation between y and the x variables, and, very importantly, when the coefficients $A_1, \ldots, A_J$ are well chosen. If these coefficients are badly chosen, the variance shown in equation (6.3.5) can be considerably greater than that of the π estimator. The derivation of optimal A_j values is considered in Section 6.8.

6.4. Introducing the Regression Estimator

In this section, we define the regression estimator for the general case with J auxiliary variables. As in Section 6.3, the value of the auxiliary variable vector for the kth element is denoted by

$$\mathbf{x}_k = (x_{1k}, \ldots, x_{Jk})'$$

The objective is to estimate the unknown y total

$$t_y = \textstyle\sum_U y_k$$

when we have observed $(y_k, \mathbf{x}_k)$ for $k \in s$, and when $\mathbf{x}_k$ is also known for $k \in U - s$.

As a starting point, look again at the difference estimator written in the form of equation (6.3.9). If the coefficients A_j in the approximate linear relation shown in (6.3.2) are no longer known beforehand, we can try the same type of estimator, but with estimated coefficients. Regression analysis is a natural tool that can be used to arrive at suitable coefficient estimators; these are denoted $\hat{B}_1, \ldots, \hat{B}_J$. (We use $\hat{B}_j$ rather than $\hat{A}_j$, because $\hat{B}_j$ is the customary notation in regression analysis.)

The general regression estimator, denoted $\hat{t}_{yr}$, is now formally defined as

$$\hat{t}_{yr} = \hat{t}_{y\pi} + \sum_{j=1}^{J} \hat{B}_j(t_{x_j} - \hat{t}_{x_j\pi}) \tag{6.4.1}$$

where

$$\hat{t}_{y\pi} = \textstyle\sum_s y_k/\pi_k = \sum_s \check{y}_k$$

is the π estimator of $t_y = \sum_U y_k$,

$$\hat{t}_{x_j\pi} = \textstyle\sum_s x_{jk}/\pi_k = \sum_s \check{x}_{jk}$$

is the π estimator of the (known) x_j total

$$t_{x_j} = \textstyle\sum_U x_{jk}$$

and $\hat{B}_1, \ldots, \hat{B}_J$ are components of the J-vector

$$\hat{\mathbf{B}} = (\hat{B}_1, \ldots, \hat{B}_J)' = \left(\textstyle\sum_s \mathbf{x}_k\mathbf{x}_k'/\sigma_k^2\pi_k\right)^{-1} \sum_s \mathbf{x}_k y_k/\sigma_k^2\pi_k \tag{6.4.2}$$

The justification, in terms of regression theory, for this particular form of the $\hat{B}_j$ is given later in the section.

Similarly as in the case of the difference estimator, we can view $\hat{t}_{yr}$ as an attempt to improve over the basic π estimator $\hat{t}_{y\pi}$. Explicitly, the regression estimator is equal to the π estimator plus an adjustment term. When the regression estimator works well, the adjustment term will often be negatively correlated with the error of the π estimator. For samples in which the π estimator alone gives a large error, the adjustment term will be about equally large as this error, but of the opposite sign, when the sample is fairly large and the linear relationship strong. Thus, $\hat{t}_{yr}$ will have a smaller error than $\hat{t}_{y\pi}$.

Now let us specify which coefficient estimators $\hat{B}_j$ should be used. The choice is based on an assumption about the shape of the finite population point scatter

$$\{(y_k, x_{1k}, \ldots, x_{Jk}) : k = 1, \ldots, N\} \tag{6.4.3}$$

This assumption is expressed in terms of a *model*. We thus assume that the scatter of the N points given in (6.4.3) looks *as if* it had been generated by a linear regression model, called ξ, with y as the regressor and $x_1, \ldots, x_J$ as regressands.

More precisely, the regression model ξ will have the following features:

i. $y_1, \ldots, y_N$ are assumed to be realized values of independent random variables $Y_1, \ldots, Y_N$,

ii. $E_\xi(Y_k) = \sum_{j=1}^{J} \beta_j x_{jk}$ $(k = 1, \ldots, N)$; and (6.4.4)

iii. $V_\xi(Y_k) = \sigma_k^2$ $(k = 1, \ldots, N)$,

where E_ξ and V_ξ denote expected value and variance with respect to the model ξ, and where $\beta_1, \ldots, \beta_J$ and $\sigma_1^2, \ldots, \sigma_N^2$ are model parameters. In what follows, we will no longer distinguish Y, the random variable, from y, the realized value. This kind of model statement is familiar from regression analysis.

Here, ξ introduces a new type of randomness. The only randomness considered so far is that different sample outcomes s can occur under the given sampling design. The new randomness exists by assumption alone and has nothing to do with the randomness created by sample selection through the given sampling design.

The model's variance parameters σ_k^2 make it possible to describe an assumed variance pattern in the point scatter defined by (6.4.3), for example, that the variability of y tends to increase with increasing x-values. We recall again that the auxiliary variable values x_{jk} $(j = 1, \ldots, J;\ k = 1, \ldots, N)$ are treated as known constants.

Example 6.4.1. Two examples of models involving a single x variable are

$$\begin{cases} E_\xi(y_k) = \beta x_k \\ V_\xi(y_k) = \sigma^2 x_k \end{cases} \tag{6.4.5}$$

(where $x_1, \ldots, x_N$ are assumed > 0), and

$$\begin{cases} E_\xi(y_k) = \beta_1 + \beta_2 x_k \\ V_\xi(y_k) = \sigma^2 \end{cases} \tag{6.4.6}$$

These models, and the estimators they generate under various sampling designs, will be discussed further in this and other chapters. In both (6.4.5) and (6.4.6), $y_1, \ldots, y_N$ are assumed independent. Without explicitly saying so, we assume in many of the model statements given later that the y_k are independent.

Remark 6.4.1. Let us discuss the role of the model ξ. The role of the model ξ is to describe the finite population point scatter. We hope that the model ξ fits the population reasonably well. We think that the finite population looks as if it might have been generated in accordance with the model ξ. However, the assumption is never made that the population was really generated by the model ξ. Our conclusions about the finite population parameters are therefore independent of model assumptions.

Ultimately the model serves as the vehicle for finding an appropriate $\hat{\mathbf{B}}$ to put into the regression estimator formula. The efficiency of the regression estimator, as compared to the π estimator, will depend on the goodness of the fit. The basic properties (approximate unbiasedness, validity of the variance formulas, etc.), however, are not dependent on whether the model ξ holds or not. Our procedures are thus *model assisted*, but they are not model dependent. Certain other estimation procedures in the sampling literature are indeed model dependent, a term coined by Hansen, Madow, and Tepping (1983). Some of these procedures are discussed in Section 14.5.

We now examine how the particular form of $\hat{\mathbf{B}}$ shown in equation (6.4.2) is motivated by the model ξ. First consider a hypothetical complete enumeration of the population (a census), where we observe y_k and $\mathbf{x}_k$ for all $k \in U$. In that case, the weighted least-squares estimator of

$$\boldsymbol{\beta} = (\beta_1, \ldots, \beta_J)'$$

under the model ξ would have been

$$\begin{aligned} \mathbf{B} &= (B_1, \ldots, B_J)' \\ &= \left(\textstyle\sum_U \mathbf{x}_k \mathbf{x}_k'/\sigma_k^2\right)^{-1} \textstyle\sum_U \mathbf{x}_k y_k/\sigma_k^2 \end{aligned} \tag{6.4.7}$$

In notation more familiar from regression analysis,

$$\mathbf{B} = (\mathbf{X}\boldsymbol{\Sigma}^{-1}\mathbf{X}')^{-1}\mathbf{X}\boldsymbol{\Sigma}^{-1}\mathbf{Y}$$

where $\mathbf{X}$ is a matrix of dimension $J \times N$ whose N columns are the J-dimensional vectors $\mathbf{x}_1, \ldots, \mathbf{x}_N$,

$$\mathbf{Y} = (y_1, \ldots, y_N)'$$

and $\boldsymbol{\Sigma}$ is the $N \times N$ diagonal matrix

$$\boldsymbol{\Sigma} = \begin{pmatrix} \sigma_1^2 & \cdots & 0 \\ \vdots & & \vdots \\ 0 & \cdots & \sigma_N^2 \end{pmatrix}$$

This is assumed known from regression theory, where it is also shown that **B** is the best linear unbiased estimator of $\boldsymbol{\beta}$ under the model. Note the model weight, $1/\sigma_k^2$, attached to the kth observation. We say that **B** corresponds to a hypothetical *population fit* of the model ξ to the population data points $(\mathbf{x}_k, y_k)$, $k = 1, \ldots, N$. Note that **B** is a finite population characteristic unknown to us. But we can estimate **B** using sample data, by applying the π estimation principle, as in Section 5.10. We thereby obtain a *sample fit* of the model. Let us write the unknown **B** as

$$\mathbf{B} = \mathbf{T}^{-1}\mathbf{t} \tag{6.4.8}$$

where

$$\mathbf{T} = \sum\nolimits_U \frac{\mathbf{x}_k\mathbf{x}_k'}{\sigma_k^2}; \qquad \mathbf{t} = \sum\nolimits_U \frac{\mathbf{x}_k y_k}{\sigma_k^2} \tag{6.4.9}$$

Here, **T** is a symmetric $J \times J$ matrix and **t** is a J-vector. The respective typical elements of **T** and of **t** are population product totals that we denote

$$t_{jj'} = \sum\nolimits_U \frac{x_{jk}x_{j'k}}{\sigma_k^2} = t_{j'j}; \qquad t_{j0} = \sum\nolimits_U \frac{x_{jk}y_k}{\sigma_k^2} \tag{6.4.10}$$

The π estimators (thus unbiased) for **T** and **t**, respectively, are

$$\hat{\mathbf{T}} = \sum\nolimits_s \frac{\mathbf{x}_k\mathbf{x}_k'}{\sigma_k^2\pi_k}; \qquad \hat{\mathbf{t}} = \sum\nolimits_s \frac{\mathbf{x}_k y_k}{\sigma_k^2\pi_k} \tag{6.4.11}$$

with the typical elements

$$\begin{aligned} \hat{t}_{jj',\pi} &= \sum\nolimits_s \frac{x_{jk}x_{j'k}}{\sigma_k^2\pi_k} = \hat{t}_{j'j,\pi} \\ \hat{t}_{j0,\pi} &= \sum\nolimits_s \frac{x_{jk}y_k}{\sigma_k^2\pi_k} \end{aligned} \tag{6.4.12}$$

These are unbiased, respectively, for $t_{jj'}$ and t_{j0}. By the π estimation principle of Chapter 5, the population parameter **B** is now estimated by

$$\begin{aligned} \hat{\mathbf{B}} &= (\hat{B}_1, \ldots, \hat{B}_J)' \\ &= \hat{\mathbf{T}}^{-1}\hat{\mathbf{t}} \\ &= \left(\sum\nolimits_s \frac{\mathbf{x}_k\mathbf{x}_k'}{\sigma_k^2\pi_k}\right)^{-1} \sum\nolimits_s \frac{\mathbf{x}_k y_k}{\sigma_k^2\pi_k} \end{aligned} \tag{6.4.13}$$

which is the expression proposed in (6.4.2). Thus, in brief, $\hat{\mathbf{B}}$ estimates, but with a certain bias, the population parameter **B**, and **B** in turn would have estimated the model parameter $\boldsymbol{\beta}$ under a hypothetical complete enumeration.

EXAMPLE 6.4.2. It is not difficult to derive the regression estimators generated by the models given by (6.4.5) and (6.4.6). The reader is encouraged to carry out the details of the derivation. For the model (6.4.5), we get

$$\hat{B} = \frac{\sum_s \check{y}_k}{\sum_s \check{x}_k}$$

and, consequently,

$$\hat{t}_{yr} = \sum_U x_k \frac{\sum_s \check{y}_k}{\sum_s \check{x}_k} \tag{6.4.14}$$

This is the ratio estimator (seen already in Section 5.6, although the auxiliary variable there was called z, not x). For the model with intercept described by (6.4.6), we have $\mathbf{x}_k = (1, x_k)'$, $\mathbf{B} = (B_1, B_2)'$ and

$$\hat{\mathbf{B}} = \begin{pmatrix} \hat{B}_1 \\ \hat{B}_2 \end{pmatrix} = \begin{pmatrix} \tilde{y}_s - \hat{B}_2 \tilde{x}_s \\ \hat{B}_2 \end{pmatrix}$$

where $\hat{B}_2$ has the expression familiar from Section 5.10,

$$\hat{B}_2 = \frac{\sum_s (x_k - \tilde{x}_s)(y_k - \tilde{y}_s)/\pi_k}{\sum_s (x_k - \tilde{x}_s)^2/\pi_k} \tag{6.4.15}$$

with

$$\tilde{x}_s = \frac{\sum_s \check{x}_k}{\hat{N}}; \qquad \tilde{y}_s = \frac{\sum_s \check{y}_k}{\hat{N}}; \qquad \hat{N} = \sum_s \frac{1}{\pi_k}$$

Consequently,

$$\hat{t}_{yr} = N[\tilde{y}_s + \hat{B}_2(\bar{x}_U - \tilde{x}_s)] \tag{6.4.16}$$

Here, the information required for the whole population is N and $N\bar{x}_U = \sum_U x_k$, the totals of the two x variables. (The first x variable is constant at unity.)

Remark 6.4.2.

(a) It is assumed that the matrices $\sum_U \mathbf{x}_k \mathbf{x}_k'/\sigma_k^2$ and $\sum_s \mathbf{x}_k \mathbf{x}_k'/\sigma_k^2 \pi_k$ are nonsingular, so their inverses exist.

(b) Note that $\hat{\mathbf{B}}$ given by (6.4.2) is not a design-unbiased estimator of $\mathbf{B}$ given by (6.4.7). The expectation of the inverse of a random matrix is not equal to the inverse of its expectation.

(c) It is necessary that, given a sample, we can compute $\hat{B}$ from equation (6.4.2). Then $\hat{B}$ must depend on known quantities only, and therefore $\sigma_1^2, \ldots, \sigma_N^2$ must conform to certain requirements. We could require that all the σ_k^2 are known, or that $\sigma_k^2 = v_k \sigma^2$ with σ^2 unknown and $v_1, \ldots, v_N$ known. (A special case is when $v_k = 1$ for all $k \in U$.) There are other important cases where the variance structure depends on several unknown parameters that vanish when $\hat{B}$ is derived. Examples will be seen in Chapter 7.

(d) Since the auxiliary variables are assumed known for all elements in the population, the matrix $\sum_U \mathbf{x}_k \mathbf{x}_k'/\sigma_k^2$ could be calculated exactly if the σ_k^2 are known. However, it is preferable to use the estimate $\sum_s \mathbf{x}_k \mathbf{x}_k'/\sigma_k^2 \pi_k$.

Remark 6.4.3. To compute the regression estimator shown in equation (6.4.1) it suffices to know the auxiliary variable totals t_{x_j} $(j = 1, \ldots, J)$, and the individual auxiliary values $\mathbf{x}_k$ only for the sampled elements $k \in s$. We do not need individual $\mathbf{x}_k$-values for $k \in U - s$, the nonsample set. In other words, if the correct totals t_{x_j} can be obtained from some other source, we can still use the regression estimator.

If erroneous totals are used, the estimator is biased. Assume that y_k and $\mathbf{x}_k$ are observed for elements in the sample, and y_k and $\mathbf{x}_k$ are unknown for elements not in the sample. The totals t_y and t_{x_j} $(j = 1, \ldots, J)$ are also unknown. But suppose that we have approximations to the unknown x-totals, obtained from some external source. We may, for example, have population totals $t^*_{x_j}$ $(j = 1, \ldots, J)$ from a previous census, or we may have estimates $\hat{t}^*_{x_j}$ $(j = 1, \ldots, J)$ from some other sample survey. Here, $t^*_{x_j}$ is ordinarily not identical to the desired $t_{x_j} = \sum_U x_{jk}$, because $t^*_{x_j}$ refers to a different point in time. The definition of the x variable may be somewhat different at the two points in time. Or the population over which the sum $t^*_{x_j}$ is formed may not be exactly identical to U. Regression estimators formed with $t^*_{x_j}$ or $\hat{t}^*_{x_j}$ instead of the desired t_{x_j} should be treated with caution. They can be severely biased.

Remark 6.4.4. It is not necessary to build the regression estimator, as we have done, around the π estimators in $y, x_1, \ldots, x_J$. Other unbiased estimators could have been used, for example, the respective pwr estimators, if sampling had been done with replacement.

The general regression estimation technique described here has evolved over the last 15 years or so. Important references, which we attempt to synthesize in this chapter, include Cassel, Särndal, and Wretman (1976), Särndal (1980, 1982), Isaki and Fuller (1982), and Wright (1983).

6.5. Alternative Expressions for the Regression Estimator

The regression estimator defined in equation (6.4.1) can be written in several different forms. The different forms that we examine now facilitate the understanding of the estimator. The definition (6.4.1) drew attention to the fact that the regression estimator consists of the π estimator plus an adjustment term, negatively correlated with the π estimator. Therefore, the regression estimator $\hat{t}_{yr}$ is usually more precise than the π estimator itself.

Let us now consider a second suggestive expression for the regression estimator. Given the sample s, the sample fit of the model ξ produces $\hat{\mathbf{B}}$, and, for $k = 1, \ldots, N$, the fitted values

$$\hat{y}_k = \mathbf{x}_k'\hat{\mathbf{B}} = \sum_{j=1}^{J} \hat{B}_j x_{jk} \tag{6.5.1}$$

as well as, for $k \in s$, the sample fit residuals

$$e_{ks} = y_k - \hat{y}_k \tag{6.5.2}$$

Note that $\hat{y}_k$ and e_{ks} depend on s through $\hat{\mathbf{B}}$. Note also that $\hat{y}_k$ is available for all $k = 1, \ldots, N$, since $\mathbf{x}_1, \ldots, \mathbf{x}_N$ are known. By contrast, the residuals can only be computed for $k \in s$.

The regression estimator can now be expressed as

$$\begin{aligned} \hat{t}_{yr} &= \sum_U \hat{y}_k + \sum_s e_{ks}/\pi_k \\ &= \sum_U \hat{y}_k + \sum_s \check{e}_{ks} \end{aligned} \tag{6.5.3}$$

This form, reminiscent of equation (6.3.4) for the difference estimator, shows the regression estimator as the sum of a population total of fitted values,

$$\sum_U \hat{y}_k = \left(\sum_U \mathbf{x}_k\right)'\hat{\mathbf{B}} \tag{6.5.4}$$

and an adjustment term, $\sum_s \check{e}_{ks}$. Only in the term (6.5.4) is auxiliary information needed at the level of the population; the adjustment term is calculated from the sample alone.

If the linear relationship is perfect, that is,

$$y_k = \sum_{j=1}^{J} B_j x_{jk}$$

for $k = 1, \ldots, N$, the term $\sum_s \check{e}_{ks}$ is, of course, zero, and $\hat{t}_{yr} = t_y$, which is an error-free estimate. The term $\sum_s \check{e}_{ks}$ vanishes in many applications even if the linear relationship is less than perfect, so that the estimator takes the attractively simple form

$$\hat{t}_{yr} = \sum_U \hat{y}_k \tag{6.5.5}$$

This simple form is obtained when the model has the structure given in the following result, where $\mathscr{S}_0$ denotes the set of all samples that can arise under a design with fixed inclusion probabilities $\pi_1, \ldots, \pi_N$.

Result 6.5.1. *A sufficient condition that*

$$\sum_s \check{e}_{ks} = 0 \tag{6.5.6}$$

for all $s \in \mathscr{S}_0$ is that there exists a constant column vector $\boldsymbol{\lambda}$ (of dimension J and not depending on k) such that, for all $k \in U$,

$$\sigma_k^2 = \boldsymbol{\lambda}'\mathbf{x}_k \tag{6.5.7}$$

Proof. We have, for any sample s,

$$\sum_s \check{e}_{ks} = \sum_s y_k/\pi_k - \left(\sum_s \mathbf{x}_k'/\pi_k\right)\hat{\mathbf{B}}$$

Now, utilizing the definition (6.4.2) of $\hat{\mathbf{B}}$ and the fact that $\sigma_k^2 = \boldsymbol{\lambda}'\mathbf{x}_k$ for all $k \in U$,

$$\begin{aligned}(\textstyle\sum_s \mathbf{x}_k'/\pi_k)\hat{\mathbf{B}} &= (\textstyle\sum_s \boldsymbol{\lambda}'\mathbf{x}_k\mathbf{x}_k'/\sigma_k^2\pi_k)\hat{\mathbf{B}} \\ &= \boldsymbol{\lambda}'(\textstyle\sum_s \mathbf{x}_k\mathbf{x}_k'/\sigma_k^2\pi_k)\hat{\mathbf{B}} \\ &= \boldsymbol{\lambda}'(\textstyle\sum_s \mathbf{x}_k y_k/\sigma_k^2\pi_k) \\ &= \textstyle\sum_s \boldsymbol{\lambda}'\mathbf{x}_k y_k/\sigma_k^2\pi_k \\ &= \textstyle\sum_s y_k/\pi_k\end{aligned}$$

It follows that $\sum_s \check{e}_{ks} = 0$.

EXAMPLE 6.5.1. Simple examples of variance structures satisfying equation (6.5.7) are as follows:

i. σ_k^2 is constant, and the regression model has an intercept. That is, $\sigma_k^2 = \sigma^2$ and $x_{1k} = 1$ for all $k \in U$.
ii. σ_k^2 is proportional to one of the x variables, that is, for some variable x_j $(j = 1, \ldots, J)$,

$$\sigma_k^2 \propto x_{jk}$$

for all k.
iii. σ_k^2 is proportional to a linear combination of the x variables, that is,

$$\sigma_k^2 \propto \sum_{j=1}^{J} a_j x_{jk}$$

for all $k \in U$ and some constants $a_1, \ldots, a_J$.

Many of the applications considered later conform to one of these variance specifications. Examples include the two models (6.4.5) and (6.4.6).

Our third and fourth expressions for $\hat{t}_{yr}$ use matrix algebra. We introduce the J-vector $\mathbf{t}_x = (t_{x_1}, \ldots, t_{x_J})'$ and its π estimator, $\hat{\mathbf{t}}_{x\pi} = (\hat{t}_{x_1\pi}, \ldots, \hat{t}_{x_J\pi})'$. With this notation we can write

$$\hat{t}_{yr} = \hat{t}_{y\pi} + (\mathbf{t}_x - \hat{\mathbf{t}}_{x\pi})'\hat{\mathbf{B}} \tag{6.5.8}$$

Expressing $\hat{\mathbf{B}}$ by means of (6.4.11) and (6.4.13), we obtain

$$\begin{aligned}\hat{t}_{y\pi} + (\mathbf{t}_x - \hat{\mathbf{t}}_{x\pi})'\hat{\mathbf{B}} &= \textstyle\sum_s \check{y}_k + (\mathbf{t}_x - \hat{\mathbf{t}}_{x\pi})'\hat{\mathbf{T}}^{-1}\textstyle\sum_s \mathbf{x}_k\check{y}_k/\sigma_k^2 \\ &= \textstyle\sum_s [1 + (\mathbf{t}_x - \hat{\mathbf{t}}_{x\pi})'\hat{\mathbf{T}}^{-1}\mathbf{x}_k/\sigma_k^2]\check{y}_k\end{aligned}$$

In other words, the regression estimator can be expressed as a linear funtion of the π expanded values $\check{y}_k$ for $k \in s$,

$$\hat{t}_{yr} = \textstyle\sum_s g_{ks}\check{y}_k = \textstyle\sum_s g_{ks} y_k/\pi_k \tag{6.5.9}$$

with the sample dependent weights

$$g_{ks} = 1 + (\mathbf{t}_x - \hat{\mathbf{t}}_{x\pi})'\hat{\mathbf{T}}^{-1}\mathbf{x}_k/\sigma_k^2 \tag{6.5.10}$$

where $\hat{\mathbf{T}}$ is given by (6.4.11). The total weight given to the observed value y_k

in (6.5.9) is the product of two weights, the sampling weight $1/\pi_k$ and the g weight g_{ks} given by equation (6.5.10). The form shown in (6.5.9) reminds us of the π estimator, but since the g_{ks} depend on the sample, there is an essential difference.

EXAMPLE 6.5.2. Returning to the single x variable models (6.4.5) and (6.4.6), we find after some calculation that for the model (6.4.5),

$$g_{ks} = \frac{\sum_U x_k}{\sum_s \check{x}_k} = \frac{N\bar{x}_U}{\hat{N}\tilde{x}_s}$$
$$= \frac{N}{\hat{N}}\left[1 + \frac{\bar{x}_U - \tilde{x}_s}{\tilde{x}_s}\right] \tag{6.5.11}$$

In this case, the g weight is constant for all $k \in s$, but it varies with s.

For the model (6.4.6) we obtain the g weight

$$g_{ks} = (N/\hat{N})[1 + a_s(x_k - \tilde{x}_s)] \tag{6.5.12}$$

with

$$a_s = \frac{\bar{x}_U - \tilde{x}_s}{\tilde{S}^2_{xs}}$$

where

$$\tilde{S}^2_{xs} = \frac{\sum_s (x_k - \tilde{x}_s)^2/\pi_k}{\sum_s 1/\pi_k}$$

is the π-weighted sample variance. All other notation is as in Example 6.4.2.

Our final expression for the regression estimator relates to the hypothetical population fit of the model ξ, which produces $\mathbf{B}$ given by equation (6.4.7), and, for $k = 1, \ldots, N$, the fitted values

$$y_k^0 = \mathbf{x}_k'\mathbf{B} \tag{6.5.13}$$

and the population fit residuals

$$E_k = y_k - y_k^0 \tag{6.5.14}$$

Since $y_k = y_k^0 + E_k$, we have, from (6.5.9),

$$\hat{t}_{yr} = \sum_s g_{ks}(\check{y}_k^0 + \check{E}_k) \tag{6.5.15}$$

We also notice that

$$\sum_s g_{ks}\check{\mathbf{x}}_k' = \sum_s \left[1 + (\mathbf{t}_x - \hat{\mathbf{t}}_{x\pi})'\hat{\mathbf{T}}^{-1}\frac{\mathbf{x}_k}{\sigma_k^2}\right]\frac{\mathbf{x}_k'}{\pi_k}$$
$$= \sum_s \check{\mathbf{x}}_k' + (\mathbf{t}_x - \hat{\mathbf{t}}_{x\pi})'\hat{\mathbf{T}}^{-1}\hat{\mathbf{T}}$$
$$= \hat{\mathbf{t}}_{x\pi}' + \mathbf{t}_x' - \hat{\mathbf{t}}_{x\pi}'$$
$$= \sum_U \mathbf{x}_k' \tag{6.5.16}$$

Multiplying both sides of equation (6.5.16) by the vector $\mathbf{B}$, we get

$$\textstyle\sum_s g_{ks}\check{y}_k^0 = (\sum_s g_{ks}\check{\mathbf{x}}_k')\mathbf{B} = (\sum_U \mathbf{x}_k')\mathbf{B} = \sum_U y_k^0 \tag{6.5.17}$$

It now follows from (6.5.15) and (6.5.17) that

$$\hat{t}_{yr} = \textstyle\sum_U y_k^0 + \sum_s g_{ks}\check{E}_k \tag{6.5.18}$$

This final expression will be useful when we discuss variance estimation in the following section.

We conclude by listing the various expressions for the regression estimator seen in this section.

$$\begin{aligned}
\hat{t}_{yr} &= \hat{t}_{y\pi} + \sum_{j=1}^{J} \hat{B}_j(t_{x_j} - \hat{t}_{x_j\pi}) && \\
&= \hat{t}_{y\pi} + (\mathbf{t}_x - \hat{\mathbf{t}}_{x\pi})'\hat{\mathbf{B}} && \text{compare (6.5.8)}\\
&= \textstyle\sum_s g_{ks}\check{y}_k && \text{compare (6.5.9)}\\
&= \textstyle\sum_U \hat{y}_k + \sum_s \check{e}_{ks} && \text{compare (6.5.3)}\\
&= \textstyle\sum_U y_k^0 + \sum_s g_{ks}\check{E}_k && \text{compare (6.5.18)}
\end{aligned}$$

Moreover, when the model satisfies $\sigma_k^2 = \boldsymbol{\lambda}'\mathbf{x}_k$ for all k, as in Result 6.5.1, then

$$\begin{aligned}
\hat{t}_{yr} &= \textstyle\sum_U \hat{y}_k && \text{compare (6.5.5)}\\
&= \textstyle\sum_s g_{ks}\check{y}_k && \text{compare (6.5.9)}
\end{aligned}$$

with the simplified g weights

$$g_{ks} = (\textstyle\sum_U \mathbf{x}_k)'\hat{\mathbf{T}}^{-1}\mathbf{x}_k/\sigma_k^2$$

Remark 6.5.1. Interestingly, when the weights g_{ks} are applied, in the manner of equation (6.5.9), to any one of the π-weighted explanatory variables, the result is an error-free estimate of that variable's population total. We have, for $j = 1, \ldots, J$,

$$\textstyle\sum_s g_{ks}\check{x}_{jk} = \sum_U x_{jk} = t_{x_j}$$

as is immediately clear from (6.5.16). Intuitively then, since the weights g_{ks} give perfect estimates for every one of the J explanatory variables, they must be nearly perfect for a regressor variable y expressible, except for minor error, as a linear combination of these explanatory variables.

6.6. The Variance of the Regression Estimator

We judge estimators by their design-based qualities, such as design expectation and design variance, under repeated sampling with a given sampling design from the fixed finite population. The regression estimator is no exception. We are thus interested in finding the statistical properties of the estimator (6.5.3) with respect to the sampling design. Usually, this cannot be done exactly, because of the complex nature of the regression estimator; we rely on approximate techniques, as in Chapter 5.

The regression estimator is not unbiased. For large samples, however, it is approximately unbiased. This statement is supported by Taylor linearization, which also provides the basis for obtaining an approximate variance, as well as a variance estimator. Result 6.6.1, which follows, summarizes a highly general regression estimation procedure with a wide range of practical applications.

Result 6.6.1. *The regression estimator $\hat{t}_{yr}$, given by equation* (6.4.1) *or equivalently by equation* (6.5.3), *is approximated through Taylor linearization by*

$$\hat{t}_{yr0} = \hat{t}_{y\pi} + (\mathbf{t}_x - \hat{\mathbf{t}}_{x\pi})'\mathbf{B} \tag{6.6.1}$$

$$= \sum_U y_k^0 + \sum_s \check{E}_k \tag{6.6.2}$$

where $\check{E}_k = E_k/\pi_k$, with E_k given by (6.5.14). *The estimator $\hat{t}_{yr}$ is approximately unbiased for $t = \sum_U y_k$ with the approximate variance*

$$AV(\hat{t}_{yr}) = \sum\sum_U \Delta_{kl}\check{E}_k\check{E}_l \tag{6.6.3}$$

and the variance estimator

$$\hat{V}(\hat{t}_{yr}) = \sum\sum_s \check{\Delta}_{kl}(g_{ks}\check{e}_{ks})(g_{ls}\check{e}_{ls}) \tag{6.6.4}$$

where $\check{e}_{ks} = e_{ks}/\pi_k$ and e_{ks} and g_{ks} are given by (6.5.2) *and* (6.5.10), *respectively.*

An approximately $100(1-\alpha)\%$ confidence interval is created as usual with the help of the normal approximation, that is,

$$\hat{t}_{yr} \pm z_{1-\alpha/2}[\hat{V}(\hat{t}_{yr})]^{1/2}$$

where $z_{1-\alpha/2}$ is the $(1-\alpha/2)$th quantile of the standardized normal distribution.

We shall verify Result 6.6.1, but it is illustrative to first apply the result in the case of our two single x variable models. The reader is encouraged to verify the details of the following two examples.

Example 6.6.1. We return to the ratio model (6.4.5). This model generates the ratio estimator given by (6.4.14). The population fit residuals applicable in (6.6.3) are

$$E_k = y_k - Bx_k$$

where

$$B = \frac{\sum_U y_k}{\sum_U x_k}$$

The sample fit residuals necessary for calculating the variance estimator (6.6.4) are

$$e_{ks} = y_k - \hat{B}x_k$$

where $\hat{B} = \sum_s \check{y}_k / \sum_s \check{x}_k$. For every $k \in s$, the g weight is given by (6.5.11). From (6.6.4), we obtain

$$\hat{V}(\hat{t}_{yr}) = \left(\frac{N\bar{x}_U}{\hat{N}\tilde{x}_s}\right)^2 \sum\sum_s \check{\Delta}_{kl}\check{e}_{ks}\check{e}_{ls}$$

In practice, this expression is worked out for each design of interest. For example, under the SI design, the variance estimator becomes

$$\hat{V}(\hat{t}_{yr}) = N^2\left(\frac{\bar{x}_U}{\bar{x}_s}\right)^2 \frac{1-f}{n} \frac{\sum_s (y_k - \hat{B}x_k)^2}{n-1}$$

where $\hat{B} = \bar{y}_s/\bar{x}_s$, and $n = fN$ is the fixed sample size.

Example 6.6.2. For the homoscedastic model with intercept given by (6.4.6), the sample fit residuals are

$$e_{ks} = y_k - \tilde{y}_s - \hat{B}_2(x_k - \tilde{x}_s)$$

with $\hat{B}_2$ given by (6.4.15). Noting also that the g weights are given by (6.5.12), we have all the necessary material for calculating the variance estimator (6.6.4).

Let us now verify Result 6.6.1.

Proof. Let us identify the different π estimators on which $\hat{t}_{yr}$ depends. We can write

$$\begin{aligned}\hat{t}_{yr} &= \hat{t}_{y\pi} + (\mathbf{t}_x - \hat{\mathbf{t}}_{x\pi})'\hat{\mathbf{B}}\\ &= \hat{t}_{y\pi} + (\mathbf{t}_x - \hat{\mathbf{t}}_{x\pi})'\hat{\mathbf{T}}^{-1}\hat{\mathbf{t}}\\ &= f(\hat{t}_{y\pi}, \hat{\mathbf{t}}_{x\pi}, \hat{\mathbf{T}}, \hat{\mathbf{t}})\end{aligned}$$

Thus $\hat{t}_{yr}$ is a nonlinear function of the following π estimators: $\hat{t}_{y\pi}$, the J π estimators $\hat{t}_{xj\pi}$ that are components of $\hat{\mathbf{t}}_{x\pi}$, and the $J(J+1)/2 + J$ different π estimators $\hat{t}_{jj',\pi}$ and $\hat{t}_{j0,\pi}$ given by (6.4.12) that appear as components of $\hat{\mathbf{T}}$ and $\hat{\mathbf{t}}$. Using the Taylor linearization we approximate this function by a linear one. The following partial derivatives are needed (see, for example, Graybill, 1983, for the rules of differentiation)

$$\frac{\partial f}{\partial \hat{t}_{y\pi}} = 1 \tag{6.6.5}$$

$$\frac{\partial f}{\partial \hat{t}_{xj\pi}} = -\hat{B}_j; \qquad j = 1, \ldots, J \tag{6.6.6}$$

$$\frac{\partial f}{\partial \hat{t}_{jj',\pi}} = (\mathbf{t}_x - \hat{\mathbf{t}}_{x\pi})'(-\hat{\mathbf{T}}^{-1}\boldsymbol{\Lambda}_{jj'}\hat{\mathbf{T}}^{-1})\hat{\mathbf{t}}; \qquad j \le j' = 1, \ldots, J, \tag{6.6.7}$$

$$\frac{\partial f}{\partial \hat{t}_{j0,\pi}} = (\mathbf{t}_x - \hat{\mathbf{t}}_{x\pi})'\hat{\mathbf{T}}^{-1}\boldsymbol{\lambda}_j; \qquad j = 1, \ldots, J \tag{6.6.8}$$

where $\boldsymbol{\lambda}_j$ is a J-vector with the jth component equal to one, and zeros elsewhere; and $\boldsymbol{\Lambda}_{jj'}$ is a $J \times J$ matrix with the value 1 in positions (j, j') and (j', j), and the value 0 everywhere else.

Proceeding as in the proof of Result 5.10.1, we evaluate these partial derivatives at the expected value point where $\hat{t}_{y\pi} = t_y$, $\hat{\mathbf{t}}_{x\pi} = \mathbf{t}_x$, $\hat{\mathbf{T}} = \mathbf{T}$, and $\hat{\mathbf{t}} = \mathbf{t}$. The partial derivatives given by (6.6.7) and (6.6.8) conveniently vanish at this point, and we obtain simply

$$\begin{aligned} \hat{t}_{yr} \doteq \hat{t}_{yr0} &= t_y + 1(\hat{t}_{y\pi} - t_y) - \sum_{j=1}^{J} B_j(\hat{t}_{x_j\pi} - t_{x_j}) \\ &= \hat{t}_{y\pi} + (\mathbf{t}_x - \hat{\mathbf{t}}_{x\pi})'\mathbf{B} \\ &= \textstyle\sum_U y_k^0 + \sum_s \check{E}_k \end{aligned} \tag{6.6.9}$$

that is, a constant term plus a π estimator in E_k. Consequently,

$$E(\hat{t}_{yr}) \doteq E(\hat{t}_{yr0}) = \textstyle\sum_U y_k^0 + E(\sum_s \check{E}_k) = t_y$$

which indicates that the regression estimator is approximately unbiased. Furthermore,

$$\begin{aligned} AV(\hat{t}_{yr}) &= V(\hat{t}_{yr0}) \\ &= V(\textstyle\sum_s \check{E}_k) \\ &= \textstyle\sum\sum_U \Delta_{kl}\check{E}_k\check{E}_l \end{aligned}$$

The variance estimator $\hat{V}(\hat{t}_{yr})$ is motivated by the exact expression given in (6.5.18) for $\hat{t}_{yr}$,

$$\hat{t}_{yr} = \textstyle\sum_U y_k^0 + \sum_s g_{ks}\check{E}_k$$

from which it follows that

$$V(\hat{t}_{yr}) = V(\textstyle\sum_s g_{ks}\check{E}_k) \tag{6.6.10}$$

This resembles the variance of a π estimator, with the difference that the sample dependent weights g_{ks} are now attached to the $\check{E}_k$. This prevents a direct application of π estimation principles for variance estimation. However, disregarding that the weights are sample dependent, and inserting $\check{e}_{ks}$ for $\check{E}_k$, we obtain the variance estimator given in equation (6.6.4). □

Remark 6.6.1. If π-estimation principles for variance estimation are applied on

$$AV(\hat{t}_{yr}) = \textstyle\sum\sum_U \Delta_{kl}\check{E}_k\check{E}_l$$

and, in addition, the unobservable $\check{E}_k = E_k/\pi_k$ is replaced by the observable $\check{e}_{ks} = e_{ks}/\pi_k$, we obtain the alternative variance estimator

$$\widehat{AV}(\hat{t}_{yr}) = \textstyle\sum\sum_s \check{\Delta}_{kl}\check{e}_{ks}\check{e}_{ls} \tag{6.6.11}$$

Both (6.6.4) and (6.6.11) are based on large sample approximations. Both formulas serve an important goal in the design-based inference. For a given nominal level $1 - \alpha$, the usual confidence interval based on the normal approximation and either of the two variance estimators gives approximately $100(1 - \alpha)\%$ coverage rate in repeated large samples. However, in a number of cases, the available evidence suggests that (6.6.4) is preferable. This formula was proposed in Särndal (1982), and its properties with respect to the design and with respect to the model were studied in Särndal, Swensson and Wretman (1989), who show that the estimator (6.6.4) is design consistent, as well as approximately unbiased with respect to the underlying model. In the applications of Result 6.6.1 given in Chapters 7 and 8, we usually present only the variance estimator $\hat{V}(\hat{t}_{yr})$ given by (6.6.4). A variance estimator somewhat similar in spirit to that in equation (6.6.4) was proposed by Kott (1990). His point of departure is to create a variance estimator that is unbiased with respect to the model but is still design consistent. The objective is achieved by attaching a ratio adjustment to the estimator (6.6.11).

A detailed discussion of asymptotic properties of the regression estimator $\hat{t}_{yr}$, such as consistency and asymptotic unbiasedness, is given in Robinson and Särndal (1983) and Leblond (1990).

Remark 6.6.2. It is true under mild conditions for the regression estimator $\hat{t}_{yr}$ given by (6.5.3) that both the bias and the variance are of order n^{-1}, where n is the fixed-sample size. It follows that the MSE and the variance of $\hat{t}_{yr}$ are the same, to the order of approximation n^{-1}. Moreover, the bias ratio (the bias divided by the standard error) tends to zero as $n^{-1/2}$. The bias ratio is often numerically small even for modest sample sizes, an important property of the regression estimator. As remarked in Section 5.2, the bias ratio affects the validity of confidence intervals. Since the bias ratio is small for the regression estimator, the confidence interval

$$\hat{t}_{yr} \pm z_{1-\alpha/2}[\hat{V}(\hat{t}_{yr})]^{1/2}$$

will be approximately valid even for quite modest size samples. For an estimator to be consistent, it suffices that both the bias and the variance tend to zero. Thus, it is possible for an estimator to be consistent and have a bias ratio that does not tend to zero, or that tends to zero at a rate slower than $n^{-1/2}$. Consistent estimators of this kind are considered in Little (1983a).

6.7. Comments on the Role of the Model

Let us summarize the role of the underlying model ξ for determining the properties of the regression estimator. The role of the model may be examined from two points of view, (i) with respect to bias, and (ii) with respect to variance. We have seen that the regression estimator $\hat{t}_{yr}$ behaves approximately as the linearized random variable $\hat{t}_{yr0}$ under repeated sampling from a given population with fixed values on all variables. Since $\hat{t}_{yr0}$ is unbiased for t_y irrespective of the shape of the finite population point scatter, it follows that the regression estimator $\hat{t}_{yr}$ is approximately unbiased for t_y, irrespective of whether the assumptions of the model ξ are true or false.

On the other hand, the appropriateness of the model ξ is crucial to achieve a small variance, as the following argument indicates. The more the population scatter conforms to a linear pattern $y_k \doteq \mathbf{x}_k'\mathbf{B}$, the smaller the population fit residuals $E_k = y_k - \mathbf{x}_k'\mathbf{B}$, and the smaller the variance of the regression estimator. This may be seen by regarding the expression (6.6.10) for the variance. In the extreme case when all E_k are zero, the variance is also zero. For an ill-fitting model, the residuals E_k can be substantial, and the variance (6.6.10) can be large.

Note that the x-variables to be utilized in the regression estimator have to be chosen from the set of auxiliary variables that are available. The crucial question for the efficiency of the regression estimator is whether an $\mathbf{x}$-vector with a known total $\sum_U \mathbf{x}_k$ is also a powerful explanatory vector for the y variable.

We do not require that the model be "true" in the sense of correctly depicting some process by which the population data may have been generated. We only believe that the population data can be fairly well described by the model. If the population data are well described by the assumed model, the regression estimator normally will bring about a large variance reduction, as compared to the π estimator. If the population is not well described by the model, the improvement on the π estimator may be modest, but the regression estimator still guarantees approximate unbiasedness. For these reasons, the regression estimator is said to be model assisted, but not model dependent.

A well specified variance structure, that is, the set of quantities $\sigma_1^2, \ldots, \sigma_N^2$, is also important, but the importance lies above all in the planning of the survey. The model variances σ_k^2, have a bearing on the optimal choice of sampling design. We show in Section 12.2 that a design is optimal, in a given sense, if the inclusion probabilities satisfy $\pi_k \propto \sigma_k$. That is, one should assign a high inclusion probability to an element about which the uncertainty σ_k is high.

6.8. Optimal Coefficients for the Difference Estimator

Section 6.3 presented some important conclusions about the difference estimator $\hat{t}_{y,\text{dif}}$ given by (6.3.4) or, alterntively, (6.3.9). However, we did not discuss the optimal choice of the constants $A_1, \ldots, A_J$. Rather, we required that these constants be independent of sample data and determined, in principle, before sampling. We now show that theoretically there is a best choice of A_1, $\ldots$, A_J, that is, a choice that minimizes the variance of the difference estimator. The result is due to Montanari (1987). The optimal A_j-values depend on unknown population characteristics, so the optimal estimator cannot be used. In the absence of good a priori knowledge of these characteristics, we have two options: One is to use the regression estimation approach given in Section 6.4. This approach plays a central role in this book. Another is to replace the optimal A_j-values by sample-based estimates, as illustrated in an example later in this section.

The variance of the difference estimator (6.3.9) can alternatively be expressed as

$$V(\hat{t}_{y,\mathrm{dif}}) = V(\hat{t}_{y\pi}) + \sum_{j=1}^{J}\sum_{j'=1}^{J} A_j A_{j'} C(\hat{t}_{x_j\pi}, \hat{t}_{x_{j'}\pi}) - 2\sum_{j=1}^{J} A_j C(\hat{t}_{y\pi}, \hat{t}_{x_j\pi}) \tag{6.8.1}$$

For compactness we introduce the following vector and matrix notation:

$$\mathbf{A} = (A_1, \ldots, A_J)'$$
$$\mathbf{\Gamma} = (\gamma_1, \ldots, \gamma_J)'$$
$$\mathbf{\Lambda} = (\lambda_{jj'})$$

where the A_j are the constants of the difference estimator,

$$\gamma_j = C(\hat{t}_{y\pi}, \hat{t}_{x_j\pi}) \qquad (j = 1, \ldots, J)$$

and the typical element of the symmetric $J \times J$ matrix $\mathbf{\Lambda}$ is

$$\lambda_{jj'} = C(\hat{t}_{x_j\pi}, \hat{t}_{x_{j'}\pi}) \qquad (j, j' = 1, \ldots, J)$$

The variance shown in (6.8.1) can now be written compactly as

$$V(\hat{t}_{y,\mathrm{dif}}) = V(\hat{t}_{y\pi}) + \mathbf{A}'\mathbf{\Lambda}\mathbf{A} - 2\mathbf{\Gamma}'\mathbf{A} \tag{6.8.2}$$

Assuming $\mathbf{\Lambda}$ to be nonsingular, we can rewrite equation (6.8.2) as

$$V(\hat{t}_{y,\mathrm{dif}}) = V(\hat{t}_{y\pi}) - \mathbf{\Gamma}'\mathbf{\Lambda}^{-1}\mathbf{\Gamma} + (\mathbf{A} - \mathbf{\Lambda}^{-1}\mathbf{\Gamma})'\mathbf{\Lambda}(\mathbf{A} - \mathbf{\Lambda}^{-1}\mathbf{\Gamma}) \tag{6.8.3}$$

The first term of (6.8.3) is not affected by the choice of $\mathbf{A}$, whereas the second term is greater than or equal to zero, with equality if and only if $\mathbf{A} = \mathbf{\Lambda}^{-1}\mathbf{\Gamma}$. Thus, provided that $\mathbf{\Lambda}$ is nonsingular, the variance of the difference estimator is minimized if we choose $\mathbf{A}$ equal to

$$\mathbf{A}_{\mathrm{opt}} = \mathbf{\Lambda}^{-1}\mathbf{\Gamma} \tag{6.8.4}$$

The resulting minimum variance is

$$V_{\min} = V(\hat{t}_{y\pi}) - \mathbf{\Gamma}'\mathbf{\Lambda}^{-1}\mathbf{\Gamma} \tag{6.8.5}$$

Equation (6.8.5) is of interest if we see the difference estimator as a way to improve over the simple π estimator. Specifically, $\mathbf{\Gamma}'\mathbf{\Lambda}^{-1}\mathbf{\Gamma}$ is the amount by which the variance is reduced, if we use the difference estimator with optimal weights instead of the π estimator. In practice, one is ordinarily not free to choose $\mathbf{A}$ equal to $\mathbf{A}_{\text{opt}}$, because the required matrices $\mathbf{\Gamma}$ and $\mathbf{\Lambda}$ are unknown. A vector $\mathbf{A}$ not too far from $\mathbf{A}_{\text{opt}}$ might, however, bring about some efficiency gain over the π estimator. Note the implicit assumption in equations (6.8.4) and (6.8.5) that $\mathbf{\Lambda}^{-1}$ exists.

Remark 6.8.1. To avoid singularity in $\mathbf{\Lambda}$, none of the J components of the adjustment term of equation (6.3.9) is allowed to be identically zero. An identically zero term might arise in the following case. Suppose one of the auxiliary vector values (say x_{k1}) is constant, $x_{k1} = 1$ for all k, corresponding to an intercept term in the regression. Then

$$\hat{t}_{x_1\pi} = \textstyle\sum_s 1/\pi_k = \hat{N} \text{ and } t_{x_1} = N$$

Under designs such as *SI* and *STSI*, $\hat{N} = N$ for all samples, so that

$$A_1(t_{x_1} - \hat{t}_{x_1\pi}) = 0$$

for all s. The term is then naturally excluded from the estimator (6.3.9), to avoid singularity in $\mathbf{\Lambda}$.

EXAMPLE 6.8.1. We consider the situation with a single auxiliary variable x. Here $J = 1$, and we have $\mathbf{\Gamma} = \gamma$ and $\mathbf{\Lambda} = \lambda$, with

$$\gamma = C(\hat{t}_{y\pi}, \hat{t}_{x\pi}) = \textstyle\sum\sum_U \Delta_{kl} \check{x}_k \check{y}_l$$

$$\lambda = V(\hat{t}_{x\pi}) = \textstyle\sum\sum_U \Delta_{kl} \check{x}_k \check{x}_l$$

The optimum value of A is then

$$A_{\text{opt}} = \gamma/\lambda$$

(provided $\lambda > 0$), with minimum variance

$$\begin{aligned} V_{\min} &= V(\hat{t}_{y\pi}) - \gamma^2/\lambda \\ &= V(\hat{t}_{y\pi})(1 - \rho^2) \end{aligned}$$

where

$$\rho = \rho(\hat{t}_{x\pi}, \hat{t}_{y\pi})$$

is the correlation between the π estimators $\hat{t}_{x\pi}$ and $\hat{t}_{y\pi}$. Thus, the gain of precision relative to the π estimator is $[V(\hat{t}_{y\pi}) - V_{\min}]/V(\hat{t}_{y\pi}) = \rho^2$, which increases with ρ.

Since γ and λ are ordinarily unknown, a practical use of the result would require some form of estimation of $A_{\text{opt}} = \gamma/\lambda$. If we make separate estimates of γ and λ, by the usual principles for variance estimation of a π estimator, we obtain

$$\hat{t} = \hat{t}_{y\pi} + \hat{A}_{\text{opt}}(t_x - \hat{t}_{x\pi}) \tag{6.8.6}$$

with

$$\hat{A}_{\text{opt}} = \frac{\sum\sum_s \check{\Delta}_{kl}\check{y}_k\check{x}_l}{\sum\sum_s \check{\Delta}_{kl}\check{x}_k\check{x}_l}$$

No model has been used to arrive at this estimator.

We may compare the estimator (6.8.6) with the regression estimator (6.4.16), which is generated by the simple regression model (6.4.6). These two estimators agree for the *SI* design; both are then given by

$$\hat{t} = N[\bar{y}_s + \hat{B}(\bar{x}_U - \bar{x}_s)]$$

where

$$\hat{B} = \frac{S_{xys}}{S_{xs}^2} = \frac{\sum_s (x_k - \bar{x}_s)(y_k - \bar{y}_s)}{\sum_s (x_k - \bar{x}_s)^2}$$

But in general the two estimators differ; not even for the *STSI* design do they agree.

Exercises

6.1. According to a reliable source, the yearly total population increase for the CO124 population is between 1% and 2%. Propose a difference estimator for the total of the variable P83 ($= y$) with P80 ($= x$) as an auxiliary variable. Calculate the unbiased difference estimate for t_y, an unbiased variance estimate, and the *cve*, given that an *SI* sample of size $n = 40$ drawn from CO124 produced the following results:

$$\textstyle\sum_s y_k = 1{,}695.6; \quad \sum_s y_k^2 = 1{,}090{,}720.82; \quad \sum_s y_k x_k = 1{,}067{,}197.7$$

$$\textstyle\sum_s x_k = 1{,}635.6; \quad \sum_s x_k^2 = 1{,}044{,}258.52; \quad \sum_U x_k = 4{,}308.1$$

6.2. Suppose it is known that the Social Democrats have the support of roughly 50% of the Swedish voters. Use this information to construct a difference estimator of the total of the variable SS82 ($= y$) for the MU284 population. The auxiliary variable is S82 ($= x$). Calculate the difference estimate, an unbiased variance estimate, and the *cve*, using that a *BE* design with $\pi = 0.2$ resulted in a sample of size $n_s = 54$, and

$$\textstyle\sum_s y_k = 1{,}223; \quad \sum_s y_k^2 = 31{,}077; \quad \sum_s y_k x_k = 62{,}561$$

$$\textstyle\sum_s x_k = 2{,}628; \quad \sum_s x_k^2 = 133{,}390; \quad \sum_U x_k = 13{,}500$$

6.3. An estimate is required of the total of the variable P85 ($= y$) for the MU281 population. The plan is to use the auxiliary variables CS82 ($= x_1$) and SS82 ($= x_2$) in a difference estimator with $y_k^0 = 3x_{1k} + 2x_{2k}$. It can be assumed in the light of earlier studies that y_k^0 approximates y_k fairly well. The known totals of CS82 and SS82 are 2,508 and 6,153, respectively. The following information was calculated from an *SI* sample of $n = 35$ municipalities:

$$\textstyle\sum_s y_k = 1{,}019; \quad S_{ys}^2 = 742.34; \quad S_{yx_1s} = 82.24; \quad S_{yx_2s} = 152.86$$

$$\sum_s x_{1k} = 321; \quad \sum_s x_{2k} = 803; \quad S^2_{x_1s} = 20.44; \quad S^2_{x_2s} = 60.41; \quad S_{x_1x_2s} = 8.22$$

Calculate a confidence interval for t_y at the approximate level 95%.

6.4. Consider *SI* sampling of n from N, and the difference estimator

$$\hat{t}_{y,\text{dif}} = \hat{t}_{y\pi} + A(t_x - \hat{t}_{x\pi}) = N[\bar{y}_s + A(\bar{x}_U - \bar{x}_s)]$$

(a) Show that $A_{\text{opt}} = S_{yxU}/S^2_{xU}$, where A_{opt} is the value of A that minimizes the variance. (b) Show that if the correlation between y and x is positive, then $\hat{t}_{y,\text{dif}}$ has a smaller variance than $\hat{t}_{y\pi}$ if $0 < A < 2A_{\text{opt}}$. (c) For the population in Exercise 6.1, we have $S_{yxU} = 13{,}039.82$, $S^2_{xU} = 12{,}479.69$ and $S^2_{yU} = 13{,}642.42$. Calculate $V(\hat{t}_{y,\text{dif}})/V(\hat{t}_{y\pi})$ for $A = 0, 0.25, 0.5, 0.75, \ldots, 2.25$.

6.5. As an alternative to the difference estimator shown in equation (6.3.4), one might consider

$$\hat{t}_{y,\text{alt}} = \sum_s y_k + \sum_{U-s} y_k^0 = \sum_U y_k^0 + \sum_s D_k$$

where $D_k = y_k - y_k^0$. Compare the estimator (6.3.4) with the new estimator $\hat{t}_{y,\text{alt}}$ with respect to bias, variance, and mean square error. Assume *SI* sampling with $f = n/N$.

6.6. Verify the result of Example 6.3.1 stating that under the *SI* design with $f = n/N$ and with $y_k^0 = x_k$,

$$V(\hat{t}_{y,\text{dif}}) = N^2[(1-f)/n](S^2_{yU} + S^2_{xU} - 2S_{xyU})$$

6.7. Show that $V(\hat{t}_{y,\text{dif}}) < V(\hat{t}_\pi)$ if and only if

$$\rho(\hat{t}_{y^0\pi}, \hat{t}_{y\pi}) > \tfrac{1}{2}[V(\hat{t}_{y^0\pi})/V(\hat{t}_{y\pi})]^{1/2}$$

where $\hat{t}_{y^0\pi} = \sum_s y_k^0/\pi_k$.

6.8. Show that the vector $\mathbf{B}$ given by (6.4.7) can be expressed alternatively as

$$\begin{aligned}\mathbf{B} &= (\mathbf{X}\boldsymbol{\Sigma}^{-1}\mathbf{X}')^{-1}\mathbf{X}\boldsymbol{\Sigma}^{-1}\mathbf{Y} \\ &= \mathbf{T}^{-1}\mathbf{t}\end{aligned}$$

where $\mathbf{X}$, $\boldsymbol{\Sigma}$, $\mathbf{Y}$, $\mathbf{T}$, and $\mathbf{t}$ are defined in Section 6.4.

6.9. Show that $\hat{\mathbf{T}}$ and $\hat{\mathbf{t}}$ given by (6.4.11) are unbiased estimators of $\mathbf{T}$ and $\mathbf{t}$ given by (6.4.9).

6.10. For the model (6.4.5), show in detail that

$$\hat{B} = \sum_s \check{y}_k / \sum_s \check{x}_k$$

and hence, as stated in Example 6.4.3, that the regression estimator generated by this model is equal to the ratio estimator given by (6.4.14).

6.11. For the model (6.4.6), show in detail that $\hat{\mathbf{B}} = (\tilde{y}_s - \hat{B}_2\tilde{x}_s, \hat{B}_2)'$ where $\hat{B}_2$ is given by (6.4.15), and hence, as stated in Example 6.4.2, that the corresponding regression estimator is

$$\hat{t}_{yr} = N[\tilde{y}_s + \hat{B}_2(\bar{x}_U - \tilde{x}_s)]$$

6.12. Show that the regression estimator

$$\hat{t}_{yr} = \hat{t}_{y\pi} + \sum_{j=1}^{J} \hat{B}_j(t_{x_j} - \hat{t}_{x_j\pi})$$

can be expressed in the following three alternative ways,

$$\begin{aligned}\hat{t}_{yr} &= \textstyle\sum_s g_{ks}\check{y}_k \\ &= \textstyle\sum_U \hat{y}_k + \sum_s \check{e}_{ks} \\ &= \textstyle\sum_U y_k^0 + \sum_s g_{ks}\check{E}_k\end{aligned}$$

6.13. Show that the variance structures (i), (ii), and (iii) of Example 6.5.1 satisfy equation (6.5.7), and hence that $\sum_s \check{e}_{ks} = 0$ in these cases.

6.14. Show that it is always true that

$$\textstyle\sum_s g_{ks}\check{e}_{ks} = \sum_s \check{e}_{ks}$$

6.15. Use equation (6.5.10) to obtain the g weights (6.5.11) and (6.5.12) corresponding to the models (6.4.5) and (6.4.6), respectively.

CHAPTER 7

Regression Estimators for Element Sampling Designs

7.1 Introduction

This chapter presents a spectrum of specific applications of regression estimation, stated in the general form in the preceding chapter. We examine some linear models that are natural candidates for describing finite populations. We derive and discuss for element sampling designs the regression estimators that are generated by these models. These include the ratio estimator, simple and multiple regression estimators, and estimators for poststratification and raking. Regression estimators for cluster sampling and sampling in two stages are treated in Chapter 8, since the auxiliary information is usually of a different nature when such designs become necessary.

This chapter does not cover all interesting models and designs. However, it should put the reader in a good position to apply the reasoning in other situations, with auxiliary data and sampling designs other than those exposed in this chapter. The technique of regression estimation is a general tool with wide applicability.

7.2. Preliminary Considerations

We present applications of the regression estimation technique for some important models and designs. For easy reference, let us summarize the most important formulas from Chapter 6. Recall that the regression estimator results from fitting a linear model. The model is used as a means of describing our idea of how the finite population point scatter looks. The general model states that

$$\begin{cases} E_\xi(y_k) = \sum_{j=1}^{J} \beta_j x_{jk} = \mathbf{x}_k'\boldsymbol{\beta} & (7.2.1) \\ V_\xi(y_k) = \sigma_k^2 & (7.2.2) \end{cases}$$

and that the y_k are independent. Here, $\mathbf{x}_1, \ldots, \mathbf{x}_N$ are the values of the auxiliary vector $\mathbf{x} = (x_1, \ldots, x_J)'$. The hypothetical population fit of the model would result in estimating $\boldsymbol{\beta}$ by

$$\mathbf{B} = (\textstyle\sum_U \mathbf{x}_k\mathbf{x}_k'/\sigma_k^2)^{-1} \sum_U \mathbf{x}_k y_k/\sigma_k^2 \qquad (7.2.3)$$

The corresponding residuals are

$$E_k = y_k - \mathbf{x}_k'\mathbf{B} \qquad (7.2.4)$$

In practice, a sample-based fit is carried out, and $\mathbf{B}$ is estimated by

$$\hat{\mathbf{B}} = (\textstyle\sum_s \mathbf{x}_k\mathbf{x}_k'/\sigma_k^2\pi_k)^{-1} \sum_s \mathbf{x}_k y_k/\sigma_k^2\pi_k \qquad (7.2.5)$$

The fitted y-values,

$$\hat{y}_k = \mathbf{x}_k'\hat{\mathbf{B}} \qquad (7.2.6)$$

can be obtained for $k = 1, \ldots, N$, and the residuals

$$e_{ks} = y_k - \hat{y}_k \qquad (7.2.7)$$

can be calculated for $k \in s$.

The regression estimator is

$$\hat{t}_{yr} = (\textstyle\sum_U \mathbf{x}_k)'\hat{\mathbf{B}} + \sum_s \check{e}_{ks} = \sum_s g_{ks}\check{y}_k \qquad (7.2.8)$$

where $\check{e}_{ks} = e_{ks}/\pi_k$, $\check{y}_k = y_k/\pi_k$, and

$$g_{ks} = 1 + (\textstyle\sum_U \mathbf{x}_k - \sum_s \check{\mathbf{x}}_k)'(\sum_s \mathbf{x}_k\mathbf{x}_k'/\sigma_k^2\pi_k)^{-1}\mathbf{x}_k/\sigma_k^2 \qquad (7.2.9)$$

The estimator (7.2.8) may be used (a) if $\mathbf{x}_1, \ldots, \mathbf{x}_N$ are known values specified in the frame and y_k is observed for $k \in s$ or (b) the total $\sum_U \mathbf{x}_k$ is known from a reliable external source, for example, an administrative data file, and y_k and $\mathbf{x}_k$ are observed for $k \in s$. The approximate variance is

$$AV(\hat{t}_{yr}) = \textstyle\sum\sum_U \Delta_{kl}\check{E}_k\check{E}_l \qquad (7.2.10)$$

where $\check{E}_k = E_k/\pi_k = (y_k - \mathbf{x}_k'\mathbf{B})/\pi_k$, and the suggested variance estimator is

$$\hat{V}(\hat{t}_{yr}) = \textstyle\sum\sum_s \check{\Delta}_{kl}(g_{ks}\check{e}_{ks})(g_{ls}\check{e}_{ls}) \qquad (7.2.11)$$

A computationally simpler variance estimator is obtained by letting $g_{ks} = 1$ for all k and s.

Recall also that simplifications occur if the variance structure satisfies

$$\sigma_k^2 = \boldsymbol{\lambda}'\mathbf{x}_k \qquad (7.2.12)$$

for all k and some constant vector $\boldsymbol{\lambda}$ not depending on k (see Result 6.5.1). Then

$$\sum_s \check{e}_{ks} = 0$$

and consequently the regression estimator is

$$\hat{t}_{yr} = \sum_U \hat{y}_k$$

For any given model, the steps involved in deriving the regression estimator are summarized as follows:

a. Derive $\hat{\mathbf{B}}$, $\hat{y}_k$, and e_{ks} to find $\hat{t}_{yr}$;
b. Identify the g_{ks} needed to obtain the variance estimator $\hat{V}(\hat{t}_{yr})$;
c. Find $\mathbf{B}$ and E_k, which are required for the approximate variance $AV(\hat{t}_{yr})$.

For each of the models considered, we first derive the regression estimator in its general form, that is, with arbitrary π_k and π_{kl}. Then, we discuss the properties of the estimator for specified sampling designs, in particular *SI*, *STSI*, and *BE*.

7.3. The Common Ratio Model and the Ratio Estimator

We consider the case of a single auxiliary variable, x, having known, positive values, $x_1, \ldots, x_k, \ldots, x_N$. Suppose the population elements are hospitals and that x_k is a known measure of the size of the kth hospital, for example, the number of available beds. Let y_k be the number of patients admitted to the kth hospital in a given observation period. In this case, it is not unrealistic to assume that the ratio y_k/x_k is more or less constant for all hospitals. If this is true, a plot of y_k/x_k against x_k for the N hospitals in the population will reveal a scatter around a horizontal straight line. Alternatively, when y_k is plotted against x_k, the scatter will conform roughly to a straight line through the origin.

The assumption may be less appropriate for some other y variable. For example, if y_k were the number of patients admitted to the kth hospital for a specific disease, a roughly constant value of y_k/x_k may not be a realistic assumption. Different hospitals may have different specializations. Sound judgment must underlie the modeling.

A regression model assuming that y_k/x_k is constant (on the average for any fixed x_k) is called a *common ratio model* or simply a *ratio model*. This model states that, for any given x_k,

$$E_\xi(y_k) = \beta x_k$$

This regression is a straight line through the origin. The model is not complete without a variance structure. We first assume that the variance around the line increases proportionally to x. The regression estimator $\hat{t}_{yr}$ is particularly simple in this case. Other variance structures are considered in Section 7.3.3. The complete model is thus

$$\begin{cases} E_\xi(y_k) = \beta x_k \\ V_\xi(y_k) = \sigma^2 x_k \end{cases} \tag{7.3.1}$$

where the parameters β and σ^2 are unknown.

The regression estimator generated by the model described in (7.3.1) is the highly important *ratio estimator*, discussed in Examples 6.4.1, 6.4.2, and 6.6.1. Its properties are summarized in Result 7.3.1, which follows.

Result 7.3.1. *The regression estimator generated by the ratio model* (7.3.1) *is the π weighted ratio estimator*

$$\hat{t}_{yr} = (\textstyle\sum_U x_k)\frac{\sum_s \check{y}_k}{\sum_s \check{x}_k} = (\sum_U x_k)\hat{B} \tag{7.3.2}$$

The approximate variance is obtained from (7.2.10) *by setting*

$$E_k = y_k - Bx_k$$

where

$$B = \frac{\sum_U y_k}{\sum_U x_k}$$

The variance estimator is obtained from equation (7.2.11) *by setting*

$$e_{ks} = y_k - \hat{B}x_k$$

and, for all $k \in s$,

$$g_{ks} = \frac{\sum_U x_k}{\sum_s \check{x}_k}$$

No proof is required. The result summarizes Examples 6.4.2, 6.5.2, and 6.6.1. Note that the model satisfies the requirement of Result 6.5.1: With $\lambda = \sigma^2$, we have $\sigma_k^2 = \sigma^2 x_k = \lambda x_k$. Consequently, the term $\sum_s \check{e}_{ks}$ vanishes, and the simple form $\hat{t}_{yr} = \sum_U \hat{y}_k$ applies.

The π weighted ratio estimator (7.3.2) is sometimes attributed to Hájek (1971). This estimator has a number of points in its favor. The form is simple and easy to accept on intuitive grounds only. Over the history of survey sampling, most of the considerable experience gathered with this estimator points to excellent performance under a variety of conditions.

Dividing (7.3.2) by N leads to the ratio estimator of the population mean, namely,

$$\hat{\bar{y}}_U = \bar{x}_U \frac{\sum_s \check{y}_k}{\sum_s \check{x}_k}$$

To obtain the AV and the variance estimator, divide the corresponding expressions for (7.3.2) by N^2.

Let us denote the ratio estimator (7.3.2) as $\hat{t}_{yra}$. We devote the rest of this section to an examination of its properties under a few important sampling designs.

7.3.1. The Ratio Estimator under *SI* Sampling

Traditional sampling texts and many journal articles analyze the ratio estimator for the *SI* design (although this design is probably used in a minority of all practical applications.) Under the *SI* design, (7.3.2) becomes the classical ratio estimator,

$$\hat{t}_{yra} = \sum_U x_k \frac{\sum_s y_k}{\sum_s x_k} = N\bar{x}_U \frac{\bar{y}_s}{\bar{x}_s} \tag{7.3.3}$$

The approximate variance is

$$AV_{SI}(\hat{t}_{yra}) = N^2 \frac{1-f}{n} S_{EU}^2 = N^2 \frac{1-f}{n} \frac{\sum_U (y_k - Bx_k)^2}{N-1} \tag{7.3.4}$$

with $f = n/N$ and $B = \sum_U y_k / \sum_U x_k$. The variance estimator is

$$\hat{V}_{SI}(\hat{t}_{yra}) = \left(\frac{\bar{x}_U}{\bar{x}_s}\right)^2 \hat{V}_0 \tag{7.3.5}$$

where

$$\hat{V}_0 = N^2 \frac{1-f}{n} S_{es}^2 = N^2 \frac{1-f}{n} \frac{\sum_s (y_k - \hat{B}x_k)^2}{n-1} \tag{7.3.6}$$

with $\hat{B} = \sum_s y_k / \sum_s x_k$. It is left as an exercise to obtain these formulas from Result 7.3.1.

A simple development shows that the approximate variance given by (7.3.4) can be written as

$$AV_{SI}(\hat{t}_{yra}) = N^2 \frac{1-f}{n} (S_{yU}^2 + B^2 S_{xU}^2 - 2BS_{xyU}) \tag{7.3.7}$$

where S_{yU}^2, S_{xU}^2 are population variances and S_{xyU} the population covariance. Correspondingly, the variance estimator (7.3.5) can be written

$$\hat{V}_{SI}(\hat{t}_{yra}) = \left(\frac{\bar{x}_U}{\bar{x}_s}\right)^2 N^2 \frac{1-f}{n} (S_{ys}^2 + \hat{B}^2 S_{xs}^2 - 2\hat{B}S_{xys}) \tag{7.3.8}$$

where $\hat{B}$, S_{ys}^2, S_{xs}^2, and S_{xys} are the sample analogs of B, S_{yU}^2, S_{xU}^2, and S_{xyU}, respectively.

Let us further examine three important aspects of the classical ratio estimator: (a) efficiency, (b) variance estimation, (c) small sample bias, and attempts to reduce such bias.

i. *Efficiency*

The ratio estimator is often, but not always, more efficient than the π estimator,

$$\hat{t}_{y\pi} = N\bar{y}_s$$

Using (7.3.7), one can easily show that $AV_{SI}(\hat{t}_{yra}) \leq V_{SI}(\hat{t}_{y\pi})$ if and only if

$$r \geq \frac{1}{2}\left(\frac{cv_{xU}}{cv_{yU}}\right) \tag{7.3.9}$$

where $r = S_{xyU}/S_{xU}S_{yU}$ is the correlation coefficient between x and y, and $cv_{xU} = S_{xU}/\bar{x}_U$; $cv_{yU} = S_{yU}/\bar{y}_U$. It follows from (7.3.9) that if the cv ratio, cv_{xU}/cv_{yU}, is not too far from unity, the correlation r needs to be approximately 0.5 or more in order to conclude that the ratio estimator is more efficient. There are cases where the ratio estimator is not as good as the expansion estimator, as when the regression of y on x (both variables assumed always positive) has a distinctly positive intercept. Then cv_{xU}/cv_{yU} may be considerably greater than unity; as a result, r may not reach the required limit shown in (7.3.9). The model fitted should then include an intercept, as discussed in Section 7.8.

On the other hand, for a population tightly scattered around $y = Bx$, the ratio estimator often yields dramatic efficiency gains compared to the π estimator.

ii. *Variance Estimation*

Because of its popularity and simplicity, the ratio estimator under SI sampling has been studied from a variety of angles. Variance estimation is one of these aspects. For example, Wu and Deng (1983) carried out an extensive empirical study of variance estimators, using a host of finite populations with different characteristics. Three of their variance estimators were

$$\hat{V}_0,\ \hat{V}_1 = (\bar{x}_U/\bar{x}_s)\hat{V}_0, \quad \text{and} \quad \hat{V}_2 = (\bar{x}_U/\bar{x}_s)^2\hat{V}_0$$

where $\hat{V}_0$ is given by (7.3.6). Here, $\hat{V}_2$ is the formula (7.3.5) resulting from g-weighting of the residuals, $\hat{V}_0$ is the simplified estimator obtained by letting $g_{ks} = 1$ for all k, and $\hat{V}_1$ is an intermediate case. Other variance estimation techniques, including the jackknife (see Chapter 11), were also examined in the Wu and Deng study.

Their study shows that $\hat{V}_0$, $\hat{V}_1$, and $\hat{V}_2$ function quite well in the basic task of yielding confidence intervals whose empirical coverage rate (in repeated SI sampling) will agree closely with the desired nominal $1 - \alpha$ rate.

It is of interest to study the shape of the sampling distribution of the "studentized statistic"

$$\frac{\hat{t}_{yra} - t_y}{\hat{V}^{1/2}} \tag{7.3.10}$$

where $\hat{V}$ is a variance estimator, for example, one of $\hat{V}_0$, $\hat{V}_1$, and $\hat{V}_2$. If $\hat{V}$ is a good variance estimator, the sampling distribution of the statistic (7.3.10) should conform closely to that of a unit-normal variate (or a Student t variable, if n is small). This will lend support to the confidence interval procedure based on the normal (or t) approximation. The Wu and Deng (1983) study indicates that out of the variance estimators $\hat{V}_0$, $\hat{V}_1$, and $\hat{V}_2$ it is, for most populations, $\hat{V}_2$ that works best for approximating the normal distribution.

Other performance criteria that have been examined include the bias and the variance of different estimators of the variance of $\hat{t}_{yra}$. Leblond (1990) compares $\hat{V}_0$, $\hat{V}_2$, and the jackknife variance estimator (see Chapter 11 for the definition). Generally, $\hat{V}_0$ and $\hat{V}_2$ underestimate, that is, they estimate $V(\hat{t}_{yra})$ with a negative bias. Usually, $\hat{V}_0$ underestimates more than $\hat{V}_2$, so in this regard $\hat{V}_2$ is preferable. On the other hand, the jackknife variance estimator is usually positively biased, so it leads to conservative confidence intervals. In very small samples, the bias of the variance estimators may be a factor of concern, because the bias may distort the confidence intervals. However, the bias rapidly approaches zero as the sample size increases. The variability of the different variance estimators is another important factor to consider. The jackknife variance estimator tends to have a higher variance than $\hat{V}_0$ and $\hat{V}_2$.

iii. *Bias*

The bias of the ratio estimator, that is, the quantity $E(\hat{t}_{yra}) - t_y$, is often small. However, in small samples, the bias may be important enough not to be ignored. Under *SI* sampling, one can easily show that the bias is of order $1/n$ and the bias ratio (see Section 5.6) of order $1/n^{1/2}$. For the validity of the confidence intervals, the bias may be a factor of some concern in very small samples, but for samples of size 20 or more, the consequence of the bias is usually negligible.

Some authors have proposed to modify either the estimator or the sampling design to reduce or eliminate the bias of $\hat{t}_{yra}$. Modified ratio estimators with reduced bias or zero bias under *SI* sampling were proposed by Quenouille (1956), Tin (1965), and Mickey (1959). Another estimator of this kind, due to Pasqual (1961), is

$$\hat{t}_y = N\bar{x}_U \frac{\bar{y}_s}{\bar{x}_s} + N \frac{\bar{y}_s - \bar{r}_s \bar{x}_s}{n-1} \tag{7.3.11}$$

where $\bar{r}_s = \sum_s r_k/n$ with $r_k = y_k/x_k$. Whereas the ratio estimator $N\bar{x}_U\bar{y}_s/\bar{x}_s$ has a bias of order $1/n$, the added term in the estimator (7.3.11) has the effect of reducing the bias to order $1/n^2$.

It was observed by Lahiri (1951), Midzuno (1952), and Sen (1953) that the ratio estimator $N\bar{x}_U\bar{y}_s/\bar{x}_s$ is unbiased if the sample s is drawn with probability

$$p(s) \propto \textstyle\sum_s x_k$$

A relatively simple procedure for implementing this design is to draw a first element with probability proportional to x, then to draw an SI sample of $n-1$ elements from the $N-1$ that remain. That is, in the first draw, the probability of drawing the element k is

$$p_k = x_k / \sum_U x_k$$

For each of the modified ratio estimation techniques that we have mentioned, an appropriate variance estimator can be worked out. The variance estimator (7.3.5) no longer applies. For a review of the many variations of the ratio-estimation technique, see P. Rao (1988).

7.3.2. The Ratio Estimator under Other Designs

i. *The BE Design*

The BE and SI designs are similar in that all inclusion probabilities are equal and dissimilar in that BE is of random sample size whereas SI is of fixed size. Using Result 7.3.1, it is shown easily that the ratio estimator for the design BE (with the inclusion probability $\pi = n/N = E_{BE}(n_s)/N$ for all N elements) is given by equation (7.3.3), just as for the SI design. Moreover, apart from a factor very near unity, its approximate variance is given by the same expression, (7.3.4), that is obtained under the SI design. That is, there is no appreciable penalty for variable sample size in this case. The variance estimator shown in equation (7.2.11) becomes

$$\hat{V}_{BE}(\hat{t}_{yra}) = \left(\frac{\bar{x}_U}{\bar{x}_s}\right)^2 N^2 \frac{1-\pi}{n_s}\left(1 - \frac{1}{n_s}\right) S_{es}^2 \tag{7.3.12}$$

where S_{es}^2 is the sample variance of the residuals e_{ks},

$$S_{es}^2 = \frac{\sum_s (y_k - \hat{B}x_k)^2}{n_s - 1}$$

with $\hat{B} = (\sum_s y_k)/(\sum_s x_k)$. If the underlying model is a good description of the population, the estimated variance (7.3.12) tends to be lower (a) if the realized sample size n_s is large or (b) if the sample mean of x, $\bar{x}_s = \sum_s x_k / n_s$, is large. Both of these tendencies are reasonable. When a sample with a large mean $\bar{x}_s$ has been realized, relatively many of the large elements are in the sample. For such a sample, it is consistent with the idea of a regression through the origin and a residual variance that increases with x to obtain a shorter confidence interval than if many of the small elements had been selected.

ii. *The STSI Design*

Another important application of the ratio estimator occurs under the design $STSI$. From the general formula (7.3.2) we obtain

$$\hat{t}_{yra} = \sum_U x_k \frac{\sum_{h=1}^{H} N_h \bar{y}_{s_h}}{\sum_{h=1}^{H} N_h \bar{x}_{s_h}} \quad (7.3.13)$$

which is known as the *combined ratio estimator*, since a combination of stratum sample means appears in both the numerator and in the denominator. It follows from Result 7.3.1 with some algebraic development that the approximate variance of the combined ratio estimator is

$$AV_{STSI}(\hat{t}_{yra}) = \sum_{h=1}^{H} N_h^2 \left(\frac{1}{n_h} - \frac{1}{N_h}\right)(S_{yU_h}^2 + B^2 S_{xU_h}^2 - 2BS_{xyU_h}) \quad (7.3.14)$$

where $B = \bar{y}_U/\bar{x}_U$.

If the stratified sampling is with proportional allocation, it can be shown that a substantial reduction compared to the *SI* sampling variance in (7.3.4) is obtained only if considerable differences exist between the stratum means of the residuals, $\bar{E}_{U_h} = \bar{y}_{U_h} - B\bar{x}_{U_h}$.

iii. *πps Sampling*

Taking x as a size variable, the πps inclusion probabilities are

$$\pi_k = nx_k/N\bar{x}_U$$

where $n = E(n_s)$ is the expected sample size. The ratio estimator (7.3.2) is now

$$\hat{t}_{yra} = \sum_U x_k \frac{\sum_s (y_k/x_k)}{n_s} \quad (7.3.15)$$

which is sometimes called the mean of the ratios estimator. If the variance structure is strongly heteroscedastic, so that the variance of the population scatter increases with x at least as strongly as $V_\xi(y_k) = \sigma^2 x_k^2$, then the estimator (7.3.15) is highly efficient. But the reduction in variance compared to *SI* sampling combined with $\hat{t}_{yra} = N\bar{x}_U(\bar{y}_s/\bar{x}_s)$ depends on the form of the finite population and may often be modest. Cassel, Särndal and Wretman (1977, Chapter 7) review a number of empirical studies of the performance of different ratio estimation strategies for finite populations of various shapes.

7.3.3. Optimal Sampling Design for the π Weighted Ratio Estimator

It is natural to ask which sampling design will minimize the variance of the ratio estimator given in equation (7.3.2). Is there an optimal choice of the π_k? The answer depends on the residual variance around the regression line. Optimal design is discussed in Chapter 12, where the criterion to minimize is the anticipated variance, that is, the model-expected value of the AV expression. Here we use Result 12.2.1, which implies that if the ratio model (7.3.1) holds, the optimal inclusion probabilities, that is, those that minimize the anticipated variance, are

$$\pi_k \propto x_k^{1/2} \tag{7.3.16}$$

If n denotes the expected sample size, we should thus use a design with

$$\pi_k = n x_k^{1/2}/(\textstyle\sum_U x_k^{1/2})$$

Let us assume that no x_k is so large that $n x_k^{1/2} > \sum_U x_k^{1/2}$.

In other words, with the ratio estimator

$$\hat{t}_{yra} = \textstyle\sum_U x_k \dfrac{\sum_s \check{y}_k}{\sum_s \check{x}_k}$$

it is possible to realize gains in efficiency by taking $\pi_k \propto x_k^{1/2}$ ($\pi p\sqrt{x}$ sampling) rather than $\pi_k =$ constant (as in *SI* sampling) or $\pi_k \propto x_k$ (as in πps sampling). The requirement is that the residual variances really obey $V_\xi(y_k) = \sigma^2 x_k$ or very nearly so. That is, to realize an efficiency gain (relative to, say *SI* sampling) is contingent on having a strong a priori knowledge about the variation of the finite population scatter. An ill-founded guess that $V_\xi(y_k) = \sigma^2 x_k$, and a subsequent decision to use $\pi p\sqrt{x}$ sampling, may lead to less efficient estimation than with *SI* sampling. The efficiency gain of $\pi p\sqrt{x}$ sampling over *SI* sampling depends on the shape of the finite population.

To construct a fixed-size design satisfying (7.3.16) is not easy, as Section 3.6 has shown. However, a simple procedure to obtain inclusion probabilities that approximate (7.3.16) is as follows: Order the quantities $x_k^{1/2}$ $k = 1, \ldots, N$, from the smallest to the largest:

$$x_1^{1/2} < x_2^{1/2} < \cdots < x_k^{1/2} < \cdots < x_N^{1/2} \tag{7.3.17}$$

Here, k denotes the kth element of this ordering. Create strata, $U_1, \ldots, U_H$, as follows. The first stratum U_1 consists of the elements k with the N_1 smallest values $x_k^{1/2}$, up to the point where (coming as close as possible)

$$\textstyle\sum_{U_1} x_k^{1/2} = (1/H) \sum_U x_k^{1/2}$$

the second stratum U_2 is composed of the following N_2 elements in the ordering given by (7.3.17), up to the point where (coming as close as possible)

$$\textstyle\sum_{U_2} x_k^{1/2} = (1/H) \sum_U x_k^{1/2}$$

and so on. In other words, each of the H strata accounts for $1/H$ of the total $\sum_U x_k^{1/2}$. In each of these strata, one then selects an *SI* sample of equal size, $n_h = n/H$, $h = 1, \ldots, H$. In this procedure, the inclusion probability, for any element in the stratum U_h, is

$$\pi_k = \frac{n_h}{N_h} = \frac{n}{HN_h} = \frac{n}{N} \frac{\sum_{U_h} x_k^{1/2}/N_h}{\sum_U x_k^{1/2}/N}$$

Because π_k is proportional to the stratum mean of $x_k^{1/2}$, it is approximately proportional to $x_k^{1/2}$. With ten or more strata, the approximation is good enough. There is no great loss of efficiency compared to the design for which (7.3.16) holds exactly. The simplification that has been gained is that *STSI* sampling is easy to carry out. The theoretical underpinnings of this *model-based stratification* method, due to Wright (1983), are given in Section 12.4.

7.3.4. Alternative Ratio Models

A ratio model with a more general variance structure is

$$\begin{cases} E_\xi(y_k) = \beta x_k \\ V_\xi(y_k) = \sigma^2 w_k = \sigma^2 x_k^\gamma \end{cases} \tag{7.3.18}$$

where β and σ^2 are unknown and γ is a known positive constant. The previous ratio model shown in (7.3.1) corresponds to $\gamma = 1$.

Assume for a moment that we can plot all N points (y_k, x_k). If $E_\xi(y_k) = \beta x_k$ is an adequate description of the finite population, then the points (y_k, x_k) will scatter around the line $y = \beta x$, while the points $(y_k/x_k, x_k)$ scatter around a horizontal line placed at the level β. This is seen in Figure 7.1(a and b) corresponding to $\gamma = 1$ and $\gamma = 2$. The variance portion of the model (7.3.18) expresses how the points (y_k, x_k) are scattered around the line $y = \beta x$. A value $\gamma > 0$ means that the scatter widens as x increases. The greater the value of γ, the stronger the tendency of the points to fan out from $y = \beta x$. This is also seen in Figure 7.1(a and b). Note that the case $\gamma = 2$ implies a constant variance for the scatter of the points $(y_k/x_k, x_k)$. For many survey populations, a (y_k, x_k) plot will show a scatter that widens with x; many actual populations correspond to values of γ in the interval $1 \le \gamma \le 2$ (Brewer, 1963b).

We assume here that γ is a fixed constant, known from experience or from a pilot study. If γ is treated as an unknown, a possibility (not discussed here) is to estimate γ as well as β and σ^2 from the sample. For work along these lines see Wright (1983).

Result 7.3.2. *Under the ratio model shown in* (7.3.18), *the regression estimator of t_y is given by*

$$\hat{t}_{yr} = (\textstyle\sum_U x_k)\hat{B} + \sum_s \dfrac{y_k - \hat{B}x_k}{\pi_k} \tag{7.3.19}$$

where

$$\hat{B} = \frac{\sum_s x_k y_k / w_k \pi_k}{\sum_s x_k^2 / w_k \pi_k}$$

with $w_k = x_k^\gamma$. The approximate variance is given by (7.2.10) *by setting*

$$E_k = y_k - Bx_k$$

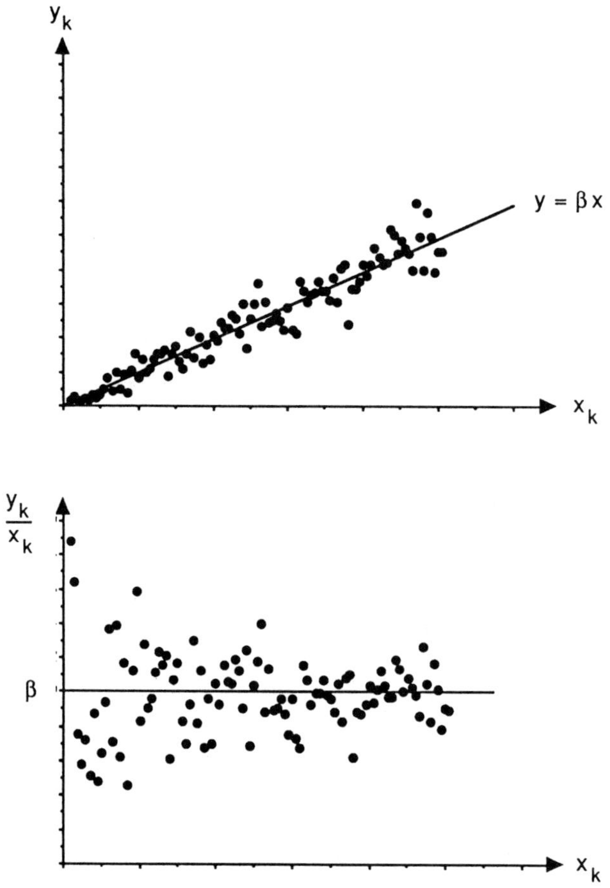

Figure 7.1(a). The case $\gamma = 1$.

with

$$B = \frac{\sum_U x_k y_k / w_k}{\sum_U x_k^2 / w_k}$$

The variance estimator is given by (7.2.11) *by setting*

$$e_{ks} = y_k - \hat{B} x_k$$

and

$$g_{ks} = 1 + \left(\sum_U x_k - \sum_s \frac{x_k}{\pi_k}\right)\left(\sum_s \frac{x_k^2}{w_k \pi_k}\right)^{-1} \frac{x_k}{w_k}$$

The conclusions are obvious, and we omit the proof. Note that the term $\sum_s \check{e}_{ks}$ does not vanish (unless $\gamma = 1$). It follows from Result 12.2.1 that an

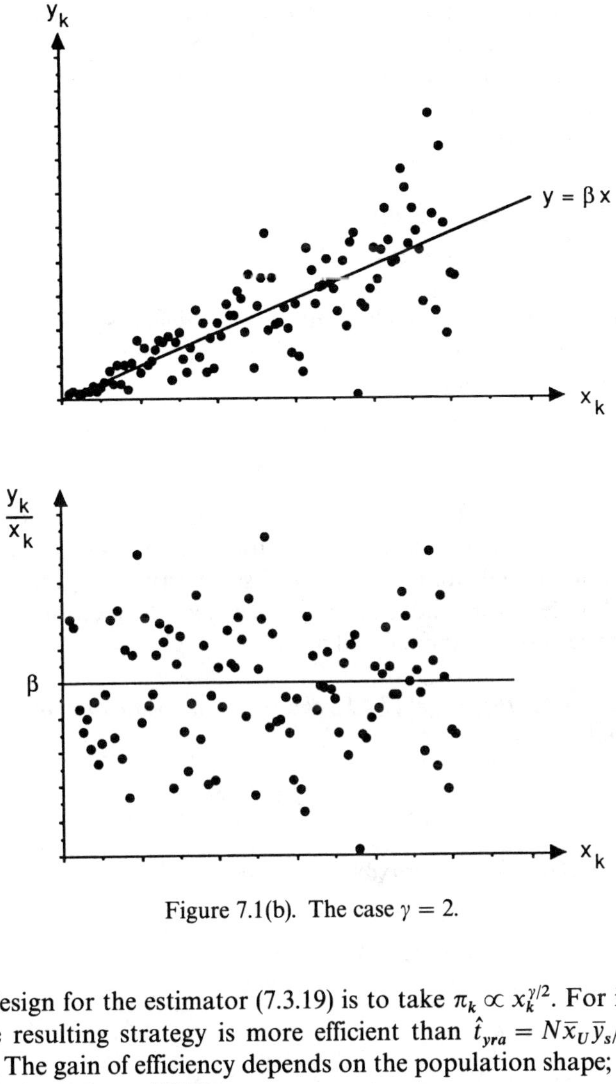

Figure 7.1(b). The case $\gamma = 2$.

optimal design for the estimator (7.3.19) is to take $\pi_k \propto x_k^{\gamma/2}$. For instance, if $\gamma = 2$, the resulting strategy is more efficient than $\hat{t}_{yra} = N\bar{x}_U\bar{y}_s/\bar{x}_s$ and *SI* sampling. The gain of efficiency depends on the population shape; see Cassel, Särndal and Wretman (1976).

For a given sampling design, which of the estimators (7.3.2) or (7.3.19) has the smaller variance in large samples? Both are derived from ratio models. To each estimator corresponds a set of residuals, $E_1, \ldots, E_N$, defined in Results 7.3.1 and 7.3.2. The answer to the question depends on which of the two sets of residuals yields the smaller value of the approximate variance, $\sum\sum_U \Delta_{kl}\check{E}_k\check{E}_l$. For example, if the *SI* design is used with the sample size *n*, the approximate variance is

$$N^2[(1-f)/n]S_{EU}^2$$

where $S_{EU}^2 = \sum_U (E_k - \bar{E}_U)^2/(N-1)$, so in this case the variance of the residuals decides the question. For many populations, the two approximate variances are nearly equal, since the residuals differ only in regard to the weighting applied in the expression for B. The ratio estimator shown in (7.3.2) has the advantage of a simpler form, and the weight given to an observation y_k can never be negative.

7.4. The Common Mean Model

A seemingly trivial but nevertheless important special case of the ratio model shown in (7.3.1) arises when $x_k = 1$ for all k. The result is the *common mean model* stating that, for all $k \in U$,

$$\begin{cases} E_\xi(y_k) = \beta \\ V_\xi(y_k) = \sigma^2 \end{cases} \tag{7.4.1}$$

The always constant x variable appears to contain no information by which one can improve over the basic π estimator. This impression is deceptive. Except in some special situations, the regression estimator generated by the model (7.4.1) differs from and is usually better than the π estimator.

The main results are summarized as follows.

Result 7.4.1. *Under the model* (7.4.1), *the regression estimator of t_y is the expanded sample mean,*

$$\hat{t}_{yr} = N\tilde{y}_s = N\frac{\sum_s \check{y}_k}{\hat{N}} \tag{7.4.2}$$

where $\hat{N} = \sum_s 1/\pi_k$. The approximate variance is obtained from the general formula (7.2.10) *by taking*

$$E_k = y_k - \bar{y}_U; \qquad k = 1, \ldots, N \tag{7.4.3}$$

where $\bar{y}_U = \sum_U y_k/N$. The variance estimator is obtained from the general formula (7.2.11) *by taking*

$$e_{ks} = y_k - \tilde{y}_s; \qquad k \in s \tag{7.4.4}$$

$$g_{ks} = N/\hat{N}; \quad all \quad k \in s \tag{7.4.5}$$

The resulting variance estimator is

$$\hat{V}(N\tilde{y}_s) = (N/\hat{N})^2 \sum\sum_s \check{\Delta}_{kl}\check{e}_{ks}\check{e}_{ls} \tag{7.4.6}$$

PROOF. The conclusions follow immediately from Result 7.3.1 by setting $x_k = 1$ for all k, noting that $B = \sum_U y_k/N = \bar{y}_U$ and that $\hat{B} = \sum_s \check{y}_k/\hat{N} = \tilde{y}_s$. □

The only a priori knowledge required for the estimator (7.4.2) is the population size N, which is the population total of the x variable; $\sum_U x_k = \sum_U 1 = N$.

For some element sampling designs, such as *SI* and *STSI*, $\hat{N} = N$; so (7.4.2) agrees with the π estimator, $\hat{t}_y = \sum_s \check{y}_k$. But for variable size designs, the estimator (7.4.2) is usually preferred. It avoids the extreme estimates sometimes obtained with the π estimator, as noted in Example 5.7.3. For further illustrations, see Example 7.4.1 and Section 8.7.

An estimator of the population mean

$$\bar{y}_U = \frac{t_y}{N} = \frac{\sum_U y_k}{N}$$

is easily found by dividing (7.4.2) by N, namely,

$$\hat{\bar{y}}_U = \tilde{y}_s = \frac{\sum_s \check{y}_k}{\hat{N}} = \frac{\sum_s y_k/\pi_k}{\sum_s 1/\pi_k} \tag{7.4.7}$$

This is the weighted sample mean derived in Section 5.7 by the π estimator substitution method. By dividing (7.4.6) by N^2, we confirm the variance estimator given in Result 5.7.1.

Example 7.4.1 (The *BE* design). If π is the constant inclusion probability for all elements $k \in U$, the sample size n_s is random with the expected value $N\pi$, which we denote by n. The expanded sample mean estimator (7.4.2) is now

$$\hat{t}_{yr} = N \frac{\sum_s y_k}{n_s} = N\bar{y}_s$$

From Result 7.4.1,

$$AV_{BE}(N\bar{y}_s) = N^2 \frac{1-\pi}{n}\left(1 - \frac{1}{N}\right) S_{yU}^2$$

$$\doteq N^2 \frac{1-\pi}{n} S_{yU}^2 \tag{7.4.8}$$

and the variance estimator is

$$\hat{V}_{BE}(N\bar{y}_s) = N^2 \frac{1-\pi}{n_s}\left(1 - \frac{1}{n_s}\right) S_{ys}^2 \tag{7.4.9}$$

where

$$S_{ys}^2 = \frac{1}{n_s - 1} \sum_s (y_k - \bar{y}_s)^2$$

If n_s is fixed (and ≥ 2), S_{ys}^2 is a conditionally unbiased estimator of S_{yU}^2; the proof is left as an exercise. It follows that

$$E_{BE}[\hat{V}_{BE}(N\bar{y}_s) | n_s] = N^2 \frac{1-\pi}{n_s}\left(1 - \frac{1}{n_s}\right) S_{yU}^2$$

This implies that, on the average, the confidence intervals calculated from equation (7.4.9) will be shorter when a large n_s is realized than when a small

n_s is realized, which is reasonable. With the variance estimator (7.4.9), we obtain what is called a conditionally valid confidence interval. This important notion is further explained in Section 7.10, where it is also shown that a better approximation to the variance than (7.4.8) is

$$AV_{BE}(N\bar{y}_s) = N^2 \frac{1-\pi}{n}\left(1 + \frac{1}{n}\right) S_{yU}^2$$

This expression shows that, to the order of approximation n^{-2}, the variance under BE sampling with expected sample size $n = N\pi$ is marginally greater, by a factor of $1 + 1/n$ only, than the variance under SI sampling, which is

$$V_{SI}(N\bar{y}_s) = N^2 \frac{1 - n/N}{n} S_{yU}^2$$

In other words, the price paid for random sample size is negligible provided the estimator $N\bar{y}_s$ is used. By comparison, the π estimator is less efficient. From equation (3.2.5), its variance is greater than the approximation in (7.4.8) by a factor of roughly $1 + (cv_{yU})^{-2}$.

7.5. Models Involving Population Groups

We now introduce an important type of model that relies on a partitioning of the population U into subpopulations $U_1, \ldots, U_g, \ldots, U_G$, according to a given classification principle. The subsets will be called population groups, or simply groups. They are *not* strata, except in the special case where they are used for stratified sample selection.

EXAMPLE 7.5.1. Suppose that the population elements are business establishments and that industry groups are formed: retail firms, manufacturing firms, mining companies, transportation companies, and so on. For many variables, firms in the same industry tend to resemble each other. The homogeneity within groups can be exploited to obtain improved estimates, via regression estimation. Auxiliary information will be necessary, for example, the sizes of the population groups, or the group totals of an auxiliary variable.

Group models have one or more parameters associated with each individual group. Once formulated and fitted, a group model leads to a particular form of regression estimator. Applications are discussed in Sections 7.6 and 7.7.

In Example 7.5.1, one possibility is that the industry groups can be identified prior to sampling. The groups may then serve as strata. That is, in each group (stratum), a probability sample is drawn. The union of these group samples constitutes a stratified sample from the whole population. This special case was covered by Section 3.7.

Another possibility (one that is of principal interest in Sections 7.6 and 7.7) is that the industry groups do not function as strata. Instead, a probability sample is drawn directly from the (ungrouped) population, and group membership is established, for sampled elements only, after the sampling has taken place. With an often-used term, the groups are then called *poststrata*. Poststratification is an important practice that often produces significant gains in efficiency, as we will see.

Denote the groups by $U_1, \ldots, U_g, \ldots, U_G$, and let N_g be the size of U_g. We then have the partitioning equations

$$\bigcup_{g=1}^{G} U_g = U; \qquad \sum_{g=1}^{G} N_g = N \tag{7.5.1}$$

If s is the sample, drawn from U with the given sampling design, there is a corresponding partitioning of s into subsets $s_1, \ldots, s_g, \ldots, s_G$, where s_g is the part of s that falls in U_g, that is,

$$s_g = U_g \cap s$$

The sample counterparts of equation (7.5.1) are then

$$\bigcup_{g=1}^{G} s_g = s; \qquad \sum_{g=1}^{G} n_{s_g} = n_s \tag{7.5.2}$$

where n_{s_g} is the size of s_g. In applications, n_{s_g} will usually be a random variable, as when an *SI* sample is drawn from the whole population and groups (poststrata) are formed after sampling.

Our first group model assumes a mean and a variance shared by all elements within the same group. More formally, for $g = 1, \ldots, G$,

$$\begin{cases} E_\xi(y_k) = \beta_g \\ V_\xi(y_k) = \sigma_{0g}^2 \end{cases} \tag{7.5.3}$$

for all $k \in U_g$, which we call the *group mean model*. In common statistical language, it is a one-way analysis of variance (one-way ANOVA) model. Here the β_g and the σ_{0g}^2 are unknown parameters.

In Example 7.5.1, if the business firms are relatively homogeneous within industry groupings, and if considerable between-group differences exist, then the model (7.5.3) will explain a good part of the variation in y, and a regression estimator derived from this model will be highly efficient. The regression estimator is obtained by first fitting the model (7.5.3), which is a special case of the general linear model expressed by equations (7.2.1) and (7.2.2); thus the general procedure outlined in Section 7.2 can be used. Readers familiar with the analysis of linear models know that a one-way ANOVA model such as the one shown in (7.5.3) can be cast in the form of a general linear model by identifying a set of dummy variables that code the group membership of each population element.

Specifically, we cast the expected value portion of (7.5.3) in the form of the general linear model formulation

$$E_\xi(y_k) = \boldsymbol{\delta}_k'\boldsymbol{\beta}$$

by letting $\boldsymbol{\beta} = (\beta_1, \ldots, \beta_g, \ldots, \beta_G)'$ and by taking

$$\boldsymbol{\delta}_k = (\delta_{1k}, \ldots, \delta_{gk}, \ldots, \delta_{Gk})' \tag{7.5.4}$$

where δ_{gk} is the dummy variable value defined by

$$\delta_{gk} = \begin{cases} 1 & \text{if } k \in U_g \\ 0 & \text{if not} \end{cases} \tag{7.5.5}$$

There are G qualitative auxiliary variables, $\delta_1, \ldots, \delta_g, \ldots, \delta_G$, such that the value of the gth variable for the kth element is given by (7.5.5). The *group membership vector* $\boldsymbol{\delta}_k$ consists of $G - 1$ entries "zero," and a single entry "one," which identifies the group to which the element k belongs. The population total of $\boldsymbol{\delta}_k$, which must be known, is the vector of group counts $(N_1, \ldots, N_G)'$. The group mean model shown in (7.5.3) is now easily fitted by the general procedure in Section 7.2. Before stating the results, we discuss an extension of the model (7.5.3) involving a known quantitative variable x related to y. We assume that $x_1, \ldots, x_k, \ldots, x_N$ are known, positive values.

The *group ratio model* is defined as follows. For $g = 1, \ldots, G$,

$$\begin{cases} E_\xi(y_k) = \beta_g x_k \\ V_\xi(y_k) = \sigma_{0g}^2 x_k \end{cases} \tag{7.5.6}$$

for $k \in U_g$. Here the β_g and the σ_{0g}^2 are unknown parameters. The ANOVA model (7.5.3) is the special case of (7.5.6) corresponding to $x_k = 1$ for all $k \in U$.

EXAMPLE 7.5.2. Suppose that the elements are business firms and that x_k represents the known value of the variable "Gross Business Income" in a given year for the kth firm, $k = 1, \ldots, N$. Let the study variable value y_k be the value of the variable "Wages and Salaries" paid by the firm during the year. The ratio y_k/x_k may be fairly constant for all firms k in a given industry group, but very different for firms in two different industries. This may be the case if one industry is labor intensive and the other is not. A constant ratio model is then appropriate for each group. Allowing the slopes to differ between groups leads to the model formulation (7.5.6).

EXAMPLE 7.5.3. We now consider another example where the model (7.5.6) may fit well. Suppose the population elements are individuals, grouped by age and sex into, say, six groups, corresponding to three age brackets crossed with the two sexes. If y_k is the (unknown) amount in dollars spent by the kth individual on a given leisure activity, say, vacationing at ski resorts, and if x_k is his or her (known) salary, it may again be appropriate to assume that the ratio y_k/x_k remains roughly constant for individuals k in a fixed age/sex group, and that different ratios apply in different groups.

The expected value part of (7.5.6) fits the general linear model (7.2.1) if we use dummy variables. For $g = 1, \ldots, G$, let δ_{gk} be given by (7.5.5). For the gth group, we create a special predictor variable x_g whose value for the kth element is

$$x_{gk} = \delta_{gk} x_k = \begin{cases} x_k & \text{if } k \in U_g \\ 0 & \text{if not} \end{cases} \tag{7.5.7}$$

That is, the gth of these new x variables is either x_k or zero, depending on whether the element k is in the gth group or not. Now the expected-value part of the model shown in (7.5.6) can be written

$$E_\xi(y_k) = \mathbf{x}_k'\boldsymbol{\beta}$$

with $\boldsymbol{\beta} = (\beta_1, \ldots, \beta_g, \ldots, \beta_G)'$ and

$$\mathbf{x}_k = (x_{1k}, \ldots, x_{gk}, \ldots, x_{Gk})' \tag{7.5.8}$$

where x_{gk} is defined by (7.5.7). The population total of $\mathbf{x}_k$, which must be known, is the vector composed of the G group totals of the variable x, $\sum_{U_g} x_k$, $g = 1, \ldots, G$.

The fit of the group ratio model shown in (7.5.6) and its special case, the group mean model (7.5.3), is a simple exercise in matrix algebra. The results are as follows.

Result 7.5.1. *In the population fit of the group ratio model* (7.5.6), β_g *is estimated by*

$$B_g = \frac{\sum_{U_g} y_k}{\sum_{U_g} x_k}; \qquad g = 1, \ldots, G \tag{7.5.9}$$

From the sample fit of the same model, the estimator of B_g *is obtained as*

$$\hat{B}_g = \frac{\sum_{s_g} \check{y}_k}{\sum_{s_g} \check{x}_k}; \qquad g = 1, \ldots, G \tag{7.5.10}$$

with $\check{y}_k = y_k/\pi_k$; $\check{x}_k = x_k/\pi_k$.

For the group mean model (7.5.3), *which implies* $x_k = 1$ *for all* k, *the corresponding results are*

$$B_g = \bar{y}_{U_g} = \frac{\sum_{U_g} y_k}{N_g}; \qquad g = 1, \ldots, G \tag{7.5.11}$$

$$\hat{B}_g = \tilde{y}_{s_g} = \frac{\sum_{s_g} \check{y}_k}{\hat{N}_g}; \qquad g = 1, \ldots, G \tag{7.5.12}$$

where $\hat{N}_g = \sum_{s_g} 1/\pi_k$.

PROOF. It suffices to show the result for $\hat{\mathbf{B}} = (\hat{B}_1, \ldots, \hat{B}_g, \ldots, \hat{B}_G)'$ since $\mathbf{B} = (B_1, \ldots, B_g, \ldots, B_G)'$ can be obtained from $\hat{\mathbf{B}}$ by setting $s = U$ (implying $s_g = U_g$ for all g) and $\pi_k = 1$ for all $k \in U$.

We apply the general formulas in Section 7.2. Equation (7.2.5) shows that the two parts that determine $\hat{B}$ are the symmetric $G \times G$ matrix $\sum_s \mathbf{x}_k\mathbf{x}_k'/\sigma_k^2\pi_k$

and the $G \times 1$ vector $\sum_s (\mathbf{x}_k y_k/\sigma_k^2 \pi_k)$, where $\mathbf{x}_k$ is defined by (7.5.8). A calculation shows that the element with index g, g' of the $G \times G$ matrix is

$$\sum_s (x_{gk} x_{g'k}/\sigma_k^2 \pi_k) = \begin{cases} \sigma_{0g}^{-2} \sum_{s_g} x_k/\pi_k & \text{if } g = g' \\ 0 & \text{if } g \neq g' \end{cases}$$

for g and $g' = 1, \ldots, G$; therefore, letting $A_g = \sum_{s_g} x_k/\pi_k$, we can write $\sum_s \mathbf{x}_k \mathbf{x}_k'/\sigma_k^2 \pi_k = \text{diag}\{A_g/\sigma_{0g}^2\}$ where $\text{diag}\{\cdot\}$ indicates the diagonal matrix. That is, all off-diagonal elements are zero, and the term within curly brackets specifies the typical diagonal element. The inversion yields a new diagonal $G \times G$ matrix

$$\left(\sum_s \mathbf{x}_k \mathbf{x}_k'/\sigma_k^2 \pi_k\right)^{-1} = \text{diag}\{\sigma_{0g}^2/A_g\} \tag{7.5.13}$$

Next we calculate

$$\sum_s \mathbf{x}_k y_k/\sigma_k^2 \pi_k = (C_1, \ldots, C_g, \ldots, C_G)' \tag{7.5.14}$$

where

$$C_g = \sigma_{0g}^{-2} \sum_{s_g} y_k/\pi_k$$

Multiplying the matrix (7.5.13) by the vector (7.5.14), the unknown model parameters σ_{0g}^2 conveniently cancel, leaving the desired result. The conclusions for the group mean model follow easily by setting $x_k = 1$ for all k. □

7.6. The Group Mean Model and the Poststratified Estimator

The regression estimator generated by the group mean model is described in the following result.

Result 7.6.1. *Under the group mean model* (7.5.3), *the regression estimator of* t_y *is*

$$\hat{t}_{yr} = \sum_{g=1}^{G} N_g \tilde{y}_{s_g} \tag{7.6.1}$$

where $\tilde{y}_{s_g} = \sum_{s_g} \check{y}_k/\hat{N}_g$ *with* $\hat{N}_g = \sum_{s_g} 1/\pi_k$. *The approximate variance is obtained from the general expression* (7.2.10) *by setting, for* $g = 1, \ldots, G$,

$$E_k = y_k - \bar{y}_{U_g} \quad \textit{for} \quad k \in U_g$$

where $\bar{y}_{U_g} = \sum_{U_g} y_k/N_g$. *The variance estimator is obtained from the general expression* (7.2.11) *by setting, for* $g = 1, \ldots, G$,

$$e_{ks} = y_k - \tilde{y}_{s_g} \quad \textit{for} \quad k \in s_g$$

$$g_{ks} = N_g/\hat{N}_g \quad \textit{for all} \quad k \in s_g$$

PROOF. It is easily seen that for the ANOVA model (7.5.3), the term $\sum_s \check{e}_{ks}$ equals zero. From Result 7.5.1, we have $\hat{y}_k = \hat{B}_g = \tilde{y}_{s_g}$ whenever $k \in U_g$, and thus the regression estimator is simply

$$\hat{t}_{yr} = \sum_U \hat{y}_k = \sum_{g=1}^{G} \sum_{U_g} \hat{y}_k = \sum_{g=1}^{G} N_g \tilde{y}_{s_g} \tag{7.6.2}$$

The required residuals E_k and e_{ks} follow immediately from equations (7.5.11) and (7.5.12) in Result 7.5.1. Finally, by identifying the coefficient of y_k/π_k in (7.6.2), we find that the g weights are $g_{ks} = N_g/\hat{N}_g$ for all $k \in s_g$. □

To estimate the population mean, divide (7.6.1) by N and the variance estimator by N^2. Here, N is known, since all N_g are required to be known. There are several practical cases where Result 7.6.1 is applicable.

a. The group membership [that is, the vector $\boldsymbol{\delta}_k$ defined by (7.5.4)] is known for all $k \in U$. The groups can be identified before sampling and serve as strata for a stratified sampling design. If groups and strata are identical, and if the *STSI* design is used, then (7.6.1) is simply the ordinary π estimator discussed in Section 3.7, and $n_{s_g} = n_g$ is fixed in advance for all strata.
b. The group membership is given in the frame for all elements $k \in U$, as in case (a), but we abstain from using the groups as strata. Practical considerations may favor some other (perhaps simpler or less costly) design, such as *SI* or *SIC*. After sample selection, y_k is observed for the elements $k \in s$, and the sampling frame is used to establish the group membership $\boldsymbol{\delta}_k$ for each sample element. In this case, the group membership information is thus utilized at the estimation stage, not at the design stage.
c. The group membership is not known beforehand for the N population elements. However, from external sources (such as a census or other reliable register), we have accurate information about the population group sizes $N_g = \sum_U \delta_{gk}$. After some convenient form of sampling we observe not only y_k but also the group membership $\boldsymbol{\delta}_k$ for each element $k \in s$. Including $\boldsymbol{\delta}_k$ among the variables to observe adds to the respondent burden, which must be considered when planning the survey. High respondent burden may cause increased nonresponse. Note also the importance of using accurate group sizes N_g in the estimator formula. Obsolete group counts will introduce bias in the poststratification estimator (7.6.1).

Here, cases (b) and (c) represent poststratification, that is, the groups are identified after sampling for sampled elements only. In comparing cases (a) and (b), an interesting question arises. Is precision lost in case (b) by forgoing the group membership information at the design stage and by using it "only" for poststratification? The answer depends, as one may suspect, on the exact nature of the sampling. If *STSI* sampling with proportional allocation is used in case (a), and if *SI* sampling (with the same sample size) is used in case (b), it is a fact that the two methods are almost equally efficient, as will be seen later in this section. But if cluster sampling was used in case (b), additional precision may be lost due to positive cluster homogeneity.

Poststratification is examined, from various points of view by Holt and Smith (1979), Doss, Hartley and Somayajulu (1979), and Jagers, Odén and Trulsson (1985). Poststratification is related to adjustment group weighting for nonresponse, as discussed in Section 15.6; see also Jagers (1986), and Bethlehem and Kersten (1985).

Let us apply Result 7.6.1 to a few specific sampling designs. We denote the estimator (7.6.1) as $\hat{t}_{ypos}$, where pos indicates poststratification.

The SI Design

This is the classical application of poststratification. Using Result 7.6.1 with $\pi_k = n/N = f$ for all k and $\pi_{kl} = f(n-1)/(N-1)$ for all $k \neq l$, we easily obtain the expressions (7.6.3) to (7.6.5) below. The estimator is

$$\hat{t}_{ypos} = \sum_{g=1}^{G} N_g \bar{y}_{s_g} \tag{7.6.3}$$

where $\bar{y}_{s_g}$ is the straight mean of the sample from group g, that is, $\bar{y}_{s_g} = \sum_{s_g} y_k / n_{s_g}$. The group count n_{s_g} is random, but the sum,

$$\sum_{g=1}^{G} n_{s_g} = n$$

is fixed. A simple derivation shows that

$$AV_{SI}(\hat{t}_{ypos}) = N^2 \frac{1-f}{n} \sum_{g=1}^{G} W'_g S^2_{yU_g} \tag{7.6.4}$$

with $W'_g = (N_g - 1)/(N-1)$ and $S^2_{yU_g}$ is the group variance,

$$S^2_{yU_g} = \frac{1}{N_g - 1} \sum\nolimits_{U_g} (y_k - \bar{y}_{U_g})^2$$

The variance estimator takes the form

$$\hat{V}_{SI}(\hat{t}_{ypos}) = (1-f) \sum_{g=1}^{G} N_g^2 \frac{S^2_{ys_g}}{n_{s_g}} \tag{7.6.5}$$

where $S^2_{ys_g}$ is the variance of the sample from the gth group,

$$S^2_{ys_g} = \frac{1}{n_{s_g} - 1} \sum\nolimits_{s_g} (y_k - \bar{y}_{s_g})^2$$

In deriving (7.6.5) from (7.2.11), we have approximated $n(n_{s_g} - 1)/(n-1)n_{s_g}$ by unity. Note that $\bar{E}_{U_g} = \bar{e}_{s_g} = 0$, which simplifies the AV and $\hat{V}$ expressions. A few comments are in order.

Equation (7.6.3) does not apply if the event

$$n_{s_g} = 0 \text{ for some } g = 1, \ldots, G$$

occurs. If the total sample size n is substantial, and if no group accounts for a very small portion of the whole population, then the event in question has near-zero probability, and no practical problem need arise. If some of the planned groups look as if they may be too rare, merging of groups should take place at the planning stage to eliminate the problem with zero counts.

All group counts n_{s_g} should be of respectable size to avoid unstable group mean estimates $\bar{y}_{s_g}$. If all n_{s_g} are at least 20, one should be on the safe side.

A better approximation to the variance than (7.6.4) (see Section 7.10 for a derivation) is given, with $W_g = N_g/N$, by

$$V_{SI}(\hat{t}_{y\text{pos}}) \doteq N^2 \frac{1-f}{n} \sum_{g=1}^{G} W_g S_{yU_g}^2 + N^2 \frac{1-f}{n^2} \sum_{g=1}^{G} (1 - W_g) S_{yU_g}^2 \quad (7.6.6)$$

In this formula, the first term, of order n^{-1}, agrees very nearly with the AV given by (7.6.4); the additional term is of order n^{-2}.

For SI sampling, the n sample observations distribute themselves randomly over the G groups, with an expected count in the gth group of

$$n_g = E(n_{s_g}) = nN_g/N = nW_g$$

That is, the expected group counts agree with a proportional allocation to the groups. We know from Section 3.7 that if groups were strata and $STSI$ sampling with proportional allocation was used, the variance of the π estimator would be given by the first term on the right-hand side of (7.6.6). The second term of (7.6.6) represents the increase (to order n^{-2}) caused by group counts that are not exactly, but only on the average, proportionally allocated. We conclude that SI sampling with poststratification is essentially as efficient as $STSI$ sampling with proportional allocation, unless the sample is very small.

SI sampling with poststratification is often much more efficient than SI sampling without poststratification. To see this, consider the usual ANOVA decomposition of the total sum of squares of y:

$$(N-1)S_{yU}^2 = \sum_{g=1}^{G} (N_g - 1)S_{yU_g}^2 + \sum_{g=1}^{G} N_g(\bar{y}_{U_g} - \bar{y}_U)^2$$

Divide by $(N-1)S_{yU}^2$ and set

$$R^2 = 1 - \frac{\sum_{g=1}^{G} (N_g - 1)S_{yU_g}^2}{(N-1)S_{yU}^2} = \frac{\sum_{g=1}^{G} N_g(\bar{y}_{U_g} - \bar{y}_U)^2}{(N-1)S_{yU}^2}$$

which measures the degree to which the ANOVA model (7.5.3) explains the variation in y. Now, under SI sampling without poststratification,

$$V_{SI}(N\bar{y}_s) = N^2 \frac{1-f}{n} S_{yU}^2$$

so that, from (7.6.4),

$$\frac{AV_{SI}\left(\sum_{g=1}^{G} N_g \bar{y}_{s_g}\right)}{V_{SI}(N\bar{y}_s)} \doteq 1 - R^2$$

When the between group variation is large (that is, when $1 - R^2$ is close to zero), poststratification greatly reduces the variance.

A strong incentive to poststratify (rather than to employ the groups as strata for an *STSI* design) arises in multipurpose surveys, that is, surveys in which many y variables are studied. (Surveys are very often multipurpose in character.) If an *STSI* design was used, the strata would be fixed once and for all. These strata may reduce variance for one or a few of the y variables but could be inefficient for many other y variables. Therefore, using *SI* or some similarly simple design together with the poststratified estimator given by (7.6.3) will often improve overall efficiency. This opens the field for different poststratifications for different y variables. For instance, certain y variables may be well explained by an age/sex grouping, others by an occupational grouping, and so forth. Here, the statistician's knowledge, intuition, and judgment concerning relationships among variables will serve in specifying efficient ANOVA models (thus different poststratifications) for different subsets of y variables.

The STSI Design

At least four cases fall under this heading, depending on how the strata and the groups are related:

i. The strata are identical to the groups of the model shown in (7.5.3). The sample sizes in the groups (= strata) are then fixed, and the regression estimator (7.6.1) is identical to the π estimator already discussed in Section 3.7.
ii. The strata cut across the model groups. For example, in a population of individuals, convenient strata ($h = 1, \ldots, H$) may be formed by a geographic classification, whereas G intersecting groups are formed by age/sex categories. Assume that the age/sex groups are a major factor in explaining y. Suppose that the geographic strata have little explanatory power; their existence rests more on practical or administrative grounds. In this situation, the sample s is thus stratified geographically and poststratified by age/sex groups. Let N_{gh} denote the population frequency in the cell gh, and denote the marginal frequencies by $N_{g\cdot} = \sum_{h=1}^{H} N_{gh}$ and $N_{\cdot h} = \sum_{g=1}^{G} N_{gh}$. In stratum h, $n_{\cdot h}$ elements are sampled from $N_{\cdot h}$. The part of the sample from stratum h that falls in group g is s_{gh}, and its size is denoted n_{gh}. The estimator shown in equation (7.6.1) now becomes

$$\hat{t}_{ypos} = \sum_{g=1}^{G} N_{g\cdot} \frac{\sum_{h=1}^{H} N_{\cdot h} n_{gh} \bar{y}_{s_{gh}} / n_{\cdot h}}{\sum_{h=1}^{H} N_{\cdot h} n_{gh} / n_{\cdot h}} \tag{7.6.7}$$

where $\bar{y}_{s_{gh}}$ is the straight mean of the n_{gh} sample elements in the ghth sample cell s_{gh}. Here, n_{gh} is random.

iii. Suppose again a situation with H geographic strata that cut across G age/sex groups. An alternative to case (ii) is to admit model parameters specific to each of the $G \times H$ cells formed by the intersection of the strata with the groups. The model is then

$$\begin{cases} E_\xi(y_k) = \beta_{gh} \\ V_\xi(y_k) = \sigma^2_{gh} \end{cases} \tag{7.6.8}$$

for elements k in the population cell U_{hg}, for $h = 1, \ldots, H$; $g = 1, \ldots, G$. Although the strata may not be seen as a primary factor in explaining y, they are taken into account in the model. For example, with six age/sex categories and eight geographic strata, the model (7.6.8) contains 48 model group means to estimate. By comparison, the model in case (ii) had only six model means requiring estimation. It is given as an exercise to show that the regression estimator generated by (7.6.8) is

$$\hat{t}_{y\mathrm{pos}} = \sum\sum N_{gh} \bar{y}_{s_{gh}} \tag{7.6.9}$$

where $\sum\sum$ is over $g = 1, \ldots, G$ and $h = 1, \ldots, H$. A drawback of (7.6.9) is the risk of empty or near-empty sample cells s_{gh}, which means that $\bar{y}_{s_{gh}}$ is either undefined or unstable as an estimate of the cell mean $\bar{y}_{U_{gh}} = \sum_{U_{gh}} y_k / N_{gh}$. For this reason, the estimator (7.6.7), which combines the strata, may be preferred; it is virtually as efficient as (7.6.9) when the strata contribute little toward explaining the y variable.

iv. Nested arrangements. One example is the subdivision of a stratum into groups that are not necessarily identical in each stratum. In a survey of households, one may, for example, stratify by household type. Groups within a stratum may then be formed by other household characteristics, with unequal numbers of groups in the different strata.

7.7. The Group Ratio Model and the Separate Ratio Estimator

The group ratio model is suitable when a known positive auxiliary variable, x, is available, and the ratio y_k/x_k is deemed roughly constant for elements in the same group. The ratio may differ considerably for elements in different groups. Examples 7.5.2 and 7.5.3 showed some practical applications of the model. If the G groups can be identified prior to sampling, they can serve as strata. Otherwise, the groups are poststrata, that is, identified after sampling, for sampled elements only. Result 7.5.1 contains the material necessary for constructing the ratio estimator.

Result 7.7.1. *Under the group ratio model given by* (7.5.6), *the regression estimator of* $t_y = \sum_U y_k$ *is*

$$\hat{t}_{yr} = \sum_{g=1}^{G} t_{xg} \hat{B}_g \tag{7.7.1}$$

where $\hat{B}_g$ *is given by* (7.5.10) *and*

$$t_{xg} = \sum_{U_g} x_k$$

The approximate variance is obtained from the general equation (7.2.10) *by setting, for* $g = 1, \ldots, G$,

$$E_k = y_k - B_g x_k \quad \text{for} \quad k \in U_g$$

where B_g *is given by* (7.5.9). *The variance estimator is obtained from the general equation* (7.2.11) *by setting, for* $g = 1, \ldots, G$,

$$e_{ks} = y_k - \hat{B}_g x_k \quad \textit{for} \quad k \in s_g$$

$$g_{ks} = t_{xg}/\hat{t}_{xg\pi} \quad \textit{for all} \quad k \in s_g$$

with

$$\hat{t}_{xg\pi} = \sum_{s_g} \check{x}_k$$

The proof involves a simple utilization of equations (7.5.9) and (7.5.10) in Result 7.5.1; no details need be given here. Setting $x_k = 1$ for all k yields Result 7.6.1.

The estimator (7.7.1) is called the *poststratified ratio estimator* or the *separate ratio estimator*. The latter term is used in particular when the groups are identified before sampling and used for stratified sampling. The group totals t_{xg} in (7.7.1) must be derived from the frame or from a reliable external source. Care should be exercised not to allow the groups to be too small. If a group sample count n_{s_g} is too small, $t_{xg}\hat{B}_g$ may have a nonnegligible bias as an estimator of the group total $t_{yg} = \sum_{U_g} y_k$. Although the bias from a single group may be modest, the accumulated bias over all groups may become considerable. A rule of thumb is to keep the number of groups sufficiently small so that no group sample count is less than 20. Let us apply Result 7.7.1 to some specific designs.

The SI Design

The realized sample is poststratified. The estimator given by (7.7.1) now takes the form

$$\hat{t}_{yr} = \sum_{g=1}^{G} t_{xg} \frac{\sum_{s_g} y_k}{\sum_{s_g} x_k} \tag{7.7.2}$$

with the approximate variance

$$AV_{SI}(\hat{t}_{yr}) = N^2 \frac{1-f}{n} \sum_{g=1}^{G} W_g' S_{EU_g}^2 \tag{7.7.3}$$

where $W_g' = (N_g - 1)/(N - 1)$, $f = n/N$ and

$$S_{EU_g}^2 = \frac{1}{N_g - 1} \sum\nolimits_{U_g} (y_k - B_g x_k)^2 \tag{7.7.4}$$

with $B_g = \sum_{U_g} y_k / \sum_{U_g} x_k$. To obtain the variance estimator, we let $\bar{x}_{U_g} = \sum_{U_g} x_k / N_g$, $\bar{x}_{s_g} = \sum_{s_g} x_k / n_{s_g}$, and

$$S_{es_g}^2 = \frac{1}{n_{s_g} - 1} \sum\nolimits_{s_g} (y_k - \hat{B}_g x_k)^2 \tag{7.7.5}$$

with $\hat{B}_g = \sum_{s_g} y_k / \sum_{s_g} x_k$. The variance estimator is then

$$\hat{V}_{SI}(\hat{t}_{yr}) = (1 - f) \sum_{g=1}^{G} (\bar{x}_{U_g} / \bar{x}_{s_g})^2 N_g^2 S_{es_g}^2 / n_{s_g} \tag{7.7.6}$$

where we have approximated $(n_{s_g} - 1)n/n_{s_g}(n - 1)$ by unity.

These expressions follow by easy manipulations of the general formulas in Result 7.7.1. The fact that $\bar{E}_{U_g} = \bar{e}_{s_g} = 0$ brings some simplification.

The STSI Design

We consider only the case where groups and strata are identical. The separate regression estimator of t_y is then

$$\hat{t}_{yr} = \sum_{g=1}^{G} t_{xg} \frac{\sum_{s_g} y_k}{\sum_{s_g} x_k} \tag{7.7.7}$$

thus identical in form to the estimator (7.7.2) obtained under the *SI* design. (The sampling distribution is different in the two cases.) The approximate variance is

$$AV_{STSI}(\hat{t}_{yr}) = \sum_{g=1}^{G} N_g^2 \frac{1 - f_g}{n_g} S_{EU_g}^2 \tag{7.7.8}$$

where the stratum sample size n_g is now fixed in advance, and $S_{EU_g}^2$ is given by (7.7.4). The variance estimator is

$$\hat{V}_{STSI}(\hat{t}_{yr}) = \sum_{g=1}^{G} \left(\frac{\bar{x}_{U_g}}{\bar{x}_{s_g}} \right)^2 N_g^2 \frac{1 - f_g}{n_g} S_{es_g}^2 \tag{7.7.9}$$

where $S_{es_g}^2$ is given by (7.7.5).

The approximate variances given by (7.7.3) and (7.7.8) are very nearly equivalent if the stratified sample is allocated by the proportional rule $n_g = nN_g/N$; $g = 1, \ldots, G$. The allocation that minimizes (7.7.8) is

$$n_g \propto N_g S_{EU_g}$$

To use this rule requires information on the group residual variances, something that may not be readily available.

7.8. Simple Regression Models and Simple Regression Estimators

In many populations where a strong linear relationship exists between the study variable y and a single auxiliary variable x, the population regression line will intercept the y axis at some distance from the origin. A model with an intercept term will then give a better regression estimator than the common ratio model discussed in Section 7.4. With only one x variable, we have a simple regression model, as opposed to the multiple regression models in Section 7.9.

The *simple regression model* of the finite population scatter states that, for $k \in U$,

$$\begin{cases} E_\xi(y_k) = \alpha + \beta x_k \\ V_\xi(y_k) = \sigma^2 \end{cases} \tag{7.8.1}$$

where α, β, and σ^2 are unknown parameters and $x_1, \ldots, x_N$ are known but not necessarily positive values of an auxiliary variable x. We can also call (7.8.1) a *common simple regression model*, since the same model is assumed for all elements in the population. The fit of this model to the whole finite population (see Examples 6.4.1, 6.4.2, and 6.5.2) leads to estimating α and β by, respectively,

$$A = \bar{y}_U - B\bar{x}_U$$

and

$$B = \frac{\sum_U (x_k - \bar{x}_U)(y_k - \bar{y}_U)}{\sum_U (x_k - \bar{x}_U)^2} \tag{7.8.2}$$

with $\bar{y}_U = \sum_U y_k/N$; $\bar{x}_U = \sum_U x_k/N$. The sample fit of the same model gives

$$\hat{A} = \tilde{y}_s - \hat{B}\tilde{x}_s$$

$$\hat{B} = \frac{\sum_s (x_k - \tilde{x}_s)(y_k - \tilde{y}_s)/\pi_k}{\sum_s (x_k - \tilde{x}_s)^2/\pi_k} \tag{7.8.3}$$

where

$$\tilde{y}_s = \textstyle\sum_s \check{y}_k/\hat{N}; \qquad \tilde{x}_s = \sum_s \check{x}_k/\hat{N}; \qquad \hat{N} = \sum_s 1/\pi_k \tag{7.8.4}$$

We now have the following result.

Result 7.8.1. *Under the model* (7.8.1), *the regression estimator of* $t_y = \sum_U y_k$ *is*

$$\hat{t}_{yr} = N[\tilde{y}_s + \hat{B}(\bar{x}_U - \tilde{x}_s)] \tag{7.8.5}$$

where $\hat{B}$ *is given by* (7.8.3). *The approximate variance is obtained from the general equation* (7.2.10) *by setting*

$$E_k = y_k - \bar{y}_U - B(x_k - \bar{x}_U) \tag{7.8.6}$$

where B is given by (7.8.2). *The variance estimator is obtained from* (7.2.11) *by taking*

$$e_{ks} = y_k - \tilde{y}_s - \hat{B}(x_k - \tilde{x}_s) \tag{7.8.7}$$

$$g_{ks} = \frac{N}{\hat{N}}\{1 + a_s(x_k - \tilde{x}_s)\} \tag{7.8.8}$$

with $\tilde{y}_s$, $\tilde{x}_s$, *and* $\hat{N}$ *given by equation* (7.8.4), *and*

$$a_s = \frac{\bar{x}_U - \tilde{x}_s}{\tilde{S}^2_{xs}}$$

where

$$\tilde{S}^2_{xs} = \frac{\sum_s (x_k - \tilde{x}_s)^2/\pi_k}{\sum_s 1/\pi_k}$$

The result summarizes conclusions from Examples 6.4.2, 6.5.2, and 6.6.2. Note that $\sum_s \check{e}_{ks} = 0$. We can proceed to discuss the estimator and its properties for various specific designs. This time we do so only briefly, since the reader is now familiar with such analyses.

Remark 7.8.1. The estimator of the population mean implied by Result 7.8.1 is

$$\hat{\bar{y}}_U = \tilde{y}_s + \hat{B}(\bar{x}_U - \tilde{x}_s)$$

To obtain the approximate variance and the variance estimator, divide the corresponding expressions in Result 7.8.1 by N^2.

In the following we denote the estimator given by (7.8.5) as $\hat{t}_{yreg}$. In particular, under the *SI* design, we obtain

$$\hat{t}_{yreg} = N[\bar{y}_s + \hat{B}(\bar{x}_U - \bar{x}_s)] \tag{7.8.9}$$

where

$$\hat{B} = \frac{\sum_s (x_k - \bar{x}_s)(y_k - \bar{y}_s)}{\sum_s (x_k - \bar{x}_s)^2}$$

and $\bar{x}_s$, $\bar{y}_s$ are straight sample means. The approximate variance is

$$AV_{SI}(\hat{t}_{yreg}) = N^2 \frac{1-f}{n} S^2_{yU}(1 - r^2) \tag{7.8.10}$$

where

$$r = \frac{S_{xyU}}{S_{xU}S_{yU}}$$

is the finite population correlation coefficient.

The variance estimator is

$$\hat{V}(\hat{t}_{yreg}) = N^2 \frac{1-f}{n} \frac{1}{n-1} \sum_s [1 + a_s(x_k - \bar{x}_s)]^2 e_{ks}^2$$

with $e_{ks} = y_k - \bar{y}_s - \hat{B}(x_k - \bar{x}_s)$ and $a_s = n(\bar{x}_U - \bar{x}_s)/\sum_s (x_k - \bar{x}_s)^2$. Under *SI* sampling, the regression estimator given by (7.8.9) will ordinarily perform better than both the expansion estimator $N\bar{y}_s$ and the ratio estimator $\hat{t}_{yra} = N\bar{x}_U(\bar{y}_s/\bar{x}_s)$. We have, from (7.8.10),

$$\frac{AV_{SI}(\hat{t}_{yreg})}{V_{SI}(N\bar{y}_s)} = 1 - r^2$$

Consequently, in large samples, the regression estimator is an improvement over the simple expansion estimator as soon as $r \neq 0$, and if $r > 0.8$, say, there is a very important improvement.

The comparison with the ratio estimator gives (see Exercise 7.23)

$$AV_{SI}(\hat{t}_{yreg}) \leq AV_{SI}(\hat{t}_{yra}) \tag{7.8.11}$$

with equality if and only if

$$B = \frac{\sum_U (x_k - \bar{x}_U)(y_k - \bar{y}_U)}{\sum_U (x_k - \bar{x}_U)^2} = \frac{\bar{y}_U}{\bar{x}_U}$$

Thus the regression estimator is better than the ratio estimator as soon as $B \neq \bar{y}_U/\bar{x}_U$. Note the meaning of "better" here: The conclusions are based on the AV expressions; thus they certainly hold in large samples, but they can also be counted on to hold in many cases with medium or even modest sample sizes.

The ratio estimator $\hat{t}_{yra}$ is preferred in some surveys to the regression estimator $\hat{t}_{yreg}$, even though the latter estimator would yield nonnegligible gains in efficiency. There are several reasons for this. The ratio estimator has a simpler structure, and it has advantages in cases where estimates are required for both the y-total $\sum_U y_k$ and the ratio of totals, $\sum_U y_k/\sum_U x_k$.

In very small samples, $\hat{t}_{yra}$ may have a smaller variance than $\hat{t}_{yreg}$, despite the large sample conclusion (7.8.11). Empirical studies on small populations, with *SI* samples of size 12 or less, have shown that the mean square error, that is, variance plus squared bias, may be smaller for the ratio estimator than for the regression estimator $\hat{t}_{yreg}$. These studies suggest that the lower efficiency of the regression estimator is sometimes due not to a higher bias, but to a higher variance. Thus very small samples may be a factor in favor of the ratio estimator. The form of the population is also a factor of importance. Nevertheless, failure to use the regression estimator (7.8.5) when the conditions for this estimator are fulfilled may cause considerable loss of precision. The regression estimator is better than the ratio estimator when an intercept in the regression leads to a more complete explanation of the y variable.

Under the *STSI* design, the estimator (7.8.5) takes the form

$$\hat{t}_{yreg} = N[\bar{y}_{st} + \hat{B}(\bar{x}_U - \bar{x}_{st})] \tag{7.8.12}$$

where $\bar{y}_{st} = \sum_{h=1}^{H} (N_h/N)\bar{y}_{s_h}$, $\bar{x}_{st}$ is analogous, and

$$\hat{B} = \frac{\sum_{h=1}^{H} (N_h/n_h) \sum_{s_h} (x_k - \bar{x}_{st})(y_k - \bar{y}_{st})}{\sum_{h=1}^{H} (N_h/n_h) \sum_{s_h} (x_k - \bar{x}_{st})^2}$$

The formula is known as the *combined regression estimator.* It is left as an exercise to determine the AV expression and the variance estimator in this case.

Some practical situations may call for fitting a *group regression model.* That is, a simple regression is fitted separately in each of a number of population groups with a priori known sizes, $N_1, \ldots, N_g, \ldots, N_G$. In this case, the regression estimator (7.2.8) takes the form

$$\hat{t}_{yr} = \sum_{g=1}^{G} N_g[\tilde{y}_{s_g} + \hat{B}_g(\bar{x}_{U_g} - \tilde{x}_{s_g})] \tag{7.8.13}$$

with

$$\hat{B}_g = \frac{\sum_{s_g} (x_k - \tilde{x}_{s_g})(y_k - \tilde{y}_{s_g})/\pi_k}{\sum_{s_g} (x_k - \tilde{x}_{s_g})^2/\pi_k}$$

and $\sim$ indicates the π-weighted means, that is,

$$\tilde{y}_{s_g} = \frac{\sum_{s_g} y_k/\pi_k}{\sum_{s_g} 1/\pi_k}$$

and $\tilde{x}_{s_g}$ is analogous to $\tilde{y}_{s_g}$.

If group membership is established after sampling, for the selected elements, (7.8.13) may be termed the *poststratified regression estimator.* To work out the variance and the variance estimator of (7.8.13) is left as an exercise. In the special case where each group is a stratum and $STSI$ sampling is used, (7.8.13) is often called the *separate regression estimator*, because a regression is fitted in each stratum.

7.9. Estimators Based on Multiple Regression Models

In this section, we examine applications of regression estimation when the model contains multiple explanatory variables. The section has two parts, dealing with the following cases: (i) the underlying model is a multiple regression with two or more quantitative x variables; (ii) the underlying model is an analysis of variance model with two or more explanatory factors, each at a number of levels. In these applications, we work directly with the general matrix forms from equations (7.2.8) and (7.2.9); there are usually no simplified expressions for the regression estimator, in contrast to the models treated earlier in this chapter.

Recall that the precision attained by the general regression estimator (7.2.8) depends on two factors: (a) the model that is fitted to obtain the estimator and (b) the sampling design. The fitted model determines the estimator formula, thereby influencing the variance. The sampling design determines the sampling distribution of the estimator and thereby its variance. The realized variance is in part due to a design effect and in part to a model fit effect. The two effects interact (see, Särndal 1981).

The AV-expression (7.2.10) is a function of the regression residuals E_k. Smaller residuals generally lead to a smaller variance. This is particularly evident under the SI design, where (7.2.10) can be written

$$AV(\hat{t}) = N^2 \frac{1-f}{n} S_{yU}^2 (1 - R^2) \tag{7.9.1}$$

where

$$R^2 = 1 - \frac{\sum_U E_k^2}{\sum_U (y_k - \bar{y}_U)^2} = 1 - \frac{\sum_U E_k^2}{(N-1)S_{yU}^2}$$

Here, R^2 is the coefficient of multiple determination for the population fit. A basic result in regression analysis states that R^2 will increase (or stay the same) when a new x variable is added to the regression.

In the case of SI sampling, if a clear-cut improvement in R^2 is perceived by adding a given x variable, there is strong incentive to use this variable in the regression estimator. If the improvement is deemed marginal at best, a sensible rule is not to further complicate the estimator by including such an x variable. To fit a simple model is better than to fit a more complicated one, unless the latter pays clear dividends in terms of efficiency gains.

Under designs other than SI, the conclusions are less obvious, since the realized variance is no longer a simple function of R^2. There is then an interaction between the effect of the model fit and the effect of the sampling design.

A factor to consider for regression estimators is the cost of the "informed expert" who will decide (y variable by y variable if the survey is multipurpose) what x variables to include in the fitted model. Another, although minor, factor is the cost of computation, which may be higher for complex regression estimators.

7.9.1. Multiple Regression Models

If a regression estimator is built from a model with two or more x variables, the hope is that the x variables jointly explain the y variable well, with increased efficiency as a result. To illustrate the performance of several regression estimators of the total $t_y = \sum_U y_k$ and to study the effect on variance that one or more auxiliary variables can have, a Monte Carlo simulation was carried out in which 5,000 repeated SI samples, each with $n = 100$, were drawn from the MU281 population of Swedish municipalities (see Appendix B).

Monte Carlo simulation is often used when an exact description of the sampling distribution of a given estimator is difficult to obtain. To obtain the exact distribution, one would have to consider all samples s that are possible under the given design. For every s, one must know the probability $p(s)$ of drawing s, and the value of the estimator, $\hat{t} = \hat{t}(s)$, of the population total t_y. It would then be possible to calculate the exact values of the expected value, the bias, and the variance of $\hat{t}$. However, this is ordinarily an impossible undertaking, since the number of possible samples is much too large. This is why simulation is frequently used to study the statistical properties of estimators used in surveys.

A simulation aimed at studying one or several estimators is typically carried out as follows. The finite population and the design are held fixed. A large number of samples is drawn from the given population according to the given design. Once drawn, a sample is replaced before the next one is drawn, so that it is always the same population that is sampled. The number of samples is denoted K, where K is a large number. When several estimators are studied in the same simulation, a large amount of calculation may be involved, but this is not a problem with modern computers. For every realized sample, the estimate $\hat{t}$ and the variance estimate $\hat{V}(\hat{t})$ are calculated. If K is large enough, the distribution of the K estimates, which may be termed the empirical sampling distribution, will closely approximate the exact sampling distribution that we cannot easily obtain. If we let $\hat{t}_j$ and $\hat{V}(\hat{t})_j$ denote the results obtained for the jth sample, we can calculate

$$\bar{\hat{t}} = \frac{1}{K} \sum_{j=1}^{K} \hat{t}_j$$

which is an estimate of the expected value $E(\hat{t})$,

$$S_{\hat{t}}^2 = \frac{1}{K-1} \sum_{j=1}^{K} (\hat{t}_j - \bar{\hat{t}})^2$$

which is an estimate of the variance $V(\hat{t})$, and finally

$$\bar{\hat{V}} = \frac{1}{K} \sum_{j=1}^{K} \hat{V}(\hat{t})_j$$

which is an estimate of the expected value of the variance estimator, that is, $E[\hat{V}(\hat{t})]$. If for each sample we also calculate the confidence interval at the approximate 95% level,

$$\hat{t} \pm 1.96[\hat{V}(\hat{t})]^{1/2}$$

and then count the number of intervals, R, say, that contain the true value of the total t, then R/K is an estimate of the actual confidence level. The actual confidence level may differ from 95%, because $(\hat{t} - t)/[\hat{V}(\hat{t})]^{1/2}$ follows only approximately the unit normal distribution.

Table 7.1. Regression Analysis on the Population of $N = 281$ Swedish Municipalities.

	Regression coefficient $\times 10^3$			
Regression	B_0	B_1	B_2	$1 - R^2$
y on x_1 only	-6.43	2.84	—	56.8%
y on x_2 only	-21.29	—	1.82	57.7%
y on x_1 and x_2	-38.30	2.48	1.59	25.3%

The simulation reported here used the following variables for the MU281 population in Appendix B. The study variable y was RMT85 $\times 10^{-4}$, where RMT85 is municipal tax receipts in 1985. Two auxiliary variables, x_1 and x_2, were used, where x_1 is CS82, which is the number of Conservative Party seats in the municipal council, and x_2 is SS82, which is the number of Social Democrat Party seats in the municipal council. For municipality k, the respective values of the three variables are denoted y_k, x_{1k} and x_{2k}, $k = 1, \ldots, 281$. Table 7.1 shows the results from several regression analyses based on all 281 data points. Each x variable by itself leaves approximately 57% of the y variation unexplained; if both x variables are included in the regression, only 25.3% of the variation in y remains unexplained. A regression estimator with the two x variables as auxiliaries should thus outperform one that uses only one of the x variables.

Moreover, the intercept is important in all three regressions. We should expect a simple regression estimator to perform better than a ratio estimator.

The following estimators were studied:

a. The π estimator

$$\hat{t}_1 = \hat{t}_\pi = N\bar{y}_s \tag{7.9.2}$$

b. Two ratio estimators, the first with x_1, the second with x_2 as the only auxiliary variable

$$\hat{t}_2 = \hat{t}_{yra}(x_1) = \sum_U x_{1k} \frac{\sum_s y_k}{\sum_s x_{1k}} \tag{7.9.3}$$

and

$$\hat{t}_3 = \hat{t}_{yra}(x_2) = \sum_U x_{2k} \frac{\sum_s y_k}{\sum_s x_{2k}} \tag{7.9.4}$$

c. Two simple regression estimators, the first with x_1, the second with x_2 as the only auxiliary variable

$$\hat{t}_4 = \hat{t}_{yreg}(x_1) = N[\bar{y}_s + \hat{B}_1(\bar{x}_{1U} - \bar{x}_{1s})] \tag{7.9.5}$$

and

$$\hat{t}_5 = \hat{t}_{yreg}(x_2) = N[\bar{y}_s + \hat{B}_2(\bar{x}_{2U} - \bar{x}_{2s})] \tag{7.9.6}$$

with

$$\hat{B}_j = \frac{\sum_s (x_{jk} - \bar{x}_{js})(y_k - \bar{y}_s)}{\sum_s (x_{jk} - \bar{x}_{js})^2}$$

where $j = 1$ for (7.9.5); $j = 2$ for (7.9.6).

d. The multiple regression estimator using both x_1 and x_2 as auxiliary variables

$$\begin{aligned}\hat{t}_6 = \hat{t}_{yr}(x_1, x_2) &= \sum_U (\hat{B}_0 + \hat{B}_1 x_{1k} + \hat{B}_2 x_{2k}) \\ &= N[\bar{y}_s + \hat{B}_1(\bar{x}_{1U} - \bar{x}_s) + \hat{B}_2(\bar{x}_{2U} - \bar{x}_{2s})]\end{aligned} \tag{7.9.7}$$

where

$$(\hat{B}_0, \hat{B}_1, \hat{B}_2)' = (\textstyle\sum_s \mathbf{x}_k \mathbf{x}_k')^{-1} \sum_s \mathbf{x}_k y_k$$

with

$$\mathbf{x}_k = (1, x_{1k}, x_{2k})'$$

For each of the 5,000 samples, the six estimates shown in equations (7.9.2) to (7.9.7) were calculated. For each estimate, two different variance estimates were calculated, namely, the g-weighted formula given by (7.2.11), which is

$$\hat{V}_g = \hat{V}_g(\hat{t}) = N^2 \frac{1-f}{n} \frac{\sum_s g_{ks}^2 e_{ks}^2}{n-1} \tag{7.9.8}$$

and the simplified variance estimate obtained by letting all $g_{ks} = 1$,

$$\hat{V}_{\text{sim}} = \hat{V}_{\text{sim}}(\hat{t}) = N^2 \frac{1-f}{n} \frac{\sum_s e_{ks}^2}{n-1} \tag{7.9.9}$$

Here, e_{ks} and g_{ks} are the expressions appropriate for each particular regression estimator, $\hat{t}_2$ to $\hat{t}_6$. For the π estimator, equation (7.9.9) applies, with $e_{ks} = y_k - \bar{y}_s$. For each of the 5,000 estimates obtained by an estimator $\hat{t}$, we let the computer calculate the two confidence intervals

$$\hat{t} \pm 1.96\hat{V}_g^{1/2} \quad \text{and} \quad \hat{t} \pm 1.96\hat{V}_{\text{sim}}^{1/2} \tag{7.9.10}$$

For each of the two intervals, it was verified whether the true population total t_y was contained in the interval or not.

Summary measures over the 5,000 samples were calculated. The results are displayed in Table 7.2, where $\bar{\hat{t}}$ and $S_{\hat{t}}^2$ are the mean and variance of the 5,000 estimates $\hat{t}$; $\bar{\hat{V}}_g$ and $\bar{\hat{V}}_{\text{sim}}$ are the means of the 5,000 variance estimates, $\hat{V}_g$ and $\hat{V}_{\text{sim}}$, respectively; and ECR_g and ECR_{sim} are the respective coverage rates, in percent, obtained, respectively, from the intervals given by (7.9.10). For instance, ECR_g is 100 times $R/5{,}000$, where R is the number of intervals found to contain the true value t_y, when the g-weighted interval was used. The final

Table 7.2. Results of a Simulation Involving 5,000 *SI* Samples of Size $n = 100$ Each; the Population Total Is $t_y = 5.315$.

Estimator	$\bar{\hat{t}}$	$S_{\hat{t}}^2$	$\bar{\hat{V}}_g$	ECR_g	$\bar{\hat{V}}_{\text{sim}}$	ECR_{sim}	AV
$\hat{t}_1 = \hat{t}_\pi$	5.31	0.204	—	—	0.203	93.6	0.204
$\hat{t}_2 = \hat{t}_{yra}(x_1)$	5.31	0.121	0.120	93.1	0.121	93.2	0.121
$\hat{t}_3 = \hat{t}_{yra}(x_2)$	5.31	0.141	0.141	93.9	0.141	93.8	0.142
$\hat{t}_4 = \hat{t}_{yreg}(x_1)$	5.30	0.119	0.115	93.1	0.114	92.5	0.116
$\hat{t}_5 = \hat{t}_{yreg}(x_2)$	5.30	0.119	0.118	93.9	0.116	93.4	0.117
$\hat{t}_6 = \hat{t}_{yr}(x_1, x_2)$	5.31	0.054	0.052	93.2	0.050	92.5	0.052

column in Table 7.2 gives the value of the AV expression for $\hat{t}$,

$$AV(\hat{t}) = N^2 \frac{1-f}{n} \frac{\sum_U E_k^2}{N-1}$$

where E_k is the appropriate population fit residual.

For the π estimator, the value in the AV column equals the exact variance,

$$V(\hat{t}_\pi) = N^2 \frac{1-f}{n} S_{yU}^2$$

Simulation summary measures do not give an exact picture of the true underlying characteristics. For example, $\bar{\hat{t}}$ estimates the true expected value of $\hat{t}$, but only to the degree of accuracy that the limited number of 5,000 repetitions can be expected to give. The imperfection caused by the finite number of repetitions is more keenly felt in the case of a variance measure ($S_{\hat{t}}^2$ in our simulation) than in the case of measures calculated as means ($\bar{\hat{t}}$, $\bar{\hat{V}}_g$, and $\bar{\hat{V}}_{\text{sim}}$).

Table 7.2 prompts the following comments.

1. The true population total is 5.315. For each of the six estimators, the value $\bar{\hat{t}}$ comes very close to this true total, indicating negligible bias in all regression estimators, $\hat{t}_2$ to $\hat{t}_6$. This is expected when n is as large as 100.
2. One notes that the four quantities $S_{\hat{t}}^2$, $\bar{\hat{V}}_g$, $\bar{\hat{V}}_{\text{sim}}$, and AV agree closely, for each estimator. Here, $S_{\hat{t}}^2$ estimates the true variance, to the degree of precision obtained with 5,000 repetitions. Note that AV is an approximate variance, in the case of the five regression estimators. That $S_{\hat{t}}^2$ and AV are close indicates that AV accurately represents the true variance when $n = 100$. For a considerably smaller sample size, there is likely to be some discrepancy between the exact variance and AV.
3. That both $\bar{\hat{V}}_g$ and $\bar{\hat{V}}_{\text{sim}}$ agree closely with $S_{\hat{t}}^2$ is a sign that the two variance estimators are unbiased or very nearly so. Again, this is not surprising when $n = 100$. For small sample sizes, both $\hat{V}_g$ and in particular $\hat{V}_{\text{sim}}$ have a tendency to underestimate the true variance. Leblond (1990) has studied the underestimation, theoretically and empirically, in the case of the ratio estimator.

4. The empirical coverage rates ECR_g and ECR_{sim} are very close to (but somewhat short of) the nominal 95% rate aimed at by the confidence interval technique. The g-weighted confidence interval procedure is slightly better (closer to the nominal 95%) than the simplified procedure.
5. The importance of including the intercept in the fitted model comes out most clearly in comparing $\hat{t}_{yra}(x_2)$ with the clearly better alternative $\hat{t}_{yreg}(x_2)$. The intercept is seen in Table 7.1 to be especially important for the x_2-based regression.
6. As expected, given the regression analysis in Table 7.1, the multiple regression estimator $\hat{t}_6$ outperforms the two simple regression estimators $\hat{t}_4$ and $\hat{t}_5$.

We duplicated the simulation, with 5,000 repeated samples, but with $n = 50$ as the size of each sample. The results were similar, but the two ECRs fell increasingly short of the nominal 95% level (ECR_g of approximately 92%; approximately 1% lower for the ECR_{sim}). This impression, also confirmed in other studies, suggests that the normality based constant 1.96 is too small to render an empirical coverage rate of 95%. Improvement will be obtained by replacing 1.96 by the value, $t_{n-1,1-\alpha/2}$, found in a table of the t distribution with $n - 1$ degrees of freedom. With $n = 50$, $t_{49;0.975} = 2.01$, which improves the ECR slightly, compared to using 1.96.

7.9.2. Analysis of Variance Models

This section deals with additive effects models of the type used in the analysis of variance. Consider a population of individuals and a two-way classification with GJ cells, U_{gj}, $g = 1, \ldots, G$; $j = 1, \ldots, J$, formed by G age groups crossed with J occupational groups. The population cell counts N_{gj} are unknown, so poststratification on the GJ cells is impossible. But we assume that the marginal counts can be easily obtained. The two sets of marginals may come from separate sources, for example, one from a census, the other from an administrative file. That is, the marginal counts

$$N_{g\cdot} = \sum_{j=1}^{J} N_{gj}; \qquad N_{\cdot j} = \sum_{g=1}^{G} N_{gj}$$

are known for $g = 1, \ldots, G$; $j = 1, \ldots, J$. The objective is to construct a regression estimator that profits from this auxiliary information. The study variable value y_k and the cell membership is observed for the individuals k in a probability sample s. The $\mathbf{x}_k$ vector appropriate for this case is given by

$$\mathbf{x}_k = (\delta_{1\cdot}, \ldots, \delta_{G\cdot}, \delta_{\cdot 1}, \ldots, \delta_{\cdot J})' \tag{7.9.11}$$

where $\delta_{g\cdot} = 1$ if k belongs to age group g, and $\delta_{g\cdot} = 0$ otherwise, $g = 1, \ldots, G$, whereas $\delta_{\cdot j} = 1$ if k belongs to occupational group j, and $\delta_{\cdot j} = 0$ otherwise, $j = 1, \ldots, J$. For every k, $\mathbf{x}_k$ contains one entry "1" to indicate age group, another entry "1" to indicate occupational group, and the other $G + J - 2$ entries are "0". The population total of $\mathbf{x}_k$,

$$\textstyle\sum_U \mathbf{x}_k = (N_{1\cdot}, \ldots, N_{G\cdot}, N_{\cdot 1}, \ldots, N_{\cdot J})'$$

corresponds exactly to what is known about the population, namely, the $G + J$ marginal counts. For simplicity, we think in terms of a model with $\sigma_k^2 = \sigma^2$ for all k. The g weights given in general by (7.2.9) can then be written as

$$g_{ks} = 1 + \mathbf{x}_k'\boldsymbol{\mu}$$

where the vector $\boldsymbol{\mu} = (u_1, \ldots, u_G, v_1, \ldots, v_J)'$ is obtained as a solution of

$$(\textstyle\sum_s \mathbf{x}_k\mathbf{x}_k'/\pi_k)\boldsymbol{\mu} = \sum_U \mathbf{x}_k - \sum_s \mathbf{x}_k/\pi_k \tag{7.9.12}$$

When $\mathbf{x}_k$ is given by (7.9.11), this is a system of $G + J$ equations structured as follows. Denote by $s_{gj} = U_{gj} \cap s$ the part of the sample s that falls in the gjth cell, and by $\hat{N}_{gj} = \sum_{s_{gj}} 1/\pi_k$ the cell count estimate. The elements, denoted m_{ab}, of the symmetrical $(G + J) \times (G + J)$ matrix $\sum_s \mathbf{x}_k\mathbf{x}_k'/\pi_k$ are functions of the $\hat{N}_{gj}$. On the diagonal, we have, for $g = 1, \ldots, G$, $m_{gg} = \sum_{j=1}^{J} \hat{N}_{gj} = \hat{N}_{g\cdot}$ and, for $j = 1, \ldots, J$, $m_{G+j,G+j} = \sum_{g=1}^{G} \hat{N}_{gj} = \hat{N}_{\cdot j}$. The elements off the diagonal are $m_{g,G+j} = m_{G+j,g} = \hat{N}_{gj}$, for $g = 1, \ldots, G$, $j = 1, \ldots, J$. All other off-diagonal elements are zero. Furthermore, on the right-hand side of the system of equations (7.9.12),

$$\textstyle\sum_U \mathbf{x}_k - \sum_s \mathbf{x}_k/\pi_k = (N_{1\cdot} - \hat{N}_{1\cdot}, \ldots, N_{G\cdot} - \hat{N}_{G\cdot}, N_{\cdot 1} - \hat{N}_{\cdot 1}, \ldots, N_{\cdot J} - \hat{N}_{\cdot J})'$$

In this case, there is no unique solution to (7.9.12), because $\sum_s \mathbf{x}_k\mathbf{x}_k'/\pi_k$ is not of full rank. The rank is $G + J - 1$, and the inverse does not exist. To obtain a solution, fix arbitrarily one component of $\boldsymbol{\mu}$, say, $v_J = 0$, and solve the system (7.9.12) for the remaining unknowns, $u_1, \ldots, u_G, v_1, \ldots, v_{J-1}$. It can be shown that $\mathbf{x}_k'\boldsymbol{\mu}$ is invariant under the fixing of one component of $\boldsymbol{\mu}$. That is, the value of $\mathbf{x}_k'\boldsymbol{\mu}$ is the same regardless of which component is fixed, and regardless of the constant value assigned to this component. A unique set of g-weights $g_{ks} = 1 + \mathbf{x}_k'\boldsymbol{\mu}$ is thereby obtained. With these weights, the regression estimator of $t_y = \sum_U y_k$ is given as usual by

$$\hat{t}_{yr} = \textstyle\sum_s g_{ks} y_k/\pi_k \tag{7.9.13}$$

The model that corresponds to this regression estimator is the two-way ANOVA model with no interactions. (It is an unbalanced model because the N_{gj} are ordinarily not equal.) To see this, let $\boldsymbol{\beta} = (\alpha_1, \ldots, \alpha_G, \beta_1, \ldots, \beta_J)'$, where the α_g and the β_j are unknown model parameters. With $\mathbf{x}_k$ given by (7.9.11), we then have $\mathbf{x}_k'\boldsymbol{\beta} = \alpha_g + \beta_j$ whenever $k \in U_{gj}$. This implies that the underlying model ξ is such that

$$E_\xi(y_k) = \mathbf{x}_k'\boldsymbol{\beta} = \alpha_g + \beta_j; \qquad V_\xi(y_k) = \sigma^2$$

for every $k \in U_{gj}$. This is an additive effects ANOVA model.

The variance estimator is given by the usual g-weighted expression,

$$\hat{V}(\hat{t}_{yr}) = \sum\sum_s (\Delta_{kl}/\pi_{kl})(g_{ks}e_{ks}/\pi_k)(g_{ls}e_{ls}/\pi_l)$$

where we use $g_{ks} = 1 + \mathbf{x}_k'\boldsymbol{\mu}$ and the residuals e_{ks} obtained from the fit of the model. These residuals are

$$e_{ks} = y_k - \mathbf{x}_k'\hat{\mathbf{B}} = y_k - (\hat{A}_g + \hat{B}_j)$$

for $k \in s_{gj} = U_{gj} \cap s$, where $\hat{\mathbf{B}} = (\hat{A}_1, \ldots, \hat{A}_G, \hat{B}_1, \ldots, \hat{B}_J)'$ is obtained by solving system of general equations

$$(\sum_s \mathbf{x}_k\mathbf{x}_k'/\pi_k)\hat{\mathbf{B}} = \sum_s \mathbf{x}_k y_k/\pi_k$$

Again, there is the problem of no unique solution. To obtain a solution $\hat{\mathbf{B}}$, fix one component, say, $\hat{B}_J = 0$, and solve for the remaining components $\hat{A}_1, \ldots, \hat{A}_G, \hat{B}_1, \ldots, \hat{B}_{J-1}$. The residuals e_{ks} and the variance estimator $\hat{V}(\hat{t}_{yr})$ can now be computed.

The technique used in this example, with weights that are benchmarked on two known marginals, goes back to Deming (1943). The estimator shown in equation (7.9.13) is closely connected with the raking ratio estimator of Deming and Stephan (1940). The theory of regression estimation leads directly to a variance estimator; this aspect is discussed in Deville and Särndal (1991). A computer program to calculate the weights, LINWEIGHT, is described in Bethlehem and Keller (1987). This program will handle extensions to multi-way tables.

7.10. Conditional Confidence Intervals

In some situations, there are good reasons to make inference conditionally on certain observed features of the sample. To illustrate, suppose the objective is to construct a $100(1-\alpha)\%$ confidence interval for the total t_y, given that a *BE* sample has been found to be of size n_{s0}. Recall that the *BE* design, with the inclusion probability π for all elements, gives a sample of random size n_s, which is distributed binomially with mean $N\pi = n$ and variance $N\pi(1-\pi)$. Having observed that $n_s = n_{s0}$, many analysts like to construct the confidence interval so that the confidence level is $1-\alpha$, say 95%, *conditionally* on the event $n_s = n_{s0}$. That is, for 95% of repeated *BE* samples obeying $n_s = n_{s0}$, the interval should contain t_y. Such an interval is called a 95% conditional confidence interval.

It follows that if the conditional coverage rate is roughly $1-\alpha$ for any fixed sample size $n_s = n_{s0}$, the coverage rate will also be $1-\alpha$ overall, that is, with respect to *BE* samples of all sizes.

Thus, a conditional confidence interval yields the desired coverage rate over all possible repeated samples. It fulfills the basic requirement for a confidence interval. An attractive additional property of the conditional confidence interval is that it also yields the desired coverage rate over repeated samples that respect the condition.

Different aspects of conditional inference in survey sampling are discussed in Holt and Smith (1979), Rao (1985), Särndal and Hidiroglou (1989), and Särndal, Swensson and Wretman (1989).

7.10.1. Conditional Analysis for *BE* Sampling

We use the *BE* design to illustrate the concept of a conditional confidence interval. Consider the estimator

$$\hat{t}_{yr} = N\frac{\sum_s y_k}{n_s} = N\bar{y}_s \tag{7.10.1}$$

obtained from the common mean model in Example 7.4.1. The formula applies if $n_s \geq 1$. If $n_s = 0$, we take, by convention, the estimate to be zero. In most cases, $n_s = 0$ has near-zero probability. If A_1 denotes the event $n_s \geq 1$, the probability of A_1 is

$$P_{A_1} = 1 - (1 - \pi)^N \doteq 1 - e^{-n} \tag{7.10.2}$$

where $n = E(n_s) = N\pi$ is the expected sample size. Even with a small expected sample size n, A_1 is almost certain to occur. For example, for $n = 10$, we have $P_{A_1} \doteq 1 - e^{-10} = 0.99995$. The extended definition of the estimator is

$$\hat{t}'_y = \begin{cases} N\bar{y}_s & \text{if } n_s \geq 1 \\ 0 & \text{if } n_s = 0 \end{cases} \tag{7.10.3}$$

Conditionally on n_s and A_1 (which translates as: n_s has a fixed value ≥ 1) we now have

$$E_{BE}(N\bar{y}_s | n_s, A_1) = t_y \tag{7.10.4}$$

and

$$V_{BE}(N\bar{y}_s | n_s, A_1) = N^2\left(\frac{1}{n_s} - \frac{1}{N}\right)S_{yU}^2 \tag{7.10.5}$$

This follows since, given A_1 and n_s, the selection of the n_s elements is formally equivalent to an *SI* selection of n_s from N elements. The proof of this is left as an exercise. In other words, $N\bar{y}_s$ is conditionally unbiased for t_y, given n_s and A_1, with the conditional variance shown in (7.10.5).

A conditionally unbiased estimator of the variance (7.10.5) is also obtained. Let A_2 be the event $n_s \geq 2$. Given n_s and A_2 (which reads: n_s has a fixed value ≥ 2), consider

$$\hat{V}_c = N^2\left(\frac{1}{n_s} - \frac{1}{N}\right)S_{ys}^2 \tag{7.10.6}$$

with

$$S_{ys}^2 = \frac{1}{n_s - 1}\sum_s (y_k - \bar{y}_s)^2$$

To see that (7.10.6) is a conditionally unbiased variance estimator, note first that

$$E_{BE}(S_{ys}^2|n_s, A_2) = S_{yU}^2$$

Consequently,

$$E_{BE}(\hat{V}_c|n_s, A_2) = N^2\left(\frac{1}{n_s} - \frac{1}{N}\right)S_{yU}^2 = V_{BE}(N\bar{y}_s|n_s, A_2)$$

A conditional confidence interval at the approximate $1 - \alpha$ level can now be constructed, provided that n_s is not extremely small, namely,

$$N\bar{y}_s \pm z_{1-\alpha/2}\hat{V}_c^{1/2}$$

with $\hat{V}_c$ given by (7.10.6). This interval will contain the unknown $t_y = \sum_U y_k$ for a proportion of roughly $1 - \alpha$ of all BE samples having a fixed size ≥ 2. Since the coverage rate will be approximately $1 - \alpha$ for any such fixed size, the overall coverage rate (over all possible samples) is also approximately $1 - \alpha$.

Remark 7.10.1. Note that the variance estimator shown in equation (7.10.6), obtained by the conditional argument, agrees very nearly with the g-weighted variance estimator (7.4.9), which was obtained from (7.2.11) and is given by

$$\hat{V}_{BE}(\hat{t}_y) = N^2\frac{1 - \pi}{n_s}\left(1 - \frac{1}{n_s}\right)S_{ys}^2$$

This follows because $1 - 1/n_s \doteq 1$, and $1 - n_s/N$ is on the average equal to $1 - \pi$. The g-weight is $N/\hat{N}$ for all k, where $\hat{N} = \sum_s 1/\pi_k$. The g-weighted variance estimator is essentially a conditional variance estimator. By contrast, the simplified variance estimator obtained by letting all $g_{ks} = 1$ in (7.2.11) is unsuitable for conditional intervals, although it gives an overall confidence level that is correct, that is, roughly equal to $1 - \alpha$.

Remark 7.10.2. The conditional moments shown in (7.10.4) and (7.10.5) may be used to derive the corresponding unconditional moments. *Unconditional* means with respect to *all* samples, regardless of their size. Let $E_1(\cdot)$ and $V_1(\cdot)$ denote the expected value and variance with respect to the distribution of n_s given $n_s \geq 1$, namely,

$$P(n_s = j) = \binom{N}{j}\pi^j(1 - \pi)^{N-j}/P_{A_1}; \qquad j = 1, 2, \ldots, N$$

This is a binomial distribution with parameters N and π, truncated at $n_s = 0$. For the estimator shown in equation (7.10.3), we have, using (7.10.4),

$$\begin{aligned} E_{BE}(\hat{t}_y') &= (1 - P_{A_1})\cdot 0 + P_{A_1}E_{BE}(N\bar{y}_s|A_1) \\ &= P_{A_1}E_1E_{BE}(N\bar{y}_s|n_s, A_1) \\ &= P_{A_1}E_1(t_y) = P_{A_1}t_y \end{aligned}$$

An exact expression for the bias is thus

$$E_{BE}(\hat{t}'_y) - t_y = -(1 - P_{A_1})t_y$$

There is a very slight negative bias, because $P_{A_1} \doteq 1 - e^{-n}$ is very near unity. The bias is caused by the arbitrary definition $\hat{t}'_y = 0$ when $n_s = 0$. The unconditional variance is

$$V_{BE}(\hat{t}'_y) = (1 - P_{A_1})t_y^2 + P_{A_1} E_{BE}[(N\bar{y}_s - t_y)^2 | A_1]$$

where

$$\begin{aligned} E_{BE}[(N\bar{y}_s - t_y)^2 | A_1] &= E_1\{E_{BE}[(N\bar{y}_s - t_y)^2 | n_s, A_1]\} \\ &= E_1[V_{BE}(N\bar{y}_s | n_s, A_1)] \end{aligned}$$

Using (7.10.5), we obtain the unconditional variance

$$V_{BE}(\hat{t}'_y) = (1 - P_{A_1})t_y^2 + P_{A_1} N^2 \left[E_1\left(\frac{1}{n_s}\right) - \frac{1}{N}\right] S_{yU}^2 \tag{7.10.7}$$

with

$$E_1\left(\frac{1}{n_s}\right) = \sum_{j=1}^{N} \left(\frac{1}{j}\right)\binom{N}{j} \pi^j (1 - \pi)^{N-j} / P_{A_1}$$

Remark 7.10.3. Equation (7.10.7) is an exact variance expression but is lengthy to calculate. An approximation by means of the Taylor expansion is possible. Letting $\Delta_s = (n_s - n)/n$, where $n = E(n_s)$, we have

$$\frac{1}{n_s} = \frac{1}{n(1 + \Delta_s)} = \frac{1}{n}(1 - \Delta_s + \Delta_s^2 - \cdots)$$

Assuming that the probability of $n_s = 0$ is negligible,

$$E\left(\frac{1}{n_s}\right) \doteq \frac{1}{n}\left[1 + \frac{V(n_s)}{n^2}\right] \tag{7.10.8}$$

This approximation holds for any design with the expected sample size n. In particular, under BE sampling, $n = N\pi$ and $V_{BE}(n_s) = N\pi(1 - \pi)$, so

$$E_{BE}\left(\frac{1}{n_s}\right) \doteq \frac{1}{n} + \frac{1 - \pi}{n^2}$$

Inserting into (7.10.7) and approximating $P_{A_1} \doteq 1$, we get, to the order of approximation n^{-2},

$$V_{BE}(\hat{t}'_y) \doteq V_{BE}(N\bar{y}_s) \doteq N^2 \frac{1 - \pi}{n}\left(1 + \frac{1}{n}\right) S_{yU}^2$$

A comparison with SI sampling (fixed size $= n$; sampling fraction $n/N = \pi$) leads to the design effect

$$\frac{V_{BE}(N\bar{y}_s)}{V_{SI}(N\bar{y}_s)} = 1 + \frac{1}{n}$$

The penalty for the random sample size is negligible (unless n is extremely small).

7.10.2. Conditional Analysis for the Poststratification Estimator

The confidence interval derived in Section 7.10.1 is an example of the following general *conditional confidence interval procedure*. A conditional confidence interval at the approximate level $1 - \alpha$ is constructed as

$$\hat{t}_y \pm z_{1-\alpha/2}[\hat{V}_c(\hat{t}_y)]^{1/2} \tag{7.10.9}$$

where $\hat{t}_y$ is conditionally unbiased, at least approximately, for t_y, and $\hat{V}_c(\hat{t}_y)$ is an (approximately) conditionally unbiased estimator of the conditional variance $V_c(\hat{t}_y)$. Here, the index c indicates conditional. In repeated samples that obey the condition, a proportion of roughly $1 - \alpha$ of all samples will give rise to intervals that contain the unknown t_y. In the *BE* sampling example above, $\hat{t}_y$ is given by (7.10.1), and $\hat{V}_c(\hat{t}_y)$ by (7.10.6).

Let us apply the conditional reasoning to the important case of the classical poststratified estimator (see Section 7.6)

$$\hat{t}_{ypos} = \sum_{g=1}^{G} N_g \bar{y}_{s_g} = \sum_{g=1}^{G} N_g (\textstyle\sum_{s_g} y_k)/n_{s_g} \tag{7.10.10}$$

Assuming *SI* sampling (n elements drawn from N, and $f = n/N$), we have

$$n = \sum_{g=1}^{G} n_{s_g}$$

Let A_1 denote the event

$$n_{s_g} \geq 1; \qquad g = 1, \ldots, G$$

In equation (7.10.10), let us define $\bar{y}_{s_g} = 0$ when $n_{s_g} = 0$. The estimator $\hat{t}_{ypos}$ is then defined even if the event $\bar{A}_1 = \text{not } A_1$ occurs.

Let $\mathbf{n}_s = (n_{s_1}, \ldots, n_{s_G})$ be the vector of group counts. These counts are random. It can be shown that the *SI* selection of a sample s, given a fixed configuration $\mathbf{n}_s$, conforms to a selection of an *STSI* sample, with n_{s_g} elements drawn from N_g in the group U_g; $g = 1, \ldots, G$ (see Exercise 7.26).

Using Result 3.7.2, we then obtain the following conditional mean and variance:

$$E_{SI}(\hat{t}_{ypos} | \mathbf{n}_s, A_1) = \sum_{g=1}^{G} N_g E[(\textstyle\sum_{s_g} y_k/n_{s_g}) | n_{s_g} \geq 1]$$

$$= \sum_{g=1}^{G} N_g \bar{y}_{U_g} = t_y \tag{7.10.11}$$

$$V_{SI}(\hat{t}_{ypos} | \mathbf{n}_s, A_1) = \sum_{g=1}^{G} N_g^2 (1/n_{s_g} - 1/N_g) S_{yU_g}^2 \tag{7.10.12}$$

where $S_{yU_g}^2$ is the variance of y in U_g. This leads immediately to a conditional variance estimator, provided that $n_{s_g} \geq 2$ for all g, namely,

$$\hat{V}_c(\hat{t}_{ypos}) = \sum_{g=1}^{G} N_g^2 \left(\frac{1}{n_{s_g}} - \frac{1}{N_g} \right) S_{ys_g}^2 \tag{7.10.13}$$

Note that

$$S_{ys_g}^2 = \frac{1}{n_{s_g} - 1} \sum_{s_g} (y_k - \bar{y}_{s_g})^2$$

is conditionally unbiased for $S_{yU_g}^2$. A conditional confidence interval at the approximate level $1 - \alpha$ is obtained from (7.10.9) with $\hat{t}_y = \hat{t}_{ypos}$ and $\hat{V}_c(\hat{t}_y)$ given by (7.10.13).

Remark 7.10.4. The conditional variance estimator shown in equation (7.10.13) agrees closely with the g-weighted variance estimator (7.6.5) used in Section 7.6. If $1 - (n_{s_g}/N_g)$ in (7.10.13) is replaced by its average value, $1 - f$, we obtain (7.6.5). Both (7.10.3) and (7.6.5) make good sense under a conditional outlook. Suppose the gth group happens to be underrepresented in the sample, so that the observed n_{s_g} is small compared to its expectation, nN_g/N. Then the gth group will tend to contribute more to the variance estimate than if n_s was larger than expected. This is a reasonable property, as argued in Holt and Smith (1979) and Särndal, Swensson and Wretman (1989).

Remark 7.10.5. We can use the conditional expressions shown in equations (7.10.11) and (7.10.12) to derive the unconditional mean and variance and a close, computationally simple approximation of the variance. Let $E_1(\cdot)$ denote the expectation under the distribution of $\mathbf{n}_s$, which is a multivariate hypergeometric distribution, truncated so that $n_{s_g} \geq 1$ for $g = 1, \ldots, G$. Now, using (7.10.11) and (7.10.12)

$$\begin{aligned} E_{SI}(\hat{t}_{ypos} | A_1) &= E_1[E_{SI}(\hat{t}_{ypos} | \mathbf{n}_s, A_1)] \\ &= E_1(t_y) = t_y \\ E_{SI}[(\hat{t}_{ypos} - t_y)^2 | A_1] &= E_1[V_{SI}(\hat{t}_{ypos} | \mathbf{n}_s, A_1)] \\ &= E_1\left[\sum_{g=1}^{G} N_g^2 \left(\frac{1}{n_{s_g}} - \frac{1}{N_g} \right) S_{yU_g}^2 \right] \\ &= \sum_{g=1}^{G} N_g^2 \left[E_1\left(\frac{1}{n_{s_g}} \right) - \frac{1}{N_g} \right] S_{yU_g}^2 \end{aligned} \tag{7.10.14}$$

Here the expectation $E_1(1/n_{s_g})$ is taken with respect to the distribution of n_{s_g}, given that $n_{s_g} \geq 1$. We conclude that, given the event A_1, $\hat{t}_{ypos}$ is unbiased for t_y, and the variance is given by (7.10.14). There is a nonzero probability of A_1 occurring. However, in many applications this probability is so small that, for all practical purposes, $\hat{t}_{ypos}$ can be treated as unbiased in the overall sense, with the variance shown in (7.10.14).

Remark 7.10.6. One can use (7.10.14) to obtain a more refined approximate variance expression than the earlier result (7.6.4). We evaluate $E_1(1/n_{s_g})$ by means of (7.10.8). First, note that if $W_g = N_g/N$ then

$$E_1(n_{s_g}) = nW_g$$

$$V_1(n_{s_g}) = \frac{N-n}{N-1} nW_g(1-W_g)$$

assuming that $n_{s_g} = 0$ has negligible probability. Thus,

$$E_1\left(\frac{1}{n_{s_g}}\right) \doteq \frac{1}{nW_g}\left[1 + \frac{(1-f)(1-W_g)}{nW_g}\right]$$

and

$$V_{SI}(\hat{t}_{\text{ypos}}) \doteq N^2 \frac{1-f}{n}\left\{\sum_{g=1}^{G} W_g S_{yU_g}^2 + \frac{1}{n}\sum_{g=1}^{G}(1-W_g)S_{yU_g}^2\right\}$$

This is the approximation (7.6.6) mentioned in Section 7.6.

7.11. Regression Estimators for Variable-Size Sampling Designs

Bernoulli sampling (*BE*) and Poisson sampling (*PO*) are examples of sampling designs with the following characteristics:

i. We have

$$\Delta_{kl} = \pi_{kl} - \pi_k \pi_l = 0 \quad \text{for all} \quad k \neq l \tag{7.11.1}$$

ii. The sample size is random. At the planning stage, it can be predicted only within certain probable limits.

Sample selection is simple in these designs. That $\Delta_{kl} = 0$ for all $k \neq l$ is a considerable advantage for the variance estimation. It implies that the product terms in the variance estimator are zero. The double sum is reduced to a simple sum. The randomness of the sample size is sometimes seen as a disadvantage. It has the effect that the variance of the π estimator is often higher than under a comparable fixed-sample-size design. For example, Section 2.10 showed that the variance of the π estimator is greater under *BE* sampling than under a comparable *SI* sampling design by a factor of approximately $1 + (cv_{yU})^{-2}$. The regression estimator is different in this respect. In many cases it is not subject to a variance penalty for randomness of the sample size. Let us examine the reasons for this.

The approximate variance (7.2.10) of the regression estimator $\hat{t}_{yr}$ can be expressed in terms of the regression residuals $E_k = y_k - \mathbf{x}_k'\mathbf{B}$ as

$$\begin{aligned} AV(\hat{t}_{yr}) &= \textstyle\sum\sum_U \Delta_{kl} \check{E}_k \check{E}_l \\ &= \textstyle\sum_U \pi_k(1-\pi_k)\check{E}_k^2 + \sum\sum_{U,\, k\neq l} (\pi_{kl} - \pi_k\pi_l)\check{E}_k\check{E}_l = V + C \end{aligned} \tag{7.11.2}$$

The term

$$C = \sum\sum_{U}_{k \neq l} (\pi_{kl} - \pi_k \pi_l) \check{E}_k \check{E}_l$$

is zero for designs that satisfy (7.11.1). The variable size designs *BE* and *PO* are examples of this.

It happens for certain fixed-size sampling designs and certain models that the term C is negligible compared to V. Then the single term

$$V = \sum_U \pi_k (1 - \pi_k) \check{E}_k^2 = \sum_U \left[\frac{1}{\pi_k} - 1 \right] E_k^2 \tag{7.11.3}$$

will give a good approximation to the variance. The following simplified variance estimator can then be used, namely,

$$\hat{V}^*(\hat{t}_{yr}) = \sum_s \frac{1}{\pi_k} \left(\frac{1}{\pi_k} - 1 \right) (g_{ks} e_{ks})^2$$

where

$$e_{ks} = y_k - \hat{\mathbf{B}}' \mathbf{x}_k$$

and the g weight g_{ks} is given by (7.2.9). A computational advantage of this variance estimator is that the formula is a simple sum, and the π_{kl} are not needed.

Let us look at specific cases. Compare the designs *SI* and *BE*, assuming that the fixed sampling fraction $f = n/N$ under *SI* equals the expected sampling fraction $E(n_s)/N = \pi$ under *BE*. Then

$$AV_{SI}(\hat{t}_{yr}) = N^2 \frac{1 - f}{n} \frac{\sum_U (E_k - \bar{E}_U)^2}{N - 1}$$

and

$$AV_{BE}(\hat{t}_{yr}) = N^2 \frac{1 - f}{n} \frac{\sum_U E_k^2}{N}$$

where $\bar{E}_U = \sum_U E_k / N$ is the mean of the residuals from the population fit. If we disregard the factor $(N - 1)/N$, it follows that $AV_{BE}(\hat{t}_{yr}) \geq AV_{SI}(\hat{t}_{yr})$ with equality if and only if $\bar{E}_U = 0$. A sufficient condition for $\bar{E}_U = 0$ to hold is that the variance structure of the underlying model conforms to equation (7.2.12). The models shown in (7.3.1), (7.4.1), (7.5.3), (7.5.6), and (7.8.1) are cases in point. The regression estimator corresponding to any one of these models will thus have approximately the same variance under *SI* sampling as under *BE* sampling.

Under the *SI* design, we have when $\bar{E}_U = 0$ that the ratio of the terms V and C in (7.11.2) is $C/V = 1/(N - 1)$, which is an example of a case where C is negligible compared to V. A similar phenomenon occurs when we compare *STSI* sampling with *STBE* sampling (see Chapter 3 for the definitions). If, for both designs, $\pi_k = f_h = n_h/N_h$ for all k in the hth stratum U_h, then

$$AV_{STSI}(\hat{t}_{yr}) = \sum_{h=1}^{H} N_h^2 \frac{1-f_h}{n_h} \frac{\sum_{U_h} (E_k - \bar{E}_{U_h})^2}{N_h - 1}$$

and

$$AV_{STBE}(\hat{t}_{yr}) = \sum_{h=1}^{H} N_h^2 \frac{1-f_h}{n_h} \frac{\sum_{U_h} E_k^2}{N_h}$$

Under models such that $\bar{E}_{U_h} = 0$ for $h = 1, \ldots, H$ the two AV expressions are equal, if we disregard the effect of the factor $(N_h - 1)/N_h$. An example of a model of this kind is the group ratio model shown in (7.5.6). For $STSI$, the term C in (7.11.2) is again negligible compared to term V, if all strata are large.

7.12. A Class of Regression Estimators

Various applications of the regression estimator (7.2.8) have been examined in this chapter. If the underlying regression model is a close description of the finite population point scatter $(y_k, \mathbf{x}_k)$, the residuals E_k are small, and the AV given by equation (7.2.10) will be small. Nevertheless, (7.2.8) has not been shown to be an optimal estimator. There exist, in fact, other estimators of regression type that are comparable to (7.2.8) in efficiency, given the same amount of auxiliary information. We now consider such alternatives, assuming that a probability sample s is selected by a design with the positive inclusion probabilities π_k and π_{kl}.

Wright (1983) considered regression estimators of the form

$$\hat{t}_y = \sum_U \hat{y}_k + \sum_s r_k(y_k - \hat{y}_k) \tag{7.12.1}$$

where the predicted values are

$$\hat{y}_k = \mathbf{x}_k'\hat{\boldsymbol{\beta}} = \mathbf{x}_k'(\textstyle\sum_s q_k \mathbf{x}_k \mathbf{x}_k')^{-1} \sum_s q_k \mathbf{x}_k y_k \tag{7.12.2}$$

and the constants r_k and q_k are to be specified. Thus, equation (7.12.1) defines a wide class of estimators, corresponding to the various choices of r_k and q_k. The statistician will require that $\hat{t}_y$ meet certain performance criteria. One criterion that is often imposed is that the estimator must be asymptotically design unbiased (ADU). The design bias, $E(\hat{t}_y) - t_y$ must then tend to zero.

The requirement of ADUness restricts the choice of the r_k and q_k, but there is more than one choice of these constants that leads to an ADU estimator. Wright (1983) shows that the estimator (7.12.1) is ADU under any choice of the constants r_k and q_k, as long as it is possible to specify a vector $\boldsymbol{\lambda}$ such that

$$1 - r_k \pi_k = \pi_k q_k \mathbf{x}_k' \boldsymbol{\lambda} \tag{7.12.3}$$

holds for $k = 1, \ldots, N$. Here we discuss two choices of the r_k, namely, (i) $r_k = 1/\pi_k$, and (ii) $r_k = 1$ for all k.

The choice $r_k = 1/\pi_k$ is interesting, because then we can take $\boldsymbol{\lambda} = \mathbf{0}$ to satisfy equation (7.12.3). That is, when $r_k = 1/\pi_k$, the estimator (7.12.1) is ADU for *any* choice of the weights q_k in (7.12.2), including the choice $q_k = 1/(\pi_k \sigma_k^2)$ that is used in the regression estimator (7.2.8). As a consequence, different q weights can be applied in (7.12.2) without upsetting the ADU property of the estimator (7.12.1).

The choice $r_k = 1$ for all k is of interest because the estimator (7.12.1) takes the appealing form

$$\begin{aligned}\hat{t}_y &= \textstyle\sum_U \hat{y}_k + \sum_s (y_k - \hat{y}_k) \\ &= \textstyle\sum_s y_k + \sum_{U-s} \hat{y}_k \end{aligned} \tag{7.12.4}$$

It consists of the sum of the observed values, y_k, for the elements in the sample and a sum of predicted values, $\hat{y}_k$, for the elements not in the sample. Many statisticians find it natural to estimate the total $t = \sum_U y_k$ in this way, because

$$t = \textstyle\sum_U y_k = \sum_s y_k + \sum_{U-s} y_k$$

where $\sum_s y_k$ is observed and the only unknown part is $\sum_{U-s} y_k$. We replace the unknown part with $\sum_{U-s} \hat{y}_k$ and obtain the estimator (7.12.4).

Since ADUness is important, the following question arises: can the weights q_k be specified in such a way that (7.12.4) is ADU? The answer is yes. That is, the estimator can have the attractive form of equation (7.12.4) as well as the ADU property. Here we consider only the case where the variances specified in the model given by (7.2.1) and (7.2.2) satisfy

$$V_\xi(y_k) = \sigma_k^2 = \mathbf{x}_k' \boldsymbol{\lambda}; \qquad k = 1, \ldots, N \tag{7.12.5}$$

for some constant vector $\boldsymbol{\lambda}$. Then the condition (7.12.3) with $r_k = 1$ is satisfied if we let $q_k = [(1/\pi_k) - 1]/\sigma_k^2$. This leads to the following result.

Result 7.12.1. *If the model variances satisfy* (7.12.5), *an ADU estimator of* t_y *is given by*

$$\hat{t}_y = \textstyle\sum_s y_k + \sum_{U-s} \hat{y}_k \tag{7.12.6}$$

where $\hat{y}_k = \mathbf{x}_k' \hat{\boldsymbol{\beta}}$ *with*

$$\hat{\boldsymbol{\beta}} = \left[\sum_s \left(\frac{1}{\pi_k} - 1\right) \frac{\mathbf{x}_k \mathbf{x}_k'}{\sigma_k^2}\right]^{-1} \left[\sum_s \left(\frac{1}{\pi_k} - 1\right) \frac{\mathbf{x}_k y_k}{\sigma_k^2}\right]$$

When the variance structure (7.12.5) holds, we know from Result 6.5.1 that the regression estimator (7.2.8) has the compact expression

$$\hat{t}_y = \textstyle\sum_U \hat{y}_k \tag{7.12.7}$$

with $\hat{y}_k = \mathbf{x}_k' \hat{\mathbf{B}}$ and

$$\hat{\mathbf{B}} = \left(\sum_s \frac{\mathbf{x}_k \mathbf{x}_k'}{\sigma_k^2 \pi_k}\right)^{-1} \sum_s \frac{\mathbf{x}_k y_k}{\sigma_k^2 \pi_k} \tag{7.12.8}$$

Both (7.12.6) and (7.12.7) are ADU estimators when (7.12.5) holds. When calculated with the aid of data from a realized sample s, the two formulas give slightly different estimates. The difference is likely to be negligible for most practical purposes.

EXAMPLE 7.12.1. Let us apply Result 7.12.1 in the case of the common ratio model shown in (7.3.1). We conclude that an ADU estimator of t_y is given by

$$\hat{t}_y = \sum_s y_k + \left(\sum_{U-s} x_k\right)\hat{\beta} \tag{7.12.9}$$

with

$$\hat{\beta} = \frac{\sum_s \left(\frac{1}{\pi_k} - 1\right) y_k}{\sum_s \left(\frac{1}{\pi_k} - 1\right) x_k}$$

The estimator (7.12.9), derived by Brewer (1979), can also be written as

$$\hat{t}_y = \left(\sum_U x_k\right)\hat{\beta} + \sum_s \frac{y_k - \hat{\beta}x_k}{\pi_k}$$

which has the form of the regression estimator (7.2.8) but with a different slope estimator. One can show that the approximate variance is given by the usual formula

$$AV(\hat{t}_y) = \sum\sum_U \Delta_{kl} \check{E}_k \check{E}_l$$

provided we take $E_k = y_k - B_{\text{alt}} x_k$, with

$$B_{\text{alt}} = \frac{\sum_U (1 - \pi_k) y_k}{\sum_U (1 - \pi_k) x_k}$$

Here, B_{alt} is the population quantity for which $\hat{\beta}$ is an estimator. For populations that are well described by the common ratio model, the estimator given by (7.12.9) thus offers an alternative to the standard ratio estimator

$$\hat{t}_{yra} = \sum_U x_k \frac{\sum_s y_k/\pi_k}{\sum_s x_k/\pi_k}$$

The two estimators do not yield identical estimates, but the difference in efficiency is unimportant. For further discussion, see Wright (1983) and Särndal and Wright (1984).

If more than one x variable are involved, (7.12.4) can also be written as a regression estimator, if the following procedure is followed. Specify a constant vector $\boldsymbol{\lambda}$, independent of k and with the same dimension as $\mathbf{x}_k$. Define the scalar quantity $a_k = \mathbf{x}_k'\boldsymbol{\lambda}$, and choose the weights q_k in (7.12.2) as

$$q_k = \left(\frac{1}{\pi_k} - 1\right)\Big/ a_k$$

Then the estimator (7.12.4) can be written alternatively as

$$\hat{t}_y = (\sum_U \mathbf{x}_k)'\hat{\boldsymbol{\beta}} + \sum_s \frac{y_k - \hat{\boldsymbol{\beta}}'\mathbf{x}_k}{\pi_k}$$

The approximate variance of this ADU estimator is

$$AV(\hat{t}_y) = \sum\sum_U \Delta_{kl} \check{E}_k \check{E}_l$$

where the appropriate residuals are

$$E_k = y_k - \mathbf{B}'_{\text{alt}}\mathbf{x}_k$$

with

$$\mathbf{B}'_{\text{alt}} = [\sum_U (1 - \pi_k)(\mathbf{x}_k\mathbf{x}'_k/a_k)]^{-1}[\sum_U (1 - \pi_k)(\mathbf{x}_k y_k/a_k)]$$

This permits a variety of weightings q_k, corresponding to the various choices of the vector $\boldsymbol{\lambda}$. However, differences in efficiency of great practical importance are not foreseen among the estimators corresponding to these possibilities.

7.13. Regression Estimation of a Ratio of Population Totals

In Section 5.6, we estimated the ratio of two population totals

$$R = \frac{\sum_U y_k}{\sum_U z_k} = \frac{t_y}{t_z}$$

The approach was to obtain the π estimators for the numerator and for the denominator, and then to use the ratio of the two as an estimator of R.

Alternatively, regression estimators can be used in the numerator as well as in the denominator. That is, for each total, $\sum_U y_k$ and $\sum_U z_k$, we use a regression estimator expressed in terms of auxiliary variables $x_1, \ldots, x_J$. Often, if the presence of such variables improves the accuracy for one or both of the totals $\sum_U y_k$ and $\sum_U z_k$, efficiency is also gained in the estimation of R.

Predicted values $\hat{y}_k$ and $\hat{z}_k$ are derived using the auxiliary vector values $\mathbf{x}_k = (x_{1k}, \ldots, x_{jk}, \ldots, x_{Jk})'$. Let the predicted y-values be

$$\hat{y}_k = \mathbf{x}'_k \hat{\mathbf{B}}_y \tag{7.13.1}$$

with

$$\hat{\mathbf{B}}_y = \left(\sum_s \frac{\mathbf{x}_k\mathbf{x}'_k}{\sigma_k^2 \pi_k}\right)^{-1} \sum_s \frac{\mathbf{x}_k y_k}{\sigma_k^2 \pi_k} \tag{7.13.2}$$

The population analog of $\hat{\mathbf{B}}_y$ is

$$\mathbf{B}_y = \left(\sum_U \frac{\mathbf{x}_k\mathbf{x}_k'}{\sigma_k^2}\right)^{-1} \sum_U \frac{\mathbf{x}_k y_k}{\sigma_k^2} \tag{7.13.3}$$

The corresponding quantities for the z variable are

$$\hat{z}_k = \mathbf{x}_k'\hat{\mathbf{B}}_z \tag{7.13.4}$$

$$\hat{\mathbf{B}}_z = \left(\sum_s \frac{\mathbf{x}_k\mathbf{x}_k'}{\sigma_k^2\pi_k}\right)^{-1} \sum_s \frac{\mathbf{x}_k z_k}{\sigma_k^2\pi_k} \tag{7.13.5}$$

$$\mathbf{B}_z = \left(\sum_U \frac{\mathbf{x}_k\mathbf{x}_k'}{\sigma_k^2}\right)^{-1} \sum_U \frac{\mathbf{x}_k z_k}{\sigma_k^2} \tag{7.13.6}$$

Regression estimators are then created for the numerator and the denominator separately. We are led to estimate R by

$$\hat{R} = \frac{\hat{t}_{yr}}{\hat{t}_{zr}} \tag{7.13.7}$$

where

$$\hat{t}_{yr} = \sum_U \hat{y}_k + \sum_s \frac{y_k - \hat{y}_k}{\pi_k} = \sum_s \frac{g_{ks}y_k}{\pi_k}$$

$$\hat{t}_{zr} = \sum_U \hat{z}_k + \sum_s \frac{z_k - \hat{z}_k}{\pi_k} = \sum_s \frac{g_{ks}z_k}{\pi_k}$$

with $\hat{y}_k$ and $\hat{z}_k$ given by (7.13.1) and (7.13.4), respectively, and the g weights depend on the auxiliary variables only,

$$g_{ks} = 1 + \left(\sum_U \mathbf{x}_k - \sum_s \mathbf{x}_k/\pi_k\right)'\left(\sum_s \frac{\mathbf{x}_k\mathbf{x}_k'}{\sigma_k^2\pi_k}\right)^{-1} \frac{\mathbf{x}_k}{\sigma_k^2} \tag{7.13.8}$$

EXAMPLE 7.13.1. The estimator given by (7.13.7) is particularly simple when a single auxiliary variable x is available and each of y and z is explained in terms of x via a common ratio model of the form of (7.3.1). The estimator of R is then

$$\hat{R} = t_x\frac{\sum_s \check{y}_k}{\sum_s \check{x}_k}\bigg/ t_x\frac{\sum_s \check{z}_k}{\sum_s \check{x}_k}$$

where $\check{y}_k = y_k/\pi_k$ and $\check{x}_k$ and $\check{z}_k$ are analogously defined. The expression for $\hat{R}$ simplifies into

$$\hat{R} = \frac{\sum_s \check{y}_k}{\sum_s \check{z}_k}$$

We are back to the estimator examined in Section 5.6.

For the estimator in equation (7.13.7), we obtain the approximate variance as in Section 5.6,

$$V(\hat{R}) \doteq \frac{1}{t_z^2}[V(\hat{t}_{yr}) + R^2 V(\hat{t}_{zr}) - 2RC(\hat{t}_{yr}, \hat{t}_{zr})] \tag{7.13.9}$$

Now, the regression estimation approach offers approximate expressions for the variances $V(\hat{t}_{yr})$ and $V(\hat{t}_{zr})$, and the covariance $C(\hat{t}_{yr}, \hat{t}_{zr})$, such that

$$AV(\hat{t}_{yr}) = \sum\sum_U \Delta_{kl} \check{E}_{yk} \check{E}_{yl}, \qquad AV(\hat{t}_{zr}) = \sum\sum_U \Delta_{kl} \check{E}_{zk} \check{E}_{zl}$$

and

$$AC(\hat{t}_{yr}, \hat{t}_{zr}) = \sum\sum_U \Delta_{kl} \check{E}_{yk} \check{E}_{zl}$$

with

$$E_{yk} = y_k - \mathbf{x}_k' \mathbf{B}_y; \qquad E_{zk} = z_k - \mathbf{x}_k' \mathbf{B}_z.$$

Here, $\mathbf{B}_y$ and $\mathbf{B}_z$ are given, respectively, by (7.13.3) and (7.13.6).

Substituting these expressions into Equation (7.13.9), we arrive at the approximate variance

$$AV(\hat{R}) = \frac{1}{t_z^2} \sum\sum_U \Delta_{kl} \frac{E_{yk} - RE_{zk}}{\pi_k} \frac{E_{yl} - RE_{zl}}{\pi_l} \tag{7.13.10}$$

The g-weighted variance estimator is

$$\hat{V}(\hat{R}) = \frac{1}{\hat{t}_{zr}^2} \sum\sum_s \check{\Delta}_{kl} \frac{g_{ks}(e_{yks} - \hat{R}e_{zks})}{\pi_k} \frac{g_{ls}(e_{yls} - \hat{R}e_{zls})}{\pi_l}$$

with

$$e_{yks} = y_k - \mathbf{x}_k' \hat{\mathbf{B}}_y; \qquad e_{zks} = z_k - \mathbf{x}_k' \hat{\mathbf{B}}_z$$

and the weight g_{ks} is given by (7.13.8).

The key to making the AV expression given by (7.13.10) numerically small is to obtain differential residuals $E_{yk} - RE_{zk}$ that are small. Now, we can write

$$E_{yk} - RE_{zk} = D_k - D_k^0$$

where D_k is the differential variable value

$$D_k = y_k - Rz_k$$

and D_k^0 its prediction in terms of $\mathbf{x}_k$, that is,

$$D_k^0 = y_k^0 - Rz_k^0 = \mathbf{x}_k'(\mathbf{B}_y - R\mathbf{B}_z)$$

We conclude that when regression estimation is used for a composite parameter such as $R = t_y/t_z$, the variance reduction extracted from the auxiliary vector depends on the extent to which $\mathbf{x}_k$ explains the variation in the differential variable $D_k = y_k - Rz_k$.

Another example of the technique in this section is found in Elvers et al. (1985), who use auxiliary variables to obtain improved estimators of the finite population regression coefficient $B = S_{zyU}/S_{zzU}$. Rao, Kovar, and Mantel (1990) consider the use of auxiliary information to get improved estimators of finite population distribution functions and quantiles.

Exercises

7.1. We want to estimate the total of the variable ME84 ($= y$) for the MU200 population using the ratio estimator shown in equation (7.3.3) with the variable P75 ($= x$) as the auxiliary variable and with $t_x = 2{,}603$. An *SI* sample of size $n = 40$ gave the following results.

$$\textstyle\sum_s y_k = 24{,}874; \qquad \sum_s y_k^2 = 19{,}136{,}668$$

$$\textstyle\sum_s x_k = 473; \qquad \sum_s x_k^2 = 6{,}539; \qquad \sum_s x_k y_k = 348{,}071.$$

Compute an approximately 95% confidence interval for the total of ME84.

7.2. Compare the following three strategies for estimating the total of RMT85 ($= y$) for the MU281 population. In strategies (2) and (3), REV84 ($= x$) is used as an auxiliary variable. (1) *SI* sampling with $n = 40$ and the π estimator $N\bar{y}_s$. (2) *SI* sampling with $n = 40$ and the ratio estimator given by (7.3.3). (3) *STSI* sampling with a total sample size of $n = 40$, allocated equally ($n_h = 8$ for all h) on $H = 5$ strata derived from a size-ordering of $(x_k)^{1/2}$ as described in Section 7.3.3, and the combined ratio estimator (7.3.13). The following population data are given;

$$\textstyle\sum_U x_k = 757{,}246; \qquad \sum_U x_k^2 = 3{,}638{,}841{,}824$$

$$\textstyle\sum_U y_k = 53{,}151; \qquad \sum_U y_k^2 = 21{,}266{,}279$$

$$\textstyle\sum_U x_k y_k = 264{,}703{,}466$$

and

h	N_h	$S^2_{xU_h}$	$S^2_{yU_h}$	S_{xyU_h}
1	91	61,722	445	4,021
2	66	60,038	792	3,326
3	53	105,494	2,965	8,340
4	41	683,674	16,595	63,493
5	30	4,037,601	78,188	360,464

7.3. Consider the MU284 population. (a) It is required to estimate the total of the variable S82 with a relative standard error of 5% based on a *BE* sample. The relative standard error is defined as $[V(\hat{t})]^{1/2}/t$. Let n_1 be the expected sample size needed to obtain the specified precision with the π estimator, $\hat{t}_\pi = \sum_s y_k/\pi$, and let n_2 be the expected sample size needed to obtain the same precision with the alternative estimator $N\bar{y}_s = N\sum_s y_k/n_s$. Calculate $100(1 - n_2/n_1)$, which is the percentage reduction in the expected sample size when the π estimator is replaced by the more efficient estimator $N\bar{y}_s$. (b) Answer the question in (a) with the variable REV84 instead of S82. Comment on the difference between the results. Use the following population data:

Variable	$\sum_U y_k$	$\sum_U y_k^2$
S82	13,500	676,292
REV84	874,017	$9{,}063 \cdot 10^9$

7.4. The following scatterplot describes the relation between the variables IMP (= y) and GNP (= x) for the 100 smallest countries (according to the value of GNP) in the CO124 population.

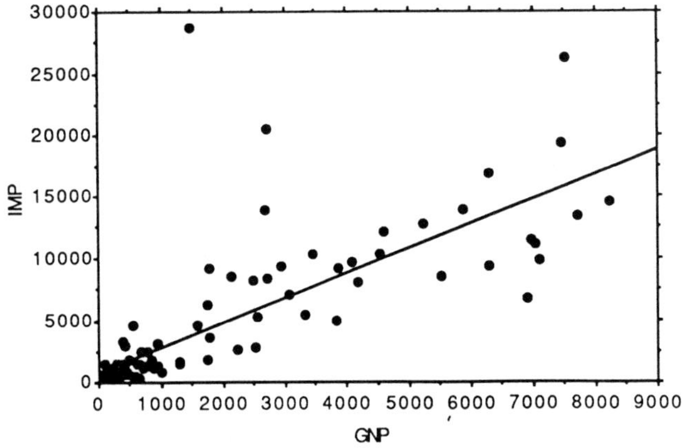

Consider the *PO* design with the expected sample size $n = 20$ and with π_k proportional to the value of GNP. Three estimators are considered for estimating the total of IMP:

$$\hat{t}_1 = \left(N/\sum_s \frac{1}{x_k}\right)\sum_s \frac{y_k}{x_k}, \quad \hat{t}_2 = \frac{t_x}{n}\sum_s \frac{y_k}{x_k}, \quad \text{and} \quad \hat{t}_3 = \frac{t_x}{n_s}\sum_s \frac{y_k}{x_k}$$

where $t_x = \sum_U x_k$. Calculations show that

$$AV(\hat{t}_1) > V(\hat{t}_2) > AV(\hat{t}_3)$$

Is this result to be expected? Explain your reasoning.

7.5. Let the MU200 population be partitioned into two groups in such a way that the 100 smallest municipalities according to the size of the variable P75 form group 1, and the remaining municipalities form group 2. For the study variable RMT85 (= y) we have

Group	N_g	$\sum_{U_g} y_k$	$\sum_{U_g} y_k^2$
1	100	5,848	371,634
2	100	13,008	1,879,106

Assume that SI sampling is used with $n = 50$. (a) Compute the approximate variance of the poststratified estimator shown in equation (7.6.1). (b) Compute the variance of the π estimator.

7.6. Let the MU200 population be partitioned into two groups as in Exercise 7.5. The study variable is RMT85 ($= y$). The following results were obtained from an SI sample of size $n = 50$:

Group	n_{s_g}	$\sum_{s_g} y_k$	$\sum_{s_g} y_k^2$
1	24	1,380	86,180
2	26	3,401	495,147

(a) Compute an approximate 95% confidence interval for the total of RMT85 based on the poststratified estimator. (b) Compute an approximate 95% confidence interval for the RMT85 total based on the π estimator.

7.7. Let the MU281 population be partitioned into two groups in such a way that the municipalities in regions 4 and 5 belong to group 1, whereas the remaining municipalities belong to group 2. Let RMT85 be the study variable ($= y$), and let P75 be the auxiliary variable ($= x$). The following population data are given

Group	N_g	$\sum_{U_g} y_k$	$\sum_{U_g} x_k$	$\sum_{U_g} y_k^2$	$\sum_{U_g} x_k^2$	$\sum_{U_g} x_k y_k$
1	92	15,212	2,079	4,562,504	81,667	607,771
2	189	37,939	4,739	16,703,775	235,041	1,967,287

Assume that SI sampling is used with $n = 90$. Compute the approximate variance (a) for the ratio estimator given by (7.3.3); (b) for the poststratified ratio estimator given by (7.7.2).

7.8. Let the MU281 population be partitioned as in Exercise 7.7. Let y and x correspond to RMT85 and P75, respectively. An SI sample of size $n = 90$ was drawn, and the following results were computed:

Group	n_{s_g}	$\sum_{s_g} y_k$	$\sum_{s_g} x_k$	$\sum_{s_g} y_k^2$	$\sum_{s_g} x_k^2$	$\sum_{s_g} x_k y_k$
1	32	5,554	761	1,746,332	31,057	232,124
2	58	11,991	1,529	5,114,681	75,439	616,606

Compute an approximate 95% confidence interval for the total of RMT85 based on the separate ratio estimator.

7.9. Consider the MU281 population. Let the study variable be ME84 ($= y$), and use S82 ($= x$) as an auxiliary variable. Compare the π estimator, the ratio estimator (7.3.3) and the regression estimator (7.8.9) of the ME84 total in terms of their approximate variances. Assume *SI* sampling with $n = 50$. Use the following population data.

$$\sum_U y_k = 388{,}134; \qquad \sum_U y_k^2 = 1{,}138{,}056{,}730$$

$$\sum_U x_k = 13{,}257; \qquad \sum_U x_k^2 = 655{,}809$$

$$\sum_U x_k y_k = 21{,}965{,}630$$

7.10. From the population considered in Exercise 7.9, an *SI* sample of size $n = 50$ was drawn. Let y and x correspond to ME84 and S82, respectively. The following results were calculated:

$$\sum_s y_k = 79{,}066; \qquad \sum_s y_k^2 = 256{,}640{,}848$$

$$\sum_s x_k = 2{,}480; \qquad \sum_s x_k^2 = 128{,}770$$

$$\sum_s x_k y_k = 4{,}666{,}496; \qquad \sum_s (g_{ks} e_k)^2 = 26{,}490{,}889$$

Compute an approximately 95% confidence interval for the ME84 total based on the regression estimator (7.8.9).

7.11. We want to estimate the total of the variable RMT85 ($= y$) for the MU281 population, using the multiple regression estimator (7.9.7) with CS82 ($= x_1$) and SS82 ($= x_2$) as auxiliary variables. An *SI* sample of size $n = 120$ was drawn, and the following data were calculated

$$\sum_s y_k = 24{,}088; \qquad \sum_s x_{1k} = 1{,}081; \qquad \sum_s x_{2k} = 2{,}718$$

$$S_{x_1 s}^2 = 19.97; \qquad S_{x_2 s}^2 = 48.21; \qquad S_{x_1 x_2 s} = 7.94$$

$$S_{x_1 y s} = 688.06; \qquad S_{x_2 y s} = 1{,}003.49; \qquad \sum_s (g_{ks} e_k)^2 = 1{,}204{,}324$$

For the population the following totals are known:

$$\sum_U x_{1k} = 2{,}508; \qquad \sum_U x_{2k} = 6{,}193$$

Compute an approximately 99% confidence interval for the RMT85 total.

7.12. Starting from Result 7.3.1, derive the results in equations (7.3.4) and (7.3.5) concerning the variance of the ratio estimator under the *SI* design.

7.13. Let $cv_{xU}/cv_{yU} > 0$, and let k be a positive constant. (a) Show that

$$V_{SI}(\hat{t}_{y\pi}) \geq k \cdot AV_{SI}(\hat{t}_{yra})$$

if and only if

$$r \geq \tfrac{1}{2}[(cv_{xU}/cv_{yU}) + (cv_{yU}/cv_{xU})(1 - 1/k)]$$

where $r = S_{xyU}/(S_{xU} S_{yU})$ is the population correlation coefficient. (b) Show that

$$V_{SI}(\hat{t}_{y\pi}) \geq AV_{SI}(\hat{t}_{yra})$$

if and only if

$$r \geq \tfrac{1}{2}(cv_{xU}/cv_{yU})$$

(c) If $cv_{xU} = cv_{yU}$, show that

$$V_{SI}(\hat{t}_{y\pi}) \geq k \cdot AV_{SI}(\hat{t}_{yra})$$

if and only if $r \geq 1 - 0.5/k$. Hence, for example, conclude that $V_{SI}(\hat{t}_{y\pi})$ is more than 20 times larger than $AV_{SI}(\hat{t}_{yra})$ if $r > 0.975$ when $cv_{xU} = cv_{yU}$.

7.14. Show that

$$\hat{t}_{yra} = N\bar{x}_U \bar{y}_s / \bar{x}_s$$

is unbiased if s is drawn with probability

$$p(s) = c\textstyle\sum_s x_k$$

where c is a constant. Determine the value of c. Show that the design in question can be executed by drawing (a) the first sample element with probability

$$p_k = x_k / N\bar{x}_U$$

and (b) $n - 1$ additional elements by *SI* sampling from the remaining $N - 1$ elements.

7.15. Consider the *BE* design with constant inclusion probability π for all elements. Use Result 7.3.1 to find the approximate variance and the variance estimator of the ratio estimator $\hat{t}_{yra}$ given by (7.3.2). Show that roughly the same variance is obtained for the ratio estimator under *SI* sampling of n from N, if $n = N\pi$, which is the expected *BE* sample size.

7.16. Consider, on the one hand, *STSI* sampling with proportional allocation ($n_h = nN_h/N$ selected from N_h elements in stratum h) and, on the other hand, *SI* sampling (n from N elements). Compare the *AV* expressions obtained for the ratio estimator

$$\hat{t}_{yra} = \left(\textstyle\sum_U x_k\right)\left(\sum_s \check{y}_k\right)/\left(\sum_s \check{x}_k\right)$$

under the two designs. Show that the *STSI* variance is smaller than the *SI* variance if differences exist between the strata means of the residuals,

$$\bar{E}_{U_h} = \textstyle\sum_{U_h} (y_k - Bx_k)/N_h = \bar{y}_{U_h} - B\bar{x}_{U_h}$$

7.17. Consider the *BE* design with the constant inclusion probability π for all N population elements. Show that the conditional probability of a *BE* sample s, given that the size is n_s (with $n_s \geq 1$) is formally equivalent to the probability of obtaining an *SI* sample of the fixed size n_s. Hence, for $n_s \geq 1$,

$$E_{BE}(\bar{y}_s | n_s) = \bar{y}_U$$

and, for $n_s \geq 2$,

$$E_{BE}(S_{ys}^2 | n_s) = S_{yU}^2$$

7.18. Consider a survey with H strata crossed with G population groups. Derive the estimator in equation (7.6.7) from the general Result 7.6.1. Derive the residuals e_{ks} and the weights g_{ks} needed for the variance estimator.

7.19. Consider a survey with H strata crossed with G population groups. $STSI$ sampling is used. Derive the poststratification estimator in equation (7.6.9), starting from the model (7.6.8). Identify the residuals e_{ks} and the weights g_{ks} needed for the variance estimator.

7.20. A population of individuals is divided into HG cells formed by a cross-classification of H geographical strata with G age groups. $STSI$ sampling from the H strata is used. Consider the estimator

$$\hat{t}_y = \sum_{g=1}^{G} \sum_{h=1}^{H} t_{xgh} \frac{\bar{y}_{s_{gh}}}{\bar{x}_{s_{gh}}}$$

where t_{xgh} is the total of x in the population cell gh, and $\bar{y}_{s_{gh}}$, $\bar{x}_{s_{gh}}$ are sample means in the cell gh. Formulate the regression model underlying the estimator $\hat{t}_y$. Find the residuals e_{ks} and the weights g_{ks} to be used in the variance estimator.

7.21. Starting from Result 7.7.1, derive the variance estimators in equations (7.7.6) and (7.7.9) for the separate ratio estimator.

7.22. Concerning the separate ratio estimator, show that the AV expressions given by (7.7.3) and (7.7.8) are very nearly equal under the proportional allocation rule $n_g \propto N_g$.

7.23. For the SI design (n from N), compare the AV expressions for the ratio estimator (7.3.3) on the one hand and the simple regression estimator (7.8.9) on the other. Show that $AV_{SI}(\hat{t}_{yreg}) \leq AV_{SI}(\hat{t}_{yra})$ with equality if and only if

$$B = S_{xyU}/S_{xU}^2 = \bar{y}_U/\bar{x}_U$$

7.24. Find the AV expression and the variance estimator for the combined regression estimator (7.8.12).

7.25. Find the AV expression and the variance estimator of the poststratified regression estimator (7.8.13), assuming SI sampling of n from N elements and poststratification of the resulting sample.

7.26. Consider the SI design (n from N). In a poststratification of the resulting sample, n_{s_g} elements are observed to fall in the gth group (poststratum); $g = 1, \ldots, G$. Let $\mathbf{n}_s = (n_{s_1}, \ldots, n_{s_G})$ be the vector of observed group counts. Show that the conditional probability of s, given $\mathbf{n}_s$, is identical to the probability of an $STSI$ sample realized by treating each group as a stratum and making independent SI selections of n_{s_g} from the N_g elements in group g; $g = 1, \ldots, G$.

CHAPTER 8

Regression Estimators for Cluster Sampling and Two-Stage Sampling

8.1. Introduction

This chapter is a continuation of Chapter 4, in the sense that cluster sampling and two-stage sampling are again considered. We now introduce regression estimators for these designs. This topic merits a separate chapter, because a variety of options arise based on the following factors:

a. The kind and the extent of the available auxiliary information.
b. The assumed form of the population point scatter.
c. The exact specification of the selection procedure, that is, the selection of the PSUs and the selection of the SSUs within the chosen PSUs.

In Section 8.2, three different cases are distinguished, depending on the auxiliary information. In each case, a linear model can be fitted and regression estimators can be obtained. Sections 8.3 through 8.7 deal with the case where the auxiliary information is about the clusters (the PSUs). Section 8.8 through 8.11 treat regression estimation for the case where the auxiliary information is about the elements (the SSUs).

We consider two-stage element sampling as described in Section 4.3.2. Results for single-stage cluster sampling are obtained as a special case. The population of elements $U = \{1, \ldots, k, \ldots, N\}$ is partitioned into clusters, $U_1, \ldots, U_i, \ldots, U_{N_I}$. They are also called primary sampling units (PSUs) when there are two stages of selection. The size of U_i is denoted N_i. We have

$$U = \bigcup_{i=1}^{N_I} U_i; \qquad N = \sum_{i=1}^{N_I} N_i$$

The notation for the population of PSUs is

$$U_{\mathrm{I}} = \{1, \ldots, i, \ldots, N_{\mathrm{I}}\}$$

At stage one, a sample of PSUs, s_{I}, is selected from U_{I} according to the design $p_{\mathrm{I}}(\cdot)$ with the inclusion probabilities $\pi_{\mathrm{I}i}$, $\pi_{\mathrm{I}ij}$.

The sampling units at the second stage (the SSUs) are the population elements, labeled $k = 1, \ldots, N$. Given that the PSU U_i is chosen at stage one, a sample s_i of size n_i is selected from U_i according to the design $p_i(\cdot)$ with the inclusion probabilities $\pi_{k|i}$ and $\pi_{kl|i}$. (For simplicity we use n_i to denote the sample size within the PSU, even though it may be random in some cases. This would deviate from our earlier notation n_{s_i}, but should not lead to confusion.) The whole sample of elements and its size are denoted

$$s = \bigcup_{i \in s_{\mathrm{I}}} s_i; \qquad n_s = \sum_{i \in s_{\mathrm{I}}} n_i \tag{8.1.1}$$

For the second-stage sampling, we assume invariance and independence, as explained in Section 4.3. The study variable value, y_k, is observed for $k \in s$.

Single-stage cluster sampling is obtained as the special case where $s_i = U_i$ with probability one, for all $i \in s_{\mathrm{I}}$. This implies observation of y_k for all elements in the selected clusters. The parameters to estimate are the population total,

$$t_y = \textstyle\sum_U y_k = \sum_{U_{\mathrm{I}}} t_{yi}$$

where $t_{yi} = \sum_{U_i} y_k$ is the ith PSU total, and the population mean,

$$\bar{y}_U = t_y/N$$

8.2. The Nature of the Auxiliary Information When Clusters of Elements Are Selected

We found in Chapters 6 and 7 that a successful use of regression estimation requires auxiliary variables capable of explaining y, the variable of interest. In surveys where selection of clusters is necessary, one does not ordinarily have access to auxiliary information as extensive as in Chapter 7. When clusters are selected, it is more likely that auxiliary variable values are known for the clusters (the PSUs) or for some but not all of the elements (SSUs).

We distinguish three different cases, according to the nature of the available auxiliary information. In case A, the information is about the clusters, in cases B and C about the elements. Auxiliary values associated with elements will be denoted by x; those associated with clusters by u.

Let $\mathbf{u} = (u_1, \ldots, u_v, \ldots, u_J)'$ be a vector of auxiliary variables taking the value $\mathbf{u}_i = (u_{1i}, \ldots, u_{vi}, \ldots, u_{Ji})'$ for the ith PSU. We then have case A.

Case A (PSU Auxiliaries). The auxiliary value $\mathbf{u}_i$ is available for all PSUs, that is, for all $i \in U_{\mathrm{I}}$.

Case A leads to regression modeling of quantities associated with the PSUs, such as the PSU totals, $t_{yi} = \sum_{U_i} y_k$, or the PSU means, t_{yi}/N_i. This kind of modeling is discussed in Sections 8.4 to 8.8.

Let $\mathbf{x} = (x_1, \ldots, x_v, \ldots, x_J)'$ be a vector whose value, $\mathbf{x}_k = (x_{1k}, \ldots, x_{vk}, \ldots, x_{Jk})'$, summarizes the auxiliary information for the kth element. We now consider cases B and C.

Case B (Complete Element Auxiliaries). The auxiliary vector value $\mathbf{x}_k$ is available for all elements k in the entire population U.

Case C (Limited Element Auxiliaries). The auxiliary vector value $\mathbf{x}_k$ is available for all elements in selected PSUs only, that is, for elements k such that

$$k \in \bigcup_{i \in s_{\mathrm{I}}} U_i$$

Cases B and C lead to regression modeling of the element values y_k (see Sections 8.9 to 8.12). Cases A and C have many practical applications because they use auxiliary information that is easily obtained in many surveys. On the other hand, if the more extensive information required by case B is available, a more efficient design than selection of clusters can usually be found. However, gathering clusters of observations has cost-saving and administrative advantages, so case B is not excluded. For example, one prefers to collect data concentrated in certain geographic areas, instead of observing a sample scattered throughout the whole territory, which would lead to increased interviewing cost.

Remark 8.2.1. In single-stage cluster sampling, one observes y_k for all elements in a selected cluster, so the cluster total t_{yi} is ascertained without error. Consequently, case C has interest only when a primary selection of clusters is followed by subsampling in the selected clusters.

Remark 8.2.2. Combinations of cases can also be considered. In a combination of cases A and C, there would be auxiliary values $\mathbf{x}_k$ available for elements in selected PSUs, as well as auxiliary values $\mathbf{u}_i$ for all PSUs $i \in U_{\mathrm{I}}$.

Some examples are useful to indicate how the various cases may arise in practice. Consider first a national survey of a certain type of establishment, say hospitals, with numerous but fairly large clusters, say counties. Here, y_k is the study variable value for the kth hospital. Due to the administrative status of a county, much information is ordinarily available to create useful county auxiliary values $\mathbf{u}_i$. These $\mathbf{u}_i$ are usually obtained from demographic sources, a census, or a current population survey, and are used in case A for explaining the cluster totals. A second possibility is to select a first-stage sample of counties and list all hospitals in the selected counties. From county or other records, it may be easy to obtain the hospital auxiliary values $\mathbf{x}_k$ necessary for case C. Finally, case B requires that the value $\mathbf{x}_k$ be available throughout the entire population of hospitals.

As a second example, consider a survey conducted in a large city, with city blocks as PSUs, and buildings as elements. The study variable measures some aspect of the kth building, such as habitable floor space or the quantity of certain equipment in the building. Suppose we have, from city registers, an up-to-date list of all buildings in the entire city, with some useful information $\mathbf{x}_k$ attached to the kth building, $k = 1, \ldots, N$. Then we are in case B. If there is no up-to-date list of buildings, it may be possible, without great cost, to obtain useful information $\mathbf{x}_k$ for the buildings k located in the *sampled* blocks. That is, following the first-stage selection of blocks, all buildings in selected blocks are listed, and, perhaps concurrently with the listing, the inexpensive information $\mathbf{x}_k$ is recorded for buildings in selected clusters (blocks). Then we are in case C. As a third possibility, suppose that city records or other sources can provide us with an auxiliary value for every block $\mathbf{u}_i$ for each $i \in U_{\text{I}}$. The values in the vector $\mathbf{u}_i$ could be measures of characteristics of the block, such as its area, the approximate population, the number, and the type of buildings (residential, high rise, etc.). We will then be in case A.

The regression estimators to be examined require a set of predicted values. Following the fit of an appropriate regression, these predicted values are calculated for a larger set than the sample. In Chapters 6 and 7, this larger set was the entire population of elements U. In this chapter, the known values $\mathbf{u}_i$ in case A will serve to calculate predicted cluster totals for all clusters $i \in U_{\text{I}}$. In case B, the auxiliary information will allow the calculation of predicted values $\hat{y}_k$ for all elements k in the whole population U. In case C, the predictions $\hat{y}_k$ are limited to elements in selected clusters, that is, they are calculated for all elements $k \in U_i$ such that $i \in s_{\text{I}}$.

Remark 8.2.3. To calculate the regression estimators in Chapters 6 and 7, it was sufficient to observe the auxiliary variables for elements in the sample and to know the population total for the auxiliary variables. Similarly, the regression estimators for two-stage sampling will work if the following less detailed information is at hand.

Case A. The value $\mathbf{u}_i$ is observed for all sampled clusters (PSUs) $i \in s_{\text{I}}$, and a correct value of $\sum_{U_{\text{I}}} \mathbf{u}_i$ is available.

Case B. The value $\mathbf{x}_k$ can be observed for all elements $k \in s$, and a correct value of $\sum_U \mathbf{x}_k$ is available.

Case C. The value $\mathbf{x}_k$ can be observed for all elements $k \in s$, and a correct value of the PSU total $\sum_{U_i} \mathbf{x}_k$ is available for each sampled PSU $i \in s_{\text{I}}$.

8.3. Comments on Variance and Variance Estimation in Two-Stage Sampling

First, recall some results from Chapter 4 about the variance, in two-stage sampling, of the π estimator, and the estimation of this variance. The π estimator is given by

$$\hat{t}_{y\pi} = \sum_s y_k/\pi_k = \sum_{s_I} \hat{t}_{yi\pi}/\pi_{Ii}$$

where

$$\hat{t}_{yi\pi} = \sum_{s_i} \check{y}_{k|i} = \sum_{s_i} y_k/\pi_{k|i}$$

is the conditional π estimator of the PSU total $t_{yi} = \sum_{U_i} y_k$. In the notation of Section 4.3, $E_{II}(\hat{t}_{yi\pi}) = t_{yi}$. To arrive easily at the variance of $\hat{t}_{y\pi}$, one can start with the following expression for the estimation error,

$$\hat{t}_{y\pi} - t_y = \underbrace{\sum_{s_I} \frac{t_{yi}}{\pi_{Ii}} - \sum_{U_I} t_{yi}}_{Q_{s_I}} + \underbrace{\sum_{s_I} \frac{1}{\pi_{Ii}}(\hat{t}_{yi\pi} - t_{yi})}_{R_s} \tag{8.3.1}$$

The only randomness in the term Q_{s_I} is due to the first stage selection, s_I. The term R_s depends on s_I as well as on the s_i; its expected value given s_I is zero.

Using the operators E_I, V_I, E_{II}, and V_{II} as in Section 4.3, we obtain from (8.3.1) the variance

$$\begin{aligned} V(\hat{t}_{y\pi}) &= V_I E_{II}(\hat{t}_{y\pi} - t_y) + E_I V_{II}(\hat{t}_{y\pi} - t_y) \\ &= V_I(Q_{s_I}) + E_I V_{II}(R_s) \\ &= V_{PSU} + V_{SSU} \end{aligned} \tag{8.3.2}$$

where

$$V_{PSU} = \sum\sum_{U_I} \Delta_{Iij} \frac{t_{yi}}{\pi_{Ii}} \frac{t_{yj}}{\pi_{Ij}}$$

and

$$V_{SSU} = \sum_{U_I} V_i/\pi_{Ii}$$

with

$$V_i = \sum\sum_{U_i} \Delta_{kl|i} \frac{y_k}{\pi_{k|i}} \frac{y_l}{\pi_{l|i}} \tag{8.3.3}$$

Estimating term by term the components of (8.3.2), we obtain the unbiased variance estimator of Result 4.3.1,

$$\hat{V}(\hat{t}_{y\pi}) = \hat{V}_{PSU} + \hat{V}_{SSU} \tag{8.3.4}$$

where

$$\hat{V}_{\text{PSU}} = \sum\sum_{s_\text{I}} \check{\Delta}_{\text{I}ij} \frac{\hat{t}_{yi\pi}}{\pi_{\text{I}i}} \frac{\hat{t}_{yj\pi}}{\pi_{\text{I}j}} - \sum_{s_\text{I}} \frac{1}{\pi_{\text{I}i}} \left(\frac{1}{\pi_{\text{I}i}} - 1 \right) \hat{V}_i \tag{8.3.5}$$

and

$$\hat{V}_{\text{SSU}} = \sum_{s_\text{I}} \hat{V}_i / \pi_{\text{I}i}^2 \tag{8.3.6}$$

with

$$\hat{V}_i = \sum\sum_{s_i} \check{\Delta}_{kl|i} \frac{y_k}{\pi_{k|i}} \frac{y_l}{\pi_{l|i}} \tag{8.3.7}$$

Single-stage cluster sampling implies that there is no subsampling in selected clusters, so we simply set $V_i = \hat{V}_i = 0$ in the above expressions. Equation (8.3.1) serves as a prototype for finding the variance and an estimator of the variance for the regression estimators now to be created for cases A, B, C.

8.4. Regression Estimators Arising Out of Modeling at the Cluster Level

We consider now case A, the case of PSU auxiliaries. For the ith cluster, there is a known auxiliary vector $\mathbf{u}_i = (u_{1i}, \ldots, u_{vi}, \ldots, u_{Ji})'$, considered a good predictor vector of the PSU total $t_{yi} = \sum_{U_i} y_k$. For example, with $J = 1$, u_i may be the number of buildings in the ith block and t_{yi} the total habitable floor space of buildings in the block.

If one has a value, say a_i, that is a suitable predictor of the PSU mean, t_{yi}/N_i, then $u_i = N_i a_i$ is, of course, a suitable predictor of t_{yi}.

Assume that the scatter of the N_I points $(t_{yi}, \mathbf{u}_i)$ is well described by

$$\begin{cases} E_\xi(t_{yi}) = \mathbf{u}_i' \boldsymbol{\beta}_\text{I} \\ V_\xi(t_{yi}) = \sigma_{\text{I}i}^2 \end{cases} \tag{8.4.1}$$

The t_{yi} are independent under the model, an assumption tacitly made throughout. If all N_I points were available, we would estimate $\boldsymbol{\beta}_\text{I}$ by

$$\mathbf{B}_\text{I} = \left(\sum_{U_\text{I}} \mathbf{u}_i \mathbf{u}_i' / \sigma_{\text{I}i}^2\right)^{-1} \sum_{U_\text{I}} \mathbf{u}_i t_{yi} / \sigma_{\text{I}i}^2 \tag{8.4.2}$$

The corresponding fitted values are, for $i \in U_\text{I}$,

$$t_{yi}^0 = \mathbf{u}_i' \mathbf{B}_\text{I} \tag{8.4.3}$$

and the residual for the ith cluster is

$$D_i = t_{yi} - t_{yi}^0 \tag{8.4.4}$$

How does one estimate $\mathbf{B}_\text{I}$, given the data from the sample $s = \bigcup_{i \in s_\text{I}} s_i$? The answer is not obvious. The first thought might be to regress the cluster totals t_{yi} on $\mathbf{u}_i$ for sampled PSUs, that is, for the clusters $i \in s_{\mathrm{I}}$. However, this approach fails when there is subsampling in selected clusters. The true t_{yi} are then unknown, and all we have are the estimates

$$t_{yi}^* = \hat{t}_{yi\pi} = \sum\nolimits_{s_i} y_k/\pi_{k|i} \tag{8.4.5}$$

With $t_{yi}^* = \hat{t}_{yi\pi}$ as an input for fitting the model (8.4.1), the sample-based estimator of the population vector $\mathbf{B}_{\mathrm{I}}$ is

$$\hat{\mathbf{B}}_{\mathrm{I}} = \left(\sum\nolimits_{s_{\mathrm{I}}} \mathbf{u}_i\mathbf{u}_i'/\sigma_{\mathrm{I}i}^2\pi_{\mathrm{I}i}\right)^{-1} \sum\nolimits_{s_{\mathrm{I}}} \mathbf{u}_i t_{yi}^*/\sigma_{\mathrm{I}i}^2\pi_{\mathrm{I}i} \tag{8.4.6}$$

Note that the dependent variable value t_{yi}^* is affected by sampling error when there is subsampling in selected clusters. The resulting predicted values are

$$\hat{t}_{yip} = \mathbf{u}_i'\hat{\mathbf{B}}_{\mathrm{I}} \tag{8.4.7}$$

for $i \in U_{\mathrm{I}}$, where the index p stands for predicted. For $i \in s_{\mathrm{I}}$, we can calculate the residual

$$d_i = t_{yi}^* - \hat{t}_{yip}$$

This technique, already used in Chapter 6, now leads to the case A regression estimator (denoted $\hat{t}_{yAr}$) of t_y, namely,

$$\hat{t}_{yAr} = \sum\nolimits_{U_{\mathrm{I}}} \hat{t}_{yip} + \sum\nolimits_{s_{\mathrm{I}}} \frac{t_{yi}^* - \hat{t}_{yip}}{\pi_{\mathrm{I}i}} \tag{8.4.8}$$

with t_{yi}^* and $\hat{t}_{yip}$ given respectively by (8.4.5) and (8.4.7). The second term of equation (8.4.8) contains the ordinary π estimator,

$$\sum\nolimits_{s_{\mathrm{I}}} \frac{t_{yi}^*}{\pi_{\mathrm{I}i}} = \sum\nolimits_{s_{\mathrm{I}}} \frac{1}{\pi_{\mathrm{I}i}} \sum\nolimits_{s_i} \frac{y_k}{\pi_{k|i}} = \sum\nolimits_{s} \frac{y_k}{\pi_k} = \sum\nolimits_{s} \check{y}_k$$

since $\pi_k = \pi_{\mathrm{I}i}\pi_{k|i}$. Thus the estimator (8.4.8) can be written as

$$\hat{t}_{yAr} = \sum\nolimits_{s} \check{y}_k + \left(\sum\nolimits_{U_{\mathrm{I}}} \mathbf{u}_i - \sum\nolimits_{s_{\mathrm{I}}} \mathbf{u}_i/\pi_{\mathrm{I}i}\right)'\hat{\mathbf{B}}_{\mathrm{I}}$$

that is, as the ordinary π estimator plus a regression adjustment term.

Remark 8.4.1. In (8.4.8), the term

$$\sum\nolimits_{s_{\mathrm{I}}} \frac{d_i}{\pi_{\mathrm{I}i}} = \sum\nolimits_{s_{\mathrm{I}}} \frac{t_{yi}^* - \hat{t}_{yip}}{\pi_{\mathrm{I}i}}$$

will vanish for many models, namely, if the variance structure satisfies

$$\sigma_{\mathrm{I}i}^2 = \boldsymbol{\lambda}'\mathbf{u}_i$$

for some constant vector $\boldsymbol{\lambda}$. This is an analog of Result 6.5.1. The case A estimator shown in equation (8.4.8) then becomes simply

$$\hat{t}_{yAr} = \sum\nolimits_{U_{\mathrm{I}}} \hat{t}_{yip}$$

Examples of this form are found in Sections 8.5 to 8.8.

Let us derive an approximate variance and a variance estimator for case A. Proceeding similarly as in Section 6.5, we write

$$\hat{t}_{yAr} = \sum_{s_{\text{I}}} \frac{g_{is_{\text{I}}} t^*_{yi}}{\pi_{\text{I}i}}$$

where the g weight for the ith PSU is

$$g_{is_{\text{I}}} = 1 + \left(\sum_{U_{\text{I}}} \mathbf{u}_i - \sum_{s_{\text{I}}} \mathbf{u}_i/\pi_{\text{I}i}\right)' \left(\sum_{s_{\text{I}}} \mathbf{u}_i \mathbf{u}_i'/\sigma^2_{\text{I}i}\pi_{\text{I}i}\right)^{-1} \mathbf{u}_i/\sigma^2_{\text{I}i}$$

After some straightforward algebra, the error of the estimator is brought onto a form analogous to equation (8.3.1), namely,

$$\hat{t}_{yAr} - t_y = \underbrace{\sum_{s_{\text{I}}} \frac{g_{is_{\text{I}}} D_i}{\pi_{\text{I}i}} - \sum_{U_{\text{I}}} D_i}_{Q_{As_{\text{I}}}} + \underbrace{\sum_{s_{\text{I}}} \frac{g_{is_{\text{I}}}}{\pi_{\text{I}i}} (t^*_{yi} - t_{yi})}_{R_{As}} \tag{8.4.9}$$

where $D_i = t_{yi} - \mathbf{u}_i' \mathbf{B}_{\text{I}}$. An essential element in deriving (8.4.9) is that

$$\sum_{s_{\text{I}}} \frac{g_{is_{\text{I}}} \mathbf{u}_i}{\pi_{\text{I}i}} = \sum_{U_{\text{I}}} \mathbf{u}_i$$

That is, the g weights give perfect estimates when applied to the auxiliary values. Now, the term $Q_{As_{\text{I}}}$ in equation (8.4.9) is a constant with respect to stage two, so $E_{\text{II}}(Q_{As_{\text{I}}}) = Q_{As_{\text{I}}}$ and $V_{\text{II}}(Q_{As_{\text{I}}}) = 0$. Moreover, given s_{I},

$$E_{\text{II}}(R_{As}) = 0; \qquad V_{\text{II}}(R_{As}) = \sum_{s_{\text{I}}} \frac{g^2_{is_{\text{I}}} V_i}{\pi^2_{\text{I}i}}$$

where

$$V_i = \sum\sum_{U_i} \Delta_{kl} \frac{y_k}{\pi_{k|i}} \frac{y_l}{\pi_{l|i}} \tag{8.4.10}$$

is the variance of $t^*_{yi} = \hat{t}_{yi\pi}$ given s_{I}. Using the technique in equation (8.3.2), we obtain

$$\begin{aligned} V(\hat{t}_{yAr}) &= V_{\text{I}}(Q_{As_{\text{I}}}) + E_{\text{I}} V_{\text{II}}(R_{As}) \\ &= V_{\text{APSU}} + V_{\text{ASSU}} \end{aligned}$$

where

$$V_{\text{APSU}} = V_{\text{I}}\left(\sum_{s_{\text{I}}} \frac{g_{is_{\text{I}}} D_i}{\pi_{\text{I}i}}\right) \tag{8.4.11}$$

and

$$V_{\text{ASSU}} = E_{\text{I}}\left(\sum_{s_{\text{I}}} \frac{g^2_{is_{\text{I}}} V_i}{\pi^2_{\text{I}i}}\right) \tag{8.4.12}$$

These are exact first- and second-stage variance components, but implicit rather than explicit, since the variance and expected value operations with respect to stage one remain to be carried out. The weights $g_{is_{\text{I}}}$ are complex, so these operations are not simple. Good approximations are obtained by using $g_{is_{\text{I}}} \doteq 1$. The conclusions are summarized in the following result.

Result 8.4.1. *In two-stage element sampling, the case A regression estimator of* $t_y = \sum_U y_k$ *is* $\hat{t}_{yAr}$ *given by equation* (8.4.8), *with*

$$t^*_{yi} = \hat{t}_{yi\pi} = \textstyle\sum_{s_i} y_k/\pi_{k|i}$$

and $\hat{t}_{yip}$ *is obtained from* (8.4.7). *The approximate variance is*

$$AV(\hat{t}_{yAr}) = AV_{\text{APSU}} + AV_{\text{ASSU}} \tag{8.4.13}$$

where

$$AV_{\text{APSU}} = \textstyle\sum\sum_{U_{\text{I}}} \Delta_{\text{I}ij} \dfrac{D_i}{\pi_{\text{I}i}} \dfrac{D_j}{\pi_{\text{I}j}} \tag{8.4.14}$$

$$AV_{\text{ASSU}} = \textstyle\sum_{U_{\text{I}}} V_i/\pi_{\text{I}i} \tag{8.4.15}$$

with D_i *and* V_i *given by* (8.4.4) *and* (8.4.10). *The estimated variance is*

$$\hat{V}(\hat{t}_{yAr}) = \hat{V}_{\text{APSU}} + \hat{V}_{\text{ASSU}} \tag{8.4.16}$$

where

$$\hat{V}_{\text{APSU}} = \textstyle\sum\sum_{s_{\text{I}}} \check{\Delta}_{\text{I}ij} \dfrac{g_{is_{\text{I}}}d_i}{\pi_{\text{I}i}} \dfrac{g_{js_{\text{I}}}d_j}{\pi_{\text{I}j}} - \sum_{s_{\text{I}}} \dfrac{1}{\pi_{\text{I}i}}\left(\dfrac{1}{\pi_{\text{I}i}} - 1\right) g^2_{is_{\text{I}}} \hat{V}_i \tag{8.4.17}$$

and

$$\hat{V}_{\text{ASSU}} = \textstyle\sum_{s_{\text{I}}} g^2_{is_{\text{I}}} \hat{V}_i/\pi^2_{\text{I}i} \tag{8.4.18}$$

with $d_i = t^*_{yi} - \mathbf{u}'_i\hat{\mathbf{B}}_{\text{I}}$ *and*

$$\hat{V}_i = \textstyle\sum\sum_{s_i} \check{\Delta}_{kl|i} \dfrac{y_k}{\pi_{k|i}} \dfrac{y_l}{\pi_{l|i}} \tag{8.4.19}$$

The result holds for single-stage cluster sampling if we set $t^*_{yi} = t_{yi} = \sum_{U_i} y_k$ *and* $V_i = \hat{V}_i = 0$, *for all i.*

DERIVATION. In equations (8.4.11) and (8.4.12), use the approximation $g_{is_{\text{I}}} \doteq 1$ for all i and s_{I}. The $V_{\text{I}}(\cdot)$ and $E_{\text{I}}(\cdot)$ operations are then simple to execute; standard results for a π estimator are involved. We obtain the AV expressions shown in (8.4.14) and (8.4.15). Turning now to the variance component estimators, we first justify $\hat{V}_{\text{APSU}}$ by assuming that the replacement of $\hat{\mathbf{B}}_{\text{I}}$ by $\mathbf{B}_{\text{I}}$ in the residuals d_i will cause only negligible approximation error,

$$\begin{aligned} E_{\text{II}}(d_i^2) &= E_{\text{II}}[(t^*_{yi} - \mathbf{u}'_i\hat{\mathbf{B}}_{\text{I}})^2] \\ &\doteq E_{\text{II}}[(t^*_{yi} - \mathbf{u}'_i\mathbf{B}_{\text{I}})^2] \\ &= [E_{\text{II}}(t^*_{yi} - \mathbf{u}'_i\mathbf{B}_{\text{I}})]^2 + V_{\text{II}}(t^*_{yi} - \mathbf{u}'_i\mathbf{B}_{\text{I}}) \\ &= D_i^2 + V_i \end{aligned}$$

Similarly, for $i \neq j$,

$$E_{\text{II}}(d_i d_j) \doteq D_i D_j$$

Noting also that $E_{\text{II}}(\hat{V}_i) = V_i$, we have

$$E_{\text{II}}(\hat{V}_{\text{APSU}}) \doteq \sum\sum_{s_{\text{I}}} \check{\Delta}_{\text{I}ij} \frac{g_{is_{\text{I}}} D_i}{\pi_{\text{I}i}} \frac{g_{js_{\text{I}}} D_j}{\pi_{\text{I}j}}$$

where the expression on the right-hand side is a g-weighted estimator of the first-stage component shown in (8.4.14); hence

$$E(\hat{V}_{\text{APSU}}) = E_{\text{I}} E_{\text{II}}(\hat{V}_{\text{APSU}}) \doteq A V_{\text{APSU}}$$

Finally,

$$E(\hat{V}_{\text{ASSU}}) = E_{\text{I}} E_{\text{II}}(\textstyle\sum_{s_{\text{I}}} g_{is_{\text{I}}}^2 \hat{V}_i / \pi_{\text{I}i}^2)$$
$$= E_{\text{I}}(\textstyle\sum_{s_{\text{I}}} g_{is_{\text{I}}}^2 V_i / \pi_{\text{I}i}^2)$$

In other words, $\hat{V}_{\text{ASSU}}$ is unbiased for (8.4.12), which is an exact expression for the second-stage variance component. Thus, (8.4.16) can be called an approximate variance estimator, but approximation is used only in the first-stage component.

When the result is applied to single-stage cluster sampling, we have $t_{yi}^* = t_{yi}$, which has no sampling error, so we simply set $V_i = \hat{V}_i = 0$. □

Remark 8.4.2. If the size of the second-stage sample is random, we know from earlier chapters that $\hat{t}_{yi\pi}$ is ordinarily more variable than $N_i \tilde{y}_{s_i}$, where

$$\tilde{y}_{s_i} = \frac{\sum_{s_i} y_k / \pi_{k|i}}{\sum_{s_i} 1/\pi_{k|i}}$$

If N_i is known, it is therefore better to build the estimator (8.4.8) with $t_{yi}^* = N_i \tilde{y}_{s_i}$ than with $t_{yi}^* = \hat{t}_{yi\pi}$. The variance is then estimated by (8.4.16) if we replace $\hat{V}_i$ in the second component, (8.4.18), by

$$\hat{V}_i' = \left(\frac{N_i}{\hat{N}_i}\right)^2 \sum\sum_{s_i} \check{\Delta}_{kl|i} \frac{e_{ks}}{\pi_{k|i}} \frac{e_{ls}}{\pi_{l|i}}$$

where $\hat{N}_i = \sum_{s_i} 1/\pi_{k|i}$ and $e_{ks} = y_k - \tilde{y}_{s_i}$. The first component, shown in equation (8.4.17), remains unchanged. See also Remark 8.6.1.

8.5. The Common Ratio Model for Cluster Totals

To illustrate Result 8.4.1, let us consider the common ratio model for *PSU* totals,

$$\begin{cases} E_\xi(t_{yi}) = \beta_{\text{I}} u_i \\ V_\xi(t_{yi}) = \sigma_{\text{I}}^2 u_i \end{cases} \tag{8.5.1}$$

for $i \in U_{\text{I}}$, where $u_i > 0$ for all i. A common slope β_{I} is assumed for the whole population of clusters. Groups of clusters, with a separate slope in each group, are considered in Section 8.8.

In many applications, the cluster total is roughly proportional to a measure of cluster size. In (8.5.1) one may then take u_i to be the known size measure for the ith cluster. With geographically defined clusters, the size measure might be the area of the cluster, or the number of elements N_i in the cluster. If the exact area or the exact number N_i is unknown, a rough indicator of size (an "eye estimate" or other good judgmental value) will often suffice to provide a powerful explanatory variable for the cluster total. The resulting regression estimator will have a greatly reduced variance, compared to an estimator that ignores information about size.

The following Result 8.5.1 is a straightforward consequence of Result 8.4.1.

Result 8.5.1. *In two-stage element sampling, and under the ratio model* (8.5.1), *the case A regression estimator of* $t_y = \sum_U y_k$ *is given by*

$$\hat{t}_{yAr} = \left(\textstyle\sum_{U_{\text{I}}} u_i\right)\hat{B}_{\text{I}} \tag{8.5.2}$$

where

$$\hat{B}_{\text{I}} = \frac{\sum_{s_{\text{I}}} t_{yi}^{*}/\pi_{\text{I}i}}{\sum_{s_{\text{I}}} u_i/\pi_{\text{I}i}}$$

with

$$t_{yi}^{*} = \hat{t}_{yi\pi} = \textstyle\sum_{s_i} y_k/\pi_{k|i}$$

The approximate variance is given by (8.4.13) *with*

$$D_i = t_{yi} - B_{\text{I}} u_i \quad \textit{for} \quad i \in U_{\text{I}}$$

where

$$B_{\text{I}} = \frac{\sum_{U_{\text{I}}} t_{yi}}{\sum_{U_{\text{I}}} u_i}$$

The variance estimator is given by (8.4.16) *with*

$$d_i = t_{yi}^{*} - \hat{B}_{\text{I}} u_i$$

and

$$g_{is_{\text{I}}} = \frac{\sum_{U_{\text{I}}} u_i}{\sum_{s_{\text{I}}} u_i/\pi_{\text{I}i}}$$

The result applies to single-stage cluster sampling by setting $t_{yi}^{*} = t_{yi}$ *and* $V_i = \hat{V}_i = 0$ *for all* i.

8.6. Estimation of the Population Mean When Clusters Are Sampled

Result 8.5.1 leads directly to a useful estimator of the population mean,

$$\bar{y}_U = \frac{1}{N} t_y = \frac{1}{N} \sum_U y_k = \frac{\sum_{U_I} t_{yi}}{\sum_{U_I} N_i}$$

namely, if we take $u_i = N_i$ in (8.5.2), which implies $\sum_{U_I} u_i = \sum_{U_I} N_i = N$. Dividing the resulting estimator (8.5.2) by N, and consequently the approximate variance and the variance estimator by N^2, we obtain Result 8.6.1 below. The estimator (8.6.1) requires no information beyond the sizes N_i of selected clusters. Note also that with $u_i = N_i$, the underlying model (8.5.1) will consider the PSU means t_{yi}/N_i as being scattered around a common mean, β_I; it is a *common mean model for cluster means.*

The conclusions drawn from Result 8.5.1 are as follows:

Result 8.6.1. *In two-stage element sampling, an approximately unbiased estimator of the population mean $\bar{y}_U = t_y/N$ is given by*

$$\hat{\bar{y}}_{Ur} = \frac{\sum_{s_I} t_{yi}^*/\pi_{Ii}}{\sum_{s_I} N_i/\pi_{Ii}} \tag{8.6.1}$$

where

$$t_{yi}^* = \hat{t}_{yi\pi} = \sum_{s_i} y_k/\pi_{k|i}$$

The approximate variance is

$$AV(\hat{\bar{y}}_{Ur}) = \frac{1}{N^2}\left(\sum\sum_{U_I} \Delta_{Iij} \frac{D_i}{\pi_{Ii}} \frac{D_j}{\pi_{Ij}} + \sum_{U_I} \frac{V_i}{\pi_{Ii}}\right) \tag{8.6.2}$$

where $D_i = N_i(\bar{y}_{U_i} - \bar{y}_U)$, with $\bar{y}_{U_i} = \sum_{U_i} y_k/N_i$; $\bar{y}_U = \sum_U y_k/N$, and V_i is given by (8.4.10). *The variance is estimated by*

$$\hat{V}(\hat{\bar{y}}_{Ur}) = \frac{1}{(\sum_{s_I} N_i/\pi_{Ii})^2}\left(\sum\sum_{s_I} \check{\Delta}_{Iij} \frac{d_i}{\pi_{Ii}} \frac{d_j}{\pi_{Ij}} + \sum_{s_I} \frac{\hat{V}_i}{\pi_{Ii}}\right) \tag{8.6.3}$$

with

$$d_i = t_{yi}^* - N_i \hat{\bar{y}}_{Ur}$$

and

$$\hat{V}_i = \sum\sum_{s_i} \check{\Delta}_{kl|i} \frac{y_k}{\pi_{k|i}} \frac{y_l}{\pi_{l|i}} \tag{8.6.4}$$

The result applies to single-stage cluster sampling by setting $t_{yi}^* = t_{yi} = \sum_{U_i} y_k$ and $V_i = \hat{V}_i = 0$ for all i.

EXAMPLE 8.6.1. Consider the *SI*, *SI* design defined in Example 4.3.1. At stage one, an *SI* sample of n_{I} PSUs is selected from N_{I}; if the ith PSU is selected, an *SI* sample of n_i SSUs is selected from N_i. The sampling fractions are denoted $f_{\mathrm{I}} = n_{\mathrm{I}}/N_{\mathrm{I}}$ and $f_i = n_i/N_i$. It follows from Result 8.6.1 that the population mean $\bar{y}_U$ is estimated by

$$\hat{\bar{y}}_{Ur} = \frac{\sum_{s_{\mathrm{I}}} N_i \bar{y}_{s_i}}{\sum_{s_{\mathrm{I}}} N_i} \tag{8.6.5}$$

where $\bar{y}_{s_i} = \sum_{s_i} y_k/n_i$. Note that $t_{yi}^* = N_i \bar{y}_{s_i}$. From equation (8.6.3), the variance is estimated by

$$\hat{V}_{SI,SI}(\hat{\bar{y}}_{Ur}) = \frac{1}{\bar{N}_{s_{\mathrm{I}}}^2}\left\{\frac{1 - f_{\mathrm{I}}}{n_{\mathrm{I}}} \frac{\sum_{s_{\mathrm{I}}} N_i^2(\bar{y}_{s_i} - \hat{\bar{y}}_{Ur})^2}{n_{\mathrm{I}} - 1} + \frac{1}{N_{\mathrm{I}} n_{\mathrm{I}}} \sum_{s_{\mathrm{I}}} N_i^2 \frac{1 - f_i}{n_i} S_{ys_i}^2\right\} \tag{8.6.6}$$

where $S_{ys_i}^2$ is the variance of y_k in s_i, and $\bar{N}_{s_{\mathrm{I}}} = \sum_{s_{\mathrm{I}}} N_i/n_I$ is the average size of the selected PSUs. Note that $d_i = N_i(\bar{y}_{s_i} - \hat{\bar{y}}_{Ur})$ and $\sum_{s_{\mathrm{I}}} d_i = 0$.

Remark 8.6.1. There exist alternatives to (8.6.1) that are sometimes preferred. One alternative estimator of the population mean is

$$\hat{\bar{y}}_U = \frac{\sum_{s_{\mathrm{I}}} \hat{t}_{yi\pi}/\pi_{\mathrm{I}i}}{\sum_{s_{\mathrm{I}}} \hat{N}_i/\pi_{\mathrm{I}i}} \tag{8.6.7}$$

with

$$\hat{N}_i = \sum_{s_i} 1/\pi_{k|i}$$

This estimator, considered in an end of chapter exercise, is of interest when estimated PSU sizes are used. Another estimator of the population mean is

$$\hat{\bar{y}}_U = \frac{\sum_{s_{\mathrm{I}}} N_i \tilde{y}_{s_i}/\pi_{\mathrm{I}i}}{\sum_{s_{\mathrm{I}}} N_i/\pi_{\mathrm{I}i}}$$

with

$$\tilde{y}_{s_i} = \left(\sum_{s_i} y_k/\pi_{k|i}\right)/\hat{N}_i \tag{8.6.8}$$

This estimator is suitable when random size subsamples are drawn within the selected clusters. We know from earlier chapters that $N_i \tilde{y}_{s_i}$ is then preferred to $\hat{t}_{yi\pi}$ as an estimator of the ith cluster total.

8.7. Design Effects for Single-Stage Cluster Sampling

We know from Section 4.2.2. that the π-estimator of the population total is often less efficient under random sampling of clusters than under direct sampling of elements. Factors that contribute to this are a positive homogeneity coefficient and a variability in the cluster sizes. The mean-per-element estimator of the population mean is also affected, but not in the same way, as we shall now see.

Consider the *SIC* design defined in Section 4.2.2. A simple random sample of n_{I} clusters is drawn from N_{I}; all elements in selected clusters are observed. Let us apply Result 8.6.1. The population mean is estimated by the sample mean per element,

$$\hat{\bar{y}}_{Ur} = \frac{\sum_{s_{\mathrm{I}}} t_{yi}}{\sum_{s_{\mathrm{I}}} N_i} \tag{8.7.1}$$

From equation (8.6.2), with $V_i = 0$, we obtain the approximate variance

$$AV_{\mathrm{SIC}}(\hat{\bar{y}}_{Ur}) = \frac{1}{\bar{N}^2}\left(\frac{1}{n_{\mathrm{I}}} - \frac{1}{N_{\mathrm{I}}}\right)\frac{\sum_{U_{\mathrm{I}}} N_i^2(\bar{y}_{U_i} - \bar{y}_U)^2}{N_{\mathrm{I}} - 1} \tag{8.7.2}$$

where

$$\bar{N} = \frac{N}{N_{\mathrm{I}}} = \frac{1}{N_{\mathrm{I}}}\sum_{U_{\mathrm{I}}} N_i$$

is the average size of the clusters in the population. The estimated variance is, from (8.6.3) or (8.6.6),

$$\hat{V}_{\mathrm{SIC}}(\hat{\bar{y}}_{Ur}) = \frac{1}{\bar{N}_{s_{\mathrm{I}}}^2}\left(\frac{1}{n_{\mathrm{I}}} - \frac{1}{N_{\mathrm{I}}}\right)\frac{\sum_{s_{\mathrm{I}}} N_i^2(\bar{y}_{U_i} - \hat{\bar{y}}_{Ur})^2}{n_{\mathrm{I}} - 1} \tag{8.7.3}$$

where $\bar{N}_{s_{\mathrm{I}}} = \sum_{s_{\mathrm{I}}} N_i/n_{\mathrm{I}}$ is the average size of the sampled clusters.

The design effect of the strategy (SIC, $\hat{\bar{y}}_{Ur}$) is given by the ratio

$$\mathrm{deff}(\mathrm{SIC}, \hat{\bar{y}}_{Ur}) = \frac{AV_{\mathrm{SIC}}(\hat{\bar{y}}_{Ur})}{V_{\mathrm{SI}}(\bar{y}_s)}$$

where $\bar{y}_s$ is the mean of a comparable *SI* sample, namely, a sample with the fixed size $n = n_{\mathrm{I}}\bar{N}$, which is the expected sample size under the SIC design. Strictly speaking, the ratio should be called a *strategy effect* in this case, since both the design and the estimator differ from their counterparts in the basic strategy (SI, $\bar{y}_s$).

The homogeneity coefficient, defined by equation (4.2.13), is

$$\delta = 1 - \frac{S_{yW}^2}{S_{yU}^2} \tag{8.7.4}$$

with

$$S_{yW}^2 = \frac{\sum_{U_{\mathrm{I}}} (N_i - 1)S_{yU_i}^2}{\sum_{U_{\mathrm{I}}} (N_i - 1)}$$

where $S_{yU_i}^2$ is the variance of y in the ith cluster. After simplification, the desired design effect can be written as

$$\text{deff}(\text{SIC}, \hat{\tilde{y}}_{Ur}) = \frac{AV_{\text{SIC}}(\hat{\tilde{y}}_{Ur})}{V_{\text{SI}}(\bar{y}_s)} = 1 + \frac{N - N_{\text{I}}}{N_{\text{I}} - 1}\delta + \frac{\text{Cov}'}{\bar{N}S_{yU}^2} \tag{8.7.5}$$

where

$$\text{Cov}' = \frac{1}{N_{\text{I}} - 1}\sum_{U_{\text{I}}} (N_i - \bar{N})D_i$$

is the covariance between N_i and $D_i = N_i(\bar{y}_{U_i} - \bar{y}_U)^2$. The details of the derivation of equation (8.7.5) are left as an exercise.

If all N_i are equal, then $\text{Cov}' = 0$, which also holds when all cluster means $\bar{y}_{U_i}$ are equal. Suppose the correlation is weak between N_i and D_i, so that the term $\text{Cov}'/\bar{N}S_{yU}^2$ can be ignored in (8.7.5). Then

$$\text{deff}(\text{SIC}, \hat{\tilde{y}}_{Ur}) \doteq 1 + (\bar{N} - 1)\delta \tag{8.7.6}$$

Experience has shown that this approximation holds with good accuracy for many populations, even if the cluster sizes N_i are quite variable. Nevertheless, the effect of varying cluster sizes merits a more detailed analysis.

To get a perspective on the design effect (8.7.5), note that an alternative estimator of $\bar{y}_U$ is obtained by dividing the π estimator $\hat{t}_{y\pi}$ by N. Since $\hat{t}_{y\pi} = (N_{\text{I}}/n_{\text{I}})\sum_{s_{\text{I}}} t_{yi} = N_{\text{I}}\bar{t}_{s_{\text{I}}}$, this alternative estimator is

$$\hat{\tilde{y}}_{U\pi} = \frac{N_{\text{I}}}{N}\bar{t}_{s_{\text{I}}} = \frac{\bar{t}_{s_{\text{I}}}}{\bar{N}}$$

In surveys that require selection of clusters, N is often unknown, and then $\hat{\tilde{y}}_{U\pi}$ cannot be computed. Moreover, even if N is known, $\hat{\tilde{y}}_{U\pi}$ is usually inferior to $\hat{\tilde{y}}_{Ur}$. This can be seen by comparing the corresponding design effects. Note first that

$$\text{deff}(\text{SIC}, \hat{t}_{y\pi}) = \text{deff}(\text{SIC}, \hat{\tilde{y}}_{U\pi}) \tag{8.7.7}$$

since dividing by the constant N does not alter the design effect. Directly from (4.2.22) we then have

$$\text{deff}(\text{SIC}, \hat{\tilde{y}}_{U\pi}) = \frac{V_{\text{SIC}}(\hat{\tilde{y}}_{U\pi})}{V_{\text{SI}}(\bar{y}_s)} = 1 + \frac{N - N_{\text{I}}}{N_{\text{I}} - 1}\delta + \frac{\text{Cov}}{\bar{N}S_{yU}^2} \tag{8.7.8}$$

where δ is the homogeneity coefficient given by (8.7.4), and Cov is the covariance between N_i and $A_i = N_i\bar{y}_{U_i}^2$, that is,

$$\text{Cov} = \frac{1}{N_{\text{I}} - 1}\sum_{U_{\text{I}}} (N_i - \bar{N})A_i$$

From (8.7.5) and (8.7.8), and after simplification, we obtain

$$\begin{aligned}\text{deff}(\text{SIC}, \hat{\tilde{y}}_{U\pi}) - \text{deff}(\text{SIC}, \hat{\tilde{y}}_{Ur}) &= \frac{\text{Cov} - \text{Cov}'}{\bar{N}S_{yU}^2} \\ &= \bar{N}\frac{cv_N(2Rcv_t - cv_N)}{(cv_{yU})^2}\end{aligned} \tag{8.7.9}$$

where the three coefficients of variation are

$$cv_N = \frac{S_{NU_\mathrm{I}}}{\bar{N}}, \qquad cv_t = \frac{S_{tU_\mathrm{I}}}{\bar{t}_{U_\mathrm{I}}}, \quad \text{and} \quad cv_{yU} = \frac{S_{yU}}{\bar{y}_U}$$

and

$$R = \frac{S_{NtU_\mathrm{I}}}{S_{NU_\mathrm{I}} S_{tU_\mathrm{I}}}$$

is the coefficient of correlation between the cluster size N_i and the cluster total $t_{yi} = \sum_{U_i} y_k$. The following definitions apply

$$S^2_{NU_\mathrm{I}} = \frac{1}{N_\mathrm{I} - 1} \sum_{U_\mathrm{I}} (N_i - \bar{N})^2; \qquad S^2_{tU_\mathrm{I}} = \frac{1}{N_\mathrm{I} - 1} \sum_{U_\mathrm{I}} (t_{yi} - \bar{t}_{U_\mathrm{I}})^2$$

$$S_{NtU_\mathrm{I}} = \frac{1}{N_\mathrm{I} - 1} \sum_{U_\mathrm{I}} (N_i - \bar{N})(t_{yi} - \bar{t}_{U_\mathrm{I}})$$

with

$$\bar{N} = \sum_{U_\mathrm{I}} N_i / N_\mathrm{I}, \qquad \bar{t}_{U_\mathrm{I}} = \sum_{U_\mathrm{I}} t_{yi} / N_\mathrm{I}$$

The difference (8.7.9) is zero if all N_i are equal. But when the cluster sizes vary, (8.7.9) is usually positive, which implies that $\hat{\bar{y}}_{Ur}$ is preferred to $\hat{\bar{y}}_{U\pi}$. A contributing factor is the often strong positive correlation R between cluster total and cluster size. To illustrate, suppose $\bar{N} = 150$, $cv_N = 0.1$, $cv_t = 0.7$, $cv_{yU} = 0.75$, and $R = 0.8$. Then there is a large difference in design effect in favor of $\hat{\bar{y}}_{Ur}$:

$$\operatorname{deff}(\mathrm{SIC}, \hat{\bar{y}}_{U\pi}) - \operatorname{deff}(\mathrm{SIC}, \hat{\bar{y}}_{Ur}) = 20.4$$

An extreme case is when all $\bar{y}_{U_i}$ are equal. The conditions are then extremely favorable for cluster sampling, and δ takes its minimal value, $\delta_{\min} = -(N_\mathrm{I} - 1)/(N - N_\mathrm{I})$. Then

$$\operatorname{deff}(\mathrm{SIC}, \hat{\bar{y}}_{Ur}) = 0; \qquad \operatorname{deff}(\mathrm{SIC}, \hat{\bar{y}}_{U\pi}) = \bar{N}(cv_N / cv_{yU})^2$$

In summary, under SIC sampling, two reasons favor $\hat{\bar{y}}_{Ur}$ over $\hat{\bar{y}}_{U\pi}$, namely, (i) $\hat{\bar{y}}_{Ur}$ and the corresponding variance estimator are calculated without knowledge of the population size N; and (ii) the design effect of $\hat{\bar{y}}_{Ur}$ is ordinarily considerably smaller.

Our discussion also shows that the design effect can be much more severe for the population total than for the population mean. If N is unknown, the estimators are $\hat{t}_{y\pi}$ and $\hat{\bar{y}}_{Ur}$. Then equation (8.7.7) holds, so the difference in design effect, that is, the difference

$$\operatorname{deff}(\mathrm{SIC}, \hat{t}_{y\pi}) - \operatorname{deff}(\mathrm{SIC}, \hat{\bar{y}}_{Ur})$$

is given by the right-hand side of (8.7.9). This difference can be large, as the numerical example showed. More is lost through sampling of clusters when the total is estimated than when the mean is estimated.

8.8. Stratified Clusters and Poststratified Clusters

In many surveys, it is natural to recognize groups of clusters. For example, clusters of households may be grouped into rural clusters, township clusters, and city clusters. Or, if a household survey is conducted in a big city, the groups may correspond to sectors of the city, each sector containing clusters of households.

When the between-group differences are deemed important enough, they should be reflected in the regression model. This is achieved by admitting model parameters for each group separately. In particular, if the groups can be identified prior to sampling, they may be used as strata in a stratified cluster sampling design.

The regression estimators generated by a cluster group model will be efficient when there is homogeneity within each group of clusters and considerable differences between groups of clusters.

More formally, we consider that the population of clusters (PSUs) U_{I} is partitioned into the groups $U_{\text{I}1}, \ldots, U_{\text{I}g}, \ldots, U_{\text{I}G}$. Let $N_{\text{I}g}$ be the size of $U_{\text{I}g}$. The *group ratio model for cluster totals* assumes that for $g = 1, \ldots, G$

$$\begin{cases} E_\xi(t_{yi}) = \beta_{\text{I}g} u_i \\ V_\xi(t_{yi}) = \sigma_{\text{I}g}^2 u_i \end{cases} \tag{8.8.1}$$

for $i \in U_{\text{I}g}$, where $u_i > 0$ for all i. Clearly, (8.8.1) contains the simple ratio model shown in (8.5.1) as the special case where $G = 1$.

In the special case where $u_i = N_i$, the model (8.8.1) implies that the cluster means per element $\bar{y}_{U_i} = t_{yi}/N_i$ are considered approximately equal within a group of clusters. On the other hand, the special case where $u_i = 1$ for all i implies that the cluster totals t_{yi} are considered roughly constant within a group.

A sample s_{I} of PSUs is drawn from U_{I}. The part of s_{I} that falls in $U_{\text{I}g}$ is denoted $s_{\text{I}g}$. That is, for $g = 1, \ldots, G$,

$$s_{\text{I}g} = s_{\text{I}} \cap U_{\text{I}g}$$

Let $n_{\text{I}g}$ denote the (possibly random) size of $s_{\text{I}g}$.

The sampling design used to draw s_{I} has the inclusion probabilities $\pi_{\text{I}i}$ and $\pi_{\text{I}ij}$. When a second stage is involved, the sample s_i from a selected PSU U_i is realized by a design whose inclusion probabilities are $\pi_{k|i}$ and $\pi_{kl|i}$. We now have the following result:

Result 8.8.1. *In two-stage element sampling and under the group ratio model shown in* (8.8.1), *the case A regression estimator of t_y is given by*

$$\hat{t}_{yAr} = \sum_{g=1}^{G} \left(\textstyle\sum_{U_{\text{I}g}} u_i\right) \hat{B}_{\text{I}g}$$

where

$$\hat{B}_{Ig} = \frac{\sum_{s_{Ig}} t_{yi}^*/\pi_{Ii}}{\sum_{s_{Ig}} u_i/\pi_{Ii}}$$

with

$$t_{yi}^* = \hat{t}_{yi\pi} = \sum_{s_i} y_k/\pi_{k|i}$$

The approximate variance is given by (8.4.13) *with*

$$D_i = t_{yi} - B_{Ig}u_i \quad \textit{for} \quad i \in U_{Ig}$$

where

$$B_{Ig} = \sum_{U_{Ig}} t_{yi}/\sum_{U_{Ig}} u_i$$

The variance estimator is given by (8.4.16) *with*

$$d_i = t_{yi}^* - \hat{B}_{Ig}u_i \quad \textit{for} \quad i \in s_{Ig}$$

and

$$g_{is_I} = (\textstyle\sum_{U_{Ig}} u_i)/(\sum_{s_{Ig}} u_i/\pi_{Ii}) \quad \textit{for} \quad i \in s_{Ig}$$

The result applies to single stage cluster sampling by setting $t_{yi}^* = t_{yi} = \sum_{U_i} y_k$ *for all i, and* $V_i = \hat{V}_i = 0$ *for all i.*

The conclusions follow by a straightforward application of the general Result 8.4.1 to the specific model shown in (8.8.1). Let us apply Result 8.8.1 in a few cases of practical interest.

i. *Stratified Cluster Sampling*

In some applications, it is a good idea to let each group form a stratum, and to select a stratified sample of clusters, possibly followed by subsampling within each of the selected clusters. A stratified cluster sampling design is appealing because of its administrative convenience, for example, when strata are formed on a geographic basis. Another reason favoring this design is that estimates may be wanted separately for each stratum.

Thus we have G strata of clusters, $U_{I1}, \ldots, U_{Ig}, \ldots, U_{IG}$. Let us limit the discussion here to the important case where the sample s_{Ig} is drawn from the stratum U_{Ig} by the *SI* design. Let $f_{Ig} = n_{Ig}/N_{Ig}$ be the sampling fraction. Moreover, suppose that an *SI* sample s_i of n_i elements is drawn from the N_i elements in U_i, if this cluster was selected at the first stage. Let $f_i = n_i/N_i$ be the sampling fraction in U_i. Our abbreviation for this design is *STSI, SI*.

The estimator obtained from Result 8.8.1 is

$$\hat{t}_{yAr} = \sum_{g=1}^{G} (\textstyle\sum_{U_{Ig}} u_i)\hat{B}_{Ig} \tag{8.8.2}$$

where

$$\hat{B}_{\mathrm{I}g} = \sum_{s_{\mathrm{I}g}} N_i \bar{y}_{s_i} / \sum_{s_{\mathrm{I}g}} u_i$$

The residuals for the variance estimator are

$$d_i = N_i \bar{y}_{s_i} - \hat{B}_{\mathrm{I}g} u_i \tag{8.8.3}$$

for clusters $i \in U_{\mathrm{I}g}$.

In particular, if $u_i = N_i$, and if we let

$$\bar{N}_{U_{\mathrm{I}g}} = \sum_{U_{\mathrm{I}g}} N_i / N_{\mathrm{I}g}; \qquad \bar{N}_{s_{\mathrm{I}g}} = \sum_{s_{\mathrm{I}g}} N_i / n_{\mathrm{I}g}$$

denote average *PSU* sizes in the gth stratum, we obtain the variance estimator

$$\hat{V}(\hat{t}_{yAr}) = \sum_{g=1}^{G} \left(\frac{\bar{N}_{U_{\mathrm{I}g}}}{\bar{N}_{s_{\mathrm{I}g}}} \right)^2 \left[N_{\mathrm{I}g}^2 \left(\frac{1}{n_{\mathrm{I}g}} - \frac{1}{N_{\mathrm{I}g}} \right) S_g^2 + \frac{N_{\mathrm{I}g}}{n_{\mathrm{I}g}} \sum_{s_{\mathrm{I}g}} \hat{V}_i \right] \tag{8.8.4}$$

where

$$\hat{V}_i = N_i^2 \left(\frac{1}{n_i} - \frac{1}{N_i} \right) \frac{\sum_{s_i} (y_k - \bar{y}_{s_i})^2}{n_i - 1}$$

and

$$S_g^2 = \frac{1}{n_{\mathrm{I}g} - 1} \sum_{s_{\mathrm{I}g}} N_i^2 (\bar{y}_{s_i} - \hat{\bar{y}}_g)^2$$

with

$$\hat{\bar{y}}_g = \sum_{s_{\mathrm{I}g}} N_i \bar{y}_{s_i} / \sum_{s_{\mathrm{I}g}} N_i$$

Note the tendency in (8.8.4) toward a smaller variance estimate if larger-than-average clusters happen to be drawn in a group ($\bar{N}_{s_{\mathrm{I}g}} > \bar{N}_{U_{\mathrm{I}g}}$); this tendency is intuitively sound.

ii. *Poststratified Clusters*

In other applications, there may be no convenient stratification of clusters. Suppose instead that the sampler selects an *SI* sample of n_{I} clusters from N_{I}, and forms poststrata of selected clusters, $s_{\mathrm{I}1}, \ldots, s_{\mathrm{I}g}, \ldots, s_{\mathrm{I}G}$. The size $n_{\mathrm{I}g}$ is now random, with expected value $n_{\mathrm{I}} N_{\mathrm{I}g} / N_{\mathrm{I}}$.

The estimator produced under this design by Result 8.8.1 is identical in form to (8.8.2), but the sampling distribution of the estimator is different. The residuals shown in equation (8.8.3) still apply. If $u_i = N_i$, and *SI* subsampling is used with the sampling fraction $f_i = n_i / N_i$, the variance estimator of (8.8.2) is

$$\hat{V}(\hat{t}_{yAr}) = \left(1 - \frac{n_{\mathrm{I}}}{N_{\mathrm{I}}}\right) \sum_{g=1}^{G} \left(\frac{\bar{N}_{U_{\mathrm{I}g}}}{\bar{N}_{s_{\mathrm{I}g}}} \right)^2 N_{\mathrm{I}g}^2 \frac{S_g^2}{n_{\mathrm{I}g}} + \frac{N_{\mathrm{I}}}{n_{\mathrm{I}}} \sum_{g=1}^{G} \left(\frac{\bar{N}_{U_{\mathrm{I}g}}}{\bar{N}_{s_{\mathrm{I}g}}} \right)^2 \sum_{s_{\mathrm{I}g}} \hat{V}_i \tag{8.8.5}$$

where $\bar{N}_{U_{\text{I}g}}$, $\bar{N}_{s_{\text{I}g}}$, S_g^2, and $\hat{V}_i$ are as in (8.8.4). The approximation $(1 - n_{\text{I}g}^{-1})/(1 - n_{\text{I}}^{-1}) \doteq 1$ has been used in arriving at (8.8.5). For an application to single-stage cluster sampling, set $\hat{V}_i = 0$ in equations (8.8.4) and (8.8.5).

8.9. Regression Estimators Arising Out of Modeling at the Element Level

We now derive the regression estimators appropriate for cases B and C, as defined in Section 8.2. In case B (the case of complete element auxiliaries), the auxiliary value $\mathbf{x}_k$ is known for all elements $k \in U$. In case C (the case of limited element auxiliaries), $\mathbf{x}_k$ is known "only" for the elements k such that

$$k \in \bigcup_{i \in s_{\text{I}}} U_i$$

We assume that subsampling of selected clusters is carried out.

To describe the population point scatter $(y_k, \mathbf{x}_k)$, we adopt the general linear model stating that

$$\begin{cases} E_\xi(y_k) = \mathbf{x}_k'\boldsymbol{\beta} \\ V_\xi(y_k) = \sigma_k^2 \end{cases} \tag{8.9.1}$$

and that the y_k are uncorrelated under the model. In a hypothetical population fit of the model (8.9.1) to all N points $(y_k, \mathbf{x}_k)$, $k = 1, \ldots, N$, the model regression vector $\boldsymbol{\beta}$ would be estimated, as in Section 6.4, by

$$\mathbf{B} = \left(\textstyle\sum_U \mathbf{x}_k\mathbf{x}_k'/\sigma_k^2\right)^{-1} \textstyle\sum_U \mathbf{x}_k y_k/\sigma_k^2 \tag{8.9.2}$$

and the associated residuals are

$$E_k = y_k - y_k^0 = y_k - \mathbf{x}_k'\mathbf{B} \tag{8.9.3}$$

The sample fit of the model (8.9.1) is based on the data points $(y_k, \mathbf{x}_k)$ corresponding to elements $k \in s$, where s is the total sample,

$$s = \bigcup_{i \in s_{\text{I}}} s_i$$

The number of such points is

$$n_s = \textstyle\sum_{s_{\text{I}}} n_i$$

which is ordinarily a random number. Our estimate of $\mathbf{B}$ is

$$\hat{\mathbf{B}} = \left(\textstyle\sum_s \mathbf{x}_k\mathbf{x}_k'/\sigma_k^2\pi_k\right)^{-1} \textstyle\sum_s \mathbf{x}_k y_k/\sigma_k^2\pi_k$$

where the sampling weights are

$$1/\pi_k = 1/\pi_{\text{I}i}\pi_{k|i}$$

The fitted values and the residuals are

$$\hat{y}_k = \mathbf{x}_k'\hat{\mathbf{B}}; \qquad e_{ks} = y_k - \hat{y}_k \tag{8.9.4}$$

The values $\hat{y}_k$ can be calculated in case B for all $k \in U$; in case C for elements k in the selected clusters. The residuals $e_{ks} = y_k - \mathbf{x}_k'\hat{\mathbf{B}}$ can be computed in both cases for $k \in s$ only.

In case B, the regression estimator is

$$\hat{t}_{yBr} = \sum_U \hat{y}_k + \sum_s \frac{e_{ks}}{\pi_k}$$
$$= \sum_{U_I} \sum_{U_i} \hat{y}_k + \sum_{s_I} \frac{1}{\pi_{Ii}} \sum_{s_i} \frac{y_k - \hat{y}_k}{\pi_{k|i}} \tag{8.9.5}$$

The prediction term, $\sum_U \hat{y}_k$, extends over the entire population.

In case C, the auxiliary information is less extensive than in case B but suffices to form a regression estimator for the total of each selected PSU, namely,

$$\hat{t}_{yir} = \sum_{U_i} \hat{y}_k + \sum_{s_i} \frac{y_k - \hat{y}_k}{\pi_{k|i}} \tag{8.9.6}$$

We assume that the sample sizes within the PSUs are not too small, so that the estimator (8.9.6) is essentially without bias for the PSU total t_{yi}.

The first-stage sampling weight $1/\pi_{Ii}$ is now applied to $\hat{t}_{yir}$. The resulting estimator of t_y is

$$\hat{t}_{yCr} = \sum_{s_I} \frac{\hat{t}_{yir}}{\pi_{Ii}} \tag{8.9.7}$$

Inserting the expression for $\hat{t}_{yir}$ given in (8.9.6), we have alternatively

$$\hat{t}_{yCr} = \sum_{s_I} \frac{t_{\hat{y}i}}{\pi_{Ii}} + \sum_{s_I} \frac{1}{\pi_{Ii}} \sum_{s_i} \frac{y_k - \hat{y}_k}{\pi_{k|i}}$$
$$= \sum_{s_I} \frac{t_{\hat{y}i}}{\pi_{Ii}} + \sum_s \frac{y_k - \hat{y}_k}{\pi_k} \tag{8.9.8}$$

where

$$t_{\hat{y}i} = \sum_{U_i} \hat{y}_k$$

Comparing (8.9.5) and (8.9.8), we see that $\hat{t}_{yBr}$ and $\hat{t}_{yCr}$ share the same adjustment term, $\sum_s e_{ks}/\pi_k$. This term vanishes for many models, as a consequence of Result 6.5.1.

As we now set out to find the approximate variances and the variance estimators of $\hat{t}_{yBr}$ and $\hat{t}_{yCr}$, it is helpful to express the estimation error in each case on a form that mimics the decomposition (8.3.1). To this end, set

$$\mathbf{t}_{xi} = \sum_{U_i} \mathbf{x}_k; \qquad \hat{\mathbf{t}}_{xi\pi} = \sum_{s_i} \mathbf{x}_k/\pi_{k|i}$$

and

$$\hat{\mathbf{T}} = \sum_s \mathbf{x}_k \mathbf{x}_k'/\sigma_k^2 \pi_k$$

The two new regression estimators can alternatively be written as g-weighted expressions. We have

$$\hat{t}_{yBr} = \sum_s \frac{g_{ksB} y_k}{\pi_k} \tag{8.9.9}$$

with $\pi_k = \pi_{Ii}\pi_{k|i}$, and the g weights

$$g_{ksB} = 1 + \left(\sum_{U_I} \mathbf{t}_{xi} - \sum_{s_I} \hat{\mathbf{t}}_{xi\pi}/\pi_{Ii}\right)' \hat{\mathbf{T}}^{-1} \mathbf{x}_k / \sigma_k^2 \tag{8.9.10}$$

Correspondingly, we have for case C

$$\hat{t}_{yCr} = \sum_s \frac{g_{ksC} y_k}{\pi_k} \tag{8.9.11}$$

with $\pi_k = \pi_{Ii}\pi_{k|i}$, and the g weights are

$$g_{ksC} = 1 + \left[\sum_{s_I} (\mathbf{t}_{xi} - \hat{\mathbf{t}}_{xi\pi})/\pi_{Ii}\right]' \hat{\mathbf{T}}^{-1} \mathbf{x}_k / \sigma_k^2 \tag{8.9.12}$$

The estimation errors can now be expressed in the manner of equation (8.3.1),

$$\hat{t}_{yBr} - t_y = \underbrace{\sum_{s_I} \frac{t_{Ei}}{\pi_{Ii}} - \sum_{U_I} t_{Ei}}_{Q_{Bs_I}} + \underbrace{\sum_{s_I} \frac{1}{\pi_{Ii}} \left(\sum_{s_i} \frac{g_{ksB} E_k}{\pi_{k|i}} - \sum_{U_i} E_k \right)}_{R_{Bs}} \tag{8.9.13}$$

and

$$\hat{t}_{yCr} - t_y = \underbrace{\sum_{s_I} \frac{t_{yi}}{\pi_{Ii}} - \sum_{U_I} t_{yi}}_{Q_{Cs_I}} + \underbrace{\sum_{s_I} \frac{1}{\pi_{Ii}} \left(\sum_{s_i} \frac{g_{ksC} E_k}{\pi_{k|i}} - \sum_{U_i} E_k \right)}_{R_{Cs}} \tag{8.9.14}$$

where E_k is the population fit residual given by (8.9.3), and $t_{Ei} = \sum_{U_i} E_k$. The Q term of equation (8.9.13) is expressed in terms of residuals, so ordinarily it contributes much less variance than the Q term of (8.9.14), which is expressed in the raw y_k scores.

The verification of equations (8.9.13) and (8.9.14), which is left as an exercise, uses the following important properties of the g weights:

$$\sum_s g_{ksB} \mathbf{x}_k / \pi_k = \sum_{U_I} \mathbf{t}_{xi} = \sum_U \mathbf{x}_k \tag{8.9.15a}$$

and

$$\sum_s g_{ksC} \mathbf{x}_k / \pi_k = \sum_{s_I} \mathbf{t}_{xi} / \pi_{Ii} \tag{8.9.15b}$$

To derive the approximate variances from (8.9.13) and (8.9.14), we approximate $g_{ksB} \doteq 1$ for all k, and $g_{ksC} \doteq 1$ for all k. This simplifies the second terms, R_{Bs} and R_{Cs}. Linear expressions are obtained. We can copy the argument for the π estimator that led from the estimation error (8.3.1) to the variance (8.3.2) and to the variance estimator (8.3.4). The variance components are now approximations and will be denoted AV.

Our summary for case B is shown in the following result.

Result 8.9.1. *In two-stage element sampling, the case B regression estimator of* t_y *is* $\hat{t}_{yBr}$ *given by equation* (8.9.5). *The approximate variance is*

$$AV(\hat{t}_{yBr}) = AV_{BPSU} + AV_{BSSU} \tag{8.9.16}$$

where

$$AV_{BPSU} = \sum\sum_{U_I} \Delta_{Iij} \frac{t_{Ei}}{\pi_{Ii}} \frac{t_{Ej}}{\pi_{Ij}} \tag{8.9.17}$$

and

$$AV_{BSSU} = \sum_{U_I} V_{Ei}/\pi_{Ii} \tag{8.9.18}$$

Here $t_{Ei} = \sum_{U_i} E_k$, *with* E_k *given by* (8.9.3), *and*

$$V_{Ei} = \sum\sum_{U_i} \Delta_{kl|i} \frac{E_k}{\pi_{k|i}} \frac{E_l}{\pi_{l|i}} \tag{8.9.19}$$

The estimated variance is

$$\hat{V}(\hat{t}_{yBr}) = \hat{V}_{BPSU} + \hat{V}_{BSSU} \tag{8.9.20}$$

with

$$\hat{V}_{BPSU} = \sum\sum_{s_I} \check{\Delta}_{Iij} \frac{\hat{t}_{Ei}}{\pi_{Ii}} \frac{\hat{t}_{Ej}}{\pi_{Ij}} - \sum_{s_I} \frac{1}{\pi_{Ii}} \left(\frac{1}{\pi_{Ii}} - 1 \right) \hat{V}_{BEi} \tag{8.9.21}$$

and

$$\hat{V}_{BSSU} = \sum_{s_I} \hat{V}_{BEi}/\pi_{Ii}^2 \tag{8.9.22}$$

Here,

$$\hat{t}_{Ei} = \sum_{s_i} g_{ksB} e_{ks}/\pi_{k|i}$$

with e_{ks} *given by* (8.9.4), *and*

$$\hat{V}_{BEi} = \sum\sum_{s_i} \check{\Delta}_{kl|i} \frac{g_{ksB} e_{ks}}{\pi_{k|i}} \frac{g_{lsB} e_{ls}}{\pi_{l|i}}$$

with g_{ksB} *given by* (8.9.10). *The sum of* (8.9.21) *and* (8.9.22) *simplifies into*

$$\hat{V}(\hat{t}_{yBr}) = \sum\sum_{s_I} \check{\Delta}_{Iij} \frac{\hat{t}_{Ei}}{\pi_{Ii}} \frac{\hat{t}_{Ej}}{\pi_{Ij}} + \sum_{s_I} \frac{\hat{V}_{BEi}}{\pi_{Ii}} \tag{8.9.23}$$

DERIVATION. Having approximated $g_{ksB} \doteq 1$ in (8.9.13), we have an expression of the same form as (8.3.1). The results given by (8.9.17) and (8.9.18) now follow directly by an analogy with the terms V_{PSU} and V_{SSU} in equation (8.3.2). To obtain estimators that correspond to V_{BPSU} and V_{BSSU}, we must substitute estimators for t_{Ei} and V_{Ei}. Here the appropriate substitutions are $\hat{t}_{Ei}$ and $\hat{V}_{BEi}$, which leads to (8.9.21) and (8.9.22). □

Computationally somewhat simpler variance component estimators may be obtained by letting $g_{ksB} = 1$ for all k and s in equations (8.9.21) and (8.9.22).

In case C, we can summarize as follows.

Result 8.9.2. *In two-stage element sampling, the case C regression estimator of* t_y *is* $\hat{t}_{yCr}$ *given by equation* (8.9.7). *The approximate variance is*

$$AV(\hat{t}_{yCr}) = AV_{\text{CPSU}} + AV_{\text{CSSU}} \tag{8.9.24}$$

where

$$AV_{\text{CPSU}} = \sum\sum_{U_{\text{I}}} \Delta_{\text{I}ij} \frac{t_{yi}}{\pi_{\text{I}i}} \frac{t_{yj}}{\pi_{\text{I}j}} \tag{8.9.25}$$

$$AV_{\text{CSSU}} = \sum_{U_{\text{I}}} V_{Ei}/\pi_{\text{I}i} \tag{8.9.26}$$

with $t_{yi} = \sum_{U_i} y_k$, *and* V_{Ei} *is given by* (8.9.19). *The estimated variance is*

$$\hat{V}(\hat{t}_{yCr}) = \hat{V}_{\text{CPSU}} + \hat{V}_{\text{CSSU}} \tag{8.9.27}$$

where

$$\hat{V}_{\text{CPSU}} = \sum\sum_{s_{\text{I}}} \check{\Delta}_{\text{I}ij} \frac{\hat{t}_{yi\pi}}{\pi_{\text{I}i}} \frac{\hat{t}_{yj\pi}}{\pi_{\text{I}j}} - \sum_{s_{\text{I}}} \frac{1}{\pi_{\text{I}i}} \left(\frac{1}{\pi_{\text{I}i}} - 1 \right) \hat{V}_i$$

$$\hat{V}_{\text{CSSU}} = \sum_{s_{\text{I}}} \hat{V}_{CEi}/\pi_{\text{I}i}^2$$

Here, $\hat{t}_{yi\pi} = \sum_{s_i} y_k/\pi_{k|i}$,

$$\hat{V}_i = \sum\sum_{s_i} \check{\Delta}_{kl|i} \frac{y_k}{\pi_{k|i}} \frac{y_l}{\pi_{l|i}}$$

and

$$\hat{V}_{CEi} = \sum\sum_{s_i} \check{\Delta}_{kl|i} \frac{g_{ksC} e_{ks}}{\pi_{k|i}} \frac{g_{lsC} e_{ls}}{\pi_{l|i}}$$

with g_{ksC} *given by* (8.9.12), *and* e_{ks} *by* (8.9.4).

Derivation. Once we have applied the approximation $g_{ksC} \doteq 1$ in (8.9.14), the AV expressions follow by a simple analogy with equation (8.3.1). As for the variance component estimators, note that $\hat{V}_{\text{CPSU}}$ is unbiased for its counterpart given in (8.9.25). Finally, $\hat{V}_{\text{CSSU}}$ is obtained after substituting $\hat{V}_{CEi}$ for V_{Ei}. □

Remark 8.9.1. A comparison of equations (8.3.2) and (8.9.24) shows that $\hat{t}_{yCr}$ and $\hat{t}_{y\pi}$ have identical first variance components. This is to be expected since the auxiliary information in case C is limited to elements within selected clusters. This information will not reduce the variance due to sampling at stage one. Comparing (8.9.16) and (8.9.24), we see that $\hat{t}_{yBr}$ and $\hat{t}_{yCr}$ share the same second variance component. This is also as it should be, because case B and case C benefit from the same auxiliary information concerning the elements in selected PSUs.

Remark 8.9.2. Alternative regression estimators for case C are

$$\hat{t}^{*}_{yCr} = N_{\mathrm{I}} \frac{\sum_{s_{\mathrm{I}}} \hat{t}_{yir}/\pi_{\mathrm{I}i}}{\sum_{s_{\mathrm{I}}} 1/\pi_{\mathrm{I}i}} \tag{8.9.28}$$

for a case where the cluster totals are considered roughly constant and

$$\hat{t}^{**}_{yCr} = N \frac{\sum_{s_{\mathrm{I}}} \hat{t}_{yir}/\pi_{\mathrm{I}i}}{\sum_{s_{\mathrm{I}}} N_i/\pi_{\mathrm{I}i}} \tag{8.9.29}$$

for a case where $N = \sum_{U_{\mathrm{I}}} N_i$ is known and the cluster totals are deemed roughly proportional to N_i. With $\hat{t}_{yir}$ given by equation (8.9.6), these estimators are often preferable to $\hat{t}_{yCr}$ given by (8.9.7) if the sample size at the first stage is random. Another reason to consider the estimator (8.9.29) is that division by N yields

$$\hat{\bar{y}}_U = \frac{\hat{t}^{**}_{yCr}}{N} = \frac{\sum_{s_{\mathrm{I}}} \hat{t}_{yir}/\pi_{\mathrm{I}i}}{\sum_{s_{\mathrm{I}}} N_i/\pi_{\mathrm{I}i}} \tag{8.9.30}$$

which is a suitable estimator of the population mean $\bar{y}_U$ under the conditions of case C. It generalizes the estimator discussed in Result 8.6.1.

8.10. Ratio Models for Elements

Ratio models are highly useful for survey work. We now examine ratio models appropriate for cases B and C and the corresponding regression estimators. The auxiliary information consists of a known positive scalar value x_k attached to the kth element. For example, if the elements are buildings, the x_k may be inexpensive measures for the buildings, whereas the values y_k of the study variable can be obtained only by a more costly process of measurement.

Three ratio models are examined:

1. The common ratio model, which assumes a common slope β for all elements throughout the population.
2. The group ratio model, in which the same slope is shared by all elements in a given population group. The slopes differ between the groups. A population group may correspond to a set of PSUs or it may cut across the PSUs. This model is considered in Section 8.11.
3. The single PSU ratio model, in which the same slope applies to all elements in one PSU. The slopes differ from one PSU to another. This model is considered in Section 8.12.

This represents a progression toward more and more detailed or parameter-rich models. In each case, it is assumed that sufficient data are available to support a stable estimate for each parameter.

The common ratio model (7.3.1) was discussed in Section 7.3. The ratios y_k/x_k are assumed to scatter around a horizontal line at the constant level β. To fit this model, we use Result 7.3.1. The slope estimator obtained from the population fit is

$$B = \frac{\sum_U y_k}{\sum_U x_k} = \frac{\sum_{U_I} t_{yi}}{\sum_{U_I} t_{xi}} \tag{8.10.1}$$

Its counterpart, obtained from the sample fit, is

$$\hat{B} = \frac{\sum_s \check{y}_k}{\sum_s \check{x}_k} = \frac{\sum_{s_I} \hat{t}_{yi\pi}/\pi_{Ii}}{\sum_{s_I} \hat{t}_{xi\pi}/\pi_{Ii}} \tag{8.10.2}$$

where $\hat{t}_{yi\pi} = \sum_{s_i} y_k/\pi_{k|i}$ is the conditional π estimator, given s_I, of the cluster total $t_{yi} = \sum_{U_i} y_k$, and $\hat{t}_{xi\pi}$ is the analogous estimator of t_{xi}.

The main conclusions for the regression estimators $\hat{t}_{yBr}$ and $\hat{t}_{yCr}$ follow from Results 8.9.1 and 8.9.2 and are summarized in the following result.

Result 8.10.1. *In two-stage element sampling, and under the common ratio model* (7.3.1), *the following regression estimators are obtained*:

$$\textit{Case B:}\ \hat{t}_{yBr} = \left(\textstyle\sum_{U_I} t_{xi}\right)\hat{B} \tag{8.10.3}$$

and

$$\textit{Case C:}\ \hat{t}_{yCr} = \left(\textstyle\sum_{s_I} t_{xi}/\pi_{Ii}\right)\hat{B} \tag{8.10.4}$$

where $\hat{B}$ *is given by* (8.10.2). *The approximate variances are given by* (8.9.16) *in case B and by* (8.9.24) *in case C with*

$$E_k = y_k - Bx_k; \qquad k = 1, \ldots, N$$

where B is given by (8.10.1). *The variance estimators are given by* (8.9.20) *in case B and by* (8.9.27) *in case C, with*

$$e_{ks} = y_k - \hat{B}x_k$$

and, for all k,

$$g_{ksB} = \frac{\sum_{U_I} t_{xi}}{\sum_{s_I} \hat{t}_{xi\pi}/\pi_{Ii}}; \qquad g_{ksC} = \frac{\sum_{s_I} t_{xi}/\pi_{Ii}}{\sum_{s_I} \hat{t}_{xi\pi}/\pi_{Ii}}$$

In particular, if $x_k = 1$ for all k, $\hat{t}_{yBr}/N$ is the population mean estimator shown in equation (8.6.7). Result 8.10.1 is a compact summary of all that is required to derive the estimator, the variance estimator, and the confidence interval. But the application of the result to specific designs will usually entail long and complex expressions. To illustrate, consider the *SI*, *SI* design as in Example 4.3.1. At stage one, n_I PSUs are drawn by *SI* from N_I; at stage two, n_i SSUs are drawn by *SI* from N_i, for $i \in s_I$. Let $f_I = n_I/N_I$; $f_i = n_i/N_i$, and set $\bar{y}_{s_i} = \sum_{s_i} y_k/n_i$; $\bar{x}_{s_i} = \sum_{s_i} x_k/n_i$; $t_{xi} = \sum_{U_i} x_k = N_i\bar{x}_{U_i}$.

Result 8.10.1 leads to the following conclusions.

Case B

The estimator of t_y is

$$\hat{t}_{yBr} = (\textstyle\sum_{U_I} N_i \bar{x}_{U_i})\hat{B}$$

where

$$\hat{B} = \frac{\sum_{s_I} N_i \bar{y}_{s_i}}{\sum_{s_I} N_i \bar{x}_{s_i}} \tag{8.10.5}$$

The variance estimator is

$$\hat{V}_{SI,SI}(\hat{t}_{yBr}) = \hat{V}_{BPSU} + \hat{V}_{BSSU} \tag{8.10.6}$$

where

$$\hat{V}_{BPSU} = N_I^2 \frac{1-f_I}{n_I} g_B^2 \left(S_{Ds_I}^2 - \frac{1}{n_I} \textstyle\sum_{s_I} N_i^2 \frac{1-f_i}{n_i} S_{es_i}^2 \right)$$

$$\hat{V}_{BSSU} = \left(\frac{N_I}{n_I}\right)^2 g_B^2 \textstyle\sum_{s_I} N_i^2 \frac{1-f_i}{n_i} S_{es_i}^2$$

with

$$S_{Ds_I}^2 = \frac{1}{n_I - 1} \textstyle\sum_{s_I} N_i^2 (\bar{y}_{s_i} - \hat{B}\bar{x}_{s_i})^2$$

which is the variance in s_I of the quantities $D_i = N_i(\bar{y}_{s_i} - \hat{B}\bar{x}_{s_i})$,

$$S_{es_i}^2 = \frac{1}{n_i - 1} \textstyle\sum_{s_i} [y_k - \bar{y}_{s_i} - \hat{B}(x_k - \bar{x}_{s_i})]^2 \tag{8.10.7}$$

and

$$g_B = \frac{\sum_{U_I} N_i \bar{x}_{U_i}/N_I}{\sum_{s_I} N_i \bar{x}_{s_i}/n_I}$$

Case C

The estimator of t_y is

$$\hat{t}_{yCr} = (N_I/n_I)(\textstyle\sum_{s_I} N_i \bar{x}_{U_i})\hat{B}$$

The variance estimator is

$$\hat{V}_{SI,SI}(\hat{t}_{yCr}) = \hat{V}_{CPSU} + \hat{V}_{CSSU}$$

with

$$\hat{V}_{CPSU} = N_I^2 \frac{1-f_I}{n_I}\left(S_{\hat{t}s_I}^2 - \frac{1}{n_I} \textstyle\sum_{s_I} \hat{V}_i \right)$$

and

$$\hat{V}_{\text{CSSU}} = \left(\frac{N_{\text{I}}}{n_{\text{I}}}\right)^2 g_C^2 \sum_{s_{\text{I}}} N_i^2 \frac{1 - f_i}{n_i} S_{es_i}^2$$

where $S_{\hat{t}s_{\text{I}}}^2$ is the variance over s_{I} of the n_{I} PSU total estimates $\hat{t}_{yi\pi} = N_i \bar{y}_{s_i}$,

$$\hat{V}_i = N_i^2 \frac{1 - f_i}{n_i} \frac{\sum_{s_i} (y_k - \bar{y}_{s_i})^2}{n_i - 1}$$

Moreover, $S_{es_i}^2$ is given by (8.10.7), and

$$g_C = \frac{\sum_{s_{\text{I}}} N_i \bar{x}_{U_i}}{\sum_{s_{\text{I}}} N_i \bar{x}_{s_i}}$$

8.11. The Group Ratio Model for Elements

In some applications, it is reasonable to expect that elements in a given population group show similarity, as discussed in Section 7.5. The group ratio model shown in (7.5.6) or its special case, the group mean model given in (7.5.3), may then be appropriate. We assume that there is a partitioning of U into G groups of elements, U_g, $g = 1, \ldots, G$. The model shown in (7.5.6) assumes that the ratio y_k/x_k is roughly constant for elements in the group U_g, but the ratio may be considerably different for elements in different groups. Then the grouping itself is an important factor in explaining the variation of the study variable, y.

In two-stage sampling, the group ratio model can be useful when

a. A model group corresponds to the elements in a given set of PSUs.
b. The model groups cut across the PSUs.

As an example of (a), suppose the elements are households, and the PSUs city blocks in a large city. A model group is defined as the households belonging to a set of (perhaps contiguous) PSUs. Households within groups formed in this way are likely to be relatively homogeneous. The group ratio model is then appropriate if the ratio y_k/x_k (say, the ratio of household savings to household income) is roughly constant for households in the same group, and if the total t_{xi} is known at least for the selected PSUs, which is required in case C.

As an example of (b), suppose the elements are individuals. The model groups consist of age/sex categories, and y_k/x_k is deemed roughly constant for individuals in the same category. On the other hand, the PSUs are formed by a geographic subdivision that is convenient for two-stage sampling, but has little effect in explaining the y variable. Here, the PSUs and the model groups are crossed.

We consider the case where each group corresponds to a set of PSUs. For $g = 1, \ldots, G$, the model groups are composed as

$$U_g = \bigcup_{i \in U_{\mathrm{I}g}} U_i \tag{8.11.1}$$

where $U_{\mathrm{I}1}, \ldots, U_{\mathrm{I}g}, \ldots, U_{\mathrm{I}G}$ is a partitioning of the population of clusters U_{I}.

With minor changes, we can use the results of Section 7.5 concerning the fit of the group ratio model. Let $s_{\mathrm{I}g}$, $g = 1, \ldots, G$ be the set of selected PSUs clusters from the group $U_{\mathrm{I}g}$, $g = 1, \ldots, G$. The set of elements sampled from group g is then

$$s_g = \bigcup_{i \in s_{\mathrm{I}g}} s_i \tag{8.11.2}$$

where s_i is the sample of elements from the ith PSU. The population group slope is

$$B_g = \frac{\sum_{U_g} y_k}{\sum_{U_g} x_k} = \frac{\sum_{U_{\mathrm{I}g}} t_{yi}}{\sum_{U_{\mathrm{I}g}} t_{xi}} \tag{8.11.3}$$

which is estimated by

$$\hat{B}_g = \frac{\sum_{s_g} \check{y}_k}{\sum_{s_g} \check{x}_k} = \frac{\sum_{s_{\mathrm{I}g}} \hat{t}_{yi\pi}/\pi_{\mathrm{I}i}}{\sum_{s_{\mathrm{I}g}} \hat{t}_{xi\pi}/\pi_{\mathrm{I}i}} \tag{8.11.4}$$

where $\hat{t}_{yi\pi} = \sum_{s_i} y_k/\pi_{k|i}$ and $\hat{t}_{xi\pi}$ is analogously defined. We assume that each group is large enough so that the probability is negligible that the set $s_{\mathrm{I}g}$ is empty.

The main conclusions are summarized in the following result.

Result 8.11.1. *In two-stage element sampling, and under the group ratio model given by* (7.5.6), *with each group defined by a set of PSUs according to equation* (8.11.2), *the following regression estimators are obtained*:

$$\text{Case B: } \hat{t}_{yBr} = \sum_{g=1}^{G} \left(\sum_{U_{\mathrm{I}g}} t_{xi}\right) \hat{B}_g$$

and

$$\text{Case C: } \hat{t}_{yCr} = \sum_{g=1}^{G} \left(\sum_{s_{\mathrm{I}g}} \frac{t_{xi}}{\pi_{\mathrm{I}i}}\right) \hat{B}_g$$

where $\hat{B}_g$ is given by (8.11.4). *The approximate variances are obtained from* (8.9.16) *in case B and from* (8.9.24) *in case C by setting*

$$E_k = y_k - B_g x_k \quad \text{for} \quad k \in U_g$$

where B_g is given by (8.11.3). *The variance estimators are obtained from* (8.9.20) *in case B and from* (8.9.27) *in case C, by setting*

$$e_{ks} = y_k - \hat{B}_g x_k \qquad \text{for} \quad k \in s_g$$

$$g_{ksB} = \frac{\sum_{U_{\mathrm{I}g}} t_{xi}}{\sum_{s_{\mathrm{I}g}} \hat{t}_{xi\pi}/\pi_{\mathrm{I}i}} \qquad \text{for all} \quad k \in s_g$$

$$g_{ksC} = \frac{\sum_{s_{\mathrm{I}g}} t_{xi}/\pi_{\mathrm{I}i}}{\sum_{s_{\mathrm{I}g}} \hat{t}_{xi\pi}/\pi_{\mathrm{I}i}} \qquad \text{for all} \quad k \in s_g$$

The two estimators in Result 8.11.1 resemble the separate ratio estimator (Section 7.6) in that, group by group, a ratio type expression estimates the group total. The estimated group totals are then simply added up to provide the global estimate.

EXAMPLE 8.11.1. We can apply Result 8.11.1 in the situation where the groups U_{Ig} in (8.11.1) are strata of clusters or poststrata of clusters, as in Section 8.8. Under the design *STSI,SI* (with the notation of Section 8.8), the estimators in cases B and C are

$$\hat{t}_{yBr} = \sum_{g=1}^{G} (\textstyle\sum_{U_{Ig}} t_{xi})\hat{B}_g$$

$$\hat{t}_{yCr} = \sum_{g=1}^{G} (N_{Ig}/n_{Ig})(\textstyle\sum_{s_{Ig}} t_{xi})\hat{B}_g$$

where

$$\hat{B}_g = \frac{\sum_{s_{Ig}} N_i \bar{y}_{s_i}}{\sum_{s_{Ig}} N_i \bar{x}_{s_i}}$$

8.12. The Ratio Model Applied within a Single PSU

An important question concerning the choice of the model is, how detailed (in terms of the number of parameters) the model should be. In the case of a ratio model, how many slope parameters should be introduced? We recall that the common ratio model used in Section 8.10 contained a single slope parameter for the whole population. The group ratio model used in Section 8.11 had one slope parameter for each group, for example, a group of PSUs. If a group is defined as a single PSU, there will be one slope parameter for each PSU, so the number of slopes to estimate may be quite large.

In principle, nothing prevents having a parameter-rich model. It will pay off, however, only if the groups are so different that significantly better estimates of the population total can be made by admitting an individual parameter for each group, and if sample sizes in each group are large enough to yield stable group parameter estimates.

The modeler may want to incorporate a separate effect for each PSU. This is done by the special case of the group ratio model given in (7.5.6) such that each group corresponds to exactly one PSU. The single PSU ratio model states that, for elements k in a selected PSU, U_i,

$$\begin{cases} E_\xi(y_k) = \beta_i x_k \\ V_\xi(y_k) = \sigma_i^2 x_k \end{cases} \tag{8.12.1}$$

where the x_k are assumed to be positive.

The single PSU ratio model can be fitted only for those PSUs where observations are made on y, that is, in those PSUs that happen to be selected in stage one. Thus, the model can be used in case C, but not in case B, since the predicted values $\hat{y}_k$ can be formed only for elements in selected PSUs. The single PSU model should be reserved for cases where the PSUs are large with considerable within PSU variation, and where the sample size in each PSU is sizeable. In this case, it is only the second variance component, relating to the sampling variance within PSUs, that can be reduced by the use of the x variable.

The basic results under the single PSU ratio model are derived as a special case of Result 8.11.1, in the sense that $s_{\text{I}g}$ now contains precisely one PSU for all g. The summation over g from 1 to G is replaced by $\sum_{s_\text{I}}$. We obtain the following result.

Result 8.12.1. *In two-stage element sampling, and under the single-PSU ratio model given by* (8.12.1), *the case C regression estimator is*

$$\hat{t}_{yCr} = \sum_{s_\text{I}} \frac{t_{xi}\hat{B}_i}{\pi_{\text{I}i}} \tag{8.12.2}$$

where $\hat{B}_i = \hat{t}_{yi\pi}/\hat{t}_{xi\pi}$. *The approximate variance is obtained from equation* (8.9.24) *by taking*

$$E_k = y_k - (t_{yi}/t_{xi})x_k$$

in the expression for V_{Ei}. *The variance estimator is obtained from equation* (8.9.27) *by taking*

$$e_{ks} = y_k - \hat{B}_i x_k \qquad \text{for} \quad k \in s_i$$

and

$$g_{ksC} = t_{xi}/\hat{t}_{xi\pi} \qquad \text{for all} \quad k \in s_i \tag{8.12.3}$$

in the expression for $\hat{V}_{CEi}$.

In the special case where $x_k = 1$ for all k, then (8.12.2) yields the estimator

$$\hat{t}_{yCr} = \sum_{s_\text{I}} N_i \tilde{y}_{s_i}/\pi_{\text{I}i}$$

Exercises

Remark. Some of the following exercises involve calculations based on approximate variance estimators. Keep in mind that the approximations may be crude because the two-stage samples obtainable with our populations are necessarily small. The exercises are given for pedagogic reasons and are not representative of the conditions in a real survey.

8.1. A field is divided into 10 areas, $i = 1, \ldots, N_\text{I} = 10$, with varying sizes. The objective is to estimate the mean size of the plants grown in this field, $\bar{y}_U = \sum_U y_k/N$, where y_k is the size of the kth plant, $k \in U = \{1, \dots, N\}$, where $N = 78$. To compare different estimation strategies using two-stage sampling, we assume that the value y_k is known for all $k \in U$. We have the following data.

Area i	Number of plants in the area (N_i)	Number of plants to be selected (n_i)	Plant size y_k, $k \in U_i$
1	11	4	12 11 12 10 13 14 11 12 13 14 14
2	6	2	10 9 7 9 8 10
3	5	2	6 5 7 5 4
4	7	2	7 8 7 7 6 8 7
5	11	4	10 13 13 12 12 14 11 12 13 14 14
6	10	3	14 15 13 12 13 14 16 12 16 15
7	5	2	6 7 6 5 4
8	7	2	9 10 8 9 9 10 9
9	6	2	7 10 8 9 9 10
10	10	3	12 13 12 13 12 12 16 14 13 15

(a) Compare the efficiency of the following four strategies:

Strategy 1. First stage: *SI* sampling of $n_{\text{I}} = 5$ from $N_{\text{I}} = 10$ areas. Second stage: For all selected areas, *SI* sampling is to be used, with n_i plants selected from N_i in the ith area, where n_i is given in the table above. Estimator: $\hat{\bar{y}}_{U\pi} = \hat{t}_\pi/N$, where $\hat{t}_\pi = \sum_s y_k/\pi_k$ is the π estimator of $t = \sum_U y_k$, and N is assumed to be known.

Strategy 2. The sampling design is as in Strategy 1, but the estimator is $\tilde{y}_s = \hat{t}_\pi/\hat{N}$, where $\hat{N} = \sum_s 1/\pi_k$ is the π estimator of N.

Strategy 3. First stage: *STSI* sampling with two strata, $U_{\text{I}1} = \{i: i = 1, 5, 6, 10\}$ (large areas), and $U_{\text{I}2} = \{i: i = 2, 3, 4, 7, 8, 9\}$ (small areas); in $U_{\text{I}1}$, $n_{\text{I}1} = 3$ areas are to be selected from $N_{\text{I}1} = 4$; in $U_{\text{I}2}$, $n_{\text{I}2} = 2$ are to be selected from $N_{\text{I}2} = 6$. Second stage: For all selected areas, *SI* sampling is to be used, with n_i plants selected from N_i, where n_i is given in the table above. The estimator is $\hat{\bar{y}}_{U\pi} = \hat{t}_\pi/N$, assuming N to be known.

Strategy 4. The sampling design is as in Strategy 3, but the estimator is $\tilde{y}_s = \hat{t}_\pi/\hat{N}$.

For each strategy, calculate the two variance components and their sum, which is the total variance of the estimator. The estimators of Strategies 1 and 3 are unbiased, and exact variance components are available. For the approximately unbiased estimators of Strategies 2 and 4, base your calculations on the approximate variance formulas given in the chapter. (Because the sample sizes are not large, one must be aware that the approximation may be crude.) Is it preferable to replace N by $\hat{N}$ even if N may be known? Does the stratification improve the estimates?

(b) Denote as Field 1 the field described by the data in (a). Another field, Field 2, is also divided in 10 areas, and the same four strategies as in (a) are to be compared. The data for Field 2 are given in the following table.

Area i	Number of plants in the area (N_i)	Number of plants to be selected (n_i)	Plant size y_k, $k \in U_i$
1	11	4	10 9 8 10 11 8 10 9 8 11 10
2	6	2	8 12 10 11 11 10
3	5	2	11 12 10 10 10
4	7	2	7 10 11 11 11 12 10
5	11	4	8 9 8 10 12 10 11 12 11 11 10
6	10	3	13 10 9 9 10 11 12 11 11 10
7	5	2	10 11 10 11 9
8	7	2	8 11 11 10 12 10 9
9	6	2	11 10 12 9 8 11
10	10	3	11 11 12 12 7 10 10 9 9 10

i. For each strategy, calculate the two variance components and the total variance, with exact or approximate expressions. ii. We note immediately by comparing the two data sets that the area means of y vary much more in Field 1 than in Field 2. Is this fact reflected in one or both of the variance components? For which of the two fields is the gain from stratification more pronounced?

8.2. Compare the strategies

$$(\text{SIC}, \hat{t}_{y\pi}), (\text{SIC}, \hat{t}_{yAr}), (\text{BEC}, \hat{t}_{y\pi}) \text{ and } (\text{BEC}, \hat{t}_{yAr})$$

for estimating the total of the variable RMT85 ($= y$) for the MU284 population using single-stage cluster sampling from the $N_I = 50$ clusters in Appendix C. Compute the (approximate) variances of the estimators assuming *SIC* sampling of size $n_I = 25$, and *BEC* sampling (that is, single-stage *BE* sampling of clusters) with the expected size 25. For the regression estimator, use the ratio model (8.5.1) with TP75 ($= u$) as the auxiliary variable. Relevant population data are:

$$\textstyle\sum_{U_I} t_{yi} = 69{,}605; \qquad \sum_{U_I} t_{yi}^2 = 207{,}411{,}399$$

$$\textstyle\sum_{U_I} u_i = 8{,}182; \qquad \sum_{U_I} u_i^2 = 2{,}338{,}656$$

$$\textstyle\sum_{U_I} t_{yi} u_i = 21{,}473{,}979$$

8.3. The objective is to estimate the total of the variable ME84 ($= y$) for the MU284 population using *SIC* sampling of size $n_I = 20$ from the $N_I = 50$ clusters in Appendix C and the ratio estimator shown in equation (8.5.2) with TP75 ($= u$) as the auxiliary variable. Use the following data

$$\textstyle\sum_{s_I} t_{yi} = 221{,}006; \qquad \sum_{s_I} t_{yi}^2 = 4{,}869{,}110{,}166$$

$$\textstyle\sum_{s_I} u_i = 3{,}290; \qquad \sum_{s_I} u_i^2 = 799{,}192$$

$$\textstyle\sum_{U_I} u_i = 8{,}182; \qquad \sum_{s_I} t_{yi} u_i = 60{,}333{,}403$$

to compute the desired estimate. Also, compute the associated *cve*.

8.4. We want an estimate of the percentage increase from 1975 to 1985 in total Swedish population as defined by the MU281 population, that is, for the parameter

$$\theta = 100\left(\frac{t_y}{t_z} - 1\right)$$

where t_y is the total of the variable P85 (= y), and t_z is the total of the variable P75 (= z). A small *POC* sample (single-stage Poisson sample of clusters) of expected size $n_{\text{I}} = 16$ was drawn from the $N_{\text{I}} = 50$ clusters defined in Appendix C (exclusive of the three largest municipalities) with inclusion probabilities proportional to t_{zi}. Estimate θ and compute the *cve* using the estimator

$$\hat{\theta} = 100\left(\frac{\hat{t}_{yAr}}{t_z} - 1\right)$$

where $\hat{t}_{yAr}$ is given by (8.5.2). Use the following sample data:

$$n_{s_{\text{I}}} = 15, \sum_{s_{\text{I}}} \frac{t_{yi}}{t_{zi}} = 15.766, \quad \text{and} \quad \sum_{s_{\text{I}}}(1 - \pi_{\text{I}i})\left(\frac{t_{yi}}{t_{zi}} - \frac{1}{n_{s_{\text{I}}}}\sum_{s_{\text{I}}}\frac{t_{yi}}{t_{zi}}\right)^2 = 0.0636$$

8.5. As a numerical illustration of the effects of different uses of auxiliary information in two-stage element sampling, let us consider the MU284 population and the two variables RMT85 (= y) and P75 (= x). There are $N_{\text{I}} = 16$ PSUs formed as follows: For $i = 1, \ldots, 12$, let

$$U_i = \{k: k = 18(i - 1) + j; j = 1, \ldots, 18\}$$

and for $i = 13, \ldots, 16$,

$$U_i = \{k: k = 216 + 17(i - 13) + j; j = 1, \ldots, 17\}$$

To estimate the total of RMT85, consider two-stage element sampling with *SI* sampling of $n_{\text{I}} = 10$ PSUs from the $N_{\text{I}} = 16$ PSUs, followed by *SI* sampling of $n_i = 10$ elements from every i in the first-stage sample. Consider the following four estimators:

(a)
$$\hat{t}_{y\pi} = \frac{N_{\text{I}}}{n_{\text{I}}}\sum_{s_{\text{I}}}\frac{N_i}{n_i}\sum_{s_i} y_k$$

corresponding to the case of no auxiliary information.

(b)
$$\hat{t}_{yAr} = \left(\sum_{U_{\text{I}}} t_{xi}\right)\frac{\sum_{s_{\text{I}}}\frac{N_i}{n_i}\sum_{s_i} y_k}{\sum_{s_{\text{I}}} t_{xi}}$$

based on the common ratio model for PSU totals given by (8.5.1) with $u_i = t_{xi}$, corresponding to the case of auxiliary information about PSUs.

(c)
$$\hat{t}_{yBr} = \left(\sum_{U_{\text{I}}} t_{xi}\right)\frac{\sum_{s_{\text{I}}}\frac{N_i}{n_i}\sum_{s_i} y_k}{\sum_{s_{\text{I}}}\frac{N_i}{n_i}\sum_{s_i} x_k}$$

based on the common ratio model (7.3.1), corresponding to the case of complete element auxiliary information

(d)
$$\hat{t}_{yCr} = \left(\frac{N_I}{n_I}\sum_{s_I} t_{xi}\right)\frac{\sum_{s_I}\frac{N_i}{n_i}\sum_{s_i} y_k}{\sum_{s_I}\frac{N_i}{n_i}\sum_{s_i} x_k}$$

based on the common ratio model (7.3.1), corresponding to the case of limited element auxiliary information, available for every element $k \in \bigcup_{i \in s_I} U_i$. Compute the (approximate) variances of the four estimators above, as well as the variance components corresponding to the two stages of selection. Use the following population data:

$\sum_{U_I} t_{yi} = 69{,}605$; $\sum_{U_I} t_{yi}^2 = 428{,}672{,}075$; $\sum_{U_I} t_{xi} = 8{,}182$

$\sum_{U_I} t_{xi}^2 = 5{,}224{,}748$; $\sum_{U_I} t_{xi}t_{yi} = 46{,}718{,}091$

and

i	$\sum_{U_i} y_k$	$\sum_{U_i} y_k^2$	$t_{yi} - \frac{t_y}{t_x} t_{xi}$	$\sum_{U_i}\left(y_k - \frac{t_y}{t_x} x_k\right)^2$
1	12,043	41,722,589	1,171	351,468
2	4,774	2,597,424	−126	25,183
3	4,259	2,679,009	−343	13,140
4	3,135	1,166,853	−659	33,324
5	3,434	1,095,082	−471	68,951
6	2,509	526,077	−520	31,556
7	7,602	13,814,724	782	1,911,233
8	9,292	45,774,818	2,818	8,562,807
9	3,030	1,030,676	−500	31,409
10	1,987	329,989	−455	17,433
11	2,251	560,209	−361	10,508
12	3,882	2,283,866	−261	6,748
13	2,414	547,610	−189	4,206
14	4,207	1,779,487	−523	23,255
15	1,294	258,136	−203	3,349
16	3,492	1,531,140	−106	24,032

8.6. Consider the following three estimators for the population mean:

$$\hat{\bar{y}}_{U1} = \frac{\sum_{s_I} \hat{t}_{yi\pi}/\pi_{Ii}}{\sum_{s_I} N_i/\pi_{Ii}}$$

$$\hat{\bar{y}}_{U2} = \frac{\sum_{s_I} N_i \tilde{y}_{s_i}/\pi_{Ii}}{\sum_{s_I} N_i/\pi_{Ii}}$$

$$\hat{\bar{y}}_{U3} = \frac{\sum_{s_I} \hat{t}_{yi\pi}/\pi_{Ii}}{\sum_{s_I} \hat{N}_i/\pi_{Ii}}$$

with

$$\tilde{y}_{s_i} = \hat{t}_{yi\pi}/\hat{N}_i = \left(\sum_{s_i} y_k/\pi_{k|i}\right)/\left(\sum_{s_i} 1/\pi_{k|i}\right)$$

Discuss the conditions under which two or all three estimators agree. In cases where they do not agree, discuss the advantages that one estimator may have over the other two.

8.7. Let u_i be a known auxiliary value for the ith PSU, $i = 1, \ldots, N_{\mathrm{I}}$, such that u_i is roughly proportional to the cluster total $t_{yi} = \sum_{U_i} y_k$. A rather large *SI* sample of n_{I} PSUs is selected. Within the ith selected PSU, an *SI* sample of n_i elements is selected from N_i. The population total $t_y = \sum_{U_{\mathrm{I}}} t_{yi}$ is estimated by

$$\hat{t}_y = (\textstyle\sum_{U_{\mathrm{I}}} u_i)\dfrac{\sum_{s_{\mathrm{I}}} \hat{t}_{yi\pi}}{\sum_{s_{\mathrm{I}}} u_i}$$

with $\hat{t}_{yi\pi} = N\bar{y}_{s_i}$. Find expressions for the estimator's approximate variance and for a variance estimator.

8.8. Consider the *SIC* design with n_{I} clusters drawn from N_{I}. In the selected clusters, all elements are observed. We estimate the population mean per element $\bar{y}_U = \sum_U y_k/N$ by the sample mean per element $\hat{\bar{y}}_{Ur} = \sum_{s_{\mathrm{I}}} t_i/\sum_{s_{\mathrm{I}}} N_i$. (a) Apply Result 8.6.1 to verify equation (8.7.2) for $AV_{SIC}(\hat{\bar{y}}_{Ur})$. (b) Compare with an *SI* sample with the fixed size $n = n_{\mathrm{I}}\bar{N}$, which is the expected size of the *SIC* sample. Verify the expression (8.7.5) for the design effect

$$\text{deff}(SIC, \hat{\bar{y}}_{Ur}) = AV_{SIC}(\hat{\bar{y}}_{Ur})/V_{SI}(\bar{y}_s)$$

8.9. Consider *SIC* sampling with n_{I} clusters selected from N_{I}. Verify equation (8.7.9) for the difference between the design effects deff$(SIC, \hat{\bar{y}}_{U\pi})$ and deff$(SIC, \hat{\bar{y}}_{Ur})$.

8.10. Consider the special case of the common mean model (7.4.1). Show that the case B estimator $\hat{t}_{yBr}$ divided by N gives the following population mean estimator:

$$\hat{\bar{y}}_U = (\textstyle\sum_{s_{\mathrm{I}}} \hat{t}_{yi\pi}/\pi_{\mathrm{I}i})/(\sum_{s_{\mathrm{I}}} \hat{N}_i/\pi_{\mathrm{I}i})$$

where $\hat{t}_{yi\pi}$ and $\hat{N}_i$ are π estimators. Derive the corresponding variance estimator.

8.11. Consider the two-stage design *SI*, *SI*. The population mean $\bar{y}_U$ is estimated by

$$\hat{\bar{y}}_U = \frac{\sum_{s_{\mathrm{I}}} N_i\bar{y}_{s_i}}{\sum_{s_{\mathrm{I}}} N_i}$$

The sizes N_i for $i \in s_{\mathrm{I}}$ are assumed known. In the first stage, an *SI* sample of PSUs is selected. The sampling fraction is $f_{\mathrm{I}} = n_{\mathrm{I}}/N_{\mathrm{I}}$. Two schemes are considered for the *SI* sampling within selected PSUs:

i. The sampling fraction is constant with $n_i/N_i = f_{\mathrm{II}}$ for all $i \in s_{\mathrm{I}}$.
ii. A constant sample size c is applied for all $i \in s_{\mathrm{I}}$, where c is determined so that the expectation of $n_s = \sum_{s_{\mathrm{I}}} n_i$ in (i) equals cn_{I}.

Compare the *AV* expressions to see if one scheme is preferred to the other under the simplifying assumption that all within-PSU variance $S^2_{yU_i}$ are equal.

8.12. The estimation errors for $\hat{t}_{yBr}$ and $\hat{t}_{yCr}$ are given by equations (8.9.13) and (8.9.14). Verify these expressions using the g weights defined in (8.9.10) and (8.9.12).

8.13. Consider the stratified cluster design $STSIC$. We draw n_{Ih} clusters from N_{Ih} in the stratum U_{Ih}. All elements in selected clusters are observed with respect to y and z. The objective is to estimate the ratio of totals given by

$$R = \frac{t_y}{t_z} = \frac{\sum_U y_k}{\sum_U z_k}$$

Specify a suitable estimator and a corresponding variance estimator.

PART III

Further Questions in Design and Analysis of Surveys

CHAPTER 9

Two-Phase Sampling

9.1. Introduction

The theory and methods presented so far form a set of basic tools in survey sampling. But many surveys require more advanced techniques. Some of these techniques are presented in Chapters 9 to 13. The topic of this chapter is two-phase sampling, or sampling followed by subsampling. The technique, originally proposed by Neyman (1938), is presented here in a general form with arbitrary sampling designs in each of the two phases. Two-phase sampling is a good compromise for surveys in which there is little or no prior knowledge about the population.

We have seen that highly efficient estimation strategies require strong auxiliary information. For example, the combined ratio estimator (see Section 7.3)

$$\hat{t} = \sum_U x_k \frac{\sum_{h=1}^{H} N_h \bar{y}_{s_h}}{\sum_{h=1}^{H} N_h \bar{x}_{s_h}}$$

is often a vast improvement on the simple strategy consisting of the *SI* design and the π estimator $N\bar{y}_s$. But the combined ratio estimator requires that all elements can be stratified, and $\sum_U x_k$ must be known. In many surveys, such extensive information is unavailable.

When the frame contains little or no useful information about the population elements, it may seem that we have only the following two options:

1. To use a very simple design like *SI* or *SIC*, combined with the π estimator. Here, acceptable precision in the estimates may be achieved only through a very large sample size, which may entail prohibitive costs.

2. To gather information about the population, use it to construct a new, highly informative frame and then select an efficient combination of sampling design and regression estimator. A much smaller sample may suffice to obtain the desired precision. However, the total survey cost may still be unacceptable due to large costs for constructing the new frame.

However, a third option can be identified. It is a compromise between the options listed above and consists of sample selection in two phases, as follows:

a. In the first phase, select a rather large sample of elements s_a by a simple sampling design $p_a(\cdot)$. For the elements in s_a, gather inexpensive information on one or more auxiliary variables.
b. With the aid of the auxiliary information collected in the first phase, select a second-phase sample s from s_a by the design $p(\cdot|s_a)$. This s is a subsample. The study variable y is then observed for the elements in the second phase sample.

The technique is called *two-phase* sampling (or sometimes *double* sampling). Extensions to more than two phases are called *multiphase* sampling. A key to successful two-phase sampling is the creation of a highly informative frame, not for the whole population (this may be too expensive) but for the part of the population from which the subsample is drawn.

An additional reason for studying two-phase sampling is that the theory is useful for estimation in the presence of nonresponse. In a survey with nonresponse, the selection of a probability sample can be seen as the first phase, and the respondents are viewed as a subselection. Therefore, the theory in this chapter (in particular Section 9.8) will find application in Chapter 15 which deals with nonresponse.

Sampling at successive occasions is a technique that relies on sampling in two or more phases. We consider this topic in Section 9.9. Two-stage sampling is a particular case of two-phase sampling. The class of two-stage designs considered in Chapter 4 was narrowed by the requirements of invariance and independence specified in Section 4.3.1. The theory for two-phase sampling in this chapter also provides a more flexible framework for two-stage sampling.

To illustrate, suppose the objective is to draw a sample s in two stages so that s will have the a priori fixed size n. From the N_{I} PSUs, of varying sizes, an *SI* sample s_{I} of n_{I} PSUs is drawn. The number of elements contained in the set of selected PSUs is

$$N_{s_{\text{I}}} = \sum_{s_{\text{I}}} N_i$$

Here, $N_{s_{\text{I}}}$ will vary from one first-stage sample to another. For the second-stage design, we retain the requirement of independence but give up the invariance (see Section 4.3.1). In a selected PSU, an *SI* subsample s_i of n_{s_i} elements is drawn, where n_{s_i} is proportional to the size N_i of the PSU, that is,

$$n_{s_i} = \left(\frac{N_i}{N_{s_I}}\right) n \tag{9.1.1}$$

The total sample size is clearly n. But since n_{s_i} depends on the outcome of the first stage, we no longer have the invariance property. The design is self-weighting (see Remark 4.3.7) in the sense that the element inclusion probability

$$\pi_k = \pi_{Ii}\pi_{k|i} = \frac{n_I n}{N_I N_{s_I}}$$

is constant for all k. Note, however, that it depends on the outcome of the first-stage sampling. This simple two-stage design is not covered by the results in Chapter 4 but can be treated with the two-phase theory presented in this chapter.

9.2. Notation and Choice of Estimator

The subscript a denotes phase one. The first-phase sample s_a of size n_{s_a} is drawn according to a sampling design $p_a(\cdot)$ such that $p_a(s_a)$ is the probability that s_a is chosen. The corresponding inclusion probabilities are denoted

$$\pi_{ak} = \sum_{s_a \ni k} p_a(s_a) \tag{9.2.1}$$

and

$$\pi_{akl} = \sum_{s_a \ni k \& l} p_a(s_a) \tag{9.2.2}$$

for elements k and $l \in U$.

Given s_a, the second-phase sample s, of size n_s, is drawn according to the design $p(\cdot|s_a)$ such that $p(s|s_a)$ is the conditional probability of choosing s. The inclusion probabilities under this design are denoted

$$\pi_{k|s_a} = \sum_{s \ni k} p(s|s_a) \tag{9.2.3}$$

and

$$\pi_{kl|s_a} = \sum_{s \ni k \& l} p(s|s_a) \tag{9.2.4}$$

for elements k and $l \in s_a$.

Our next task is to find a simple unbiased estimator for the population total $t = \sum_U y_k$. (To simplify the notation in this chapter, we write t for the y total instead of t_y as in some of the earlier chapters.) A natural candidate is the π estimator

$$\hat{t}_\pi = \sum_s \check{y}_k = \sum_s y_k/\pi_k \tag{9.2.5}$$

where $\pi_k = \Pr(k \in s)$ is the inclusion probability of the kth element. Obviously, the estimator (9.2.5) requires that the π_k can be calculated, but this is not always possible in two-phase sampling. We have

$$\pi_k = \sum_{s \ni k} p(s)$$

where $p(s)$ is the probability (jointly over the two phases) of drawing the sample s. Now

$$p(s) = \sum_{s_a \supset s} p_a(s_a) p(s|s_a)$$

Hence

$$\pi_k = \sum_{s \ni k} \sum_{s_a \supset s} p_a(s_a) p(s|s_a) = \sum_{s_a \ni k} \sum_{\substack{s \subset s_a \\ s \ni k}} p_a(s_a) p(s|s_a)$$

$$= \sum_{s_a \ni k} p_a(s_a) \left[\sum_{\substack{s \subset s_a \\ s \ni k}} p(s|s_a) \right] = \sum_{s_a \ni k} p_a(s_a) \pi_{k|s_a} \tag{9.2.6}$$

Thus, to determine π_k in practice, we must know the probabilities $p_a(s_a)$ for every s_a (which we ordinarily do) and we must also know the $\pi_{k|s_a}$ for every s_a (which we ordinarily do not, because $\pi_{k|s_a}$ may depend on the outcome of phase one).

EXAMPLE 9.2.1. Suppose that the first phase calls for drawing an *SI* sample of size n_a. Then

$$p_a(s_a) = 1 \bigg/ \binom{N}{n_a}$$

for every s_a of the fixed size n_a (and $p_a(s_a) = 0$ otherwise), and $\pi_{ak} = n_a/N$ for all $k \in s_a$. Furthermore, suppose that, for every $k \in s_a$, we record an auxiliary variable x believed to be roughly proportional to the study variable y. Suppose the second phase is carried out by Poisson sampling, in such a way that the conditional inclusion probability of element k is proportional to x, that is,

$$\pi_{k|s_a} = n x_k / \textstyle\sum_{s_a} x_k$$

(For simplicity, $n x_k < \sum_{s_a} x_k$ for all k is assumed). Clearly, $p_a(s_a)$ is known for every s_a. However, $\pi_{k|s_a}$ depends on s_a and is thus known only for the first-phase sample s_a actually realized. Consequently, the π_k required for the estimator (9.2.5) cannot be obtained.

Our discussion shows that the π estimator cannot always be used in practice. Consequently, we seek an unbiased estimator that uses weighting in a more practical manner. To this end, let

$$\textstyle\sum_{s_a} \check{y}_{ak} = \sum_{s_a} y_k / \pi_{ak}$$

be the π estimator of $t = \sum_U y_k$ that could be formed if y_k and π_{ak} were known for every $k \in s_a$. But y_k is observed only for $k \in s$. Now, given s_a, $\sum_{s_a} \check{y}_{ak}$ is unbiasedly estimated by the conditional π estimator

$$\sum_s \frac{\check{y}_{ak}}{\pi_{k|s_a}} = \sum_s \frac{y_k}{\pi_{ak}\pi_{k|s_a}} \tag{9.2.7}$$

This estimator will work, for the $\pi_{k|s_a}$ are known for the sample s_a realized in phase one. Introducing the quantity

$$\pi_k^* = \pi_{ak}\pi_{k|s_a} \tag{9.2.8}$$

and noting that the weight attached to y_k in the new estimator (9.2.7) is $1/\pi_k^*$, we may say that (9.2.7) is obtained by "π^* expansion" of the y_k values in the second phase sample. Denote the π^*-expanded y-value by

$$\check{\check{y}}_k = \frac{\check{y}_{ak}}{\pi_{k|s_a}} = \frac{y_k}{\pi_{ak}\pi_{k|s_a}} = \frac{y_k}{\pi_k^*} \tag{9.2.9}$$

where $\check{\check{}}$ reminds us that a two-fold expansion (one for each phase) is used. We call (9.2.7) the π^* *estimator* and denote it by $\hat{t}_{\pi^*}$; in other words

$$\hat{t}_{\pi^*} = \sum_s y_k/\pi_k^* = \sum_s \check{\check{y}}_k \tag{9.2.10}$$

where π_k^* is given by equation (9.2.8).

Note that $\hat{t}_{\pi^*}$ does not in general agree with the π estimator $\hat{t}_\pi$ given by equation (9.2.5). This is because $\pi_k \neq \pi_k^*$ (see Remark 9.2.1 below), except in rare cases. The π^* estimator is examined in the next section, and examples of its use in practice will be seen in Section 9.4.

Remark 9.2.1. Note that there are subtle differences between the various probabilities. For example, for $k \in s_a$,

$$\Pr(k \in s | k \in s_a) = \frac{\Pr(k \in s)}{\Pr(k \in s_a)} = \pi_k/\pi_{ak} \tag{9.2.11}$$

which will in general differ from $\pi_{k|s_a}$. Here, $\pi_{k|s_a}$ is the probability of including k in the second-phase sample s under the particular design used when s_a is realized in phase one. By contrast, π_k/π_{ak} is the conditional probability of selecting k in the second phase, given that k was present in s_a. Now, using equation (9.2.11), we conclude that π_k and π_k^* are not in general equal, because

$$\begin{aligned}\pi_k &= \Pr(k \in s) = \Pr(k \in s_a)\Pr(k \in s | k \in s_a) \\ &= \pi_{ak}(\pi_k/\pi_{ak}) \neq \pi_{ak}\pi_{k|s_a} = \pi_k^*\end{aligned}$$

9.3. The π^* Estimator

In this section, we derive the essential results concerning the π^* estimator. This requires further notation. Let

$$\pi_{kl}^* = \pi_{akl}\pi_{kl|s_a} \tag{9.3.1}$$

$$\Delta_{akl} = \pi_{akl} - \pi_{ak}\pi_{al} \tag{9.3.2}$$

and

$$\Delta_{kl|s_a} = \pi_{kl|s_a} - \pi_{k|s_a}\pi_{l|s_a} \tag{9.3.3}$$

It is helpful to express the total estimation error of the π^* estimator given by (9.2.10) as a sum of two components,

$$\hat{t}_{\pi^*} - t = \underbrace{(\textstyle\sum_{s_a} \check{y}_{ak} - \sum_U y_k)}_{Q_{s_a}} + \underbrace{(\textstyle\sum_s \check{\check{y}}_k - \sum_{s_a} \check{y}_{ak})}_{R_s} \tag{9.3.4}$$

where Q_{s_a} may be called the error due to the first phase of sampling and R_s the error due to the second phase. Recall that the π^* estimator is unbiased, given s_a, for the estimate $\sum_{s_a} \check{y}_k$ that would be formed if there was just one phase of sampling. This implies that the component R_s in (9.3.4) has expected value zero, conditionally on s_a. This important property is used in the derivations in this section.

We now easily obtain the following result.

Result 9.3.1. *In two-phase sampling, the population total* $t = \sum_U y_k$ *is estimated unbiasedly by the* π^* *estimator*

$$\hat{t}_{\pi^*} = \textstyle\sum_s \check{\check{y}}_k = \sum_s y_k/\pi_k^* \tag{9.3.5}$$

The variance of $\hat{t}_{\pi^*}$ *is given by*

$$V(\hat{t}_{\pi^*}) = \textstyle\sum\sum_U \Delta_{akl}\check{y}_{ak}\check{y}_{al} + E_{p_a}(\sum\sum_{s_a} \Delta_{kl|s_a}\check{\check{y}}_k\check{\check{y}}_l) \tag{9.3.6}$$

where $\check{y}_{ak} = y_k/\pi_{ak}$, $\check{\check{y}}_k = y_k/\pi_k^*$, *and the* Δ *quantities are given by equations* (9.3.2) *and* (9.3.3). *An unbiased estimator of the variance is given by*

$$\hat{V}(\hat{t}_{\pi^*}) = \sum\sum_s \frac{\Delta_{akl}}{\pi_{kl}^*}\check{y}_{ak}\check{y}_{al} + \sum\sum_s \frac{\Delta_{kl|s_a}}{\pi_{kl|s_a}}\check{\check{y}}_k\check{\check{y}}_l \tag{9.3.7}$$

Each component of (9.3.7) *is unbiased for its counterpart in* (9.3.6).

Note that the variance (9.3.6) is not stated explicitly but as an expected value over the first phase design. This is necessary because the $\Delta_{kl|s_a}$ may depend on the actually realized sample s_a. Fortunately, this causes no problem for the variance estimation. We see that the variance estimator (9.3.7) is stated in an explicit form, making direct calculation possible.

PROOF. The proof of Result 9.3.1 is simple if we invoke the standard conditioning approach (see Section 4.3.2). In this case, it is appropriate to condition on the sample s_a realized in phase one. For the expected value of $\hat{t}_{\pi^*}$, we then have

$$E(\hat{t}_{\pi^*}) = E_{p_a}E(\hat{t}_{\pi^*}|s_a) = E_{p_a}(\textstyle\sum_{s_a} \check{y}_{ak}) = t$$

That is, the π^* estimator is unbiased. Turning to the variance, we have

$$V(\hat{t}_{\pi^*}) = V_{p_a}E(\hat{t}_{\pi^*}|s_a) + E_{p_a}V(\hat{t}_{\pi^*}|s_a)$$

reflecting the variation due to each of the two phases of sampling. But from the decomposition (9.3.4)

$$V_{p_a} E(\hat{t}_{\pi^*}|s_a) = V_{p_a}(Q_{s_a}) = \sum\sum_U \Delta_{akl} \check{y}_{ak} \check{y}_{al}$$

and

$$E_{p_a} V(\hat{t}_{\pi^*}|s_a) = E_{p_a} V(R_s|s_a) = E_{p_a}(\sum\sum_{s_a} \Delta_{kl|s_a} \check{\check{y}}_k \check{\check{y}}_l)$$

using known properties of the π estimator, and the unconditional variance (9.3.6) follows. For the variance estimator, we use $E(\cdot) = E_{p_a} E(\cdot|s_a)$, and obtain, for the first component,

$$E\left(\sum\sum_s \frac{\Delta_{akl}}{\pi^*_{kl}} \check{y}_{ak} \check{y}_{al}\right) = E_{p_a}\left(\sum\sum_{s_a} \frac{\Delta_{akl}}{\pi_{akl}} \check{y}_{ak} \check{y}_{al}\right)$$
$$= \sum\sum_U \Delta_{akl} \check{y}_{ak} \check{y}_{al}$$

and, for the second component,

$$E\left(\sum\sum_s \frac{\Delta_{kl|s_a}}{\pi_{kl|s_a}} \check{\check{y}}_k \check{\check{y}}_l\right) = E_{p_a}(\sum\sum_{s_a} \Delta_{kl|s_a} \check{\check{y}}_{ak} \check{\check{y}}_{al})$$

This shows the unbiasedness of each component of the variance estimator. It follows that

$$E[\hat{V}(\hat{t}_{\pi^*})] = V(\hat{t}_{\pi^*})$$

as claimed in the result. □

Remark 9.3.1. The variance estimator shown in equation (9.3.7) can also be written as a compact, single-term expression,

$$\hat{V}(\hat{t}_{\pi^*}) = \sum\sum_s \check{\check{\Delta}}^*_{kl} \check{\check{y}}_k \check{\check{y}}_l \tag{9.3.8}$$

where

$$\Delta^*_{kl} = \pi^*_{kl} - \pi^*_k \pi^*_l \tag{9.3.9}$$

and

$$\check{\check{\Delta}}^*_{kl} = \Delta^*_{kl}/\pi^*_{kl} \tag{9.3.10}$$

It is left as an exercise to verify equation (9.3.8). The simplicity of (9.3.8) is somewhat deceptive. When the variance estimator is worked out for cases of practical interest, the resulting expressions can be complex. Example 9.4.2 will illustrate this.

EXAMPLE 9.3.1. Two-stage sampling can be regarded as a special case of two-phase sampling. Let us in particular consider the two-stage design SI, SI defined in Example 4.3.1. Using SI sampling in each stage, n_I PSUs are drawn from N_I in the first stage, and n_{s_i} elements are subsampled from N_i, provided that the ith PSU was selected. We assume that the sizes n_{s_i} are not necessarily fixed prior to sampling, but that they may depend on the set of PSUs s_{I} obtained in stage one. For example, the n_{s_i} may be determined by equation (9.1.1). Let $f_{s_i} = n_{s_i}/N_i$ and $\hat{t}_i = N_i \bar{y}_{s_i}$. By Result 9.3.1, the π^* estimator is now given by

$$\hat{t}_{\pi^*} = (N_{\mathrm{I}}/n_{\mathrm{I}}) \textstyle\sum_{s_{\mathrm{I}}} \hat{t}_i \tag{9.3.11}$$

and its variance by

$$V(\hat{t}_{\pi^*}) = N_{\mathrm{I}}^2 \frac{1-f_{\mathrm{I}}}{n_{\mathrm{I}}} S_{tU_{\mathrm{I}}}^2 + E_{SI}\left(\frac{N_{\mathrm{I}}^2}{n_{\mathrm{I}}^2} \textstyle\sum_{s_{\mathrm{I}}} N_i^2 \frac{1-f_{s_i}}{n_{s_i}} S_{yU_i}^2\right) \tag{9.3.12}$$

where $S_{tU_{\mathrm{I}}}^2$ and $S_{yU_i}^2$ are defined as in Example 4.3.1. The second variance component is given nonexplicitly as an expectation with respect to the *SI* selection at stage one. The unbiased variance estimator is

$$\hat{V}(\hat{t}_{\pi^*}) = N_{\mathrm{I}}^2 \frac{1-f_{\mathrm{I}}}{n_{\mathrm{I}}} S_{\hat{t}s_{\mathrm{I}}}^2 + \frac{N_{\mathrm{I}}}{n_{\mathrm{I}}} \textstyle\sum_{s_{\mathrm{I}}} N_i^2 \frac{1-f_{s_i}}{n_{s_i}} S_{ys_i}^2 \tag{9.3.13}$$

where $S_{\hat{t}s_{\mathrm{I}}}^2$ and $S_{ys_i}^2$ are defined as in Example 4.3.1. The conclusions (9.3.12) and (9.3.13) are more generally valid than their counterparts in Example 4.3.1, since it is now permitted to fix the n_{s_i} conveniently after the sample of PSUs has been obtained. If the n_{s_i} are set by an invariant rule (as in Example 4.3.1), then equation (9.3.13) confirms the earlier variance estimator (4.3.23).

9.4. Two-Phase Sampling for Stratification

Stratified sampling is a strongly variance-reducing technique, provided that the strata are well constructed. Powerful stratification variables are needed to obtain good strata. If no such information is available at the outset, we can use two-phase sampling in the following way. A large first-phase sample is selected and stratified with the aid of auxiliary characteristics observed, at low cost, for the elements in the first-phase sample. The second phase is then carried out as stratified sampling, with a considerably smaller sample size, and the study variable y is observed for this small stratified sample. The procedure is called *two-phase sampling for stratification.* We examine it here, assuming that the π^* estimator is used.

More specifically, the two-phase design is defined in the following way.

1. In the first phase, a large sample s_a of size n_{s_a} is drawn according to a given design $p_a(\cdot)$. For the elements in s_a, information is recorded that will permit a stratification.

2a. The information is used to stratify s_a into H_{s_a} strata denoted s_{ah} ($h = 1, \ldots, H_{s_a}$), with n_{ah} elements in stratum h. Thus

$$s_a = \bigcup_{h=1}^{H_{s_a}} s_{ah}; \qquad n_{s_a} = \sum_{h=1}^{H_{s_a}} n_{ah}$$

2b. From stratum h, a sample s_h (where $s_h \subset s_{ah}$) of size n_h is drawn according to the design $p_h(\cdot|s_a)$. Subsampling is carried out independently from one stratum to another. For the total subsample s, the decomposition is

$$s = \bigcup_{h=1}^{H_{s_a}} s_h; \qquad n_s = \sum_{h=1}^{H_{s_a}} n_h$$

Remark 9.4.1. Note that the present notation, for reasons of simplicity, deviates slightly from our earlier norm. We write n_{ah} instead of $n_{s_{ah}}$ and n_h instead of n_{s_h}, although the n_{ah} and the n_h are ordinarily random.

Remark 9.4.2. The frequency interpretation of the procedure defined by (1), (2a), and (2b) is as follows. There is a long series of experiments, in which each experiment consists of the realization of a first-phase sample s_a, followed by the realization of a second-phase subsample, according to the sampling designs specified for each phase. When s_a is given, the strata are also given. In the second phase, exactly the same stratified design is used every time that a given s_a is realized. However, for two nonidentical first-phase samples, the stratifications may be different. The number of strata may be different. This is recognized in our notation H_{s_a}, a number which may vary from one s_a to another. Ideally, the strata are defined by characteristics that are inexpensive to observe yet powerful as explanatory factors for the study variable y.

The principal results for the π^* estimator are summarized in the following result.

Result 9.4.1. *In two-phase sampling for stratification, the π^* estimator for the population total t can be written*

$$\hat{t}_{\pi^*} = \sum_{h=1}^{H_{s_a}} \sum\nolimits_{s_h} \check{\check{y}}_k \tag{9.4.1}$$

The variance is

$$V(\hat{t}_{\pi^*}) = \sum\sum\nolimits_U \Delta_{akl}\check{y}_{ak}\check{y}_{al} + E_{p_a}\left(\sum_{h=1}^{H_{s_a}} \sum\sum\nolimits_{s_{ah}} \Delta_{kl|s_a}\check{\check{y}}_k\check{\check{y}}_l\right) \tag{9.4.2}$$

where $\check{y}_{ak} = y_k/\pi_{ak}$ and $\check{\check{y}}_k = y_k/\pi_k^*$. An unbiased variance estimator is given by

$$\hat{V}(\hat{t}_{\pi^*}) = \sum\sum\nolimits_s \frac{\Delta_{akl}}{\pi_{kl}^*}\check{y}_{ak}\check{y}_{al} + \sum_{h=1}^{H_{s_a}} \sum\sum\nolimits_{s_h} \frac{\Delta_{kl|s_a}}{\pi_{kl|s_a}}\check{\check{y}}_k\check{\check{y}}_l \tag{9.4.3}$$

Each component of the variance estimator (9.4.3) *is unbiased for its counterpart in equation* (9.4.2).

The expressions given in (9.4.2) and (9.4.3) are simple and compact, because they are stated in general terms. When applied to specific designs, the formulas may, however, become rather lengthy, as illustrated by the following two examples.

EXAMPLE 9.4.1. Let us apply Result 9.4.1, specifying that the second phase design is *STSI*, whereas the first-phase design remains general. Then

$$\pi_{k|s_a} = \frac{n_h}{n_{ah}} = f_h \quad \text{for} \quad k \in s_{ah}$$

and

$$\pi_{kl|s_a} = \begin{cases} f_h & \text{for} \quad k = l \in s_{ah} \\ f_h \dfrac{n_h - 1}{n_{ah} - 1} & \text{for} \quad k \in s_{ah}, l \in s_{ah}, k \neq l \\ f_h f_{h'} & \text{for} \quad k \in s_{ah}, l \in s_{ah'}, h \neq h' \end{cases} \tag{9.4.4}$$

From Result 9.4.1, the π^* estimator is now

$$\hat{t}_{\pi^*} = \sum_{h=1}^{H_{s_a}} n_{ah} \bar{\check{y}}_{s_h} \tag{9.4.5}$$

where

$$\bar{\check{y}}_{s_h} = \frac{1}{n_h} \sum\nolimits_{s_h} \check{y}_{ak} = \frac{1}{n_h} \sum\nolimits_{s_h} \frac{y_k}{\pi_{ak}} \tag{9.4.6}$$

Its variance can be written as

$$V(\hat{t}_{\pi^*}) = \sum\sum\nolimits_U \Delta_{akl} \check{y}_{ak} \check{y}_{al} + E_{p_a}\left(\sum_{h=1}^{H_{s_a}} n_{ah}^2 \frac{1 - f_h}{n_h} S^2_{\check{y}s_{ah}} \right) \tag{9.4.7}$$

where $S^2_{\check{y}s_{ah}}$ is the variance in stratum h of the expanded values $\check{y}_{ak} = y_k/\pi_{ak}$, that is,

$$S^2_{\check{y}s_{ah}} = \frac{1}{n_{ah} - 1} \sum\nolimits_{s_{ah}} (\check{y}_{ak} - \bar{\check{y}}_{s_{ah}})^2$$

with

$$\bar{\check{y}}_{s_{ah}} = \sum\nolimits_{s_{ah}} \check{y}_{ak}/n_{ah}$$

The second component of equation (9.4.7), which contains a familiar stratified form, must be left as an expected value, since n_{ah}, n_h, and H_{s_a} may be determined as a function of the first-phase sample s_a actually realized. The unbiased variance estimator is given by

$$\hat{V}(\hat{t}_{\pi^*}) = \sum\sum\nolimits_s \frac{\Delta_{akl}}{\pi^*_{kl}} \check{y}_{ak} \check{y}_{al} + \sum_{h=1}^{H_{s_a}} n_{ah}^2 \frac{1 - f_h}{n_h} S^2_{\check{y}s_h} \tag{9.4.8}$$

where $\pi^*_{kl} = \pi_{akl} \pi_{kl|s_a}$ with $\pi_{kl|s_a}$ given by (9.4.4) and

$$S^2_{\check{y}s_h} = \frac{1}{n_h - 1} \sum\nolimits_{s_h} (\check{y}_{ak} - \bar{\check{y}}_{s_h})^2 \tag{9.4.9}$$

with $\bar{\check{y}}_{s_h}$ given by (9.4.6). Note that if $U = s_a$ with probability one (that is, if the entire population is selected in phase one), then (9.4.7) and (9.4.8) are reduced to the well-known formulas for *STSI* sampling given in Result 3.7.2.

EXAMPLE 9.4.2. Let us continue the preceding example by further specifying that the first phase calls for drawing an *SI* sample of n_a elements from N, whereas *STSI* still applies in the second phase. Set $f_a = n_a/N$, which is the phase-one sampling fraction, $w_{ah} = n_{ah}/n_a$, which is the relative size of stratum h, and $f_h = n_h/n_{ah}$, which is the phase-two sampling fraction in stratum h. Result 9.4.1 now leads to the following conclusions. The π^* estimator can be written

$$\hat{t}_{\pi^*} = N \sum_{h=1}^{H_{s_a}} w_{ah} \bar{y}_{s_h} = N \hat{\bar{y}}_U \tag{9.4.10}$$

The variance is

$$V(\hat{t}_{\pi^*}) = N^2 \frac{1-f_a}{n_a} S_{yU}^2 + E_{SI}\left(N^2 \sum_{h=1}^{H_{s_a}} w_{ah}^2 \frac{1-f_h}{n_h} S_{ys_{ah}}^2 \right) = V_1 + V_2 \tag{9.4.11}$$

Here, S_{yU}^2 and $S_{ys_{ah}}^2$ denote the variance of y in U and in s_{ah}. From equation (9.4.8) we obtain the following unbiased variance component estimators

$$\hat{V}_1 = N^2 \frac{1-f_a}{n_a}\left[\sum_{h=1}^{H_{s_a}} w_{ah}(1-\delta_h) S_{ys_h}^2 + \frac{n_a}{n_a - 1} \sum_{h=1}^{H_{s_a}} w_{ah}(\bar{y}_{s_h} - \hat{\bar{y}}_U)^2 \right] \tag{9.4.12}$$

and

$$\hat{V}_2 = N^2 \sum_{h=1}^{H_{s_a}} w_{ah}^2 \frac{1-f_h}{n_h} S_{ys_h}^2 \tag{9.4.13}$$

where $S_{ys_h}^2$ is the variance of y in s_h and

$$\delta_h = (1/n_h)[(n_a - n_{ah})/(n_a - 1)]$$

Adding the components (9.4.12) and (9.4.13) we obtain, after some algebra, the unbiased variance estimator

$$\hat{V}(\hat{t}_{\pi^*}) = \hat{V}_1 + \hat{V}_2 = N(N-1) \sum_{h=1}^{H_{s_a}} \left(\frac{n_{ah}-1}{n_a - 1} - \frac{n_h - 1}{N-1} \right) \frac{w_{ah} S_{ys_h}^2}{n_h}$$
$$+ \frac{N(N-n_a)}{n_a - 1} \sum_{h=1}^{H_{s_a}} w_{ah}(\bar{y}_{s_h} - \hat{\bar{y}}_U)^2 \tag{9.4.14}$$

Since $n_h \geq 2$ is required to calculate $S_{ys_h}^2$, the second-phase design should be chosen with this in mind.

When N is much greater than n_a and $(n_{ah} - 1)/(n_a - 1) \doteq w_{ah}$, we can approximate the right hand side of (9.4.14), which leads to

$$\hat{V}(\hat{t}_{\pi^*}) \doteq N^2 \sum_{h=1}^{H_{s_a}} \frac{w_{ah}^2 S_{ys_h}^2}{n_h} + \frac{N^2}{n_a} \sum_{h=1}^{H_{s_a}} w_{ah}(\bar{y}_{s_h} - \hat{\bar{y}}_U)^2$$

Here the first term resembles the variance estimator (7.6.5) used in poststratified sampling. The second term adds little if the phase-one sample is considerably greater than the phase-two sample.

Remark 9.4.3. In Example 9.4.2, the strata were defined after examination of the first-phase sample s_a. Different stratifications are permitted for different samples s_a. A somewhat different situation arises when the strata are set once and for all, at the level of the whole population. That is, there exist H predetermined strata, $U_1, \ldots, U_h, \ldots, U_H$, reflecting a subdivision of the population by a given principle, for example, a fixed number of age-sex groups. These strata are not identified prior to the first phase of sampling, but every sample s_a can be stratified by the predetermined principle. Under these conditions, let us consider the following two-phase selection.

a. Let n_a be the size of the *SI* sample s_a drawn in the first phase. Let n_{ah} be the number of elements in s_a that belong to U_h.
b. The second phase is carried out by the *STSI* design. The size n_h of the subsample s_h drawn from $s_{ah} = s_a \cap U_h$, is determined by

$$n_h = v_h n_{ah}; \qquad 0 < v_h \leq 1$$

where the v_h are a priori fixed constants, $h = 1, \ldots, H$.

There is now a nonzero probability that one or more of the sets s_{ah} is empty. If n_a is very large, however, this event has a negligible probability. On this assumption, we obtain for the π^* estimator (9.4.10) the variance

$$V(\hat{t}_{\pi^*}) = N^2 \frac{1 - f_a}{n_a} S_{yU}^2 + N^2 \sum_{h=1}^{H} \frac{W_h S_{yU_h}^2}{n_a} \left(\frac{1}{v_h} - 1 \right) \tag{9.4.15}$$

where $W_h = N_h/N$ is the relative size of the stratum U_h, and $S_{yU_h}^2$ denotes the variance of y in U_h. Equation (9.4.14) is still the appropriate variance estimator, assuming that $\Pr(n_h \geq 2)$ is very near unity for all h. These results were obtained by Rao (1973).

9.5. Auxiliary Variables for Selection in Two Phases

The π^* estimator shown in equation (9.3.5) relies exclusively on weighting of elements. The inclusion probabilities of the second-phase design may depend on information gathered in the first phase, but the auxiliary variables do not appear in the formula for the π^* estimator. We now consider explicit use of auxiliary variables, in the form of difference estimators and regression estimators for two-phase sampling. Auxiliary values may now be of two kinds: (i) values obtained by observing the elements in the first-phase sample s_a, that is, values that appear in the frame used for the second phase; (ii) values available at the outset for all N elements of the population U, that is, values given in the initial frame. As mentioned, a central idea in two-phase sampling is that significant auxiliary information of type (i) can be obtained. In many applications, there is no relevant information of type (ii).

The notation will be as follows:

a. Let $\mathbf{x}_k$ be a vector of J auxiliary values available for all $k \in s_a$.
b. Let $\mathbf{x}_{1k}$ be the vector of J_1 auxiliary values available for all k in the population U.

We assume that $\mathbf{x}_k$ contains variable values known beforehand for all of U as well as variable values known for $k \in s_a$ only. In other words, we can write

$$\mathbf{x}_k = (\mathbf{x}'_{1k}, \mathbf{x}'_{2k})'.$$

where $\mathbf{x}_{1k}$ is the vector of J_1 values known for all of U, if at all available, and $\mathbf{x}_{2k}$ is the vector of $J_2 = J - J_1$ values recorded by (relatively inexpensive) observation of elements k in the first-phase sample s_a only. In the regression estimators to be presented, $\mathbf{x}_k$ serves to get predicted y-values, $\hat{y}_k$ from s to s_a, and $\mathbf{x}_{1k}$ is used to obtain predictions $\hat{y}_{1k}$ from s_a to U. Quantities known, predicted, or observed can now be summarized as follows.

Set of elements	Known auxiliary vector values	Observed study variable values	Predicted study variable values
Population, U	$\mathbf{x}_{1k}$	—	$\hat{y}_{1k}$
First-phase sample, s_a	$\mathbf{x}_k = (\mathbf{x}'_{1k}, \mathbf{x}'_{2k})'$	—	$\hat{y}_k$, $\hat{y}_{1k}$
Second-phase sample, s	$\mathbf{x}_k = (\mathbf{x}'_{1k}, \mathbf{x}'_{2k})'$	y_k	$\hat{y}_k$

The goal in the following sections is to improve on the π^* estimator, assuming that the designs in each of the two phases are fixed. The extent to which the variance is reduced by the use of regression estimation will depend on (a) the strength of $\mathbf{x}_2$ as a predictor of y, and (b) the strength of $\mathbf{x}_1$ as a predictor of y. Different cases can be distinguished.

The typical case in a survey that uses two-phase sampling is that $\mathbf{x}_2$ is a strong predictor of y, whereas $\mathbf{x}_1$ is a weak predictor of y. At the planning stage, the survey statistician uses his judgement to identify x variables that are likely to explain y well, and these variables will form the vector $\mathbf{x}_2$, whose value, $\mathbf{x}_{2k}$, is then observed for the elements $k \in s_a$.

When the first phase consists of a sample selection and the second phase corresponds to a subselection imposed by nonresponse, it is also important to distinguish the two categories of auxiliary values, $\mathbf{x}_{1k}$ and $\mathbf{x}_{2k}$. The vector $\mathbf{x}_{2k}$ then contains the x variable values that are known or made available for all elements designated for the sample, respondents as well as nonrespondents. Only respondents provide the value y_k. The $\mathbf{x}_{1k}$ are values known for the whole population. For reasons that become clear in Chapter 15, it is important that $\mathbf{x}_2$ in particular, but preferably also $\mathbf{x}_1$, are strong predictors of y. Special effort and expense may be required to obtain the values $\mathbf{x}_{2k}$, but their presence improves the inference significantly.

We start with some simply derived results on difference estimators in two-phase sampling (Section 9.6); they provide natural stepping stones to the regression estimators (Sections 9.7 and 9.8).

9.6. Difference Estimators

Let $\mathbf{x}_k$ (available for $k \in s_a$) and $\mathbf{x}_{1k}$ (available for $k \in U$) be the predictors defined in the preceding section. Suppose that the approximate linear relationships

$$y_k \doteq \mathbf{x}_k'\mathbf{A} = y_k^0 \tag{9.6.1}$$

and

$$y_k \doteq \mathbf{x}_{1k}'\mathbf{A}_1 = y_{1k}^0 \tag{9.6.2}$$

hold, where $\mathbf{A}$ (of dimension J) and $\mathbf{A}_1$ (of dimension J_1) are known vectors. Here, y_k^0 and y_{1k}^0 may be called *proxy values* for the element k. Form the corresponding differences

$$D_k = y_k - y_k^0 \tag{9.6.3a}$$

which are defined for $k \in s_a$, and

$$D_{1k} = y_k - y_{1k}^0 \tag{9.6.3b}$$

which are defined for $k \in U$. We expect these differences to be numerically small, in particular the D_k, because $\mathbf{x}_k$ is assumed to be a stronger predictor than $\mathbf{x}_{1k}$. Both sets of differences can contribute significantly toward improving on the weighting-only estimator t_{π^*} given by equation (9.3.5). The differences are used as follows to construct a new estimator of $t = \sum_U y_k$. Consider first

$$\sum_U y_{1k}^0 + \sum_{s_a} \frac{y_k - y_{1k}^0}{\pi_{ak}} = \sum_U y_{1k}^0 + \sum_{s_a} \frac{D_{1k}}{\pi_{ak}} \tag{9.6.4}$$

This unbiased difference estimator could be used if the y_k were observed for all $k \in s_a$. But this is not the case. For the unknown term $\sum_{s_a} y_k/\pi_{ak}$ in (9.6.4), we therefore substitute the conditionally unbiased difference estimator

$$\sum_{s_a} \frac{y_k^0}{\pi_{ak}} + \sum_s \frac{y_k - y_k^0}{\pi_k^*} = \sum_{s_a} \frac{y_k^0}{\pi_{ak}} + \sum_s \frac{D_k}{\pi_k^*} \tag{9.6.5}$$

which makes good use of the known proxies y_k^0. The result, obtained by two rounds of differencing, is

$$\hat{t}_{\text{dif}} = \sum_U y_{1k}^0 + \sum_{s_a} \frac{y_k^0 - y_{1k}^0}{\pi_{ak}} + \sum_s \frac{y_k - y_k^0}{\pi_k^*}$$

$$= \sum_U y_{1k}^0 + \sum_{s_a} \frac{D_{1k} - D_k}{\pi_{ak}} + \sum_s \frac{D_k}{\pi_k^*} \tag{9.6.6}$$

As is easily seen, the estimation error of this difference estimator can be written

$$\hat{t}_{\text{dif}} - t = \underbrace{\left(\sum_{s_a} \frac{D_{1k}}{\pi_{ak}} - \sum_U D_{1k}\right)}_{Q_{Ds_a}} + \underbrace{\left(\sum_s \frac{D_k}{\pi_k^*} - \sum_{s_a} \frac{D_k}{\pi_{ak}}\right)}_{R_{Ds}} \tag{9.6.7}$$

Compare this expression with its analog (9.3.4) for the simple π^* estimator. Whereas both error terms in equation (9.3.4) are in terms of raw scores y_k, the two error components in equation (9.6.7) are expressed in the usually less-variable quantities D_{1k} and D_k, respectively. A two-faceted improvement over $\hat{t}_{\pi^*}$ is thus expected. We arrive by analogy at the following result.

Result 9.6.1. *In two-phase sampling, the difference estimator*

$$\hat{t}_{\text{dif}} = \sum_U y_{1k}^0 + \sum_{s_a} \frac{y_k^0 - y_{1k}^0}{\pi_{ak}} + \sum_s \frac{y_k - y_k^0}{\pi_k^*} \tag{9.6.8}$$

is unbiased for the population total $t = \sum_U y_k$. Its variance is given by

$$V(\hat{t}_{\text{dif}}) = \sum\sum_U \Delta_{akl} \frac{D_{1k}}{\pi_{ak}} \frac{D_{1l}}{\pi_{al}} + E_{p_a}\left(\sum\sum_{s_a} \Delta_{kl|s_a} \check{\check{D}}_k \check{\check{D}}_l\right) \tag{9.6.9}$$

where $\check{\check{D}}_k = D_k/\pi_k^*$. The variance is unbiasedly estimated by

$$\hat{V}(\hat{t}_{\text{dif}}) = \sum\sum_s \frac{\Delta_{akl}}{\pi_{kl}^*} \frac{D_{1k}}{\pi_{ak}} \frac{D_{1l}}{\pi_{al}} + \sum\sum_s \frac{\Delta_{kl|s_a}}{\pi_{kl|s_a}} \check{\check{D}}_k \check{\check{D}}_l \tag{9.6.10}$$

Each component of (9.6.10) *is unbiased for its counterpart in* (9.6.9).

The proof is very similar to the proof of Result 9.3.1. We have from equation (9.6.7)

$$E(\hat{t}_{\text{dif}} - t) = E_{p_a} E(\hat{t}_{\text{dif}} - t | s_a)$$
$$= E_{p_a}(Q_{Ds_a}) = 0$$

which establishes the unbiasedness. For the variance we have

$$V(\hat{t}_{\text{dif}}) = V(\hat{t}_{\text{dif}} - t) = V_{p_a} E(\hat{t}_{\text{dif}} - t | s_a) + E_{p_a} V(\hat{t}_{\text{dif}} - t | s_a)$$
$$= V_{p_a}(Q_{Ds_a}) + E_{p_a} V(R_{Ds} | s_a)$$

and the unconditional variance (9.6.9) follows from familiar π estimation principles. Verification of the unbiasedness of the variance estimator (9.6.10) is left as an exercise.

Remark 9.6.1. Note that $\hat{t}_{\text{dif}}$ also can be written

$$\hat{t}_{\text{dif}} = \sum_s \frac{y_k}{\pi_k^*} + \left(\sum_U y_{1k}^0 - \sum_{s_a} \frac{y_{1k}^0}{\pi_{ak}}\right) + \left(\sum_{s_a} \frac{y_k^0}{\pi_{ak}} - \sum_s \frac{y_k^0}{\pi_k^*}\right) \tag{9.6.11}$$

That is, $\hat{t}_{\text{dif}}$ equals the unbiased π^* estimator, $\hat{t}_{\pi^*} = \sum_s y_k/\pi_k^*$, plus two unbiased estimators of zero, corresponding to the two levels of auxiliary information. When the procedure works well, each zero-estimator will have a strong negative correlation with $\hat{t}_{\pi^*}$. The effect of the zero-estimators is usually a reduction in the variance of $\hat{t}_{\pi^*}$.

Although the difference estimator shown in equation (9.6.8) was conceived for situations with two sources of auxiliary information ($\mathbf{x}_{1k}$ for $k \in U$ on the one hand and $\mathbf{x}_{2k}$ for $k \in s_a$ on the other), it also applies to the special cases that occur when one of these sources of information is absent.

Case 1. Suppose that the auxiliary information $\mathbf{x}_{2k}$ gathered for $k \in s_a$ is used to full advantage in the sampling design for the second phase. Then there is little point in having the estimator also depend on $\mathbf{x}_{2k}$. However, the values $\mathbf{x}_{1k}$ available for $k \in U$ may bring some improvement. In Result 9.6.1 we then have

$$\mathbf{x}_k = (\mathbf{x}_{1k}', \mathbf{x}_{2k}')' = \mathbf{x}_{1k}$$

and

$$y_k^0 = y_{1k}^0; \qquad D_k = D_{1k}$$

The difference estimator (9.6.8) reduces to

$$\hat{t}_{\text{dif}1} = \sum_U y_{1k}^0 + \sum_s \frac{y_k - y_{1k}^0}{\pi_k^*} \tag{9.6.12}$$

The differences D_{1k} will replace the D_k in the expressions (9.6.9) and (9.6.10).

Case 2. Suppose that there is no useful $\mathbf{x}_{1k}$ information, but that the information on $\mathbf{x}_{2k}$ gathered for the elements $k \in s_a$ will bring some improvement. With $\mathbf{x}_{1k}$ missing, we set

$$\mathbf{x}_k = (\mathbf{x}_{1k}', \mathbf{x}_{2k}')' = \mathbf{x}_{2k}$$

in Result 9.6.1. In equation (9.6.11), the first zero-estimator vanishes, and the difference estimator takes the form

$$\hat{t}_{\text{dif}2} = \sum_{s_a} \frac{y_k^0}{\pi_{ak}} + \sum_s \frac{y_k - y_k^0}{\pi_k^*} \tag{9.6.13}$$

It is not hard to verify that the estimation error of $\hat{t}_{\text{dif}2}$ is now given by

$$\hat{t}_{\text{dif}2} - t = \left(\sum_{s_a} \frac{y_k}{\pi_{ak}} - \sum_U y_k\right) + \left(\sum_s \frac{D_k}{\pi_k^*} - \sum_{s_a} \frac{D_k}{\pi_{ak}}\right) \tag{9.6.14}$$

with the variance

$$V(\hat{t}_{\text{dif}2}) = \sum\sum_U \Delta_{akl}\check{y}_{ak}\check{y}_{al} + E_{p_a}(\sum\sum_{s_a} \Delta_{kl|s_a}\check{\check{D}}_k\check{\check{D}}_l) \tag{9.6.15}$$

and the unbiased variance estimator

$$\hat{V}(\hat{t}_{\text{dif}2}) = \sum\sum_s \frac{\Delta_{akl}}{\pi_{kl}^*}\check{y}_{ak}\check{y}_{al} + \sum\sum_s \frac{\Delta_{kl|s_a}}{\pi_{kl|s_a}}\check{\check{D}}_k\check{\check{D}}_l \tag{9.6.16}$$

The first-phase component remains undifferenced in this case.

Two simple examples will illustrate the difference estimator in case 2.

EXAMPLE 9.6.1. There is a single auxiliary variable x. That is, $J_2 = 1$. Let $y_k^0 = Ax_k$ and $D_k = y_k - Ax_k$, where A is a known constant. The difference estimator (9.6.13) can be written

$$\begin{aligned}\hat{t}_{\text{dif}2} &= \sum_s \frac{y_k}{\pi_k^*} + A\left(\sum_{s_a} \frac{x_k}{\pi_{ak}} - \sum_s \frac{x_k}{\pi_k^*}\right) \\ &= \hat{t}_{\pi^*} + A[\hat{t}_{x\pi_a} - \hat{t}_{x\pi^*}] \end{aligned} \tag{9.6.17}$$

If the first-phase design is SI, with n_a elements drawn from N, and if the second-phase design is SI, with n elements drawn from n_a, then

$$\hat{t}_{\text{dif}2} = N[\bar{y}_s + A(\bar{x}_{s_a} - \bar{x}_s)] \tag{9.6.18}$$

EXAMPLE 9.6.2. There is a single auxiliary variable x. Let $y_k^0 = A_1 + A_2x_k$ and $D_k = y_k - A_1 - A_2x_k$, where the constants A_1 and A_2 are known. In this case the difference estimator (9.6.13) can be written

$$\hat{t}_{\text{dif}2} = \hat{t}_{\pi^*} + A_1(\hat{N}_{\pi_a} - \hat{N}_{\pi^*}) + A_2(\hat{t}_{x\pi_a} - \hat{t}_{x\pi^*}) \tag{9.6.19}$$

where

$$\hat{N}_{\pi_a} = \sum_{s_a} 1/\pi_{ak}; \qquad \hat{N}_{\pi^*} = \sum_s 1/\pi_k^*$$

With SI sampling in both phases, as in Example 9.6.1, the estimator (9.6.19) is simplified to $\hat{t}_{\text{dif}2} = N[\bar{y}_s + A_2(\bar{x}_{s_a} - \bar{x}_s)]$.

9.7. Regression Estimators for Two-Phase Sampling

We present a general approach to regression estimation for two-phase sampling following Särndal and Swensson (1987). The regression estimators bear close structural resemblance to the difference estimators seen in the preceding section. As usual, the relationship between the study variable and the auxiliary variables is described by a regression model, one for each level of auxiliary information. The model parameters are then estimated from current sample data, and the resulting predicted values become essential elements in building the regression estimator.

Our starting point is as in Section 9.2, with a general sampling design in each of the two phases. We consider two sources of auxiliary values. Let $\mathbf{x}_{1k}$ be known for all $k \in U$, and let $\mathbf{x}_{2k}$ be a value obtained by observation for elements k in the first-phase sample s_a. For an element $k \in s_a$, the complete information is thus summarized in the vector

$$\mathbf{x}_k' = (\mathbf{x}_{1k}', \mathbf{x}_{2k}') \tag{9.7.1}$$

although in practice the complete vector $(\mathbf{x}_{1k}', \mathbf{x}_{2k}')$ may not be used for prediction. The study variable value y_k is observed for the elements k in the second-phase sample s.

The difference estimator given by (9.6.8) of Result 9.6.1 suggests the regression estimator

$$\hat{t}_r = \sum_U \hat{y}_{1k} + \sum_{s_a} \frac{\hat{y}_k - \hat{y}_{1k}}{\pi_{ak}} + \sum_s \frac{y_k - \hat{y}_k}{\pi_k^*} \tag{9.7.2}$$

where $\hat{y}_k$ and $\hat{y}_{1k}$ are predicted values obtained from appropriate regression fits.

Our immediate task is to specify the necessary predictions, starting at the bottom level with $\hat{y}_k$, then proceeding to the top level with $\hat{y}_{1k}$.

Bottom Level

The bottom level predictions $\hat{y}_k$ will be calculated for $k \in s_a$ and are based on the predictor vector $\mathbf{x}_k$ available for $k \in s_a$. A model, denoted ξ, describes the point scatter $(y_k, \mathbf{x}_k)$ in the finite population in the following way,

$$\begin{cases} E_\xi(y_k) = \mathbf{x}_k'\boldsymbol{\beta} \\ V_\xi(y_k) = \sigma_k^2 \end{cases} \tag{9.7.3}$$

If the y_k-values were known for the whole set s_a, an estimator of the unknown $\boldsymbol{\beta}$ vector could be formed, at the level of s_a, namely,

$$\mathbf{B}_{s_a} = \left(\sum_{s_a} \frac{\mathbf{x}_k\mathbf{x}_k'}{\sigma_k^2 \pi_{ak}}\right)^{-1} \sum_{s_a} \frac{\mathbf{x}_k y_k}{\sigma_k^2 \pi_{ak}} \tag{9.7.4}$$

and one could obtain the residuals

$$E_k = y_k - \mathbf{x}_k'\mathbf{B}_{s_a}, \qquad k \in s_a \tag{9.7.5}$$

What we can actually calculate from the available data is the regression coefficient vector

$$\hat{\mathbf{B}}_s = \left(\sum_s \frac{\mathbf{x}_k\mathbf{x}_k'}{\sigma_k^2 \pi_k^*}\right)^{-1} \sum_s \frac{\mathbf{x}_k y_k}{\sigma_k^2 \pi_k^*} \tag{9.7.6}$$

the predicted values

$$\hat{y}_k = \mathbf{x}_k'\hat{\mathbf{B}}_s \qquad \text{for} \quad k \in s_a \tag{9.7.7}$$

and the residuals

$$e_{ks} = y_k - \hat{y}_k \qquad \text{for} \quad k \in s \tag{9.7.8}$$

Top Level

The top level predictions $\hat{y}_{1k}$ are calculated for $k \in U$. As input, use the information $\mathbf{x}_{1k}$ for $k \in U$, and y_k for $k \in s$. Through a new model, denoted ξ_1, we try to capture the essence of the point scatter $(y_k, \mathbf{x}_{1k})$, $k = 1, \ldots, N$. This new model assumes that

$$\begin{cases} E_{\xi_1}(y_k) = \mathbf{x}'_{1k}\boldsymbol{\beta}_1 \\ V_{\xi_1}(y_k) = \sigma^2_{1k} \end{cases} \tag{9.7.9}$$

The hypothetical population fit of this model would lead to the $\boldsymbol{\beta}_1$ estimator

$$\mathbf{B}_1 = \left(\sum\nolimits_U \frac{\mathbf{x}_{1k}\mathbf{x}'_{1k}}{\sigma^2_{1k}}\right)^{-1} \sum\nolimits_U \frac{\mathbf{x}_{1k}y_k}{\sigma^2_{1k}} \tag{9.7.10}$$

and the residuals

$$E_{1k} = y_k - \mathbf{x}'_{1k}\mathbf{B}_1 \tag{9.7.11}$$

The counterpart to $\mathbf{B}_1$, at the level of the first-phase sample, is

$$\hat{\mathbf{B}}_{1s_a} = \left(\sum\nolimits_{s_a} \frac{\mathbf{x}_{1k}\mathbf{x}'_{1k}}{\sigma^2_{1k}\pi_{ak}}\right)^{-1} \sum\nolimits_{s_a} \frac{\mathbf{x}_{1k}y_k}{\sigma^2_{1k}\pi_{ak}} \tag{9.7.12}$$

which, however, is impossible to compute, since the y_k values are known in s only. Stepping down to the level of s, we obtain quantities that can actually be calculated, namely, the regression coefficient vector

$$\hat{\mathbf{B}}_{1s} = \left(\sum\nolimits_{s} \frac{\mathbf{x}_{1k}\mathbf{x}'_{1k}}{\sigma^2_{1k}\pi^*_k}\right)^{-1} \sum\nolimits_{s} \frac{\mathbf{x}_{1k}y_k}{\sigma^2_{1k}\pi^*_k} \tag{9.7.13}$$

the predictions

$$\hat{y}_{1k} = \mathbf{x}'_{1k}\hat{\mathbf{B}}_{1s} \qquad \text{for} \quad k \in U \tag{9.7.14}$$

and the residuals

$$e_{1ks} = y_k - \hat{y}_{1k} \qquad \text{for} \quad k \in s \tag{9.7.15}$$

Remark 9.7.1. We can view (9.7.13) as a conditional estimator, given s_a, of the regression coefficient (9.7.12), which in turn estimates (9.7.10). An alternative estimator of (9.7.10) is proposed in Remark 9.7.2 below.

The specification of the regression estimator (9.7.2) is thus formally completed by saying that the predictions $\hat{y}_k$ and $\hat{y}_{1k}$ are calculated according to equations (9.7.7) and (9.7.14), respectively. The resulting estimator is not unbiased (only approximately so), and its variance will be given as an approximation. The estimator can alternatively be expressed in terms of g weights. There is one set of g weights for each phase. They are convenient for discussing the variance estimation. We define

$$g_{ks} = 1 + \left(\sum_{s_a} \frac{\mathbf{x}_k}{\pi_{ak}} - \sum_s \frac{\mathbf{x}_k}{\pi_k^*}\right)' \left(\sum_s \frac{\mathbf{x}_k \mathbf{x}_k'}{\sigma_k^2 \pi_k^*}\right)^{-1} \frac{\mathbf{x}_k}{\sigma_k^2} \tag{9.7.16}$$

for $k \in s$, and

$$g_{1ks_a} = 1 + \left(\sum_U \mathbf{x}_{1k} - \sum_{s_a} \frac{\mathbf{x}_{1k}}{\pi_{ak}}\right)' \left(\sum_{s_a} \frac{\mathbf{x}_{1k} \mathbf{x}_{1k}'}{\sigma_{1k}^2 \pi_{ak}}\right)^{-1} \frac{\mathbf{x}_{1k}}{\sigma_{1k}^2} \tag{9.7.17}$$

for $k \in s_a$.

One can then show that the estimation error of the estimator determined by (9.7.2), (9.7.7), and (9.7.14) is

$$\hat{t}_r - t = \left(\sum_{s_a} \frac{g_{1ks_a} y_k}{\pi_{ak}} - \sum_U y_k\right) + \left(\sum_s \frac{g_{ks} y_k}{\pi_k^*} - \sum_{s_a} \frac{y_k}{\pi_{ak}}\right) + \Delta \tag{9.7.18}$$

where

$$\Delta = \left(\sum_U \mathbf{x}_{1k} - \sum_{s_a} \mathbf{x}_{1k}/\pi_{ak}\right)'(\hat{\mathbf{B}}_{1s} - \hat{\mathbf{B}}_{1s_a}) \tag{9.7.19}$$

The essential properties of the new regression estimator are stated in the following result.

Result 9.7.1. *In two-phase sampling, an approximately unbiased estimator of the population total $t = \sum_U y_k$ is given by*

$$\hat{t}_r = \sum_U \hat{y}_{1k} + \sum_{s_a} \frac{\hat{y}_k - \hat{y}_{1k}}{\pi_{ak}} + \sum_s \frac{y_k - \hat{y}_k}{\pi_k^*} \tag{9.7.20}$$

where $\hat{y}_k$ and $\hat{y}_{1k}$ are determined, respectively, by equations (9.7.7) *and* (9.7.14). *The approximate variance is given by*

$$AV(\hat{t}_r) = \sum\sum_U \Delta_{akl} \check{E}_{1k} \check{E}_{1l} + E_{p_a}\left\{\sum\sum_{s_a} \Delta_{kl|s_a} \check{\check{E}}_k \check{\check{E}}_l\right\} \tag{9.7.21}$$

where $\check{E}_{1k} = E_{1k}/\pi_{ak}$ and $\check{\check{E}}_k = E_k/\pi_k^*$. The residuals E_k and E_{1k} are defined by (9.7.5) *and* (9.7.11), *respectively. The variance is estimated by*

$$\hat{V}(\hat{t}_r) = \sum\sum_s \frac{\Delta_{akl}}{\pi_{kl}^*} g_{1ks_a} \check{e}_{1ks} g_{1ls_a} \check{e}_{1ls} + \sum\sum_s \frac{\Delta_{kl|s_a}}{\pi_{kl|s_a}} g_{ks} \check{\check{e}}_{ks} g_{ls} \check{\check{e}}_{ls} \tag{9.7.22}$$

where $\check{e}_{1ks} = e_{1ks}/\pi_{ak}$ and $\check{\check{e}}_{ks} = e_{ks}/\pi_k^*$ and e_{1ks}, g_{1ks_a}, e_{ks}, and g_{ks} are given, respectively, by (9.7.15), (9.7.17), (9.7.8), *and* (9.7.16).

A simplified variance estimator is obtained by setting the g weights equal to unity for all k.

Proof. The g weights shown in equations (9.7.16) and (9.7.17) have the properties

$$\sum_s \frac{g_{ks}\mathbf{x}_k'}{\pi_k^*} = \sum_{s_a} \frac{\mathbf{x}_k'}{\pi_{ak}}$$

and

$$\sum_{s_a} \frac{g_{1ks_a}\mathbf{x}_{1k}'}{\pi_{ak}} = \sum_U \mathbf{x}_{1k}'$$

We obtain, from (9.7.18), the following expression for the error of the estimator,

$$\hat{t}_r - t = \left(\sum_{s_a} \frac{g_{1ks_a}E_{1k}}{\pi_{ak}} - \sum_U E_{1k}\right) + \left(\sum_s \frac{g_{ks}E_k}{\pi_k^*} - \sum_{s_a} \frac{E_k}{\pi_{ak}}\right) + \Delta$$

where E_k and E_{1k} are given, respectively, by (9.7.5) and (9.7.11). Approximating $g_{ks} \doteq 1$, $g_{1ks_a} \doteq 1$, and dropping the term Δ (which is small compared to the two terms that precede), we get

$$\hat{t}_r - t \doteq \underbrace{\left(\sum_{s_a} \frac{E_{1k}}{\pi_{ak}} - \sum_U E_{1k}\right)}_{Q_{Es_a}} + \underbrace{\left(\sum_s \frac{E_k}{\pi_k^*} - \sum_{s_a} \frac{E_k}{\pi_{ak}}\right)}_{R_{Es}} \qquad (9.7.23)$$

The right-hand side of (9.7.23), which equals $Q_{Es_a} + R_{Es}$, has the same structure as the decomposition (9.6.7), obtained earlier for the difference estimator, but with E_{1k} replacing D_{1k} and E_k replacing D_k. It follows from Result 9.6.1 that the linear random variable on the right-hand side of (9.7.23) has the expected value zero and the variance (9.7.21), which mimics (9.6.9). We now make the assumption that the nonlinear random variable $\hat{t}_r - t$ on the left-hand side of (9.7.23) behaves approximately as the linear variable on the right-hand side. It follows that $\hat{t}_r$ is approximately unbiased and that the variance is given approximately by (9.7.21). The variance estimator given by (9.7.22) is obtained by replacing E_k and E_{1k} by the sample-based counterparts, e_{ks} and e_{1ks}, and by applying the g weights to these residuals. □

Remark 9.7.2. The regression estimator (9.7.20) can be modified by replacing $\hat{\mathbf{B}}_{1s}$ given by (9.7.13) by an alternative estimator of $\hat{\mathbf{B}}_{1s_a}$,

$$\hat{\mathbf{B}}_{1s}^* = \left(\sum_{s_a} \frac{\mathbf{x}_{1k}\mathbf{x}_{1k}'}{\sigma_{1k}^2\pi_{ak}}\right)^{-1} \left[\sum_{s_a} \frac{\mathbf{x}_{1k}\hat{y}_k}{\sigma_{1k}^2\pi_{ak}} + \sum_s \frac{\mathbf{x}_{1k}(y_k - \hat{y}_k)}{\sigma_{1k}^2\pi_k^*}\right]$$

where $\hat{y}_k$ is the predictive value given by (9.7.7). In this approach, the predictions $\hat{y}_k$ arising out of the bottom level fit are used to estimate the vector

$$\sum_{s_a} \frac{\mathbf{x}_{1k}y_k}{\sigma_{1k}^2\pi_{ak}}$$

which appears in the vector $\hat{\mathbf{B}}_{1s_a}$ given by (9.7.12). Result 9.7.1 continues to hold if we change the predicted values and the residuals in accordance with the following:

$$\hat{y}_{1k} = \mathbf{x}'_{1k}\hat{\mathbf{B}}^*_{1s}$$

and

$$e_{1ks} = y_k - \mathbf{x}'_{1k}\hat{\mathbf{B}}^*_{1s}$$

while g_{1ks_a} is as before. The efficiency is expected to be about the same whether $\hat{\mathbf{B}}_{1s}$ or $\hat{\mathbf{B}}^*_{1s}$ is used.

Before giving a few examples, let us discuss two special cases of the regression estimator shown in equation (9.7.20).

Case 1. No new x variables are recorded for $k \in s_a$. This case corresponds to Case 1 in the discussion of the difference estimator, Section 9.6. There is no vector $\mathbf{x}_{2k}$ to be used in the estimator. In equations (9.7.4) to (9.7.8), which summarize the bottom level fit, we then have

$$\mathbf{x}_k = (\mathbf{x}'_{1k}, \mathbf{x}'_{2k})' = \mathbf{x}_{1k}$$

Assuming that $\sigma_k^2 = \sigma_{1k}^2$, it follows that $\hat{y}_k = \hat{y}_{1k}$, so the middle term of (9.7.20) vanishes, leaving simply

$$\hat{t}_{r1} = \sum_U \hat{y}_{1k} + \sum_s \frac{y_k - \hat{y}_{1k}}{\pi_k^*} \tag{9.7.24}$$

The approximate variance given by (9.7.21) and the variance estimator given by (9.7.22) continue to apply if we set $\mathbf{x}_k = \mathbf{x}_{1k}$ in the expressions for E_k, e_{ks}, and g_{ks}.

Case 2. Here we suppose, as in case 2 for the difference estimator (see Section 9.6), that no useful $\mathbf{x}_{1k}$ information exists, making it impossible to compute the predictions $\hat{y}_{1k}$. This is ordinarily the case when a two-phase sampling design is contemplated. The fit of the model (9.7.3), summarized by equations (9.7.4) to (9.7.8), is now based solely on

$$\mathbf{x}_k = (\mathbf{x}'_{1k}, \mathbf{x}'_{2k})' = \mathbf{x}_{2k}$$

where $\mathbf{x}_{2k}$ is the information gathered for $k \in s_a$. The predictions $\hat{y}_{1k}$ drop out of the estimator formula; that is, the estimator (9.7.20) becomes

$$\hat{t}_{r2} = \sum_{s_a} \frac{\hat{y}_k}{\pi_{ak}} + \sum_s \frac{y_k - \hat{y}_k}{\pi_k^*} \tag{9.7.25}$$

One can show that the estimation error of t_{r2} can be written

$$\hat{t}_{r2} - t = \left(\sum_{s_a} \frac{y_k}{\pi_{ak}} - \sum_U y_k\right) + \left(\sum_s g_{ks}\frac{E_k}{\pi_k^*} - \sum_{s_a} \frac{E_k}{\pi_{ak}}\right) \tag{9.7.26}$$

which leads to the approximate variance

$$AV(\hat{t}_{r2}) = \sum\sum_U \Delta_{akl}\check{y}_{ak}\check{y}_{al} + E_{p_a}(\sum\sum_{s_a} \Delta_{kl|s_a}\check{\check{E}}_k\check{\check{E}}_l) \tag{9.7.27}$$

and the variance estimator

$$\hat{V}(\hat{t}_{r2}) = \sum\sum_s \frac{\Delta_{akl}}{\pi_{kl}^*} \check{y}_{ak}\check{y}_{al} + \sum\sum_s \frac{\Delta_{kl|s_a}}{\pi_{kl|s_a}} g_{ks}\check{e}_{ks} g_{ls}\check{e}_{ls} \tag{9.7.28}$$

This resembles the case of limited auxiliary information in regression estimation for two-stage sampling, treated as case C in Section 8.9.

EXAMPLE 9.7.1. Regression through the origin. Assume that y is well explained by a single x variable through the model ξ such that

$$\begin{cases} E_\xi(y_k) = \beta x_k \\ V_\xi(y_k) = \sigma^2 x_k \end{cases} \tag{9.7.29}$$

The bottom level predictions given by (9.7.7) are now

$$\hat{y}_k = \hat{B}_s x_k$$

with

$$\hat{B}_s = \frac{\sum_s \check{\check{y}}_k}{\sum_s \check{\check{x}}_k} \tag{9.7.30}$$

where $\check{\check{y}}_k = y_k/\pi_k^*$, and analogously for $\check{\check{x}}_k$. The residuals are

$$e_{ks} = y_k - \hat{B}_s x_k \tag{9.7.31}$$

and the g weights are given by

$$g_{ks} = \frac{\sum_{s_a} \check{x}_{ak}}{\sum_s \check{\check{x}}_k} \tag{9.7.32}$$

with $\check{x}_{ak} = x_k/\pi_{ak}$. When x_k is observed for $k \in s_a$ (and no further auxiliary information exists), we have Case 2. From equation (9.7.25), the estimator of $t = \sum_U y_k$ is

$$\hat{t}_{r2} = \left(\sum_{s_a} \check{x}_{ak}\right) \frac{\sum_s \check{\check{y}}_k}{\sum_s \check{\check{x}}_k} = \left(\sum_{s_a} \check{x}_{ak}\right) \hat{B}_s \tag{9.7.33}$$

and the variance estimator is given by (9.7.28) with the residuals shown in (9.7.31) and the g weights shown in (9.7.32).

In particular, if the SI design is used in both phases, with n_a elements selected from N in phase one, and n from n_a in phase two, then

$$\hat{t}_{r2} = N\bar{x}_{s_a}(\bar{y}_s/\bar{x}_s)$$

and the variance estimator (9.7.28) can be written

$$\hat{V}(\hat{t}_{r2}) = N^2\left(1 - \frac{n_a}{N}\right)\frac{S_{ys}^2}{n_a} + N^2\left(1 - \frac{n}{n_a}\right)\left(\frac{\bar{x}_{s_a}}{\bar{x}_s}\right)^2 \frac{S_{es}^2}{n}$$

where

$$S_{es}^2 = \frac{1}{n-1}\sum_s \left(y_k - \frac{\bar{y}_s}{\bar{x}_s}x_k\right)^2; \qquad S_{ys}^2 = \frac{1}{n-1}\sum_s (y_k - \bar{y}_s)^2$$

EXAMPLE 9.7.2. Returning to the preceding example, suppose that the model (9.7.29) adequately describes the relation between y and x. In addition, we invoke a simple top level model, namely,

$$\begin{cases} E_{\xi_1}(y_k) = \beta_1 \\ V_{\xi_1}(y_k) = \sigma_1^2 \end{cases}$$

corresponding to $x_{1k} = 1$ for all k. Result 9.7.1 gives

$$\hat{t}_r = (N - \hat{N}_{\pi_a})\tilde{y}_s + \hat{t}_{r2}$$

where $\hat{t}_{r2}$ is given by equation (9.7.33) and

$$\hat{N}_{\pi_a} = \sum_{s_a} 1/\pi_{ak}; \qquad \tilde{y}_s = \frac{\sum_s \check{y}_k}{\sum_s 1/\pi_k^*}$$

This estimator requires that N be known, but for first-phase designs such that $\hat{N}_{\pi_a} = N$ (for example, the *SI* design), we have $\hat{t}_r = \hat{t}_{r2}$. Under the alternative approach of Remark 9.7.2, the estimator is

$$\hat{t}_r = N\tilde{x}_{s_a}\hat{B}_s \tag{9.7.34}$$

with

$$\tilde{x}_{s_a} = \frac{\sum_{s_a} x_k/\pi_{ak}}{\sum_{s_a} 1/\pi_{ak}}$$

Dividing equation (9.7.34) by N leads to an estimator of the population mean $\bar{y}_U$, namely,

$$\hat{\bar{y}}_{Ur} = \tilde{x}_{s_a}\hat{B}_s$$

For the variance estimator (9.7.22), we need (9.7.31) and (9.7.32), as well as

$$e_{1ks} = y_k - \hat{\bar{y}}_{Ur}$$

and

$$g_{1ks_a} = \frac{N}{\hat{N}_{\pi_a}}$$

9.8. Stratified Bernoulli Sampling in Phase Two

We now consider stratified Bernoulli sampling in phase two. One reason to examine this design is its connection with the theory of nonresponse. In stratified Bernoulli sampling, the inclusion probabilities are constant for all elements in a stratum. We do not require that they be known. In a survey with nonresponse, this corresponds to response probabilities that are unknown but assumed constant within groups. The results in this section will be used later in Chapter 15.

As in Example 9.4.1, a first-phase sample s_a (drawn by an arbitrary design) is stratified into H_{s_a} strata s_{ah}, $h = 1, \ldots, H_{s_a}$. Stratified Bernoulli sampling is then used to draw the subsample. That is, for each element $k \in s_{ah}$, a Bernoulli experiment is carried out to decide whether the element is to be included or not in the second-phase sample s_h. The probability of inclusion is fixed at θ_h for every element k in the stratum h. The Bernoulli experiments are independent, and the probabilities θ_h may vary between strata. For elements in stratum s_{ah} we have

$$\pi_{k|s_a} = \theta_h$$

and $\Delta_{kl|s_a} = 0$, whether k and l $(k \neq l)$ belong to different strata s_{ah} or not.

From Result 9.4.1 we obtain the π^* estimator

$$\hat{t}_{\pi^*} = \sum_{h=1}^{H_{s_a}} \theta_h^{-1} \sum\nolimits_{s_h} \check{y}_{ak} \tag{9.8.1}$$

with the variance

$$V(\hat{t}_{\pi^*}) = \sum\sum\nolimits_U \Delta_{akl} \check{y}_{ak} \check{y}_{al} + E_{p_a}\left[\sum_{h=1}^{H_{s_a}} (\theta_h^{-1} - 1) \sum\nolimits_{s_{ah}} \check{y}_{ak}^2\right]$$

for which we have the unbiased estimator

$$\hat{V}(\hat{t}_{\pi^*}) = \sum\sum\nolimits_s \frac{\Delta_{akl}}{\pi_{kl}^*} \check{y}_{ak} \check{y}_{al} + \sum_{h=1}^{H_{s_a}} \theta_h^{-1}(\theta_h^{-1} - 1) \sum\nolimits_{s_h} \check{y}_{ak}^2$$

However, we know from Section 3.2 that π expansion is not particularly efficient when the sample size is random, as is the case here. One can thus expect the variance contribution of stratum h, given s_a,

$$(\theta_h^{-1} - 1) \sum\nolimits_{s_{ah}} \check{y}_{ak}^2$$

to be large, and we circumvent this problem in the following approach which uses estimates of the θ_h rather than the θ_h themselves. This approach is of great value when the second phase is a selection caused by nonresponse, as in Chapter 15. In that case, the θ_h correspond to unknown response probabilities, so they must be estimated. Given s_h and $n_h \geq 1$,

$$\hat{t}_{hc} = \frac{n_{ah}}{n_h} \sum\nolimits_{s_h} \check{y}_{ak} \tag{9.8.2}$$

is conditionally unbiased for $\sum_{s_{ah}} \check{y}_{ak}$. The reason is that, given s_a and a fixed vector $\mathbf{n} = (n_1, \ldots, n_h, \ldots, n_{H_{s_a}})$ with $n_h \geq 1$ for all h, the second-phase sampling is equivalent to an *STSI* selection with n_h elements chosen from n_{ah} in stratum h. It is left as an exercise to prove this.

Equation (9.8.2) applies as long as $n_h \geq 1$. To be completely covered, a separate value of $\hat{t}_{hc}$ must be defined when $n_h = 0$. For example, we can define $\hat{t}_{hc}$ as 0 if $n_h = 0$. Or a sample dependent value may be specified, for example, the sample mean in a stratum deemed similar to stratum h. In practice, the probability of $n_h = 0$ will ordinarily be very small, and it is of little consequence how the definition is made. The resulting bias is negligible.

In the following derivations we act as if the sample s_a were so large that the probability, given s_a, of the event

$$\bar{A}_1 = \{n_h = 0 \text{ for some } h = 1, \ldots, H_{s_a}\}$$

is negligible. The estimator

$$\hat{t}_c = \sum_{h=1}^{H_{s_a}} \hat{t}_{hc} = \sum_{h=1}^{H_{s_a}} \frac{n_{ah}}{n_h} \sum\nolimits_{s_h} \check{y}_{ak} \tag{9.8.3}$$

is then unbiased for t, because

$$E(\hat{t}_c) = E_{p_a} E_{\mathbf{n}} E_c(\hat{t}_c) = E_{p_a} E_{\mathbf{n}} \left[\sum_{h=1}^{H_{s_a}} E_c \left(\frac{n_{ah}}{n_h} \sum\nolimits_{s_h} \check{y}_{ak} \right) \right]$$

$$= E_{p_a} E_{\mathbf{n}} \left(\sum_{h=1}^{H_{s_a}} \sum\nolimits_{s_{ah}} \check{y}_{ak} \right) = E_{p_a}(\textstyle\sum_{s_a} \check{y}_{ak}) = \sum_U y_k = t$$

where $E_c(\cdot)$ denotes conditional expectation given s_a and $\mathbf{n}$, and $E_{\mathbf{n}}(\cdot)$ denotes conditional expectation, given s_a, over all realizations $\mathbf{n}$ obeying $\sum_{h=1}^{H_{s_a}} n_h = n_s$.

Moreover, we can use the results in Example 9.4.1 to obtain the variance

$$V(\hat{t}_c) = V_{p_a} E_{\mathbf{n}} E_c(\hat{t}_c) + E_{p_a} V_{\mathbf{n}} E_c(\hat{t}_c) + E_{p_a} E_{\mathbf{n}} V_c(\hat{t}_c)$$

$$= V_{p_a}(\textstyle\sum_{s_a} \check{y}_{ak}) + 0 + E_{p_a} E_{\mathbf{n}} \left(\displaystyle\sum_{h=1}^{H_{s_a}} n_{ah}^2 \frac{1 - f_h}{n_h} S_{\check{y}s_{ah}}^2 \right)$$

$$= \sum\sum\nolimits_U \Delta_{akl} \check{y}_{ak} \check{y}_{al} + E_{p_a} E_{\mathbf{n}} \left(\sum_{h=1}^{H_{s_a}} n_{ah}^2 \frac{1 - f_h}{n_h} S_{\check{y}s_{ah}}^2 \right) \tag{9.8.4}$$

where $V_c(\cdot)$ denotes variance conditionally on s_a and $\mathbf{n}$. In addition, suppose that the event

$$\bar{A}_2 = \{n_h \leq 1 \text{ for some } h = 1, \ldots, H_{s_a}\}$$

has negligible probability, given s_a. An unbiased variance estimator is then given by

$$\hat{V}(\hat{t}_c) = \sum\sum\nolimits_s \frac{\Delta_{akl}}{\pi_{kl}^*} \check{y}_{ak} \check{y}_{al} + \sum_{h=1}^{H_{s_a}} n_{ah}^2 \frac{1 - f_h}{n_h} S_{\check{y}s_h}^2 \tag{9.8.5}$$

where $\pi_{kl}^* = \pi_{ak} \pi_{kl|s_a}$ with $\pi_{kl|s_a}$ given by (9.4.4). Note that (9.8.5) agrees with the variance estimator for $\hat{t}_{\pi^*}$ given by equation (9.4.8) in Example 9.4.1.

9.9. Sampling on Two Occasions

In many surveys, the same population is sampled repeatedly and the same study variable is measured at each occasion, so that development over time can be followed. For example, in many countries, labor-force surveys are conducted monthly to estimate the number of employed and the rate of unemployment. Other examples are monthly surveys in which data on price of goods are collected to determine a consumer price index, and political opinion surveys conducted at regular intervals to measure voter preferences. Special techniques are used for repeated surveys. Here, we examine sampling on two occasions. A key issue is the extent to which elements sampled at a previous occasion should be retained in the sample selected at the current occasion. The optimal "matching proportion" depends on the parameter under estimation. The matching problem has been studied by a number of authors; an early reference is Patterson (1950). Our treatment is in terms of general sampling designs.

We consider sampling on two occasions from a finite population $U = \{1, \ldots, k, \ldots, N\}$ assumed to be composed of the *same* elements at two different occasions. The study variable, for example, unemployment or net household income, is observed at each occasion, but not necessarily for the same set of elements. The study variable will be denoted z at the first occasion and y at the second.

At the first occasion, a sample s_a is drawn by the sampling design $p_a(\cdot)$, and the variable z is measured for all elements in s_a. The inclusion probabilities associated with the design are denoted π_{ak} and π_{akl}. We set $\Delta_{akl} = \pi_{akl} - \pi_{ak}\pi_{al}$. We assume here that the π estimator

$$\hat{t}_{zs_a} = \sum_{s_a} z_k/\pi_{ak}$$

is used as an estimator of the total $t_z = \sum_U z_k$. To the sample s_a drawn at the first occasion corresponds a complement sample, $s_a^c = U - s_a$. The complement sample is not surveyed at the first occasion, but we need the probabilities of inclusion in the complement sample induced by the design $p_a(\cdot)$. We denote by π_{ak}^c the probability that k is an element of s_a^c and by π_{akl}^c the probability that both k and l are elements of s_a^c. Also, set $\Delta_{akl}^c = \pi_{akl}^c - \pi_{ak}^c\pi_{al}^c$. Then

$$\pi_{ak}^c = 1 - \pi_{ak}$$

$$\pi_{akl}^c = 1 - \pi_{ak} - \pi_{al} + \pi_{akl}$$

$$\Delta_{akl}^c = \Delta_{akl}$$

What sampling design should be chosen at the second occasion? In single-occasion surveys the choice depends on several factors, such as the parameters to estimate, the available auxiliary information, cost and measurement considerations, and so on. When the same study variable is observed in repeated surveys, the interest usually lies in estimating both *parameters of level* and *parameters of change*. A number of parameters of interest are of the form

$$t = \phi t_z + \psi t_y$$

where $t_z = \sum_U z_k$, $t_y = \sum_U y_k$, and ϕ and ψ are constants. For example, (a) $\phi = 0$, $\psi = 1$ leads to $t = t_y$, *the current total*, which is a parameter of level; (b) $\phi = -1$, $\psi = 1$ leads to $t = t_y - t_z$, *the absolute change*; (c) $\phi = -1/t_z$ $\psi = 1/t_z$ leads to $t = (t_y - t_z)/t_z$, *the relative change*; and (d) $\phi = 1$, $\psi = 1$ leads to $t = t_y + t_z$, which is *the sum of the totals* over the two occasions for the characteristic under study.

In choosing a design for the second occasion, we have more information than at the first occasion: For every $k \in s_a$, we know the value z_k. For the new design, we can consider no overlap, complete overlap, or partial overlap with the first sample s_a. As this section shows, different parameters have different optimal sampling designs at the second occasion. It is intuitively clear that there are cases in which the information from the first occasion may be used to improve the estimation. Hence, we opt for partial overlap. At the second occasion, two independent samples are drawn, one *matched sample* and one *unmatched sample*. The matched sample, denoted s_m, is drawn from s_a by the design $p_m(\cdot|s_a)$. The unmatched sample, denoted s_u, is drawn from s_a^c according to the design $p_u(\cdot|s_a^c)$ and is independent of s_m. The quantities

$$\pi_{k|s_a}, \quad \pi_{kl|s_a}, \quad \Delta_{kl|s_a} = \pi_{kl|s_a} - \pi_{k|s_a}\pi_{l|s_a}$$

are associated with $p_m(\cdot|s_a)$, and

$$\pi_{k|s_a^c}, \quad \pi_{kl|s_a^c}, \quad \Delta_{kl|s_a^c} = \pi_{kl|s_a^c} - \pi_{k|s_a^c}\pi_{l|s_a^c}$$

are the analogous quantities for $p_u(\cdot|s_a^c)$. The variable y is observed for all elements in s_m and s_u. The total sample at the second occasion is thus $s = s_m \cup s_u$.

9.9.1. Estimating the Current Total

In many cases, there is good reason to assume that y_k is well approximated by $y_k^0 = Kz_k$, where K is a known constant. The value of K may be suggested by a preceding study or by subject matter theory. Using the first sample s_a, the matched sample s_m and the differences $D_k = y_k - y_k^0$, we can form an unbiased difference estimator of the current total, t_y, namely,

$$\hat{t}_1 = \hat{t}_{y^0 s_a} + \hat{t}_{D s_m} \tag{9.9.1}$$

where

$$\hat{t}_{y^0 s_a} = \sum_{s_a} \frac{y_k^0}{\pi_{ak}} \quad \text{and} \quad \hat{t}_{D s_m} = \sum_{s_m} \frac{D_k}{\pi_{ak}\pi_{k|s_a}}$$

A second unbiased estimator of the current total can be formed from the unmatched sample, namely

$$\hat{t}_2 = \hat{t}_{y s_u} = \sum_{s_u} \frac{y_k}{\pi_{ak}^c \pi_{k|s_a^c}} \tag{9.9.2}$$

It is easy to show that both $\hat{t}_1$ and $\hat{t}_2$ are unbiased for the current total. By linear combination we obtain the new unbiased estimator

$$\hat{t}_y = w_1 \hat{t}_1 + w_2 \hat{t}_2 \tag{9.9.3}$$

where w_1 and w_2 are nonnegative constant weights to be determined and such that $w_1 + w_2 = 1$. We call $\hat{t}_y$ a *composite estimator*; it combines the matched sample estimator with the unmatched sample estimator. The optimal choice of $w_1 = 1 - w_2$ will be considered later. First, we give the following result, where $V_1 = V(\hat{t}_1)$, $V_2 = V(\hat{t}_2)$, and $C = C(\hat{t}_1, \hat{t}_2)$.

Result 9.9.1. *The variance of the composite estimator* (9.9.3) *is given by*

$$V(\hat{t}_y) = w_1^2 V_1 + w_2^2 V_2 + 2w_1 w_2 C \tag{9.9.4}$$

where

$$V_1 = \sum\sum_U \Delta_{akl} \frac{y_k}{\pi_{ak}} \frac{y_l}{\pi_{al}} + E\left(\sum\sum_{s_a} \Delta_{kl|s_a} \frac{D_k}{\pi_{ak}\pi_{k|s_a}} \frac{D_l}{\pi_{al}\pi_{l|s_a}}\right) \tag{9.9.5}$$

$$V_2 = \sum\sum_U \Delta_{akl}^c \frac{y_k}{\pi_{ak}^c} \frac{y_l}{\pi_{al}^c} + E\left(\sum\sum_{s_a^c} \Delta_{kl|s_a^c} \frac{y_k}{\pi_{ak}^c\pi_{k|s_a^c}} \frac{y_l}{\pi_{al}^c\pi_{l|s_a^c}}\right) \tag{9.9.6}$$

and

$$C = -\sum\sum_U \Delta_{akl} \frac{y_k}{\pi_{ak}} \frac{y_l}{\pi_{al}^c} \tag{9.9.7}$$

Here, the expectation E in equations (9.9.5) and (9.9.6) is with respect to $p_a(\cdot)$. The proof is left as an exercise.

EXAMPLE 9.9.1. The expressions in Result 9.9.1 take simple forms when simple random selection is used throughout. Assume that s_a is an *SI* sample of size n from U, which implies that the complement s_a^c is an *SI* sample of size $N - n$ from U. Let $f = n/N$. Furthermore, assume that s_m is an *SI* sample of size $m = \mu n$ from s_a, and that s_u is an *SI* sample of size $u = n - m = (1 - \mu)n = \nu n$ from s_a^c. Here, the quantity $\mu = 1 - \nu$ is called the *matching proportion*. Then

$$\hat{t}_1 = N\bar{y}_{s_a}^0 + N\bar{D}_{s_m} = N[\bar{y}_{s_m} + K(\bar{z}_{s_a} - \bar{z}_{s_m})] \tag{9.9.8}$$

whereas

$$\hat{t}_2 = N\bar{y}_{s_u} \tag{9.9.9}$$

It follows easily that

$$V_1 = N^2\left(\frac{1-f}{n} S_{yU}^2 + \frac{1-\mu}{\mu n} S_{DU}^2\right) \tag{9.9.10}$$

$$V_2 = N^2 \frac{1-\nu f}{\nu n} S_{yU}^2 \tag{9.9.11}$$

and

$$C = -NS_{yU}^2 \tag{9.9.12}$$

The optimal matching proportion and the optimal value of K will be determined later.

We now determine $w_1 = 1 - w_2$ and K in the composite estimator (9.9.3) so as to minimize its variance. We obtain

$$\begin{aligned} V(\hat{t}_y) &= w_1^2 V_1 + w_2^2 V_2 + 2w_1 w_2 C \\ &= (V_1 + V_2 - 2C)\left\{w_1 - \frac{V_2 - C}{V_1 + V_2 - 2C}\right\}^2 + \frac{V_1 V_2 - C^2}{V_1 + V_2 - 2C} \\ &\geq \frac{V_1 V_2 - C^2}{V_1 + V_2 - 2C} = V(\hat{t}_y)_{\min} \end{aligned} \tag{9.9.13}$$

because $V_1 + V_2 - 2C = V(\hat{t}_1 - \hat{t}_2) > 0$. Equality holds if and only if

$$w_1 = 1 - w_2 = \frac{V_2 - C}{V_1 + V_2 - 2C} \tag{9.9.14}$$

The minimal variance

$$V(\hat{t}_y)_{\min} = \frac{V_1 V_2 - C^2}{V_1 + V_2 - 2C} \tag{9.9.15}$$

is a strictly increasing function of V_1, and V_1 is the only component of (9.9.15) that depends on K. To minimize (9.9.15) is equivalent to finding the K-value which minimizes V_1. Let

$$\hat{t}_{ys_m} = \sum\nolimits_{s_m} \frac{y_k}{\pi_{ak}\pi_{k|s_a}}, \quad \hat{t}_{zs_m} = \sum\nolimits_{s_m} \frac{z_k}{\pi_{ak}\pi_{k|s_a}} \quad \text{and} \quad \hat{t}_{zs_a} = \sum\nolimits_{s_a} \frac{z_k}{\pi_{ak}}$$

Then $\hat{t}_1$ can be written

$$\hat{t}_1 = \hat{t}_{ys_m} + K(\hat{t}_{zs_a} - \hat{t}_{zs_m}) \tag{9.9.16}$$

and an alternative expression for its variance is

$$\begin{aligned} V(\hat{t}_1) &= V(\hat{t}_{ys_m}) + K^2 V(\hat{t}_{zs_a} - \hat{t}_{zs_m}) + 2KC(\hat{t}_{ys_m}, \hat{t}_{zs_a} - \hat{t}_{zs_m}) \\ &= V(\hat{t}_{ys_m}) + K^2 EV(\hat{t}_{zs_m}|s_a) - 2KEC(\hat{t}_{ys_m}, \hat{t}_{zs_m}|s_a) \\ &= V(\hat{t}_{ys_m}) + EV(\hat{t}_{zs_m}|s_a)\left\{K - \frac{EC(\hat{t}_{ys_m}, \hat{t}_{zs_m}|s_a)}{EV(\hat{t}_{zs_m}|s_a)}\right\}^2 - \frac{[EC(\hat{t}_{ys_m}, \hat{t}_{zs_m}|s_a)]^2}{EV(\hat{t}_{zs_m}|s_a)} \\ &\geq V(\hat{t}_{ys_m}) - \frac{[EC(\hat{t}_{ys_m}, \hat{t}_{zs_m}|s_a)]^2}{EV(\hat{t}_{zs_m}|s_a)} = V(\hat{t}_1)_{\min} \end{aligned} \tag{9.9.17}$$

Here, the operators V and C are associated with $p_m(\cdot|s_a)$, and the expectation E is associated with $p_a(\cdot)$. Thus, $EV(\cdot|s_a)$ is shorthand for the quantities written earlier as $E_{p_a}V(\cdot|s_a)$.

Equality holds if and only if

$$K = \frac{EC(\hat{t}_{ys_m}, \hat{t}_{zs_m}|s_a)}{EV(\hat{t}_{zs_m}|s_a)} = K_{\text{opt}} \tag{9.9.18}$$

This is a complex expression in general, but, in special cases, it can be easily evaluated, as in the following example.

EXAMPLE 9.9.2. Let us return to Example 9.9.1. Let $r = r_{yzU} = S_{yzU}/(S_{yU}S_{zU})$, which is the correlation coefficient between y and z in the population U. Then

$$K_{\text{opt}} = rS_{yU}/S_{zU} \tag{9.9.19}$$

and with $K = K_{\text{opt}}$, equation (9.9.10) becomes

$$V_1 = N^2 \frac{S_{yU}^2}{\mu n}[(1 - r^2) + \mu(r^2 - f)] \tag{9.9.20}$$

Furthermore,

$$V_2 = N^2 \frac{S_{yU}^2}{\mu n} \frac{\mu(1 - \nu f)}{\nu} \tag{9.9.21}$$

and

$$C = N^2 \frac{S_{yU}^2}{\mu n}(-\mu f) \tag{9.9.22}$$

The minimal variance given by (9.9.15) can, after some tedious algebra, be written

$$V_{\min} = V(\hat{t}_y)_{\min} = N^2 \frac{S_{yU}^2}{n}\left\{\frac{1 - \nu r^2}{1 - \nu^2 r^2} - f\right\} \tag{9.9.23}$$

It is simple in this case to obtain the optimal matching proportion $\mu = 1 - \nu$. Differentiating $V_{\min}$ with respect to ν and equating to zero, we get the following optimal proportions:

$$\nu_{\text{opt}} = 1/[1 + (1 - r^2)^{1/2}] \tag{9.9.24}$$

and

$$\mu_{\text{opt}} = \frac{(1 - r^2)^{1/2}}{1 + (1 - r^2)^{1/2}} \tag{9.9.25}$$

If V_{opt} denotes the value of $V_{\min}$ when $\nu = \nu_{\text{opt}}$, we have

$$V_{\text{opt}} = V(\hat{t}_y)_{\text{opt}} = N^2 \frac{S_{yU}^2}{n}\left(\frac{1 - \nu_{\text{opt}} r^2}{1 - \nu_{\text{opt}}^2 r^2} - f\right) \tag{9.9.26}$$

$$= N^2 \frac{S_{yU}^2}{n}\left[\frac{1}{2\nu_{\text{opt}}} - f\right] = N^2 \frac{S_{yU}^2}{n}\left[\frac{1 + (1 - r^2)^{1/2}}{2} - f\right] \tag{9.9.27}$$

Naturally, we want to know if matching produces a gain in precision. If there is no matching and the π estimator is used for the current total, the variance is

$$V(\hat{t}_\pi) = N^2 \frac{S_{yU}^2}{n}(1 - f)$$

assuming an *SI* sample of n from N. The relative variance reduction due to matching (which is a measure of the gain in precision) is expressed by

Table 9.1. Relative Variance Reduction Due to Matching, and Optimal Matching Proportion for Selected Values of r^2 and $f = n/N$.

r^2	Optimum matching proportion, %	$100[1 - V_{\text{opt}}/V(\hat{t}_\pi)]$ for $f =$				
		0.4	0.2	0.1	0.01	0
0.5	41	24	18	16	15	15
0.6	39	31	23	20	19	18
0.7	35	38	28	25	23	23
0.8	31	46	35	31	28	28
0.9	24	57	43	38	35	34
0.95	18	65	49	43	39	39
0.99	9	75	56	50	45	45
0.999	3	81	61	54	49	48

$$1 - \frac{V_{\text{opt}}}{V(\hat{t}_\pi)} = 1 - \frac{(1/2v_{\text{opt}}) - f}{1 - f} = \frac{1 - (1/2v_{\text{opt}})}{1 - f}$$

Table 9.1. shows this relative variance reduction, as well as the optimum matching proportion, for selected values of r^2 and $f = n/N$. For the values of r^2 in Table 9.1 the optimum matching proportion never exceeds approximately 40%, and it drops markedly for values of r^2 close to unity. For sampling fractions f of 10% or less, the reduction in variance lies roughly in the range 15% to 50%. For larger values of f, the reduction can be much greater.

Remark 9.9.1. It is interesting to note (but far from trivial to verify) that an alternative derivation of the optimal estimator of the current total is obtained by the following argument. Start with a linear combination of four unbiased estimators of population totals,

$$\hat{t}_y = \alpha\hat{t}_{zs_a} + \beta\hat{t}_{zs_m} + \gamma\hat{t}_{ys_m} + \delta\hat{t}_{ys_u} \tag{9.9.28}$$

where α, β, γ, and δ are constants to be determined. If the estimator (9.9.28) is to be unbiased for the current total t_y we must have $\beta = -\alpha$ and $\delta = 1 - \gamma$, that is,

$$\hat{t}_y = \alpha(\hat{t}_{zs_a} - \hat{t}_{zs_m}) + \gamma(\hat{t}_{ys_m} - \hat{t}_{ys_u}) + \hat{t}_{ys_u} \tag{9.9.29}$$

Next, find the variance $V(\hat{t}_y)$. It is left as an exercise to show that the equations $\partial V(\hat{t}_y)/\partial\alpha = 0$ and $\partial V(\hat{t}_y)/\partial\gamma = 0$ lead to the optimal values

$$\alpha_{\text{opt}} = \gamma_{\text{opt}} \frac{EC(\hat{t}_{ys_m}, \hat{t}_{zs_m} | s_a)}{EV(\hat{t}_{zs_m} | s_a)} \tag{9.9.30}$$

and

$$\gamma_{\text{opt}} = \frac{V(\hat{t}_{ys_u}) - C(\hat{t}_{ys_a}, \hat{t}_{ys_a^c})}{V(\hat{t}_{ys_m} - \hat{t}_{ys_u}) - \dfrac{[EC(\hat{t}_{ys_m}, \hat{t}_{zs_m} | s_a)]^2}{EV(\hat{t}_{zs_m} | s_a)}} \tag{9.9.31}$$

where

$$\hat{t}_{ys_a} = \sum_{s_a} \frac{y_k}{\pi_{ak}} \quad \text{and} \quad \hat{t}_{ys_a^c} = \sum_{s_a^c} \frac{y_k}{\pi_{ak}^c}$$

It is also left as an exercise to show that *SI* sampling, as in Examples 9.9.1 and 9.9.2, yields

$$\alpha_{\text{opt}} = r \frac{S_{yU}}{S_{zU}} \frac{\mu}{1 - v^2 r^2} \quad \text{and} \quad \gamma_{\text{opt}} = \frac{\mu}{1 - v^2 r^2}$$

When these values are inserted into $V(\hat{t}_y)$, we finally confirm the results given in Example 9.9.2.

If a good K-value cannot be specified in advance in the difference estimator in equation (9.9.1), a regression estimator can be used instead. Suppose that the model

$$\begin{cases} E_\xi(y_k) = \alpha + \beta z_k \\ V_\xi(y_k) = \sigma^2 \end{cases}$$

for $k \in U$ is a good description of the relation between y and z. From the matched sample, construct the regression estimator

$$\hat{t}_{1r} = \hat{t}_{ys_m} + \hat{A}(\hat{N}_{s_a} - \hat{N}_{s_m}) + \hat{B}(\hat{t}_{zs_a} - \hat{t}_{zs_m}) \tag{9.9.32}$$

where

$$\hat{N}_{s_a} = \sum_{s_a} \frac{1}{\pi_{ak}}, \quad \hat{N}_{s_m} = \sum_{s_m} \frac{1}{\pi_{ak}\pi_{k|s_a}}, \quad \hat{A} = \tilde{y}_{s_m} - \hat{B}\tilde{z}_{s_m}$$

and

$$\hat{B} = \frac{\sum_{s_m} (z_k - \tilde{z}_{s_m})(y_k - \tilde{y}_{s_m})/\pi_{ak}\pi_{k|s_a}}{\sum_{s_m} (z_k - \tilde{z}_{s_m})^2/\pi_{ak}\pi_{k|s_a}}$$

with

$$\tilde{z}_{s_m} = \hat{t}_{zs_m}/\hat{N}_{s_m} \quad \text{and} \quad \tilde{y}_{s_m} = \hat{t}_{ys_m}/\hat{N}_{s_m}$$

Now, $\hat{t}_{1r}$ is approximately unbiased for t_y, and its approximate variance is given by

$$AV_1 = \sum\sum_U \Delta_{akl} \frac{y_k}{\pi_{ak}} \frac{y_l}{\pi_{al}} + E\left\{\sum\sum_{s_a} \Delta_{kl|s_a} \frac{E_k}{\pi_{ak}\pi_{k|s_a}} \frac{E_l}{\pi_{al}\pi_{l|s_a}}\right\} \tag{9.9.33}$$

where

$$E_k = y_k - \tilde{y}_{s_a} - \hat{B}_{s_a}(z_k - \tilde{z}_{s_a}), \quad \tilde{y}_{s_a} = \hat{t}_{ys_a}/\hat{N}_{s_a}, \quad \tilde{z}_{s_a} = \hat{t}_{zs_a}/\hat{N}_{s_a}$$

and

$$\hat{B}_{s_a} = \frac{\sum_{s_a} (z_k - \tilde{z}_{s_a})(y_k - \tilde{y}_{s_a})/\pi_{ak}}{\sum_{s_a} (z_k - \tilde{z}_{s_a})^2/\pi_{ak}}$$

The unmatched sample is utilized as before. That is, we keep the unbiased estimator $\hat{t}_2 = \hat{t}_{ys_u}$ given by (9.9.2). Linear combination of $\hat{t}_{1r}$ and $\hat{t}_2$ leads to

$$\hat{t}_{yr} = \omega_1 \hat{t}_{1r} + \omega_2 \hat{t}_2 \tag{9.9.34}$$

where $\omega_1 + \omega_2 = 1$. This new estimator is approximately unbiased with the approximate variance

$$AV(\hat{t}_{yr}) = \omega_1^2 AV_1 + \omega_2^2 V_2 + 2\omega_1\omega_2 AC \tag{9.9.35}$$

where the expressions for AV_1, $V_2 = V(\hat{t}_2)$, and $AC = AC(\hat{t}_{1r}, \hat{t}_2) = C$ are given by equations (9.9.33), (9.9.6), and (9.9.7), respectively.

A derivation analogous to (9.9.13) shows that the approximate variance in (9.9.35) is minimized by

$$\omega_1 = 1 - \omega_2 = \frac{V_2 - AC}{AV_1 + V_2 - 2AC} \tag{9.9.36}$$

which gives the minimum approximate variance

$$AV(\hat{t}_{yr})_{\min} = \frac{(AV_1)V_2 - (AC)^2}{AV_1 + V_2 - 2AC} \tag{9.9.37}$$

EXAMPLE 9.9.3. Suppose that SI sampling is used as in Examples 9.9.1 and 9.9.2. Then $\hat{t}_{1r}$ in the estimator (9.9.34) is given by

$$\hat{t}_{1r} = N[\bar{y}_{s_m} + \hat{B}(\bar{z}_{s_a} - \bar{z}_{s_m})]$$

where

$$\hat{B} = S_{zys_m}/S^2_{zs_m}$$

and $AV_1 = V_1$, V_2, and $AC = C$ are given by (9.9.20) to (9.9.22). This leads to

$$AV(\hat{t}_{yr})_{\min} = N^2 \frac{S^2_{yU}}{n}\left\{\frac{1 - vr^2}{1 - v^2r^2} - f\right\} \tag{9.9.37a}$$

This approximate variance is the same as the variance shown in equation (9.9.23). Thus, the optimal matching proportion is given in this case, too, by (9.9.25), and Table 9.1 shows the variance reduction.

9.9.2. Estimating the Previous Total

The data collected for the sample at the second occasion can also be used to improve on the original estimator of the previous total t_{zU}. Instead of the π estimator $\hat{t}_{z\pi}$, consider

$$\hat{t}_z = \alpha\hat{t}_{zs_a} + \beta\hat{t}_{zs_m} + \gamma\hat{t}_{ys_m} + \delta\hat{t}_{ys_u} \tag{9.9.38}$$

where α, β, γ, and δ are constants to be determined. If the estimator (9.9.38) is to be unbiased for the previous total t_z we must have $\alpha = 1 - \beta$ and $\gamma = -\delta$, that is,

$$\hat{t}_z = \hat{t}_{zs_a} + \beta(\hat{t}_{zs_m} - \hat{t}_{zs_a}) + \delta(\hat{t}_{ys_u} - \hat{t}_{ys_m}) \tag{9.9.39}$$

The optimal values of β and δ are found by equating the partial derivatives of $V(\hat{t}_z)$ with respect to β and δ to zero. This leads to

$$\beta_{\text{opt}} = \delta_{\text{opt}} \frac{EC(\hat{t}_{ys_m}, \hat{t}_{zs_m} | s_a)}{EV(\hat{t}_{zs_m} | s_a)} \tag{9.9.40}$$

and

$$\delta_{\text{opt}} = \frac{C(\hat{t}_{zs_a}, \hat{t}_{ys_a}) - C(\hat{t}_{zs_a}, \hat{t}_{ys_a^c})}{V(\hat{t}_{ys_m} - \hat{t}_{ys_u}) - \dfrac{[EC(\hat{t}_{ys_m}, \hat{t}_{zs_m} | s_a)]^2}{EV(\hat{t}_{zs_m} | s_a)}} \tag{9.9.41}$$

EXAMPLE 9.9.4. Using *SI* sampling as in Example 9.9.1, we get

$$\beta_{\text{opt}} = \frac{\mu v r^2}{1 - v^2 r^2} \quad \text{and} \quad \delta_{\text{opt}} = \frac{S_{zU}}{S_{yU}} \frac{\mu v r}{1 - v^2 r^2} \tag{9.9.42}$$

and

$$V(\hat{t}_z)_{\min} = N^2 \frac{S_{zU}^2}{n} \left\{ \frac{1 - v r^2}{1 - v^2 r^2} - f \right\} \tag{9.9.43}$$

The expression within curly brackets is the same as in equation (9.9.23). The optimal matching proportion is therefore the same as for estimating the current total.

9.9.3. Estimating the Absolute Change and the Sum of the Totals

To estimate the absolute change $\Delta = t_y - t_z$, we can form the two estimators

$$\hat{\Delta}_1 = \hat{t}_{1r} - \hat{t}_{zs_a} \tag{9.9.44}$$

and

$$\hat{\Delta}_2 = \hat{t}_2 - \hat{t}_{zs_a} \tag{9.9.45}$$

where $\hat{t}_{1r}$ is the regression estimator of the current total given by (9.9.32) and where $\hat{t}_2 = \hat{t}_{s_u}$ is given by (9.9.2). Linear combination of $\hat{\Delta}_1$ and $\hat{\Delta}_2$ gives

$$\hat{\Delta} = \delta_1 \hat{\Delta}_1 + \delta_2 \hat{\Delta}_2 \tag{9.9.46}$$

where $\delta_1 + \delta_2 = 1$. This is an approximately unbiased estimator of the absolute change Δ. The approximate variance is given by

$$AV(\hat{\Delta}) = \delta_1^2 AV(\hat{\Delta}_1) + \delta_2^2 V(\hat{\Delta}_2) + 2AC(\hat{\Delta}_1, \hat{\Delta}_2) \tag{9.9.47}$$

The reasoning used earlier in this section leads directly to

$$\delta_{1\text{opt}} = 1 - \delta_{2\text{opt}} = \frac{V(\hat{\Delta}_2) - AC(\hat{\Delta}_1, \hat{\Delta}_2)}{AV(\hat{\Delta}_1) + V(\hat{\Delta}_2) - 2AC(\hat{\Delta}_1, \hat{\Delta}_2)} \tag{9.9.48}$$

and we have

$$AV(\hat{\Delta})_{\min} = \frac{AV(\hat{\Delta}_1)V(\hat{\Delta}_2) - [AC(\hat{\Delta}_1, \hat{\Delta}_2)]^2}{AV(\hat{\Delta}_1) + V(\hat{\Delta}_2) - 2AC(\hat{\Delta}_1, \hat{\Delta}_2)} \tag{9.9.49}$$

EXAMPLE 9.9.5. Again, consider SI sampling as in Example 9.9.1. Then

$$\hat{\Delta}_1 = N[\bar{y}_{s_m} + \hat{B}(\bar{z}_{s_a} - \bar{z}_{s_m}) - \bar{z}_{s_a}] \quad \text{and} \quad \hat{\Delta}_2 = N(\bar{y}_{s_u} - \bar{z}_{s_a})$$

and

$$AV(\hat{\Delta}_1) = \frac{N^2}{n}\left[\frac{S_{yU}^2}{\mu}(1 - \mu f - \nu r^2) + (1 - f)S_{zU}^2 - 2(1 - f)S_{zyU}\right]$$

$$V(\hat{\Delta}_2) = \frac{N^2}{n}\left[\frac{1 - \nu f}{\nu} S_{yU}^2 + (1 - f)S_{zU}^2 + 2fS_{zyU}\right]$$

and

$$AC(\hat{\Delta}_1, \hat{\Delta}_2) = \frac{N^2}{n}[(1 - f)(S_{zU}^2 - S_{zyU}) - f(S_{yU}^2 - S_{zyU})]$$

A simplifying assumption, often realistic, is that $S_{yU}^2 = S_{zU}^2 = S^2$. Then we get

$$AV(\hat{\Delta}_1) = \frac{N^2S^2}{n}\left[\frac{1}{\mu}(1 - \nu r^2) + 1 - 2r - 2f(1 - r)\right]$$

$$V(\hat{\Delta}_2) = \frac{N^2S^2}{n}\left[\frac{1}{\nu} + 1 - 2f(1 - r)\right]$$

and

$$AC(\hat{\Delta}_1, \hat{\Delta}_2) = \frac{N^2S^2}{n}[1 - r - 2f(1 - r)]$$

From (9.9.49), we now get

$$AV(\hat{\Delta})_{\min} = 2\frac{N^2S^2}{n}(1 - r)\left\{\frac{1}{1 - \nu r} - f\right\} \tag{9.9.50}$$

which is an increasing function of ν for every r such that $0 < r < 1$. Hence, if $0 < r < 1$, the optimum matching proportion is 100%; that is, the best policy for estimating the absolute change is to use the same sample at both occasions. By contrast, for estimating level (the total of y, for example), the optimal matching proportion seldom exceeds 40%, as Table 9.1 shows.

To estimate the sum of the totals, $T = t_y + t_z$, we form the estimators

$$\hat{T}_1 = \hat{t}_{1r} + \hat{t}_{zs_a} \tag{9.9.51}$$

and

$$\hat{T}_2 = \hat{t}_2 + \hat{t}_{zs_a} \tag{9.9.52}$$

where $\hat{t}_{1r}$ is the regression estimator of the current total t_y given by (9.9.32) and where $\hat{t}_2 = \hat{t}_{s_u}$ is the π estimator shown in equation (9.9.2).

The derivation of the best linear weights is left as an exercise. One can show that the approximate minimum variance is

$$AV_{\min} = 2\frac{N^2S^2}{n}\frac{1+r}{1+vr} \tag{9.9.53}$$

if we assume *SI* sampling, $S_{yU}^2 = S_{zU}^2 = S^2$ and $f = 0$. This implies that the optimum matching proportion is zero when $r > 0$. That is, the best policy for estimating the sum of the two totals is to draw a completely new sample at the second occasion.

Remark 9.9.2. A number of large-scale surveys are designed to measure population changes over time. Well-known examples are the labor force surveys conducted in many countries at regular, often monthly, intervals. These frequently use sample overlap in some form. Design and estimation for such surveys may require special methods, for example, the use of time-series analysis combined with design-based survey-sampling tools. We do not enter into this topic here; for recent overviews, the reader is referred to Duncan and Kalton (1987) and Binder and Hidiroglou (1988).

Exercises

9.1. Two-phase sampling from the MU284 population was carried out as follows to estimate the total of the variable REV84 ($= y$). (i) In the first phase, an *SI* sample s_a of size $n_a = 150$ was drawn. The 1975 population variable P75 ($= x$) was recorded for every municipality in s_a. (ii) After inspection of the data from the first phase, it was decided to draw a Poisson sample s in the second phase. The expected size was $n = 10$, and the inclusion probabilities were proportional to the value of P75. This led to the selection of 11 municipalities for which the recorded values of REV84 and P75 were as follows:

REV84	P75
2,653	33
17,949	247
1,060	12
1,324	12
2,223	18
2,553	30
2,216	20
13,205	138
3,475	35
7,072	62
4,623	47

(a) Compute an unbiased estimate of the total of REV84. (The total of P75 for the 150 municipalities in the first phase sample was 4,060.) (b) Show that an unbiased variance estimator is given by

$$\hat{V}(\hat{t}_{\pi*}) = \left(\frac{N}{n}\bar{x}_{s_a}\right)^2 \frac{1}{n_a - 1}\left[(n_a - f_a)\sum_s \left(\frac{y_k}{x_k}\right)^2 - (1 - f_a)\left(\sum_s \frac{y_k}{x_k}\right)^2\right]$$

$$- \frac{N}{n}\bar{x}_{s_a}\sum_s \frac{y_k^2}{x_k}$$

where

$$\bar{x}_{s_a} = \frac{1}{n_a}\sum_{s_a} x_k \quad \text{and} \quad f_a = \frac{n_a}{N}$$

(c) Use the expression in (b) to compute a variance estimate.

9.2. Two-phase sampling from the MU284 population was carried out as follows to estimate the total of the variable SS82 ($= y$). (i) In the first phase, an *SI* sample s_a of size $n_a = 160$ was drawn. The variable S82 ($= x$) was observed for every municipality in s_a. (ii) The selected 160 municipalities were partitioned into four strata of equal size, so that stratum 1 contained the 40 smallest municipalities (according to the value of the variable S82), stratum 2 the 40 smallest of the remaining 120 municipalities, and so on. From each stratum, an *SI* sample of size 20 was finally drawn, and the study variable SS82 was observed. The following results were obtained:

Stratum h	$\bar{y}_{s_h}$	$S^2_{ys_h}$
1	17.05	19.945
2	19.75	24.197
3	22.40	28.359
4	31.25	42.829

(a) Compute an unbiased estimate of the total of SS82. (b) Compute unbiased estimates of the two variance components V_1 and V_2 in equation (9.4.11). (c) Compute an approximately 95% confidence interval for the total of SS82. (d) Estimate the variance that would be obtained with a single-phase *SI* sample of size $n = 80$.

9.3. (Continuation of Exercise 9.2.). Suppose that the value of the variable S82 is known for every municipality in the MU284 population. Two-phase sampling is to be used as follows for the estimation of the total of the variable SS82. The municipalities are size-ordered according to the value of S82. The 71 smallest municipalities are placed in stratum 1, the 71 smallest of the remaining 213 municipalities in stratum 2, and so on. (i) In the first phase, an *SI* sample s_a of size n_a is to be drawn; that is, the information about S82 is disregarded in the first-phase sampling. (ii) Let s_{ah} be the set of elements in s_a which belong to stratum h, and let n_{ah} be the size of s_{ah}; that is, stratum membership is ascertained for every element in the realized first-phase sample. An *SI* sample of size

$$n_h = \nu_h n_{ah} = 0.5 n_{ah} \; (h = 1, 2, 3, 4)$$

is to be drawn. Determine the first-phase sample size n_a if the objective is to obtain an approximately 95% confidence interval for the total of SS82 with a width of 600, based on the π^* estimator shown in equation (9.4.10). Use the following assumptions (planning values) concerning the variances: $S_{yU}^2 = 55$, and

Stratum h	$S_{yU_h}^2$
1	20
2	20
3	25
4	50

9.4. The following two-phase procedure for difference estimation was used to estimate the total of the variable P85 ($= y$) for the MU200 population. (i) An *SI* sample s_a of $n_a = 150$ municipalities was drawn in the first phase, and the variable P75 ($= x$) was observed. (ii) An *SI* subsample s of size $n = 30$ was drawn, and the study variable P85 was observed. The following results were obtained:

$$\textstyle\sum_{s_a} x_k = 1{,}945; \qquad \sum_s x_k = 414; \qquad \sum_s y_k = 422$$
$$S_{ys}^2 = 52.8; \qquad S_{Ds}^2 = 3.5$$

where S_{Ds}^2 is the variance in s of $D_k = y_k - x_k$. (a) Derive an unbiased estimator of the variance of the two-phase difference estimator given by equation (9.6.17) with $A = 1$. (b) Use the result in (a) to compute an approximately 95% confidence interval for the total of P85.

9.5. Use the two-phase regression estimator given by (9.7.33) to obtain an approximately 95% confidence interval for the total of the variable RMT85 ($= y$) from the following sample data, based on *SI* samples in both phases, with $n_a = 100$ and $n = 50$ from the MU281 population. The auxiliary variable is P75 ($= x$).

$$\textstyle\sum_{s_a} x_k = 2{,}619; \qquad \sum_s x_k = 1{,}230; \qquad \sum_s y_k = 9{,}594$$
$$\textstyle\sum_s x_k y_k = 520{,}753; \qquad \sum_s x_k^2 = 64{,}078; \qquad \sum_s y_k^2 = 4{,}272{,}462$$

9.6. (Continuation of Exercise 9.5). After the calculation of the confidence interval in Exercise 9.5, suppose you find out that an additional auxiliary variable, CS82 ($= x_{1k}$), is available. The value of this variable is known for every municipality in the population. Improve on the already calculated confidence interval by the following approach. Use the information available about P75 ($= x$) and CS82 ($= x_{1k}$), assuming that

$$E_\xi(y_k) = \beta x_{1k}$$
$$V_\xi(y_k) = \sigma^2 x_{1k}$$

is a good description of the scatter of the points (y_k, x_{1k}). Use the following data

$$\textstyle\sum_U x_{1k} = 2{,}508; \qquad \sum_{s_a} x_{1k} = 907; \qquad \sum_s x_{1k} = 421$$

$$\textstyle\sum_s x_{1k} y_k = 115{,}897; \qquad \sum_s x_{1k}^2 = 4{,}791;$$

9.7. To study important properties of the estimator (9.7.33) under two-phase sampling with the *SI* design in both phases, a simulation study was carried out with repeated sampling from the MU200 population using RMT85 ($= y$) as the study variable and P75 ($= x$) as the auxiliary variable.

(i) From the $N = 200$ population elements, a first-phase *SI* sample s_a of size $n_a = 90$ was drawn. Data on x was collected for all elements $k \in s_a$. In the second phase, an *SI* sample s of size $n = 30$ was drawn from s_a, and data on y was collected for all elements $k \in s$. From the sample data, the following results were calculated: The point estimate $\hat{t}_{r2}$, the variance estimate $\hat{V}_1 = \hat{V}(\hat{t}_{r2})$ using the appropriate g weight, the simple variance estimate $\hat{V}_2 = \hat{V}(\hat{t}_{r2})$ based on $g_{ks} = 1$ for every $k \in s$, the upper (U) and lower (L) approximately 95% confidence limits

$$U_i = \hat{t}_{r2} + 1.96\hat{V}_i^{1/2} \quad \text{and} \quad L_i = \hat{t}_{r2} - 1.96\hat{V}_i^{1/2}; \qquad i = 1, 2$$

and, finally, the length $D_i = U_i - L_i$ of the corresponding confidence interval. Furthermore, it was observed whether, or not, the intervals covered the known population total $t_y = \sum_U y_k = 18{,}856$.

(ii) The procedure in (i) was repeated independently until $u = 5{,}000$ two-phase samples had been drawn. The following results were calculated from the simulation study.

$$\bar{\hat{t}}_{r2} = \frac{1}{u}\sum_{j=1}^{u} \hat{t}_{r2j} = 18{,}831$$

$$S^2(\hat{t}_{r2}) = \frac{1}{u-1}\sum_{j=1}^{u} (\hat{t}_{r2j} - \bar{\hat{t}}_{r2})^2 = 8.11 \cdot 10^5$$

$$\bar{\hat{V}}_1 = \frac{1}{u}\sum_{j=1}^{u} \hat{V}_{1j} = 7.84 \cdot 10^5 \quad \text{and} \quad \bar{\hat{V}}_2 = \frac{1}{u}\sum_{j=1}^{u} \hat{V}_{2j} = 7.86 \cdot 10^5$$

$$\bar{D}_1 = \frac{1}{u}\sum_{j=1}^{u} D_{1j} = 3{,}428 = \frac{1}{u}\sum_{j=1}^{u} D_{2j} = \bar{D}_2$$

$$S_{D_1} = 544 \quad \text{and} \quad S_{D_2} = 568$$

where S_{D_1} and S_{D_2} are the standard deviations of the $u = 5{,}000$ confidence interval lengths D_{1j} and D_{2j}, respectively. The coverage rate of the confidence intervals based on $\hat{V}_1$ and $\hat{V}_2$ was 93.1% and 93.0%, respectively. The approximate variance $AV(\hat{t}_{r2})$ computed from known population data is $7.92 \cdot 10^5$. Discuss the results of the simulation study.

9.8. Assume that *SI* sampling is used as in Examples 9.9.1 and 9.9.2, and that the current total is estimated by

$$\hat{t}_{yr} = \omega_1 \hat{t}_{1r} + \omega_2 \hat{t}_2$$

where $\hat{t}_{1r}$ is given in Example 9.9.3, and where $\hat{t}_2$ is given by equation (9.9.9). (a) Show that under the optimal choice of v, v_{opt}, the best weights ω_1 and ω_2 are always given by $\omega_1 = \omega_2 = 1/2$. (b) Because r^2 is unknown, the proportion of unmatched elements must in practice be determined with the aid of a "planning value" denoted r_p, which leads to the unmatched proportion

$$v_p = \frac{1}{1 + (1 - r_p^2)^{1/2}}$$

Using $\omega_1 = \omega_2 = 1/2$ and $v = v_p$, first show that

$$AV(\hat{t}_{yr}) = N^2 \frac{S_{yU}^2}{n} \left[\frac{1}{4} \cdot \frac{1 - v_p^2 r^2}{v_p(1 - v_p)} - f \right]$$

The percentage increase in the approximate variance of $\hat{t}_{yr}$ caused by the use of a nonoptimal value v_p is

$$100 \left[\frac{AV(\hat{t}_{yr})}{AV_{\text{opt}}} - 1 \right]$$

Confirm the values of this quantity given in the following table for the case $f = n/N = 0$:

	r_p^2					
r^2	0.60	0.70	0.80	0.90	0.95	0.99
0.70	0.4	0	0.7	5.5	15.2	64.8
0.80	1.9	0.6	0	1.9	7.7	41.7
0.90	6.0	3.7	1.5	0	1.5	17.8
0.95	10.8	7.8	4.6	1.1	0	6.2
0.99	20.4	16.6	12.3	6.7	3.1	0

(c) Consider the MU200 population. The total of the variable P75 ($= z$) was estimated from an *SI* sample s_a of size $n = 100$ drawn in 1975. Using the π estimator, the estimate was $200 \cdot 12.85 = 2{,}570$. To estimate the total of the variable P85 ($= y$) in 1985, it was decided to use the estimator in (b) based on an *SI* sample s_m of m municipalities from s_a, and an *SI* sample s_u of u municipalities from $U - s_a$, with $m + u = n = 100$. The unmatched proportion was determined using the planning value $r_p^2 = 0.9$. Compute an approximately 95% confidence interval for the total of P85 from the following data:

$$\bar{z}_{s_m} = 11.54; \qquad \bar{y}_{s_m} = 11.79; \qquad \bar{y}_{s_u} = 13.91$$

$$S_{zs_m}^2 = 24.26; \qquad S_{zys_m} = 22.99;$$

$$S_{ys}^2 = 43.58; \qquad S_{es_m}^2 = 1.33$$

where S_{ys}^2 is the variance of y_k in $s = s_m \cup s_u$, and $S_{es_m}^2$ is the variance in s_m of the residuals $e_k = y_k - \bar{y}_{s_m} - \hat{B}(z_k - \bar{z}_{s_m})$ with $\hat{B} = S_{zys_m}/S_{ys_m}^2$. Verify also that s is realized with the same probability as that of a simple random sample of $n = 100$ from U.

(d) As an alternative to the strategy in (c), consider estimating the 1985 current total by the strategy based on an *SI* sample s_u of size $n = 100$ from U and the simple expansion estimator $N\bar{y}_s$. Use data in (c) to estimate the length of an approximately 95% confidence interval for t_y under the alternative strategy.

9.9. Verify that the unbiased variance estimator for the π^* estimator can be expressed by the single-term equation (9.3.8).

9.10. Verify the expressions for the variance estimators shown by equations (9.4.12) to (9.4.14) in Example 9.4.2.

9.11. Prove the unbiasedness of the variance estimator shown in (9.6.10) in two-phase sampling for difference estimation.

9.12. Verify that the estimation error of $\hat{t}_{\text{dif}2}$ is given by equation (9.6.14).

9.13. Verify that the estimation error of $\hat{t}_{r2}$ is given by equation (9.7.26).

9.14. Consider two-phase sampling with stratified Bernoulli sampling in phase two, as described in Section 9.8. Prove that, given s_a and a fixed vector $\mathbf{n} = (n_1, \ldots, n_h, \ldots, n_{H_{s_a}})$ with $n_h \geq 1$ for all h, the second-phase sampling is equivalent to an *STSI* selection with n_h elements chosen from n_{ah} in stratum h.

9.15. Verify the variance and covariance equations (9.9.10) to (9.9.12).

9.16. Prove that the minimal variance (9.9.15) in the context of Example 9.9.2 can be written in the form of equation (9.9.23) and that the optimal matching proportion is given by (9.9.25).

9.17. Deduce the optimal values of α and γ given in Remark 9.9.1 by equations (9.9.30) and (9.9.31). Also, confirm the expressions α_{opt} and γ_{opt} for *SI* sampling given in the same remark.

9.18. Show that the optimal values of β and δ in the estimator (9.9.38) are as given by equations (9.9.40) and (9.9.41). Also, verify the expressions β_{opt} and δ_{opt} for *SI* sampling in Example 9.9.4.

9.19. Derive step by step the approximate minimum variance shown in equation (9.9.50).

9.20. Derive step by step the approximate minimum variance shown in equation (9.9.53).

9.21. To simplify the discussion in Section 9.9, the difference estimator (9.9.1) of the current total was formed with the aid of a single auxiliary variable z. To generalize, suppose that several auxiliary variables, $z_1, \ldots, z_J$ are available for elements in the first sample s_a and that y_k is well approximated by

$$y_k^0 = \sum_{j=1}^{J} A_j z_{jk} = \mathbf{A}'\mathbf{z}_k$$

where $\mathbf{A}$ is a vector of *known* constants. We form the unbiased estimator

$$\hat{t}_1 = \hat{t}_{y^0 s_a} + \hat{t}_{D s_m} = \hat{t}_{y s_m} + \sum_{j=1}^{J} A_j(\hat{t}_{z_j s_a} - \hat{t}_{z_j s_m})$$

where

$$\hat{t}_{z_j s_a} = \sum_{s_a} \frac{z_{jk}}{\pi_{ak}} \quad \text{and} \quad \hat{t}_{z_j s_m} = \sum_{s_m} \frac{z_{jk}}{\pi_{ak}\pi_{k|s_a}}$$

and combine it linearly—as in the simpler case—with the unbiased estimator $\hat{t}_2 = \hat{t}_{ys_u}$. Let $V_1 = V(\hat{t}_1)$, $V_2 = V(\hat{t}_2)$, and $C = C(\hat{t}_1, \hat{t}_2)$. (a) Prove that the composite estimator

$$\hat{t}_y = w_1 \hat{t}_1 + w_2 \hat{t}_2$$

is unbiased for the current total, and that its variance is given by equations (9.9.4) to (9.9.7). (b) Prove that the optimum choices of $w_1 = 1 - w_2$ and $\mathbf{A}$ are given by

$$w_{1\text{opt}} = 1 - w_{2\text{opt}} = \frac{V_2 - C}{V_{1\min} + V_2 - 2C}$$

and

$$\mathbf{A}_{\text{opt}} = \mathbf{\Lambda}^{-1}\mathbf{\Gamma}$$

where

$$V_{1\min} = V(\hat{t}_{ys_m}) - \mathbf{\Gamma}'\mathbf{\Lambda}^{-1}\mathbf{\Gamma}$$

leading to the minimum variance

$$V(\hat{t}_y)_{\min} = \frac{V_{1\min} V_2 - C^2}{V_{1\min} + V_2 - 2C}$$

Here, $\mathbf{\Gamma}$ is a vector of J components with typical element

$$\gamma_j = EC(\hat{t}_{ys_m}, \hat{t}_{z_j s_m} | s_a)$$

while $\mathbf{\Lambda}$ is a $J \times J$ matrix with typical element

$$\lambda_{ij} = EC(t_{z_i s_m}, t_{z_j s_m} | s_a).$$

CHAPTER 10

Estimation for Domains

10.1. Introduction

It is the rule rather than the exception in a survey that estimates are needed not only for the population as a whole, but also for various subpopulations. For example, in a household survey, the survey statistician may be asked to provide separate estimates for each of a number of household types, defined, say, on the basis of household size (one-member households, two-member households, etc.). In a labor-force survey conducted in a given country, estimates of unemployment are usually required not only at the national level, but also at provincial and local levels.

Often, the data base (the sample size) for the whole population is substantial, so that estimates of excellent precision can be obtained nationally, whereas when the total sample is broken down at a detailed local level, some areas will have so few observations (or none at all) that estimates of acceptable precision become impossible. On the other hand, other subpopulations pose no data shortage problem. In an *SI* sample of $n = 1{,}000$ from a population of individuals, around 500 observations are expected in each of the subpopulations "men" and "women," a solid base for a separate estimate for each sex.

Subpopulations for which separate point estimates and confidence intervals are required are called *domains*. Sometimes the longer term *domain of study* is used, in particular about a subpopulation designated at the planning stage as one for which a separate estimate is required. Adequate resources can then be reserved to guarantee that the domain is sampled sufficiently to obtain acceptable precision in the estimates. More generally, however, a domain is any subpopulation for which a separate estimate may be requested, before or after the planning stage. Often, the need for estimates for certain domains manifests itself only after the sampling design has been decided or after the sampling and the field work have been completed. To respond to these unplanned requests, the statistician must make the best of the data at hand. The number of observations that happen to fall in the domain is ordinarily random and sometimes very small. These are the features that give domain estimation its particular flavor.

A *small domain* is one that accounts for only a minor fraction of the whole population. In such a domain, the statistician is likely to have very few observations (or none at all). This complication creates "the small domain estimation problem," for which an extensive recent literature has emerged. *Small area estimation* is another term often heard in this connection. It refers primarily to the case where the small domains are geographically defined. A source of information on recent developments in small-domain estimation is the volume edited by Platek et al., (1986). It is beyond the scope of this chapter to give a detailed account of the many special techniques that have been proposed for the small area problem.

The aim of this chapter is to familiarize the reader with the basic tools for estimation of domain totals and means. Most of these techniques assume that the number of observations from the domain is modest but not extremely small. Some issues in domain estimation were brought up in Sections 3.3 and 5.8.

10.2. The Background for Domain Estimation

Consider a partitioning of the population $U = \{1, \ldots, k, \ldots, N\}$ into D subsets, $U_1, \ldots, U_d, \ldots, U_D$, called domains. Let N_d be the size of U_d. We have the partitioning equations

$$U = \bigcup_{d=1}^{D} U_d; \qquad N = \sum_{d=1}^{D} N_d \tag{10.2.1}$$

The objective is to estimate the domain totals

$$t_d = \textstyle\sum_{U_d} y_k; \qquad d = 1, \ldots, D$$

or the domain means

$$\bar{y}_{U_d} = t_d/N_d; \qquad d = 1, \ldots, D$$

Remark 10.2.1. In this chapter, we use t_d for the domain total and $\bar{y}_{U_d}$ for the domain mean. The general notation for estimators of these parameters is $\hat{t}_d$ and $\hat{\bar{y}}_{U_d}$, respectively.

If N_d is unknown (often the case in practice), then the parameter $\bar{y}_{U_d} = t_d/N_d$ is a ratio of two unknowns. Note that the estimator of $\bar{y}_{U_d}$ considered in Section 5.8 resulted from viewing $\bar{y}_{U_d}$ as a ratio of two unknown totals.

Remark 10.2.2. The division into domains represents a new type of partitioning of the population into subsets. Important partitionings encountered in earlier chapters are: (a) a set of strata (for purposes of stratified sampling); (b) a set of PSUs (for two-stage or multistage sampling); (c) a set of groups (for group models in regression estimation).

If the elements in U_d (say, a geographic area) can be identified beforehand and listed, the statistician can select a sample directly from the domain, according to a suitable design. The domain is then in effect designated as a stratum, and the sample selection in the domain is controlled by the choice of design in the stratum. However, even if U_d could be singled out as a stratum, one does not always choose to do so in practice. If the number of domains D is large, a high cost may be associated with controlled selection in every domain. This chapter is primarily about the case where it is impossible (for example, for lack of frame) or impractical (for example, for cost reasons) to single out each domain as a stratum.

We assume that a survey is carried out on the population U. That is, a probability sample s of size n_s is drawn from U according to a specified sampling design $p(\cdot)$ with the inclusion probabilities π_k, π_{kl}, and as usual we set $\Delta_{kl} = \pi_{kl} - \pi_k \pi_l$ and $\check{\Delta}_{kl} = \Delta_{kl}/\pi_{kl}$.

Let s_d denote the part of s that happens to fall in U_d, that is,

$$s_d = s \cap U_d$$

Denote by n_{s_d} the size of s_d. The sample analogs of equations (10.2.1) are then

$$s = \bigcup_{d=1}^{D} s_d; \qquad n_s = \sum_{d=1}^{D} n_{s_d} \tag{10.2.2}$$

Here, n_{s_d} is random and possibly quite small. With this random number of observations, the statistician's task is to make the best possible estimate for the domain.

It is useful, as in Section 5.8, to work with domain indicator variables. For the dth domain, define

$$z_{dk} = \begin{cases} 1 & \text{if } k \in U_d \\ 0 & \text{otherwise} \end{cases} \tag{10.2.3}$$

The vector

$$\mathbf{z}_k = (z_{1k}, \ldots, z_{dk}, \ldots, z_{Dk})$$

consists of $D - 1$ entries "0" and a single entry "1" specifying the domain of the kth element. The z_{dk} are constants, not random variables, and are often unknown before sampling.

The domain size can be expressed as

$$N_d = \textstyle\sum_U z_{dk}$$

The π estimator of N_d (thus unbiased) is given by

$$\hat{N}_d = \sum_s z_{dk}/\pi_k = \sum_{s_d} 1/\pi_k$$

Moreover, the domain sample size can be represented as

$$n_{s_d} = \sum_U z_{dk} I_k = \sum_{U_d} I_k$$

where I_k is the usual sample membership indicator, that is, $I_k = 1$ if $k \in s$ and $I_k = 0$ if not. Under the given sampling design $p(\cdot)$, the expected domain sample size is therefore

$$E(n_{s_d}) = \sum_U z_{dk} \pi_k = \sum_{U_d} \pi_k \tag{10.2.4}$$

Thus, the design, or more specifically the π_k induced by the design, determines whether a good or a poor sample size can be expected in the domain.

In particular, under the *SI* design, with n elements chosen from N,

$$E_{SI}(n_{s_d}) = N_d n/N = f P_d N \tag{10.2.5}$$

where $f = n/N$ and

$$P_d = N_d/N$$

is the relative size of the domain. In equation (10.2.5), two usually small fractions, f and P_d, are multiplied into a usually large number, the population size N. Seemingly minor changes in the three numbers f, P_d, and N can cause considerable change in $E_{SI}(n_{s_d})$. For example, with $P_d = 0.3\%$, $f = 1\%$, and $N = 100\,000$, we have

$$E_{SI}(n_{s_d}) = 3$$

which obviously offers a poor outlook for a precise estimate. With some increase in each of the three numbers, the prospects are much brighter. For $P_d = 0.9\%$, $f = 2\%$, and $N = 400{,}000$, we have

$$E_{SI}(n_{s_d}) = 72$$

Especially if some strong auxiliary information can be exploited, approximately 70 observations may give a decent domain estimate.

Purcell and Kish (1979) point out that the choice of estimation method depends on the situation, including the relative size of the domain, P_d. They introduce four types of domains qualified as *major*, *minor*, *mini*, and *rare* depending on whether the relative size $P_d = N_d/N$ satisfies, respectively, $P_d \geq 0.1$, $0.01 \leq P_d < 0.1$, $0.0001 \leq P_d < 0.01$, and $P_d < 0.0001$. For a rare or mini domain, even a very large overall sample may not produce a single observation from the domain. Special estimation methods (not considered here) may then have to be used.

Remark 10.2.3. The domain estimation problem appears by coincidence in connection with imperfect frames. Suppose that the frame population U_F has overcoverage but no other imperfections. Then the target population U is a subset (a domain) of U_F. For instance, for a population of business establishments, the frame may be somewhat outdated, and the overcoverage $U_F - U$

corresponds to firms having gone out of business since the frame was set up. Estimation in the presence of overcoverage is discussed in Section 14.7.2.

Remark 10.2.4. If the domain proportions P_d $(d = 1, \ldots, D)$ are highly variable, the domain sample sizes $n_{s_1}, \ldots, n_{s_D}$ may also be considerably different. For many domains, estimates of unnecessarily high precision may result, whereas in the smallest domains, very poor estimates may be obtained. This suggests a two-phase sampling approach. Once the large first-phase sample s has been divided up into its domain parts s_d, subsampling may take place, with small subsampling rates in the large domains, and up to 100% subsampling in the smallest domains.

10.3. The Basic Estimation Methods for Domains

Let us first examine the basic techniques for estimating the domain total

$$t_d = \sum_{U_d} y_k$$

the domain mean

$$\bar{y}_{U_d} = \frac{t_d}{N_d} = \frac{1}{N_d} \sum_{U_d} y_k$$

and, at the end of the section, the difference between two domain means. Let

$$y_{dk} = z_{dk} y_k = \begin{cases} y_k & \text{if } k \in U_d \\ 0 & \text{otherwise} \end{cases} \tag{10.3.1}$$

where z_{dk} is the indicator quantity defined by (10.2.3). Then the domain total is

$$t_d = \sum_U y_{dk}$$

and the corresponding π estimator is

$$\hat{t}_{d\pi} = \sum_s \check{y}_{dk} = \sum_{s_d} \check{y}_k$$

When N_d is known, an ordinarily better estimator of t_d is obtained using Result 5.8.1. Consider the π-weighted domain sample mean

$$\tilde{y}_{s_d} = \frac{1}{\hat{N}_d} \sum_{s_d} \check{y}_k \tag{10.3.2}$$

with

$$\hat{N}_d = \sum_{s_d} 1/\pi_k$$

Multiplying $\tilde{y}_{s_d}$ by N_d gives an alternative estimator of t_d which we denote $\tilde{t}_d$. That is,

$$\tilde{t}_d = N_d \tilde{y}_{s_d}$$

This requires $n_{s_d} \geq 1$.

In summary, the basic techniques for domain estimation are:

1. To estimate the domain total when N_d is unknown, use the π estimator

$$\hat{t}_{d\pi} = \sum_{s_d} \check{y}_k$$

Its properties are given in Result 10.3.1.

2. To estimate the domain total when N_d is known, use

$$\tilde{t}_d = N_d \tilde{y}_{s_d} = \frac{N_d}{\hat{N}_d} \sum_{s_d} \check{y}_k$$

Result 10.3.1 gives the fundamental properties.

3. To estimate the domain mean, whether N_d is known or not, use

$$\hat{\bar{y}}_{U_d} = \tilde{y}_{s_d}$$

as described in Result 5.8.1.

Result 10.3.1. *The π estimator of the domain total $t_d = \sum_{U_d} y_k$ is*

$$\hat{t}_{d\pi} = \sum_{s_d} \check{y}_k \tag{10.3.3}$$

with the variance

$$V(\hat{t}_{d\pi}) = \sum\sum_{U_d} \Delta_{kl} \check{y}_k \check{y}_l \tag{10.3.4}$$

An unbiased variance estimator is

$$\hat{V}(\hat{t}_{d\pi}) = \sum\sum_{s_d} \check{\Delta}_{kl} \check{y}_k \check{y}_l \tag{10.3.5}$$

When the domain size N_d is known, an ordinarily preferred estimator is

$$\tilde{t}_d = N_d \tilde{y}_{s_d} \tag{10.3.6}$$

where $\tilde{y}_{s_d}$ is the π-weighted mean given by (10.3.2). The approximate variance is

$$AV(\tilde{t}_d) = \sum\sum_{U_d} \Delta_{kl} \left(\frac{y_k - \bar{y}_{U_d}}{\pi_k}\right)\left(\frac{y_l - \bar{y}_{U_d}}{\pi_l}\right) \tag{10.3.7}$$

A variance estimator is given by

$$\hat{V}(\tilde{t}_d) = \left(\frac{N_d}{\hat{N}_d}\right)^2 \sum\sum_{s_d} \check{\Delta}_{kl} \left(\frac{y_k - \tilde{y}_{s_d}}{\pi_k}\right)\left(\frac{y_l - \tilde{y}_{s_d}}{\pi_l}\right) \tag{10.3.8}$$

A confidence interval constructed in the usual manner with the aid of (10.3.5) or (10.3.8) gives approximately $100(1-\alpha)\%$ coverage in repeated draws of samples s from the whole population U, with the given design $p(\cdot)$.

Here, the expressions (10.3.4) and (10.3.5) follow easily if we apply the general Result 2.8.1 to the domain variable y_d whose typical value y_{dk} is defined by (10.3.1). The conclusions (10.3.7) and (10.3.8) follow from (5.8.4) and (5.8.5) by simply multiplying by N_d^2. Note that (10.3.7) is expressed by means of deviations from the domain mean, $y_k - \bar{y}_{U_d}$, whereas (10.3.4) involves the raw

y_k-scores. As a consequence, the variance of $\tilde{t}_d$ is ordinarily smaller than that of $\hat{t}_{d\pi}$. The improvement becomes more obvious if we focus on a specific design, as in the following example.

EXAMPLE 10.3.1. Under the *SI* design with n elements drawn from N, the π estimator of t_d is

$$\hat{t}_{d\pi} = (N/n) \textstyle\sum_{s_d} y_k$$

Letting $P_d = N_d/N$ and $Q_d = 1 - P_d$, the variance shown in equation (10.3.4) takes the form

$$\begin{aligned} V_{SI}(\hat{t}_{d\pi}) &= N^2 \frac{1-f}{n} \frac{(N_d - 1)S^2_{yU_d} + N_d Q_d \bar{y}^2_{U_d}}{N-1} \\ &\doteq N^2 \frac{1-f}{n} P_d(S^2_{yU_d} + Q_d \bar{y}^2_{U_d}) \end{aligned} \qquad (10.3.9)$$

where

$$S^2_{yU_d} = \frac{1}{N_d - 1} \textstyle\sum_{U_d} (y_k - \bar{y}_{U_d})^2$$

is the domain variance. The variance estimator shown in (10.3.5) takes an analogous form,

$$\begin{aligned} \hat{V}_{SI}(\hat{t}_{d\pi}) &= N^2 \frac{1-f}{n} \frac{(n_d - 1)S^2_{ys_d} + n_d q_d \bar{y}^2_{s_d}}{n-1} \\ &\doteq N^2 \frac{1-f}{n} p_d(S^2_{ys_d} + q_d \bar{y}^2_{s_d}) \end{aligned}$$

where

$$S^2_{ys_d} = \frac{1}{n_{s_d} - 1} \textstyle\sum_{s_d} (y_k - \bar{y}_{s_d})^2 \qquad (10.3.10)$$

$$\bar{y}_{s_d} = \textstyle\sum_{s_d} y_k / n_{s_d}, \qquad p_d = n_{s_d}/n, \quad \text{and} \quad q_d = 1 - p_d$$

Turning now to the estimator $\tilde{t}_d$, which requires knowledge of the domain size, we obtain from equation (10.3.7)

$$\begin{aligned} AV_{SI}(\tilde{t}_d) &= N^2 \frac{1-f}{n} \frac{(N_d - 1)S^2_{yU_d}}{N-1} \\ &\doteq N^2 \frac{1-f}{n} P_d S^2_{yU_d} \end{aligned} \qquad (10.3.11)$$

whereas the variance estimator given by (10.3.8) takes the form

$$\hat{V}_{SI}(\tilde{t}_d) = \left(\frac{N_d}{\hat{N}_d}\right)^2 N^2 \frac{1-f}{n} \frac{(n_{s_d}-1)S^2_{ys_d}}{n-1}$$

$$\doteq N_d^2\left(\frac{1}{n_{s_d}} - \frac{1}{\hat{N}_d}\right)S^2_{ys_d} \tag{10.3.12}$$

with $\hat{N}_d = Nn_{s_d}/n$. Obviously, the calculation of (10.3.12) requires $n_{s_d} \geq 2$.

The increase in variance due to not knowing N_d can be expressed through the ratio of equations (10.3.9) and (10.3.11),

$$\frac{V_{SI}(\hat{t}_{d\pi})}{AV_{SI}(\tilde{t}_d)} \doteq 1 + \frac{Q_d}{(cv_{yU_d})^2} \tag{10.3.13}$$

where $cv_{yU_d} = S_{yU_d}/\bar{y}_{U_d}$ is the coefficient of variation of y in the dth domain. For example, if $cv_{yU_d} = 0.5$, the variance of $\hat{t}_{d\pi}$ is roughly five times that of $\tilde{t}_d$ when the domain accounts for only a few percent of the population (Q_d near unity). If the domain accounts for 50% of the population (with cv_{yU_d} unchanged at 0.5) the disadvantage of $\hat{t}_{d\pi}$ is less pronounced, but still considerable; the variance ratio drops to roughly three.

Example 10.3.2. Suppose the domain can be identified in advance and sampled in a controlled manner. Would this lead to a better estimator than (10.3.6)? Intuitively one would expect so; for a more complete answer in one specific case, let us consider the *SI* design. If the domain can be identified in advance, the statistician may select an *SI* sample of n_d (now a fixed number) from the N_d domain elements. The unbiased estimator of t_d is then

$$N_d\bar{y}_{s_d} = N_d \textstyle\sum_{s_d} y_k/n_d$$

with the exact variance

$$V'_{SI}(N_d\bar{y}_{s_d}) = N_d^2\left(\frac{1}{n_d} - \frac{1}{N_d}\right)S^2_{yU_d} \tag{10.3.14}$$

Compare this with the uncontrolled domain estimation approach. The AV expression shown in (10.3.11) can be written

$$AV_{SI}(\tilde{t}_d) \doteq N_d^2\left(\frac{1}{n_d^0} - \frac{1}{N_d}\right)S^2_{yU_d} \tag{10.3.15}$$

where

$$n_d^0 = nN_d/N$$

is the expected domain sample count.

That is, if the sample size n_d under controlled conditions is equal to the expected domain sample count $E(n_{s_d}) = n_d^0$ under uncontrolled conditions, the two variances will be roughly the same, which is easy to accept on intuitive grounds. Thus, to the degree of approximation used here, there is no loss of precision through the domain estimation approach, provided N_d is known. For a closer approximation to $AV_{SI}(\tilde{t}_d)$, showing that a minor loss of precision is actually incurred, see Section 10.4.

EXAMPLE 10.3.3. Consider the STSI design, with H strata that cut across the D domains. The domain mean estimator shown by equation (10.3.2) is then

$$\hat{\bar{y}}_{U_d} = \tilde{y}_{s_d} = \sum_{h=1}^{H} \frac{N_h}{n_h} \sum\nolimits_{s_{dh}} y_k \Big/ \sum_{h=1}^{H} \frac{N_h}{n_h} n_{s_{dh}}$$

where N_h/n_h is the inverse of the sampling fraction in stratum h, s_{dh} is the intersection of the domain U_d and the simple random sample drawn in stratum h, and $n_{s_{dh}}$ is the random size of this intersection. The estimated variance (10.3.8) divided by N_d^2 can after some algebra be written as

$$\hat{V}_{STSI}(\hat{\bar{y}}_{U_d}) = \frac{1}{\hat{N}_d^2} \sum_{h=1}^{H} N_h^2 \frac{1 - f_h}{n_h} \frac{\sum_{s_{dh}} (y_k - \bar{y}_{s_{dh}})^2 + n_{s_{dh}}(1 - p_{dh})(\bar{y}_{s_{dh}} - \hat{\bar{y}}_{U_d})^2}{n_h - 1}$$

where

$$p_{dh} = n_{s_{dh}}/n_h, \qquad \hat{N}_d = \sum_{h=1}^{H} N_h(n_{s_{dh}}/n_h)$$

and $\bar{y}_{s_{dh}}$ is the straight mean of y in the cell dh.

A nuisance contribution to this expression are the terms in $(\bar{y}_{s_{dh}} - \hat{\bar{y}}_{U_d})^2$, due to differences between the strata means in the domain. A stratification that is highly efficient when estimates are made for the entire population or for a large domain may lose most of its advantage when estimates are made for a small domain, as observed by Durbin (1958).

In many applications, it is of interest to find out if two domains differ. If the sampled population consists of individuals, we may wish to compare men with women, or the over 65 age group with the 30 to 50 age group, or residents of a rural area with those of an urban area. Of particular importance is the comparison of two domain means. The social scientist or the decision-maker will often want not only a point estimate of the difference between the two domain means, but also an indication of whether the difference is significantly large. It is thus of interest to set a confidence interval for the unknown difference of the two means. If the value zero is not contained in an interval at the 95% level, there is strong indication that the domain means really differ.

If we take the two domains to be compared as U_1 and U_2 (that is, $d = 1$ and $d = 2$), the difference to estimate is

$$D = \bar{y}_{U_1} - \bar{y}_{U_2} = \frac{t_1}{N_1} - \frac{t_2}{N_2}$$

An obvious estimator is the difference of the corresponding π-weighted sample means,

$$\hat{D} = \tilde{y}_{s_1} - \tilde{y}_{s_2}$$

with $\tilde{y}_{s_d}$ given by (10.3.2) for $d = 1, 2$. The variance is

$$V(\hat{D}) = V(\tilde{y}_{s_1}) + V(\tilde{y}_{s_2}) - 2C(\tilde{y}_{s_1}, \tilde{y}_{s_2})$$

Approximations to the two variances on the right-hand side are obtained by dividing the expression in equation (10.3.7) by N_d^2, $d = 1, 2$, whereas an approximate expression for the covariance is

$$AC(\tilde{y}_{s_1}, \tilde{y}_{s_2}) = \frac{1}{N_1 N_2} \sum_{k \in U_1} \sum_{l \in U_2} \Delta_{kl} \frac{y_k - \bar{y}_{U_1}}{\pi_k} \frac{y_l - \bar{y}_{U_2}}{\pi_l} \tag{10.3.16}$$

The resulting approximate variance of $\hat{D}$ can be expressed as

$$AV(\hat{D}) = \sum\sum_{U_1 \cup U_2} \Delta_{kl} \check{A}_k \check{A}_l \tag{10.3.17}$$

with $\check{A}_k = A_k/\pi_k$, where

$$A_k = \frac{z_{1k}(y_k - \bar{y}_{U_1})}{N_1} - \frac{z_{2k}(y_k - \bar{y}_{U_2})}{N_2}$$

The corresponding variance estimator, needed for the confidence interval calculation, is

$$\hat{V}(\hat{D}) = \sum\sum_{s_1 \cup s_2} \check{\Delta}_{kl} \check{a}_k \check{a}_l \tag{10.3.18}$$

with $\check{a}_k = a_k/\pi_k$, where

$$a_k = \frac{z_{1k}(y_k - \tilde{y}_{s_1})}{\hat{N}_1} - \frac{z_{2k}(y_k - \tilde{y}_{s_2})}{\hat{N}_2}$$

EXAMPLE 10.3.4. The approximate covariance shown in (10.3.16) is seen to be zero for a design such that π_k is constant for all k and Δ_{kl} is constant for all $k \neq l$. A case in mind is the *SI* design (n elements from N), in which case

$$\hat{D} = \bar{y}_{s_1} - \bar{y}_{s_2}$$

with $\bar{y}_{s_d} = \sum_{s_d} y_k / n_{s_d}$, $d = 1, 2$. In the absence of a covariance term, we have from (10.3.15)

$$AV_{SI}(\hat{D}) \doteq \sum_{d=1}^{2} \left(\frac{1}{n_d^0} - \frac{1}{N_d} \right) S_{yU_d}^2$$

and the variance estimator shown in (10.3.18) is correspondingly additive,

$$\hat{V}_{SI}(\hat{D}) = \sum_{d=1}^{2} \left(\frac{1}{n_{s_d}} - \frac{1}{\hat{N}_d} \right) S_{ys_d}^2$$

where $\hat{N}_d = N n_{s_d}/n$, and we have approximated $n(n_{s_d} - 1)/n_{s_d}(n - 1)$ by unity.

In Section 10.9, we use auxiliary information to create improved estimators for the difference of two domain means.

10.4. Conditioning on the Domain Sample Size

Despite the finding in Example 10.3.2 that the domain estimation approach, with *SI* sampling from the whole population and $\tilde{t}_d = N_d \tilde{y}_{s_d}$, is essentially as efficient as direct *SI* sampling within the a priori identified domain, it is expected that some price is still paid for the uncontrolled domain sample size. A better variance appoximation than (10.3.11) is required to reveal this limited loss of efficiency. To this end, conditioning on n_{s_d} is helpful. Similar reasoning was used in Section 7.10.

Denote by A_d the event $n_{s_d} \geq 1$. If n is very large, $\Pr(A_d)$ is near unity, even if the relative domain size P_d is quite small. For a fixed value of n_{s_d} such that $n_{s_d} \geq 1$, the domain part of the sample, $s_d = s \cap U_d$, behaves as a simple random selection of n_{s_d} from N_d. For $\tilde{t}_d = N_d \tilde{y}_{s_d}$ we thus have

$$\begin{cases} E_{SI}(\tilde{t}_d | A_d, n_{s_d}) = t_d & (10.4.1) \\ V_{SI}(\tilde{t}_d | A_d, n_{s_d}) = N_d^2 \left(\dfrac{1}{n_{s_d}} - \dfrac{1}{N_d} \right) S_{yU_d}^2 & (10.4.2) \end{cases}$$

Thus, $\tilde{t}_d$ is conditionally unbiased, given any domain sample size such that $n_{s_d} \geq 1$, and the conditional variance is shown by equation (10.4.2). Averaging over all values $n_{s_d} \geq 1$,

$$\begin{cases} E_{SI}(\tilde{t}_d | A_d) = t_d & (10.4.3) \\ V_{SI}(\tilde{t}_d | A_d) = N_d^2 \left[E_{SI}\left(\dfrac{1}{n_{s_d}} | A_d \right) - \dfrac{1}{N_d} \right] S_{yU_d}^2 & (10.4.4) \end{cases}$$

To obtain (10.4.4), we have used

$$V[E_{SI}(\tilde{t}_d | A_d, n_{s_d})] = V(t_d | A_d) = 0$$

where $V(\cdot)$ denotes variance with respect to the distribution of n_{s_d}. In other words, given that the domain contains at least one observation, $\tilde{t}_d$ is unbiased for t_d under *SI* sampling from U.

The moments (10.4.3) and (10.4.4) still have a condition, namely, the event A_d. Now, supposing that n is so large that A_d will almost certainly occur, we conclude from (10.4.3) and (10.4.4) that, for all practical purposes, $\tilde{t}_d$ is unbiased for t_d with the unconditional variance

$$V_{SI}(\tilde{t}_d) = N_d^2 \left[E_{SI}\left(\frac{1}{n_{s_d}} \right) - \frac{1}{N_d} \right] S_{yU_d}^2 \qquad (10.4.5)$$

To put an equality sign here is to stretch the truth slightly; we obtain equation (10.4.5) on the assumption that $n_{s_d} = 0$ has negligible probability.

By Taylor expansion, using equation (7.10.8), we obtain

$$E_{SI}\left(\frac{1}{n_{s_d}} \right) \doteq \frac{1}{n_d^0} + \frac{(1-f)(1-P_d)}{(n_d^0)^2} \qquad (10.4.6)$$

where $f = n/N$, and

$$n_d^0 = E(n_{s_d}) = nN_d/N = nP_d$$

From (10.4.5) and (10.4.6), and with $Q_d = 1 - P_d$

$$V_{SI}(\tilde{t}_d) = N_d^2\left(\frac{1}{n_d^0} - \frac{1}{N_d}\right)\left(1 + \frac{Q_d}{n_d^0}\right)S_{yU_d}^2 \tag{10.4.7}$$

Comparing the uncontrolled sampling variance (10.4.7) with the controlled sampling variance (10.3.14) we obtain

$$\frac{V_{SI}(\tilde{t}_d)}{V'_{SI}(N_d\bar{y}_{s_d})} = 1 + \frac{Q_d}{n_d^0}$$

which is roughly $1 + 1/n_d^0$ if n is much larger than $n_d^0 = nP_d$. For example, if the expected domain sample size is $n_d^0 = 10$ elements, the variance is increased by roughly 10%. In other words, there is a nonnegligible loss of precision caused by the lack of control of the domain sample size.

Remark 10.4.1. The conditional variance shown in (10.4.2) is estimated unbiasedly (given n_{s_d} such that $n_{s_d} \geq 2$) by

$$\hat{V}_{SI}^* = N_d^2\left(\frac{1}{n_{s_d}} - \frac{1}{N_d}\right)S_{ys_d}^2$$

where $S_{ys_d}^2$ is given by (10.3.10). That is, in repeated SI samples from the whole population, considering only those samples in which the domain sample count n_{s_d} has a fixed value, the average value of $\hat{V}_{SI}^*$ is as shown in equation (10.4.2). This conditional variance estimator essentially agrees with the expression (10.3.12) flowing from the principal approach in this book. The very minor difference, $1/\hat{N}_d$ instead of $1/N_d$, is of negligible consequence in practice.

10.5. Regression Estimators for Domains

Improved domain estimators can be obtained when further auxiliary information (beyond the knowledge of the N_d) is available. We now examine this option, using the regression estimation approach. This avenue is particularly important as a means of offsetting the scarcity of the sample data in small domains.

Let $\mathbf{x}_k$ denote the value of the kth element of the auxiliary vector $\mathbf{x}$, of dimension J, say. When estimates are required for a number of small domains, it is particularly important that the practitioner seek out strong auxiliary variables from registers and other available sources. As in Chapter 6, we fit a regression model describing the presumed relationship between the study variable and the auxiliary variables. The fitted values $\hat{y}_k$ are used to build a suitable regression estimator.

Let us describe the population scatter by the model stating that, for $k \in U$,

$$\begin{cases} E_\xi(y_k) = \mathbf{x}_k'\boldsymbol{\beta} \\ V_\xi(y_k) = \sigma_k^2 \end{cases} \tag{10.5.1}$$

As before, the fit of this model leads to the estimated regression coefficient vector

$$\hat{\mathbf{B}} = \left(\sum\nolimits_s \frac{\mathbf{x}_k\mathbf{x}_k'}{\sigma_k^2\pi_k}\right)^{-1} \sum\nolimits_s \frac{\mathbf{x}_k y_k}{\sigma_k^2\pi_k}$$

The resulting predicted values are

$$\hat{y}_k = \mathbf{x}_k'\hat{\mathbf{B}}$$

and the residuals are given by

$$e_{ks} = y_k - \hat{y}_k = y_k - \mathbf{x}_k'\hat{\mathbf{B}}$$

Let us consider some simple examples of this general model formulation that might be of interest when domains are involved. In (a), (b), and (c) below, a single, always positive auxiliary variable x is assumed.

a. The model stating that

$$\begin{cases} E_\xi(y_k) = \beta x_k \\ V_\xi(y_k) = \sigma^2 x_k \end{cases} \tag{10.5.2}$$

for $k \in U$ assumes that the ratio y_k/x_k may be considered roughly constant throughout the population. There is a single slope parameter β to estimate, using the combined data from all domains.

b. In some cases it is more realistic to assume that slopes are not equal in all domains. A formulation in which each domain has its own slope parameter is

$$\begin{cases} E_\xi(y_k) = \beta_d x_k \\ V_\xi(y_k) = \sigma_d^2 x_k \end{cases} \tag{10.5.3}$$

for $k \in U_d$, $d = 1, \ldots, D$. Using domain dummy variables, the model (10.5.3) may be included under the umbrella of the general linear form shown in (10.5.1). An advantage of (10.5.3) is that differences between the domains have a chance to be reflected. If the dth domain has enough sample points, the model will generate a good regression estimator. A drawback of (10.5.3) is that when the domains are numerous, say several hundred, some of them may contain very few observations or none at all. Consequently, weak estimates or none at all will result for the domain slopes.

c. The models (10.5.2) and (10.5.3) represent two extremes: one single slope parameter and a slope for each domain. An intermediate case would be of interest in many applications involving smallish sample sizes in individual domains. The domains are divided into a limited number (say ten or less) of groups so that the slope (that is, the y/x ratio) is considered to be the same for all elements in a given group of domains. A group of domains, for example, may correspond to a geographic area. With the number of slopes to estimate reduced to ten or less, it is more likely that the data will support stable estimates of group slopes. Here, as in the single-slope model (10.5.2), strength is borrowed from data outside the domain itself.

For the domain total,

$$t_d = \sum_{U_d} y_k$$

we construct an estimator (in the general manner of Chapter 6) as a sum of regression-predicted values plus an adjustment term involving regression residuals.

Two alternatives are as follows:

a. $$\hat{t}_{dr} = \sum_{U_d} \hat{y}_k + (N_d/\hat{N}_d) \sum_{s_d} e_{ks}/\pi_k \qquad (10.5.4)$$

where $\hat{N}_d = \sum_{s_d} 1/\pi_k$ and N_d is known, and

b. $$\hat{t}'_{dr} = \sum_{U_d} \hat{y}_k + \sum_{s_d} e_{ks}/\pi_k \qquad (10.5.5)$$

if N_d is unknown.

For some model formulations [for example, the models given by (10.6.1) and (10.7.3) below], $\sum_{s_d} e_{ks}/\pi_k = 0$, so the estimators (10.5.4) and (10.5.5) agree and we have

$$\hat{t}_{dr} = \hat{t}'_{dr} = \sum_{U_d} \hat{y}_k$$

If, on the other hand, $\sum_{s_d} e_{ks}/\pi_k$ is not identically equal to zero, then (10.5.4) is usually preferred to (10.5.5) when N_d is known. The reason is that the size of s_d is random. The term $(N_d/\hat{N}_d)\sum_{s_d} e_{ks}/\pi_k$ in (10.5.4) tends to be less variable than the corresponding term $\sum_{s_d} e_{ks}/\pi_k$ in (10.5.5). In the following, interest will be focused on the estimator (10.5.4), which was examined in Särndal and Hidiroglou (1989). The estimator (10.5.5) was analyzed in Särndal (1984).

The first term of (10.5.4) and (10.5.5), that is,

$$\sum_{U_d} \hat{y}_k = \left(\sum_{U_d} \mathbf{x}_k\right)' \hat{\mathbf{B}}$$

can be considered as an estimator of t_d in its own right. It is simply the sum of the predicted values for the domain and is called the *synthetic estimator* corrresponding to the fitted regression model. If the adjustment term is nonzero, this estimator is biased, but its variance usually is very small, and it sometimes is used for estimation in extremely small domains. We examine the reasons for synthetic estimation in Section 10.8.

Remark 10.5.1. Summing the domain estimators (10.5.5) over all D domains, we obtain

$$\sum_{d=1}^{D} \hat{t}'_{dr} = \sum_U \hat{y}_k + \sum_s e_{ks}/\pi_k$$

which is the regression estimator for the population as a whole. This attractive property (often required by survey statisticians) states that the estimates made for the different subpopulations add up to the estimate made (by the same estimation method) for the population as a whole. In published tables of estimates, such internal consistency is desirable. Many users of official statistics look for consistency in a set of domain estimates. The domain estimates obtained by (10.5.4) do not have this property, except under special circumstances, for example, if the term $\sum_{s_d} e_{ks}/\pi_k = 0$ for all s_d and all domains. If additivity is essential, (10.5.5) should be used for domain totals, despite some loss of efficiency compared to (10.5.4).

Remark 10.5.2. The computation of the prediction term

$$\sum_{U_d} \hat{y}_k = \left(\sum_{U_d} \mathbf{x}_k\right)' \hat{\mathbf{B}}$$

in the estimators (10.5.4) and (10.5.5) requires a priori knowledge of the domain total of the auxiliary vector,

$$\sum_{U_d} \mathbf{x}_k = \sum_U z_{dk} \mathbf{x}_k \tag{10.5.6}$$

where z_{dk} is the domain indicator given by (10.2.3). Situations where these totals are known include the following: (a) the domain membership z_{dk}, as well as the auxiliary value $\mathbf{x}_k$, is known (given in the sampling frame), for every element $k \in U$; (b) the vector of domain totals

$$\sum_{U_d} \mathbf{x}_k' = \left(\sum_{U_d} x_{1k}, \ldots, \sum_{U_d} x_{Jk}\right)$$

is known from an accurate source.

To illustrate the requirement in situation (b), suppose that the elements are individuals, the domains are geographic areas, and $\mathbf{x}_k'$ is the J-vector composed of $J-1$ entries "0" and a single "1" indicating the age/sex group of the kth individual. The vector (10.5.6) that must be known is then given by

$$(N_{d1}, \ldots, N_{dj}, \ldots, N_{dJ})$$

where N_{dj} is the population count in the dth region and the jth age/sex group. In some applications, these counts may be taken from the latest census figures. (See the discussion of group model for domains in Section 10.7.)

Remark 10.5.3. Accurate statistics for smaller geographic domains may be available for certain variables in years when a census is conducted. However, censuses are infrequent, and rather few variables are ordinarily included in a census. For planning purposes, small-area statistics are often required for between-census years or for variables other than those covered by the census. Suppose that limited surveys of the population are undertaken in the years between censuses. In such surveys, minor geographic areas may be represented by rather few data points. It may then be possible to use census-year data as the auxiliary information for improved estimates, through the regression technique.

The domain estimator (10.5.4) can be written as a weighted sum of the π-expanded y-values,

$$\hat{t}_{dr} = \sum_s g_{dks} \check{y}_k$$

where the g weights (which depend on the domain d, the whole sample s, and the element k) are

$$g_{dks} = \frac{N_d}{\hat{N}_d} z_{dk} + \left(\sum_{U_d} \mathbf{x}_k - \frac{N_d}{\hat{N}_d} \sum_{s_d} \check{\mathbf{x}}_k\right)' \left(\sum_s \frac{\mathbf{x}_k \mathbf{x}_k'}{\sigma_k^2 \pi_k}\right)^{-1} \frac{\mathbf{x}_k}{\sigma_k^2} \tag{10.5.7}$$

An appealing property of the weights is that

$$\sum_s g_{dks} \check{\mathbf{x}}_k = \sum_U z_{dk} \mathbf{x}_k = \sum_{U_d} \mathbf{x}_k \tag{10.5.8}$$

That is, the g weights are calibrated in such a way that the g-weighted sample sum of the auxiliary values equals the known domain total of these values.

To discuss the variance properties, we need the population fit residuals

$$E_k = y_k - \mathbf{x}_k' \mathbf{B} \tag{10.5.9}$$

where

$$\mathbf{B} = \left(\sum_U \frac{\mathbf{x}_k \mathbf{x}_k'}{\sigma_k^2}\right)^{-1} \sum_U \frac{\mathbf{x}_k y_k}{\sigma_k^2}$$

The fundamental properties of the regression estimator $\hat{t}_{dr}$ are now expressed in the following result:

Result 10.5.1. *Assuming that the domain size N_d is known, the regression estimator of the domain total $t_d = \sum_{U_d} y_k$ is*

$$\hat{t}_{dr} = \sum_{U_d} \hat{y}_k + (N_d/\hat{N}_d) \sum_{s_d} \check{e}_{ks} = \sum_s g_{dks} \check{y}_k \tag{10.5.10}$$

where

$$\check{e}_{ks} = e_{ks}/\pi_k = (y_k - \hat{y}_k)/\pi_k, \qquad \check{y}_k = y_k/\pi_k$$

and g_{dks} is given by equation (10.5.7). The approximate variance is

$$AV(\hat{t}_{dr}) = \sum\sum_{U_d} \Delta_{kl} \left(\frac{E_k - \bar{E}_{U_d}}{\pi_k}\right)\left(\frac{E_l - \bar{E}_{U_d}}{\pi_l}\right) \tag{10.5.11}$$

where E_k is given by (10.5.9) and $\bar{E}_{U_d} = \sum_{U_d} E_k/N_d$. An estimator of the variance is given by

$$\hat{V}(\hat{t}_{dr}) = \sum\sum_s \check{\Delta}_{kl} \frac{g_{dks} e_{ks}}{\pi_k} \frac{g_{dls} e_{ls}}{\pi_l} \tag{10.5.12}$$

Result 10.5.1 derives from writing the error of $\hat{t}_{dr}$ as

$$\hat{t}_{dr} - t_d = N_d[(\tilde{E}_{s_d} - \bar{E}_{U_d}) - (\hat{\mathbf{B}} - \mathbf{B})'(\tilde{\mathbf{x}}_{s_d} - \bar{\mathbf{x}}_{U_d})]$$

where

$$\tilde{E}_{s_d} = \frac{1}{\hat{N}_d} \sum_{s_d} \frac{E_k}{\pi_k}; \qquad \bar{E}_{U_d} = \frac{1}{N_d} \sum_{U_d} E_k$$

and $\tilde{\mathbf{x}}_{s_d}$ and $\bar{\mathbf{x}}_{U_d}$ are the analogously defined means of the vector $\mathbf{x}$. The term with a minus sign within the square brackets is a product of two "small" components (both $\hat{\mathbf{B}} - \mathbf{B}$ and $\tilde{\mathbf{x}}_{s_d} - \bar{\mathbf{x}}_{U_d}$ are near the zero vector in probability). Dropping this term,

$$\hat{t}_{dr} - t_d \doteq N_d(\tilde{E}_{s_d} - \bar{E}_{U_d})$$

Here, $N_d \tilde{E}_{s_d} = (N_d/\hat{N}_d) \sum_{s_d} \check{E}_k$ has the same structure as $(N_d/\hat{N}_d) \sum_{s_d} \check{y}_k = N_d \tilde{t}_d$, which was considered in Result 10.3.1. The former is expressed in the residuals $E_k = y_k - \mathbf{x}_k'\mathbf{B}$, the latter in the raw scores y_k. Replacing y_k by E_k in equation (10.3.7) gives the desired AV result shown in (10.5.11).

The estimated variance shown in (10.5.12) is justified by the identity

$$\hat{t}_{dr} - t_d = \sum_s g_{dks} \check{E}_k - \sum_U z_{dk} E_k$$

which suggests the g-weighted variance estimator

$$\hat{V} = \sum\sum_s \check{\Delta}_{kl} (g_{dks} \check{E}_k)(g_{dls} \check{E}_l)$$

Replacing the unknown E_k by e_{ks} leads to (10.5.12).

The double sum in (10.5.12) is over the entire sample s, because data from other domains than U_d may have been used in estimating the model. Simplification occurs for models [such as (10.6.1) and (10.7.3) below] for which g_{dks} has the structure

$$g_{dks} = z_{dk} h_{dks} \tag{10.5.13}$$

where z_{dk} is the domain indicator for element k, and the weight h_{dks} may depend on d, k, and s. In such cases, the g weights equal zero outside the domain, and (10.5.12) reduces to a double sum over the domain sample s_d alone,

$$\hat{V}(\hat{t}_{dr}) = \sum\sum_{s_d} \check{\Delta}_{kl} \frac{h_{dks} e_{ks}}{\pi_k} \frac{h_{dls} e_{ls}}{\pi_l} \tag{10.5.14}$$

For cases where the sample s is drawn from U by SI sampling, Särndal and Hidiroglou (1989) suggest the variance estimator

$$\hat{V}^*(\hat{t}_{dr}) = N_d^2 \left(\frac{1}{n_{s_d}} - \frac{1}{N_d} \right) \frac{\sum_{s_d} (e_{ks} - \bar{e}_{s_d})^2}{n_{s_d} - 1}$$

where n_{s_d} is the size of $s_d = s \cap U_d$ and $\bar{e}_{s_d}$ is the mean of e_{ks} in s_d. For other designs, and if g_{dks} is not of the form of (10.5.13), the computation may be simplified by using the alternative estimator

$$\hat{V}^*(\hat{t}_{dr}) = \left(\frac{N_d}{\hat{N}_d} \right)^2 \sum\sum_{s_d} \check{\Delta}_{kl} \frac{e_{ks} - \tilde{e}_{s_d}}{\pi_k} \frac{e_{ls} - \tilde{e}_{s_d}}{\pi_l}$$

instead of (10.5.12).

Sections 10.6 and 10.7 illustrate the use of Result 10.5.1. When Result 10.5.1 is used to estimate the domain mean $\bar{y}_{U_d} = t_d/N_d$, divide the estimator (10.5.10) by the known N_d, and divide (10.5.11) and (10.5.12) by N_d^2.

10.6. A Ratio Model for Each Domain

In a first example of the regression technique for domains, our description of the population is in terms of a separate ratio model for each domain. That is, for $d = 1, \ldots, D$, the model is such that

$$\begin{cases} E_\xi(y_k) = \beta_d x_k \\ V_\xi(y_k) = \sigma_d^2 x_k \end{cases} \tag{10.6.1}$$

for $k \in U_d$, where all $x_k > 0$. The slopes β_d may vary with the domains. The model (10.6.1) can be written in the general form given by (10.5.1), if we set

$$\boldsymbol{\beta} = (\beta_1, \ldots, \beta_d, \ldots, \beta_D)'$$

and

$$\mathbf{x}_k = (z_{1k}x_k, \ldots, z_{dk}x_k, \ldots, z_{Dk}x_k)'$$

where z_{dk} is the domain indicator (10.2.3). Here, $\mathbf{x}_k$ is a D vector with $D - 1$ zeroes and a single nonzero entry, $z_{dk}x_k = x_k$, if k belongs to the dth domain.

Applying Result 10.5.1, we obtain after the necessary derivations the *domain ratio estimator*,

$$\hat{t}_{dra} = \sum_{U_d} x_k \frac{\sum_{s_d} \check{y}_k}{\sum_{s_d} \check{x}_k} = \left(\sum_{U_d} x_k\right)\hat{B}_d \tag{10.6.2}$$

which requires that the domain total of x be known. The residuals and the g weights are obtained, respectively, as

$$e_{ks} = y_k - \hat{B}_d x_k$$

and

$$g_{dks} = \left(\frac{\sum_{U_d} x_k}{\sum_{s_d} \check{x}_k}\right) z_{dk}$$

From (10.5.13) and (10.5.14), the estimated variance is

$$\hat{V}(\hat{t}_{dra}) = \left(\frac{\sum_{U_d} x_k}{\sum_{s_d} \check{x}_k}\right)^2 \sum\sum_{s_d} \check{\Delta}_{kl} \check{e}_{ks} \check{e}_{ls} \tag{10.6.3}$$

This is a domain analog of the variance estimator (7.3.2) that applies for the ratio estimator of the entire population total.

EXAMPLE 10.6.1. Under the *SI* design with n elements chosen from N, equation (10.6.2) yields

$$\hat{t}_{dra} = (\textstyle\sum_{U_d} x_k)\hat{B}_d \tag{10.6.4}$$

with

$$\hat{B}_d = \frac{\sum_{s_d} y_k}{\sum_{s_d} x_k}$$

Its approximate variance, from equation (10.5.11), is

$$AV(\hat{t}_{dra}) = N^2 \frac{1-f}{n} \frac{(N_d - 1)S^2_{EU_d}}{N-1}$$

where

$$S^2_{EU_d} = \frac{1}{N_d - 1} \textstyle\sum_{U_d} (y_k - B_d x_k)^2$$

with $B_d = (\sum_{U_d} y_k)/(\sum_{U_d} x_k)$. The variance estimator (10.6.3) takes the form

$$\hat{V}(\hat{t}_{dra}) = \left\{\frac{n}{n-1} \frac{n_{s_d} - 1}{n_{s_d}}\right\} \left(\frac{\bar{x}_{U_d}}{\bar{x}_{s_d}}\right)^2 N_d^2 \left(\frac{1}{n_{s_d}} - \frac{1}{\hat{N}_d}\right) S^2_{es_d}$$

$$\doteq \left(\frac{\bar{x}_{U_d}}{\bar{x}_{s_d}}\right)^2 N_d^2 \left(\frac{1}{n_{s_d}} - \frac{1}{\hat{N}_d}\right) S^2_{es_d} \tag{10.6.5}$$

where $\hat{N}_d = N n_{s_d}/n$, $\bar{x}_{U_d}$ and $\bar{x}_{s_d}$ are straight means, and

$$S^2_{es_d} = \frac{1}{n_{s_d} - 1} \textstyle\sum_{s_d} (y_k - \hat{B}_d x_k)^2$$

Monte Carlo experiments have been carried out, for example, in Leblond (1990), showing that even with a modest number of observations in the domain, valid conditional confidence intervals are obtained with the aid of the point estimator (10.6.4) and the variance estimator (10.6.5). That is, if repeated SI samples of size n are drawn from the population U, and if we consider only those samples for which the domain sample size n_{s_d} has a fixed value, we find that the estimator $\hat{t}_{dra}$ has very little bias and the conditional coverage rate of the interval computed in the usual way from (10.6.4) and (10.6.5) is very near the nominal confidence level aimed at.

Remark 10.6.1. Suppose that $x_k = 1$ for all $k \in U$ in the model (10.6.1). The resulting model is a one-way ANOVA model with domain effects,

$$\begin{cases} E_\xi(y_k) = \beta_d \\ V_\xi(y_k) = \sigma_d^2 \end{cases} \tag{10.6.6}$$

for $k \in U_d$, $d = 1, \ldots, D$. The corresponding regression estimator (10.6.2) then takes the form $N_d \tilde{y}_{s_d}$. That is, the ANOVA model shown in (10.6.6) provides the formal justification for the expanded domain sample mean proposed in Result 10.2.1.

Remark 10.6.2. One way to avoid the difficulty caused by domain counts that are too small is to replace the model (10.6.1) by a ratio model with fewer than D slope parameters. For example, one may consider a model in which the same slope is shared by all elements in a suitably large group of domains. To work out the regression estimator (10.5.10), the AV and the variance estimator for this case presents no particular difficulty. The form of $\hat{t}_{dr}$ is complicated somewhat by the fact that $\sum_{s_d} e_{ks}/\pi_k$ does not vanish.

10.7. Group Models for Domains

Another useful device in regression estimation for domains is to model in terms of population groups of elements perceived to be homogeneous. This leads to a poststratified domain estimator. The idea here is that the groups are a powerful factor for explaining the variance of the y variable, whereas the domains themselves are perhaps not. For example, age/sex groups will often explain a good part of individual variation, but beyond that, a partitioning into geographic domains may be comparatively weak explanatory factor.

Partition the population into G groups denoted $U_{\cdot 1}, \ldots, U_{\cdot g}, \ldots, U_{\cdot G}$. The sizes, usually unknown, are denoted $N_{\cdot 1}, \ldots, N_{\cdot g}, \ldots, N_{\cdot G}$. The case considered here occurs when the G population groups cut across the D domains to form a grid of DG population cells denoted U_{dg}, $d = 1, \ldots, D$; $g = 1, \ldots, G$. Let N_{dg} be the size of U_{dg}. For added clarity, the domains are now denoted $U_{d\cdot}$, $d = 1, \ldots, D$, and their respective sizes $N_{d\cdot}$, $d = 1, \ldots, D$.

The population and its size are now partitioned according to the equations

$$U = \bigcup_{d=1}^{D} U_{d\cdot} = \bigcup_{g=1}^{G} U_{\cdot g} = \bigcup_{d=1}^{D} \bigcup_{g=1}^{G} U_{dg} \tag{10.7.1a}$$

and

$$N = \sum_{d=1}^{D} N_{d\cdot} = \sum_{g=1}^{G} N_{\cdot g} = \sum_{d=1}^{D} \sum_{g=1}^{G} N_{dg} \tag{10.7.1b}$$

The sample s is analogously partitioned into domain subsamples $s_{d\cdot}$, group subsamples $s_{\cdot g}$, and cell subsamples s_{dg}. Denoting their respective sizes by $n_{s_{d\cdot}}$, $n_{s_{\cdot g}}$, and $n_{s_{dg}}$, we can write the sample analogs of the equations (10.7.1a) and (10.7.1b) as

$$s = \bigcup_{d=1}^{D} s_{d\cdot} = \bigcup_{g=1}^{G} s_{\cdot g} = \bigcup_{d=1}^{D} \bigcup_{g=1}^{G} s_{dg} \tag{10.7.2a}$$

and

$$n_s = \sum_{d=1}^{D} n_{s_{d\cdot}} = \sum_{g=1}^{G} n_{s_{\cdot g}} = \sum_{d=1}^{D} \sum_{g=1}^{G} n_{s_{dg}} \tag{10.7.2b}$$

The $n_{s_{dg}}$ are assumed to be random. Ordinarily, $n_{s_{d\cdot}}$ and $n_{s_{\cdot g}}$ are also random but, for example, $n_{s_{\cdot g}}$ would be fixed if the gth group is a stratum from which a fixed number of elements is drawn. In many applications the domains are numerous, say, a hundred or more. By contrast, one would ordinarily work with only a limited number of groups, say 10 or less. From the point of view of efficiency, extremely fine grouping does not ordinarily pay. The variance reductions would often be marginal by extending the number of groups beyond 10, say.

EXAMPLE 10.7.1. To illustrate the practical constraints, suppose we have an *SI* sample s of size $n = 3{,}000$ split on $D = 200$ domains of varying relative sizes $P_d = N_{d\cdot}/N$. Although an average domain contains 15 sample points, it is likely that a number of domain sample sizes $n_{s_{d\cdot}}$ are five or less, and some may be zero.

Assuming $G = 8$ roughly equal-sized groups, the group sample sizes $n_{s_{\cdot g}}$ are approximately 400, that is, fairly large. As for the cells, they number as many as $DG = 1{,}600$ in this example. Clearly, we can expect many of them to be empty, that is, $n_{s_{dg}} = 0$. In such cells, one cannot compute a sample cell mean, for example.

No matter what estimation technique is used, in this case it is obviously difficult to obtain precise estimates (narrow confidence intervals) for all 200 domain totals or means. Ultimately, the domain sample size $n_{s_{d\cdot}}$ is the decisive factor for the precision, even though the precision can be improved by the auxiliary information. However, for the larger domains, especially if the groups are a strong explanatory factor, good precision may be obtained by the estimators (10.7.4) and (10.8.2) presented below. The estimator (10.8.2) avoids the difficulty with zero cell sample counts.

The group model for domains states that

$$\begin{cases} E_\xi(y_k) = \beta_{dg} \\ V_\xi(y_k) = \sigma_{dg}^2 \end{cases} \tag{10.7.3}$$

for $k \in U_{dg}$. For each of the DG cells, there is a parameter, β_{dg}, to estimate. It is not hard to show that the population fit leads to

$$B_{dg} = \textstyle\sum_{U_{dg}} y_k/N_{dg} = \bar{y}_{U_{dg}}$$

and that the sample fit (provided s_{dg} is nonempty) gives

$$\hat{B}_{dg} = \textstyle\sum_{s_{dg}} \check{y}_k/\hat{N}_{dg} = \tilde{y}_{s_{dg}}$$

with

$$\hat{N}_{dg} = \textstyle\sum_{s_{dg}} 1/\pi_k$$

Result 10.5.1 now produces the poststratified domain estimator,

$$\hat{t}_{d\text{pos}} = \sum_{g=1}^{G} N_{dg}\tilde{y}_{s_{dg}} \tag{10.7.4}$$

where the population cell counts N_{dg} must be known. The approximate variance is given by (10.5.11) with

$$E_k = y_k - \bar{y}_{U_{dg}} \tag{10.7.5}$$

for $k \in U_{dg}$. The variance estimator is obtained from equation (10.5.12). The residuals and the g weights are

$$e_{ks} = y_k - \tilde{y}_{s_{dg}}$$

and

$$g_{dks} = (N_{dg}/\hat{N}_{dg})z_{dk}$$

for $k \in s_{dg}$. Since g_{dks} is of the form shown in equation (10.5.13), the variance estimator is a double sum extending over $s_{d\cdot}$ only, rather than over all of s. That is, the expression (10.5.14) can be used.

EXAMPLE 10.7.2. For the SI design with $f = n/N$, further simplification of the variance estimator for $\hat{t}_{d\text{pos}}$ results from the fact that $\check{\Delta}_{kl}$ is constant for all $k \neq l$. In this case, the variance estimator (10.5.14) reduces to

$$\begin{aligned} \hat{V}_{SI}(\hat{t}_{d\text{pos}}) &= \sum_{g=1}^{G} \left(\frac{n}{n-1} \frac{n_{s_{dg}} - 1}{n_{s_{dg}}} \right) N_{dg}^2 \left(\frac{1}{n_{dg}} - \frac{1}{\hat{N}_{dg}} \right) S_{ys_{dg}}^2 \\ &\doteq \sum_{g=1}^{G} N_{dg}^2 \left(\frac{1}{n_{s_{dg}}} - \frac{1}{\hat{N}_{dg}} \right) S_{ys_{dg}}^2 \end{aligned} \tag{10.7.6}$$

where $\hat{N}_{dg} = Nn_{s_{dg}}/n$ and

$$S_{ys_{dg}}^2 = \frac{1}{n_{s_{dg}} - 1} \sum\nolimits_{s_{dg}} (y_k - \bar{y}_{s_{dg}})^2 \tag{10.7.7}$$

is the sample variance in the cell; $\bar{y}_{s_{dg}}$ now denotes the straight cell mean.

Remark 10.7.1. The estimator (10.7.4), which is generated by the model (10.7.3), is a logical extension to a domain of the classical poststratified estimator (7.6.1). Here, it is the domain part of the sample, $s_{d\cdot} = s \cap U_{d\cdot}$, that is poststratified with the aid of the G groups; the result is the sample cells $s_{d1}, \ldots, s_{dG}$. The estimator (10.7.4) is constructed as the weighted sum of the sample cell means, with the known cell counts N_{dg} as weights.

Remark 10.7.2. The poststratified estimator (10.7.4) supposes that none of the cell sample counts $n_{s_{d1}}, \ldots, n_{s_{dG}}$ is extremely small. When all cell counts are of respectable size, (10.7.4) will yield good estimates, and all the more so if the groups explain much of the variance in y. A prudent rule is to require at least 10 observations in each cell. Obviously, in practice, the conditions are often less favorable. In Example 10.7.1, which involves $n = 3{,}000$ observations and $DG = 200 \times 8 = 1{,}600$ cells, the probability is considerable for an average size domain that some cell counts $n_{s_{d1}}, \ldots, n_{s_{dG}}$ are zero or extremely small (one or two observations). For such a domain, the poststratified domain estimator

(10.7.4) should not (or cannot) be used. The component $\tilde{y}_{s_{dg}}$ is likely to contribute both bias and high variance if the cell count $n_{s_{dg}}$ is extremely small.

In the large-scale computerized production of statistical tables by means of the estimator (10.7.4), the possibility of zero or near-zero cell counts may be a serious problem, since the computer inflexibly follows a given routine. An estimator that avoids the difficulty without appreciable loss of efficiency is needed; one such procedure is given in the next section.

Remark 10.7.3. The domain size $N_{d\cdot} = \sum_{g=1}^{G} N_{dg}$ is a known quantity in applications where (10.7.4) is used. The estimator of the domain mean $\bar{y}_{U_d} = t_d/N_{d\cdot}$ generated by the model (10.7.3) is therefore

$$\hat{\bar{y}}_{U_d} = \sum_{g=1}^{G} P_{dg}\tilde{y}_{s_{dg}}$$

where $P_{dg} = N_{dg}/N_{d\cdot}$ is the known cell proportion within the domain d.

10.8. Problems Arising for Small Domains; Synthetic Estimation

The domain/group model (10.7.3) in the previous section contained DG β parameters to estimate, possibly a very large number. Remark 10.7.2 warned that this may cause difficulties, including the impossibility to compute the estimator (10.7.4) when some cell counts $n_{s_{dg}}$ $(g = 1, \ldots, G)$ are zero. It is often wise to reduce the number of model parameters; one option is to work with a model expressed in terms of group effects only. This stabilizes the parameter estimates and causes little loss of efficiency if the groups (rather than the domains) are the principal factor in explaining the variation in y. As mentioned before, age/sex groups may be a dominant factor for explaining variation in individuals, whereas geographically defined domains may be unimportant. Such a reduction of the model meets with the objective to limit as much as possible (yet without loss of precious auxiliary information) the number of parameters in the model used to generate the estimator of the domain mean or total.

Consider therefore the group model such that, for $g = 1, \ldots, G$,

$$\begin{cases} E_\xi(y_k) = \beta_g \\ V_\xi(y_k) = \sigma_g^2 \end{cases} \tag{10.8.1}$$

for all $k \in U_{\cdot g}$. In this one-way ANOVA model, $U_{\cdot 1}, \ldots, U_{\cdot g}$ are the population groups introduced in the preceding section, from which we also borrow other notation used below.

From Result 10.5.1, the regression estimator of the domain total t_d generated by this more parsimonious model is

$$\hat{t}_{dr} = \sum_{g=1}^{G} N_{dg}\tilde{y}_{s_{\cdot g}} + (N_{d\cdot}/\hat{N}_{d\cdot}) \sum_{g=1}^{G} \hat{N}_{dg}(\tilde{y}_{s_{dg}} - \tilde{y}_{s_{\cdot g}}) \tag{10.8.2}$$

an approximately unbiased estimator, with

$$\tilde{y}_{s_{\cdot g}} = \textstyle\sum_{s_{\cdot g}} \check{y}_k/\hat{N}_{\cdot g}; \qquad \tilde{y}_{s_{dg}} = \sum_{s_{dg}} \check{y}_k/\hat{N}_{dg}$$

and

$$\hat{N}_{\cdot g} = \textstyle\sum_{s_{\cdot g}} 1/\pi_k; \qquad \hat{N}_{dg} = \sum_{s_{dg}} 1/\pi_k; \qquad \hat{N}_{d\cdot} = \sum_{s_{d\cdot}} 1/\pi_k$$

The necessary quantities N_{dg} must be available from accurate sources (registers, a census, etc.). Note that the term

$$\sum_{g=1}^{G} \hat{N}_{dg}\tilde{y}_{s_{dg}}/\hat{N}_{d\cdot} = \textstyle\sum_{s_{d\cdot}} \check{y}_k/\hat{N}_{d\cdot} = \tilde{y}_{s_{d\cdot}}$$

causes no computational difficulty as long as the domain has at least one observation. We can thus write the estimator (10.8.2) as

$$\hat{t}_{dr} = N_{d\cdot}\left[\tilde{y}_{s_{d\cdot}} + \sum_{g=1}^{G}\left(\frac{N_{dg}}{N_{d\cdot}} - \frac{\hat{N}_{dg}}{\hat{N}_{d\cdot}}\right)\tilde{y}_{s_{\cdot g}}\right]$$

The appropriate residuals for the AV formula shown in equation (10.5.11) are in this case

$$E_k = y_k - \bar{y}_{U_{\cdot g}} \tag{10.8.3}$$

for $k \in U_{dg}$, where $\bar{y}_{U_{\cdot g}}$ is the y-mean in $U_{\cdot g}$. The quantities required for the variance estimator (10.5.12) are, for $k \in s_{dg}$, given by

$$e_{ks} = y_k - \tilde{y}_{s_{\cdot g}}$$

and

$$g_{dks} = N_{d\cdot}\left\{\frac{z_{dk}}{\hat{N}_{d\cdot}} + \left(\frac{N_{dg}}{N_{d\cdot}} - \frac{\hat{N}_{dg}}{\hat{N}_{d\cdot}}\right)\frac{1}{\hat{N}_{\cdot g}}\right\}$$

Remark 10.8.1. It is of interest to compare the performance of the estimator (10.8.2) with that of the poststratified estimator given by (10.7.4). The latter should have a smaller variance if data are abundant and there are domain differences as well as group differences. The residual (10.8.3) that determines the approximate variance of the estimator (10.8.2) can be written as

$$E_k = E_k^* + (\bar{y}_{U_{dg}} - \bar{y}_{U_{\cdot g}})$$

where the overbar denotes a straight mean over indicated sets, and

$$E_k^* = y_k - \bar{y}_{U_{dg}}$$

is the residual associated with the poststratified estimator (10.7.4). A case of special interest to examine is when

$$\bar{y}_{U_{dg}} = \bar{y}_{U_{\cdot g}} \tag{10.8.4}$$

for all groups $g = 1, \ldots, G$. This states that, in each group, the group mean for the domain equals the group mean for the population as a whole, that is, in a sense, the domain is like the population. We then have equality of the residuals; $E_k = E_k^*$ for all k. Consequently, the approximate variance of the estimators (10.7.4) and (10.8.2) are equal, and both are nearly unbiased. In survey practice (10.8.4) cannot be counted on to hold exactly. But in the presence of strong group effects and no pronounced domain effects, little loss of efficiency is suffered by using the estimator derived from the much simpler model shown in (10.8.1). In addition, the estimator (10.8.2) enjoys the advantage that calculation is possible despite some zero cell counts. In rare instances, (10.8.2) may yield impossible estimates. If y is an always nonnegative variable, only nonnegative estimates of t_d can be tolerated, but a remote possibility exists for very small domains that (10.8.2) overcorrects and turns out a negative estimate.

The first term of the estimator (10.8.2),

$$\hat{t}_{dsy} = \sum_{g=1}^{G} N_{dg} \tilde{y}_{s \cdot g} \tag{10.8.5}$$

can be considered an estimator in its own right; however, it is biased. It is called a *synthetic estimator* and has been studied extensively in the recent literature, especially for the *SI* design, in which case $\tilde{y}_{s \cdot g}$ becomes $\bar{y}_{s \cdot g}$, the unweighted group mean. Early references are National Center for Health Statistics (1968) and Gonzalez (1973).

We have

$$E(\hat{t}_{dsy}) \doteq \sum_{g=1}^{G} N_{dg} \bar{y}_{U \cdot g}$$

so the bias of $\hat{t}_{dsy}$ as an estimator of t_d is approximately

$$B(\hat{t}_{dsy}) = E(\hat{t}_{dsy}) - t_d \doteq -\sum_{g=1}^{G} N_{dg}(\bar{y}_{U_{dg}} - \bar{y}_{U \cdot g})$$

If equation (10.8.4) holds for $g = 1, \ldots, G$, then $\hat{t}_{dsy}$ is unbiased. For (10.8.4) to hold in a given practical setting would require an extreme stroke of luck; normally, the unknown bias is nonzero, perhaps considerable, and a confidence interval constructed around the point estimate would be off-center and invalid. Why then would the statistician be interested in the estimator (10.8.5)? The answer is that the variance of $\hat{t}_{dsy}$ is often extremely small, and in most cases much smaller than that of the nearly unbiased estimator shown in equation (10.8.2). This is not surprising since the group means $\tilde{y}_{s \cdot g}$ in (10.8.5) are determined with high precision as a result of substantial group sample sizes and perhaps small within-group variances. On the other hand, the bias correction term

$$(N_{d \cdot}/\hat{N}_{d \cdot}) \sum_{g=1}^{G} \hat{N}_{dg}(\tilde{y}_{s_{dg}} - \tilde{y}_{s \cdot g})$$

will contribute a large amount of variance to the estimator (10.8.2). Thus $\hat{t}_{dsy}$ will be highly precise (low variance), but may be highly inaccurate (considerable bias and MSE). To use $\hat{t}_{dsy}$ is a gamble; the statistician hopes for a small or negligible bias. If this assumption holds, the synthetic estimator is attractive. Thus the synthetic estimator (10.8.5) cannot be recommended unless the statistician is willing to run the risk of bias and invalid confidence statements.

Remark 10.8.2. More generally, a synthetic estimator is defined by the prediction term of the regression estimator (10.5.10),

$$\hat{t}_{dsy} = \sum_{U_d} \hat{y}_k = (\sum_{U_d} \mathbf{x}_k)'\hat{\mathbf{B}}$$

where strength has been borrowed from outside the domain in deriving the predictions $\hat{y}_k$. For example, a synthetic estimator obtained by imposing the simple one-parameter ratio model (10.5.2) on the domain d is given by

$$\hat{t}_{dsy} = (\sum_{U_d} x_k)\hat{B} \tag{10.8.6}$$

where

$$\hat{B} = \frac{\sum_s \check{y}_k}{\sum_s \check{x}_k}$$

Here, $\hat{B}$ estimates a slope parameter

$$B = \frac{\sum_U y_k}{\sum_U x_k} \tag{10.8.7}$$

for the population as a whole. The variance of $\hat{B}$ is often very small because the estimation uses the full sample s. Consequently, the variance of the estimator (10.8.6) is small, but the bias is substantial if the domain ratio

$$B_d = \frac{\sum_{U_d} y_k}{\sum_{U_d} x_k}$$

differs considerably from the overall population ratio shown in equation (10.8.7).

A number of estimators have been constructed as a weighted linear combination of a biased, low variance synthetic term and an unbiased, high-variance standard-estimator term, such as the weighted domain sample mean. Different statistical principles may be invoked in constructing the weights. One technique is the Empirical Bayes approach examined by Fay and Herriot (1979). Other combined estimators have been proposed by Schaible (1979), Drew, Singh and Choudhry (1982), Battese and Fuller (1984), and others. A spectrum of techniques is presented in the book edited by Platek et al. (1987).

The combined estimators usually have nonnegligible bias in small to modest samples, so the MSE contains a fairly important squared-bias term. Therefore, it is usually impossible to obtain a valid confidence interval. However, the MSE may be smaller than for an unbiased estimator. Moreover, the weighting of the terms may be done in such a way that the bias tends to zero as the sample size increases. Considerable weight is given to the synthetic term when the domain sample size is very small, and, as this size increases, the weight is gradually shifted onto the standard estimator, so that the combined estimator is consistent.

10.9. More on the Comparison of Two Domains

Let us return to the important question of comparing two domain means. If the two domains are U_1 and U_2 (corresponding to $d = 1$ and $d = 2$), the difference to estimate is

$$D = \bar{y}_{U_1} - \bar{y}_{U_2} = \frac{t_1}{N_1} - \frac{t_2}{N_2}$$

An obvious estimator, the difference of the two π-weighted sample means, was examined in Section 10.3. The regression approach is generally more efficient, provided that suitable auxiliary information is available. For $d = 1, 2$, suppose that the domain mean $\bar{y}_{U_d}$ is estimated by $\hat{\bar{y}}_{U_d r} = \hat{t}_{dr}/N_d$, where $\hat{t}_{dr}$ is the regression estimator (10.5.10) and N_d is one of the auxiliaries assumed to be known in this section. To estimate D, we then use

$$\begin{aligned} \hat{D}_r &= \hat{\bar{y}}_{U_1 r} - \hat{\bar{y}}_{U_2 r} \\ &= \textstyle\sum_s g^*_{ks} \check{y}_k \end{aligned} \tag{10.9.1}$$

with the g weight

$$g^*_{ks} = \frac{g_{1ks}}{N_1} - \frac{g_{2ks}}{N_2} \tag{10.9.2}$$

where, for $d = 1, 2$, g_{dks} is given by (10.5.7).

The approximate variance is obtained as

$$AV(\hat{D}_r) = \textstyle\sum\sum_{U_1 \cup U_2} \Delta_{kl} \check{A}_k \check{A}_l \tag{10.9.3}$$

with $\check{A}_k = A_k/\pi_k$, where

$$A_k = \frac{z_{1k}(E_k - \bar{E}_{U_1})}{N_1} - \frac{z_{2k}(E_k - \bar{E}_{U_2})}{N_2}$$

Here, z_{dk} is the domain indicator (10.2.3). The variance estimator is given by (10.5.12) if g_{dks} is replaced by g^*_{ks}, which is given by (10.9.2).

Remark 10.9.1. We can write (10.9.3) as

$$AV(\hat{D}_r) = AV(\hat{\bar{y}}_{U_1 r}) + AV(\hat{\bar{y}}_{U_2 r}) - 2AC(\hat{\bar{y}}_{U_1 r}, \hat{\bar{y}}_{U_2 r})$$

where the approximate variance $AV(\hat{\bar{y}}_{U_d r})$, $d = 1, 2$, is given by (10.5.11) divided by N_d^2, and the approximate covariance term is

$$AC(\hat{\tilde{y}}_{U_1 r}, \hat{\tilde{y}}_{U_2 r}) = \frac{1}{N_1 N_2} \sum_{k \in U_1} \sum_{l \in U_2} \Delta_{kl} \frac{E_k - \bar{E}_{U_1}}{\pi_k} \frac{E_l - \bar{E}_{U_2}}{\pi_l}$$

Example 10.9.1. For the domain and group model (10.7.3), the estimator (10.9.1) is a difference between two poststratified domain estimators,

$$\hat{D}_r = \sum_{g=1}^{G} P_{1g} \tilde{y}_{s_{1g}} - \sum_{g=1}^{G} P_{2g} \tilde{y}_{s_{2g}}$$

where $P_{dg} = N_{dg}/N_{d\cdot}$; see Remark 10.7.3. The weights necessary for the variance estimator are

$$g_{ks}^* = \frac{z_{1k} P_{1g}}{\hat{N}_{1g}} - \frac{z_{2k} P_{2g}}{\hat{N}_{2g}}$$

if $k \in s_{1g} \cup s_{2g}$, $g = 1, \ldots, G$, and $g_{ks}^* = 0$ for all other k. In particular, for the *SI* design with n elements chosen from N, the variance estimator takes the simple additive form

$$\hat{V}_{SI}(\hat{D}) = \sum_{d=1}^{2} \sum_{g=1}^{G} P_{dg}^2 \left(\frac{1}{n_{s_{dg}}} - \frac{1}{\hat{N}_{dg}} \right) S_{ys_{dg}}^2$$

where $\hat{N}_{dg} = N n_{s_{dg}}/n$ and $S_{ys_{dg}}^2$ is given by (10.7.7); moreover, we have replaced $n(n_{s_{dg}} - 1)/(n - 1)n_{s_{dg}}$ by unity.

Exercises

Remark. All of the numerical exercises in this chapter concern estimation for three domains corresponding to a grouping of the 284 Swedish municipalities into three major regions. These were obtained by merging the 8 smaller regions indicated in Appendix B as follows:

Major region 1 (roughly East and Central Sweden). All municipalities in regions 1, 2, and 3.

Major region 2 (roughly South and West Sweden). All municipalities in regions 4 and 5.

Major region 3 (roughly North Sweden). All municipalities in regions 6, 7, and 8.

10.1. Consider the MU284 population, the three major Swedish regions, and the study variable CS82 ($= y$). (a) Assuming that an *SI* sample of $n = 150$ municipalities is to be drawn from the whole MU284 population, compute, for each of the three major Swedish regions, the (exact or approximate) variance of the estimators

$$\hat{t}_{d\pi} = \sum_{s_d} \check{y}_k \quad \text{and} \quad \tilde{t}_d = N_d \tilde{y}_{s_d} = N_d \hat{t}_{d\pi}/\hat{N}_d$$

Use the following population data:

Major region d	N_d	$\sum_{U_d} y_k$	$\sum_{U_d} y_k^2$
1	105	1,093	14,685
2	94	1,005	12,213
3	85	485	3,491

(b) An *SI* sample of size $n = 150$ was drawn from the MU284 population, and the following data were computed from the sample:

Major region d	n_{s_d}	$\sum_{s_d} y_k$	$\sum_{s_d} y_k^2$
1	63	691	9,869
2	55	587	6,993
3	32	171	1,005

For each of the two estimators in (a), compute approximately 95% confidence intervals for the three major region totals of the variable CS82.

10.2. Use the data in Exercise 10.1 to compute an approximately 95% confidence interval for the difference between the population means of region 2 and region 3.

10.3. Suppose that the *SI* design in Exercise 10.1 is replaced by the following *STSI* design. Two strata are created using the variable S82, so that stratum 1 consists of the 142 smallest municipalities according to the value of S82 and stratum 2 consists of the other 142 municipalities. Estimates are required of the major region totals of the variable CS82 ($= y$). (a) Verify that the (approximate) variances of the two estimators

$$\hat{t}_{d\pi} = \sum_{s_d} \check{y}_k \quad \text{and} \quad \tilde{t}_d = N_d \tilde{y}_{s_d} = N_d \hat{t}_{d\pi}/\hat{N}_d$$

are given by

$$V(\hat{t}_{d\pi}) = \sum_{h=1}^{2} N_h^2 \frac{1-f_h}{n_h} \frac{1}{N_h - 1}\left\{(N_{dh} - 1)S_{yU_{dh}}^2 + N_{dh}\left(1 - \frac{N_{dh}}{N_h}\right)\bar{y}_{U_{dh}}^2\right\}$$

$$AV(\tilde{t}_d) = \sum_{h=1}^{2} N_h^2 \frac{1-f_h}{n_h} \frac{1}{N_h - 1}\left\{(N_{dh} - 1)S_{yU_{dh}}^2 + N_{dh}\left(1 - \frac{N_{dh}}{N_h}\right)(\bar{y}_{U_{dh}} - \bar{y}_{U_d})^2\right\}$$

(b) Assuming that $n_1 = n_2 = 75$, compute the variances in (a) using the following population data

Major region d	Stratum h 1	2
1	(47; 349; 3,375)	(58; 744; 11,310)
2	(50; 437; 4,159)	(44; 568; 8,054)
3	(45; 194; 1,000)	(40; 291; 2,491)

where the components of the three-dimensional vectors are N_{dh}, $\sum_{U_{dh}} y_k$, and $\sum_{U_{dh}} y_k^2$, respectively

(c) Using the *STSI* design described above (with $n_1 = n_2 = 75$), a sample was drawn from the MU284 population, and the following data were computed from the sample:

Major region d	Stratum h 1	2
1	(28; 191; 1,595)	(26; 334; 4,920)
2	(23; 208; 2,058)	(21; 292; 4,516)
3	(24; 98; 486)	(28; 207; 1,821)

where the components of the three-dimensional vectors are $n_{s_{dh}}$, $\sum_{s_{dh}} y_k$, and $\sum_{s_{dh}} y_k^2$, respectively. Using $\tilde{t}_d$, compute point estimates of the three regional totals of CS82. Also, compute the corresponding *cve*'s.

10.4. In a survey of the MU284 population, the study variable is REV84 ($= y$), and P75 ($= x$) is an auxiliary variable. The goal is to estimate the major region totals of REV84. (a) Assuming that an *SI* sample of $n = 160$ municipalities is to be drawn, compute the approximate variances of the ratio estimators $\hat{t}_{dra1}$ and $\hat{t}_{dra2}$ generated by the models (10.5.2) and (10.5.3), respectively. Use the following population data:

Major region d	N_d	$\sum_{U_d} y_k$	$\sum_{U_d} y_k^2$	$\sum_{U_d} x_k$	$\sum_{U_d} x_k^2$	$\sum_{U_d} y_k x_k$
1	105	382,906	5,377,345,182	3,654	616,568	56,907,240
2	94	281,392	2,729,354,250	2,772	341,592	29,922,838
3	85	209,719	956,277,147	1,756	68,714	7,550,583

(b) An *SI* sample of size $n = 160$ was drawn from the MU284 population, and the following data were computed from the sample:

Major region d	n_{s_d}	$\sum_{s_d} y_k$	$\sum_{s_d} y_k^2$	$\sum_{s_d} x_k$	$\sum_{s_d} x_k^2$	$\sum_{s_d} y_k x_k$
1	58	258,287	4,762,200,163	2,456	555,030	50,951,390
2	55	169,268	2,047,462,576	1,684	248,648	22,212,708
3	47	124,662	564,098,390	999	38,703	4,190,969

Use $\hat{t}_{dra2}$ to compute point estimates of the major region totals. Also, compute the corresponding variance estimates.

10.5. A domain, a stratum, a poststratum, and a primary sampling unit are four conceptually different types of subset of a finite population. Give precise verbal definitions of each of these concepts.

10.6. A *BE* sample is drawn from U with the constant inclusion probability $\pi_k = \pi$ for all $k \in U$. For the mean $\bar{y}_{U_d}$ of a domain U_d of U, two estimators are considered:

$$\hat{\bar{y}}_{U_d 1} = (1/N_d \pi) \sum_{s_d} y_k \quad \text{and} \quad \hat{\bar{y}}_{U_d 2} = (1/n_{s_d}) \sum_{s_d} y_k$$

where $s_d = s \cap U_d$, n_{s_d} is the size of s_d and N_d the known size of U_d. Compare the variances of the two estimators (for $\hat{\bar{y}}_{U_d 2}$ use the *AV* expression).

10.7. Apply the general Result 10.5.1 to the domain ratio model shown in (10.6.1) and verify in detail that this model leads to equations (10.6.2) and (10.6.3) for the domain ratio estimator and its variance estimator.

10.8. In a survey design, there are H strata that cut across D domains. The cell counts N_{dh} are unknown, but the marginal counts $N_{d\cdot} = \sum_{h=1}^{H} N_{dh}$ and $N_{\cdot h} = \sum_{d=1}^{D} N_{dh}$ are known, $d = 1, \ldots, D$; $h = 1, \ldots, H$. The domain total t_d is estimated by

$$\hat{t}_d = N_{d\cdot} \frac{\sum_{h=1}^{H} f_h^{-1} \sum_{s_{dh}} y_k}{\sum_{h=1}^{H} f_h^{-1} n_{dh}}$$

where $f_h = n_{\cdot h}/N_{\cdot h}$ is the *STSI* sampling fraction in stratum h, $s_{dh} = s_h \cap U_d$ is the part of the sample from stratum h that falls in the dth domain, and n_{dh} is the size of s_{dh}. Formulate the model underlying the estimator. Find the approximate variance and the corresponding variance estimator. Durbin (1958) points out: "If the proportion in the domain of study is small most of the advantage of the stratification has been lost, while only if the proportion is close to unity has the advantage been retained." Can this be seen from the expression for the approximate variance?

10.9. Consider the *SI* design with $f = n/N$. The estimator $\hat{t}_{dpos}$ given by equation (10.7.4) takes the form

$$\hat{t}_{dpos} = \sum_{g=1}^{G} N_{dg} \bar{y}_{s_{dg}}$$

where $\bar{y}_{s_{dg}}$ is the mean of y in the sample cell $s_{dg} = U_{dg} \cap s$. Verify that the variance estimator obtained from Result 10.5.1 is

$$\hat{V}(\hat{t}_{dpos}) = (1 - f) \sum_{g=1}^{G} a_{dg} N_{dg}^2 S_{ys_{dg}}^2 / n_{s_{dg}}$$

where

$$S^2_{ys_{dg}} = \sum_{s_{dg}} (y_k - \bar{y}_{s_{dg}})^2/(n_{s_{dg}} - 1)$$

and

$$a_{dg} = n(n_{s_{dg}} - 1)/(n - 1)n_{s_{dg}} \doteq 1$$

Use the conditional argument to obtain a very nearly equivalent variance estimator. Condition on the sample cell counts $n_{s_{dg}}$.

10.10. Consider a two-stage element sampling design according to the general description in Section 4.3. Let

$$\hat{\bar{y}}_{U_d} = N_d \tilde{y}_{s_d}$$

be the estimator of the domain mean $\bar{y}_{U_d}$ where

$$\tilde{y}_{s_d} = (\sum_{s_d} y_k/\pi_k)/(\sum_{s_d} 1/\pi_k)$$

with $\pi_k = \pi_{Ii}\pi_{k|i}$. Find the AV expression for $\hat{\bar{y}}_{U_d}$.

10.11. From a population U of N individuals, an *SI* sample s of size n is drawn. Let U_1 and U_2 be the subpopulations "men" and "women," respectively. Let s_1 be the set of men in the sample and s_2 the set of women in the sample. Denote the sizes by n_{s_1} and n_{s_2} $(n = n_{s_1} + n_{s_2})$. Show that the probability of obtaining s_1, s_2, given n_{s_1}, n_{s_2}, is the same as the probability of obtaining an *STSI* sample with n_{s_1} men and n_{s_2} women.

10.12. In a survey, a comparison is wanted between two domains (areas) U_A and U_B of the sampled population U. For area A, define the ratio of totals $R_A = \sum_{U_A} y_k/\sum_{U_A} z_k$ (for example, food expenditures per income dollar, or the prevalence of a certain disease); let R_B be analogously defined for area B. An *SI* sample s with $f = n/N$ is drawn from U. (a) To estimate the difference $R_A - R_B$, form the estimator $\hat{R}_A - \hat{R}_B$, where $\hat{R}_A = \sum_{s_A} y_k/\sum_{s_A} z_k$, and $s_A = U_A \cap s$ denotes the part of the sample s that comes from area A; $\hat{R}_B$ is analogous. Find the approximate variance of $\hat{R}_A - \hat{R}_B$ and the corresponding variance estimator. (b) To estimate the ratio R_A/R_B, form the estimator $\hat{R}_A/\hat{R}_B$. Find the approximate variance of $\hat{R}_A/\hat{R}_B$ and the corresponding variance estimator.

CHAPTER 11

Variance Estimation

11.1. Introduction

The precision of an estimator $\hat{\theta}$ is usually discussed in terms of the variance $V(\hat{\theta})$. Usually the exact value of the variance is unknown, because it depends on unknown population quantities. After survey data have been obtained, however, an estimate of the variance can be calculated. When survey results are presented, it is good practice to provide variance estimates $\hat{V}(\hat{\theta})$ for the estimators $\hat{\theta}$ used in the survey. A variance estimate is often used to construct a confidence interval,

$$\hat{\theta} \pm \text{constant} \times [\hat{V}(\hat{\theta})]^{1/2}$$

assuming that the sampling distribution of $\hat{\theta}$ is approximately normal.

This chapter presents some special variance estimation methods in addition to those already familiar from earlier chapters. Three general variance estimation methods have been dealt with so far.

1. Chapter 2 considered the case where $\hat{\theta}$ is a π estimator. Two unbiased estimators of $V(\hat{\theta})$ were given, the Horvitz-Thompson estimator and the Yates-Grundy-Sen estimator; see Section 2.8. Both estimators can also be modified and used when $\hat{\theta}$ is a linear function of several π estimators.
2. When $\hat{\theta}$ is a nonlinear function of π estimators, an exact expression for the variance $V(\hat{\theta})$ cannot usually be found. In Chapter 5, the Taylor linearization was used to obtain an approximate variance, $AV(\hat{\theta})$. Section 5.5 dealt with the estimation of this approximate variance.
3. Chapter 6 presented a weighted residual variance estimator for the regression estimator. This variance estimator has advantages over the Taylor linearization estimator.

This chapter presents some approximate techniques for variance estimation, including random groups, balanced half-samples, jackknife, and bootstrap. These techniques are able to handle both complex sampling designs and complex estimators. Thus, they can be used in cases where methods (1) to (3) above are not easily applicable. Since they usually require extensive computation, they are sometimes called computer intensive.

These variance estimators are theoretically complicated, because it is hard to derive general results on their statistical properties, for example, their bias. The few theoretical results of this chapter are limited to the case in which the variance of a π estimator is estimated. These techniques are primarily used with more complex estimators, however. In these cases, no theoretical results are available, and our knowledge about the statistical properties of the variance estimators is limited to conclusions drawn from simulation studies or other sources of empirical evidence.

Throughout this chapter, we assume sampling without replacement. Section 11.2 presents a computationally simplified variance estimator for the π estimator, which is calculated as if the sample selection had been *with* replacement although the actual selection is *without* replacement. The subsequent sections introduce the special techniques of random groups, balanced half-samples, jackknife, and bootstrap.

A more detailed account of these methods (and of the Taylor linearization method) is given by Wolter (1985). The following presentation owes much to that source. For a brief review of variance estimation techniques, the reader is also referred to Rust (1985).

A variance estimator $\hat{V}(\hat{\theta})$ should be:

a. Unbiased, or nearly so, that is, it should satisfy

$$E[\hat{V}(\hat{\theta})] \doteq V(\hat{\theta})$$

b. Stable, that is, the variance of the variance estimator should be small.
c. Nonnegative, that is, always take nonnegative values.
d. Produce confidence intervals that cover θ with a probability approximately equal to the stated confidence level.

Now, to what extent do the variance estimators presented in this chapter fulfill the above requirements? They are all nonnegative. When $\hat{\theta}$ is a π estimator, they often have a positive bias. Thus, they will overestimate $V(\hat{\theta})$. When $\hat{\theta}$ is a nonlinear function of π estimators, exact results do not exist. However, it is often hoped that these variance estimators will continue to work well. As for stability and confidence level, we refer to simulation studies (see, for example, those mentioned in Section 11.7).

We now present a result that is used several times later in this chapter. Suppose $\hat{\theta}$ is an estimator of θ, based on data from a probability sample. In the techniques that follow, we often consider a number (say A) of subsets of the original sample, and a separate estimator of θ calculated from each subset,

$$\hat{\theta}_1, \ldots, \hat{\theta}_a, \ldots, \hat{\theta}_A$$

An alternative full-sample estimator of θ can now be formed by averaging the $\hat{\theta}_a$,

$$\hat{\theta}^* = \frac{1}{A} \sum_{a=1}^{A} \hat{\theta}_a$$

Sometimes, $\hat{\theta}^*$ and $\hat{\theta}$ can be defined so that they will be identical.

Our prime interest here is in estimating the variance of $\hat{\theta}^*$, that is, $V(\hat{\theta}^*)$. We consider the following two expressions,

$$\hat{V}_1 = \frac{1}{A(A-1)} \sum_{a=1}^{A} (\hat{\theta}_a - \hat{\theta}^*)^2 \tag{11.1.1}$$

and

$$\hat{V}_2 = \frac{1}{A(A-1)} \sum_{a=1}^{A} (\hat{\theta}_a - \hat{\theta})^2 \tag{11.1.2}$$

The use of the expression (11.1.1) as an estimator of $V(\hat{\theta}^*)$ is justified by the following simple result

$$E(\hat{V}_1) = V(\hat{\theta}^*) - \frac{1}{A(A-1)} \sum_{a=1}^{A} \sum_{\substack{b=1 \\ a \neq b}}^{A} C(\hat{\theta}_a, \hat{\theta}_b) + \frac{1}{A(A-1)} \sum_{a=1}^{A} [E(\hat{\theta}_a) - E(\hat{\theta}^*)]^2 \tag{11.1.3}$$

It follows from (11.1.3) that, if $\hat{\theta}_1, \ldots, \hat{\theta}_A$ were uncorrelated and had the same expected value, then $\hat{V}_1$ given by (11.1.1) would be unbiased for $V(\hat{\theta}^*)$. To verify equation (11.1.3) is left as an exercise.

The use of the expression (11.1.2) as an estimator of $V(\hat{\theta}^*)$ is justified by the identity

$$\sum_{a=1}^{A} (\hat{\theta}_a - \hat{\theta})^2 = \sum_{a=1}^{A} (\hat{\theta}_a - \hat{\theta}^*)^2 + A(\hat{\theta}^* - \hat{\theta})^2 \tag{11.1.4}$$

from which it follows that

$$\hat{V}_2 \geq \hat{V}_1$$

In fact, both $\hat{V}_2$ and $\hat{V}_1$ are also used to estimate the variance of $\hat{\theta}$, $V(\hat{\theta})$, under the assumption that $V(\hat{\theta}^*)$ and $V(\hat{\theta})$ are approximately equal.

Most of the approximate variance estimators examined in this chapter are of the form shown in equations (11.1.1) and (11.1.2). In the cases that we consider below, $\hat{\theta}_1, \ldots, \hat{\theta}_A$ will usually be correlated, and both $\hat{V}_1$ and $\hat{V}_2$ will be biased estimators of $V(\hat{\theta}^*)$ and of $V(\hat{\theta})$.

11.2. A Simplified Variance Estimator under Sampling without Replacement

We examine one technique, which consists of taking a variance estimator appropriate for sampling with replacement and using it for sampling without replacement. A simple example illustrates the principle.

EXAMPLE 11.2.1. Under *SI* sampling of n elements from N ($f - n/N$), the π estimator of the population total $t = \sum_U y_k$ is

$$\hat{t}_\pi = N\bar{y}_s$$

Its variance

$$V = V(\hat{t}_\pi) = N^2(1 - f)S^2_{yU}/n \tag{11.2.1}$$

is estimated unbiasedly by

$$\hat{V} = N^2(1 - f)S^2_{ys}/n \tag{11.2.2}$$

Here S^2_{yU} and S^2_{ys} have their usual meanings. When the sampling fraction f is negligible, a new, marginally simpler, and somewhat biased variance estimator, $\hat{V}_0$, is obtained by omitting the finite population correction factor $1 - f$,

$$\hat{V}_0 = N^2 S^2_{ys}/n \tag{11.2.3}$$

Note that $\hat{V}_0$ is the variance estimator appropriate for the mean of n with replacement draws (see Section 3.3.2). The relative bias,

$$\frac{E(\hat{V}_0) - V}{V} = \frac{f}{1 - f}$$

is positive and unimportant if f is small.

It is possible to generalize the previous example to an arbitrary design. The following results were given by Durbin (1953) and are discussed by Wolter (1985).

We seek to estimate the total $t = \sum_U y_k$. A sample s of fixed size n is drawn by an arbitrary design with inclusion probabilities π_k and π_{kl} ($k, l = 1, \ldots, N$). The π estimator of t is

$$\hat{t}_\pi = \sum_s y_k/\pi_k$$

with variance

$$V = -\tfrac{1}{2}\sum\sum_U \Delta_{kl}(\check{y}_k - \check{y}_l)^2 \tag{11.2.4}$$

where $\check{y}_k = y_k/\pi_k$. The usual unbiased Yates-Grundy-Sen estimator of V is

$$\hat{V} = -\tfrac{1}{2}\sum\sum_s \check{\Delta}_{kl}(\check{y}_k - \check{y}_l)^2 \tag{11.2.5}$$

This formula has the computational drawback of a double sum, and it also requires that all Δ_{kl} are calculated for k and $l \in s$. To simplify, let us instead consider the variance estimator

$$\hat{V}_0 = \frac{1}{n(n-1)} \sum_s (y_k/p_k - \hat{t}_\pi)^2 \tag{11.2.6}$$

where p_k is taken as $p_k = \pi_k/n$ for $k \in s$. Note that $\hat{V}_0$ has the form of the variance estimator that applies for the pwr estimator under sampling with replacement (see Section 2.9). That is, when $\hat{V}_0$ is adopted as a variance estimator for $\hat{t}_\pi$, the following procedure is used. We calculate the variance estimator as if sampling had been performed with replacement, whereas in actual fact it was without replacement. The computational simplification that results from getting rid of the π_{kl} and the double sum is often considerable. We commented briefly on this technique in Section 3.6.3.

Remark 11.2.1. Under the *SI* design, equations (11.2.4) to (11.2.6) are identical with (11.2.1) to (11.2.3), respectively.

The trade off for the simplification is that $\hat{V}_0$ is no longer unbiased for V. In many cases, however, the bias is positive, so that $\hat{V}_0$ will be "on the safe side," as in the introductory Example 11.2.1. It is an easy exercise to prove that under an arbitrary fixed size design,

$$E(\hat{V}_0) - V = \frac{n}{n-1}(V_0 - V) \tag{11.2.7}$$

where V_0 is the variance of the pwr estimator of t (see Section 2.9) corresponding to a hypothetical with-replacement design, such that the draw probabilities are $p_k = \pi_k/n$, $k = 1, \dots, N$, in each of n independent draws. That is,

$$V_0 = \frac{1}{n} \sum_U p_k(y_k/p_k - t)^2$$

as shown in Section 2.9. Now, the bias (11.2.7) of $\hat{V}_0$ is positive in cases where sampling without replacement (n elements; inclusion probabilities π_k; π estimator) is more efficient than sampling with replacement (n draws; draw probabilities $p_k = \pi_k/n$; pwr estimator). The result stated in equation (11.2.7) will be of interest later in the chapter. We shall see that, under certain conditions, the bias of other variance estimators is related in a simple manner to the bias of $\hat{V}_0$.

Remark 11.2.2. Under stratified sampling, the principle expressed in (11.2.6) should be applied in each stratum separately. Thus, supposing that

$$\hat{t}_\pi = \sum_{h=1}^{H} \hat{t}_{\pi h}$$

with

$$\hat{t}_{\pi h} = \sum_{s_h} y_k/\pi_k$$

we define the simplified variance estimator as

$$\hat{V}_0 = \sum_{h=1}^{H} \frac{1}{n_h(n_h - 1)} \sum_{s_h} (y_k/p_k - \hat{t}_{\pi h})^2$$

with $p_k = \pi_k/n_h$ for $k \in s_h$. For example, with *SI* sampling in each stratum, the expression is

$$\hat{V}_0 = \sum_{h=1}^{H} N_h^2 S_{ys_h}^2/n_h$$

Remark 11.2.3. Now we state a result analogous to (11.2.7) for multistage sampling. We assume that a first-stage sample s_I of n_I PSUs is drawn from the set U_I containing N_I PSUs. Following subsampling of the selected PSUs (with a known but arbitrary subsampling design in any number of stages), we calculate estimated PSU totals $\hat{t}_i$, for $i \in s_I$. The population total $t = \sum_U y_k$ is then estimated by

$$\hat{t} = \sum_{s_I} \hat{t}_i/\pi_{Ii}.$$

By analogy with equation (11.2.6), we define a simplified estimator of $V = V(\hat{t})$ as

$$\hat{V}_0 = \frac{1}{n_I(n_I - 1)} \sum_{s_I} (\hat{t}_i/p_i - \hat{t})^2 \tag{11.2.8}$$

with $p_i = \pi_{Ii}/n_I$. (This is the variance estimator v introduced in Section 4.6.) A result similar to (11.2.7) can be shown for multistage sampling, namely,

$$E(\hat{V}_0) - V = \frac{n_I}{n_I - 1}(V_0 - V) \tag{11.2.9}$$

where V_0 is the variance of the estimator of t described in Section 4.5 under with-replacement sampling of PSUs. Actually, (11.2.7) can be seen as a special case of (11.2.9), if we assume $\hat{t}_i = t_i$ for all $i \in s_I$.

11.3. The Random Groups Technique

11.3.1. Independent Random Groups

The independent random groups technique is a special technique for drawing a sample that leads to simple variance estimation. The idea of independent random groups has its origin in work by Mahalanobis (1939, 1944, 1946) and Deming (1956). The terminology in this area varies somewhat. Interpenetrating samples (Malahanobis) and replicated samples (Deming) are alternative terms for what we call random groups.

Instead of drawing one large sample s of size n, we draw A independent samples, $s_1, \ldots, s_A$, of equal size $m = n/A$. We assume that the first sample s_1 is drawn, then replaced into the population. The second sample s_2 is then drawn, by sampling again from the whole population, by the same design that produced s_1, and independently of s_1. Then s_2 is replaced into the population, a third independent sample s_3 is drawn by the same design that produced s_1 and s_2, and so on, until A samples have been drawn. The common design by which each sample s_a is drawn may be without replacement. What is important, however, is replacement of each sample s_a before the next sample s_{a+1} is drawn.

Under these assumptions, the samples, or random groups as they are also called, $s_1, \ldots, s_A$, can be considered outcomes of A independent repetitions of the same random experiment, namely, drawing m elements from a given population by a fixed sampling design. The A samples are not necessarily disjoint. (By contrast, if the samples were not replaced into the population, then they would be disjoint, but no longer independent.)

For each $a = 1, \ldots, A$, an estimator $\hat{\theta}_a$ of θ is calculated on data from s_a only. The same estimator formula applies throughout. The average of the $\hat{\theta}_a$, here denoted $\hat{\theta}_{\mathrm{IRG}}$, is a natural candidate for estimating θ,

$$\hat{\theta}_{\mathrm{IRG}} = \frac{1}{A} \sum_{a=1}^{A} \hat{\theta}_a \tag{11.3.1}$$

As an estimator of $V(\hat{\theta}_{\mathrm{IRG}})$, the variance of $\hat{\theta}_{\mathrm{IRG}}$, let us consider the independent random groups variance estimator,

$$\hat{V}_{\mathrm{IRG1}} = \frac{1}{A(A-1)} \sum_{a=1}^{A} (\hat{\theta}_a - \hat{\theta}_{\mathrm{IRG}})^2 \tag{11.3.2}$$

Under the assumptions of independent random groups, that is, the A samples are drawn independently and by the same design, the estimators $\hat{\theta}_1, \ldots, \hat{\theta}_A$ are independent and identically distributed random varibles. It follows immediately from (11.1.3) that $\hat{V}_{\mathrm{IRG1}}$ is unbiased,

$$E(\hat{V}_{\mathrm{IRG1}}) = V(\hat{\theta}_{\mathrm{IRG}}) \tag{11.3.3}$$

Remark 11.3.1. For (11.3.3) to hold, it is not required that the $\hat{\theta}_a$ be identically distributed. We can see from equation (11.1.3) that (11.3.3) holds under the somewhat weaker assumption that $\hat{\theta}_1, \ldots, \hat{\theta}_a, \ldots, \hat{\theta}_A$ have the same expected value. Thus, they need not have the same variance, nor need they be unbiased for θ, as long as they are independent and have the same expected value.

The variance estimator shown in equation (11.3.2) is strikingly simple to calculate. Once $\hat{\theta}_1, \ldots, \hat{\theta}_a, \ldots, \hat{\theta}_A$ have been obtained, very little work remains to arrive at $\hat{V}_{\mathrm{IRG1}}$. Another nicety is that the unbiasedness of (11.3.3) holds regardless of the complexity of the estimator $\hat{\theta}_a$ and regardless of the complexity of the sampling design used to draw s_a. For example, $\hat{\theta}_a$ may be a correlation coefficient estimator, and the design may be one that does not allow unbiased variance estimation by traditional techniques, such as systematic sampling with a single random start.

However, certain problems are also associated with the independent random groups technique:

1. Selection and data collection for a series of independent samples may turn out more costly and cumbersome than drawing and observing a single large sample. Moreover, one must be careful not to create an undesired dependence between the $\hat{\theta}_a$. Such dependence could be introduced through the interviewers, or at the data handling stage by the processing staff. These ideas will be further discussed in Chapter 16.
2. To give a stable variance estimator, the number A of independent samples should be large. In practice, however, A can usually not be so large, which makes the variance estimator unstable. For example, Mahalanobis proposed to use as few as four groups, whereas Deming suggests ten.

These weaknesses may explain why the independent random groups technique is not often used in practice.

The estimator $\hat{\theta}_{\text{IRG}}$ was obtained as the average of the estimates for the different independent samples $s_1, \ldots, s_A$. Alternatively, θ could obviously be estimated from the pooled sample. Applying the estimator formula to the data corresponding to

$$s = \bigcup_{a=1}^{A} s_a$$

we obtain a pooled sample estimator, denoted $\hat{\theta}$. Clearly, $\hat{\theta}$ may differ from $\hat{\theta}_{\text{IRG}}$. For example, a correlation coefficient estimate based on the pooled s is not in general equal to the average of the A correlation coefficient estimates obtained from the different subsets $s_1, \ldots, s_A$.

A related question is: How should the variance of $\hat{\theta}$ be estimated? Often, $\hat{V}_{\text{IRG1}}$ defined by (11.3.2) is in fact also used for estimating $V(\hat{\theta})$, even though $\hat{V}_{\text{IRG1}}$ is not unbiased for $V(\hat{\theta})$ when $\hat{\theta}$ and $\hat{\theta}_{\text{IRG}}$ are nonidentical.

A sometimes used alternative estimator of $V(\hat{\theta})$ is

$$\hat{V}_{\text{IRG2}} = \frac{1}{A(A-1)} \sum_{a=1}^{A} (\hat{\theta}_a - \hat{\theta})^2 \tag{11.3.4}$$

Little seems to be known about whether $\hat{V}_{\text{IRG1}}$ or $\hat{V}_{\text{IRG2}}$ should be preferred as an estimator of $V(\hat{\theta})$. However, from (11.1.4)

$$\hat{V}_{\text{IRG2}} \geq \hat{V}_{\text{IRG1}} \tag{11.3.5}$$

so $\hat{V}_{\text{IRG2}}$ is more conservative than $\hat{V}_{\text{IRG1}}$. When it comes to estimating $V(\hat{\theta}_{\text{IRG}})$, it follows from (11.3.5) that $\hat{V}_{\text{IRG2}}$ will have a positive bias.

11.3.2. Dependent Random Groups

The dependent random groups technique is an attempt to adapt the independent random groups technique to a sample that does not fulfill the requirements of independent random groups. Assume that we first draw one large sample from the whole population by a probability sampling design. After this sample has been obtained, a random mechanism is used to divide it into a number of disjoint subsamples, the random groups. These will not be independent, but a significant feature of the procedure to be described is that they are treated as if they were independent. These ideas were first described by Hansen, Hurwitz, and Madow (1953).

Let s be the sample drawn from the population U; we call it the full sample. Now, s is divided into A disjoint random groups, $s_1, \ldots, s_a, \ldots, s_A$, that is,

$$s = \bigcup_{a=1}^{A} s_a$$

Let s be of fixed size n, and assume for simplicity that the groups are of equal size, $m = n/A$. We assume that s is divided into groups by a randomization device so that "each random group has essentially the same sampling design as the parent sample" (Wolter, 1985). It is not always self-evident how this can be accomplished. In some cases, there are several possible ways as will be seen later in this subsection.

Let $\hat{\theta}_1, \ldots, \hat{\theta}_a, \ldots, \hat{\theta}_A$ be estimators of θ, where $\hat{\theta}_a$ is based on data from the group s_a only; $a = 1, \ldots, A$. The $\hat{\theta}_a$ are assumed unbiased, or nearly so, but need not be π estimators. We can now consider two alternative full-sample estimators of θ: (a) the estimator $\hat{\theta}_{\text{DRG}}$ formed by averaging,

$$\hat{\theta}_{\text{DRG}} = \frac{1}{A} \sum_{a=1}^{A} \hat{\theta}_a \tag{11.3.6}$$

and (b) the estimator $\hat{\theta}$ based on data from the full sample s, disregarding the division into random groups. In some cases $\hat{\theta}_{\text{DRG}}$ and $\hat{\theta}$ can be defined so that they will be identical. As earlier in the chapter, we can form two alternative variance estimators,

$$\hat{V}_{\text{DRG1}} = \frac{1}{A(A-1)} \sum_{a=1}^{A} (\hat{\theta}_a - \hat{\theta}_{\text{DRG}})^2 \tag{11.3.7}$$

and

$$\hat{V}_{\text{DRG2}} = \frac{1}{A(A-1)} \sum_{a=1}^{A} (\hat{\theta}_a - \hat{\theta})^2 \tag{11.3.8}$$

Both of these are possible estimators of $V(\hat{\theta}_{\text{DRG}})$ as well as of $V(\hat{\theta})$, but neither is unbiased for any of these two purposes.

Some immediate conclusions are:

1. It follows from (11.1.3) that if the random groups are formed so that

$\hat{\theta}_1, \ldots, \hat{\theta}_A$ have the same expected value, not necessarily equal to θ, then the bias of $\hat{V}_{\text{DRG1}}$ as an estimator of $V(\hat{\theta}_{\text{DRG}})$ is given by

$$E(\hat{V}_{\text{DRG1}}) - V(\hat{\theta}_{\text{DRG}}) = -\frac{1}{A(A-1)} \sum_{a=1}^{A} \sum_{\substack{b=1 \\ a \neq b}}^{A} C(\hat{\theta}_a, \hat{\theta}_b)$$

In particular, if all pairs $\hat{\theta}_a$ and $\hat{\theta}_b$ have the same covariance γ, then

$$E(\hat{V}_{\text{DRG1}}) - V(\hat{\theta}_{\text{DRG}}) = -\gamma \tag{11.3.9}$$

2. It follows from (11.1.4) that $\hat{V}_{\text{DRG2}} \geq \hat{V}_{\text{DRG1}}$.

Example 11.3.1. Let a sample s of size $n = Am$ be drawn from a population U of size N by the *SI* design. Then, divide s at random into A nonoverlapping random groups $s_1, \ldots, s_A$ of equal size m, in such a way that each s_a is an *SI* sample from U. This can be achieved as follows. First, draw s_1 as an *SI* sample of size m from s. Then, draw s_2 as an *SI* sample from the remaining $n - m$ elements of s, that is, from $s - s_1$. Then, draw s_3 from $s - s_1 - s_2$, and so on. Supposing that $\theta = t = \sum_U y_k$ is to be estimated, we let $\hat{\theta} = N\bar{y}_s$ and $\hat{\theta}_a = N\bar{y}_{s_a}$. In this case, $\hat{\theta}$ and $\hat{\theta}_{\text{DRG}}$ are identical,

$$\hat{\theta} = \hat{\theta}_{\text{DRG}} = N\bar{y}_s$$

with variance

$$V(N\bar{y}_s) = N^2(1-f)S_{yU}^2/n$$

The random groups variance estimator is therefore

$$\hat{V}_{\text{DRG1}} = \hat{V}_{\text{DRG2}} = N^2 \sum_{a=1}^{A} (\bar{y}_{s_a} - \bar{y}_s)^2 / A(A-1)$$

which we denote simply $\hat{V}_{\text{DRG}}$. Since, for $a \neq b$,

$$C(\bar{y}_{s_a}, \bar{y}_{s_b}) = V(\bar{y}_s) - S_{yU}^2/n$$

it follows from (11.3.9) that the relative bias of $\hat{V}_{\text{DRG}}$ is

$$\frac{E(\hat{V}_{\text{DRG}}) - V(N\bar{y}_s)}{V(N\bar{y}_s)} = \frac{f}{1-f}$$

which is unimportant when f is small, and equal to the relative bias of the variance estimator $\hat{V}_0$ defined in (11.2.3). In the extreme case, when $A = n$, that is, $m = 1$, then $\hat{V}_{\text{DRG}}$ and $\hat{V}_0$ are identical. In this example, an easy correction for bias is obtained by multiplying $\hat{V}_{\text{DRG}}$ with the finite population correction factor $(1 - f)$.

Example 11.3.2. Consider πps sampling for estimating the population total $\theta = t = \sum_U y_k$. A sample s of fixed size $n = Am$ is selected by πps sampling with inclusion probabilities π_k, $k = 1, \ldots, N$ (see Section 3.6.2). The resulting sample s is divided at random into A groups $s_1, \ldots, s_A$ of equal size m, as in

Example 11.3.1. That is, s_1 is an *SI* sample from s, s_2 an *SI* sample from $s - s_1$, s_3 from $s - s_1 - s_2$, and so on. Consequently, each s_a behaves as if it had been realized by a design with the inclusion probabilities

$$\pi_{k,a} = (m/n)\pi_k$$

for $k = 1, \ldots, N$. Here $\pi_{k,a}$ is the probability that the element k will finally be included in s_a. We assume $\hat{\theta}$ and $\hat{\theta}_a$ to be π estimators, that is,

$$\hat{\theta} = \sum_s y_k/\pi_k = \hat{t}_\pi$$

and

$$\hat{\theta}_a = \sum_{s_a} y_k/\pi_{k,a} = (n/m) \sum_{s_a} y_k/\pi_k = \hat{t}_a$$

Then

$$\hat{\theta}_{\text{DRG}} = \sum_{a=1}^{A} \hat{t}_a/A = \hat{t}_\pi = \hat{\theta}$$

Thus, $\hat{\theta}$ and $\hat{\theta}_{\text{DRG}}$ are identical, and so are the two variance estimators,

$$\hat{V}_{\text{DRG1}} = \hat{V}_{\text{DRG2}} = \frac{1}{A(A-1)} \sum_{a=1}^{A} (\hat{t}_a - \hat{t}_\pi)^2 = \hat{V}_{\text{DRG}} \tag{11.3.10}$$

It is an easy exercise to show that, for the π estimator,

$$E(\hat{V}_{\text{DRG}}) = E(\hat{V}_0) \tag{11.3.11}$$

where $\hat{V}_0$ is the simplified variance estimator defined by equation (11.2.6). Thus, $\hat{V}_0$ and the random groups estimator of the present example share the same bias. (Again, if $A = n$, then $\hat{V}_{\text{DRG}}$ and $\hat{V}_0$ are identical.) From (11.2.7), the bias is

$$E(\hat{V}_{\text{DRG}}) - V(\hat{t}_\pi) = \frac{n}{n-1}[V_0 - V(\hat{t}_\pi)]$$

Note that Example 11.3.1 is just a special case of the present example. Also note that the results of this example extend to linear functions of several π estimators. For nonlinear functions of π estimators, the simplicity of the results in the present example will no longer hold.

EXAMPLE 11.3.3. When the sample s is obtained by a stratified sampling design, there are two fundamentally different ways to apply the dependent random groups technique. These yield two different variance estimators, and it is at present not known which of the two is preferable. To keep things simple, we assume that the population total

$$\theta = t = \sum_{h=1}^{H} t_h$$

is estimated by

$$\hat{\theta} = \hat{t} = \sum_{h=1}^{H} \hat{t}_h$$

where $\hat{t}_h$ is an estimator of the stratum total $t_h = \sum_{U_h} y_k$. The first approach is to start with the variance

$$V(\hat{t}) = \sum_{h=1}^{H} V(\hat{t}_h)$$

and to apply the dependent random groups technique within each stratum separately. That is, each $V(\hat{t}_h)$ $(h = 1, \ldots, H)$ is estimated by $\hat{V}_{h,\text{DRG}}$, which can be of the type shown in (11.3.7) or the type shown in (11.3.8), and the variance of $\hat{t}$ is finally estimated by

$$\hat{V}_{\text{DRG}} = \sum_{h=1}^{H} \hat{V}_{h,\text{DRG}}$$

The second approach is to form random groups across the strata, so that each s_a is a stratified sample from the whole population. If the full sample s has n_h elements from the hth stratum $(h = 1, \ldots, H)$, then the random group s_a $(a = 1, \ldots, A)$ will have $m_h = n_h/A$ elements from the hth stratum. This results in A dependent random groups of equal size $m = \sum_{h=1}^{H} m_h$, each one allocated to strata in the same way as the full sample. As an estimator of the variance of $\hat{\theta} = \hat{t}$ (or of $\hat{\theta}_{\text{DRG}}$), we then take either (11.3.7) or (11.3.8). Further detail on these approaches is found in Wolter (1985), especially for the case in which n_h is not an integer multiple of A. Little is known about which of the two approaches is better.

EXAMPLE 11.3.4. When the design involves two or more stages of sampling, a widespread practice is to apply the random groups technique to the PSUs only. Thus, the sample s_I of PSUs is divided into random groups $s_{\text{I}a}$ $(a = 1, \ldots, A)$. We assume that these groups of PSUs are formed in the same way as groups of elements were formed in Examples 11.3.1, 11.3.2, and 11.3.3. For example, consider πps sampling of PSUs. For the population total $\theta = t$, consider the estimator given by (4.4.1),

$$\hat{\theta} = \hat{t} = \sum_{s_\text{I}} \hat{t}_i / \pi_{\text{I}i}$$

where $\pi_{\text{I}i}$ is the inclusion probability of the ith PSU and $\hat{t}_i$ is an estimator of the ith PSU total t_i, based on elements subsampled within this PSU in one or more stages. The subsampling design is arbitrary. The set s_I of n_I selected PSUs is now divided at random as in Examples 11.3.1 and 11.3.2 into A random groups $s_{\text{I}1}, \ldots, s_{\text{I}a}, \ldots, s_{\text{I}A}$ of equal size, $m_\text{I} = n_\text{I}/A$. It then follows that, for a given s_I, $s_{\text{I}a}$ behaves as an SI sample from s_I. The estimator of t based on $s_{\text{I}a}$ is, in analogy with Example 11.3.2,

$$\hat{\theta}_a = (n_\text{I}/m_\text{I}) \sum_{s_{\text{I}a}} \hat{t}_i / \pi_{\text{I}i} = \hat{t}_a$$

Again, $\hat{\theta}$ and $\hat{\theta}_{\text{DRG}}$ coincide,

$$\hat{\theta} = \hat{\theta}_{\text{DRG}} = \frac{1}{A} \sum_{a=1}^{A} \hat{t}_a = \hat{t}$$

and the variance estimator is

$$\hat{V}_{\text{DRG}} = \frac{1}{A(A-1)} \sum_{a=1}^{A} (\hat{t}_a - \hat{t})^2$$

It is left as an exercise to show that this variance estimator has the same bias as the simplified variance estimator $\hat{V}_0$ defined by equation (11.2.8). Especially, when $A = n_{\text{I}}$ (and consequently $m_{\text{I}} = 1$), then $\hat{V}_{\text{DRG}}$ and $\hat{V}_0$ are identical. As for the bias of $\hat{V}_{\text{DRG}}$, one can reach conclusions similar to those in Section 11.2.

Remark 11.3.2. The variance estimators (11.3.7) and (11.3.8) may suffer from instability in that they are ordinarily based on a small number of groups. Norlén and Waller (1979) suggest an improved procedure, namely, to repeat the whole random groups procedure independently a number of times, say R times, on the same sample s. Thus, we obtain a sequence of R independent variance estimators $\hat{V}_1, \ldots, \hat{V}_r, \ldots, \hat{V}_R$, where each $\hat{V}_r$ is a dependent random groups variance estimator. Averaging gives the final variance estimator

$$\hat{V} = \frac{1}{R} \sum_{r=1}^{R} \hat{V}_r$$

This repeated random groups procedure will obviously give a variance estimator with reduced variability. Let

$$V(\hat{V}_r) = c_1 + c_2$$

where $c_1 = E[V(\hat{V}_r|s)]$ and $c_2 = V[E(\hat{V}_r|s)]$. Then,

$$V(\hat{V}) = (c_1/R) + c_2 < c_1 + c_2$$

Note that the repeated random groups are formed from one and the same sample, so the only additional cost, usually modest, is for the increased computational effort.

11.4. Balanced Half-Samples

The method of balanced half-samples was originally developed for the case with a large number of strata and a sample composed of only two elements (or two PSUs) per stratum. In this case, the dependent random groups technique will have poor stability, because only two disjoint groups can be formed. The idea of using half-samples for variance estimation was introduced at the United States Bureau of the Census around 1960. The balanced half-sample technique that we shall now describe goes back to McCarthy (1966, 1969). Other names for the technique are balanced repeated replication (BRR) and pseudoreplication.

We describe the technique as used when the variance of a π estimator is to be estimated. Several variance estimation methods have already been given for the π estimator. However, the π estimator illustrates well the way the technique works. The case where $\hat{\theta}$ is a more complex estimator is of more interest in practice and will be discussed at the end of the section.

Let a population U with N elements be divided into H strata $U_1, \ldots, U_H$ of sizes $N_1, \ldots, N_H$, respectively. Let a sample s_h of fixed size $n_h = 2$ be drawn independently from each stratum ($h = 1, \ldots, H$) by a design with the inclusion probabilities π_k and π_{kl}. The finite population total can be written as

$$t = \sum_{h=1}^{H} t_h = \sum_{h=1}^{H} \sum\nolimits_{U_h} y_k$$

The corresponding π estimator is given by

$$\hat{t}_\pi = \sum_{h=1}^{H} \hat{t}_{\pi h} = \sum_{h=1}^{H} \sum\nolimits_{s_h} y_k/\pi_k \tag{11.4.1}$$

We recall the simplified variance estimator given in Remark 11.2.2, which here (with $n_h = 2$) is

$$\hat{V}_0 = \tfrac{1}{2} \sum_{h=1}^{H} \sum\nolimits_{s_h} (y_k/p_k - \hat{t}_{\pi h})^2 \tag{11.4.2}$$

The balanced half-sample technique offers an alternative way of computing the variance estimator (11.4.2), as we now show. Consider a given sample $s = \bigcup_{h=1}^{H} s_h$, where each s_h consists of exactly two elements. A half-sample is a set consisting of exactly one of the two elements from each s_h. There are thus 2^H possible half-samples.

The basic idea of the balanced half-sample technique is to select a set of half-samples from the set of all 2^H half-samples, to calculate an estimate $\hat{t}_a$ of t for each selected half-sample, and then use these $\hat{t}_a$ values to estimate $V(\hat{t}_\pi)$. We shall see that the selection of the half-samples in an ingenious way (namely, as a "balanced set") makes it possible to calculate $\hat{V}_0$ given by (11.4.2) from the observed $\hat{t}_a$ values.

We need additional notation to describe the technique. For each s_h, let its two elements be temporarily relabeled (at random) as $h1$ and $h2$. Thus, a half-sample consists of only one of the two elements $h1$ and $h2$ from each stratum h. For each possible half-sample ($a = 1, \ldots, 2^H$), we further define indicator variables δ_{ah} and ε_{ah} ($h = 1, \ldots, H$) such that

$$\delta_{ah} = \begin{cases} 1 & \text{if the } a\text{th half-sample contains element } h1 \\ 0 & \text{if the } a\text{th half-sample contains element } h2 \end{cases}$$

and

$$\varepsilon_{ah} = \begin{cases} 1 & \text{if the } a\text{th half-sample contains element } h1 \\ -1 & \text{if the } a\text{th half-sample contains element } h2 \end{cases}$$

Thus, each possible half-sample ($a = 1, \ldots, 2^H$) is characterized by the vector $(\delta_{a1}, \ldots, \delta_{aH})$, or, alternatively, by the vector $(\varepsilon_{a1}, \ldots, \varepsilon_{aH})$. Clearly, $\varepsilon_{ah} = 2\delta_{ah} - 1$.

A set of A half-samples (labeled $a = 1, \ldots, A$) is said to be *balanced* if

$$\sum_{a=1}^{A} \varepsilon_{ah}\varepsilon_{ah'} = 0 \tag{11.4.3}$$

for all pairs of strata h and h' $(h \neq h')$. A set of half-samples that is balanced and also satisfies

$$\sum_{a=1}^{A} \varepsilon_{ah} = 0 \tag{11.4.4}$$

for each $h = 1, \ldots, H$ is said to be in *full orthogonal balance*.

The balanced half-sample technique of variance estimation can now be described in the following way.

i. Obtain a balanced set of half-samples s_a $(a = 1, \ldots, A)$ from s.
ii. For each half-sample $(a = 1, \ldots, A)$, calculate

$$\hat{t}_a = \sum_{h=1}^{H} [\delta_{ah} y_{h1}/p_{h1} + (1 - \delta_{ah}) y_{h2}/p_{h2}] \tag{11.4.5}$$

where $p_{hi} = \pi_{hi}/n_h = \pi_{hi}/2$ $(i = 1, 2)$.
iii. The balanced half-sample variance estimator $\hat{V}_{BH}$ is obtained as

$$\hat{V}_{BH} = \frac{1}{A} \sum_{a=1}^{A} (\hat{t}_a - \hat{t}_\pi)^2 \tag{11.4.6}$$

where $\hat{t}_\pi$ is the π estimator shown in (11.4.1).

This variance estimator has the following important property: if the set of half-samples is balanced, then

$$\hat{V}_{BH} = \hat{V}_0 \tag{11.4.7}$$

where $\hat{V}_0$ is defined by (11.4.2). Thus, the balanced half-sample procedure is just an alternative way of calculating $\hat{V}_0$. The proof is left as an exercise. Furthermore, if the set of half-samples is in full orthogonal balance, then the mean of the $\hat{t}_a$ values is equal to $\hat{t}_\pi$, that is,

$$\hat{t}_{BH} = \hat{t}_\pi \tag{11.4.8}$$

where $\hat{t}_{BH} = \sum_{a=1}^{A} \hat{t}_a/A$. The verification of equation (11.4.8) is proposed as an exercise. So, under full orthogonal balance, the variance estimator $\hat{V}_{BH}$ can be alternatively obtained as

$$\frac{1}{A} \sum_{a=1}^{A} (\hat{t}_a - \hat{t}_{BH})^2$$

EXAMPLE 11.4.1. Consider the *STSI* design with two elements sampled from each stratum. Then, $\pi_{h1} = \pi_{h2} = 2p_{h1} = 2p_{h2} = 2/N_h$,

$$\hat{t}_\pi = \sum_{h=1}^{H} N_h \bar{y}_{s_h}$$

and

$$V(\hat{t}_\pi) = \tfrac{1}{2} \sum_{h=1}^{H} N_h^2 (1 - f_h) S_{yU_h}^2$$

Suppose we have a balanced set of half-samples. Then,

$$\hat{t}_a = \sum_{h=1}^{H} N_h [\delta_{ah} y_{h1} + (1 - \delta_{ah}) y_{h2}]$$

and it is easily verified that (11.4.7) holds, that is,

$$\hat{V}_{BH} = \hat{V}_0$$

with $\hat{V}_{BH}$ given by (11.4.6) and

$$\hat{V}_0 = \tfrac{1}{2} \sum_{h=1}^{H} N_h^2 S_{ys_h}^2$$

If the set of half-samples is in full orthogonal balance, it is also easily verified that $\hat{t}_{BH} = \hat{t}_\pi$.

We now discuss how one can obtain a balanced set of half-samples. The set consisting of all possible 2^H half-samples is balanced and also in full orthogonal balance (see Exercise 11.13). However, for computational reasons, we want a much smaller number of half-samples. In all cases of practical interest, it turns out that we can find a set consisting of not more than $H + 3$ balanced half-samples, for reasons that are discussed later.

It follows from equation (11.4.3) that finding a balanced set of A half-samples is the same as finding an $A \times H$ matrix with elements ε_{ah} equal to either $+1$ or -1 and with all the columns pairwise orthogonal. Such matrices of order $c \times c$, originally intended for use in experimental design, were given by Plackett and Burman (1946) for $c = 2, 4, 8, 12, 16, \ldots, 200$. Wolter (1985) also presents such matrices for $c = 2, 4, 8, 12, 16, \ldots, 100$. The matrices presented by these authors are not unique, that is, for a given c, there exist other $c \times c$ matrices of ε_{ah} values that also satisfy the balancing condition. One should note that the matrices of Plackett and Burman and of Wolter do not quite meet the condition of full orthogonal balance. Their matrices have exactly one column of only $+1$ or only -1, so (11.4.4) is not verified. The remaining columns in each matrix, however, satisfy the condition of full orthogonal balance. The following example shows how these matrices can be used.

Example 11.4.2. Construction of a set of balanced half-samples. Suppose there are $H = 8$ strata. Then we can use the following 8×8 matrix given by Wolter (1985):

$$\begin{bmatrix} +1 & +1 & +1 & +1 & +1 & +1 & +1 & +1 \\ +1 & -1 & +1 & -1 & +1 & -1 & +1 & -1 \\ +1 & -1 & -1 & +1 & +1 & -1 & -1 & +1 \\ +1 & +1 & -1 & -1 & +1 & +1 & -1 & -1 \\ +1 & +1 & +1 & +1 & -1 & -1 & -1 & -1 \\ +1 & -1 & +1 & -1 & -1 & +1 & -1 & +1 \\ +1 & -1 & -1 & +1 & -1 & +1 & +1 & -1 \\ +1 & +1 & -1 & -1 & -1 & -1 & +1 & +1 \end{bmatrix}$$

Let each row define a half-sample. Thus, the first half-sample consists of the first element from each stratum. The second half-sample consists of the first element from strata 1, 3, 5, and 7 (since $\varepsilon_{2h} = +1$ for $h = 1, 3, 5, 7$), and the second element from strata 2, 4, 6, and 8 (since $\varepsilon_{2h} = -1$ for $h = 2, 4, 6, 8$). Proceeding similarly with the remaining six rows, we get a total set of $A = 8$ half-samples. The set is balanced because the matrix is orthogonal by construction. Now suppose $H = 7$. Then we can use any set of seven columns from the 8×8 matrix to obtain eight balanced half-samples. Similarly, when $H = 6$ we can use any set of six columns, and when $H = 5$ any set of five columns, to obtain $A = 8$ balanced half-samples.

The principle illustrated by the preceding example is that a given $c \times c$ matrix (where c is a multiple of 4) can be used to obtain a set of $A = c$ balanced half-samples when $H = c, c - 1, c - 2$, or $c - 3$. In other words, for $H = 5, 6, 7$, or 8, we get $A = 8$ half-samples from the 8×8 table, for $H = 9$, 10, 11, or 12, we get $A = 12$ half-samples from the 12×12 table, and so on. Thus, the number of balanced half-samples need not be greater than $H + 3$.

If full orthogonal balance is required, the procedure in the following example will work.

Example 11.4.3. Construction of a set of half-samples in full orthogonal balance. Consider again the 8×8 matrix of Example 11.4.2. If the first column is omitted, the remaining seven columns satisfy the condition of full orthogonal balance, by construction. That is, these seven columns are pairwise orthogonal, and each column sum is zero. Thus, for $H = 7$ we obtain a set of $A = 8$ half-samples in full orthogonal balance by omitting the first column. For $H = 6$ strata, we obtain $A = 8$ half-samples in full orthogonal balance by omitting the first column and any one of the remaining seven columns. For $H = 5$ strata, we omit the first column and any two of the remaining columns. Note, though, that for $H = 8$ strata, the full 8×8 matrix in Example 11.4.2 will not serve our purpose. Instead, we should use a 12×12 matrix. Deletion of the first column and three of the remaining 11 columns in such a matrix yields $A = 12$ half-samples in full orthogonal balance.

The principle just illustrated is that a given $c \times c$ matrix (where c is a multiple of 4) can be used to obtain $A = c$ half-samples in full orthogonal balance when $H = c - 1, c - 2, c - 3$, or $c - 4$. Thus, for $H = 4, 5, 6$, or 7, we get $A = 8$ half-samples from the 8×8 table, for $H = 8, 9, 10$, or 11, we get $A = 12$ half-samples from the 12×12 table, and so on.

Remark 11.4.1. The balanced half-sample technique can be extended to multistage sampling with stratified sampling of two PSUs from each stratum of PSUs. The technique is then applied to the PSUs. Consider estimation of the population total t by the unbiased estimator

$$\hat{t} = \sum_{h=1}^{H} (\textstyle\sum_{s_{\mathrm{I}h}} \hat{t}_i)/\pi_{\mathrm{I}i}$$

where $s_{\mathrm{I}h}$ is the sample of two PSUs from the hth stratum of PSUs, $\pi_{\mathrm{I}i}$ is the inclusion probability of the ith PSU, and $\hat{t}_i$ is an unbiased estimator of the ith PSU total t_i, based on subsampling within the PSU by an arbitrary design in one or more stages. We can proceed as in single-stage sampling. The two selected PSUs from the hth stratum are relabeled as $h1$ and $h2$. As in Example 11.4.2, a balanced set of half-samples of PSUs is chosen. For each half-sample ($a = 1, \ldots, A$) we calculate

$$\hat{t}_a = \sum_{h=1}^{H} [\delta_{\mathrm{I}ah}\hat{t}_{h1}/p_{\mathrm{I}h1} + (1 - \delta_{\mathrm{I}ah})\hat{t}_{h2}/p_{\mathrm{I}h2}]$$

where $\delta_{\mathrm{I}ah} = 1$ if the ath half-sample contains the PSU labeled $h1$, (and $\delta_{\mathrm{I}ah} = 0$ otherwise), $p_{\mathrm{I}hj} = \pi_{\mathrm{I}hj}/2$ for $j = 1, 2$, and $\hat{t}_{hj}$ is the estimator of t_{hj} based on data from subsampling within PSU hj. The following variance estimator can now be calculated,

$$\hat{V}_{BH} = \frac{1}{A} \sum_{a=1}^{A} (\hat{t}_a - \hat{t})^2$$

For a balanced set of half-samples, this coincides with the variance estimator (11.2.8).

Remark 11.4.2. A limitation of the balanced half-sample technique described above is the requirement that exactly two elements (or two PSUs) are drawn from each stratum. Modifications of the technique have been suggested for cases where the stratum sample sizes n_h exceed two. For a review, see Wolter (1985).

Remark 11.4.3. The balanced half-sample technique described here yields an approximate variance estimator identical with (11.4.2), through an alternative calculation method. The resulting variance estimator would have been appropriate under sampling with replacement. The approach neglects that our sample was in fact drawn without replacement, but can be modified to take into account the without-replacement feature, as is now described. For each half-sample ($a = 1, \ldots, A$) we calculate

$$\hat{t}_a^* = \hat{t} + \sum_{h=1}^{H} \left(\frac{\pi_{h1}\pi_{h2}}{\pi_{h1,h2}} - 1\right)^{1/2} \left[\frac{\delta_{ah} 2y_{h1}}{\pi_{h1}} + \frac{(1-\delta_{ah})2y_{h2}}{\pi_{h2}} - \hat{t}_h\right]$$

where $\hat{t} = \sum_{h=1}^{H} \hat{t}_h$ and $\hat{t}_h = y_{h1}/\pi_{h1} + y_{h2}/\pi_{h2}$. Clearly, this requires known second-order inclusion probabilities $\pi_{h1,h2}$ such that

$$\pi_{h1}\pi_{h2}/\pi_{h1,h2} - 1 \geq 0$$

for $h = 1, \ldots, H$. A modified variance estimator is now defined as

$$\hat{V}_{BH}^* = \frac{1}{A} \sum_{a=1}^{A} (\hat{t}_a^* - \hat{t})^2 \tag{11.4.9}$$

It is left as an exercise to verify that if the set of half-samples is in full orthogonal balance, then $\hat{V}_{BH}^*$ is identical with the Yates-Grundy variance estimator, which in this case (with all $n_h = 2$) is

$$\hat{V} = \sum_{h=1}^{H} \left(\frac{\pi_{h1}\pi_{h2}}{\pi_{h1,h2}} - 1\right)\left(\frac{y_{h1}}{\pi_{h1}} - \frac{y_{h2}}{\pi_{h2}}\right)^2$$

For example, with *SI* sampling in each stratum, we have

$$\pi_{h1} = \pi_{h2} = \frac{2}{N_h}$$

$$\pi_{h1,h2} = \frac{2}{N_h(N_h - 1)}$$

and

$$\frac{\pi_{h1}\pi_{h2}}{\pi_{h1,h2}} - 1 = \frac{N_h - 2}{N_h} > 0$$

so the modified method works and gives the unbiased estimator

$$\hat{V}_{BH}^* = \hat{V} = \tfrac{1}{2} \sum_{h=1}^{H} N_h^2 (1 - f_h) S_{ys_h}^2$$

Remark 11.4.4. In our discussion above, we have used balanced half-samples to estimate the variance of the π estimator $\hat{t}_\pi$. In this case, however, the more standard methods discussed in Chapter 2 and in Section 11.2 can be easily applied, and the balanced half-sample technique offers no particular advantage. Its main interest lies in applications to variance estimation for more complex parameters, for example, finite population regression and correlation coefficients. In such cases, we have no exact results for the properties (such as the bias) of the balanced half-sample variance estimator. But as the technique is known to perform well for the simple π estimator, the tacit assumption is that it continues to give satisfactory results, even if the situation is more complex. A considerable body of empirical evidence is available to support this assumption. (See, for example, references given in Section 11.7.)

When the parameter θ and the estimator $\hat{\theta}$ are complex, the literature suggests several alternative balanced half-sample estimators for the variance $V(\hat{\theta})$. We continue to assume stratified sampling of two elements per stratum, but not necessarily *SI* sampling within strata. Let $\hat{\theta}$ be a full-sample estimator of θ. Let $\hat{\theta}_a$ be an estimator of θ based on data from the ath half-sample and of the same structure as $\hat{\theta}$. Let $\hat{\theta}_a^c$ be an estimator of θ based on the complement of the ath half-sample, that is, on those sample elements that do not belong to the ath half-sample. Four variance estimators that have been suggested in this situation are

$$\hat{V}_{\text{BH1}} = \frac{1}{A} \sum_{a=1}^{A} (\hat{\theta}_a - \hat{\theta})^2 \tag{11.4.10}$$

$$\hat{V}_{\text{BH2}} = \frac{1}{A} \sum_{a=1}^{A} (\hat{\theta}_a^c - \hat{\theta})^2 \tag{11.4.11}$$

$$\hat{V}_{\text{BH3}} = \tfrac{1}{2}(\hat{V}_{\text{BH1}} + \hat{V}_{\text{BH2}}) \tag{11.4.12}$$

and

$$\hat{V}_{\text{BH4}} = \frac{1}{4A} \sum_{a=1}^{A} (\hat{\theta}_a - \hat{\theta}_a^c)^2 \tag{11.4.13}$$

Substituting

$$\hat{\theta}_{\text{BH}} = \frac{1}{A} \sum_{a=1}^{A} \hat{\theta}_a$$

for $\hat{\theta}$ in (11.4.10) to (11.4.13) leads to even more formulas. Theoretical comparisons between these alternative variance estimators are lacking. We note that, if $\hat{\theta}$ is a π estimator or a linear combination of π estimators and if the set of half-samples is balanced, then (11.4.10) to (11.4.13) are identical. These four estimators are included in a Monte Carlo study by Frankel (1971). There are no dramatic differences in the performance (bias, MSE, coverage rate) between the four estimators. From a computational point of view, $\hat{V}_{\text{BH1}}$ and $\hat{V}_{\text{BH2}}$ are slightly simpler than the other two estimators.

11.5. The Jackknife Technique

The jackknife technique originated outside the field of survey sampling. The first idea, developed by Quenouille (1949, 1956), was to use jackknifing to reduce the bias of an estimator, in an infinite-population context. Tukey (1958) subsequently suggested that the technique might also be used to produce variance estimates. For finite populations, the jackknife technique was first considered by Durbin (1959). Here, we give a review of the jackknife technique as it is commonly used to estimate a variance in survey sampling. A more detailed account is given in Wolter (1985).

Let s be a sample (the full sample) of n elements obtained by an arbitrary nonstratified element sampling design. (Stratified sampling will be discussed later.) Let the population parameter θ be estimated by $\hat{\theta}$, an estimator based on data from the full sample s. The aim is to estimate $V(\hat{\theta})$.

The jackknife technique starts with partitioning the sample s into A dependent random groups of equal size m $(= n/A)$, as described in Section 11.3.2. We assume that, for any given s, each group is an SI sample from s, even if s itself is a non-SI sample. Next, for each group ($a = 1, \ldots, A$), we calculate $\hat{\theta}_{(a)}$, an estimator of θ of the same functional form as $\hat{\theta}$, but based only on the data that remain after omitting the ath group. For $a = 1, \ldots, A$, we then define

$$\hat{\theta}_a = A\hat{\theta} - (A - 1)\hat{\theta}_{(a)} \tag{11.5.1}$$

sometimes called the ath pseudovalue. The jackknife estimator of θ (an alternative to the estimator $\hat{\theta}$) is

$$\hat{\theta}_{\text{JK}} = \frac{1}{A} \sum_{a=1}^{A} \hat{\theta}_a \tag{11.5.2}$$

and the jackknife variance estimator is defined as

$$\hat{V}_{\text{JK}1} = \frac{1}{A(A-1)} \sum_{a=1}^{A} (\hat{\theta}_a - \hat{\theta}_{\text{JK}})^2 \tag{11.5.3}$$

In practice, $\hat{V}_{\text{JK}1}$ is used as an estimator of $V(\hat{\theta})$ as well as of $V(\hat{\theta}_{\text{JK}})$. A sometimes used alternative to $\hat{V}_{\text{JK}1}$ is

$$\hat{V}_{\text{JK}2} = \frac{1}{A(A-1)} \sum_{a=1}^{A} (\hat{\theta}_a - \hat{\theta})^2 \tag{11.5.4}$$

We note that $\hat{V}_{\text{JK}2} \geq \hat{V}_{\text{JK}1}$, which follows from (11.1.4).

Moreover, it follows from (11.1.3) that if the $\hat{\theta}_a$ ($a = 1, \ldots, A$) were uncorrelated random variables with the same expected value, then $\hat{V}_{\text{JK}1}$ would be unbiased for $V(\hat{\theta}_{\text{JK}})$. However, the $\hat{\theta}_a$ are in general correlated, so unbiasedness does not hold. There are no exact (finite sample size) results for the properties (bias, etc.) of the jackknife variance estimators when $\hat{\theta}$ is more complex than a π estimator. In the simple case where θ is a population total and $\hat{\theta}$ the corresponding π estimator, it will be shown in the following example that the jackknife technique works well, as far as bias is concerned. This result, together with empirical evidence, seems to be the main justification for using the jackknife technique in more complex situations.

Example 11.5.1. Let θ be the population total, $\theta = t = \sum_U y_k$. Suppose the sample s, of fixed size n, is obtained by πps sampling, without replacement, with inclusion probabilities $\pi_1, \ldots, \pi_N$. Let $\hat{\theta}$ be the π estimator based on the full sample s, that is,

$$\hat{\theta} = \hat{t}_\pi = \sum_s y_k/\pi_k$$

Let s be divided into A equal-sized random groups as described above. We define

$$\hat{\theta}_{(a)} = \hat{t}_{\pi(a)} = [A/(A-1)] \sum_{s-s_a} y_k/\pi_k \tag{11.5.5}$$

It follows from (11.5.1), (11.5.2), and (11.5.5) that

$$\hat{\theta}_a = \hat{t}_a = A\hat{t} - (A-1)\hat{t}_{\pi(a)} = A \sum_{s_a} y_k/\pi_k$$

and

$$\hat{\theta}_{\text{JK}} = \hat{t}_{\text{JK}} = \frac{1}{A}\sum_{a=1}^{A} \hat{t}_a = \hat{t}_\pi$$

Thus, the two variance estimators (11.5.3) and (11.5.4) coincide,

$$\hat{V}_{\text{JK}1} = \hat{V}_{\text{JK}2} = \frac{1}{A(A-1)}\sum_{a=1}^{A} (\hat{t}_a - \hat{t}_\pi)^2 = \hat{V}_{\text{JK}} \tag{11.5.6}$$

We note that (11.5.6) is identical with the random groups variance estimator for the same sampling design and random grouping, namely, $\hat{V}_{\text{DRG}}$ given by (11.3.10). That is, $\hat{V}_{\text{JK}1} = \hat{V}_{\text{JK}2} = \hat{V}_{\text{DRG}}$. In other words, when the π estimator is involved, the jackknife technique is an alternative way to compute $\hat{V}_{\text{DRG}}$. It follows from (11.3.11) that for the π estimator,

$$E(\hat{V}_{\text{JK}}) = E(\hat{V}_0)$$

where $\hat{V}_0$ is the simplified variance estimator shown in (11.2.6). The following familiar expression is obtained for the bias of $\hat{V}_{\text{JK}}$,

$$E(\hat{V}_{\text{JK}}) - V(\hat{t}) = \frac{n}{n-1}[V_0 - V(\hat{t})]$$

In the special case where $A = n$ and $m = 1$ (common in practice), it can be shown that the jackknife variance estimator is equal to $\hat{V}_0$ not only in expectation, but identically,

$$\hat{V}_{\text{JK}} = \hat{V}_0$$

so in that case the jackknife is an alternative way of computing $\hat{V}_0$.

EXAMPLE 11.5.2. In case of *SI* sampling, the results of the preceding example are simplified as follows.

$$\hat{t}_\pi = N\bar{y}_s$$

$$\hat{t}_{\pi(a)} = \frac{N}{n-m}\sum_{s-s_a} y_k = N\bar{y}_{s-s_a}$$

$$\hat{t}_a = \frac{N}{m}\sum_{s_a} y_k = N\bar{y}_{s_a}$$

$$\hat{t}_{\text{JK}} = \hat{t}_\pi = N\bar{y}_s$$

$$\hat{V}_{\text{JK}} = \frac{N^2}{A(A-1)} \sum_{a=1}^{A} (\bar{y}_{s_a} - \bar{y}_s)^2$$

$$E(\hat{V}_{\text{JK}}) = E(N^2 S_{ys}^2/n)$$

$$E(\hat{V}_{\text{JK}}) - V(N\bar{y}_s) = N S_{yU}^2$$

It follows that in the case of *SI* sampling, we can remove the bias of the jackknife estimator simply by multiplying by $1 - f = 1 - n/N$. In particular, if $A = n$ and $m = 1$ we obtain

$$\hat{V}_{\text{JK}} = N^2 S_{ys}^2/n$$

and

$$(1-f)\hat{V}_{\text{JK}} = N^2(1-f)S_{ys}^2/n$$

which is the ordinary unbiased variance estimator.

Remark 11.5.1. When the jackknife technique is applied to a stratified sample, one will commonly use other variance estimators than (11.5.3) or (11.5.4). According to Wolter (1985, p. 175), one should be "especially careful not to apply the classical jackknife estimators... to stratified sampling problems." Assume that the sample from stratum h ($h = 1, \ldots, H$) is partitioned at random into A_h groups, for a total of $A = \sum_{h=1}^{H} A_h$ groups. As before, let $\hat{\theta}$ be the full-sample estimator of θ. Let $\hat{\theta}_{(ha)}$ be the estimator of θ based on what remains from the stratum h sample after omitting the ath group. One estimator that has been suggested for $V(\hat{\theta})$ is (see Rust 1985)

$$\hat{V}_{\text{JK3}} = \sum_{h=1}^{H} [(A_h - 1)/A_h] \sum_{a=1}^{A_h} [\hat{\theta}_{(ha)} - \hat{\theta}]^2 \tag{11.5.7}$$

which would be unbiased if sample selection were with replacement and if $\hat{\theta}$ were the π estimator of the population total $\theta = t = \sum_U y_k$.

In particular, suppose *SI* sampling is used in each stratum. Let

$$\hat{\theta} = \sum_{h=1}^{H} N_h \bar{y}_{s_h}$$

and

$$\hat{\theta}_{(ha)} = N_1 \bar{y}_{s_1} + \cdots + N_{h-1}\bar{y}_{s_{h-1}} + N_h \bar{y}_{s_h - s_{ha}} + N_{h+1}\bar{y}_{s_{h+1}} + \cdots + N_H \bar{y}_{s_H}$$

We then obtain

$$\hat{V}_{\text{JK3}} = \sum_{h=1}^{H} \frac{N_h^2}{A_h(A_h - 1)} \sum_{a=1}^{A_h} (\bar{y}_{s_{ha}} - \bar{y}_{s_h})^2$$

Under the further assumption that $A_h = n_h$, we get simply

$$\hat{V}_{\text{JK3}} = \sum_{h=1}^{H} N_h^2 S_{ys_h}^2 / n_h$$

which is the traditional variance estimator, with the finite population correction omitted. Other variance estimators for the stratified sampling case are discussed by Wolter (1985).

Remark 11.5.2. In multistage sampling, the jackknife technique is usually applied at the PSU level. Suppose a first-stage sample s_I containing n_I PSUs is drawn from the set U_I composed of N_I PSUs. Let s_I be partitioned at random into A groups of PSUs, with m PSUs in each group. Let $\hat{\theta}$ be the full-sample estimator of θ, and let $\hat{\theta}_{(a)}$ denote the θ estimator that is based on the data remaining after deletion of the ath group of PSUs. Using equations (11.5.1) to (11.5.4) in this new context, we obtain

$$\hat{\theta}_a = A\hat{\theta} - (A-1)\hat{\theta}_{(a)}; \qquad a = 1, \dots, A$$

$$\hat{\theta}_{JK} = \frac{1}{A}\sum_{a=1}^{A}\hat{\theta}_a$$

and the variance estimators are

$$\hat{V}_{JK1} = \frac{1}{A(A-1)}\sum_{a=1}^{A}(\hat{\theta}_a - \hat{\theta}_{JK})^2$$

$$\hat{V}_{JK2} = \frac{1}{A(A-1)}\sum_{a=1}^{A}(\hat{\theta}_a - \hat{\theta})^2$$

which serve the dual purpose of estimating $V(\hat{\theta})$ as well as $V(\hat{\theta}_{JK})$. For example, let $\theta = t = \sum_{U_I} t_i$, and

$$\hat{\theta} = \hat{t} = \sum_{s_I} \hat{t}_i/\pi_{Ii}$$

where $\hat{t}_i$ is an estimator of the ith PSU total t_i based on subsampling within the ith PSU and π_{Ii} is the first-stage inclusion probability of this PSU. Analogous with Example 11.5.1, we let

$$\hat{\theta}_{(a)} = \hat{t}_{(a)} = [A/(A-1)]\sum_{s_I - s_{Ia}} \hat{t}_i/\pi_{Ii}$$

which gives

$$\hat{\theta}_a = \hat{t}_a = A\sum_{s_{Ia}} \hat{t}_i/\pi_{Ii}$$

$$\hat{\theta}_{JK} = \hat{t}_{JK} = \hat{t}$$

and the variance estimator

$$\hat{V}_{JK1} = \hat{V}_{JK2} = \frac{1}{A(A-1)}\sum_{a=1}^{A}(\hat{t}_a - \hat{t})^2 = \hat{V}_{JK}$$

In stratified sampling of PSUs, the technique described in Remark 11.5.1 could be applied to the PSUs.

Remark 11.5.3. For the jackknife, we need to fix a suitable number of groups A. For good accuracy in the resulting variance estimator, one would like to have as many groups as possible, which means $A = n$, that is, $m = 1$. On the other hand, for computational reasons, one would prefer as few groups as possible, the most extreme choice being $A = 2$, and $m = n/2$. In practice, the most frequent choice appears to be $A = n$, or, when computation cost is a serious consideration, a compromise between the two extremes $A = n$ and $A = 2$.

Remark 11.5.4. Kovar, Rao, and Wu (1988) found in an empirical study that the jackknife method performed poorly for estimating the variance of estimators of population quantiles.

11.6. The Bootstrap

Like the jackknife, the bootstrap technique was introduced outside survey sampling as a means of obtaining approximate variance estimates and confidence intervals. The originator was Efron (1979, 1981, 1982).

Gradually, the bootstrap technique began to attract the attention of survey samplers, as an alternative to the other approximate variance estimation techniques described in this chapter. So far, the technique is somewhat unexplored for survey sampling. The bootstrap technique was originally designed for use with independent observations, the standard assumption of traditional statistical theory. One basic problem, not yet definitely answered, is how the technique should be correctly modified to accomodate the special features of survey sampling, including the nonindependence arising in sampling without replacement and other complexities of designs and estimators.

The modest aim in this section is to indicate one simple use of the bootstrap idea in survey sampling. For further reading on bootstrap in survey sampling, the reader is referred to Bickel and Freedman (1984), Bondesson and Holm (1985), Gross (1980), Kovar, Rao, and Wu (1988), McCarthy and Snowden (1985), and Rao and Wu (1984, 1987).

Suppose a probability sample s is drawn from a population U by an arbitrary sampling design without replacement. The population parameter θ is estimated by $\hat{\theta}$, and we seek an estimate of $V(\hat{\theta})$. The following is a brief description of how the bootstrap technique works.

i. Using the sample data, construct an artificial population U^*, assumed to mimic the real, but unknown, population U.
ii. Draw a series of independent samples, "resamples," or "bootstrap samples," from U^* by a design identical to the one by which s was drawn from U. Independence implies that each bootstrap sample must be replaced into U^* before the next one is drawn. For each bootstrap sample, calculate an estimate $\hat{\theta}_a^*$ ($a = 1, \ldots, A$) in the same way as $\hat{\theta}$ was calculated.
iii. The observed distribution of $\hat{\theta}_1^*, \ldots, \hat{\theta}_A^*$ is considered an "estimate" of the sampling distribution of the estimator $\hat{\theta}$, and $V(\hat{\theta})$ is estimated by

$$\hat{V}_{BS} = \frac{1}{A-1} \sum_{a=1}^{A} (\hat{\theta}_a^* - \hat{\theta}^*)^2$$

where

$$\hat{\theta}^* = \frac{1}{A} \sum_{a=1}^{A} \hat{\theta}_a^*$$

Remark 11.6.1. In some applications of the bootstrap technique, when a confidence interval is the primary goal, a variance estimate $\hat{V}_{BS}$ is not calculated, but a confidence interval is obtained directly from the observed distribution of $\hat{\theta}_1^*, \ldots, \hat{\theta}_A^*$. For example, for a 95% confidence interval one takes the interval between the lower and upper 2.5% points of that distribution.

The following example shows how the boostrap principles can be used to estimate a variance $V(\hat{\theta})$, when $\hat{\theta}$ is the π estimator of the population total.

EXAMPLE 11.6.1. Let $\theta = t = \sum_U y_k$ and $\hat{\theta} = \hat{t}_\pi = \sum_s y_k/\pi_k$, where the π_k are the inclusion probabilities under the design used to draw s. For simplicity, suppose the design has a fixed sample size n. We seek to estimate $V(\hat{\theta}) = V$.

The first step is to construct the artificial population U^*. One possibility is to let U^* be composed of replicates of the elements in s. For each $k \in s$ we create $1/\pi_k$ artificial elements of U^*, all sharing the same y-value, namely, y_k and the same π-value, π_k. For simplicity, we assume that $1/\pi_k$ is an integer for all $k \in s$. Let us use the subscript i to label the artificial elements of U^*. For the ith element, let y_i^* and π_i^* be the respective values of y and π. Thus, to each $k \in s$ correspond $1/\pi_k$ elements in U^*, all of which have $y_i^* = y_k$ and $\pi_i^* = \pi_k$. By this construction, U^* will be of size

$$N^* = \sum_s 1/\pi_k = \hat{N}$$

and the U^* population total is

$$t_{U^*} = \sum_{U^*} y_i^* = \sum_s (1/\pi_k) y_k = \hat{t}_\pi$$

The next step is to draw resamples from U^*. To simplify, we let each resample be a *pps* sample of n draws with replacement (as in Section 2.9), with draw probabilities $p_i^* = \pi_i^*/n$ for each $i \in U^*$. Thus, the resamples are with replacement, although the original sample s was without replacement. Note that $\sum_{U^*} p_i^* = 1$.

For each bootstrap sample ($a = 1, \ldots, A$) we calculate the pwr estimator defined in Section 2.9,

$$\hat{\theta}_a^* = \hat{t}_a^* = \frac{1}{n} \sum_{j=1}^{n} y_{i_j(a)}^* / p_{i_j(a)}^*$$

where the y-value $y_{i_j(a)}^*$ is associated with the element from U^* that was obtained in the jth draw for the ath bootstrap sample. Letting

$$\hat{t}^* = \frac{1}{A} \sum_{a=1}^{A} \hat{t}_a^*$$

we finally define the variance estimator as

$$\hat{V}_{BS} = \frac{1}{A-1} \sum_{a=1}^{A} (\hat{t}_a^* - \hat{t}^*)^2$$

Its bias is easily obtained. Given s, $t_1^*, \ldots, \hat{t}_A^*$ are independent and identically distributed random variables. Using equation (2.9.8), we have

$$\begin{aligned} E(\hat{V}_{BS}|s) &= V(\hat{t}_a^*|s) \\ &= (1/n) \textstyle\sum_{U^*} p_i^*(y_i^*/p_i^* - t_{U^*})^2 \\ &= (1/n^2) \textstyle\sum_s (y_k/p_k - \hat{t}_\pi)^2 \\ &= [(n-1)/n] \hat{V}_0 \end{aligned}$$

where $\hat{V}_0$ is the simplified estimator of $V(\hat{\theta})$ defined by (11.2.6). Hence, for the π estimator,

$$E(\hat{V}_{BS}) = E[E(\hat{V}_{BS}|s)] = \frac{n-1}{n} E(\hat{V}_0)$$

In other words, apart from the factor $(n-1)/n$, the bootstrap technique, as used here, gives a variance estimator with the same expected value (and the same bias) as several of the techniques seen earlier in this chapter.

11.7. Concluding Remarks

The variety of variance estimation methods in this chapter may leave the reader wondering what technique one should actually use in a given survey application. The literature to date offers little guidance in this choice. By and large, all of the methods mentioned give satisfactory results and differences are often minor.

Some theoretical results, with emphasis on asymptotics, are found in Krewski and Rao (1981) and Rao and Wu (1984, 1987).

For finite sample size properties of the various techniques, one must rely on empirical evidence based on simulation studies from known (real or artificial) populations. A number of such studies have been carried out in the last 20 years. Different variance estimators have been compared in terms of bias, mean square error, and coverage rate of the resulting confidence intervals. Pioneering studies were done by Frankel (1971) and Kish and Frankel (1974).

Wolter (1985) summarizes findings from five different studies: Frankel (1971), Bean (1975), Mulry and Wolter (1981), Dippo and Wolter (1984), and a preliminary version of Deng and Wu (1987). He concludes, "...we feel that it may be warranted to conclude that the TS [Taylor series] method is good, perhaps best in some circumstances, in terms of the MSE and bias criteria, but the BHS [balanced half-samples] method in particular, and secondarily the RG [random groups] and J [jackknife] methods, are preferable from the point of view of confidence interval coverage probabilities" (Wolter, 1985, p. 316). Wolter (1985) also compares the variance estimation techniques from the point of view of flexibility and cost. Flexibility is defined as the capacity of a technique to handle various types of estimators $\hat{\theta}$ and various sampling designs likely to occur in survey practice. By and large, the four above mentioned techniques are considered equally flexible. Nevertheless, the random groups technique may have a slight edge in terms of versatility, because it can handle a larger variety of sampling designs. As for computing cost, the jackknife is more expensive than the other techniques, which are roughly equal as far as cost is concerned.

A review of variance estimation techniques, including bootstrap, as well as empirical results from a simulation study are given by Kovar, Rao, and Wu (1988). A comparative summary of variance estimation techniques is also found in Rust (1985). Computer programs for these variance estimation techniques (and modifications of them) are available, either as separate programs or as routines in statistical packages. For comparative studies of such programs, one may consult Francis (1981), Cohen, Burt, and Jones (1986), and Cohen, Xanthopoulos, and Jones (1988).

Exercises

11.1. Let $\hat{V} = \hat{V}(\hat{\theta})$ be an unbiased estimator of the variance $V = V(\hat{\theta})$. Prove that $\hat{V}^{1/2}$ underestimates the standard deviation $V^{1/2}$, that is,

$$E(\hat{V}^{1/2}) \leq V^{1/2}$$

11.2. Verify equation (11.1.3), namely, that

$$E(\hat{V}_1) = V(\hat{\theta}^*) - \sum\sum_{a \neq b}^{A} C(\hat{\theta}_a, \hat{\theta}_b)/A(A-1)$$
$$+ \sum_{a=1}^{A} [E(\hat{\theta}_a) - E(\hat{\theta}^*)]^2/A(A-1)$$

11.3. Verify the statement of Remark 11.2.1, namely, that under the *SI* design, equations (11.2.4) to (11.2.6) are identical to equations (11.2.1) to (11.2.3), respectively.

11.4. Verify equation (11.2.7), namely, that under an arbitrary fixed-size design,

$$E(\hat{V}_0) - V = [n/(n-1)](V_0 - V)$$

where $\hat{V}_0$ is defined by (11.2.6).

11.5. Verify equation (11.2.9) of Remark 11.2.3, namely, that for multistage sampling,

$$E(\hat{V}_0) - V = [n_{\mathrm{I}}/(n_{\mathrm{I}} - 1)](V_0 - V)$$

where $\hat{V}_0$ is defined by (11.2.8).

11.6. Show that for dependent random groups, under the *SI* design as described in

Example 11.3.1, the covariance of two group means is

$$C(\bar{y}_{s_a}, \bar{y}_{s_b}) = V(\bar{y}_s) - S^2_{yU}/n = -S^2_{yU}/N$$

11.7. Verify equation (11.3.11), namely, that for dependent random groups, under πps sampling as described in Example 11.3.2, the variance estimator $\hat{V}_{\text{DRG}}$ based on π estimators has an expected value satisfying

$$E(\hat{V}_{\text{DRG}}) = E(\hat{V}_0)$$

where $\hat{V}_0$ is given by (11.2.6).

11.8. Show that under multistage sampling, with dependent random groups of PSUs as described in Example 11.3.4, the variance estimator $\hat{V}_{\text{DRG}}$ in this example has an expected value satisfying

$$E(\hat{V}_{\text{DRG}}) = E(\hat{V}_0)$$

where $\hat{V}_0$ is given by (11.2.8).

11.9. Verify the statement made in Remark 11.3.2, namely, that the variance of the repeated random groups variance estimator $\hat{V}$ in the remark is given by

$$V(\hat{V}) = (c_1/R) + c_2$$

11.10. Verify equation (11.4.7), namely, that if the set of half-samples is balanced, then

$$\hat{V}_{\text{BH}} = \hat{V}_0$$

where $\hat{V}_0$ is defined by (11.4.2).

11.11. Verify equation (11.4.8), namely, that if the set of half-samples is in full orthogonal balance, then

$$\hat{t}_{\text{BH}} = \hat{t}_\pi$$

11.12. Verify the details of Example 11.4.1.

11.13. Show that the set of all 2^H possible half-samples is always in full orthogonal balance.

11.14. Show that under multistage sampling, the variance estimator $\hat{V}_{\text{BH}}$ in Remark 11.4.1 is identical to the simplified variance estimator $\hat{V}_0$ defined by (11.2.8), provided the set of half-samples is balanced.

11.15. Show that the modified half-samples variance estimator $\hat{V}^*_{\text{BH}}$ defined by (11.4.9) is identical to the Yates-Grundy variance estimator, provided there is full orthogonal balance.

11.16. Show that if $\hat{\theta}$ is a π estimator and the set of half-samples is balanced, then the four variance estimators $\hat{V}_{\text{BH1}}$ to $\hat{V}_{\text{BH4}}$, defined by equations (11.4.10) to (11.4.13), are identical.

11.17. Prove that under the assumptions of Example 11.5.1:

a. $E[\hat{\theta}_{(a)}|s] = \sum_s y_k/\pi_k$, and hence $E[\hat{\theta}_{(a)}] = t$
b. $\hat{\theta}_a = A\sum_{s_a} y_k/\pi_k$
c. $\hat{\theta}_{\text{JK}} = \sum_s y_k/\pi_k$
d. When $A = n$ (that is, $m = 1$), $\hat{V}_{\text{JK}} = \hat{V}_0$

11.18. Verify the details of Remark 11.5.2.

CHAPTER 12

Searching for Optimal Sampling Designs

12.1. Introduction

A number of optimization problems arise in survey design. At the planning stage of a survey, an efficient (optimal or near optimal) sampling design must be specified. The search for the best design can in some cases be given a precise mathematical formulation. For example, what is the best sampling design in a specified class of designs if the objective is to minimize variance or to minimize variance for a given cost of sampling. Some problems of this kind are formulated and solved in this chapter.

Sections 12.2 to 12.5 deal with optimal design in the presence of univariate or multivariate auxiliary information. Earlier chapters have shown that the regression estimator uses auxiliary information in an efficient manner. The question considered in this chapter is: What sampling design is optimal in combination with the regression estimator? What are the best inclusion probabilities, for a given auxiliary information? Section 12.2 shows that the optimal π_k are determined by the variance structure of the underlying regression model. More specifically, it is optimal under the variance criterion adopted in Section 12.2 to take

$$\pi_k \propto [V_\xi(y_k)]^{1/2}$$

where $V_\xi(y_k)$ is the variance specified for y_k by the model ξ. A simplified design whose inclusion probabilities come close to this optimum is obtained by the technique called model-based stratification, which is examined in Sections 12.4 to 12.5.

A second type of optimality problem concerns the optimum allocation of a sample. Limited sample resources are to be distributed in the best possible way among subsets of a finite population, for example, strata or PSUs. A classic example was seen in Section 3.7, namely, the allocation of an *STSI* sample of a given size to a fixed number of strata. Many other sample allocation problems arise in connection with the standard survey designs. A few of these are considered in Sections 12.7 to 12.9. For example, for two-stage sampling with *SI* selection in each stage, we want to find the optimal sampling fraction for the first stage, and the optimal subsampling fractions within PSUs at the second stage. The criterion may be to minimize variance for a fixed cost or to minimize cost for a fixed variance.

Many optimality problems in sampling have been considered and solved. We do not touch on all in this chapter, but reviews are found in Rao (1979b) and Bellhouse (1984). For example, Bellhouse reviews six different types of optimality problem.

12.2. Model-Based Optimal Design for the General Regression Estimator

In a survey, the efficiency of the design/estimator combination (the strategy) depends on how well the auxiliary information can be utilized. The regression estimator $\hat{t}_{yr}$ presented in Chapter 6 is an explicit function of known auxiliary variable values. One approach is to use the regression estimator together with the *SI* design, which is simple to implement. Then the auxiliary information is used only at the estimation stage. However, this approach may not fully exhaust the information content of the auxiliary variables. Additional mileage can often be obtained by using the auxiliary variables at the design stage as well.

In this section we ask: What sampling design is optimal for the regression estimator, given the underlying regression model? How should the inclusion probabilities be determined, given our knowledge of the auxiliary variables?

Suppose that $\mathbf{x}_1, \ldots, \mathbf{x}_N$ are known auxiliary vector values. We consider the regression estimator of $t = \sum_U y_k$,

$$\hat{t}_r = \sum_U \hat{y}_k + \sum_s \frac{e_{ks}}{\pi_k} \tag{12.2.1}$$

where

$$e_{ks} = y_k - \hat{y}_k$$

is the sample fit residual for the kth element, and

$$\hat{y}_k = \mathbf{x}_k' \hat{\mathbf{B}}$$

is the fitted value under the weighted least-squares regression fit, with $\hat{\mathbf{B}}$ given by

$$\hat{\mathbf{B}} = \left(\sum_s \mathbf{x}_k \mathbf{x}_k' / a_k \pi_k\right)^{-1} \sum_s \mathbf{x}_k y_k / a_k \pi_k \tag{12.2.2}$$

Here, $1/a_k$ is a suitable weight for the kth element. The approximate variance of $\hat{t}_r$,

$$AV(\hat{t}_r) = \sum\sum_U \Delta_{kl} \check{E}_k \check{E}_l \tag{12.2.3}$$

depends on the unknown y_k-values through the census fit residuals

$$E_k = y_k - \mathbf{x}_k' \mathbf{B}$$

where

$$\mathbf{B} - \left(\sum_U \mathbf{x}_k \mathbf{x}_k'/a_k\right)^{-1} \sum_U \mathbf{x}_k y_k/a_k$$

The regression assumed to underly the estimator (12.2.1) can be written as

$$y_k = \mathbf{x}_k' \boldsymbol{\beta} + \varepsilon_k \tag{12.2.4}$$

where ε_k is an error whose expected value is zero with respect to the model for every k.

Earlier chapters have shown situations in which the estimator (12.2.1) is considerably more efficient than the simple π estimator $\hat{t}_\pi = \sum_s \check{y}_k$. Large gains in efficiency are realized when the residuals E_k are small, and this happens when the available x variables explain y well.

A desirable further goal is to find the design (the inclusion probabilities) for which the approximate variance (12.2.3) is minimum. But the direct minimization of (12.2.3) is impossible, since the census-fit residuals E_k are unknown. The problem is more tractable if we derive a measure of variance by also taking into account the assumed statistical properties of the model errors ε_k. The variance pattern of the ε_k is particularly important. This approach leads to an expression that can be easily minimized to find optimal inclusion probabilities. Thus, in this approach it is not the real variance that is minimized, but a simplified substitute.

To this end, we specify that the model errors in (12.2.4) satisfy

$$\begin{cases} E_\xi(\varepsilon_k) = 0 \\ V_\xi(\varepsilon_k) = \sigma_k^2 \\ E_\xi(\varepsilon_k \varepsilon_l) = 0; \quad k \neq l \end{cases} \tag{12.2.5}$$

We assume, to start with, that $\sigma_1^2, \ldots, \sigma_N^2$ are known up to a constant multiplier. (It is possible to relax this assumption, see Section 12.5.) Often

$$\sigma_k^2 = c\, h(\mathbf{x}_k)$$

where $h(\mathbf{x}_k)$ is a known function of the known auxiliary value $\mathbf{x}_k$ and c is a possibly unknown constant. If there is just one x variable, we may work with

$$\sigma_k^2 = c x_k^\gamma \tag{12.2.6}$$

where x_k is the known x-value for the element k and c and γ are ordinarily unknown constants. We shall see that the secret in constructing an efficient design is to make a correct specification of the variance structure of the model

(12.2.5). In particular, if the specification (12.2.6) is used, an accurate guess is required about the value of γ.

Under the model ξ consisting of (12.2.4) and (12.2.5), the study variable values $y_1, \ldots, y_N$ are random quantities. Thus the total $t = \sum_U y_k$ is random. Moreover, if $\hat{t}$ is an estimator of t, the estimation error $\hat{t} - t$ is an important random quantity whose statistical properties can be examined jointly under the sampling design $p(\cdot)$ and the model ξ. In particular, the variance of $\hat{t} - t$ under the design and the model is called the *anticipated variance*, a fitting term coined by Isaki and Fuller (1982). That is, the anticipated variance of $\hat{t} - t$ is

$$E_\xi E_p[(\hat{t} - t)^2] - [E_\xi E_p(\hat{t} - t)]^2$$

For emphasis, we write $E_p(\cdot)$ to denote the expectation with respect to the sampling design $p(\cdot)$ and $E_\xi(\cdot)$ to denote the expectation with respect to the model composed of (12.2.4) and (12.2.5). The model is the survey sampler's conceptualization of a *superpopulation* and is often an expression of prior knowledge or belief. The sampler is willing to consider the vector of finite population values $(y_1, \ldots, y_N)$ as a realization of this superpopulation. The anticipated variance is a measure of average squared error computed with respect to both the superpopulation and the sampling design. The variance anticipated at the time the survey is designed is conceptually different from the design variance, which is the variance for a particular finite population vector $(y_1, \ldots, y_N)$ and a given randomized selection scheme. This design variance, $V_p(\hat{t}) = E_p(\hat{t}^2) - [E_p(\hat{t})]^2$, has been used extensively in the preceding chapters.

If $E_\xi E_p(\hat{t} - t) = 0$, the anticipated variance becomes simply

$$E_\xi E_p[(\hat{t} - t)^2] \tag{12.2.7}$$

We are particularly interested in deriving the anticipated variance (12.2.7) when $\hat{t}$ is the regression estimator $\hat{t}_r$ given by (12.2.1), and (12.2.5) is assumed to hold. In the following, we obtain an approximate expression for the anticipated variance, which depends not on the unknown y_k but (assuming that σ_k^2 is a specified function of $\mathbf{x}_k$) only on the known auxiliary values $\mathbf{x}_1, \ldots, \mathbf{x}_N$. A nearly optimum design, expressed in terms of known entities, can thus be specified.

It is helpful to consider an analog to $\hat{t}_r$, expressed in the model errors ε_k rather than in the sample fit residuals e_k,

$$\hat{t}_r^* = \textstyle\sum_U y_k^* + \sum_s \check{\varepsilon}_k \tag{12.2.8}$$

where

$$\check{\varepsilon}_k = \varepsilon_k/\pi_k = (y_k - y_k^*)/\pi_k$$

and

$$y_k^* = \mathbf{x}_k'\boldsymbol{\beta}$$

Here, the random variable (12.2.8) is a tool only; obviously, it is not an estimator, because the ε_k are not observed. The variance of (12.2.8) with respect to the sampling design only is given exactly by (12.2.3) with ε_k substituted for E_k, that is, by

$$V_p(\hat{t}_r^*) = \sum\sum_U \Delta_{kl}\check{\varepsilon}_k\check{\varepsilon}_l \tag{12.2.9}$$

The derivations that follow are considerably simplified by the fact that $\hat{t}_r$ can be approximated by $\hat{t}_r^*$ when statistical properties such as bias and variance are evaluated jointly with respect to the model ξ and the sampling design $p(\cdot)$. This can be seen as follows. Write the error of $\hat{t}_r$ as

$$\hat{t}_r - t = \hat{t}_r^* - t + R \tag{12.2.10}$$

where

$$\hat{t}_r^* - t = \sum_s \check{\varepsilon}_k - \sum_U \varepsilon_k$$

is the error of $\hat{t}_r^*$, and the residual term R is given by

$$\begin{aligned} R = \hat{t}_r - \hat{t}_r^* &= (\hat{\mathbf{B}} - \boldsymbol{\beta})'(\textstyle\sum_s \check{\mathbf{x}}_k - \sum_U \mathbf{x}_k) \\ &= \sum_{j=1}^{J} (\hat{B}_j - \beta_j)(\textstyle\sum_s \check{x}_{jk} - \sum_U x_{jk}) \end{aligned}$$

where $\hat{\mathbf{B}}$ is given by (12.2.2). Here, R is a sum of J products. The terms $\sum_s \check{x}_{jk} - \sum_U x_{jk} = D_s(x_j)$ and $\hat{t}_r^* - t = \sum_s \check{\varepsilon}_k - \sum_U \varepsilon_k = D_s(\varepsilon)$ have expected value zero and are of the same order of importance, because both are defined in terms of π estimators. Now, $D_s(x_j)$ is multiplied by another factor that is near zero in probability with respect to ξ and $p(\cdot)$ jointly, namely, the factor $\hat{B}_j - \beta_j$. Note that $E_\xi E_p(\hat{B}_j) = E_p E_\xi(\hat{B}_j) = E_p(\beta_j) = \beta_j$. The remainder term R on the right-hand side of (12.2.10) is therefore of a lower order of importance than the term $\hat{t}_r^* - t = D_s(\varepsilon)$. For a more complete discussion of these matters, including the regularity conditions that are required, see Robinson and Särndal (1983), and Wright (1983). Ignoring R in equation (12.2.10) leads to the approximation

$$\hat{t}_r - t \doteq \hat{t}_r^* - t$$

We thus approximate the anticipated variance of $\hat{t}_r$ as follows

$$\begin{aligned} E_\xi E_p[(\hat{t}_r - t)^2] &\doteq E_\xi E_p[(\hat{t}_r^* - t)^2] \\ &= E_\xi(\textstyle\sum\sum_U \Delta_{kl}\check{\varepsilon}_k\check{\varepsilon}_l) = \sum\sum_U \Delta_{kl} E_\xi(\check{\varepsilon}_k\check{\varepsilon}_l) \end{aligned} \tag{12.2.11}$$

Under the model (12.2.5),

$$E_\xi(\check{\varepsilon}_k\check{\varepsilon}_l) = \begin{cases} \sigma_k^2/\pi_k^2 & \text{if } k = l \\ 0 & \text{if } k \neq l \end{cases}$$

That is, only terms with $k = l$ give a nonzero contribution to the double sum (12.2.11), which is a considerable simplification. We now obtain

$$E_\xi E_p[(\hat{t}_r - t)^2] \doteq \sum_U \left(\frac{1}{\pi_k} - 1\right)\sigma_k^2 \tag{12.2.12}$$

The expression on the right-hand side of (12.2.12), denoted $ANV(\hat{t}_r)$, is an approximation of the anticipated variance of $\hat{t}_r$. That is, we set

$$ANV(\hat{t}_r) = \sum_U \left(\frac{1}{\pi_k} - 1\right)\sigma_k^2 \tag{12.2.13}$$

We now state the main result of this section, which specifies the inclusion probabilities π_k that minimize (12.2.13). A design with such inclusion probabilities is called an *optimal design.*

Result 12.2.1. *Let $p(\cdot)$ be a sampling design such that the expected sample size satisfies*

$$E(n_s) = n$$

for some given value n. Suppose that $t = \sum_U y_k$ is estimated by the regression estimator (12.2.1). *An optimal design, that is, a design for which the approximate anticipated variance* (12.2.13) *is minimized, is such that the first-order inclusion probabilities are determined by*

$$\pi_k = \pi_{0k} = \frac{n\sigma_k}{\sum_U \sigma_k} \tag{12.2.14}$$

where it is assumed that $\pi_{0k} \leq 1$ for all k. The minimum value of the approximate anticipated variance is then

$$ANV_0(\hat{t}_r) = \frac{1}{n}\left(\sum_U \sigma_k\right)^2 - \sum_U \sigma_k^2 \tag{12.2.15}$$

Proof. The Cauchy-Schwartz inequality can be used to minimize (12.2.13) under the constraint

$$E_p(n_s) = \Sigma_U \pi_k = n \tag{12.2.16}$$

We get

$$\left(\sum_U \sigma_k^2/\pi_k\right)\left(\sum_U \pi_k\right) \geq \left(\sum_U \sigma_k\right)^2$$

where equality holds if and only if

$$\pi_k \propto \sigma_k; \quad k = 1, \ldots, N$$

Using equation (12.2.16), the results (12.2.14) and (12.2.15) follow. □

Remark 12.2.1. The qualification "optimal design" in Result 12.2.1 should be interpreted with the following in mind: (a) an approximation to the anticipated variance is minimized; the optimality is thus at best approximate for any finite expected sample size n; (b) more than one design can be optimal, since the π_{0k} given by equation (12.2.14) do not uniquely determine the design; (c) the optimality of (12.2.14) is derived assuming a specific variance structure, so the extent to which the result is of value in practice depends on how well the assumed variance structure reflects the true variance structure.

Note that, for practical use of Result 12.2.1, it is sufficient to know the σ_k up to a constant, unknown multiplier. The multiplier vanishes when the π_{0k} are calculated.

Remark 12.2.2. Godambe and Joshi (1965) showed an interesting inequality concerning the anticipated variance, in which the expression on the right-hand side of equation (12.2.13) appears as a lower bound. Let $\hat{t}$ be any estimator of t satisfying

$$E_\xi E_p(\hat{t} - t) = 0 \tag{12.2.17}$$

For such an estimator, Godambe and Joshi showed that

$$E_\xi E_p(\hat{t} - t)^2 \geq \sum_U \left(\frac{1}{\pi_k} - 1\right)\sigma_k^2$$

Since the lower bound on the right-hand side coincides with the expression in (12.2.13), we conclude that the regression estimator given in (12.2.1), which satisfies (12.2.17), has the important property of attaining approximately the Godambe and Joshi lower bound.

The approximation (12.2.12) to the anticipated variance holds for a variety of weightings $1/a_k$ in (12.2.2). It is practical to use a "simple form weighting" defined by

$$a_k = \boldsymbol{\lambda}'\mathbf{x}_k$$

where $\boldsymbol{\lambda}$ is a vector independent of k, since the estimator (12.2.1) then simplifies to

$$\hat{t}_r = \sum_U \hat{y}_k$$

An optimum design for this estimator is then chosen in agreement with (12.2.14).

Example 12.2.1. Let us examine the ratio model

$$y_k = \beta x_k + \varepsilon_k \tag{12.2.18}$$

where $E_\xi(\varepsilon_k) = 0$. In this case, the regression estimator (12.2.1) is

$$\hat{t}_r = \left(\sum_U x_k\right)\hat{B} + \sum_s \frac{y_k - \hat{B}x_k}{\pi_k}$$

with

$$\hat{B} = \frac{\sum_s x_k y_k / a_k \pi_k}{\sum_s x_k^2 / a_k \pi_k}$$

where $1/a_k$ is the weight given to the data point (y_k, x_k).

In simple form weighting, a_k is set proportional to x_k, which gets rid of the term $\sum_s (y_k - \hat{B}x_k)/\pi_k$. We obtain simply

$$\hat{t}_r = \hat{t}_{ra} = \Sigma_U x_k \frac{\sum_s \check{y}_k}{\sum_s \check{x}_k} \tag{12.2.19}$$

The residuals that determine the approximate variance in equation (12.2.3) are

$$E_k = y_k - (\textstyle\sum_U y_k / \sum_U x_k) x_k \tag{12.2.20}$$

As pointed out in Section 7.3.4, when the sampling design is fixed, the weighting $1/a_k$ used in $\hat{B}$ has a minor effect on the variance of $\hat{t}_r$, as long as extreme weightings are avoided. There is no great incentive to complicate the estimator by choosing a weighting other than the simple form weighting $a_k \propto x_k$. No significant reduction in variance would be realized by some other choice. For instance, little is gained by taking $a_k = cx_k^2$ and obtaining the more complex estimator

$$\hat{t}_r = (\textstyle\sum_U x_k)\hat{B} + \sum_s (y_k - \hat{B}x_k)/\pi_k$$

where

$$\hat{B} = \frac{\sum_s y_k / x_k \pi_k}{\sum_s 1/\pi_k}$$

In this case, the residuals E_k that apply in the AV expression (12.2.3) are

$$E_k = y_k - N^{-1}(\textstyle\sum_U y_k / x_k) x_k$$

Although these residuals might for some populations lead to a smaller approximate variance than the residuals given in (12.2.20), the reduction is likely to be small, assuming that the sampling design is the same in both cases.

Let us now specify a variance structure for the model errors in (12.2.18). Assume that

$$V_\xi(\varepsilon_k) = cx_k^\gamma \tag{12.2.21}$$

where γ is a specified constant. From Result 12.2.1, an optimal design is one that satisfies

$$\pi_k = \pi_{0k} = \frac{nx_k^{\gamma/2}}{\sum_U x_k^{\gamma/2}} \tag{12.2.22}$$

In summary, an efficient estimation strategy under the model composed of (12.2.18) and (12.2.21) is to use the ratio estimator (12.2.19), and, if a good guess can be made about γ, to choose a design that satisfies (12.2.22). A simplified design with inclusion probabilities that approximate (12.2.22) is presented in Section 12.4. Exercise 12.1 illustrates that the ratio estimator (12.2.19) can have a considerably smaller variance if the inclusion probabilities are proportional to $x_k^{1/2}$ or to x_k, rather than constant, as is the case in *SI* sampling.

12.3. Model-Based Optimal Design for the Group Mean Model

The example in this section illustrates Result 12.2.1 in the important case of the group mean model introduced in Section 7.5. Let $U_1, \ldots, U_g, \ldots, U_G$ be a partitioning of the population U into groups. Consider the single factor ANOVA model stating that

$$y_k = \beta_g + \varepsilon_k \tag{12.3.1}$$

for $k \in U_g$, or, equivalently,

$$y_k = \mathbf{x}_k' \boldsymbol{\beta} + \varepsilon_k$$

with $\mathbf{x}_k' = (0, \ldots, 1, \ldots, 0)$, where the single entry "1" indicates the group to which k belongs, and $\boldsymbol{\beta} = (\beta_1, \ldots, \beta_g, \ldots, \beta_G)'$. If the weights in equation (12.2.2) are set to $1/a_k = A_g$, a constant for all $k \in U_g$, the predicted values are

$$\hat{y}_k = \hat{B}_g \quad \text{for all} \quad k \in U_g$$

where

$$\hat{B}_g = \tilde{y}_{s_g} = \textstyle\sum_{s_g} \check{y}_k / \hat{N}_g$$

with $\check{y}_k = y_k/\pi_k$ and $\hat{N}_g = \sum_{s_g} 1/\pi_k$. We know from Result 7.6.1 that (12.2.1) is then the poststratified estimator

$$\hat{t}_r = \sum_{g=1}^{G} N_g \tilde{y}_{s_g} \tag{12.3.2}$$

The optimal design for this estimator depends on our assumption about the variance structure. The following examples are illustrative.

Homoscedastic Variance throughout the Population

Homoscedastic variance is assumed throughout the population, so that

$$V_\xi(\varepsilon_k) = \sigma^2 \tag{12.3.3}$$

for all $k \in U$. By Result 12.2.1, an equal probability design is then optimal, that is, a design with

$$\pi_k = \pi_{0k} = n/N$$

for all $k \in U$, where $n = E(n_s)$ is the expected sample size. An example is the *SI* design, in which case (12.3.2) is the classical poststratified estimator (see Section 7.6). In other words, the ANOVA model shown in (12.3.1) together with the homoscedastic variance structure in (12.3.3) lead to the classical poststratification strategy. For an equal probability design other than *SI*, the estimator (12.3.2) will have roughly the same variance as under *SI* sampling.

Homoscedastic Variance within Each Group

Homoscedastic variance is assumed within each group. For $g = 1, \ldots, G$,

$$V_\xi(\varepsilon_k) = \sigma_{0g}^2 \tag{12.3.4}$$

for all $k \in U_g$. The σ_{0g}^2 may differ from group to group. An optimal design, according to Result 12.2.1, is one with

$$\pi_k = \pi_{0k} = \frac{n\sigma_{0g}}{\sum_{g=1}^{G} N_g \sigma_{0g}}$$

for all $k \in U_g$. These inclusion probabilities are achieved, for example, by the use of *STSI* sampling such that (a) each group U_g of the model (12.3.1) is a stratum; (b) in each group (stratum), an *SI* sample is drawn; (c) the sample size in stratum g is

$$n_g = n\frac{N_g \sigma_{0g}}{\sum_{g=1}^{G} N_g \sigma_{0g}} \tag{12.3.5}$$

where n is the predetermined total sample size.

This resembles Neyman's optimal allocation rule discussed in Section 3.7.4. The difference is that the rule is expressed here in terms of the group standard deviations σ_{0g} of the model. A calculation of the optimal n_g requires that we first quantify the σ_{0g}, through an educated guess or with the aid of prior knowledge. In summary, we can say that the classic strategy calling for sample selection by *STSI*, sample allocation by Neyman's rule, and estimation by the wellknown stratified formula

$$\hat{t} = \sum_{g=1}^{G} N_g \bar{y}_{s_g}$$

is a strategy generated by the ANOVA model (12.3.1) with model variances assumed to be equal for all elements in the same group, as expressed by equation (12.3.4).

12.4. Model-Based Stratified Sampling

Model-based stratified sampling, proposed by Wright (1983), is a simple selection procedure by which one can obtain inclusion probabilities that come near those of the optimum design of Result 12.2.1. The procedure has the simplicity characteristic of *STSI* sampling. At the same time, the loss of efficiency is negligible, compared to a design that exactly satisfies equation (12.2.14). The procedure uses a stratification of the population according to the value of σ_k, so that each stratum consists of elements with neighboring values of σ_k. The sampling fraction in a stratum is taken as proportional to the average σ_k within a stratum. Let us consider the background for this procedure.

Assume that $\sigma_1^2, \ldots, \sigma_N^2$ are specified up to a constant multiplier. We can then calculate the optimal design inclusion probabilities in Result 12.2.1, namely,

$$\pi_k = \frac{n\sigma_k}{\sum_U \sigma_k} = \pi_{0k}; \qquad k = 1, \ldots, N \tag{12.4.1}$$

where $n = E(n_s)$ is a given expected sample size. For simplicity, we assume that all $\pi_{0k} < 1$. If not, elements with $\pi_{0k} \geq 1$ would be chosen with certainty.

Simplicity of execution is an important consideration in choosing the sampling design. If a random-size design is acceptable, a simple solution is to use Poisson sampling. That is, a list-sequential scheme would be used, consisting of independent Bernoulli trials such that the kth element is given the probability of selection as in (12.4.1). However, some samplers do not like the uncontrolled sample size that is a part of Poisson sampling. If we insist on a fixed-sample-size design, the construction of a design respecting equation (12.4.1) meets with the difficulties discussed in Section 3.6.2. We know from that section that it is hard to specify a fixed-size design that (a) is easy to execute; (b) has first-order inclusion probabilities proportional to N given numbers; and (c) has second-order inclusion probabilities π_{kl} that are simple to calculate. One approximate solution is Sunter's scheme given in Section 3.6.2.

We therefore seek an alternative design with inclusion probabilities that are good approximations of the optimal π_k given by (12.4.1), but with no great loss of efficiency. A solution is model-based stratification, used together with the equal aggregate σ rule. We now present the reasoning that leads to this technique.

For an optimal design, the inclusion probabilites π_k are given by (12.4.1). We seek an alternative design which is simple to execute and nearly is as efficient as the optimal design. Here, efficiency is measured as follows. Suppose that to attain an a priori fixed value of the ANV defined by (12.2.13) we need n^* observations with the optimal design and n observations ($n \geq n^*$) with the alternative design. Define the *model-based efficiency* of the alternative design (relative to the optimal design) as

$$\frac{n^*}{n} = \text{eff} \tag{12.4.2}$$

Using equations (12.2.13) and (12.2.15), we obtain an equation that expresses the equality of the ANVs under the two designs, namely,

$$\textstyle\sum_U \sigma_k^2/\pi_k - \sum_U \sigma_k^2 = (\sum_U \sigma_k)^2/n^* - \sum_U \sigma_k^2$$

Since $\sum_U \pi_k = n$, it follows that

$$\text{eff} = \frac{n^*}{n} = \frac{(\sum_U \sigma_k)^2}{(\sum_U \pi_k)(\sum_U \sigma_k^2/\pi_k)} \tag{12.4.3}$$

We want an alternative design that is easy to execute and whose efficiency is near unity. *Model-based STSI sampling* satisfies these requirements, provided the strata are constructed in a particular way. The procedure is defined as follows. Let the population U be partitioned into the strata $U_1, \ldots, U_h, \ldots, U_H$. Let N_h be the size of U_h. From U_h, draw an *SI* sample of n_h elements, where

$$n_h = n\frac{\sum_{U_h} \sigma_k}{\sum_U \sigma_k} = n\frac{N_h\bar{\sigma}_h}{N\bar{\sigma}} \tag{12.4.4}$$

with

$$\bar{\sigma}_h = \frac{1}{N_h}\sum_{U_h} \sigma_k; \qquad \bar{\sigma} = \frac{1}{N}\sum_U \sigma_k \tag{12.4.5}$$

We now give a simple lower bound for the efficiency realized by model-based *STSI* sampling. Define

$$cv_{\sigma h} = \frac{S'_{\sigma h}}{\bar{\sigma}_h}$$

where

$$(S'_{\sigma h})^2 = \frac{1}{N_h}\sum_{U_h} (\sigma_k - \bar{\sigma}_h)^2$$

(Here, there is a slight advantage to dividing by N_h rather than $N_h - 1$ in defining the variance.) That is, $cv_{\sigma h}$ is he coefficient of variation of σ_k in stratum h. Also, we define the quantity

$$\text{MCV} = \max_{h=1,\ldots,H} cv_{\sigma h} \tag{12.4.6}$$

We then have the following result.

Result 12.4.1. *The efficiency as defined by equation* (12.4.3) *of model-based stratified sampling satisfies*

$$\text{eff} \geq \frac{1}{1 + (\text{MCV})^2}$$

Result 12.4.1 states that the efficiency is close to unity if the strata U_h are constructed so that MCV is near zero. Note that unity is an upper bound to *eff*. The obvious next step is to give a practical procedure for constructing the strata such that MCV is small. That is, we want the largest $cv_{\sigma h}$ to be small. First, however, let us prove Result 12.4.1.

PROOF. Given *STSI* sampling, it follows that $\pi_k = n_h/N_h$ for all $k \in U_h$. Then, using (12.4.4), we have

$$\pi_k = \frac{n\bar{\sigma}_h}{N\bar{\sigma}} \quad \text{for all} \quad k \in U_h$$

With these π_k, we obtain

$$n \sum_U \frac{\sigma_k^2}{\pi_k} = N\bar{\sigma} \sum_{h=1}^{H} \left(\sum_{U_h} \sigma_k^2\right)/\bar{\sigma}_h$$

$$= N\bar{\sigma} \sum_{h=1}^{H} N_h \bar{\sigma}_h (1 + cv_{\sigma h}^2)$$

Inserting this expression into equation (12.4.3), and noting that $\sum_U \pi_k = n$, we get

$$\text{eff} = \frac{(N\bar{\sigma})^2}{n \sum_U \sigma_k^2/\pi_k} = \frac{N\bar{\sigma}}{\sum_{h=1}^{H} N_h \bar{\sigma}_h (1 + cv_{\sigma h}^2)}$$

From (12.4.6), we now see that

$$\text{eff} \geq \frac{N\bar{\sigma}}{[1 + (\text{MCV})^2] \sum_{h=1}^{H} N_h \bar{\sigma}_h} = \frac{1}{1 + (\text{MCV})^2}$$

which proves Result 12.4.1. □

Remark 12.4.1. The value σ_k is a measure of our uncertainty about the value y_k. We may call $\sum_U \sigma_k$ "the uncertainty of the whole finite population," and $\sum_{U_h} \sigma_k$ "the uncertainty of the hth stratum." In model-based *STSI* sampling, the stratum sample proportion, n_h/n, is taken as equal to the proportion of the total uncertainty accounted for by the stratum.

In many cases, a small value of MCV is obtained by the *equal aggregate σ rule*. The rule is defined through the following steps. Recall that the σ_k are known up to a constant multiplier.

Step 1. Order the values σ_k in increasing magnitude:

$$\sigma_1 \leq \sigma_2 \leq \cdots \leq \sigma_k \leq \cdots \leq \sigma_N \tag{12.4.7}$$

Here, we let k denote the index of the element with the kth smallest value σ_k.

Step 2. Specify the number of strata wanted, H say, and calculate $\sum_U \sigma_k = N\bar{\sigma}$. In the first stratum, U_1, include the first N_1 elements ordered as in (12.4.7), up to the point where, as closely as possible,

$$\sum_{U_1} \sigma_k = \sum_U \sigma_k/H$$

In the second stratum, U_2, include the next N_2 elements ordered as in (12.4.7), to the point where, as closely as possible,

$$\sum_{U_2} \sigma_k = \sum_U \sigma_k/H$$

and so on. That is, every statum is such that it accounts for (very nearly) one Hth of the total, $\sum_U \sigma_k = N\bar{\sigma}$, so that

$$\sum_{U_h} \sigma_k = N_h \bar{\sigma}_h = N\bar{\sigma}/H; \qquad h = 1, \ldots, H \tag{12.4.8}$$

Step 3. Allocate equally (or as closely as possible) the n elements into the strata, that is, take

$$n_h = n/H; \qquad h = 1, \ldots, H \tag{12.4.9}$$

Finally, select, by SI sampling, n_h among the N_h elements in U_h, $h = 1, \ldots, H$.

Remark 12.4.2. If σ_k is modeled according to (12.2.6), step 1 of the procedure is equivalent to a size-ordering of the N positive values $x_k^{\gamma/2}$. (Assuming $\gamma > 0$, this in turn is equivalent to a size-ordering of the N values x_k.)

We now have the following consequence of Result 12.4.1.

Result 12.4.2. *Suppose that the strata are determined by the equal aggregate σ rule for model-based stratification, as specified by steps* 1 *to* 3 *above. The model-based efficiency, eff, then satisfies the inequality in Result* 12.4.1.

PROOF. Combining equations (12.4.8) and (12.4.9) we see that the stratum sample sizes n_h of the equal aggregate σ rule satisfy (12.4.4). The rule is thus a special case of model-based $STSI$ sampling. Result 12.4.2 follows from Result 12.4.1.

Thus, the key to successful results with the equal aggregate σ rule is to make sure that MCV is small. For many populations, it has been observed (Wright, 1983) that the values $cv_{\sigma 1}, \ldots, cv_{\sigma H}$ obtained under the rule are fairly equal for a given H, and that MCV decreases as H increases. However, it is the structure of the particular population that determines whether or not the maximum $cv_{\sigma h}$ has a tendency to drop as H increases. It is a good idea in practice to take the precaution of examining the series of $cv_{\sigma h}$ values, to ensure that not one of them (for instance, the value corresponding to the stratum of the smallest σ_k) is systematically greater than the rest.

Remark 12.4.3. Results 12.4.1 and 12.4.2 state that the anticipated variance obtained by stratified sampling according to the equal aggregate σ rule comes close to the minimum anticipated variance. Note the difference between the anticipated variance and the ordinary variance, which is the variance with respect to the sampling design alone. This latter variance is approximated by (12.2.3). If the statistician can specify a correct or nearly correct model for the σ_k^2, we can expect that model-based $STSI$ sampling is highly efficient also in the sense that the design variance is small.

12.5. Applications of Model-Based Stratification

Suppose that a single, always positive x variable is available, so that $x_1, \ldots, x_k, \ldots, x_N$ are known positive numbers. Consider the model stating that, for $k \in U$,

$$\begin{cases} E_\xi(y_k) = \beta x_k & (12.5.1) \\ V_\xi(y_k) = \sigma_0^2 x_k^\gamma & (12.5.2) \end{cases}$$

where σ_0 is an unknown multiplier and γ is a known positive value. For many populations, the scatter around the regression line will widen as x increases, which corresponds to a positive value of γ. The greater γ is, the more pronounced the heteroscedasticity is. Experience has shown that many finite populations conform well to a variance structure like the one shown in equation (12.5.2) with a value of γ in the interval (0, 2) or the narrower interval (1, 2); see Brewer (1963b). In regularly repeated surveys, one may have a good a priori idea of the value of γ. If γ is unknown, one may estimate γ from the sample, as discussed in Godfrey, Roshwalb and Wright (1984) and Särndal and Wright (1984).

To estimate $t = \sum_U y_k$ under the model consisting of (12.5.1) and (12.5.2), we can use SI sampling and the ratio estimator

$$\hat{t} = N\bar{x}_U \frac{\sum_s y_k}{\sum_s x_k} \tag{12.5.3}$$

Further utility can be extracted from the x variable by using it to advantage in the sampling design. According to Result 12.2.1, an optimal design is one for which the inclusion probabilities are given by

$$\pi_k = n x_k^{\gamma/2} / \textstyle\sum_U x_k^{\gamma/2} \tag{12.5.4}$$

where n is the expected sample size. We know from Section 3.6.2 that it is not simple to construct a fixed-size πps sampling scheme with π_k exactly proportional to x_k. To obtain a design that is simple to execute, one can use the approximately optimal solution consisting of model-based stratification and the equal aggregate σ rule. That is, the estimator is the combined ratio estimator

$$\hat{t} = N\bar{x}_U \frac{\sum_{h=1}^{H} N_h \bar{y}_{s_h}}{\sum_{h=1}^{H} N_h \bar{x}_{s_h}} \tag{12.5.5}$$

where the strata are formed according to steps 1 to 3 of the preceding section, with $\sigma_k = \sigma_0 x_k^{\gamma/2}$. For each stratum U_h, we should have, as closely as possible,

$$\textstyle\sum_{U_h} x_k^{\gamma/2} = \frac{1}{H} \sum_U x_k^{\gamma/2} \tag{12.5.6}$$

$h = 1, \ldots, H$. The sample is allocated to the strata by the "equal parts rule," that is, $n_h = n/H$ in every stratum. Exercise 12.1 gives a numerical illustration of the use of the estimator (12.5.5) and the rule (12.5.6).

The following special cases are noted:

1. If $\gamma = 0$ is assumed, then equation (12.5.6) tells us that the strata should be of equal size, that is, for $h = 1, \ldots, H$,

$$N_h = N/H \tag{12.5.7}$$

2. If $\gamma = 1$ is assumed, then (12.5.6) yields strata based on equal aggregate square root size: For $h = 1, \ldots, H$,

$$\textstyle\sum_{U_h} x_k^{1/2} = \sum_U x_k^{1/2}/H \tag{12.5.8}$$

3. If $\gamma = 2$ is assumed, then equation (12.5.6) states that the strata should obey the equal aggregate size rule: For $h = 1, \ldots, H$,

$$\textstyle\sum_{U_h} x_k = \sum_U x_k/H \tag{12.5.9}$$

The three rules expressed in (12.5.7) to (12.5.9) are used in practice. In particular, (12.5.9) was suggested by Mahalanobis (1952).

For several theoretical distributions, Raj (1964) compared the variance obtained with the equal aggregate size rule (12.5.9) and $n_h = n/H$ for all h with the variance obtained by optimal (minimum variance) strata limits and $n_h = n/H$ for all h. The optimal limits must be calculated from one assumed distribution to another and are less practical than the easily obtained equal aggregate size limits. The equal aggregate size limits lead to some efficiency loss compared to optimal limits. The rule expressed in (12.5.8), which corresponds to $\gamma = 1$, is a compromise that will serve well for many finite populations.

12.6. Other Approaches to Efficient Stratification

The equal aggregate σ rule (steps 1 to 3 of Section 12.4) simultaneously resolves two tasks: (a) the stratification of the population and (b) the allocation of a sample (of a given size) to the strata. Earlier approaches often separate these two tasks.

There is a considerable literature on optimal (that is, minimum variance) stratification for the usual unbiased estimator

$$\hat{t} = \sum_{h=1}^{H} N_h \bar{y}_{s_h} \tag{12.6.1}$$

Suppose that the *STSI* sample is allocated by the Neyman rule,

$$n_h \propto N_h S_{yU_h} \tag{12.6.2}$$

where $S_{yU_h}^2$ is the variance of y in the stratum U_h. Given the estimator (12.6.1), and the optimal allocation (12.6.2), Dalenius (1957) and Dalenius and Hodges

(1959) examined the question of optimal stratification of the population. For a given number H of strata, they sought the stratum boundaries that minimize the variance of the estimator (12.6.1), when the sample is allocated by (12.6.2). This variance is

$$V_0(\hat{t}) = \frac{1}{n}\left(\sum_{h=1}^{H} N_h S_{yU_h}\right)^2 - \sum_{h=1}^{H} N_h S_{yU_h}^2 \qquad (12.6.3)$$

If the finite population correction is ignored, minimizing the expression on the right hand side of (12.6.3) is equivalent to minimizing

$$\sum_{h=1}^{H} W_h S_{yU_h} \qquad (12.6.4)$$

where $W_h = N_h/N$.

The minimization of (12.6.4) was studied by assuming that y follows a theoretical distribution of specified shape, for instance, the exponential distribution, $f(y) = e^{-y}$ for $y > 0$. For a given density for y and a given H, one can then calculate the stratum boundaries that minimize the continuous variable equivalent of (12.6.4), as done in Dalenius (1957). The derivation of the exact optimum is cumbersome for most densities. Various approximate rules have been suggested, the best known being the *cum* $\sqrt{\mathrm{f}}$ *rule* of Dalenius and Hodges (1959), which we now examine.

Note that in practice, the y variable is unknown, so a stratification of the y distribution is impossible. However, if x is a completely known auxiliary variable, strongly correlated with y, we can stratify the distribution of the N x-values, $x_1, \ldots, x_N$. We describe how this is done with the cum $\sqrt{\mathrm{f}}$ technique. Let $f_X(x)$ be the assumed density function of x, where $f_X(x) > 0$ for $A_L \leq x \leq A_U$. Define $a_0 = A_L$ and $a_H = A_U$, where H is the given number of strata to be formed. We seek $H - 1$ stratum boundaries $a_1, \ldots, a_{H-1}$ in the interior of the interval (A_L, A_U), such that

$$A_L = a_0 < a_1 < a_2 < \cdots < a_{H-1} < a_H = A_U$$

The cum $\sqrt{\mathrm{f}}$ rule creates the H strata in such a way that, for $h = 1, \ldots, H$,

$$\int_{a_{h-1}}^{a_h} [f_X(x)]^{1/2}\, dx = \frac{1}{H}\int_{A_L}^{A_U} [f_X(x)]^{1/2}\, dx \qquad (12.6.5)$$

In other words the strata limits are set so that each stratum accounts for $1/H$ of the total integral of $[f_X(x)]^{1/2}$.

The actual population is finite, not infinite, so the smooth density function is an unattainable ideal. As an approximation, we can work with a histogram based on the N values $x_1, \ldots, x_N$. Consider such a histogram based on J equally wide classes, where J is considerably larger than the prescribed number of strata H, for instance, $J = 30$ and $H = 6$. Using the histogram, we can approximate the cum $\sqrt{\mathrm{f}}$ rule as follows: Calculate $f_j^{1/2}$, the square root of f_j, where f_j is the frequency of x_k-values in the jth interval; $j = 1, \ldots, J$. Starting at the left tail of the histogram, neighboring classes are then joined together to form strata in such a way that each of the H strata contributes approximately $1/H$ of the accumulated scores $f_j^{1/2}$. That is, the sum of $f_j^{1/2}$ over the classes in each stratum should be as close as possible to

$$\frac{1}{H}\sum_{j=1}^{J} f_j^{1/2}$$

Once these strata have been determined, the stratum x-variance

$$S_{xU_h}^2 = \frac{1}{N_h - 1}\sum\nolimits_{U_h}(x_k - \bar{x}_{U_h})^2$$

is easily calculated, and an allocation of the sample based on "the Neyman rule on x",

$$n_h = n\frac{N_h S_{xU_h}}{\sum_{h=1}^{H} N_h S_{xU_h}}; \qquad h = 1, \ldots, H \tag{12.6.6}$$

is easily derived.

We know that the best allocation for the given strata is determined by "the Neyman rule on y",

$$n_h = n\frac{N_h S_{yU_h}}{\sum_{h=1}^{H} N_h S_{yU_h}}; \qquad h = 1, \ldots, H \tag{12.6.7}$$

But in most applications, the standard deviatons S_{yU_h} are unknown. The S_{xU_h} are reasonably good substitutes if the correlation between x and y is strong. No general rules of thumb can be given, but in many cases, the Neyman rule on x may work well if there is a correlation of 0.90 or more between x and y, whereas if the correlation drops to 0.80 or less, the efficiency of the allocation (12.6.6) compared to that of (12.6.7) starts to decline considerably.

When $x_1, \ldots, x_k, \ldots, x_N$ are known positive auxiliary values, several options exist for the combination of stratification, sample allocation and estimation. Two contenders are the following.

Strategy A. Use the combined ratio estimator (12.5.5) and *STSI* sampling based on model-based stratification according to the equal aggregate σ rule, with σ_k modeled as $\sigma_k = \sigma_0 x_k^{\gamma/2}$, where γ is specified constant. Equal allocation, $n_h = n/H$, is used in all strata. This is the procedure summarized by (12.5.5) and (12.5.6).

Strategy B. Use the unbiased estimator (12.6.1) and *STSI* sampling, where the strata are created by the cum $\sqrt{\text{f}}$ rule applied to the x-value distribution, and the sample allocation is by the Neyman rule on x given by (12.6.6).

For many populations and for a range of γ-values, strategy A will often yield a lower variance than strategy B. Around the point of the minimum, the variance is fairly insensitive to changes in the value of γ. Provided one is successful in choosing a good γ-value, strategy A is thus often preferable to strategy B. This is perhaps not surprising, since strategy A utilizes the known x-values in both the design and estimation stages.

Needless to say, smaller variances would be obtained with Strategy B if the strata could be created by the cum $\sqrt{\text{f}}$ rule applied to the y distribution itself and if the sample could be allocated by the Neyman rule on y given by (12.6.7). However if this much prior knowledge were available about the study variable y, there would hardly be a need to carry out a survey. Studies of model-based and other stratification rules are reported in Godfrey, Roshwalb and Wright (1984) and in Bethel (1989b).

Remark 12.6.1. For highly skewed populations (such as those encountered in many business surveys) it is often necessary to include the largest elements with probability one. These elements define what is called a self-representing stratum, or a "take-all" stratum. Suppose that *SI* sampling is used in each of the other strata, the "take-some" strata. The question of optimal stratum boundaries (cut-off point for the take-all stratum; boundaries for the take-some strata) was examined by Glasser (1962), Hidiroglou (1986a), Lavallée and Hidiroglou (1987). The last mentioned paper assumes that the sample is power-allocated (see Section 12.7), and gives an algorithm for finding the boundaries.

Remark 12.6.2. In some surveys, several categoric variables are available for stratification, and a considerable number of strata is formed by cross classification. Suppose there are H_i classes of the ith stratification variable, $i = 1, \ldots, K$. Cross classification leads to a total number of strata equal to $H = \prod_{i=1}^{K} H_i$. Kish and Anderson (1978) consider the efficiency gains resulting from this type of multivariate stratification.

12.7. Allocation Problems in Stratified Random Sampling

Sections 12.7 to 12.10 deal with optimal allocation of limited sampling resources to subsets of the finite population, such as strata or PSUs. A design of a given type is assumed, for instance, an *STSI* design or a two-stage design *SI*, *SI*. The problem is to specify the best sampling fractions to use with such a design.

A classic problem of this type concerning the *STSI* design was treated in Results 3.7.3 and 3.7.4. The π estimator of the total $\sum_U y_k$,

$$\hat{t}_\pi = \sum_{h=1}^{H} N_h \bar{y}_{s_h}$$

has the variance

$$V = V_{STSI}(\hat{t}_\pi) = B + \sum_{h=1}^{H} A_h/n_h \tag{12.7.1}$$

where

$$B = -\sum_{h=1}^{H} N_h S_h^2; \qquad A_h = N_h^2 S_h^2$$

Here we use the simplified notation S_h^2 for the stratum variance

$$S_{yU_h}^2 = \frac{1}{N_h - 1} \sum_{U_h} (y_k - \bar{y}_{U_h})^2$$

The cost of sampling is assumed to be

$$C = c_0 + \sum_{h=1}^{H} n_h c_h \tag{12.7.2}$$

where c_0 is the fixed cost, and $c_h > 0$ the per-element cost of sampling in stratum h.

In this case, the optimal allocation problem is to determine the n_h that minimize the variance (12.7.1) subject to a fixed cost C, or, conversely, to determine the n_h that minimize the cost C for a given variance V. The n_h are subject to the obvious constraints

$$n_h \leq N_h; \qquad h = 1, \ldots, H \tag{12.7.3}$$

The solution given in Result 3.7.3 is to take

$$n_h \propto (A_h/c_h)^{1/2} \tag{12.7.4}$$

More explicitly, (12.7.4) implies that the optimum allocation is given by

$$n_h = \left(\frac{A_h}{c_h}\right)^{1/2} \frac{C - c_0}{\sum_{h=1}^{H} (A_h c_h)^{1/2}} = \frac{N_h S_h}{c_h^{1/2}} \frac{C - c_0}{\sum_{h=1}^{H} N_h S_h c_h^{1/2}} \tag{12.7.5}$$

if C is fixed, and by

$$n_h = \left(\frac{A_h}{c_h}\right)^{1/2} \frac{\sum_{h=1}^{H} (A_h c_h)^{1/2}}{V - B} = \frac{N_h S_h}{c_h^{1/2}} \frac{\sum_{h=1}^{H} N_h S_h c_h^{1/2}}{V - B} \tag{12.7.6}$$

if V is fixed.

Remark 12.7.1. If one or more of the constraints $n_h \leq N_h$ are violated, the solution is to take $n_h = N_h$ for strata in which the value obtained by (12.7.5) or (12.7.6) exceeds the restriction N_h, and to readjust the remaining stratum sample sizes. To illustrate, take the case where V is fixed and C is minimized. Suppose that, after a reordering of the strata, the number n_h in (12.7.6) exceeds the limit N_h for $h = 1, \ldots, p$ $(p \leq H)$. Then let

$$\begin{cases} n_h = N_h; & h = 1, \ldots, p \\ n_h = K(A_h/c_h)^{1/2}; & h = p + 1, \ldots, H \end{cases}$$

with the constant K determined by

$$K = \sum_{h=p+1}^{H} (A_h c_h)^{1/2} \bigg/ \left(V - B - \sum_{h=1}^{p} A_h/N_h \right)$$

If all $n_h \leq N_h$, this represents the solution; if not, repeat the process until all $n_h \leq N_h$.

Remark 12.7.2. To be completely covered, one ought to include also the constraints

$$n_h \geq 1; \qquad h = 1, \ldots, H$$

which are required to calculate a stratum mean, or

$$n_h \geq 2; \qquad h = 1, \ldots, H$$

which are required to calculate a stratum variance. A more general optimization problem that includes such constraints is considered below. Algorithms for optimal allocation under different types of constraints are discussed, for *STSI* and other designs, in Hughes and Rao (1979).

More complex sample allocation issues arise in stratified sampling when there are several variables of study, $y_1, \ldots, y_i, \ldots, y_I$ $(I \geq 2)$. The totals $t_i = \sum_U y_{ik}$, $i = 1, \ldots, I$, are to be estimated. The multicharacter *STSI* allocation problem can be formulated as the minimization of cost under a constraint on the variance of each of the I estimates.

Assuming *STSI* sampling, the ith total is estimated by

$$\hat{t}_{i\pi} = \sum_{h=1}^{H} N_h \bar{y}_{is_h} \tag{12.7.7}$$

where $\bar{y}_{is_h}$ is the mean of y_i in the sample s_h taken from stratum h. The ith variance is given by

$$V_i = V_{STSI}(\hat{t}_{i\pi}) = B_i + \sum_{h=1}^{H} \frac{A_{ih}}{n_h} \tag{12.7.8}$$

with

$$B_i = -\sum_{h=1}^{H} N_h S_{ih}^2$$

and

$$A_{ih} = N_h^2 S_{ih}^2$$

where S_{ih}^2 is the variance of the variable y_i in stratum U_h.

The requirement of a stated minimal precision, V_{i0}, for the ith estimated total gives the constraints

$$V_i \leq V_{0i}; \qquad i = 1, \ldots, I \tag{12.7.9}$$

The stratum sample sizes are subject to the constraints

$$n_h \leq N_h; \qquad h = 1, \ldots, H \tag{12.7.10}$$

and we may also wish to include the constraints

$$n_h \geq 1; \qquad h = 1, \ldots, H \tag{12.7.11}$$

or

$$n_h \geq 2; \qquad h = 1, \ldots, H \tag{12.7.12}$$

Let the cost function again be given by equation (12.7.2).

The multicharacter STSI allocation problem consists of minimizing the cost (12.7.2) subject to the precision constraints given in (12.7.9), as well as to the sample size constraints (12.7.10), and possibly also (12.7.11) or (12.7.12).

This fits the description of the following general optimization problem, which is also of interest for several other allocation problems in sampling.

Problem I. Minimize the objective function

$$\sum_{j=1}^{J} \frac{q_j}{y_j} \tag{12.7.13}$$

subject to the constraints

$$\sum_{j=1}^{J} Q_{ij} y_j \leq Q_{i0}; \qquad i = 1, \ldots, I \tag{12.7.14}$$

and

$$y_{j0} \leq y_j; \qquad j = 1, \ldots, J \tag{12.7.15}$$

Here q_j, Q_{i0}, Q_{ij}, and y_{j0} are constants such that $q_j > 0$ and $y_{j0} \geq 0$ for all j, $Q_{i0} > 0$ for all i, and $Q_{ij} \geq 0$ for all i and j.

To see that the multicharacter *STSI* allocation problem fits the mold of problem I, let the quantities J, j, q_j, y_j, Q_{ij}, and Q_{i0} correspond, respectively, to H, h, c_h, $1/n_h$, $A_{ih} > 0$ and $V_{0i} - B_i > 0$. Moreover, the sample size constraints (12.7.10) are of the type given in (12.7.15) with y_{j0} corresponding to $1/N_h$. The constraints (12.7.11) and (12.7.12) translate into additional constraints of the type shown in (12.7.14).

In problem I, the y_j are treated as continuous variables and the minimum of (12.7.13) is sought over all real vectors $(y_1, \ldots, y_J)$ in the convex set formed by (12.7.14) and (12.7.15). Treating $1/n_h$ as equivalent to one of the continuous variables y_j is not quite correct, as only integer solutions n_h are of interest.

However, if y_j is obtained by solving problem I, the practical approach is to set n_h equal to the integer nearest to $1/y_j$. The approximation that results is without consequence for most practical purposes.

Problem I can be described as the minimization of the convex objective function shown in (12.7.13) subject to a number of linear constraints. Thus problem I defines a convex mathematical programming problem. Several questions now arise. Does there exist a solution to problem I? If one does exist how is it calculated? Is it possible to state an analytic solution, or must one resort to a numerical solution? These questions are discussed in Danielsson (1975), who shows that problem I has a solution. This follows from the Kuhn-Tucker theorem in optimization theory. Moreover, the solution can be stated analytically although in a complex form, expressed by a lengthy algorithm that we do not show here. It is a "complex analytic solution," as opposed to the "simple analytic solution" obtained in the case of $I = 1$. Heavy computational work is normally required to carry out the steps of this analytic solution. Ordinarily, problem I is instead solved by one of the efficient computer routines available for convex mathematical programming. The computer will use a search technique to arrive at a solution arbitrarily close to the optimum. Thus, it is a routine matter to solve problem I numerically.

A computer program of this kind is discussed in Huddleston, Claypool and Hocking (1970). They compare the optimum solution (obtained through a convex programming algorithm) with a compromise solution based on averaging the stratum sampling fractions obtained by applying the Neyman allocation formula for each variable separately. They find that the convex programming solution is superior to the compromise solution. It pays to search for the real optimum, as opposed to relying on an intuitive ad hoc solution. Bethel (1989a) expresses the optimal multicharacter *STSI* allocation as a closed expression in terms of normalized Lagrangian multipliers. The optimization problem still remains; programming considerations for obtaining a solution are discussed in Bethel's paper.

Remark 12.7.3. A simplified approach to allocation for multicharacter *STSI* sampling is to specify an importance weight, H_i, for the ith variable, $i = 1, \ldots, I$, and then use these weights to form a linear combination of the variances V_i given by (12.7.8)

$$V_{\text{lin}} = \sum_{i=1}^{I} H_i V_i$$

The expression V_{lin} is minimized subject to the fixed sampling cost given by (12.7.2), or, conversely, the cost is minimized subject to a fixed V_{lin}. This removes the multicharacter complication from the problem. We are back to the case of $I = 1$. It now easily follows that the n_h should be chosen as

$$n_h \propto \frac{N_h}{c_h^{1/2}} \left(\sum_{i=1}^{I} H_i S_{ih}^2 \right)^{1/2}$$

See Rao (1979b) for a discussion. The weakness of this compromise is the arbitrariness in the choice of the importance weights H_i.

Another application of problem I occurs in *STSI* sampling with a single study variable, when a desired precision is specified not only for the overall estimate, $\hat{t}_\pi = \sum_{h=1}^{H} N_h \bar{y}_{s_h}$, but also for each stratum total estimate, $\hat{t}_{h\pi} = N_h \bar{y}_{s_h}$, $h = 1, \ldots, H$. For example, an estimate of fixed precision is wanted for each province of a country, as well as for the country as a whole. The problem may be formulated as the minimization of the total sampling cost (12.7.2) subject to the constraints

$$V \leq V_0$$

and

$$N_h^2 \left(\frac{1}{n_h} - \frac{1}{N_h} \right) S_h^2 \leq V_{oh}; \qquad h = 1, \ldots, H$$

where $S_h^2 = S_{yU_h}^2$ is the variance of y in stratum h, V is given by (12.7.1) and $V_0, V_{01}, \ldots, V_{0H}$ are specified quantities. These constraints can be recast as

$$\sum_{h=1}^{H} \frac{N_h^2 S_h^2}{n_h} \leq V_0 + \sum_{h=1}^{H} N_h S_h^2 \qquad (12.7.16)$$

and

$$\frac{1}{n_h} \leq \frac{V_{0h}}{N_h^2 S_h^2} + \frac{1}{N_h}; \qquad h = 1, \ldots, H \qquad (12.7.17)$$

The problem can be put into the form of problem I if we let J, j, and y_j correspond to H, h, and $1/n_h$, respectively. We have $H + 1$ constraints that can be expressed in the form of (12.7.14). The inequality (12.7.16) is the first of these constraints, with Q_{1j} and Q_{10} corresponding to $N_h^2 S_h^2$ and $V_0 + \sum_{h=1}^{H} N_h S_h^2$, respectively. The other H constraints come from the H inequalities (12.7.17), with Q_{i0} corresponding to $(V_{0h}/N_h^2 S_h^2) + N_h^{-1}$. For a given index $i \geq 2$, all except one of the Q_{ij} are zero. The techniques available for solving problem I can be used, for example, a convex linear programming routine.

In an *STSI* sampling design where estimates are required for strata totals as well as for the whole population total, the Neyman allocation

$$n_h \propto N_h S_h$$

may give excellent ("unnecessarily good") precision for the overall population estimate and for the large strata estimates, whereas insufficient precision may result for the small strata.

To avoid underrepresentation in small geographical strata when large discrepancies exist in the strata sizes, Bankier (1988) suggests to "power allocate" the sample. For a given constant power a, he minimizes the criterion

$$\sum_{h=1}^{H} [(t_{xh})^a CV(\hat{t}_{yh})]^2$$

subject to $\sum_{h=1}^{H} n_h = n$, where $CV(\hat{t}_{yh}) = [V(\hat{t}_{yh})]^{1/2}/t_{yh}$ is the coefficient of variation for the estimator $\hat{t}_{yh}$ of the stratum total $t_{yh} = \sum_{U_h} y_k$, and t_{xh} is a measure of size or importance of stratum h. Assuming *SI* sampling in each stratum and $\hat{t}_{yh} = N_h \bar{y}_{s_h}$ leads to the allocation

$$n_h = n \frac{(t_{xh})^a cv_{yh}}{\sum_{h=1}^{H} (t_{xh})^a cv_{yh}} \tag{12.7.18}$$

where $cv_{yh} = S_h/\bar{y}_{U_h} = S_{yU_h}/\bar{y}_{U_h}$ is the coefficient of variation of y in stratum h. The constant a is called the power of the allocation. The Neyman allocation is the special case obtained by setting $a = 1$ and $t_{xh} = t_{yh}$. On the other hand, setting $a = 0$ results in an allocation where the $CV(\hat{t}_{yh})$ are almost equal from stratum to stratum, assuming that the coefficients of variation cv_{yh} vary little between the strata and that the strata sampling fractions are small.

Note also that if cv_{yh} is constant from stratum to stratum, then the power allocation (12.7.18) can be written as

$$n_h \propto (N_h \bar{x}_{U_h})^a$$

if $t_{xh} = \sum_{U_h} x_k = N_h \bar{x}_{U_h}$, where x_k is a size measure of element k. In practice, a suitable choice of the power a may be 1/2 or 1/3. Such a choice can be viewed as a compromise between the Neyman allocation and the allocation that yields practically constant precision for the strata estimates $\hat{t}_{yh}$, as measured by $CV(\hat{t}_{yh})$. Power allocation often makes it possible to increase considerably the precision for the estimates in the small strata, with only a slight reduction of precision in the estimates for the large strata and for the population as a whole, as compared to the Neyman allocation.

12.8. Allocation Problems in Two-Stage Sampling

12.8.1. The π Estimator of the Population Total

Let us consider the two-stage sampling design *SI*, *SI*, as in Example 4.3.1, from where the notation is borrowed. Our problem is to determine optimally the first-stage sampling fraction, $n_{\mathrm{I}}/N_{\mathrm{I}} = f_{\mathrm{I}}$, as well as the various subsampling fractions, $n_i/N_i = f_i$.

The π estimator is given by (4.3.21) and its variance by (4.3.22), which can also be expressed as

$$V_{SI,SI}(\hat{t}_\pi) = A_0 + \frac{A_1}{f_{\mathrm{I}}} + \frac{1}{f_{\mathrm{I}}} \sum_{U_I} \frac{A_{2i}}{f_i} \tag{12.8.1}$$

if we let

$$A_0 = -N_{\mathrm{I}} S^2_{tU_{\mathrm{I}}}; \qquad A_1 = N_{\mathrm{I}} G; \qquad A_{2i} = N_i S_i^2 \tag{12.8.2}$$

with

$$G = S^2_{tU_{\mathrm{I}}} - \frac{1}{N_{\mathrm{I}}} \sum_{U_{\mathrm{I}}} N_i S_i^2 \tag{12.8.3}$$

where

$$S^2_{tU_{\mathrm{I}}} = \frac{1}{N_{\mathrm{I}} - 1} \sum_{U_{\mathrm{I}}} (t_i - \bar{t}_{U_{\mathrm{I}}})^2$$

and

$$S_i^2 = S^2_{yU_i} = \frac{1}{N_i - 1} \sum_{U_i} (y_k - \bar{y}_{U_i})^2 \tag{12.8.4}$$

The aim is to minimize (12.8.1) for a fixed cost (or conversely), so we need a suitable cost function. Consider

$$c_0 + n_{\mathrm{I}} c_u + (\textstyle\sum_{s_{\mathrm{I}}} N_i) c_l + \sum_{s_{\mathrm{I}}} n_i c_{2i} \tag{12.8.5}$$

where c_0 is a fixed cost, and the remaining three terms represent variable costs; c_u is the cost of selecting one PSU, c_l is the listing cost per element in a selected cluster, and c_{2i} is the cost of obtaining the desired information from a selected element, including interviewing cost.

Now, the cost (12.8.5) is impractical for our purposes, since it is a random variable. It depends on the random set s_{I}. *Expected cost* is a more suitable criterion. If EVC denotes expected variable cost, that is, the expected value of the last three terms of (12.8.5), we have

$$\mathrm{EVC} = n_{\mathrm{I}} c_u + \frac{n_{\mathrm{I}}}{N_{\mathrm{I}}} (\textstyle\sum_{U_{\mathrm{I}}} N_i) c_l + \frac{n_{\mathrm{I}}}{N_{\mathrm{I}}} \sum_{U_{\mathrm{I}}} n_i c_{2i}$$

or, equivalently,

$$\mathrm{EVC} = a_1 f_{\mathrm{I}} + f_{\mathrm{I}} (\textstyle\sum_{U_{\mathrm{I}}} a_{2i} f_i) \tag{12.8.6}$$

where

$$a_1 = N_{\mathrm{I}} c_1; \qquad a_{2i} = N_i c_{2i} \tag{12.8.7}$$

with

$$c_1 = c_u + \bar{N} c_l \tag{12.8.8}$$

Here c_1 represents an average per cluster cost. It includes the cost of selecting a cluster and listing the elements in a selected cluster. By contrast, c_{2i} is a per element cost. It includes the cost of observing a selected element in a selected cluster.

We now seek to determine f_{I} and f_i subject to the constraints

$$0 < f_i \leq 1; \qquad i \in U_{\mathrm{I}} \tag{12.8.9}$$

$$0 < f_{\mathrm{I}} \leq 1 \tag{12.8.10}$$

It helps to distinguish the following two cases:

Simple Case. Here, $G \geq 0$ (thus $A_1 \geq 0$) and a simple technique (Cauchy-Schwartz inequality or the Lagrange multiplier method) provides a solution. A negative value of A_1 is possible, although rather exceptional in practice, as discussed below.

General Case. The simple case does not necessarily apply; a numerical solution may then be found by mathematical programming methods. The general case is discussed in Section 12.10.

The simple case has two versions that can be treated simultaneously:

a. Minimize the variance (12.8.1) under the expected variable cost constraint

$$\mathrm{EVC} = C_0 \tag{12.8.11}$$

where EVC is given by (12.8.6).

b. Minimize the expected variable cost (12.8.6) under the constraint of a fixe. variance,

$$V_{SI,SI}(\hat{t}_\pi) = V_0 \tag{12.8.12}$$

where $V_{SI,SI}(\hat{t}_\pi)$ is given by (12.8.1).

Ignoring for the moment (12.8.9) and (12.8.10), it is easy to show (invoking either the Cauchy-Schwarz inequality or the Lagrange multiplier method) that the optimum subsampling fractions for either case (a) or (b) are

$$f_i = \left(\frac{a_1}{A_1}\frac{A_{2i}}{a_{2i}}\right)^{1/2} \tag{12.8.13}$$

To complete the solution for case (a), find the optimal first-stage fraction by solving the fixed variance equation (12.8.11) for f_{I}, which gives

$$f_{\mathrm{I}} = \frac{C_0}{a_1}\left[1 + \sum_{U_{\mathrm{I}}}\left(\frac{a_{2i}A_{2i}}{a_1 A_1}\right)^{1/2}\right]^{-1} \tag{12.8.14}$$

It is left as an exercise to find the optimal f_{I} for case (b), by means of the fixed variance equation (12.8.12).

We substitute into equations (12.8.13) and (12.8.14) the expressions for A_1, A_{2i}, a_1, and a_{2i} given by (12.8.2) and (12.8.7). This leads to the following result.

Result 12.8.1. *Let the expected variable cost be given by* (12.8.6), *and let G and S_i be defined, respectively, by equations* (12.8.3) *and* (12.8.4). *The minimization of $V_{SI,SI}(\hat{t}_\pi)$ subject to $EVC = C_0$ leads to the following optimal sampling fractions*

$$f_{\mathrm{I}} = \frac{n_{\mathrm{I}}}{N_{\mathrm{I}}} = \frac{C_0}{c_1}\left[N_{\mathrm{I}} + \frac{1}{(c_1 G)^{1/2}} \sum_{U_{\mathrm{I}}} N_i S_i (c_{2i})^{1/2}\right]^{-1} \tag{12.8.15}$$

for the first stage and

$$f_i = \frac{n_i}{N_i} = \left(\frac{c_1}{c_{2i}}\right)^{1/2} \frac{S_i}{G^{1/2}} \tag{12.8.16}$$

for subsampling of the ith cluster, provided that $G > 0$ and that neither (12.8.9) *nor* (12.8.10) *is violated.*

If the solution in Result 12.8.1 does not obey the constraints (12.8.9) and (12.8.10), one may use the mathematical programming solution in Section 12.10.

The sign of G merits a comment. Using (4.2.19) we can write, after some algebra,

$$G = \frac{\bar{N}(N-1)\delta S_U^2}{N_{\mathrm{I}} - 1} + \frac{\sum_{U_{\mathrm{I}}}(N_i - \bar{N})Q_i}{N_{\mathrm{I}} - 1} + \frac{\sum_{U_{\mathrm{I}}}(N_i - \bar{N})S_i^2}{N - N_{\mathrm{I}}} \tag{12.8.17}$$

where

$$\bar{N} = \sum_{U_{\mathrm{I}}} N_i / N_{\mathrm{I}}, \qquad Q_i = N_i (\bar{y}_{U_i})^2$$

and

$$\delta = 1 - S_W^2 / S_U^2$$

is the homogeneity coefficient, with

$$S_U^2 = \frac{1}{N-1} \sum_U (y_k - \bar{y}_U)^2; \qquad S_W^2 = \frac{1}{N - N_I} \sum_{U_I} (N_i - 1) S_i^2$$

It can happen that $G < 0$, for example, if $N_i = \bar{N}$ for all i, and $\delta < 0$. Ordinarily, however, G is positive, becuase δ is likely to be positive and, when the N_i vary, the net contribution of the last two terms in equation (12.8.17) is often positive.

Remark 12.8.1. If (12.8.15) leads to $n_{\mathrm{I}} \geq N_{\mathrm{I}}$, Result 12.8.1 implies that *all* clusters should be selected and subsampled with the sampling fractions in (12.8.16). In other words, *STSI* sampling should be used with the clusters as strata.

Remark 12.8.2. If follows from (12.8.16) that the optimal sample size in the ith cluster satisfies

$$n_i \propto N_i S_i / c_{2i}^{1/2}$$

This is in agreement with the optimal sample allocation to stratum h in *STSI* sampling, as seen in equation (3.7.18), namely,

$$n_h \propto N_h S_h / c_h^{1/2}$$

Remark 12.8.3. In practice one may not have all the information necessary to calculate the optimum expressed by (12.8.15) and (12.8.16). Various ordinarily unknown population quantities are involved. However, we may have approximate information about the ratios

$$\frac{c_1}{c_{2i}} = \frac{\text{per cluster cost}}{\text{per element cost in cluster } i}$$

and about $S_i/G^{1/2}$. Here, $S_i/G^{1/2}$ may be hard to assess, but one notes from (12.8.17) that if $N_i = N$ for all i, then

$$\frac{N_i S_i}{G^{1/2}} = \left[\frac{(N_{\mathrm{I}} - 1)\bar{N}}{(N-1)\delta}\right]^{1/2} \frac{S_i}{S_U}$$

where S_i/S_U is the cluster-to-whole-population standard deviation ratio and δ is the homogeneity coefficient. Thus, for equal-sized clusters, if good guesses can be made about δ, about the standard deviation ratios S_i/S_U, and about the cost ratios c_{2i}/c_1, equations (12.8.15) and (12.8.16) can be used as a guide to the optimum allocation.

12.8.2 Estimation of the Population Mean

Under the two-stage sampling design SI, SI, the estimator of the population mean given in Example 8.6.1 is

$$\hat{\bar{y}}_{Ur} = \frac{\sum_{s_{\mathrm{I}}} N_i \bar{y}_{s_i}}{\sum_{s_{\mathrm{I}}} N_i} \tag{12.8.18}$$

Its approximate variance satisfies

$$N^{-2} AV_{SI,SI}(\hat{\bar{y}}_{Ur}) = N_{\mathrm{I}}^2 \left(\frac{1}{n_{\mathrm{I}}} - \frac{1}{N_{\mathrm{I}}}\right) S_{\mathrm{I}}^2 + \frac{N_{\mathrm{I}}}{n_{\mathrm{I}}} \sum_{U_{\mathrm{I}}} N_i^2 \left(\frac{1}{n_i} - \frac{1}{N_i}\right) S_i^2 \tag{12.8.19}$$

where

$$S_{\mathrm{I}}^2 = \frac{\sum_{U_{\mathrm{I}}} N_i^2 (\bar{y}_{U_i} - \bar{y}_U)^2}{N_{\mathrm{I}} - 1} \tag{12.8.20}$$

Again, we can consider two optimization problems.

a. Minimize the approximate variance shown in equation (12.8.19), subject to a fixed expected variable cost,

$$\mathrm{EVC} = C_0 \tag{12.8.21}$$

where EVC is given by (12.8.6).

b. Minimize the expected variable cost shown in equation (12.8.6), subject to the constraint of a fixed precision

$$AV_{SI,SI}(\hat{\bar{y}}_{Ur}) = V_0 \tag{12.8.22}$$

Equation (12.8.19) can be written in the form of (12.8.1), if we let $A_0 = -N_{\mathrm{I}}S_{\mathrm{I}}^2$, $A_1 = N_{\mathrm{I}}G'$, and $A_{2i} = N_i S_i^2$, where

$$G' = S_{\mathrm{I}}^2 - \frac{1}{N_{\mathrm{I}}} \sum_{U_{\mathrm{I}}} N_i S_i^2 \tag{12.8.23}$$

Alternatively, we can express G' as

$$G' = \frac{\bar{N}(N-1)\delta S_U^2}{N_{\mathrm{I}}-1} + \frac{\sum_{U_{\mathrm{I}}}(N_i - \bar{N})Q_i'}{N_{\mathrm{I}}-1} + \frac{\sum_{U_{\mathrm{I}}}(N_i - \bar{N})S_i^2}{N - N_{\mathrm{I}}} \tag{12.8.24}$$

where $Q_i' = N_i(\bar{y}_{U_i} - \bar{y}_U)^2$. Here, G' can be negative, for example, if $\delta < 0$ and all $N_i = \bar{N}$.

It follows by analogy with Result 12.8.1 that the optimal subsampling fractions are now given by

$$f_i = \frac{n_i}{N_i} = \left(\frac{c_1}{c_{2i}}\right)^{1/2} \frac{S_i}{(G')^{1/2}}$$

and f_{I} is determined either by the condition (12.8.21) or by the condition (12.8.22). If all $N_i = \bar{N}$, then $G' = G$, and the solution agrees with the one in Result 12.8.1.

An interesting special case arises when the same subsampling fraction is used in each selected cluster. Let the constant subsampling fraction be $f_i = f_{\mathrm{II}}$ for all i. The estimator (12.8.18) is now self-weighting,

$$\hat{\bar{y}}_{Ur} = \frac{\sum_{s_{\mathrm{I}}} \sum_{s_i} y_k}{\sum_{s_{\mathrm{I}}} n_i}$$

which is simply the mean of y per sampled element. The optimization problem reduces to finding two optimal sampling fractions f_{I} and f_{II}. The approximate variance satisfies

$$N^{-2} AV_{SI,SI}(\hat{\bar{y}}_{Ur}) = A_0 + \frac{A_1}{f_{\mathrm{I}}} + \frac{A_2}{f_{\mathrm{I}} f_{\mathrm{II}}} \tag{12.8.25}$$

with $A_0 = -N_{\mathrm{I}}S_{\mathrm{I}}^2$, $A_1 = N_{\mathrm{I}}S_{\mathrm{I}}^2 - \sum_{U_{\mathrm{I}}} N_i S_i^2 = N_{\mathrm{I}}G'$, and $A_2 = \sum_{U_{\mathrm{I}}} N_i S_i^2$ where S_{I}^2 and G' are given, respectively, by (12.8.20) and (12.8.23).

For further simplicity, suppose that the per element cost, c_{2i}, is the same for all clusters, so that

$$c_{2i} = c_2, \qquad i \in U_{\mathrm{I}}$$

The expected cost (12.8.6) can then be written

$$\mathrm{EVC} = a_1 f_{\mathrm{I}} + a_2 f_{\mathrm{I}} f_{\mathrm{II}} \tag{12.8.26}$$

where

$$a_1 = N_{\mathrm{I}} c_1, \qquad a_2 = N c_2$$

Here c_2 and c_1 are the per element cost and the per cluster cost, respectively.

Minimizing the approximate variance in (12.8.19) subject to a fixed value of the expected variable cost given in (12.8.26), or the converse, is a simpler version of the problem treated earlier in this section. If $A_1 > 0$ and the constraints $0 < f_{\mathrm{I}} \leq 1$, $0 < f_{\mathrm{II}} \leq 1$ are without consequence, the optimum is specified by

$$f_{\mathrm{II}} = \left(\frac{a_1 A_2}{a_2 A_1}\right)^{1/2} = \left(\frac{c_1}{c_2 G' N} \sum_{U_{\mathrm{I}}} N_i S_i^2\right)^{1/2} \tag{12.8.27}$$

The fraction f_{I} is determined from the fixed expected cost equation or from the fixed variance equation. For example, in the former case,

$$f_{\mathrm{I}} = \frac{C_0}{c_1}\left[N_{\mathrm{I}} + \left(\frac{c_2 N}{c_1 G'} \sum_{U_{\mathrm{I}}} N_i S_i^2\right)^{1/2}\right]^{-1} \tag{12.8.28}$$

where C_0 is the fixed value of EVC.

Let us examine this solution when $N_i = \bar{N}$ for all i, in which case

$$A_1 = N_{\mathrm{I}} G' = \frac{N(N-1)\delta S_U^2}{N_{\mathrm{I}} - 1}$$

$$A_2 = \sum_{U_{\mathrm{I}}} N_i S_i^2 = N(1-\delta) S_U^2$$

The number of units to be sampled is $n_{\mathrm{I}} = N_{\mathrm{I}} f_{\mathrm{I}}$ at stage one, and $n_{\mathrm{II}} = \bar{N} f_{\mathrm{II}}$ within each selected cluster at stage two. It follows from (12.8.27) and (12.8.28) that these numbers can be expressed in the following simple form

$$\begin{cases} n_{\mathrm{I}} \doteq \dfrac{C_0}{c_1}\left\{1 + \left[\dfrac{c_2(1-\delta)}{c_1 \delta}\right]^{1/2}\right\}^{-1} \\[2ex] n_{\mathrm{II}} \doteq \left[\dfrac{c_1(1-\delta)}{c_2 \delta}\right]^{1/2} \end{cases} \tag{12.8.29}$$

where $N/(N-1)$ and $N_{\mathrm{I}}/(N_{\mathrm{I}}-1)$ have been approximated by unity. If accurate figures are available for c_1, c_2, and δ, we can obtain a close indication of the optimal values of n_{I} and n_{II}. It is suggested that (12.8.29) may be used as an approximate solution even if the N_i vary modestly.

Remark 12.8.4. Other cost functions than the one given in (12.8.6) are sometimes called for. For example, if the cost of traveling between clusters is substantial, a realistic expected cost function may be

$$c_0 + c_t n_{\mathrm{I}}^{1/2} + n_{\mathrm{I}} \bar{N} c_l + \frac{n_{\mathrm{I}}}{N_{\mathrm{I}}} \sum_{U_{\mathrm{I}}} f_i c_{2i}$$

The term $c_t n_{\mathrm{I}}^{1/2}$ reflects a traveling cost. This nonlinear expected cost function will lead to different optimal allocations than those that we have found earlier in this section with a linear cost function.

12.9. Allocation in Two-Phase Sampling for Stratification

Optimization for two-phase sampling for stratification can be treated with techniques similar to those in Sections 12.7 and 12.8. As in Remark 9.4.3, consider a population with H predetermined strata. An SI sample of n_a elements is drawn from N. This sample is stratified, and STSI sampling is used in the second phase to select n_h elements from n_{ah} in stratum h. We want to find the optimal fractions $f_a = n_a/N$ for phase one and $v_h = n_h/n_{ah}$, $h = 1, \ldots, H$, for phase two. The fractions v_h are treated as constants fixed in advance and not dependent on the outcome of the first-phase sample.

The estimator of $t = \sum_U y_k$ is

$$\hat{t}_{\pi*} = N \sum_{h=1}^{H} \frac{n_{ah}}{n_a} \bar{y}_{s_h}$$

Its variance, given by (9.4.15), can be written as

$$V(\hat{t}_{\pi*}) = A_0 + \frac{A_1}{f_a} + \frac{1}{f_a} \sum_{h=1}^{H} \frac{A_{2h}}{v_h} \tag{12.9.1}$$

with

$$A_0 = -NS_U^2; \qquad A_1 = N\left(S_U^2 - \sum_{h=1}^{H} W_h S_h^2\right); \qquad A_{2h} = NW_h S_h^2 \tag{12.9.2}$$

where $W_h = N_h/N$ and

$$S_U^2 = S_{yU}^2 = \frac{1}{N-1} \sum_U (y_k - \bar{y}_U)^2; \qquad S_h^2 = S_{yU_h}^2 = \frac{1}{N_h - 1} \sum_{U_h} (y_k - \bar{y}_{U_h})^2$$

Clearly, $A_0 < 0$ and $A_{2h} > 0$ for all h. We can express A_1 as

$$A_1 = \frac{N}{N-1}\left[N \sum_{h=1}^{H} W_h(\bar{y}_{U_h} - \bar{y}_U)^2 - \sum_{h=1}^{H} (1 - W_h) S_h^2\right]$$

Usually, the term within square brackets is positive, so that $A_1 > 0$. However, $A_1 \leq 0$ can clearly occur. For many situations, a realistic variable cost function is

$$c_1 n_a + \sum_{h=1}^{H} c_{2h} n_h$$

where c_1 and c_{2h} are the per element sampling costs in phase one and phase two, respectively. The n_h are random variables, so it is better to consider the expected variable cost. Now,

$$n_h = n_a w_{ah} v_h$$

where n_a and v_h are fixed and $w_{ah} = n_{ah}/n_a$ is random, so we obtain

$$E(n_h) = n_a v_h E(w_{ah}) = n_a v_h W_h$$

The expected variable cost is

$$\text{EVC} = a_1 f_a + f_a \left(\sum_{h=1}^{H} a_{2h} v_h \right) \tag{12.9.3}$$

where

$$a_1 = Nc_1; \qquad a_{2h} = NW_h c_{2h} \tag{12.9.4}$$

The optimal fractions f_a and v_h are to be determined, subject to the constraints

$$0 < v_h \leq 1; \qquad h = 1, \ldots, H \tag{12.9.5}$$

$$0 < f_a \leq 1 \tag{12.9.6}$$

The variance (12.9.1) and the expected variable cost (12.9.3) are expressed in the same form as their counterparts, equations (12.8.1) and (12.8.6), in the two-stage problem of Section 12.8.1. Moreover, the constraints (12.9.5) and (12.9.6) have the same form as those appearing in the two-stage problem. We can directly use the results in Section 12.8.1 to determine the fractions f_a and $v_1, \ldots, v_H$ so that (a) the variance (12.9.1) is minimized for a fixed expected variable cost,

$$\text{EVC} = C_0 \tag{12.9.7}$$

where EVC is given by (12.9.3), or (b) the expected variable cost (12.9.3) is minimized for a fixed value of the variance (12.9.1),

$$V(\hat{t}_{\pi*}) = V_0 \tag{12.9.8}$$

In analogy with (12.8.13), the optimal within-stratum sampling fraction is

$$v_h = \left(\frac{a_1 A_{2h}}{A_1 a_{2h}} \right)^{1/2}$$

When the variance is minimized for a fixed expected variable cost, the optimal first-phase fraction is the analog of (12.8.14), that is,

$$f_a = \frac{C_0}{a_1} \left[1 + \sum_{h=1}^{H} \left(\frac{a_{2h} A_{2h}}{a_1 A_1} \right)^{1/2} \right]^{-1}$$

Using (12.9.2) and (12.9.4), these fractions can be expressed equivalently as in the following result.

Result 12.9.1. *In two-phase sampling for stratification, the minimization of the variance* (12.9.1) *subject to* $EVC = C_0$, *where EVC is the expected variable cost given by* (12.9.3), *leads to the following optimal sampling fractions:*

$$f_a = \frac{C_0}{Nc_1} \left[1 + \sum_{h=1}^{H} \left(\frac{c_{2h}}{c_1} \right)^{1/2} \frac{W_h S_h}{\left(S_U^2 - \sum_{h=1}^{H} W_h S_h^2 \right)^{1/2}} \right]^{-1} \tag{12.9.9}$$

for phase one, and

$$v_h = \left(\frac{c_1}{c_{2h}}\right)^{1/2} \frac{S_h}{\left(S_U^2 - \sum_{h=1}^{H} W_h S_h^2\right)^{1/2}}; \qquad h = 1, \ldots, H \qquad (12.9.10)$$

for sampling of stratum h in phase two, provided that $A_1 > 0$ *and that none of the constraints* (12.9.5) *and* (12.9.6) *are violated.*

12.10. A Further Comment on Mathematical Programming

Problem I, a convex mathematical programming problem, was formulated and used in Section 12.7. The following is an extension of problem I.

PROBLEM II. Minimize (12.7.13) subject to (12.7.14) and (12.7.15), where the constants are such that $q_j > 0$, $y_{jo} \geq 0$ for $j = 1, \ldots, J$; $Q_{io} \geq 0$ for $i = 1, \ldots, I$; and some (but not all) of the Q_{ij} may be negative.

Some of the Q_{ij} may now be negative, and some of the Q_{io} may be zero.

In an extensive study of allocation problems in surveys, Danielsson (1975) considers four different types of mathematical programming problems, including problems I and II. A somewhat different classification is given in Hughes and Rao (1979). It can be shown that problem II has a unique solution. It is hard to express this solution analytically. Consequently, one must rely on a mathematical programming algorithm to obtain a numerical solution. Since all constraints are linear, special algorithms may be used.

Problem II applies, for example, in the following situation. Reconsider the sample allocation problem for the *SI*, *SI* design, as in Section 12.8.1. The simple case solution in Result 12.8.1 may be inapplicable because $A_1 < 0$ (that is, $G < 0$) or because some of the fractions (12.8.15) and (12.8.16) do not respect the constraints (12.8.9) and (12.8.10). A solution may then be obtained by mathematical programming, using the formulation of problem II. Suppose we want to determine the sampling fractions that minimize the expected variable cost (12.8.6), subject to $V_{SI,SI}(\hat{t}_\pi) \leq V_0$, and subject to the constraints given by (12.8.9) and (12.8.10). Here, $V_{SI,SI}(\hat{t}_\pi)$ is the variance (12.8.1) and V_0 is the specified bound on the precision. This problem fits the mold of problem II with $J = N_I + 1$, namely, if y_1 corresponds to $1/f_I$, and if y_{j+1} corresponds to $1/f_I f_j$ and q_{j+1} to a_{2j} in (12.8.6), where j now is the cluster index; $j = 1, \ldots, N_I$. The precision requirement $V_{SI,SI}(\hat{t}_\pi) \leq V_0$ provides the first of the constraints (12.7.14). An additional set of N_I constraints of the form of (12.7.14) arises from the requirement $f_I f_j \leq f_I$ for all j, that is,

$$y_1 - y_{j+1} \leq 0; \qquad j = 1, \ldots, N_I$$

A constraint of the form of (12.7.15) is that $f_I \leq 1$, that is, $y_1 \geq 1$. Having formulated the problem this way, it is always possible to find a numerical solution with a convex programming algorithm.

12.11. Sampling Design and Experimental Design

There are close connections between experimental design and sampling design. Not surprisingly, certain ideas from the former field have been found useful in the latter field. An example is the following. In an *SI* sampling design with n elements chosen from N, the number of possible samples is $\binom{N}{n}$, the inclusion probabilities are $\pi_k = n/N$ for all k, and $\pi_{kl} = n(n-1)/N(N-1)$ for all $k \neq l$.

The number of possible samples under a specified design is sometimes called the *support size*. That is, for a given design $p(\cdot)$, the support size is the number of samples s with $p(s) > 0$. The question now arises whether we can construct a design with support size less than $\binom{N}{n}$ but with the same π_k and π_{kl} as under *SI* sampling. The practical reason for wanting to reduce the support size may be that some of the samples are undesirable because of considerable geographic spread of the elements and thereby a high sampling cost. It s interesting to note that Chakrabarti (1963), using the theory of experimental design, showed that there exist balanced incomplete block designs (PBIB) that yield the same π_k and π_{kl} as *SI*. If the PBIB has $b < \binom{N}{n}$ blocks, one would select with equal probability one of the blocks, and the elements in it would form the sample. That is, $p(s) = 1/b$ for each of the b samples. Since the π_k and π_{kl} determine the variance, this sampling scheme is as efficient as *SI* sampling for estimating the population mean or total. For further reading on the connections between experimental design and sampling design, the reader is referred to Hedayat (1979), Bellhouse (1984), and Fienberg and Tanur (1987).

Exercises

12.1. Twelve different strategies, 1 to 12 below, are considered for sampling the MU200 population and estimating the total $t_y = \sum_U y_k$ of the study variable ME84 ($= y$). The size (or the expected size) of the sample is 80 in each case, and P75 ($= x$) is used as an auxiliary variable for all but the first strategy. The known x total, $\sum_U x_k$, is denoted t_x. For the strategies 5 to 8, the strata are formed by the cum $\sqrt{\text{f}}$ rule applied to the x-value distribution, and the sample is allocated by the Neyman rule on x. Strategies 9 to 12 use model-based stratification, and equal parts of the sample are allocated to each stratum; strategies 9 and 11 are according to equation (12.5.8), and 10 and 12 according to equation (12.5.9).

1. *SI*; $N\bar{y}_s$
2. *SI*; $t_x(\sum_s y_k)/(\sum_s x_k)$

3. PO; $t_x(\sum_s \check{y}_k)/(\sum_s \check{x}_k)$ with $\pi_k \propto x_k^{1/2}$
4. PO; $t_x(\sum_s \check{y}_k)/(\sum_s \check{x}_k)$ with $\pi_k \propto \mathrm{x}_k$
5. $STSI$; $\sum_{h=1}^{2} N_h \bar{y}_{s_h}$
6. $STSI$; $\sum_{h=1}^{4} N_h \bar{y}_{s_h}$
7. $STSI$; $t_x(\sum_{h=1}^{2} N_h \bar{y}_{s_h})/(\sum_{h=1}^{2} N_h \bar{x}_{s_h})$
8. $STSI$; $t_x(\sum_{h=1}^{4} N_h \bar{y}_{s_h})/(\sum_{h=1}^{4} N_h \bar{x}_{s_h})$
9. $STSI$; $t_x(\sum_{h=1}^{2} N_h \bar{y}_{s_h})/(\sum_{h=1}^{2} N_h \bar{x}_{s_h})$
10. $STSI$; $t_x(\sum_{h=1}^{2} N_h \bar{y}_{s_h})/(\sum_{h=1}^{2} N_h \bar{x}_{s_h})$
11. $STSI$; $t_x(\sum_{h=1}^{4} N_h \bar{y}_{s_h})/(\sum_{h=1}^{4} N_h \bar{x}_{s_h})$
12. $STSI$; $t_x(\sum_{h=1}^{4} N_h \bar{y}_{s_h})/(\sum_{h=1}^{4} N_h \bar{x}_{s_h})$

The (exact or approximate) variances for the estimators in the twelve strategies are $39.1 \cdot 10^6$, $4.5 \cdot 10^6$, $3.4 \cdot 10^6$, $3.0 \cdot 10^6$, $11.6 \cdot 10^6$, $5.2 \cdot 10^6$, $3.2 \cdot 10^6$, $3.3 \cdot 10^6$, $3.8 \cdot 10^6$, $3.7 \cdot 10^6$, $3.6 \cdot 10^6$, and $3.2 \cdot 10^6$, respectively. Discuss and compare the strategies in light of the precision they yield.

12.2. It is proposed to estimate the total of the variable RMT85 for the MU281 population by means of $STSI$ sampling and the π estimator. The population is stratified into $H = 6$ strata as follows:

h	N_h	$S^*_{yU_h}$
1	79	15
2	65	21
3	45	26
4	54	53
5	21	85
6	17	194

where the values $S^*_{yU_h}$ are known to be good approximations of the true stratum standard deviations S_{yU_h}. (a) Use the Neyman rule and the values $S^*_{yU_h}$ to allocate a sample of total size $n = 150$. (b) Suppose that the cost of sampling is given by equation (12.7.2), where

h	1	2	3	4	5	6
c_h	200	180	160	120	120	100

Determine n_h so that the total cost is minimized under the condition that the variance of the π estimator is the same as in (a).

12.3. The three major regions in Exercise 10.4 are used as strata for an $STSI$ sampling design. Estimates are required of the three domain totals t_{yh} and for the entire population total t_y. Four allocations of the sample are considered: power allocation according to equation (12.7.18) with $a = 0$, 1/3, and 1/2, and Neyman allocation. Use the strata summary information in Exercise 10.4 to compute $CV(\hat{t}_{yh})$ and $CV(\hat{t}_y)$ for each of the four allocations. Compare and discuss the results. The total sample size is $n = 160$.

12.4. The MU284 population is partitioned into PSUs as in Exercise 8.5. To estimate the mean of the variable S82 $(=y)$, the two-stage design *SI*, *SI* will be used in such a way that n_{I} PSUs are to be selected in stage one and a constant number of municipalities, n_{II}, is to be selected from each PSU in stage two. The homogeneity coefficient for S82 is $\delta = 0.075$. The per cluster cost is $c_1 = 55$, the per element cost is $c_{2i} = c_2 = 7$ for every i, and $C_0 = \text{EVC}$ is fixed at the value 1,000. Determine approximately optimal sizes n_{I} and n_{II}, if the objective is to minimize the approximate variance of the estimator (12.8.18) of the population mean.

12.5. Consider two-phase sampling for regression estimation. In phase one, an *SI* sample of n_a elements is chosen from N, and the auxiliary value x_k is recorded for $k \in s_a$. In phase two, an *SI* subsample of size $n = \nu n_a$ is drawn, and y_k is observed for the subsample. The model $E_\xi(y_k) = \alpha + \beta x_k$; $V_\xi(y_k) = \sigma^2$ is formulated.

(a) Show that the approximate variance of the estimator $\hat{t}_{r2}$ obtained from equation (9.7.25) can be written

$$AV(\hat{t}_{r2}) = N^2 S_{yU}^2 \left(\frac{r_{xyU}^2}{n_a} + \frac{1 - r_{xyU}^2}{n} - \frac{1}{N} \right)$$

where r_{xyU} is the population correlation coefficient of x and y.

(b) Let c_a be the per element cost of observing x_k for the elements $k \in s_a$, let c be the per element cost of observing y_k for $k \in s$, and consider the cost function

$$C = c_a n_a + cn$$

Show that if C is fixed at C_0, then $AV(\hat{t}_{r2})$ is minimized by

$$\nu = \left(\frac{c_a}{c} \frac{1 - r_{xyU}^2}{r_{xyU}^2} \right)^{1/2}$$

(c) Consider the case of the MU281 population, the study variable REV84 $(=y)$, and the auxiliary variable P75 $(=x)$. Determine the approximately optimal sizes n_a and n if $C_0 = 4{,}000$, $c_a = 10$, and $c = 100$, supposing it is known that $r_{xyU} \doteq 0.9$.

(d) For the case considered in (c), suppose that single-phase *SI* sampling and the π estimator $N\bar{y}_s$ is used instead of the two-phase sampling. Compute the variance ratio $AV(\hat{t}_{r2})/V_{SI}(N\bar{y}_s)$. Use the same costs as in (c).

12.6. A sample of n elements is to be selected by *STSI* sampling from a population divided into two strata. How would you allocate the sample,

a. If you want to estimate the population mean, $\bar{y}_U$, with smallest possible variance?
b. If you want to estimate the difference, $\bar{y}_{U_1} - \bar{y}_{U_2}$, between the two strata means with smallest possible variance?
c. If you want to estimate the ratio, $\bar{y}_{U_1}/\bar{y}_{U_2}$, of the two strata means with smallest possible variance?

12.7. Under *STSI* sampling of n elements from a given stratified population, find the allocation that minimizes the variance of the estimator $\hat{t}_y$ of the population total $t_y = \sum_U y_k$,

a. When $\hat{t}_y$ is the separate ratio estimator.
b. When $\hat{t}_y$ is the combined ratio estimator.

12.8. Verify equation (12.8.13), namely, that under two-stage sampling with the design SI, SI, as described in Section 12.8, the optimum subsampling fractions are

$$f_i = [(a_1/A_1)(A_{2i}/a_{2i})]^{1/2}$$

12.9. Under the same assumptions as in Exercise 12.8, find the first-stage sampling fraction f_{I} that

a. Minimizes the variance (12.8.1) under the constraint of a fixed expected variable cost as stated in equation (12.8.11).
b. Minimizes the expected variable cost (12.8.6) under the constraint of a fixed variance as stated in equation (12.8.12).

12.10. Consider two-stage sampling with the design SI, SI. From each of the n_{I} clusters obtained in the first stage, the same number of elements is to be selected, that is, $n_i = n_0$ for all $i \in U_{\text{I}}$. (However, the clusters are not assumed to be of equal size.) The total cost of the survey is given by

$$C = n_{\text{I}} c_u + n_{\text{I}} n_0 c_{20}$$

where c_u is the per cluster cost and c_{20} is the cost associated with a selected element in a selected cluster. Find n_{I} and n_0 so that $V(\hat{t}_{y\pi})$ is minimized for given total cost, and find the minimum variance.

12.11. Consider two-stage sampling with the design SI, SI. Suppose that the subsampling fraction f_i is constant, that is, $f_i = f_0$, for all $i \in U_{\text{I}}$. The total cost of the survey is given by

$$C = n_{\text{I}} c_u + (\textstyle\sum_{s_{\text{I}}} N_i) c_l + (\sum_{s_{\text{I}}} n_i) c_{20}$$

The population total t_y is estimated by the ratio-to-size estimator

$$\hat{t}_y = N(\textstyle\sum_{s_{\text{I}}} N_i \bar{y}_{s_i})/(\sum_{s_{\text{I}}} N_i)$$

Find the sizes n_{I} and n_i that minimize $AV(\hat{t}_y)$ for a given expected variable cost.

12.12. Consider the π estimator $\hat{t}_{y\pi} = \sum_s y_k/\pi_k$ and the difference estimator

$$\hat{t}_{y,\text{dif}} = \textstyle\sum_U y_k^0 + \sum_s D_k/\pi_k$$

where $D_k = y_k - y_k^0$. Here, y_k^0 is a known value for all $k \in U$. A way to compare $\hat{t}_{y,\text{dif}}$ with $\hat{t}_{y\pi}$ is to compare the corresponding anticipated variances. Assume a model ξ under which $E_\xi(D_k) = 0$, $V_\xi(D_k) = \sigma_k^2$, $\text{Cov}_\xi(D_k, D_l) = 0$; $k \neq l$. Show that

$$E_\xi[V(\hat{t}_{y,\text{dif}})] = \textstyle\sum_U (\pi_k^{-1} - 1)\sigma_k^2$$

$$E_\xi[V(\hat{t}_{y\pi})] - E_\xi[V(\hat{t}_{y,\text{dif}})] = \textstyle\sum\sum_U \Delta_{kl}(y_k^0/\pi_k)(y_l^0/\pi_l)$$

where $\hat{t}_{y\pi}$ is the π estimator. From these results, derive the simplified expressions that apply under SI sampling with $f = n/N$ and $\sigma_k^2 = \sigma^2$ for all k.

CHAPTER 13

Further Statistical Techniques for Survey Data

13.1. Introduction

Survey data are special because the observations are usually sampled with unequal probabilities. The survey data consist of the measurements made on one or more study variables for the elements k in a probability sample s. This sample is drawn from the finite population $U = \{1, \ldots, k, \ldots, N\}$ according to a given sampling design, and the inclusion probabilities π_k are usually not equal. The data cannot be considered a sample of independent and identically distributed observations on a random variable, as is often assumed in traditional statistical theory and in many of the widely used computer packages for statistical analysis. However, in recent years, computer packages specially equipped to handle survey data have emerged. Among these are SUPER-CARP, PC-CARP, and SUDAAN.

This chapter discusses some of the issues that arise because of the special character of survey data. Sections 13.2 and 13.3 deal with design effects for the estimation of finite population parameters arising in regression and correlation analyses. Section 13.4 presents a technique for variance estimation for a vector of complex finite population parameters. Section 13.5 is devoted to hypothesis tests based on categorical survey data. Finally, in Section 13.6, a distinction is made between descriptive parameters and superpopulation model parameters, and different types of inference from survey data are discussed.

13.2. Finite Population Parameters in Multivariate Regression and Correlation Analysis

In Chapter 5, we examined estimators for certain finite population parameters that depend on two or more variables of study, such as a ratio of totals, a regression coefficient, and others. This chapter develops these ideas further. Consider a survey with a multidimensional variable of interest, $\mathbf{y} = (y_1, \ldots, y_i, \ldots, y_q)'$, whose value for the kth element is denoted

$$\mathbf{y}_k = (y_{1k}, \ldots, y_{ik}, \ldots, y_{qk})'$$

$k = 1, \ldots, N$. Let $\bar{y}_{iU} = \sum_U y_{ik}/N$ be the population mean of the ith variable. With the definitions used in Section 5.9, the finite population covariance between y_i and y_j is

$$S_{ij} = \frac{1}{N-1} \sum_U (y_{ik} - \bar{y}_{iU})(y_{jk} - \bar{y}_{jU}) \tag{13.2.1}$$

When $j = i$, we get the finite population variance of y_i,

$$S_{ii} = S_i^2 = \frac{1}{N-1} \sum_U (y_{ik} - \bar{y}_{iU})^2$$

Arranged in matrix form, these quantities define the *finite population covariance matrix*

$$\mathbf{S} = \begin{bmatrix} S_{11} & S_{12} & \cdots & S_{1q} \\ S_{12} & S_{22} & \cdots & S_{2q} \\ \vdots & \vdots & & \vdots \\ S_{1q} & S_{2q} & \cdots & S_{qq} \end{bmatrix}$$

or, for short,

$$\mathbf{S} = (S_{ij}) \tag{13.2.2}$$

The finite population correlation coefficient of y_i and y_j is calculated from the N pairs (y_{ik}, y_{jk}), $k = 1, \ldots, N$, as

$$r_{ij} = \frac{S_{ij}}{(S_{ii}S_{jj})^{1/2}} = \frac{\sum_U (y_{ik} - \bar{y}_{iU})(y_{jk} - \bar{y}_{jU})}{[\sum_U (y_{ik} - \bar{y}_{iU})^2 \sum_U (y_{jk} - \bar{y}_{jU})^2]^{1/2}} \tag{13.2.3}$$

We have $r_{ii} = 1$ for $i = 1, \ldots, q$. The finite population correlation matrix is a symmetric $q \times q$ matrix defined as

$$\mathbf{R} = (r_{ij}) \tag{13.2.4}$$

Important finite population characteristics such as multiple regression coefficients and partial and multiple correlation coefficients can be expressed as functions of the variances and covariances S_{ij}, or, alternatively, as functions of the r_{ij}. This section discusses such descriptive parameters and their estimation. They are called *descriptive* because they measure characteristics of the finite population at a given moment in time. They represent the results of the regression and correlation analyses that one would carry out if the entire finite population is observed, as in a census. To make inferences about these parameters when only a sample of the population is observed can be described as drawing conclusions about the results of a hypothetical regression or correlation analysis carried out on the full population.

First consider a multiple regression analysis with y_1 as the criterion variable and $y_2, \ldots, y_q$ as the explanatory variables. If the whole population is observed, we can determine the intercept B_1 and the regression slopes $B_2, \ldots, B_q$ so as to minimize the ordinary least squares (OLS) criterion

$$\sum_U (y_{1k} - B_1 - B_2 y_{2k} - \cdots - B_q y_{qk})^2$$

The vector $\mathbf{B} = (B_1, B_2, \ldots, B_q)'$ that minimizes this expression is

$$\mathbf{B} = \left(\sum_U \mathbf{z}_k \mathbf{z}_k'\right)^{-1} \sum_U \mathbf{z}_k y_{1k} \tag{13.2.5}$$

where $\mathbf{z}_k = (1, y_{2k}, \ldots, y_{qk})'$; $k = 1, \ldots, N$. Alternatively, equation (13.2.5) can be stated in terms of variances and covariances. Let $\mathbf{S}^{11}$ denote the $(q-1) \times (q-1)$ matrix obtained by deleting row one and column one in the covariance matrix $\mathbf{S}$ given by (13.2.2). Thus, $\mathbf{S}^{11}$ is the covariance matrix of $(y_2, \ldots, y_q)$. Moreover, let

$$\mathbf{S}_1 = (S_{12}, \ldots, S_{1q})'$$

This is the $(q-1)$-vector composed of the covariances of y_1 with $y_2, \ldots, y_q$. Standard texts on regression analysis point out that (13.2.5) can be expressed equivalently as

$$(B_2, \ldots, B_q)' = (\mathbf{S}^{11})^{-1}\mathbf{S}_1 \tag{13.2.6}$$

and

$$B_1 = \bar{y}_{1U} - B_2 \bar{y}_{2U} - \cdots - B_q \bar{y}_{qU} \tag{13.2.7}$$

The predicted value (the OLS prediction) for element k is $\hat{y}_{1k} = \hat{y}_{1 \cdot 2 \ldots q;k}$ defined by

$$\begin{aligned} \hat{y}_{1k} &= \mathbf{z}_k'\mathbf{B} = B_1 + B_2 y_{2k} + \cdots + B_q y_{qk} \\ &= \bar{y}_{1U} + \sum_{i=2}^{q} B_i(y_{ik} - \bar{y}_{iU}) \end{aligned} \tag{13.2.8}$$

and $y_{1k} - \hat{y}_{1k}$ is the regression residual for the kth element.

The traditional measure of the fit of the regression is the (*multiple*) *coefficient of determination*

$$R^2 = 1 - \frac{\text{residual sum of squares of } y_1}{\text{total sum of squares of } y_1}$$

or, more explicitly,

$$R^2 = R^2_{1(23\ldots q)} = 1 - \frac{\sum_U (y_{1k} - \hat{y}_{1k})^2}{\sum_U (y_{1k} - \bar{y}_{1U})^2}$$

$$= 1 - \frac{\sum_U y_{1k}^2 - \mathbf{B}' \sum_U \mathbf{z}_k y_{1k}}{\sum_U y_{1k}^2 - N(\bar{y}_{1U})^2} \tag{13.2.9}$$

where $\mathbf{B}$ is given by (13.2.5). One can show that equivalent expressions for R^2 are

$$R^2 = \frac{\mathbf{S}_1'(\mathbf{S}^{11})^{-1}\mathbf{S}_1}{S_{11}} \tag{13.2.10}$$

where $\mathbf{S}_1$ and $\mathbf{S}^{11}$ are as in (13.2.6), and

$$R^2 = 1 - \frac{|\mathbf{R}|}{R^{11}}$$

where $|\mathbf{R}|$ is the determinant of the correlation matrix $\mathbf{R}$ and R^{11} is the cofactor associated with the element in row 1 and column 1 of $\mathbf{R}$ given by (13.2.4). That is, R^{11} is the determinant of $\mathbf{R}^{11}$, the $(q - 1) \times (q - 1)$ matrix obtained by deleting row one and column one in $\mathbf{R}$. The finite population *multiple correlation coefficient* is defined as

$$R = \sqrt{R^2}$$

One can show that R is the finite population correlation coefficient, calculated in the manner of equation (13.2.3), between y_{1k} and the OLS prediction $\hat{y}_{1k}$ defined by (13.2.8).

The *finite population partial correlation coefficient* between y_1 and y_2, controlling for the other $q - 2$ variables, denoted $r_{12\cdot 3\ldots q}$, is the finite population correlation coefficient between $E_1 = y_1 - \hat{y}_{1\cdot 3\ldots q}$ and $E_2 = y_2 - \hat{y}_{2\cdot 3\ldots q}$, where $\hat{y}_{1\cdot 3\ldots q}$ is the OLS prediction of y_1 in terms of $y_3, \ldots, y_q$ and, analogously, $\hat{y}_{2\cdot 3\ldots q}$ is the OLS prediction of y_2 in terms of $y_3, \ldots, y_q$. In other words, $r_{12\cdot 3\ldots q}$ is calculated in the manner of equation (13.2.3) from the values (E_{1k}, E_{2k}), $k = 1, \ldots, N$, where

$$E_{1k} = y_{1k} - \hat{y}_{1\cdot 3\ldots q;k}; \qquad E_{2k} = y_{2k} - \hat{y}_{2\cdot 3\ldots q;k}$$

Alternatively, $r_{12\cdot 3..q}$ can be expressed in terms of cofactors. Let S^{ij} and R^{ij} be the cofactors associated with the ij elements of $\mathbf{S}$ and of $\mathbf{R}$, respectively. Then

$$r_{12\cdot 3\ldots q} = -\frac{S^{12}}{(S^{11}S^{22})^{1/2}} = -\frac{R^{12}}{(R^{11}R^{22})^{1/2}} \tag{13.2.11}$$

EXAMPLE 13.2.1. For $q = 3$ variables, it follows easily from the general formulas above that the coefficient of determination is

$$R^2 = R^2_{1(23)} = \frac{r_{12}^2 + r_{13}^2 - 2r_{12}r_{13}r_{23}}{1 - r_{23}^2}$$

the two regression slopes are

$$B_2 = \frac{S_{12} - (S_{13}S_{23}/S_{33})}{S_{22} - (S_{23}^2/S_{33})} = \frac{S_1}{S_2}\frac{r_{12} - r_{13}r_{23}}{1 - r_{23}^2}$$

and

$$B_3 = \frac{S_{13} - (S_{12}S_{23}/S_{22})}{S_{33} - (S_{23}^2/S_{22})} = \frac{S_1}{S_3}\frac{r_{13} - r_{12}r_{23}}{1 - r_{23}^2}$$

where $S_i = S_{ii}^{1/2}$ is the population standard deviation of y_i, for $i = 1, 2, 3$. The finite population partial correlation coefficient between y_1 and y_2, controlling for y_3 is

$$r_{12\cdot 3} = \frac{r_{12} - r_{13}r_{23}}{[(1 - r_{13}^2)(1 - r_{23}^2)]^{1/2}}$$

Let us estimate these new parameters. The vector $\mathbf{y}_k = (y_{1k}, \ldots, y_{ik}, \ldots, y_{qk})'$ is observed for the elements $k \in s$, where s is drawn by a given sampling design $p(\cdot)$ with $\pi_k > 0$ and $\pi_{kl} > 0$. For $i, j = 1, \ldots, q$, the parameter S_{ij} can be expressed as

$$S_{ij} = \frac{1}{N-1}t_{ij} - \frac{1}{N(N-1)}t_i t_j$$

with $t_{ij} = \sum_U y_{ik}y_{jk}$, $t_i = \sum_U y_{ik}$, $t_j = \sum_U y_{jk}$. Substitution of the appropriate π estimator for each total gives

$$\tilde{S}_{ij} = \frac{1}{\hat{N}-1}\hat{t}_{ij\pi} - \frac{1}{\hat{N}(\hat{N}-1)}\hat{t}_{i\pi}\hat{t}_{j\pi} \tag{13.2.12}$$

with $\hat{t}_{ij\pi} = \sum_s y_{ik}y_{jk}/\pi_k$, $\hat{t}_{i\pi} = \sum_s y_{ik}/\pi_k$, $\hat{t}_{j\pi} = \sum_s y_{jk}/\pi_k$ and $\hat{N} = \sum_s 1/\pi_k$. The statistical properties of $\tilde{S}_{ij}$ follow from the analysis in Section 5.9. Note that $\tilde{S}_{ij}$ is not unbiased, but consistent under simple assumptions. (Consistency was discussed in Section 5.3.) The estimated finite population covariance matrix is thus

$$\tilde{\mathbf{S}} = (\tilde{S}_{ij})$$

We use the $\tilde{S}_{ij}$ to obtain a consistent estimator of the correlation coefficient $r_{ij} = S_{ij}/(S_{ii}S_{jj})^{1/2}$, namely,

$$\tilde{r}_{ij} = \frac{\tilde{S}_{ij}}{(\tilde{S}_{ii}\tilde{S}_{jj})^{1/2}}$$

or, equivalently

$$\tilde{r}_{ij} = \frac{\sum_s (y_{ik} - \tilde{y}_{is})(y_{jk} - \tilde{y}_{js})/\pi_k}{\{[\sum_s (y_{ik} - \tilde{y}_{is})^2/\pi_k][\sum_s (y_{jk} - \tilde{y}_{js})^2/\pi_k]\}^{1/2}} \tag{13.2.13}$$

with $\tilde{y}_{is} = (\sum_s y_{ik}/\pi_k)/\hat{N}$. This is the π-weighted sample correlation coefficient.

It is not unbiased for R_{ij}, but consistent under general conditions. The estimated finite population correlation matrix is then

$$\tilde{\mathbf{R}} = (\tilde{r}_{ij})$$

We also obtain estimators for the multiple regression coefficients, the multiple coefficient of determination, and the partial correlation coefficient. These parameters were expressed above as functions of the variances and covariances S_{ij} or of the correlations r_{ij}. We replace S_{ij} by $\tilde{S}_{ij}$ (or, equivalently, r_{ij} by $\tilde{r}_{ij}$) and obtain consistent estimators that are sample-based copies of the corresponding parameters.

1. The estimator of the multiple regression coefficient vector shown in (13.2.5) is

$$\hat{\mathbf{B}} = (\hat{B}_1, \hat{B}_2, \ldots, \hat{B}_q)' = \left(\textstyle\sum_s \mathbf{z}_k \mathbf{z}_k'/\pi_k\right)^{-1} \textstyle\sum_s \mathbf{z}_k y_{1k}/\pi_k \qquad (13.2.14)$$

   or, equivalently,

$$(\hat{B}_2, \ldots, \hat{B}_q)' = (\tilde{\mathbf{S}}^{11})^{-1}\tilde{\mathbf{S}}_1$$

   and

$$\hat{B}_1 = \tilde{y}_{1s} - \sum_{i=2}^{q} \hat{B}_i \tilde{y}_{is}$$

   where $\tilde{\mathbf{S}}^{11}$ is obtained from $\tilde{\mathbf{S}}$ by eliminating row one and column one, and $\tilde{\mathbf{S}}_1 = (\tilde{S}_{12}, \ldots, \tilde{S}_{1q})'$.
2. The estimator of the coefficient of determination $R^2 = R^2_{1\cdot 23\ldots q}$ defined by (13.2.9) or (13.2.10) is

$$\hat{R}^2 = \frac{\tilde{\mathbf{S}}_1'(\tilde{\mathbf{S}}^{11})^{-1}\tilde{\mathbf{S}}_1}{\tilde{S}_{11}} \qquad (13.2.15)$$

   or, equivalently,

$$\hat{R}^2 = 1 - \frac{\sum_s y_{1k}^2/\pi_k - \hat{\mathbf{B}}' \sum_s \mathbf{z}_k y_{1k}/\pi_k}{\sum_s y_{1k}^2/\pi_k - \hat{N}(\tilde{y}_{1s})^2}$$

   The multiple correlation coefficient $R = R_{1(23\ldots q)}$ is estimated by $\hat{R} = \sqrt{\hat{R}^2}$.
3. The estimator of the partial correlation coefficient $r_{12\cdot 3\ldots q}$ given by (13.2.11) is

$$\hat{r}_{12\cdot 3\ldots q} = -\frac{\tilde{S}^{12}}{(\tilde{S}^{11}\tilde{S}^{22})^{1/2}} \qquad (13.2.16)$$

   where $\tilde{S}^{ij}$ $(i, j = 1, 2)$ is the cofactor associated with the ij element of $\tilde{\mathbf{S}}$.

Section 13.4 presents a method of variance estimation that is often useful for complex statistics such as (13.2.14) to (13.2.16).

Remark 13.2.1 Sample weighting of observations may be important for other statistical analyses than the ones mentioned in this section. If the aim is to carry out a factor analysis or a principal components analysis of the finite population, and the data consist of a probability sample from the finite population, appropriately sample-weighted correlation matrices should be used. Factors should be extracted from the sample-weighted correlation matrix $\tilde{\mathbf{R}}$, rather than from a matrix of unweighted sample correlation coefficients. Factor analysis is discussed in Fuller (1987), and principal components analysis in Skinner, Holmes and Smith (1986).

13.3. The Effect of Sampling Design on a Statistical Analysis

The sampling design has an effect on the conclusions drawn from survey data. The method of analysis must take into account the survey design. Many computer programs for standard statistical analyses implicitly assume independent and identically distributed observations. When the data are obtained from a sample survey, this assumption is approximately fulfilled under *SI* sampling with a small sampling fraction. But under most other sampling designs, the routine use of standard procedures can lead to erroneous variance estimates and therefore invalid conclusions.

Under many frequently used sampling designs, the variance of a given complex statistic is higher than it would have been under *SI* sampling. To proceed as if *SI* sampling had been used may lead to a misrepresentation of the actual variance. For example, a confidence interval calculated on the assumption of *SI* sampling may be on the average too short, so that the actual coverage probability falls short of the desired nominal confidence level of $1 - \alpha = 95\%$, say. This situation generally arises when clusters of elements are sampled and observed.

If the estimated variance based on the *SI* assumption overestimates the actual variance, the resulting confidence statements are still in error. However, the consequences are less harmful, in the sense that an interval calculated from the overstated variance will have a true confidence level at least equal to the $100(1 - \alpha)\%$ aimed at. The confidence interval is then conservative. This is often the case for designs that realize efficiency gains compared to the *SI* design, such as an efficient form of stratified sampling.

In the last few decades, many studies have been undertaken to examine the effect of sampling design on the common types of statistical analysis. Some early papers dealing with regression analysis are Konijn (1962), Kish and Frankel (1974), and Fuller (1975), whereas the effect of design on chi-square tests was studied by Nathan (1969, 1975), Holt, Scott and Ewings (1980), and Fellegi (1980).

The study by Kish and Frankel was one of the first to show the effect of sampling design on the results of statistical analyses. If $\hat{\theta}_s$ denotes an unbiased estimator of the population parameter θ, the design effect for $\hat{\theta}_s$ under the design $p(s)$ is

$$\text{deff}_p(\hat{\theta}_s) = \frac{V_p(\hat{\theta}_s)}{V_{SI}(\hat{\theta}_{s,SI})}$$

where $\hat{\theta}_{s,SI}$ is the expression for $\hat{\theta}_s$ under the SI design. To make the comparison fair, the same (expected) sample size is assumed under the design $p(\cdot)$ as under SI. (If the estimator $\hat{\theta}_s$ is not unbiased but consistent, we define the deff as the ratio of the approximate variances.) For a statistical analysis performed on survey data, it is important to know the effect of the design. For example, if $\text{deff}_p(\hat{\theta}_s)$ exceeds unity, $\hat{\theta}_s$ has a greater variance under $p(s)$ than under SI.

Remark 13.3.1. An alternative view of a deff value greater than unity is as follows. If the statistic $\hat{\theta}_s$ is consistent for θ, the absolute difference $|\hat{\theta}_s - \theta|$ will be less than some fixed small positive number with a probability near one, if the sample size is sufficiently large. Now if $\text{deff}(\hat{\theta}_s)$ exceeds unity, the convergence in probability to θ is slowed down, compared to SI sampling. That is, the sample may have to be considerably larger (than under SI sampling) for $|\hat{\theta}_s - \theta|$ to be small with a high probability.

To get an idea of the effect of the design, we need to evaluate the deff. This may be relatively easy for a simple statistic and a simple design. For example, the design effect under SIC sampling of the π weighted sample mean $\tilde{y}_s = (\sum_s y_k/\pi_k)/(\sum_s 1/\pi_k)$ is

$$\text{deff}_{SIC}(\tilde{y}_s) = V_{SIC}(\tilde{y}_s)/V_{SI}(\bar{y}_s) \doteq 1 + (\overline{N} - 1)\delta$$

where δ is the homogeneity coefficient (8.7.4), assuming that the clusters are of equal size, $N_i = \overline{N}$ for all i. If $\delta > 0$, there is loss of precision due to the selection of clusters.

For more complex statistics, such as the estimators (13.2.14) to (13.2.16), the evaluation of the deff is not so simple. Consider the statistic $\hat{R}^2$ given by equation (13.12.15). Under a given design $p(\cdot)$, how does $\text{deff}_p(\hat{R}^2)$ compare with $\text{deff}_p(\tilde{y}_s)$? Is $\hat{R}^2$ more or less sensitive to clustering of observations than a simple statistic such as $\tilde{y}_s$? A compact analytic answer is hard to obtain.

Partly for this reason, Kish and Frankel (1970, 1974) undertook a large-scale Monte Carlo study of the design effect and other properties of complex analytic statistics, including several of those reviewed above for regression and correlation analysis. Repeated samples were drawn with a stratified cluster sampling design. All strata contained the same number of clusters, say A. Two clusters were selected by SIC sampling from each stratum. There was no subsampling, so the inclusion probability was constant, $\pi_k = 2/A$ for all elements k. However, the π_{kl} differed depending on whether k and l belong to the same cluster or not, a factor of consequence for the variance estimation.

For each sample, the parameter estimates were calculated, and summary performance measures were calculated. For each parameter of interest, the simulated sampling distributions were studied with respect to: (a) bias; (b) approach to normality; (c) values of $(\text{deff})^{1/2}$; (d) accuracy of the estimation of the MSE; (e) coverage rates attained by the confidence intervals.

The design effect of a complex statistic $\hat{\theta}_s$ in the study was compared to the design effect of the weighted sample mean statistic $\tilde{y}_s = (\sum_s y_k/\pi_k)/\hat{N}$, where y is one of the variables used to calculate $\hat{\theta}_s$. Kish and Frankel (1974) list the following main conclusions:

a. $\text{deff}_p(\hat{\theta}_s)$ was greater than unity, indicating that a variance is greater under the cluster sampling design than under *SI* sampling.
b. $\text{deff}_p(\hat{\theta}_s)$ was usually less than $\text{deff}_p(\tilde{y}_s)$, where y is a variable used in the calculation of $\hat{\theta}_s$. That is, the design effect for $\hat{\theta}_s$ tended to be less pronounced than for the weighted sample mean statistic.
c. $\text{deff}_p(\hat{\theta}_s)$ was related to $\text{deff}_p(\tilde{y}_s)$ in such a way that for a variable y for which $\text{deff}_p(\tilde{y}_s)$ was high, $\text{deff}_p(\hat{\theta}_s)$ also tended to be high.

Although these conclusions are limited to one type of design (stratified cluster sampling), they still convey the important message that selection of clusters can lead to a severe variance increase for a variety of survey statistics, and that erroneous inferences may result if one were to proceed as if *SI* sampling had been used.

Remark 13.3.2. For complex designs and complex statistics, simplified approaches to variance estimation are of interest. Let $\hat{\theta}_s$ be a complex estimator of the parameter θ, and suppose y is a variable used in the calculation of $\hat{\theta}_s$. An estimate is wanted for $V_p(\hat{\theta}_s)$, the variance of $\hat{\theta}_s$ under the complex design $p(\cdot)$. Kish and Frankel suggest that safe overestimates of the variance $V_p(\hat{\theta}_s)$ may be calculated as

$$\hat{V}_p(\hat{\theta}_s) = \text{deff}_p(\tilde{y}_s)\hat{V}_{SI}(\hat{\theta}_{s,SI})$$

assuming that good approximate values are available for $\text{deff}_p(\tilde{y}_s)$ and $\hat{V}_{SI}(\hat{\theta}_{s,SI})$, and that $\text{deff}_p(\hat{\theta}_s) < \text{deff}_p(\tilde{y}_s)$; point (b) above suggests that this inequality sometimes holds. The method may be useful in some cases for obtaining conservative variance estimates, but it requires good prior knowledge.

Kish and Frankel (1974) go on to suggest that one might model the unknown $\text{deff}_p(\hat{\theta}_s)$ as

$$\text{deff}_p(\hat{\theta}_s) = 1 + f_{\hat{\theta}_s}[\text{deff}_p(\tilde{y}_s) - 1] \tag{13.3.1}$$

where $\text{deff}_p(\tilde{y}_s) > 1$, and the factor $f_{\hat{\theta}_s}$ is unique to the statistic $\hat{\theta}_s$. The advantage with this modeling would be that if $f_{\hat{\theta}_s}$ is taken to be a known value, for example, from experience with the same variables and similar populations, and if $\text{deff}_p(\tilde{y}_s)$ and $\hat{V}_{SI}(\hat{\theta}_{s,SI})$ are approximately known, one can obtain an estimate of the variance of $\hat{\theta}_s$ as

$$\hat{V}_p(\hat{\theta}_s) = \text{deff}_p(\hat{\theta}_s)\hat{V}_{SI}(\hat{\theta}_{s,SI})$$

where $\text{deff}_p(\hat{\theta}_s)$ is given by (13.3.1). The method requires good approximations for ordinarily unknown quantities. For a use of the approach in connection with a survey, see Kish, Groves and Krotki (1976).

13.4. Variances and Estimated Variances for Complex Analyses

The Taylor linearization method for variance estimation has been used extensively throughout the book, and other methods were proposed in Chapter 11. Binder (1983) proposed an *implicit differentiation method* that we now examine. It is a method for producing a variance estimator for a vector of complex survey statistics. The method rests on a utilization of the Taylor series expansion for a vector-valued parameter, and is particularly useful when the parameters are implicitly defined. The following simple example shows the nature of the technique.

EXAMPLE 13.4.1. For the elements $k \in s$, y_k and z_k are observed. We want to estimate the finite population covariance

$$S_{yz} = \frac{1}{N-1} \sum_U (y_k - \bar{y}_U)(z_k - \bar{z}_U)$$

where $\bar{y}_U = \sum_U y_k/N$ and $\bar{z}_U = \sum_U z_k/N$. Equivalently,

$$S_{yz} = \frac{t_{yz}}{t_1 - 1} - \frac{t_y t_z}{t_1(t_1 - 1)}$$

with

$$t_1 = \sum_U 1 = N; \qquad t_y = \sum_U y_k; \qquad t_z = \sum_U z_k; \qquad t_{yz} = \sum_U y_k z_k$$

This expresses S_{yz} as a function of four population totals that must first be estimated. The population size N is also viewed as an unknown. Now, S_{yz} is defined implicitly by the following system of equations

$$\begin{aligned} t_1 - N &= 0 \\ t_y - N\bar{y}_U &= 0 \\ t_z - N\bar{z}_U &= 0 \\ t_{yz} - N\bar{y}_U\bar{z}_U - (N-1)S_{yz} &= 0 \end{aligned} \qquad (13.4.1)$$

It is true in this case that S_{yz} has an easily derived explicit expression, but the implicit approach that we now follow has some advantages. We define an estimator of S_{yz} through the system shown in (13.4.1), but with the π estimators

$$\hat{t}_{1\pi} = \sum_s 1/\pi_k = \hat{N}; \qquad \hat{t}_{y\pi} = \sum_s y_k/\pi_k$$
$$\hat{t}_{z\pi} = \sum_s z_k/\pi_k; \qquad \hat{t}_{yz\pi} = \sum_s y_k z_k/\pi_k \tag{13.4.2}$$

replacing the population totals t_1, t_y, t_z, and t_{yz}, respectively. The resulting estimator is

$$\tilde{S}_{yz} = \frac{1}{\hat{N} - 1}\hat{t}_{yz\pi} - \frac{1}{\hat{N}(\hat{N} - 1)}\hat{t}_{y\pi}\hat{t}_{z\pi} \tag{13.4.3}$$

which agrees with the earlier estimator (5.9.7). We seek an expression for the approximate variance of $\tilde{S}_{yz}$ and an estimator of that variance. Obviously, the variance depends on the covariance matrix of $(\hat{t}_{1\pi}, \hat{t}_{y\pi}, \hat{t}_{z\pi}, \hat{t}_{yz\pi})$, the vector of π estimators used to define $\tilde{S}_{yz}$. The implicit differentiation method offers a convenient way to express this dependence.

Alternatively, supposing we want simultaneously to find estimators of N, $\bar{y}_U$, $\bar{z}_U$, and S_{yz}, we can treat $\boldsymbol{\theta} = (N, \bar{y}_U, \bar{z}_U, S_{yz})'$ as a parameter vector. This vector is uniquely defined by the system (13.4.1) in terms of the population totals N, t_y, t_z, and t_{yz}. Substitution of π estimators in (13.4.1) implies a uniquely defined estimator $\hat{\boldsymbol{\theta}}$ of $\boldsymbol{\theta}$. We want an expression for the covariance matrix of $\hat{\boldsymbol{\theta}}$, as well as an estimator of that matrix. The implicit differentiation method is useful for such a problem. We return to the present example after some general theory.

Suppose that we observe $\mathbf{y}_k = (y_{1k}, \ldots, y_{qk})'$ for the elements $k \in s$. Let $\boldsymbol{\theta} = (\theta_1, \ldots, \theta_p)'$ be an unknown p-dimensional finite population parameter vector defined as the unique solution of the system of equations

$$W_i(\boldsymbol{\theta}) = \sum_U h_i(\mathbf{y}_k, \boldsymbol{\theta}) - v_i(\boldsymbol{\theta}) = 0 \tag{13.4.4}$$

for $i = 1, \ldots, p$. Here, the unknown population total $\sum_U h_i(\mathbf{y}_k, \boldsymbol{\theta})$ may depend on $\mathbf{y}_k$ and on $\boldsymbol{\theta}$, whereas $v_i(\boldsymbol{\theta})$ does not depend on the $\mathbf{y}_k$. Let us summarize the p equations in matrix form as

$$\mathbf{W}(\boldsymbol{\theta}) = \sum_U \mathbf{h}(\mathbf{y}_k, \boldsymbol{\theta}) - \mathbf{v}(\boldsymbol{\theta}) = \mathbf{0} \tag{13.4.5}$$

With the ultimate goal of estimating $\boldsymbol{\theta}$, start by replacing the unknown population totals in (13.4.5) by suitable estimators. Here we use π estimators, although ratio estimators, for example, could be considered. Substituting $\sum_s \check{h}_i(\mathbf{y}_k, \boldsymbol{\theta}) = \sum_s h_i(\mathbf{y}_k, \boldsymbol{\theta})/\pi_k$ into (13.4.4), we get, for $i = 1, \ldots, p$, a statistic that "estimates" $W_i(\boldsymbol{\theta})$, for any fixed $\boldsymbol{\theta}$, namely,

$$\hat{W}_i(\boldsymbol{\theta}) = \sum_s \check{h}_i(\mathbf{y}_k, \boldsymbol{\theta}) - v_i(\boldsymbol{\theta}) \tag{13.4.6}$$

which we assume to be consistent and approximately normally distributed in large samples, for any fixed $\boldsymbol{\theta}$. Here, $\hat{W}_i(\boldsymbol{\theta})$ cannot be calculated, since it depends on the unknown $\boldsymbol{\theta}$. The matrix summary form of the p estimators shown in (13.4.6) is

$$\hat{\mathbf{W}}(\boldsymbol{\theta}) = \sum_s \check{\mathbf{h}}(\mathbf{y}_k, \boldsymbol{\theta}) - \mathbf{v}(\boldsymbol{\theta}) \tag{13.4.7}$$

We define our estimator $\hat{\boldsymbol{\theta}} = (\hat{\theta}_1, \ldots, \hat{\theta}_i, \ldots, \hat{\theta}_p)'$ of $\boldsymbol{\theta}$ as the solution of the equation $\hat{\mathbf{W}}(\boldsymbol{\theta}) = \mathbf{0}$, assuming that a unique solution exists. That is, $\hat{\boldsymbol{\theta}}$ satisfies

$$\hat{\mathbf{W}}(\hat{\boldsymbol{\theta}}) = \mathbf{0} \tag{13.4.8}$$

The next objective is to seek an approximation to the variance of $\hat{\boldsymbol{\theta}}$ and from there to construct a variance estimator. Let us use the Taylor series expansion of $\hat{\mathbf{W}}(\hat{\boldsymbol{\theta}})$ around the point $\hat{\boldsymbol{\theta}} = \boldsymbol{\theta}$, where $\boldsymbol{\theta}$ is the true unknown parameter value defined by the equation (13.4.5). Define

$$\hat{\mathbf{J}}(\boldsymbol{\theta}) = \partial\hat{\mathbf{W}}(\boldsymbol{\theta})/\partial\boldsymbol{\theta}$$

as the $p \times p$ matrix whose ij element is the partial derivative

$$\partial\hat{W}_i(\boldsymbol{\theta})/\partial\theta_j$$

Using (13.4.8), we have

$$\mathbf{0} = \hat{\mathbf{W}}(\hat{\boldsymbol{\theta}}) \doteq \hat{\mathbf{W}}(\boldsymbol{\theta}) + \hat{\mathbf{J}}(\boldsymbol{\theta})(\hat{\boldsymbol{\theta}} - \boldsymbol{\theta})$$

Solving for $\hat{\boldsymbol{\theta}} - \boldsymbol{\theta}$ we obtain

$$\hat{\boldsymbol{\theta}} - \boldsymbol{\theta} \doteq -\hat{\mathbf{J}}^{-1}(\boldsymbol{\theta})\hat{\mathbf{W}}(\boldsymbol{\theta})$$

supposing that the inverse, $\hat{\mathbf{J}}^{-1}(\boldsymbol{\theta})$, of $\hat{\mathbf{J}}(\boldsymbol{\theta})$ exists. This system of equations, which approximates each of the implied estimators $\hat{\theta}_1, \ldots, \hat{\theta}_p$ by a linear combination of the p variables $\hat{W}_1(\boldsymbol{\theta}), \ldots, \hat{W}_p(\boldsymbol{\theta})$, leads to an approximation of $\mathbf{V}(\hat{\boldsymbol{\theta}})$, the covariance matrix of $\hat{\boldsymbol{\theta}}$. We obtain

$$\mathbf{V}(\hat{\boldsymbol{\theta}}) \doteq [\hat{\mathbf{J}}^{-1}(\boldsymbol{\theta})]\boldsymbol{\Sigma}(\boldsymbol{\theta})[\hat{\mathbf{J}}^{-1}(\boldsymbol{\theta})]' \tag{13.4.9}$$

where $\boldsymbol{\Sigma}(\boldsymbol{\theta})$ denotes the $p \times p$ symmetric matrix whose ij element σ_{ij} is the covariance between $\hat{W}_i(\boldsymbol{\theta})$ and $\hat{W}_j(\boldsymbol{\theta})$. Equivalently, σ_{ij} is the covariance between $\sum_s \check{h}_i(\mathbf{y}_k, \boldsymbol{\theta})$ and $\sum_s \check{h}_j(\mathbf{y}_k, \boldsymbol{\theta})$. That is, with $\Delta_{kl} = \pi_{kl} - \pi_k\pi_l$ as usual, we have

$$\sigma_{ij} = \sum\sum_U \Delta_{kl}\check{h}_i(\mathbf{y}_k, \boldsymbol{\theta})\check{h}_j(\mathbf{y}_l, \boldsymbol{\theta})$$

A final step is to transform the approximate covariance matrix shown in (13.4.9) into an estimated covariance matrix. This entails replacing $\boldsymbol{\Sigma}(\boldsymbol{\theta})$ by an estimator, $\hat{\boldsymbol{\Sigma}}(\boldsymbol{\theta})$, and, because $\boldsymbol{\theta}$ is unknown, $\boldsymbol{\theta}$ is replaced by its estimator $\hat{\boldsymbol{\theta}}$, both in $\hat{\boldsymbol{\Sigma}}(\boldsymbol{\theta})$ and in $\hat{\mathbf{J}}^{-1}(\boldsymbol{\theta})$. Consequently, $\hat{\boldsymbol{\Sigma}}(\boldsymbol{\theta})$ is replaced by $\hat{\boldsymbol{\Sigma}}(\hat{\boldsymbol{\theta}})$, a $p \times p$ symmetrical matrix whose ijth element is given by

$$\hat{\sigma}_{ij} = \sum\sum_s \check{\Delta}_{kl}\check{h}_i(\mathbf{y}_k, \hat{\boldsymbol{\theta}})\check{h}_j(\mathbf{y}_l, \hat{\boldsymbol{\theta}})$$

with $\check{\Delta}_{kl} = \Delta_{kl}/\pi_{kl}$. If we assume that consistency is preserved in these operations, our end result is the following consistent estimator of $\mathbf{V}(\hat{\boldsymbol{\theta}})$

$$\hat{\mathbf{V}}(\hat{\boldsymbol{\theta}}) = [\hat{\mathbf{J}}^{-1}(\hat{\boldsymbol{\theta}})]\hat{\boldsymbol{\Sigma}}(\hat{\boldsymbol{\theta}})[\hat{\mathbf{J}}^{-1}(\hat{\boldsymbol{\theta}})]' \tag{13.4.10}$$

Binder (1983) states the regularity conditions needed to ensure asymptotic normality of the $\hat{W}_i(\boldsymbol{\theta})$ and consistency of $\hat{\mathbf{V}}(\hat{\boldsymbol{\theta}})$.

In a case where the number of parameters to estimate is large, it is time-saving to standardize the calculations. The expression (13.4.10) has the advantage that all the desired variance estimates can be obtained from $\hat{\boldsymbol{\Sigma}}(\hat{\boldsymbol{\theta}})$, which is the estimated covariance matrix of the π estimators $\sum_s \check{h}_i(\mathbf{y}_k, \boldsymbol{\theta})$ that serve in the process. If interest is focused on a particular component of $\hat{\boldsymbol{\theta}}$, say $\hat{\theta}_i$, the appropriate variance estimator is obtained by isolating the diagonal element, $\hat{V}_{ii}$, of $\hat{\mathbf{V}}(\hat{\boldsymbol{\theta}})$. Let us look at some examples of this technique.

EXAMPLE 13.4.2. Continuing Example 13.4.1, let us use the implicit differentiation method to obtain an estimated variance for $\tilde{S}_{yz}$. The system of equations (13.4.1), which defines $\boldsymbol{\theta} = (N, \bar{y}_U, \bar{z}_U, S_{yz})'$, is of the general form shown in equation (13.4.5) with

$$h_1(\mathbf{y}_k, \boldsymbol{\theta}) \equiv 1, \qquad h_2(\mathbf{y}_k, \boldsymbol{\theta}) = y_k, \qquad h_3(\mathbf{y}_k, \boldsymbol{\theta}) = z_k, \qquad h_4(\mathbf{y}_k, \boldsymbol{\theta}) = y_k z_k$$

$$v_1(\boldsymbol{\theta}) = N, \quad v_2(\boldsymbol{\theta}) = N\bar{y}_U, \quad v_3(\boldsymbol{\theta}) = N\bar{z}_U, \quad v_4(\boldsymbol{\theta}) = N\bar{y}_U\bar{z}_U + (N-1)S_{yz}$$

The equations $\hat{W}_i(\boldsymbol{\theta}) = 0$, $i = 1, 2, 3, 4$, are then $\hat{t}_{1\pi} - v_1(\boldsymbol{\theta}) = 0$; $\hat{t}_{y\pi} - v_2(\boldsymbol{\theta}) = 0$; $\hat{t}_{z\pi} - v_3(\boldsymbol{\theta}) = 0$; and $\hat{t}_{yz\pi} - v_4(\boldsymbol{\theta}) = 0$, where the π estimators are given by (13.4.2). The implied estimator of $\boldsymbol{\theta}$ is $\hat{\boldsymbol{\theta}} = (\hat{N}, \tilde{y}_s, \tilde{z}_s, \tilde{S}_{yz})$, where $\tilde{S}_{yz}$ is given by (13.4.3) and

$$\hat{N} = \textstyle\sum_s 1/\pi_k, \qquad \tilde{y}_s = (\sum_s y_k/\pi_k)/\hat{N}, \qquad \tilde{z}_s = (\sum_s z_k/\pi_k)/\hat{N}$$

We find

$$\hat{\mathbf{J}}(\boldsymbol{\theta}) = \begin{bmatrix} -1 & 0 & 0 & 0 \\ -\bar{y}_U & -N & 0 & 0 \\ -\bar{z}_U & 0 & -N & 0 \\ -\bar{y}_U\bar{z}_U - S_{yz} & -N\bar{z}_U & -N\bar{y}_U & -(N-1) \end{bmatrix}$$

We now invert $\hat{\mathbf{J}}(\boldsymbol{\theta})$ and obtain the variance estimator from (13.4.10). Looking in particular at $\tilde{S}_{yz}$, we obtain the fourth row of the inverse $\hat{\mathbf{J}}^{-1}(\hat{\boldsymbol{\theta}})$ as

$$\mathbf{a} = \frac{1}{\hat{N} - 1}(\tilde{S}_{yz} - \tilde{y}_s\tilde{z}_s, \tilde{z}_s, \tilde{y}_s, -1)$$

With $\mathbf{a}$ defined in this way, the estimated variance of $\tilde{S}_{yz}$ is therefore

$$\hat{V}(\tilde{S}_{yz}) = \mathbf{a}\hat{\boldsymbol{\Sigma}}\mathbf{a}'$$

where $\hat{\boldsymbol{\Sigma}} = \hat{\boldsymbol{\Sigma}}(\hat{\boldsymbol{\theta}})$ is the 4×4 matrix whose ijth element is $\sum\sum_s \check{\Delta}_{kl}(h_{ik}/\pi_k)(h_{jl}/\pi_l)$, with $h_{1k} \equiv 1$, $h_{2k} = y_k$, $h_{3k} = z_k$, and $h_{4k} = y_k z_k$. This result confirms the variance estimator obtained by the Tepping-Woodruff technique in Section 5.9.

EXAMPLE 13.4.3. Let use use the implicit differentiation method to estimate the coefficient of multiple determination R^2, defined by (13.2.9) or equivalently by (13.2.10), as well as a variance estimator. For the kth element, let y_k

be the value of the criterion variable of the regression, and let $\mathbf{z}_k = (z_{1k}, \ldots, z_{ik}, \ldots, z_{qk})'$ be the value of the explanatory vector, where $z_{1k} \equiv 1$. Consider $\boldsymbol{\theta} = (\bar{y}_U, \mathbf{B}', R^2)'$, a $(q + 2)$-vector of finite population parameters defined by the system

$$t_y - N\bar{y}_U = 0$$

$$\mathbf{t} - \mathbf{T}\mathbf{B} = \mathbf{0}$$

$$(t_{yy} - N\bar{y}_U^2)(R^2 - 1) + t_{yy} - \mathbf{t}'\mathbf{B} = 0$$

where the population totals are $t_y = \sum_U y_k$, $\mathbf{t} = \sum_U \mathbf{z}_k y_k$, $\mathbf{T} = \sum_U \mathbf{z}_k \mathbf{z}_k'$ and $t_{yy} = \sum_U y_k^2$, and $\mathbf{0}$ is a q-vector of zeroes. The second line of the system is shorthand for q equations. Here, $\bar{y}_U = t_y/N$ is the finite population y-mean, $\mathbf{B} = \mathbf{T}^{-1}\mathbf{t}$ is the vector of OLS regression coefficients, and R^2 is the coefficient of determination.

$$R^2 = 1 - \frac{\sum_U (y_k - \mathbf{z}_k'\mathbf{B})^2}{\sum_U (y_k - \bar{y}_U)^2} = 1 - \frac{t_{yy} - \mathbf{t}'\mathbf{B}}{t_{yy} - N(\bar{y}_U)^2}$$

We assume in this example that the population size N is known. The four unknown totals have the following π estimators

$$\hat{t}_{y\pi} = \sum_s y_k/\pi_k, \quad \hat{\mathbf{t}}_\pi = \sum_s \mathbf{z}_k y_k/\pi_k, \quad \hat{\mathbf{T}}_\pi = \sum_s \mathbf{z}_k \mathbf{z}_k'/\pi_k, \quad \hat{t}_{yy\pi} = \sum_s y_k^2/\pi_k$$

Let $v_1(\boldsymbol{\theta}) = N\bar{y}_U$ and $v_3(\boldsymbol{\theta}) = N\bar{y}_U^2(R^2 - 1)$. The equations $\hat{W}_i(\boldsymbol{\theta}) = 0$ are

$$\hat{W}_1(\boldsymbol{\theta}) = \hat{t}_{y\pi} - v_1(\boldsymbol{\theta}) = 0$$

$$\hat{\mathbf{W}}_2(\boldsymbol{\theta}) = \hat{\mathbf{t}}_\pi - \hat{\mathbf{T}}_\pi \mathbf{B} = \mathbf{0}$$

$$\hat{W}_3(\boldsymbol{\theta}) = \hat{t}_{yy\pi} R^2 - \hat{\mathbf{t}}_\pi' \mathbf{B} - v_3(\boldsymbol{\theta}) = 0$$

Here, $\hat{\mathbf{W}}_2(\boldsymbol{\theta}) = \mathbf{0}$ consists of q equations. The estimators implied by the system are

$$\hat{\bar{y}}_{U\pi} = \hat{t}_{y\pi}/N$$

$$\hat{\mathbf{B}} = \hat{\mathbf{T}}_\pi^{-1}\hat{\mathbf{t}}_\pi$$

$$\hat{R}^2 = 1 - \frac{\hat{t}_{yy\pi} - \hat{\mathbf{t}}_\pi'\hat{\mathbf{B}}}{\hat{t}_{yy\pi} - N(\hat{\bar{y}}_{U\pi})^2}$$

Letting $SY = \hat{t}_{yy\pi} - N\bar{y}_U^2$, we obtain

$$\hat{\mathbf{J}}(\boldsymbol{\theta}) = \begin{bmatrix} -N & \mathbf{0}' & 0 \\ \mathbf{0} & -\hat{\mathbf{T}}_\pi & \mathbf{0} \\ -2N\bar{y}_U(R^2 - 1) & -\hat{\mathbf{t}}_\pi' & SY \end{bmatrix}$$

Suppose we are particularly interested in obtaining the variance estimator of $\hat{R}^2$. For this we need the final row of $\hat{\mathbf{J}}^{-1}(\boldsymbol{\theta})$, evaluated at $\boldsymbol{\theta} = \hat{\boldsymbol{\theta}}$, which after some algebra is obtained as

$$\mathbf{a} = \frac{1}{\hat{t}_{yy\pi} - N(\hat{\bar{y}}_{U\pi})^2}[-2\hat{\bar{y}}_{U\pi}(\hat{R}^2 - 1), -\hat{\mathbf{B}}', 1]$$

Let $\hat{\boldsymbol{\Sigma}} = \hat{\boldsymbol{\Sigma}}(\hat{\boldsymbol{\theta}})$ be the estimated $(q + 2) \times (q + 2)$ covariance matrix of $\hat{\boldsymbol{\theta}}$. The ij element of $\hat{\boldsymbol{\Sigma}}$ is then $\hat{\sigma}_{ij} = \sum\sum_s \check{\Delta}_{kl}(h_{ik}/\pi_k)(h_{jl}/\pi_l)$ with $h_{1k} = y_k$, $h_{i+1,k} = z_{ik}(y_k - \mathbf{z}_k'\hat{\mathbf{B}})$; $i = 1, \ldots, q$, and $h_{q+2,k} = (\hat{R}^2 y_k - \mathbf{z}_k'\hat{\mathbf{B}})y_k$. With these definitions of $\mathbf{a}$ and $\hat{\boldsymbol{\Sigma}}$, the desired variance estimator of $\hat{R}^2$ is finally obtained as

$$\hat{V}(\hat{R}^2) = \mathbf{a}\hat{\boldsymbol{\Sigma}}\mathbf{a}'$$

From the example we can also obtain a variance estimator for the estimated regression coefficient vector, $\hat{\mathbf{B}} = \hat{\mathbf{T}}_\pi^{-1}\hat{\mathbf{t}}_\pi$; it confirms Result 5.10.1, namely,

$$\hat{\mathbf{V}}(\hat{\mathbf{B}}) = \hat{\mathbf{T}}_\pi^{-1}\hat{\boldsymbol{\Sigma}}_1\hat{\mathbf{T}}_\pi^{-1}$$

where $\hat{\boldsymbol{\Sigma}}_1$ is the $q \times q$ matrix whose ij-element is $\sum\sum_s \check{\Delta}_{kl}(u_{ik}/\pi_k)(u_{jl}/\pi_l)$ with $u_{ik} = z_{ik}(y_k - \mathbf{z}_k'\hat{\mathbf{B}})$; $i = 1, \ldots, q$.

In Examples 13.4.2 and 13.4.3, we could have used the Tepping-Woodruff method of Chapter 5 to obtain variance estimators, so implicit differentiation is an alternative method in these cases. However, implicit differentiation is particularly useful when the parameter estimates cannot be expressed explicitly as a function of the sample statistics, as in the following example.

Example 13.4.4. Consider the logistic regression model in which the criterion variable y is dichotomous; $y_k = 0$ or $y_k = 1$. Given $\mathbf{x}_k$, the model states that

$$\begin{aligned}\Pr(y_k = 1) &= e^{\mathbf{x}_k'\boldsymbol{\beta}}/\{1 + e^{\mathbf{x}_k'\boldsymbol{\beta}}\} \\ &= \mu(\mathbf{x}_k'\boldsymbol{\beta}) \qquad (13.4.11)\end{aligned}$$

and $\Pr(y_k = 0) = 1 - \Pr(y_k = 1) = 1 - \mu(\mathbf{x}_k'\boldsymbol{\beta})$. Here, $\boldsymbol{\beta}$ is a parameter of the model. If $(y_k, \mathbf{x}_k)$ were observed for all $k \in U$, as in a census, we could use maximum likelihood to estimate $\boldsymbol{\beta}$. The likelihood function is then

$$L(\boldsymbol{\beta}) = \prod_{k \in U_1} \mu(\mathbf{x}_k'\boldsymbol{\beta}) \prod_{k \in U_0} [1 - \mu(\mathbf{x}_k'\boldsymbol{\beta})]$$

where $U_1 = \{k : k \in U \text{ and } y_k = 1\}$ and $U_0 = U - U_1$. The likelihood equation $\partial \log L(\boldsymbol{\beta})/\partial\boldsymbol{\beta} = 0$ leads to

$$\textstyle\sum_U [y_k - \mu(\mathbf{x}_k'\boldsymbol{\beta})]\mathbf{x}_k = \mathbf{0}$$

The finite population parameter $\mathbf{B}$ is now defined as the solution for $\boldsymbol{\beta}$ of this system of equations. That is, $\mathbf{B}$ satisfies

$$\textstyle\sum_U [y_k - \mu(\mathbf{x}_k'\mathbf{B})]\mathbf{x}_k = \mathbf{0} \qquad (13.4.12)$$

Here, $\mathbf{B}$ cannot be stated explicitly.

Now, suppose only data from a sample are available. For any fixed **B**, the estimating equation (13.4.12) is in its turn estimated by

$$\sum_s [y_k - \mu(\mathbf{x}_k'\mathbf{B})]\mathbf{x}_k/\pi_k = \mathbf{0}$$

We define the estimator of **B** as the value, $\hat{\mathbf{B}}$, that satisfies this last equation. (Note that $\hat{\mathbf{B}}$ is then an estimator of the finite population parameter **B**, not of the model parameter $\boldsymbol{\beta}$.) The implicit differentiation method can now be used to obtain an estimated covariance matrix for $\hat{\mathbf{B}}$. More generally, this offers a way to estimate finite population parameters defined implicitly through a generalized linear model (GLIM) as presented in Nelder and Wedderburn (1972).

Remark 13.4.1. In the previous example, (13.4.11) is a *superpopulation model* that leads to the descriptive parameter **B**, which is implicitly defined through (13.4.12). Superpopulation models, model parameters, and the estimation of such parameters from survey data are discussed in Sections 13.6 and 14.5.

13.5. Analysis of Categorical Data for Finite Populations

13.5.1. Test of Homogeneity for Two Populations

A proper analysis of survey data requires that the sampling design be taken into account, when conclusions are wanted about the finite population. There is a considerable recent literature on the analysis of survey data. Valid methods for testing independence in contingency tables and other analyses of survey data were given by Nathan (1975) and Koch, Freeman and Freeman (1975). Moreover, Cohen (1976), Altham (1976), Fellegi (1980), Rao and Scott (1981), and others show that sampling of clusters may have a considerable effect on the standard Pearson chi-square statistic, and that this statistic may require adjustment to ensure valid conclusions. We examine some of these issues.

The test for homogeneity for two or more populations is a classic statistical procedure. Here we consider the comparison of two finite populations, U_1 and U_2, whose elements can be classified into c possible categories. On the basis of a probability sample from each population, we wish to test the null hypothesis that the two unknown vectors of population proportions are identical.

For example, we wish to compare the health status of the individuals in two different regions, say, two countries, or two provinces of the same country. The categories correspond to c different states of health. Independent health surveys are carried out, one in each region, with possibly different sampling designs. Health status is observed for sampled individuals. The null hypothesis states that health status, as measured by the vector of category proportions, is the same in the two populations.

More formally, let $U_1 = \{1, \ldots, k, \ldots, N_1\}$ and $U_2 = \{1, \ldots, k \ldots, N_2\}$ be the two finite populations and N_1 and N_2 the respective sizes. For $i = 1, 2$, consider a partition of U_i into c nonoverlapping and exhaustive categories (population cells) $U_{ij}; j = 1, \ldots, c$. The size of U_{ij} is denoted by N_{ij}. We have

$$U_i = \bigcup_{j=1}^{c} U_{ij}; \qquad N_i = \sum_{j=1}^{c} N_{ij}$$

The unknown proportion in category j of the ith population is

$$P_{ij} = \frac{N_{ij}}{N_i}; \qquad j = 1, \ldots, c; \qquad i = 1, 2$$

For $i = 1, 2$, we have a vector of proportions

$$\mathbf{P}_i = (P_{i1}, \ldots, P_{ij}, \ldots, P_{i,c-1})'; \qquad i = 1, 2$$

The null hypothesis is

$$H_0 : \mathbf{P}_1 = \mathbf{P}_2$$

We let

$$\mathbf{P} = (P_1, \ldots, P_j, \ldots, P_{c-1})'$$

denote the (unknown) common vector value under H_0. The alternative hypothesis, H_1, states that H_0 is not true.

The two populations are sampled independently, each according to its own sampling design. For $i = 1, 2$, the sample s_i, of size n_i, is drawn from U_i by the sampling design $p_i(\cdot)$, with positive inclusion probabilities π_{ik} and π_{ikl}. Category membership is observed for each element k in s_i, which is thereby partitioned into c subsets (sample cells) denoted $s_{i1}, \ldots, s_{ic}$. Let n_{ij} be the (usually random) size of $s_{ij} = s_i \cap U_{ij}, j = 1, \ldots, c; i = 1, 2$. We have

$$s_i = \bigcup_{j=1}^{c} s_{ij} \quad \text{and} \quad n_i = \sum_{j=1}^{c} n_{ij}$$

There is a vector-valued variable of interest, whose value for element k of population i is

$$\mathbf{y}_{ik} = (y_{i1k}, \ldots, y_{ijk}, \ldots, y_{i,c-1,k})'$$

where

$$y_{ijk} = \begin{cases} 1 & \text{if } k \in U_{ij} \\ 0 & \text{if not} \end{cases}$$

Thus, if k is in one of the categories $1, \ldots, c - 1$, the vector $\mathbf{y}_{ik}$ contains $c - 2$ components "0" and a single component "1". If k is in the cth category, $\mathbf{y}_{ik}$ is the zero vector.

To test H_0 against H_1 requires appropriate estimators of $\mathbf{P}_1$ and $\mathbf{P}_2$, as well as estimators of the corresponding $(c-1) \times (c-1)$ covariance matrices. A naive thought would be to estimate

$$P_{ij} = \frac{N_{ij}}{N_i} = \frac{\sum_{U_i} y_{ijk}}{N_i}$$

by the sample proportion in the jth category, n_{ij}/n_i. However, n_{ij}/n_i is not a consistent estimator of P_{ij}, unless the sampling design is self-weighting.

For a correct analysis, sample weighting is required. Some of the theory in Chapters 5 and 7 will be useful. Now, P_{ij} is simply the population mean of the ijth y variable; substituting π estimators, we get

$$\hat{P}_{ij} = \frac{\hat{N}_{ij}}{\hat{N}_i} \tag{13.5.1}$$

where

$$\hat{N}_{ij} = \sum_{s_i} y_{ijk}/\pi_{ik} \quad \text{and} \quad \hat{N}_i = \sum_{s_i} 1/\pi_{ik}$$

(Whether N_i is unknown or not, one would normally divide by $\hat{N}_i$ in defining an estimator of P_{ij}.) The resulting estimator of $\mathbf{P}_i$ is

$$\hat{\mathbf{P}}_i = (\hat{P}_{i1}, \ldots, \hat{P}_{i,c-1})'$$

Denote the $(c-1) \times (c-1)$ covariance matrix of $\hat{\mathbf{P}}_i$ under the design $p_i(\cdot)$ as $\mathbf{V}_i = \mathbf{V}_{p_i}(\hat{\mathbf{P}}_i) = (v_{ijj'})$. For the test we need a corresponding estimated covariance matrix,

$$\hat{\mathbf{V}}_i = \hat{\mathbf{V}}_{p_i}(\hat{\mathbf{P}}_i) = (\hat{v}_{ijj'}) \tag{13.5.2}$$

This is obtained by defining, for $j = 1, \ldots, c-1$ and $i = 1, 2$,

$$e_{ijk} = y_{ijk} - \frac{\hat{N}_{ij}}{\hat{N}_i} \quad \text{for} \quad k \in s_i$$

The covariance $v_{ijj'}$ between $\hat{P}_{ij}$ and $\hat{P}_{ij'}$ is estimated by

$$\hat{C}(\hat{P}_{ij}, \hat{P}_{ij'}) = \hat{N}_i^{-2} \sum\sum_{s_i} \check{\Delta}_{ikl} \check{e}_{ijk} \check{e}_{ij'l} = \hat{v}_{ijj'} \tag{13.5.3}$$

where $\check{e}_{ijk} = e_{ijk}/\pi_{ik}$ and $\check{\Delta}_{ikl} = \Delta_{ikl}/\pi_{ikl}$. When $j' = j$, (13.5.3) is the estimated variance of $\hat{P}_{ij}$. These elements determine the estimated covariance matrix defined in (13.5.2).

Let us test H_0 assuming that the two sampling designs have fixed sizes n_1 and n_2, respectively. One option is to base the test on the *Wald statistic*. In the present application, this statistic uses the difference $\hat{\mathbf{P}}_1 - \hat{\mathbf{P}}_2$ and is given by

$$X_{WH}^2 = (\hat{\mathbf{P}}_1 - \hat{\mathbf{P}}_2)'(\hat{\mathbf{V}}_1 + \hat{\mathbf{V}}_2)^{-1}(\hat{\mathbf{P}}_1 - \hat{\mathbf{P}}_2)$$

Work by Grizzle, Starmer and Koch (1969) and Koch, Freeman and Freeman (1975) was instrumental in showing how the Wald statistic can be used in hypothesis tests with survey data, in particular in linear and log-linear model analysis.

Let χ^2_m denote the chi-square random variable with m degrees of freedom. Under the conditions indicated below, and if H_0 holds, the large sample distribution of X^2_{WH} is χ^2_{c-1}. At a fixed level α, (say, 5%), an approximate (asymptotically valid) test is obtained by rejecting H_0 if X^2_{WH} exceeds $\chi^2_{c-1,1-\alpha}$, a critical value determined by

$$P(\chi^2_{c-1} \geq \chi^2_{c-1,1-\alpha}) = \alpha$$

The necessary assumptions for the Wald statistic test are as follows:

a. For $i = 1, 2$, $\hat{\mathbf{V}}_i$ is a consistent estimator of the covariance matrix $\mathbf{V}_i$.
b. A central limit theorem is in effect, for each of the two sampling designs, so that in large samples, $\hat{\mathbf{P}}_i - \mathbf{P}_i$ follows approximately a $(c - 1)$-dimensional normal distribution, $i = 1, 2$.

As for (a), the estimated covariance matrix defined by (13.5.2) and (13.5.3) is consistent under general conditions. Ordinarily, an estimated covariance matrix based on balanced repeated replication, on the jackknife, or on one of the other techniques in Chapter 11 would also work. The requirement (b) implies that, for $i = 1$ and 2, the variable $n_i^{1/2}(\hat{\mathbf{P}}_i - \mathbf{P}_i)$ must approach the $(c - 1)$-dimensional normal distribution with mean vector $\mathbf{0}$ and covariance matrix $n_i\mathbf{V}_i = n_i\mathbf{V}_{p_i}(\hat{\mathbf{P}}_i)$, as n_i increases.

There are some drawbacks with the Wald statistic. First, when the cells are numerous, it has been observed that the Wald statistic has a tendency to become unstable and to yield significance levels that are too high compared to the nominal α; see Fay (1985) and Thomas and Rao (1987). Secondly, the computation of X^2_{WH} requires the full matrices $\hat{\mathbf{V}}_1$ and $\hat{\mathbf{V}}_2$, that is, the estimated variances $\hat{v}_{ijj}$ (the diagonal elements) as well as the estimated covariances $\hat{v}_{ijj'}$ (the off-diagonal elements). As argued by Scott and Rao (1981), a researcher who carries out a secondary analysis (that is, an analysis based not on the original micro data, but on summary information released by a statistical agency) may have access only to the published variance estimates $\hat{v}_{ijj}$, but not to the covariance estimates $\hat{v}_{ijj'}$ $(j \neq j')$. It would then be convenient to have an (at least approximate) test that can be carried out with the estimated variances only; we now give such a procedure.

The classic Pearson chi-square statistic for testing the homogeneity of two independent samples is

$$\sum_{i=1}^{2} \sum_{j=1}^{c} \frac{(n_{ij} - a_{ij})^2}{a_{ij}} = \sum_{i=1}^{2} \sum_{j=1}^{c} \frac{(n_i\bar{P}_{ij} - n_i\bar{P}_{+j})^2}{n_i\bar{P}_{+j}} \tag{13.5.4}$$

where $a_{ij} = n_i(n_{1j} + n_{2j})/(n_1 + n_2)$ is the estimated ij cell frequency under H_0, $\bar{P}_{ij} = n_{ij}/n_i$, and $\bar{P}_{+j} = (n_{1j} + n_{2j})/(n_1 + n_2)$. The proof that the statistic (13.5.4) is distributed approximately as χ^2_{c-1}, traditionally given in intermediate to advanced texts on statistical theory, relies on independent multinomial distributions for each of the vectors $(n_{11}, \ldots, n_{1,c-1})$ and $(n_{21}, \ldots, n_{2,c-1})$. When finite populations are involved, this holds only under simple random sampling with replacement. The cell probability estimator $\bar{P}_{ij}$ used in (13.5.4) ignores the sampling weights and is consistent for P_{ij} only if the design is self-weighting. A cautious analyst, who is aware of the importance of sample weighting, may get the good idea to use the standard chi-square formula, but expressed in terms of the appropriately weighted estimates $\hat{P}_{ij}$ given by (13.5.1),

$$X_{PH}^2 = \sum_{i=1}^{2} \sum_{j=1}^{c} \frac{(n_i \hat{P}_{ij} - n_i \hat{P}_{+j})^2}{n_i \hat{P}_{+j}} \tag{13.5.5}$$

where $\hat{P}_{ic} = 1 - \hat{P}_{i1} - \cdots - \hat{P}_{i,c-1}$ and

$$\hat{P}_{+j} = \frac{n_1 \hat{P}_{1j} + n_2 \hat{P}_{2j}}{n_1 + n_2}; \qquad j = 1, \ldots, c \tag{13.5.6}$$

We call (13.5.5) *the modified chi-square statistic*; it reduces to (13.5.4) if both designs are self-weighting.

The naive user would perhaps proceed as if X_{PH}^2 were distributed approximately as χ_{c-1}^2. This is incorrect. The reason is that in mimicking the classic Pearson statistic (13.5.4), X_{PH}^2 will also (erroneously) assume the variance-covariance structure that applies under independent multinomial distributions for the category count vectors. Although the modified statistic (13.5.5) attempts to adjust for the unequal probability sampling, it does not go far enough. The consequence for an analyst who fixes an α, and rejects H_0 if $X_{PH}^2 > \chi_{c-1,1-\alpha}^2$, is a Type I error probability, $\Pr(X_{PH}^2 > \chi_{c-1,1-\alpha}^2)$, that is not approximately α as intended. The test is then invalid. The Type I error probability may be considerably greater than α.

The large-sample distribution of X_{PH}^2 is given in the following result from Scott and Rao (1981), which we state but do not prove.

Result 13.5.1. *Under $H_0 : \mathbf{P}_1 = \mathbf{P}_2$, the statistic X_{PH}^2 given by equation* (13.5.5) *is distributed asymptotically as $\sum_{j=1}^{c-1} \delta_j z_j^2$, where $z_1, \ldots, z_{c-1}$ are independent $N(0, 1)$ random variables and $\delta_1, \ldots, \delta_{c-1}$ are the eigenvalues of*

$$\frac{n_2 \mathbf{A}_1 + n_1 \mathbf{A}_2}{n_1 + n_2} \tag{13.5.7}$$

where $\mathbf{A}_i = \mathbf{R}^{-1} n_i \mathbf{V}_i$, with $\mathbf{R} = diag(\mathbf{P}) - \mathbf{PP}'$, and $\mathbf{V}_i = \mathbf{V}_{p_i}(\hat{\mathbf{P}}_i) = (v_{ijj'})$ is the covariance matrix of $\hat{\mathbf{P}}_i$ under the sampling design $p_i(\cdot)$.

In Result 13.5.1, diag($\mathbf{P}$) denotes the matrix whose diagonal elements are those of the vector $\mathbf{P}$, whereas all off-diagonal elements are zero. Thus, the distribution of X_{PH}^2 is approximately that of a weighted combination of χ_1^2 variables, where the weights are the eigenvalues δ_i of the matrix $(n_2 \mathbf{A}_1 + n_1 \mathbf{A}_2)/(n_1 + n_2)$, called the design effects matrix.

The special case of $n_1 \mathbf{V}_1 = n_2 \mathbf{V}_2 = \mathbf{R}$ corresponds to simple random sampling with replacement. For this design, (13.5.5) equals the classic Pearson statistic (13.5.4), and $\delta_i = 1$ for all i. Otherwise, the δ_i will vary. Denoting their mean by

$$\bar{\delta} = \sum_{j=1}^{c-1} \delta_j/(c-1)$$

we have, using Result 13.5.1 and the fact that $E(\chi_m^2) = m$,

$$E(\mathrm{X}_{PH}^2) \doteq \sum_{j=1}^{c-1} \delta_j = (c-1)\bar{\delta} \tag{13.5.8}$$

Whenever the δ_j are not all equal to one, the test that treats X_{PH}^2 as a χ_{c-1}^2 random variable is erroneous. First, suppose that $\bar{\delta} > 1$. The statistic X_{PH}^2 will then tend to "too easily" exceed the critical value $\chi_{c-1,1-\alpha}^2$, found in the χ_{c-1}^2 table for the desired α level, say, $\alpha = 5\%$. [Note that $E(\chi_{c-1}^2) = c - 1 < \bar{\delta}(c-1) \doteq E(\mathrm{X}_{PH}^2)$.] The true Type I error probability, $\Pr(\mathrm{X}_{PH}^2 > \chi_{c-1,1-\alpha}^2)$, can be considerably greater than the nominal α and, consequently, false significances may result. This typically occurs in sampling of clusters with positive homogeneity. Scott and Rao (1981) show examples where the Type I probability was as large as 40%, for a test with a nominal α of 5%. When $\bar{\delta} < 1$, the test that treats X_{PH}^2 as χ_{c-1}^2 is also incorrect, but a redeeming feature is that the test is usually conservative, that is, the Type I error probability is ordinarily less than the specified α. Under efficient *STSI* sampling, $\bar{\delta} < 1$ is normally the case.

To treat X_{PH}^2 as a χ_{c-1}^2 is wrong, but can any valid test be based on the modified chi-square statistic X_{PH}^2? There are several possibilities. The first method is to use the distribution of a weighted sum of χ_1^2 variables, for which accurate approximations have been obtained by Solomon and Stephens (1977). With the aid of tables of this distribution, the statistic X_{PH}^2 can be used to carry out a correct test of H_0.

A second method is to use a simplified, approximate test statistic suggested by Scott and Rao (1981). A "mean eigenvalue correction" is applied to X_{PH}^2. The resulting statistic is

$$\mathrm{X}_{MH}^2 = \mathrm{X}_{PH}^2/\hat{\bar{\delta}} \tag{13.5.9}$$

Here, $\hat{\bar{\delta}}$ is an estimate of the unknown eigenvalue average, $\bar{\delta}$. The null hypothesis H_0 is tested by referring X_{MH}^2 to a table of the χ_{c-1}^2 distribution. That is, H_0 is rejected if X_{MH}^2 exceeds the critical value $\chi_{c-1,1-\alpha}^2$ found in a table of the χ_{c-1}^2 distribution. The reason for mean eigenvalue correction is easy to see. From (13.5.8),

$$E(\mathrm{X}_{PH}^2/\bar{\delta}) \doteq c - 1 = E(\chi_{c-1}^2)$$

that is, $\mathrm{X}_{PH}^2/\bar{\delta}$ and χ_{c-1}^2 agree approximately in expectation. Higher order moments will not agree, but if the δ_j do not vary too much, the chi-square approximation is still reasonably good. Thus, if $\bar{\delta}$ exceeds unity, dividing by $\bar{\delta}$ deflates X_{PH}^2, and $\mathrm{X}_{PH}^2/\bar{\delta}$ is brought to the same average level as χ_{c-1}^2, under H_0.

Let us further examine $\bar{\delta}$, the mean of the eigenvalues of the design effects matrix shown in (13.5.7). If $\operatorname{tr}(\mathbf{A}_i)$ denotes the trace of $\mathbf{A}_i = \mathbf{R}^{-1} n_i \mathbf{V}_i$, $i = 1, 2$, we have

$$(c-1)\bar{\delta} = \sum_{j=1}^{c-1} \delta_j = [n_2 \operatorname{tr}(\mathbf{A}_1) + n_1 \operatorname{tr}(\mathbf{A}_2)]/(n_1 + n_2)$$

After some algebra, $\operatorname{tr}(\mathbf{A}_i)$ can be expressed as

$$\operatorname{tr}(\mathbf{A}_i) = \sum_{j=1}^{c} \frac{n_i v_{ijj}}{P_j}$$

where P_j is the null hypothesis proportion in the jth category, which gives

$$\bar{\delta} = \frac{n_1 n_2}{n_1 + n_2} \sum_{i=1}^{2} \sum_{j=1}^{c} \frac{v_{ijj}}{P_j} \tag{13.5.10}$$

Thus, $\bar{\delta}$ is a function of the variances v_{ijj} alone. An alternative expression for $\bar{\delta}$ is even more telling. Using (13.5.10), we can write

$$(c-1)\bar{\delta} = \sum_{i=1}^{2} \sum_{j=1}^{c} w_{ij} d_{ij} \tag{13.5.11}$$

where we have set

$$w_{ij} = \frac{n_1 n_2}{(c-1)(n_1 + n_2)} \frac{1 - P_j}{n_i}$$

and

$$d_{ij} = \frac{n_i v_{ijj}}{P_j(1 - P_j)}$$

In equation (13.5.11), the w_{ij} are nonnegative weights adding up to unity,

$$\sum_{i=1}^{2} \sum_{j=1}^{c} w_{ij} = 1$$

Moreover, d_{ij} can be interpreted as a design effect associated with the ijth cell. Under the ith sampling design, the (approximate) variance of $\hat{P}_{ij}$ is v_{ijj}, whereas under multinomial sampling, when H_0 holds, the variance of $\hat{P}_{ij}$ would be $P_j(1 - P_j)/n_i$, and d_{ij} is the ratio of these two quantities. Thus, (13.5.11) expresses $\bar{\delta}$ as a weighted average of the $2c$ cell design effects d_{ij}.

Now, the proposed test statistic X^2_{MH} given by equation (13.5.9) calls for an estimate, $\hat{\bar{\delta}}$, in place of the unknown $\bar{\delta}$. To this end, replace v_{ijj} and P_j in (13.5.10) by $\hat{v}_{ijj}$ and $\hat{P}_{+j}$, given by (13.5.3) (with $j' = j$) and (13.5.6); we assume that these are consistent estimators.

The estimated cell design effect for the cell ij is

$$\hat{d}_{ij} = \frac{n_i \hat{v}_{ijj}}{\hat{P}_{+j}(1 - \hat{P}_{+j})} \tag{13.5.12}$$

and we have

$$\hat{\bar{\delta}} = \sum_{i=1}^{2} (1 - f_i)\left[\sum_{j=1}^{c} (1 - \hat{P}_{+j})\hat{d}_{ij}\right]\bigg/(c - 1)$$

where $f_i = n_i/(n_1 + n_2)$. It is of interest in the analysis to calculate all the $\hat{d}_{ij}$ to see if some cells ij are particularly sensitive to the design. For example, one may want to identify cells with a particularly large $\hat{d}_{ij}$.

Clearly, when $X^2_{MH} = X^2_{PH}/\hat{\bar{\delta}}$ is treated as a χ^2_{c-1}, several approximations have been made, but if the δ_j vary little, the procedure has proved to work well, giving a Type I error probability close to the nominal α. Empirical studies of the procedure were carried out by Scott and Rao (1981) and Holt, Scott and Ewings (1980). For the populations considered, the procedure gave, with $\alpha = 5\%$ nominal level, a true Type I error probability not exceeding 0.06. Also, X^2_{MH} is computationally simpler than the Wald statistic X^2_{WH}. For X^2_{MH} we need only the diagonal elements $\hat{v}_{ijj}$ of the estimated covariance matrix $\hat{\mathbf{V}}_i$. By contrast, the Wald statistic X^2_{WH} requires the full estimated covariance matrix.

A third method for basing a test on the statistic X^2_{PH} is to use the Satterthwaite correction. The variance of X^2_{MH} is greater than $2(c - 1)$, the variance of χ^2_{c-1}, and increasingly so if the variance of the δ_j is considerable. In this situation, an improvement on the preceding method is obtained by applying the Satterthwaite (1946) correction to X^2_{PH}. This procedure uses the fact that

$$X^2_{PH}/\bar{\delta}(1 + C^2)$$

is an approximate χ^2_ν random variable with degrees of freedom $\nu = p/(1 + C^2)$, where $p = c - 1$ and C^2 is the squared coefficient of variation of the δ_j,

$$C^2 = \bar{\delta}^{-2} \sum_{j=1}^{p} (\delta_j - \bar{\delta})^2/p$$

and $\bar{\delta} = \sum_{j=1}^{p} \delta_j/p$. To make the test operational, $\bar{\delta}$ and C^2 are replaced by sample-based estimates, calculated from the covariance estimates $\hat{v}_{ijj'}$. The procedure gives improved approximation but requires the full matrix $\hat{\mathbf{V}}_i$, just as the Wald statistic.

What sample sizes n_1 and n_2 are necessary for the various χ^2 approximations considered in this section? A complete answer to this question is not available; however, sample sizes are often very large (several thousands in many surveys conducted by national statistical agencies), so lack of observations is often not a serious concern.

13.5.2. Testing Homogeneity for More than Two Finite Populations

The test of homogeneity in the previous section can be generalized to an arbitrary number r of finite populations (for example, countries). Let N_i be the size of the ith population. As in Section 13.5.1, we assume for $i = 1, \ldots, r$ that there are c categories (for example, states of health), $j = 1, \ldots, c$, which form a total of rc population cells U_{ij}. The objective is to test the equality of the r vectors of category proportions

$$\mathbf{P}_i = (P_{i1}, \ldots, P_{ij}, \ldots, P_{i,c-1})', \qquad i = 1, \ldots, r$$

where $P_{ij} = N_{ij}/N_i$ and N_{ij} is the size of U_{ij}. The formally stated null hypothesis is

$$H_0 : \mathbf{P}_1 = \mathbf{P}_2 = \cdots = \mathbf{P}_r$$

We denote by $\mathbf{P} = (P_1, \ldots, P_{c-1})'$ the common value of the $\mathbf{P}_i$ under H_0. The alternative hypothesis states that H_0 does not hold.

Suppose that independent samples are drawn, one from each population, $i = 1, \ldots, r$. The sample s_i from the ith population is of fixed size n_i and let

$$n = \sum_{i=1}^{r} n_i$$

The sampling designs $p_i(s_i)$ need not be identical. Our sample-weighted estimator of $\mathbf{P}_i$ is

$$\hat{\mathbf{P}}_i = (\hat{P}_{i1}, \ldots, \hat{P}_{ij}, \ldots, \hat{P}_{i,c-1})$$

with

$$\hat{P}_{ij} = \frac{\sum_{s_i} y_{ijk}/\pi_{ik}}{\sum_{s_i} 1/\pi_{ik}} \tag{13.5.13}$$

where $y_{ijk} = 1$ if $k \in U_{ij}$ and $y_{ijk} = 0$ if not, and π_{ik} is the inclusion probability of unit k in population i; $i = 1, \ldots, r$. The covariance matrix of $\hat{\mathbf{P}}_i$ is denoted

$$\mathbf{V}_i = \mathbf{V}(\hat{\mathbf{P}}_i) = (v_{ijj'})$$

for which we assume that a consistent estimator

$$\hat{\mathbf{V}}_i = (\hat{v}_{ijj'})$$

can be formed, for example, by using Taylor linearization covariance estimates.

As in the case $r = 2$, several possibilities exist for testing H_0. We can use the Wald statistic, or a modified chi-square statistic.

i. *The Test Based on the Wald Statistic*

Consider the $(r-1)(c-1)$-dimensional vector of estimated proportions

$$\hat{\mathbf{Q}} = [(\hat{\mathbf{P}}_1 - \hat{\mathbf{P}}_r)', \ldots, (\hat{\mathbf{P}}_i - \hat{\mathbf{P}}_r)', \ldots, (\hat{\mathbf{P}}_{r-1} - \hat{\mathbf{P}}_r)']' \tag{13.5.14}$$

The large-sample variance-covariance matrix of $\hat{\mathbf{Q}}$ can be shown to be

$$\mathbf{V}(\hat{\mathbf{Q}}) = \bigoplus_{i=1}^{r-1} \mathbf{V}_i + \mathbf{V}_r \otimes \mathbf{J} \tag{13.5.15}$$

where $\mathbf{V}_i$ is the covariance matrix of $\hat{\mathbf{P}}_i$ $(i = 1, \ldots, r)$, $\mathbf{J}$ is a $(r-1) \times (r-1)$ matrix with every element equal to unity, $\oplus$ denotes the direct sum operator, and $\otimes$ stands for the Kronecker product of matrices. Let $\hat{\mathbf{V}}(\hat{\mathbf{Q}})$ be the consistent estimator of $\mathbf{V}(\hat{\mathbf{Q}})$ obtained by replacing $\mathbf{V}_i = (v_{ijj'})$ in (13.5.15) by its estimate $\hat{\mathbf{V}}_i = (\hat{v}_{ijj'})$; $i = 1, \ldots, r$. We assume asymptotic normality for the $\hat{\mathbf{P}}_i$. Under H_0, $\hat{\mathbf{Q}}$ has a mean vector of zero, and the Wald statistic,

$$X_{WH}^2 = \hat{\mathbf{Q}}'[\hat{\mathbf{V}}(\hat{\mathbf{Q}})]^{-1}\hat{\mathbf{Q}} \tag{13.5.16}$$

has an approximate $\chi^2_{(r-1)(c-1)}$ distribution. An approximately correct test can thus be obtained by rejecting H_0 if X_{WH}^2 exceeds the critical value $\chi^2_{(r-1)(c-1),\,1-\alpha}$. The calculation of the statistic (13.5.16), which includes inversion of the $(r-1)(c-1) \times (r-1)(c-1)$ matrix $\hat{\mathbf{V}}(\hat{\mathbf{Q}})$, can be tedious. Moreover, in a secondary data analysis, the estimated covariances $\hat{v}_{ijj'}$ $(j \neq j')$ may not be available.

ii. *Tests Using the Modified Pearson Chi-Square Statistic*

For this case, the statistic is

$$X_{PH}^2 = \sum_{i=1}^{r} \sum_{j=1}^{c} \frac{(n_i \hat{P}_{ij} - n_i \hat{P}_{+j})^2}{n_i \hat{P}_{+j}} \tag{13.5.17}$$

with $\hat{P}_{ij}$ given by (13.5.13) and

$$\hat{P}_{+j} = \frac{1}{n} \sum_{i=1}^{r} n_i \hat{P}_{ij} \tag{13.5.18}$$

To generalize Result 13.5.1, the large sample distribution of X_{PH}^2 is the same as that of a weighted sum of independent χ_1^2 variables, $\sum_{j=1}^{m} \delta_j z_j^2$, where $m = (r-1)(c-1)$, and $z_1, \ldots, z_m$ are independent $N(0, 1)$ random variables. The design effects matrix (that is, the matrix having the weights δ_i as eigenvalues) is given by

$$n(\mathbf{F} \otimes \mathbf{R}^{-1})\mathbf{V}(\hat{\mathbf{Q}}) \tag{13.5.19}$$

where

$$\mathbf{F} = \text{diag}(\mathbf{f}) - \mathbf{f}\mathbf{f}'$$

with

$$\mathbf{f} = (f_1, \ldots, f_i, \ldots, f_{r-1})'; \qquad f_i = n_i/n$$

and

$$\mathbf{R} = \text{diag}(\mathbf{P}) - \mathbf{P}\mathbf{P}'$$

where $\mathbf{P}$ denotes the common vector of proportions under H_0 and $\mathbf{V}(\hat{\mathbf{Q}})$ is the covariance matrix of $\hat{\mathbf{Q}}$ given by (13.5.15).

If the π_{ik} are equal, within each population, X_{PH}^2 reduces to the classic Pearson chi-square

$$\sum_{i=1}^{r} \sum_{j=1}^{c} \frac{(n_{ij} - a_{ij})^2}{a_{ij}}$$

where n_{ij} is the count in the sample cell $s_{ij} = s_i \cap U_{ij}$ and

$$a_{ij} = n_i \sum_{i=1}^{r} n_{ij}/n$$

Under independent multinomial sampling (that is, simple random sampling with replacement) from each of the r populations, this statistic is distributed asymptotically as $\chi^2_{(r-1)(c-1)}$.

A mean eigenvalue correction leads to a test that is both valid and simple. If the δ_i do not vary too much, a good approximate test of H_0 is obtained by treating

$$X^2_{MH} = X^2_{PH}/\hat{\bar{\delta}} \tag{13.5.20}$$

as a $\chi^2_{(r-1)(c-1)}$ random variable, where $\hat{\bar{\delta}}$ is an estimate of the mean, $\bar{\delta}$, of the $(r-1)(c-1)$ eigenvalues of the design effects matrix shown in (13.5.19).

In this case, $\bar{\delta}$ can be expressed as the weighted sum of cell design effects

$$\bar{\delta} = \sum_{i=1}^{r} \sum_{j=1}^{c} w_{ij} d_{ij}$$

The weights $w_{ij} = (1 - f_i)(1 - P_j)/(r - 1)(c - 1)$ add up to unity, and

$$d_{ij} = \frac{n_i v_{ijj}}{P_j(1 - P_j)}$$

is the design effect associated with the ijth cell.

The estimate $\hat{\bar{\delta}}$ required for calculating the statistic (13.5.20) is again obtained by replacing v_{ijj} and P_j by the respective estimates $\hat{v}_{ijj}$ and $\hat{P}_{+j}$, and estimated cell design effects, $\hat{d}_{ij}$, are obtained in the manner of (13.5.12). A Sattherthwaite corrected test is also possible. It uses the full estimated variance matrix, just like the Wald statistic.

13.5.3. Discussion of Categorical Data Tests for Finite Populations

The following chi-square tests are standard procedures for analyzing categorical data:

a. Test of goodness of fit, that is, to test if a c-vector of category proportions for a sampled population is in agreement with a specific constant vector.
b. Test of homogeneity in an $r \times c$ table, that is, to test if a c-vector of category proportions is the same for each of r independently sampled populations.
c. Test of independence in an $r \times c$ table.

For these tests, there exist finite population equivalents that take the sampling design into account. The tests in category (b) were discussed in Sections 13.5.1 and 13.5.2. As an example of type (a), one may wish to test if a health status distribution estimated from current survey data in a certain country equals a specified vector, say, the known distribution in another country. For an example of type (c), one may wish to test if two characteristics, such as health status and occupational status, are independent, based on results from a current survey. Rao and Hidiroglou (1981) and Hidiroglou and Rao (1987) illustrate how the three types of test work in an application to Canada Health Survey data. The theory of these tests is discussed in Rao and Scott (1981, 1984). For each of the types (a), (b), and (c), a test may be based on the Wald statistic; this test is asymptotically valid and requires the full estimated covariance matrix.

One can also use a test based on the modified chi-square statistic, that is, the classic chi-square formula expressed in sample-weighted estimates of proportions. This statistic, denoted X^2, is distributed as $\sum_{j=1}^{p} \delta_j z_j^2$, where $z_1^2, \ldots, z_p^2$ are independent χ_1^2 random variables, and $\delta_1, \ldots, \delta_i, \ldots, \delta_p$ are the eigenvalues of the appropriate design effects matrix. This matrix takes the form $\mathbf{A}^{-1}\mathbf{B}$, where $\mathbf{A}$ and $\mathbf{B}$ are symmetric $p \times p$ matrices. There is a corresponding estimated design effect matrix, $\hat{\mathbf{A}}^{-1}\hat{\mathbf{B}}$, with eigenvalues denoted $\hat{\delta}_1, \ldots, \hat{\delta}_j, \ldots, \hat{\delta}_p$. There are at least three ways of carrying out the test.

1. Use the distribution of $\sum_{j=1}^{p} \delta_j z_j^2$. Accurate approximations can be obtained by the method of Solomon and Stephens (1977).
2. Use the Satterthwaite (1946) approximation. The method is based on the fact that

$$X_{\mathrm{SAT}}^2 = X^2/\bar{\delta}(1 + C^2)$$

   has approximately the distribution of a χ_ν^2 random variable with $\nu = p/(1 + C^2)$ degrees of freedom, where C^2 is the squared coefficient of variation of the δ_j, and $\bar{\delta}$ is their mean.
3. Use the mean eigenvalue correction. This method, which is quite accurate if the δ_j do not vary too much, rests on the idea that $X^2/\bar{\delta}$ can be considered distributed roughly as a χ_p^2 random variate.

In practice, the test statistics in the above three procedures are calculated by using the eigenvalues $\hat{\delta}_1, \ldots, \hat{\delta}_j, \ldots, \hat{\delta}_p$ of the estimated design effects matrix $\hat{\mathbf{A}}^{-1}\hat{\mathbf{B}}$, instead of the unknown eigenvalues $\delta_1, \ldots, \delta_i, \ldots, \delta_p$ of $\mathbf{A}^{-1}\mathbf{B}$.

Remark 13.5.1. (a) Other approximate test procedures have been suggested for categorical survey data. Fellegi (1980) considers a modified Wald statistic tested with the aid of an F distribution. In this procedure, the Wald statistic is multiplied by an appropriate constant and is then referred to a table of the F distribution. Fay (1985) proposes a test based on chi-square and obtained by jackknifing. (b) Computer packages such as PC-CARP, SUPER-CARP, and SUDAAN are equipped to perform most of the categorical data analyses described in this section. (c) Extensions to analysis of three and higher dimension contingency tables have been considered, as in Rao and Scott (1984) and Hidiroglou and Rao (1987). (d) Logistic regression analysis, generalized linear model (GLIM) analysis, and logit model analysis for survey data have been examined, as in Rao and Scott (1987), Kumar and Rao (1984), and Nordberg (1989). (e) There is a trade-off between the control of the significance level and the power of the test. We noted that the modified chi-square statistic X_{PH}^2 can be used for an approximate test, if it is adjusted by dividing by $\hat{\bar{\delta}}$. This is to obtain control of the significance level. The test also needs to be examined from the point of view of its power. Thomas and Rao (1987) examine the power of some of the categorical data tests discussed in this section.

Remark 13.5.2. How should a null hypothesis be formulated for finite populations? For example, if the means of two finite populations are compared, some would argue that an exact equality never occurs. There is always a difference, however small, between the means of two actual finite populations. One is sure from the outset that the null hypothesis expressing the equality of the means is false, and such a null hypothesis should not be considered. A counterargument is that in practice, nonrejection of a null hypothesis H_0 is rarely taken as an outright conclusion that H_0 is true, but rather as an indication that "H_0 cannot be rejected." In some cases, it may be more appropriate to address a null hypothesis stating that the means are not more than a small specified quantity apart.

Remark 13.5.3. Frequently, survey data are used to compare groups of elements to see whether (significant) differences exist. One must then keep in mind that comparison of groups takes a number of methodologically different forms when finite populations are involved. For example, we can have the following specifications.

a. A probability sample of individuals is drawn in a certain area, say a country; compare men with women.
b. A probability sample of individuals is drawn in area A; another probability sample is drawn, independently of the first and perhaps with a different sampling design, in area B; compare individuals in area A with individuals in area B.
c. Two independent probability samples of individuals are drawn, one in area A, one in area B; compare women in area A with women in area B.

In case (a), the comparison is between two domains (men and women) of one and the same sampled population. The domain sample "men" is ordinarily not independent of the domain sample "women." Some theory relevant for this case was given in Chapter 10. In case (b), two independently sampled entire populations are compared. In Section 13.5.1 we examined case (b) for categorical survey data. In case (c), a domain (women) of the first population (area A) is compared with the corresponding domain of the second, independently sampled population (area B). The particular features of each case must be taken into account in setting up a correct analysis.

13.6. Types of Inference When a Finite Population Is Sampled

Many texts on statistics distinguish statistics for descriptive purposes from statistics for inferential purposes. With finite populations in mind, let us discuss this distinction.

i. *Description*

For the finite population $U = \{1, \ldots, k, \ldots, N\}$, consider two variables of interest y and z. Suppose that (y_k, z_k) is observed for $k = 1, \ldots, N$. This occurs in a census with 100% response rate. With the aid of these data we can calculate quantities such as the totals

$$t_y = \sum_U y_k; \qquad t_z = \sum_U z_k \tag{13.6.1}$$

the means

$$\bar{y}_U = t_y/N; \qquad \bar{z}_U = t_z/N \tag{13.6.2}$$

the ratio of totals

$$R = \frac{\sum_U y_k}{\sum_U z_k} \tag{13.6.3}$$

and the regression slope

$$B = \frac{\sum_U (y_k - \bar{y}_U)(z_k - \bar{z}_U)}{\sum_U (z_k - \bar{z}_U)^2} \tag{13.6.4}$$

These quantities portray different aspects of the finite population, at the current moment in time. Each may provide interesting information about the population U. For example, R and B measure, in somewhat different ways, the number of units of y per unit of z in the finite population. For instance, R can be the part of a household income dollar that is saved. Other descriptive measures, based on multivariate observations, were given in Section 13.2.

If U_a is a subset of U, and if (y_k, z_k) is observed for the elements $k \in U_a$, we can describe the set U_a by quantities directly analogous to those given in (13.6.1) to (13.6.4). To $\bar{y}_U$ corresponds $\bar{y}_{U_a} = (\sum_{U_a} y_k)/N_a$, and so on, where N_a is the size of U_a. The point is that finite population data are sometimes analyzed with only one intent: to describe the population or the subpopulation. One simply calculates one or more descriptive quantities, with no attempt to generalize the conclusions to a larger universe. There is no statistical inference.

When the finite population is not completely observed, quantities such as those shown by equations (13.6.1) to (13.6.4) are unknown and may become the target of investigation. Then they are commonly referred to as *descriptive parameters.*

ii. *Inference*

To make inference, in a statistician's language, usually means to draw conclusions, with the aid of probability statements, from a sample to a larger universe. Confidence intervals and hypothesis tests are traditional tools of inference. Suppose we have data obtained by measuring the elements k in a probability sample s drawn from the finite population U. Two important types of inference are as follows:

a. Inference about the finite population U itself.
b. Inference about a model or a superpopulation thought to have generated U.

Case (a) has occupied a major part in this book. The objective is to estimate and in other ways make inferences about descriptive parameters that characterize U. This is inference about the "now," about the current state of the finite population. By contrast, case (b) poses questions about the process that underlies the finite population U. The need for (b) is shown by the following example.

EXAMPLE 13.6.1. Consider a model ξ which describes the relationship between the observable variables y and $\mathbf{z}$, where y_k may be household savings and $\mathbf{z}_k$ a vector of suitable explanatory variables for the kth household. The model ξ states that, for fixed vectors $\mathbf{z}_1, \ldots, \mathbf{z}_N$,

$$\begin{cases} 1.\ y_k = \mathbf{z}_k'\boldsymbol{\beta} + \varepsilon_k; \qquad k = 1, \ldots, N \\ 2.\ \varepsilon_1, \ldots, \varepsilon_N \text{ are independent random variables with} \\ \quad E_\xi(\varepsilon_k) = 0; \qquad V_\xi(\varepsilon_k) = \sigma^2; \qquad k = 1, \ldots, N \\ 3.\ \varepsilon_1, \ldots, \varepsilon_N \text{ are normally distributed} \end{cases} \tag{13.6.5}$$

Often ξ is referred to as a *superpopulation model*; the vector valued parameter $\boldsymbol{\beta}$ is a *model parameter.* (The term *analytic parameter* is also used sometimes; we avoid it here, because analysis is also involved when the target of interest is the finite population itself.) Inference to model parameters is often of interest in economic and sociological research based on the analysis of survey data. The model builder is interested not in the finite population U at the present moment in time, but rather in the process or the causal system relating y to $\mathbf{z}$. The question then arises: How should this be done when we observe $(y_k, \mathbf{z}_k)$ only for elements k in a probability sample s? Some options are considered in this section. Superpopulations are further discussed in Section 14.5.

Remark 13.6.1. A descriptive parameter has the property that in a census with 100% response and no measurement errors, its value can be established exactly, without error. A model parameter (for example, $\boldsymbol{\beta}$ in the regression model above) is different in this respect. It is part of a hypothetical construct, therefore, it can never be calculated exactly.

The following table, adapted with minor changes from Hartley and Sielken (1975), shows different types of inference that may be considered when a finite population is sampled.

Parameters of interest	Sampling from a fixed finite population	Two-step sampling from a superpopulation
Finite population parameters or descriptive parameters	Classic finite population sampling theory: case 1	Superpopulation theory for the finite population: case 3
Infinite superpopulation parameters or model parameters	Infeasible: case 2	Inference about superpopulation parameters from two-step sampling: case 4

Let us examine the different cases.

Case 1. The finite population values $y_1, \ldots, y_N$ are seen as fixed constants, as in most of the preceding chapters. Case 1 inference aims at knowledge about unknown descriptive parameters of the finite population U. If it had been possible to observe the variable(s) of interest for *all* N elements in U, we could have described U in terms of various quantities of interest, such as those given in (13.6.1) to (13.6.4) or any of the multivariate descriptive parameters discussed in Section 13.2. However, the observations available are limited to the sample s, hence the necessity to estimate the result of a hypothetical analysis of the whole finite population U. An analyst interested in a regression slope such as (13.6.4) implicitly considers that a linear regression model is consistent with the finite population scatter. The finite population U is the target of the inference in this case. The necessary stochastic structure is induced by the sampling design; the sample s is the random element. An estimator of the design variance can be obtained. This leads to a pure *design-based inference*, as we have encountered it throughout most of this book. In addition, error models (such as nonresponse mechanisms and measurement error models, see Chapters 15 and 16) may become necessary to complete the stochastic structure.

Case 2. The target of inference is the superpopulation and its parameters; however, the way that the finite population relates to the superpopulation is not specified, so the case becomes infeasible.

In cases 3 and 4, the finite population values $y_1, \ldots, y_N$ are considered to be the realized values of N random variables, $Y_1, \ldots, Y_N$, whose joint distribution is specified by a model ξ, for example, the model given by (13.6.5). The realization of $y_1, \ldots, y_N$ under the model ξ forms step one. The selection of a sample s from U forms step two. That is, realized values of the N random variables Y_k are observed for a sample of elements, and remain unobserved for the elements not in the sample s.

Case 3. A superpopulation model is invoked, but the inference still concerns the finite population and its parameters. Let θ denote such a parameter and $\hat{\theta}_s$ its estimator; for example, $\theta = \sum_U y_k$ and

$$\hat{\theta}_s = \textstyle\sum_s y_k/\pi_k \quad \text{or} \quad \hat{\theta}_s = (\sum_U x_k)(\sum_s y_k/\pi_k)/(\sum_s x_k/\pi_k)$$

Since two steps are involved, the evaluation of a potential estimator will ordinarily use a criterion that takes into account both the model ξ and the sampling design $p(s)$. Under one such criterion we would seek an estimator $\hat{\theta}_s$ for which

$$E_\xi E_p[(\hat{\theta}_s - \theta)^2] \tag{13.6.6}$$

is small or minimum. If $E_\xi E_p(\hat{\theta}_s - \theta) = 0$, the quantity in (13.6.6) is the anticipated variance used in Section 12.2. Note that both $\hat{\theta}_s$ and θ are now random variables. Both are functions of the y_k-values, which are random quantities under the model. The random variable $\hat{\theta}_s$ is a predictor of the random variable θ.

A more extreme position is to base the evaluation and the inference on the model alone; this is pure *model-based inference*. The sampling design $p(s)$ and its inclusion probabilities play no role in the inference; design unbiasedness is seen as unimportant. This position is taken notably in papers by Royall (1970), Royall and Herson (1973a, b), Royall and Eberhardt (1975), and Royall and Cumberland (1981a, b). In these papers, the purely model-based evaluation consists of finding the estimator to minimize the model mean square error, given s,

$$E_\xi[(\hat{\theta}_s - \theta)^2 \mid s] \tag{13.6.7}$$

under the constraint $E_\xi[(\hat{\theta}_s - \theta) \mid s] = 0$. This criterion is conditional on s. But in practice, how should s be chosen? An extreme (nonprobabilistic or purposive) course of action is to take s as the set of elements k for which the criterion (13.6.7) is minimized; see Royall (1970). Such a sample may prove unsuitable or at least risky (one cannot have complete trust in the model!). Balanced sampling is a (still nonrandomized) device suggested by Royall and Eberhardt (1975), whereby the selection of a sample is limited to those leading to inferences that are in a given sense robust to model breakdown. Balanced sampling is considered in Section 14.5. An interesting combination of randomization and balancing in sample selection is suggested by Deville, Grosbras and Roth (1988). Model-based inference is further discussed in Section 14.5.

An approach following case 3, and with selection of a balanced sample, will often lead to inferences about descriptive parameters that are not much different from what would be obtained by the design-based approach of case 1. Särndal (1978) discusses a variety of criteria for judging estimation strategies when both a model and a sampling design are considered in the inference process.

Case 4. This case is truly different from the other three, and it is important in practice (but beside the principal theme of this book). Given data from a sample survey, case 4 specifies a setup for statistical conclusions about models and causal systems, such as the one in Example 13.6.1. The goal is to estimate or draw other inferences about the superpopulation and its parameters. The following example illustrates the process.

EXAMPLE 13.6.2. Assuming the model in Example 13.6.1, let us use the method of maximum likelihood to estimate the model parameter $\boldsymbol{\beta}$. For a census, the available observations are $(y_k, \mathbf{z}_k)$; $k = 1, \ldots, N$, so the complete finite population likelihood is

$$L(\boldsymbol{\beta}) = \prod_{k=1}^{N} \left\{ \frac{1}{\sqrt{2\pi}\,\sigma} \exp[-(y_k - \mathbf{z}_k'\boldsymbol{\beta})^2/2\sigma^2] \right\}$$

The log likelihood (the "estimating function" for $\boldsymbol{\beta}$) is, apart from a constant,

$$h = h(\boldsymbol{\beta}) = \textstyle\sum_U (y_k - \mathbf{z}_k'\boldsymbol{\beta})^2$$

Maximization of $h(\boldsymbol{\beta})$ leads to the census fit estimator of $\boldsymbol{\beta}$,

$$\hat{\boldsymbol{\beta}}_U = \mathbf{B} = (\textstyle\sum_U \mathbf{z}_k\mathbf{z}_k')^{-1} \sum_U \mathbf{z}_k y_k \tag{13.6.8}$$

However, the available data are limited to a sample; we observe $(y_k, \mathbf{z}_k)$ for $k \in s$ only. The π estimator of the log likelihood $h = h(\boldsymbol{\beta})$ is, for any fixed $\boldsymbol{\beta}$,

$$\hat{h}_\pi = \textstyle\sum_s (y_k - \mathbf{z}_k'\boldsymbol{\beta})^2/\pi_k$$

Maximizing this expression we obtain the estimator

$$\hat{\boldsymbol{\beta}}_s = \hat{\mathbf{B}} = (\textstyle\sum_s \mathbf{z}_k\mathbf{z}_k'/\pi_k)^{-1} \sum_s \mathbf{z}_k y_k/\pi_k \tag{13.6.9}$$

examined in Section 5.10 and in Section 13.3 as an estimator of the finite population parameter $\mathbf{B}$. But here it appears as an estimator of the model parameter $\boldsymbol{\beta}$. The statistical properties of $\hat{\boldsymbol{\beta}}_s$ can be evaluated jointly with respect to model and sample selection as follows. Let $\mathbf{Z} = (\mathbf{z}_1, \ldots, \mathbf{z}_N)'$; $\mathbf{y} = (y_1, \ldots, y_N)'$. We use the double subscript ξp to indicate "jointly under model and design." Then, given $\mathbf{Z}$,

$$E_{\xi p}(\hat{\boldsymbol{\beta}}_s) = \boldsymbol{\beta}$$

that is, $\hat{\boldsymbol{\beta}}_s$ is unbiased with respect to the two-step approach. The steps of the derivation are

$$\begin{aligned} E_{\xi p}(\hat{\boldsymbol{\beta}}_s \mid \mathbf{Z}) &= E_\xi[E_p(\hat{\boldsymbol{\beta}}_s \mid \mathbf{y}, \mathbf{Z})] = E_\xi[\textstyle\sum_{s \in \mathscr{S}} p(s)\hat{\boldsymbol{\beta}}_s] \\ &= \textstyle\sum_{s \in \mathscr{S}} p(s) E_\xi(\hat{\boldsymbol{\beta}}_s) = \boldsymbol{\beta}[\textstyle\sum_{s \in \mathscr{S}} p(s)] = \boldsymbol{\beta} \end{aligned}$$

where $\mathscr{S}$ denotes the set of all samples. Note that the order of the expectations E_ξ and E_p may be interchanged since we are assuming that the design $p(\cdot)$ is noninformative, as specified in Remark 2.4.4.

The variance under the two-step approach has one sampling variance component (since a sample rather than the entire population is observed), and one model variance component (since the N population data points scatter according to $V_\xi(y_k \mid \mathbf{z}_k) = \sigma^2$; $k = 1, \ldots, N$).

This is seen from the decomposition

$$V_{\xi p}(\hat{\boldsymbol{\beta}}_s \mid \mathbf{Z}) = V_\xi[E_p(\hat{\boldsymbol{\beta}}_s \mid \mathbf{y}, \mathbf{Z})] + E_\xi[V_p(\hat{\boldsymbol{\beta}}_s \mid \mathbf{y}, \mathbf{Z})] \tag{13.6.10}$$

Now, $\hat{\boldsymbol{\beta}}_s$ is approximately unbiased with respect to the design. That is, $E_p(\hat{\boldsymbol{\beta}}_s \mid \mathbf{y}, \mathbf{Z}) \doteq \mathbf{B}$, so irrespective of the sampling design, the first component is roughly equal to

$$V_\xi(\mathbf{B} \mid \mathbf{Z}) = \sigma^2(\textstyle\sum_U \mathbf{z}_k \mathbf{z}_k')^{-1}$$

This represents the variance in a census fit of the model. The second term is the additional variance contributed by sampling of the finite population. This component is exactly zero in the census case. If estimates of the two components of (13.6.10) can be worked out, we can calculate a confidence interval for $\boldsymbol{\beta}$ or develop a test of the null hypothesis $\boldsymbol{\beta} = \boldsymbol{\beta}_0$.

The approach taken in Nordberg (1989) relies on the two-step sampling of case 4 (and a nonresponse selection step as well). Nordberg is concerned with regression analysis and extensions to generalized linear model analysis. The objective is to use the survey data to estimate the underlying (generalized) linear model and its parameters. The theory of *estimating functions* can be used in connection with the present example, see Godambe and Thompson (1986).

Remark 13.6.2. When the model parameter $\boldsymbol{\beta}$ is estimated from survey data, statisticians familiar with regression analysis will be faced with the question whether one should use the sample weighted estimator shown in (13.6.9) or the usual regression theory estimator

$$\boldsymbol{\beta}_{s\,\text{mod}} = (\textstyle\sum_s \mathbf{z}_k \mathbf{z}_k')^{-1} \textstyle\sum_s \mathbf{z}_k y_k \tag{13.6.11}$$

where "mod" signifies that the estimator is based on a model. Standard texts on regression theory show that (13.6.11) has commendable properties under model assumptions. If we forget about the sampling design, $\hat{\boldsymbol{\beta}}_{s\,\text{mod}}$ is the best linear unbiased estimator (BLUE) of $\boldsymbol{\beta}$, assuming (1) and (2) of the model (13.6.5) hold for the elements $k \in s$. For any constant vector $\mathbf{c}$ conformable with $\boldsymbol{\beta}$, the BLUE property implies that

$$E_\xi[\mathbf{c}'(\hat{\boldsymbol{\beta}}_{s\,\text{mod}} - \boldsymbol{\beta})^2 \mid s, \mathbf{Z}] \leq E_\xi[\mathbf{c}'(\hat{\boldsymbol{\beta}}_s - \boldsymbol{\beta})^2 \mid s, \mathbf{Z}]$$

where $\hat{\boldsymbol{\beta}}_s$ is the sample-weighted estimator shown in (13.6.9). Since this inequality holds for every s, it follows that

$$E_\xi E_p[\mathbf{c}'(\hat{\boldsymbol{\beta}}_{s\,\text{mod}} - \boldsymbol{\beta})^2 \mid s, \mathbf{Z}] \leq E_\xi E_p[\mathbf{c}'(\hat{\boldsymbol{\beta}}_s - \boldsymbol{\beta})^2 \mid s, \mathbf{Z}]$$

By this evaluation taking both ξ and $p(\cdot)$ into account, $\hat{\boldsymbol{\beta}}_{s\,\text{mod}}$ is a better estimator of $\boldsymbol{\beta}$ than $\hat{\boldsymbol{\beta}}_s$. Moreover, $\hat{\boldsymbol{\beta}}_{s\,\text{mod}}$ is the maximum likelihood estimator of $\boldsymbol{\beta}$, assuming (1), (2), and (3) of (13.6.5) hold for $k \in s$. Despite these attractive properties of $\hat{\boldsymbol{\beta}}_{s\,\text{mod}}$, there are reasons that favor the sample-weighted estimator $\hat{\boldsymbol{\beta}}_s$ in practice.

1. The estimator $\hat{\boldsymbol{\beta}}_s$ is more robust than the unweighted estimator $\hat{\boldsymbol{\beta}}_{s\,\text{mod}}$ in the sense that $\hat{\boldsymbol{\beta}}_s$ is model-unbiased for $\boldsymbol{\beta}$ if the model is true, and design-consistent for $\mathbf{B}$ whether the model holds or not. The fact that the analyst seldom has complete trust in the model speaks for $\hat{\boldsymbol{\beta}}_s$.
2. A simple regression model as shown in (13.6.5), with homoscedastic and uncorrelated errors, may be inadequate for a survey population. For example, within natural population clusters, the errors will often be positively correlated. We can easily formulate a new model incorporating this more realistic error structure. However, the intracluster correlation of the errors, as well as the exact composition of the clusters, are likely to be unknown. So although in theory there is a best $\boldsymbol{\beta}$ estimator for this more realistic model, it is impossible to compute the best estimator and its variance estimator. One may always use the easily computed $\hat{\boldsymbol{\beta}}_{s\,\text{mod}}$, perhaps without too much loss of model-based efficiency, but it is then difficult to get a correct model-based variance estimate. For a more complete discussion, see Pfeffermann and Smith (1985). Again, these considerations may speak in favor of the sample-weighted $\hat{\boldsymbol{\beta}}_s$.

Remark 13.6.3. Regression analysis is a widely used tool and not surprisingly its application to survey data has attracted the attention of many authors. To decide on the parameter of interest is extremely important. Kish and Frankel (1974), Jönrup and Rennermalm (1976), Shah, Holt and Folsom (1977) see the finite population parameter $\mathbf{B}$ given by (13.5.6) as a natural one to estimate. Others question the relevance of $\mathbf{B}$. They argue that an "analytic" finite population parameter, such as $\mathbf{B}$, has relevance only when the model associated with that parameter is true or at least well supported by available data. Clearly, if the model (13.6.5) holds, the distinction between $\boldsymbol{\beta}$ and $\mathbf{B}$ can be essentially ignored for large finite populations. Estimation of the superpopulation parameter $\boldsymbol{\beta}$ in (13.6.5), using the two step procedure discussed above, was discussed by Fuller (1975), Thomsen (1978), Pfeffermann and

Nathan (1981). Special model parameters have been considered. Konijn (1962) assumed that the finite population is divided into disjoint subpopulations, $i = 1, \ldots, M$. If β_i denotes the regression model slope for the ith subpopulation, he considered the problem of estimating the weighted parameter $\sum_{i=1}^{M} N_i\beta_i/N$, where N_i is the size of subpopulation i. Other special parameters are estimated in Porter (1973) and in Pfeffermann and Nathan (1977). Pfefferman and Holmes (1985) discuss the robustness aspect in connection with inference for regression analysis with survey data. Hidiroglou (1986b) considers sample-weighted estimation of regression coefficients with errors in the variables, that is, not only y but also $\mathbf{z}$ is affected by random error. Ten Cate (1986) studies the influence of the sampling design on the estimation of $\boldsymbol{\beta}$ when the design is informative (or endogenous), that is, when the design depends on the values y_k of the criterion variable. Other issues in regression analysis with survey data are discussed in Brewer and Mellor (1973), Holt and Smith (1976), Nathan and Holt (1980), Nathan (1981), DuMouchel and Duncan (1983), Kalton (1983), and Pfeffermann and Smith (1985). A more general source of information on statistical analysis in complex surveys is the book edited by Skinner, Holt and Smith (1989).

Remark 13.6.4. In this section, we considered (a) inference about the finite population U at the time of the survey and (b) inference about a model (superpopulation) thought to have generated U. A third type is inference about the finite population, but at a future point in time. An example of the reasoning is given in Cassel, Särndal and Wretman (1979). The sampling design is used in the estimation of the population total $t = \sum_U y_k$, and nonobserved y-values are assumed to behave according to a certain regression model, which is taken to be the same at the time of the survey as at the future point in time.

Exercises

13.1. Let the finite population correlation coefficient $r_{yz} = r_{yzU} = S_{yzU}/(S_{yU}^2 S_{zU}^2)^{1/2}$ be estimated by $\tilde{r}_{yz} = \tilde{S}_{yz}/(\tilde{S}_{yy}\tilde{S}_{zz})$, where $\tilde{S}_{yz}$, $\tilde{S}_{yy}$, and $\tilde{S}_{zz}$ are defined in the manner of equation (13.2.12). Use Result 5.5.1 to show that the u_k in the approximate variance expression are given by

$$u_k = \left[y_k^* z_k^* - \frac{r_{yzU}}{2}(y_k^{*2} + z_k^{*2}) \right] \Big/ (N - 1)$$

where

$$y_k^* = (y_k - \bar{y}_U)/S_{yU} \quad \text{and} \quad z_k^* = (z_k - \bar{z}_U)/S_{zU}$$

Also specify the quantities $\hat{u}_k$ required for the variance estimator shown in equation (5.5.13).

13.2. *STSI* sampling from the MU200 population is used to estimate the population correlation coefficient $r_{yz} = r_{yzU}$, where the two study variables are REV84 ($= y$)

and ME84 ($=z$). The design uses two strata, stratum 1 composed of the 132 smallest municipalities according to the value of the variable P75, and stratum 2 consisting of the other 68 municipalities. The strata sample sizes are $n_1 = n_2 = 30$. Compute r_{yz} and the approximate coefficient of variation for the estimator $\tilde{r}_{yz}$ using the following summary population data ($\tilde{r}_{yz}$ and u_k are as in Exercise 13.1):

Stratum h	$\sum_{U_h} y_k$	$\sum_{U_h} y_k^2$	$\sum_{U_h} z_k$	$\sum_{U_h} z_k^2$	$\sum_{U_h} y_k z_k$
1	164,404	311,741,580	62,604	33,192,750	84,898,976
2	163,181	483,526,473	73,618	85,512,874	188,159,967

Stratum h	$\sum_{U_h} u_k$	$\sum_{U_h} u_k^2$
1	-0.0422	0.0084170
2	0.0422	0.0006749

13.3. An *SI* sample of $n = 120$ municipalities was drawn from the MU281 population. Calculate a point estimate for the correlation, r_{yzU}, between the study variables CS82 ($=y$) and S82 ($=z$). Use the following sample data, where $\hat{u}_k$ is defined as in Exercise 13.1:

$$\textstyle\sum_s y_k = 1{,}057; \qquad \sum_s y_k^2 = 11{,}557$$
$$\textstyle\sum_s z_k = 5{,}624; \qquad \sum_s z_k^2 = 275{,}816$$
$$\textstyle\sum_s y_k z_k = 52{,}793; \qquad \sum_s \hat{u}_k^2 = 6.678 \cdot 10^{-4}$$

Also, compute the associated *cve*.

13.4. Verify the details of Examples 13.4.1 and 13.4.2, where the implicit differentiation method is used to estimate the covariance $S_{yz} = S_{yzU}$ and to derive a corresponding variance estimator.

13.5. Verify the details of Example 13.4.3, where the implicit differentiation method is used to estimate the finite population regression coefficient vector, $\mathbf{B}$, and the coefficient of multiple determination, R^2, and to derive the corresponding variance estimators.

13.6. Verify the result used in the derivation of equation (13.5.10), namely, that

$$\operatorname{tr}(\mathbf{A}_i) = n_i \sum_{j=1}^{c} v_{ijj}/P_j$$

13.7. For the case $c = 2$, consider testing the null hypothesis $H_0\colon \mathbf{P}_1 = \mathbf{P}_2$ against $H_1\colon \mathbf{P}_1 \neq \mathbf{P}_2$, with notation as in Section 13.5.1. In this case,

$$P_{12} = 1 - P_{11}, \qquad P_{22} = 1 - P_{21}, \qquad P_2 = 1 - P_1$$

and

$$\hat{P}_{12} = 1 - \hat{P}_{11}, \qquad \hat{P}_{22} = 1 - \hat{P}_{21}, \qquad \hat{P}_{+2} = 1 - \hat{P}_{+1}$$

(a) Show that the Wald statistic takes the form

$$X_{WH}^2 = (\hat{P}_{11} - \hat{P}_{21})^2/[\hat{V}(\hat{P}_{11}) + \hat{V}(\hat{P}_{21})]$$

(b) Show that the modified chi-square statistic in equation (13.5.5) can be written

$$X_{PH}^2 = \frac{n_1 n_2}{n_1 + n_2} \cdot \frac{(\hat{P}_{11} - \hat{P}_{21})^2}{\hat{P}_{+1}(1 - \hat{P}_{+1})}$$

(c) Show that the estimated weighted average design effect can be written

$$\hat{\bar{\delta}} = \frac{\hat{V}(\hat{P}_{11}) + \hat{V}(\hat{P}_{21})}{[(n_1 + n_2)/n_1 n_2]\hat{P}_{+1}(1 - \hat{P}_{+1})}$$

Observe that the numerator is the estimated variance of $\hat{P}_{11} - \hat{P}_{21}$, whereas the denominator is an estimate of that variance, assuming *SIR* sampling from both populations and assuming that the null hypothesis is true.

(d) Using the results in (a) to (c), deduce that $X_{WH}^2 \equiv X_{MH}^2$, where X_{MH}^2 is given by (13.5.9).

(e) Propose an alternative test statistic whose large sample distribution, under the null hypothesis, is approximately that of the standard normal variate.

13.8. To study if the political power structure is different in large and small municipalities, the MU284 population was subdivided into two populations. The first population, called "Large," consisted of the 140 largest municipalities as determined by the value of the variable P75; the second, called "Small," consisted of the remaining 144 municipalities. From each population, independent *SI* samples of 100 municipalities each were drawn. Every sampled municipality was classified into one of the following mutually exclusive categories defined by the strength of the Social Democrat Party in the municipal assembly:

Strong majority (if SS82 $> 50\% \times$ S82)

Weak majority (if max[CS82, S82-SS82-CS82] $<$ SS82 $\leq 50\% \times$ S82)

No majority

The two samples were distributed as follows:

Strength of the Social Democrat Party	Sample from population: Large	Small
Strong majority	38	32
Weak majority	42	31
No majority	20	37
Total	100	100

The null hypothesis is that the two populations do not differ with respect to the distribution of Social Democrat seats. Carry out the test of the null hypothesis using each of the test statistics X_{WH}^2, X_{PH}^2, and X_{MH}^2. Compare the conclusions.

PART IV

A Broader View of Errors in Surveys

CHAPTER 14

Nonsampling Errors and Extensions of Probability Sampling Theory

14.1. Introduction

The preceding chapters show that the probability sampling approach can be adapted to a wide variety of situations. We now examine the probability sampling approach, its logic, advantages, limitations, and possible extensions. This chapter is in large part a nontechnical overview. A brief history of the probability sampling approach is presented in Section 14.2. The importance of a measurable probability sampling design is discussed in Section 14.3. In practice, costs or other reasons may force the statistician to use sampling that does not meet all of the requirements of the probability sampling approach. Examples are discussed in Section 14.4; such methods should be used with care. Section 14.5 outlines a model-based approach to estimation. In Sections 14.7 to 14.10, we examine the survey operations and review the nonsampling errors that they may generate, for instance, frame error, measurement error, coding, editing and imputation error, and nonresponse error. We return to a more formal presentation in Chapters 15 and 16, which are devoted to estimation in the presence of nonresponse and to measurement error models.

14.2. Historic Notes: The Evolution of the Probability Sampling Approach

The idea that a partial investigation, that is, observation of a sample of limited size, could be a serious contender to complete enumeration of the population of a country goes back to the end of the 19th century. The 1891 census in

Norway used what became known as the *representative method*. It was argued that a partial investigation could give reliable results, provided that the observations form a representative picture of the whole field of study. The representative method of sample surveys is described in Kiaer (1897). The sample should reflect the target population on important characteristics. Now, representativity could be obtained in two ways: (a) by randomized selection from the entire population (a large simple random sample will automatically reflect the finite population) and (b) by purposive selection of a sample that mirrors the population on chosen variables. In the first decades of the 20th century, sample surveys were carried out by one or the other of these two representative methods, and the question of which was "better" was not settled until the early 1930s. A highly influential paper by Neyman (1934) contains a number of important conclusions. One is the optimal allocation result for a stratified random sample (see Section 3.7.4), anticipated by Tschuprow (1923). For an *STSI* sample of fixed size, the allocation to stratum h should be determined as

$$n_h \propto N_h S_{yU_h}$$

Consequently, the sampling fraction in stratum h is

$$\frac{n_h}{N_h} \propto S_{yU_h}$$

Since n_h/N_h is the inclusion probability for any element k in stratum h, Neyman had in fact shown that unequal probability selection pays when the stratum variances are unequal. Neyman (1934) goes on to discuss confidence intervals for survey estimates:

> If we are interested in a collective character X of a population π and use methods of sampling and of estimation, allowing us to ascribe to every possible sample, Σ, a confidence interval $X_1(\Sigma)$, $X_2(\Sigma)$ such that the frequency of error in the statements
>
> $$X_1(\Sigma) \leq X \leq X_2(\Sigma)$$
>
> does not exceed the limit $1 - \varepsilon$ prescribed in advance whatever the unknown properties of the population, I should call the method of sampling representative and the method of estimation consistent.

Neyman used the term "representative" with a new meaning. Stratified random sampling is representative, according to Neyman, even though sampling rates may vary widely between strata. The realization that equal probability sampling is not necessary for valid conclusions was a great step forward. Moreover, the confidence intervals are valid "whatever the unknown properties of the population." This distribution-free feature of the approach is appealing.

Neyman advocated the probability sampling approach, and was critical of the purposive sample selection approach. In the 1934 paper, he discussed in detail a purposive sampling method used by the Italian statisticians Gini and Galvani. To estimate the average rate of natural increase of the population in

Italy, Gini and Galvani had purposely selected a sample of 29 out of 214 districts in such a way that the weighted mean of a control variable in the sample was equal to the weighted mean of that same variable in the whole population. In modern terminology this procedure is called balanced sampling. Neyman concludes that

> ··· the consistency of the estimate suggested by Gini and Galvani, based on a purposely selected sample, depends upon hypotheses which it is impossible to test except by an extensive enquiry. If these hypotheses are not satisfied, which I think is a rather general case, we are not able to appreciate the accuracy of the results obtained. Thus this is not what I should call a representative method. Of course it may give sometimes perfect results, but these will be due rather to the uncontrollable intuition of the investigator and good luck than to the method itself.

Neyman's (1934) paper is a powerful rejection of purposive sampling and also sets the basis for estimation from cluster samples.

Once it was established that valid conclusions about finite populations did not require equal inclusion probabilities, a rapid development followed. It was realized that each sampling unit can be given its own individual inclusion probability. Sampling methodologists in England, India, and the United States were at the forefront of these developments. Hansen and Hurwitz (1943) introduced pps with replacement sampling in connection with multistage sampling. From there the step was not long to the general unequal probability sampling formulation (the π estimator) of Horvitz and Thompson (1952) that is used extensively in this book.

The probability sampling approach has held a dominant position in survey sampling until our time, although not always without heated debate. Hansen, Madow, and Tepping (1983) defend the approach and evoke the dangers of model-dependent estimation. The discussion at the end of their paper gives a good idea of different points of view.

For a brief history of the development of survey sampling, and in particular the probability sampling approach, see the review papers by Hansen, Dalenius and Tepping (1985) and Rao and Bellhouse (1990). Books that discuss the theoretical foundations of inference from survey data with an emphasis on the developments in recent decades are Cassel, Särndal and Wretman (1977) and Chaudhuri and Vos (1988).

14.3. Measurable Sampling Designs

One important requirement for a probability sampling design (see Section 1.4) is that

$$\pi_k > 0 \quad \text{for all} \quad k \in U \tag{14.3.1}$$

If all π_k are known, this condition is necessary and sufficient for obtaining an unbiased estimator of $t = \sum_U y_k$; for a proof, see for example Cassel, Särndal, and Wretman (1977, p. 68). If $\pi_k = 0$ for certain elements $k \in U$, we can estimate without bias the y-total only for that subset of U for which $\pi_k > 0$.

A probability sampling design for which both (14.3.1) and

$$\pi_{kl} > 0 \quad \text{for all} \quad k \neq l \in U \tag{14.3.2}$$

hold is called a *measurable probability sampling design.* For such a design, it is possible to calculate, from the sample itself, an unbiased or nearly unbiased estimate of the sampling variance of an estimator. For example, for the π estimator, $\hat{t}_\pi = \sum_s \check{y}_k = \sum_s y_k/\pi_k$, an unbiased variance estimator is

$$\hat{V}(\hat{t}_\pi) = \sum\sum_s [(\pi_{kl} - \pi_k\pi_l)/\pi_{kl}]\check{y}_k\check{y}_l \tag{14.3.3}$$

and for the regression estimator, $\hat{t}_r = \sum_s g_{ks}\check{y}_k$, an approximately unbiased variance estimator recommended in Section 7.2 is

$$\hat{V}(\hat{t}_r) = \sum\sum_s [(\pi_{kl} - \pi_k\pi_l)/\pi_{kl}](g_{ks}\check{e}_k)(g_{ls}\check{e}_l) \tag{14.3.4}$$

with g_{ks} given by (7.2.9). The variance estimators shown by equations (14.3.3) and (14.3.4) are called *design-based.* They can be used to create a *design-based confidence interval,*

$$\hat{t} \pm z_{1-\alpha/2}[\hat{V}(\hat{t})]^{1/2} \tag{14.3.5}$$

where $\Pr(Z \geq z_{1-\alpha/2}) = \alpha/2$, and Z is $N(0, 1)$.

The word "design-based" refers to the fact that, except for very small sample sizes, the frequency with which the parameter is included in the interval is roughly equal to the prescribed $1 - \alpha$ under repeated draws of samples s with the fixed sampling design. This holds independently of the shape of the finite population. Another term often used for an interval with these properties is *valid confidence interval*, and (14.3.3) and (14.3.4) are also referred to as *valid variance estimators.* A confidence interval is invalid if it does not cover the unkown parameter value roughly 100 $(1 - \alpha)$% of the time, in repeated samples, where $1 - \alpha$ is the intended confidence level.

The distribution-free property of design-based confidence intervals is attractive but often entails trade-offs. There may exist alternative, biased estimators with smaller variance and smaller mean square error (MSE) than the estimator that leads to a valid confidence interval. Nevertheless, most sampling statisticians prefer a valid confidence interval, because such an interval is an objective statement of the precision obtained from the sample, regardless of the shape of the population. If a calculated interval is wide, it is a sign that steps need to be taken to improve the estimation, for example, by increased sample size or by better use of auxiliary information.

A good illustration is estimation for small domains. If $\hat{t}_d$ is an estimator of the domain total t_d, the mean square error

$$\text{MSE}(\hat{t}_d) = V(\hat{t}_d) + \{\text{Bias}(\hat{t}_d)\}^2$$

is an average measure of how close the estimator is to the parameter. Several studies have shown that the synthetic estimator (10.8.5) is excellent for many domains when the mean square error is used as the criterion. That is, the synthetic estimator has the good property of coming closer on the average to the unknown parameter than most competing estimators. But if the bias is considerable, there is little sense in calculating a confidence interval. We can remove the bias of the synthetic estimator, in the manner of equation (10.8.2), and then obtain a valid confidence interval. This interval may be quite wide, but it states objectively the precision that can be obtained from the data.

A sampling design may be nonmeasurable for one of the following reasons:

i. A probability sampling design is used for sample selection, but the design does not respect the requirement (14.3.2).
ii. The design is not a probability sampling design.

Case (i) applies for several designs that may be of interest in practice, for example:

a. Stratified sampling with a single element drawn from one or more strata (see Section 3.7).
b. Systematic sampling with a single random start (see Section 3.4.4).
c. Selection of a single cluster of elements.

In case (ii), the probability distribution $p(s)$ of the different samples s is unknown. The inclusion probabilities needed for the confidence interval calculation are then unknown. Nonprobability sampling designs are nevertheless sometimes used in practice (see Section 14.4).

14.4. Some Nonprobability Sampling Methods

Surveys are sometimes carried out by nonprobability sampling methods, because these methods often lead to lower costs. For example, a good frame may not be available, and the cost of constructing one would be prohibitive. Let us review some of the more important nonprobabilistic methods.

Scientists often make conclusions from arbitrary or fortuitous samples. Much research in medical, biological, and physical sciences is based on experimental units that are gathered haphazardly or that "happen to be handy." The scientist will often implicitly assume that the units are typical for a larger universe about which conclusions are desired. Often such an assumption is ill founded.

An example of a simplistic nonprobability selection is a sample composed of the first 300 persons encountered in the street. There is no attempt to introduce random selection. No one would have much faith in conclusions drawn from such a sample, if the target population is the whole city or the whole country.

Expert choice sampling, a form of *judgment sampling*, is a more developed form of nonrandom selection. For example, suppose an "expert" designates five "typical cities" where data are to be collected for conclusions about the whole country. With skill and ingenuity on the part of the judgment sampler, a fairly good sample *may* result, but there is no way to be sure. The design is not measurable, so valid variance estimates cannot be calculated. The method lacks objectivity; a different expert would probably pick different cities.

More sophisticated nonprobability sampling methods include (i) quota sampling and (ii) balanced sampling. Both (i) and (ii) appeal to notions of representativity similar to those prevailing at the turn of the century, as discussed in Section 14.2. Since the inclusion probabilities are ordinarily unknown in these methods, one cannot determine the design-based properties of the estimators. Many elements may have zero inclusion probability. Accurate estimates are not excluded in either method, but it is ordinarily impossible to get an objective assessement of the precision.

Quota sampling is often used in market research. The basic principle is that the sample contains a fixed number of elements in specified population cells. Suppose that the population is divided according to three controls, sex, age group, and geographic area. With two sexes, four age groups, and six areas, we get a total of $2 \times 4 \times 6 = 48$ population cells. In each cell, the investigator fixes a number (a "quota") of elements to be included in the sample. Now the interviewer simply "fills the quotas," that is, interviews the predetermined number of persons in each of the quota cells. These may be the first persons encountered, or it may be left to the interviewer to exercise judgment in the quota selection. The method resembles stratified sampling, but the selection within strata is nonprobabilistic. Because that selection is nonprobabilistic, there is neither an unbiased point estimate nor a valid variance estimate within the cell. Refusals to answer may occur, but the desired quota in each cell can always be attained by approaching new individuals. ("Rare cells" may require considerable effort.) Thus it is relatively simple in quota sampling to eliminate nonresponse but the problem of selection bias from nonresponse remains.

Different methods are in use for fixing the quotas n_h. A simple method is to take the n_h proportional to the population cell counts N_h, assuming these latter are known. An alternative method is an adaptation of Neyman's rule, namely to take n_h proportional to $N_h\sigma_h$, if $\sigma_1, \ldots, \sigma_H$ are assumed values for the standard deviation of y in the different cells.

To estimate $t = \sum_U y_k$ from a quota sample, one can use

$$\hat{t} = \sum_{h=1}^{H} N_h \bar{y}_{s_h} \tag{14.4.1}$$

where h is the cell index, $h = 1, \ldots, H$, and $\bar{y}_{s_h}$ is the sample mean in cell h. The choice of the controls is crucial. Well-chosen controls may go a long way toward eliminating selection bias. Experienced quota samplers have a good feel for how the quota cells should be defined. Low cost is often an advantage with quota sampling. However, there is no objective method to assess the bias and the precision of quota sampling estimates. The quota sampling estimator shown in equation (14.4.1) can be justified by a model for the y-values, and the statistical properties can be worked out using that model. This is done in Example 14.5.2 below. A weakness of this argument is that the model may be false, which renders the model-based inferences invalid.

Quota sampling sometimes combines probabilistic and nonprobabilistic techniques, as when primary units (city blocks or districts, say) are first selected with a probability sampling scheme, and elements are then selected by the quota method. In each primary unit selected, the interviewer is required to meet the specified quotas.

Balanced sampling resembles quota sampling in that control variables are used to guide the sample selection. Suppose $x_1, \ldots, x_J$ are control variables, qualitative or quantitative, with known population totals. The objective then is to select a sample such that the sample mean of each control variable equals (or closely approximates) its known population counterpart, that is

$$\bar{x}_{js} \doteq \bar{x}_{jU} \tag{14.4.2}$$

for $j = 1, \ldots, J$. In practice, several samples may have to be drawn (say, with simple random sampling) and successively rejected until one is obtained that respects (14.4.2). If the x variables are strongly correlated with y, then $\bar{y}_s$ should also be close to $\bar{y}_U$, so one can simply use $N\bar{y}_s$ as an estimator of t. A more controlled procedure for obtaining a balanced sample was given by Deville, Grosbras and Roth (1988). They use a sequential probabilistic selection procedure that leads in the end to a balanced sample. The rules of probability sampling are respected and the advantages of balanced sampling are exploited at the same time.

Probability sampling requires that $\pi_k > 0$ for *all* $k \in U$. There are sampling methods in current use that employ probabilistic selection with $\pi_k > 0$ for part of the population U, whereas $\pi_k = 0$ for the remainder of U. Such methods take an intermediate position between probability sampling and nonprobabilistic selection with π_k that are unknown throughout the population. One of these techniques is *cut-off sampling*, which we now consider. In cut-off sampling there is a usually deliberate exclusion of part of the target population from sample selection. This procedure, which leads to biased estimates, is justified by the following argument: (i) that it would cost too much, in relation to a small gain in accuracy, to construct and maintain a reliable frame for the entire population; (ii) that the bias caused by the cut-off is deemed negligible. In particular, the procedure is used when the distribution of the values $y_1, \ldots, y_N$ is highly skewed, and no reliable frame exists for the small elements. Such populations are often found in business surveys. A considerable portion of the population may consist of small business enterprises whose contribution to the total of a variable of interest (for example, sales) is modest or negligible. At the other extreme, such a population often contains some giant enterprises whose inclusion in the sample is virtually mandatory in order not to risk large error in an estimated total. One may decide in such a case to cut off (exclude from the frame, thus from sample selection) the enterprises with few employees, say five or less. The procedure is not recommended if a good frame for the whole population can be constructed without excessive cost.

Let U_c denote the cut-off portion of the population; let U_0 be the rest of the population, from which we assume that a probability sample is selected in the normal way. The whole population is thus $U = U_0 \cup U_c$. Each element in the cut-off portion has zero inclusion probability; that is, $\pi_k = 0$ for all $k \in U_c$. Let $\hat{t}_0$ be an estimator of $t_0 = \sum_{U_0} y_k$, for example, $\hat{t}_0 = \sum_{s_0} y_k/\pi_k$. But we need an estimator of the whole total $t = \sum_U y_k$. How can this be achieved? Two possible courses of action are as follows.

The statistician may be willing to assume that $t_c = \sum_{U_c} y_k$ is a negligible portion of the whole total $t = \sum_U y_k$. If $\hat{t}_0$ by itself is used to estimate t, the relative bias is

$$\frac{E(\hat{t}_0) - t}{t} = -\frac{t_c}{t} = -\left(1 - \frac{t_0}{t}\right)$$

which is negative, but negligible under the assumption. We assume that y is an always positive variable.

A second approach is to use a ratio adjustment for the cut-off. Let x be an auxiliary variable, for example, the variable of interest measured for the entire population at an earlier date, or some other known variable roughly proportional to the current variable of interest y. Let

$$R_{U_0} = \frac{\sum_{U_0} y_k}{\sum_{U_0} x_k}$$

and let

$$\hat{R}_{U_0} = \frac{\sum_{s_0} y_k/\pi_k}{\sum_{s_0} x_k/\pi_k}$$

be the consistent estimator of R_{U_0}, based on the probability sample from U_0. To extend the conclusions to the whole population, an unverifiable assumption is necessary. Assume that $R_{U_0} = R_U = \sum_U y_k / \sum_U x_k$. Then, $\hat{R}_{U_0}$ can serve to estimate R_U as well, and by ratio adjustment we arrive at

$$\hat{t}_{\text{cut}} = \left(\textstyle\sum_U x_k\right)\hat{R}_{U_0}$$

as an estimator of the whole current total $t = \sum_U y_k$, assuming $\sum_U x_k$ or a close estimate is available. The relative bias is approximately

$$\frac{E(\hat{t}_{\text{cut}}) - t}{t} \doteq \frac{R_{U_0}}{R_U} - 1$$

which can be positive or negative. It is zero if the assumption $R_U = R_{U_0}$ holds. This assumption is one that the statistician may be more inclined to make than the assumption in the first approach that t_c/t is negligible.

Remark 14.4.1. Cut-off estimation should be used with great caution, especially for rapidly changing populations. The set U_c, which should contain only small elements, may in fact contain elements with large y_k-values caused by sudden growth. In such cases, neither of the above courses of action will lead to good estimates, and the available data will not in any way indicate or warn that poor estimates may result.

Remark 14.4.2. Involuntary cut-off occurs when the frame has undercoverage. The part of the target population not listed in the frame cannot be selected. An unknown segment of the population (both large and small elements) may be omitted from selection (see also Section 14.7.2).

Remark 14.4.3. In a two-stage element sampling design, the inclusion probability of the kth element is $\pi_k = \pi_{Ii}\pi_{k|i}$ where π_{Ii} is the inclusion probability of the ith primary sampling unit and $\pi_{k|i}$ the inclusion probability of k in subsampling from the ith primary sampling unit. According to the probability sampling requirements, we must have

$$\pi_k = \pi_{Ii}\pi_{k|i} > 0 \tag{14.4.3}$$

for all k in the target population. Thus $\pi_{Ii} > 0$ is required for any cluster i containing at least one element k. Suppose the elements k are households, and the primary sampling units are geographic areas labeled $i = 1, \ldots, N_I$. To respect (14.4.3), the sampling must be done so that all areas with at least one household have nonzero probability. A prohibitive cost may be associated with interviewing a few households scattered in geographically remote areas. One approach is to exclude remote-area households from the target population and make valid estimates for a somewhat reduced population of households. Another approach, considerably riskier, is to apply cut-off reasoning, and be forced to depend on unverifiable assumptions in order to obtain estimates for the entire population of households.

14.5. Model-Based Inference from Survey Samples

Model-based reasoning starts from a specification of a model for the N-dimensional distribution of $\mathbf{Y} = (Y_1, \ldots, Y_k, \ldots, Y_N)$, where Y_k is a random variable tied to the kth element. Let this model be denoted ξ. It is often called a superpopulation model. The actual finite population vector, $\mathbf{y} = (y_1, \ldots, y_k, \ldots, y_N)$, is considered to be a realization of $\mathbf{Y}$. Suppose the objective is to estimate the total $t = \sum_U y_k$. We assume that a sample s has been drawn. The realized value y_k of the random variable Y_k is observed for the elements $k \in s$, nonobserved for the elements $k \in U - s$. The recorded y data are thus $\{y_k: k \in s\}$. An estimator of t is a function of these data, that is,

$\hat{t} = \hat{t}(\{y_k: k \in s\})$. Using the specifications of ξ, the distribution of $\hat{t} - t$ can be derived *for the given sample* s. The model-based mean square error of $\hat{t} - t$ can be obtained and estimated, which leads to a model-based prediction interval for t. Examples of this procedure are given later in this section.

A model-based inference is interpreted by visualizing a long series of realizations of the finite population vector $\mathbf{Y} = (Y_1, \ldots, Y_k, \ldots, Y_N)$, for the fixed s. The inference is tied to the particular s that was realized, and not to other samples. Some statisticians see this as a methodologic advantage. However, to visualize repeated finite populations is considered artificial by others, since in reality there is only one finite population. In the following we use the symbol y_k to denote both the random variable Y_k associated with the kth element and a particular value taken by this random variable.

The estimator $\hat{t}$ is said to be *model unbiased* if, given s,

$$E_\xi[(\hat{t} - t)|s] = 0$$

A quantity that measures the variability is the *model mean square error*

$$\text{MSE}_\xi(\hat{t}) = E_\xi[(\hat{t} - t)^2|s]$$

Suppose an estimator, $\widehat{\text{MSE}}_\xi(\hat{t})$, can be obtained from the sample data. A *prediction interval* for t at the approximate level $(1 - \alpha)$, *for the particular* s, is then obtained with the aid of the model unbiased $\hat{t}$ as

$$\hat{t} \pm z_{1-(\alpha/2)}[\widehat{\text{MSE}}_\xi(\hat{t})]^{1/2} \tag{14.5.1}$$

This is a prediction interval rather than a confidence interval, since in this approach $t = \sum_U y_k$ is itself a random variable. As usual, $z_{1-\alpha/2}$ is the constant value exceeded with probability $\alpha/2$ by the unit normal variate. The procedure relies implicitly on the assumption that the central limit theorem holds for $\hat{t} - t$.

The interpretation of $1 - \alpha$ in this case is that, given s, the difference $\hat{t} - t$ is within the limits $\pm z_{1-(\alpha/2)}[\widehat{\text{MSE}}_\xi(\hat{t})]^{1/2}$ for approximately $100(1 - \alpha)\%$ of all vectors $(y_1, \ldots, y_k, \ldots, y_N)$ that can be generated under the specified N-dimensional model distribution ξ. Note that the interval is actually for the finite population total, t, and not for some parameter of the hypothetical superpopulation (c.f., Section 13.6.) The interval shown in (14.5.1) is a *model-based interval* for t. It is *not* a valid confidence interval in the design-based sense explained in Section 14.3. In contrast to (14.3.5), the interval shown in (14.5.1) does not usually guarantee coverage of the unknown parameter at the rate of $1 - \alpha$ in repeated samples, independently of the form of the population.

Proponents of model-based inference advocate randomized selection of the sample as a safeguard against selection bias, but the randomization probabilities play no role in the inference. Misspecifications of the model will lead to invalid conclusions, particularly for large samples. Although the variance diminishes as n grows, a squared bias term remains at an essentially constant level. This phenomenon is well illustrated in Hansen, Madow and Tepping (1983). Brewer (1963b), Royall (1970), and Royall and Herson (1973a, b) lay the foundations for the model-based approach. This approach has contributed greatly to the understanding of the role of modeling in survey sampling and to the development of model-assisted, design-based methods.

EXAMPLE 14.5.1. Consider the ratio estimator

$$\hat{t}_{ra} = (\textstyle\sum_U x_k)\frac{\sum_s y_k}{\sum_s x_k}$$

We ignore how the sample s was drawn. The statistical properties of $\hat{t}_{ra}$ can, however, be derived from a model. Suppose that the modeler is willing to assume that the finite population values $y_1, \ldots, y_N$ are generated by the model stating that the y_k are independent random variables with

$$\begin{cases} E_\xi(y_k) = \beta x_k \\ V_\xi(y_k) = \sigma^2 x_k \end{cases} \tag{14.5.2}$$

As is easily shown, $\hat{t}_{ra}$ is then model unbiased for $t = \sum_U y_k$, that is

$$E_\xi[(\hat{t}_{ra} - t)|s] = 0$$

It is left as an exercise to show that the model mean square error is given by

$$\text{MSE}_\xi(\hat{t}_{ra}) = \frac{\bar{x}_U \bar{x}_{U-s}}{\bar{x}_s} N^2 \frac{1-f}{n} \sigma^2 \tag{14.5.3}$$

where $\bar{x}_U$, $\bar{x}_s$, and $\bar{x}_{U-s}$ denote, respectively, the mean of x in U, s, and $U - s$ (the nonsample part of U). Here, $f = n/N$ is the sampling fraction. To arrive at an estimator of $\text{MSE}_\xi(\hat{t}_{ra})$, we only need to specify an estimator $\hat{\sigma}^2$ of σ^2. Various possibilities are open. From the theory of weighted least squares we obtain

$$\hat{\sigma}^2 = \frac{1}{n-1}\sum_s \frac{(y_s - bx_k)^2}{x_k}$$

with $b = \bar{y}_s/\bar{x}_s$. This is a model unbiased estimator of σ^2 under the model expressed by (14.5.2), that is, $E_\xi(\hat{\sigma}^2) = \sigma^2$; however a weakness is that it can be severely biased if some other model holds, so that $V_\xi(y_k)$ is no longer proportional to x_k. This led Royall and Eberhardt (1975) to seek model-based variance estimators for the ratio estimator that are more robust to misspecification of the model. They suggested

$$\hat{\sigma}^2 = \frac{1}{\bar{x}_s\{1 - (cv_{xs})^2/n\}} \frac{\sum_s (y_s - bx_k)^2}{n-1} \tag{14.5.4}$$

where $cv_{xs} = S_{xs}/\bar{x}_s$ is the coefficient of variation of x in s. This is also unbiased for σ^2 under the model (14.5.2) and has the additional advantage of being approximately unbiased under certain models in which $V_\xi(y_k)$ is not proportional to x_k. Substituting the estimator (14.5.4) for σ^2 in $\text{MSE}_\xi(\hat{t}_{ra})$, we obtain

$$\widehat{\text{MSE}}(\hat{t}_{ra}) = \frac{\bar{x}_U \bar{x}_{U-s}}{\bar{x}_s^2} N^2 \frac{1-f}{n} \frac{1}{1-(cv_{xs})^2/n} \frac{\sum_s (y_k - bx_k)^2}{n-1} \quad (14.5.5)$$

This is very nearly identical to the variance estimator (7.3.5) obtained by the probability sampling approach, in the case of *SI* sampling, if we note that $\bar{x}_{U-s} \doteq \bar{x}_U$ and $1-(cv_{xs})^2/n \doteq 1$.

For the sample selection, Royall and Eberhardt (1975) favor balancing on the x variable. That is, one should select a sample s that satisfies $\bar{x}_U \doteq \bar{x}_s$. This has the effect of controlling the model bias of $\hat{t}_{ra}$ that occurs if (14.5.2) is false and instead the true model is one of a specific class of alternative models.

The estimator shown in (14.5.5) has an interesting property which is related to a conditioning on the value of $\bar{x}_s$. Suppose we fix a narrow interval of values for $\bar{x}_s$. Denote this interval as A. Draw, say, 1,000 simple random samples, and look in particular at those samples for which $\bar{x}_s \in A$. Empirical findings of Royall and Cumberland (1981a) show that the average of (14.5.5) over the samples s satisfying $\bar{x}_s \in A$ is near the average of $(\hat{t}_{ra} - t)^2$ over samples s conditioned in the same way. This is true no matter where on the axis the narrow interval A is situated. That is, $\widehat{\text{MSE}}(\hat{t}_{ra})$ given by (14.5.5) has a property of "tracking" $\text{MSE}_\xi(\hat{t}_{ra})$ when both quantities are plotted as functions of the value $\bar{x}_s$ that one is conditioning on.

Example 14.5.2. When a quota sample is selected (Section 14.4), an often used estimator of $t = \sum_U y_k$ is

$$\hat{t}_{\text{qu}} = \sum_{h=1}^{H} N_h \bar{y}_{s_h}$$

where h refers to a quota cell, $h = 1, \ldots, H$. Since quota sampling is nonprobabilistic, we cannot derive any design-based statistical properties of $\hat{t}_{\text{qu}}$. However, we can formulate a suitable model that (i) generates the estimator $\hat{t}_{\text{qu}}$ and (ii) leads to an estimator of the model mean square error. Consider the model ξ under which the y_k are independent random variables with

$$E_\xi(y_k) = \mu_h$$

$$V_\xi(y_k) = \sigma_h^2$$

for all $k \in U_h$; $h = 1, \ldots, H$. Quota samplers may feel that this adequately reflects the quota cells that they have created. It is then easy to show that $\hat{t}_{\text{qu}}$ is model unbiased for t under the model ξ, that is, $E_\xi[(\hat{t}_{\text{qu}} - t)|s] = 0$. It is left as an exercise to show that the model mean square error is

$$\text{MSE}_\xi(\hat{t}_{\text{qu}}) = \sum_{h=1}^{H} N_h^2 \frac{1-f_h}{n_h} \sigma_h^2$$

where $f_h = n_h/N_h$ is the observed fraction in cell h. A model unbiased estimator of the cell variance is

$$\hat{\sigma}_h^2 = \frac{1}{n_h - 1} \sum_{s_h} (y_k - \bar{y}_{s_h})^2$$

When this expression is substituted for σ_h^2 in $\text{MSE}_\xi(\hat{t}_{\text{qu}})$, we obtain precisely the variance estimator (3.7.9) of the probability sampling approach. This is a formal equivalence only. There is an essential difference between the two approaches. With simple random probability sampling within the cell, valid confidence intervals are obtained regardless of whether some model about the y_k-values within the cell holds true or not.

Remark 14.5.1. The model-based approach, illustrated by the preceding two examples, offers an alternative to design-based inference for survey populations. Advocates of the model-based approach do not reject randomized selection per se. However, they do oppose the use of the randomization distribution to make statistical inference on the grounds that such inferences, although assumption free, refer to repeated draws of samples, instead of to the particular sample that was actually drawn.

14.6. Imperfections in the Survey Operations

14.6.1. Ideal Conditions for the Probability Sampling Approach

The *survey operations* identified in Section 1.7 were grouped into five phases:

1. Sample selection, including frame construction.
2. Data collection, that is, contacting and observing elements in the selected sample.
3. Data processing, including coding, editing, imputation, and outlier detection and treatment.
4. Estimation and analysis, including point estimation, variance estimation, and confidence intervals.
5. Dissemination of the survey results.

The probability sampling approach assumes ideally that we can carry out the following actions:

a. Construct a perfect frame for the target population.
b. Mechanically select a sample from that frame, in a way that obeys the probability sampling design.
c. Observe the true value for each study variable, for every element k in the sample s.

d. Process the data without introducing errors.
e. Use the processed, correct data to make valid inferences about the finite population: point estimation, variance estimation, and confidence interval calculation.

These represent *ideal conditions for the probability sampling approach.* When the ideal conditions hold, the sampling error is the only error in the survey estimates. In the absence of all nonsampling errors, the randomization distribution (the sampling design) contains the entire stochastic structure needed for statistical inference. Confidence statements can in most cases be obtained that are valid regardless of the shape of the sampled population. These methods do not require models or assumptions that are hard to verify.

In most surveys, the ideal conditions do not hold. The question that arises is: how can we take into consideration the errors introduced by imperfect survey operations in the statistical inference, which until now recognized sampling variation only? How can variance and bias components caused by nonsampling error be estimated? Because there is currently no complete theory of nonsampling errors, and because one does not ordinarily have sufficient knowledge about the distributions of the nonsampling errors, these questions can usually not be fully answered. However, as this and the following two chapters will show, useful partial answers can be given in many circumstances.

14.6.2. Extension of the Probability Sampling Approach

Statistical inference for finite populations becomes more complex when nonsampling errors are also taken into account. One reason is that the sampling distributions of the nonsampling errors are insufficiently known. In survey sampling as in most areas of statistics, the statistician must make model assumptions. In survey sampling, assumptions are necessary about the nonsampling errors and their distributions.

From now on, we use a framework for statistical inference that can be described as an *extended probability sampling approach.* In this approach, inferences about the finite population rely on the following:

a. The randomization distribution induced by the probability sampling.
b. One or several model assumptions about the different nonsampling errors.

In this framework, (a) gives structure to the sample-to-sample variation and (b) to the variation due to the different nonsampling errors. For example, a response mechanism can be assumed to describe how observations drop out due to nonresponse, and one or more measurement error models may be used. These models may contain unknown parameters that must first be estimated.

The known randomization distribution continues to play an important part in the inferences. But because (b) implies that unverifiable assumptions are also invoked, the conclusions about the finite population depend, for their validity, on the truth of those models. We no longer have the distribution-free property that pure probability sampling theory prides itself on. This will be particularly apparent in the chapters devoted to nonresponse (Chapter 15) and measurement errors (Chapter 16).

Sampling practitioners like to come as close as possible to the ideal conditions identified in Section 14.6.1. This is manifested in their insistance on near-perfect frames, on great efforts to keep the nonresponse low, on making sure that interviewers and other staff are well trained to maintain high quality of responses, and so on. Most government statistical offices carry out their surveys from carefully prepared, up-to-date frames and with considerable attention to the elimination of nonresponse and measurement errors. With nonsampling errors reduced to a minimum, the survey comes close to the ideal conditions.

Remark 14.6.1. In many surveys, the sampling error may be small compared to the nonsampling errors. In a large national survey with a sample size of 20,000 observations, the variance due to the sampling error is likely to be small for estimates made for the entire population or for large domains thereof. The nonsampling errors in such a survey could, however, be considerable, and significant attention and resources should be devoted to controlling these errors. The importance of nonsampling errors was recognized early in India, as seen in Mahalanobis (1944, 1946). Research on nonsampling errors has progressed considerably in recent years. The United States Bureau of the Census has shown leadership in the modeling of nonsampling error. Such modeling is considered in Chapter 16. Groves (1987) reviews the recent progress in nonsampling error research.

Our discussion has showed that inference in the presence of nonsampling errors must rely in part on modeling. Some have used this as an argument in favor of a *fully model-based approach* with the following components:

1. A model describing the generation of the population values. The finite population vector, $\mathbf{y} = (y_1, \ldots, y_k, \ldots, y_N)$ is viewed as a realization of the N-dimensional vector $\mathbf{Y} = (Y_1, \ldots, Y_k, \ldots, Y_N)$ and a model is formulated for the distribution of $\mathbf{Y}$.
2. One or several model assumptions about the nonsampling errors and their distributions.

The randomization distribution in step (1) of the extended probability sampling approach is replaced here by a model distribution for the y-values.

If $\hat{t}$ is a proposed estimator of $t = \sum_U y_k$, the distribution of the error $\hat{t} - t$ can be obtained, under the complete set of model assumptions, and for the actual set of respondents. Randomized selection of the sample may be used (to guard against selection bias), but randomization probabilities play no role in the inference. In this and subsequent chapters, we use the extended probability sampling approach.

14.7. Sampling Frames

14.7.1. Frame Imperfections

The frame consists of entities that we call units. The objective in a survey is to draw a sample of units from the frame, to identify the corresponding population elements (for example, individuals or farms or business firms) and to contact and observe these elements. The units in the frame are associated with the elements. It is this association that provides the observational access to the population through the frame. Using the ideas of Dalenius (1974) and Jessen (1978) we now describe more formally the relations between the sampling frame and the population.

The frame is a collection of M units,

$$F = \{F_1, \ldots, F_i, \ldots, F_M\}$$

We can think of the typical frame unit F_i as a record in a computer file or in a list. The typical element is denoted E_k. For any unit F_i and element E_k, we define an indicator quantity L_{ik} as follows:

$$L_{ik} = \begin{cases} 1 & \text{if an association exists between the frame unit} \\ & F_i \text{ and the element } E_k \\ 0 & \text{otherwise} \end{cases}$$

When $L_{ik} = 1$, we also say that a *link* exists between the frame unit F_i and the element E_k. Dalenius (1986) identifies three important types of association that can exist between the frame units and the population elements. A one-to-one relationship exists when every unit F_i is linked with one and only one element E_k and every element E_k is linked with one and only one unit F_i. A many-to-one relationship exists when each frame unit F_i is linked with one and only one element E_k, but an element E_k may have links with more than one unit F_i. An example of this occurs when the frame units are individuals and the elements are households. Another example occurs when some elements are listed more than once in the frame. A one-to-many relationship exists when each frame unit F_i is linked with several elements E_k and no element E_k is linked with more than one unit F_i. For example, the elements can be the farms in a country and the units can be administrative regions; F_i then corresponds to a cluster of farms. Jessen (1978) goes further and identifies six different types of association.

Different sets of elements need to be distinguished. The *target population*, denoted U, is the population that we wish to study, that is, the collection of elements E_k for which estimates are required. The *frame population*, denoted U_F, is the set of elements E_k with links to the frame F. Expressed in terms of links,

$$U_F = \{E_k\colon L_{ik} = 1 \text{ for some } i \text{ such that } F_i \in F\}$$

that is, each element E_k in U_F is linked with at least one unit F_i in the frame. Note that U_F may contain elements in the target population U as well as elements not in U (nonpopulation elements).

For example, suppose that the elements are individuals, and that the target population is individuals residing in Sweden. Any individual not residing in Sweden is a nonpopulation element. One reason why such a person may be in the frame population may be that at one point in time he was a resident, and he was not removed from the frame when he ceased to be one. Or the person may be deceased and was not removed from the frame. On the other hand, there may exist individuals that belong to the target population but are absent from the frame. The complete set of elements is

$$C = U \cup U_F$$

This set is decomposed into three pairwise disjoint parts as follows,

1. The set of target population elements with link to F,

$$U_{\text{link}} = U \cap U_F$$

2. The set of target population elements without link to F,

$$U_{\text{nolink}} = U - U_{\text{link}}$$

3. The set of nonpopulation elements with link to F,

$$\bar{U} = U_F - U_{\text{link}}$$

It is useful to introduce the counts

$$L_{i\cdot} = \sum_{k \in U} L_{ik}$$

which is the number of target population elements with links to the frame unit F_i and

$$L_{\cdot k} = \sum_{i \in F} L_{ik}$$

which is the number of frame units having a link with the element E_k.

Now we can describe sample selection from the frame. A sample is drawn from F by a specified sampling design. To the units in this sample are linked a set of elements s_F, where $s_F \subset U_F$. Some elements in s_F are target population elements, namely, the set $s = s_F \cap U$. The remainder, $\bar{s} = s_F - s$, consists of nonpopulation elements. The elements in $\bar{s}$ may no longer exist, or, if they do exist, they are of no interest to the survey. The different sets are illustrated in Figure 14.1.

We can distinguish the following five types of frame imperfection. The first three are easy to describe by the concept of links.

1. Some target population elements do not have a link to any unit in the frame and hence they cannot be selected for the survey. That is, U_{nolink} is nonempty, or, equivalently, $L_{\cdot k} = 0$ for some elements in U.

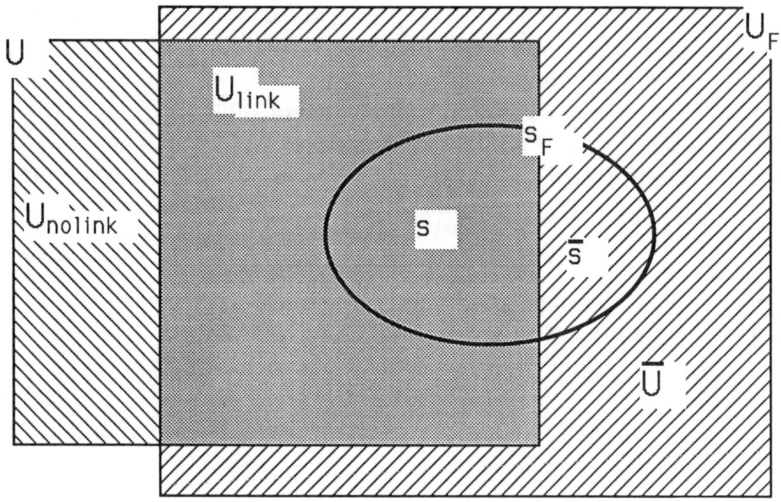

Figure 14.1. Target population and frame population.

2. Some units in the frame have links with nonpopulation elements, that is, $\overline{U}$ is nonempty, or, equivalently, $L_{\cdot k} > 0$ for some elements in $\overline{U}$.
3. Some target population elements have links to more than one frame object, that is, $L_{\cdot k} > 1$ for some elements in U.
4. The frame contains information that can be used for such purposes as stratification or selection with probability proportional to size or regression estimation, but the information is incorrect.
5. The frame information is not detailed enough or not current enough to allow access to some target population elements.

A given frame may have one or more of these imperfections. The first three are often referred to as undercoverage, overcoverage, and duplicate listings, respectively. The fourth type of imperfection may lead to some loss of precision in the survey estimates, but the validity of the inferences is usually not at stake. The confidence intervals may become wider but not invalid. The fifth type of frame imperfection is akin to undercoverage, but with the difference that it can be recognized from the frame itself that some elements cannot be accessed. Undercoverage, on the other hand, means that there are elements in the population without link to the frame. Their existence is unknown to the sampler.

Remark 14.7.1. We have distinguished the target population from the frame population. Sometimes other distinctions are made. Suppose that a social scientist wants to study a certain population of individuals. His/her perception of the target population may be unrealistic. Certain individuals are perhaps inaccessible or, on closer reflection, may not really be of interest to the study. In the planning stages, one of the statistician's responsibilities is to identify a population that is realistic to survey, that is, one for which a good frame can be constructed within given budget constraints. The term *survey population* is sometimes used roughly in the sense "population that can realistically be studied" or "the realistic target population." The survey population may thus differ from the original target population, and both may differ from the frame population that is finally obtained. Murthy (1983) discusses different concepts of population.

14.7.2. Estimation in the Presence of Frame Imperfections

Let us consider some special estimation techniques that may be used when there are frame imperfections. Some end of chapter exercises also examine such techniques. In cases (i) to (iii) below, s_F denotes a sample of elements obtained by probability sampling from the frame. We have $s_F \subset U_F$, where U_F is the frame population; the target population is U.

i. *The Frame Has Overcoverage but No Other Imperfections*

In this case, the target population U defines a domain of the frame population U_F. A probability sample s_F is selected from U_F. Since $U \subset U_F$, a positive inclusion probability π_k is guaranteed for each element in U, so that there is no conflict with the probability sampling requirements.

A first case of interest occurs when each sample element can be identified, at the time of observation, as either a member or a nonmember of the target population U. Nonmembers can be weeded out. The part of the sample belonging to U is identified as $s = U \cap s_F$, which is of random size. Valid parameter estimates can be obtained for characteristics of U. If the overcoverage is slight, so is the reduction in sample size caused by the overcoverage. The estimator $\sum_s y_k/\pi_k$ is unbiased for the total $\sum_U y_k$. The theory of domain estimation (Chapter 10) can be used to derive the properties of this estimator. Things become more complex when it is not possible to identify and eliminate nonmembers of U. Procedures for this case are reviewed by Lessler (1982).

ii. *The Frame Has Duplicate Listings but No Other Imperfections*

In this case, $L_{\cdot k} > 1$ for some target population elements. Units are sampled from the frame with probabilities that we assume positive and known (or possible to determine). If the number of links $L_{\cdot k}$ is known for every target population element E_k, it is possible in simple cases such as *SI* sampling from the frame to determine the inclusion probabilities for the target population elements, and standard estimation techniques can be used for the parameters of interest. Otherwise, the requirement of a known positive inclusion probability for every target population element will usually be violated. Special techniques must then be used. An example of such a technique is given in Exercise 14.10. If cost permits, an obvious remedy to the problem is to examine the frame before sample selection to identify and eliminate duplicate listings.

iii. *The Frame Has Undercoverage but No Other Imperfections*

In this case $U_F \subset U$. Undercoverage implies impossibility to obtain data from parts of the target population. Elements in the undercoverage portion, $U - U_F$, have $\pi_k = 0$, since they are excluded from the frame population. The implications for the properties of the survey estimates are more or less serious, depending on the parameter under estimation and on the approach used in the estimation. If s_F is a probability sample from U_F, then

$$\hat{t}_{F\pi} = \sum_{s_F} y_k/\pi_k$$

is an unbiased estimator of the total of U_F. But if $\hat{t}_{F\pi}$ is used as an estimator of the target population total, $t = \sum_U y_k$, there will clearly be a negative bias (assuming $y_k > 0$ for all k). The use of a well-chosen ratio adjustment may reduce the bias. Supposing that $\sum_U x_k$ is a known auxiliary total, we can construct the estimator

$$\hat{t}_{Fra} = \sum_U x_k \frac{\sum_{s_F} y_k/\pi_k}{\sum_{s_F} x_k/\pi_k}$$

Note the formal resemblance with the ratio adjusted estimator proposed for cut-off sampling in Section 14.4. It is easy to show that the relative bias is given approximately by

$$\frac{E(\hat{t}_{Fra}) - t}{t} \doteq \frac{R_{U_F}}{R_U} - 1 \tag{14.7.1}$$

where

$$R_{U_F} = \frac{\sum_{U_F} y_k}{\sum_{U_F} x_k}; \qquad R_U = \frac{\sum_U y_k}{\sum_U x_k}$$

The bias can be positive or negative. In either case, it is unknown. If the population missed is like the population covered in the sense that $R_{U_F} = R_U$, then $\hat{t}_{Fra}$ is approximately unbiased for t. It follows from (14.7.1) that if the unverifiable assumption $R_{U_F} = R_U$ holds, then the ratio adjustment leads to an approximately unbiased estimator. The dilemma caused in this case by the undercoverage is that valid conclusions are impossible unless the assumption happens to be correct.

14.7.3. Multiple Frames

Multiple-frame techniques assume that two or more frames exist, each with some undercoverage, but that together they give exact coverage of the target population. The theory of domain estimation (Chapter 10) is helpful here. We assume the following:

1. Every element k in the target population U is in at least one of two available frames denoted U_A and U_B
2. A probability sample is taken from each frame, s_A from U_A and s_B from U_B.
3. Frame membership can be determined for every sampled element.

We distinguish three disjoint domains of the population U,

Domain A, consisting of elements in U_A only, that is, $A = U - U_B$.
Domain B, consisting of elements in U_B only, that is, $B = U - U_A$.
Domain AB, consisting of elements in both U_A and U_B, that is, $AB = U_A \cap U_B$.

Under our assumptions, we have $U = A \cup B \cup AB$. An unbiased estimator of $t = \sum_U y_k$ is now obtained as

$$\hat{t} = \hat{t}_A + \hat{t}_B + \hat{t}_{AB}$$

where the three terms are estimators (unbiased or approximately unbiased) for the domain totals $\sum_A y_k$, $\sum_B y_k$ and $\sum_{AB} y_k$, respectively. Here, π estimators or more efficient estimators may be used. The variance and a variance estimator can be worked out using domain estimation theory. Note that the three components $\hat{t}_A$, $\hat{t}_B$ and $\hat{t}_{AB}$ are generally not independent.

14.7.4. Frame Construction and Maintenance

Once a frame is constructed, in many cases, it will serve for a variety of different surveys and for repetitions of the same survey over time. It is important at the outset to construct a high-quality frame and to maintain the frame's quality over time. Wright and Tsao (1983) list the following steps for a program of frame construction and maintenance:

Step 1, choice of frame units, should consider factors such as (i) the cost involved in establishing and maintaining the desired access to population elements and the desired extent of auxiliary information, (ii) the availability of the type of information one would like to see for the frame units, (iii) the stability of the frame units over a period of time, and (iv) the time needed to construct the frame.

Step 2, development of the frame, often consists of the construction of a data base. Considerable research may be needed to collect and organize the information required for units in the frame.

Step 3, validation of the frame, consists of examining the quality of the frame obtained through steps 1 and 2. The coverage achieved through the frame and the quality of the information in the frame should be assessed at this step.

Step 4, administration, should include steps to preserve the frame, if it is to serve a number of surveys over time. The needs of different users of the frame and modifications and updating should be considered.

Step 5, frame maintenance, involves carrying out the adjustments and the updating that is required. Duplicates and "deaths" should be removed, "births" incorporated, and auxiliary information updated. For rapidly changing populations, a frame that was established with considerable cost may quickly become out of date, unless there is continuous upkeep. Business establishments in rapidly changing industries are a good example. To maintain an up-to-date frame may involve considerable cost.

14.8. Measurement and Data Collection

Measurement error was defined in Section 1.7 as any error arising in the data collection phase of the survey operations. The idea of measurement error is relevant only under the supposition that a *true value* exists of the variable y for each population element k. This value is denoted θ_k. Let y_k be the value recorded at the end of the data collection phase. The *measurement error* is the difference $y_k - \theta_k$.

The notion of a true value is sometimes easy to accept. Consider, for example, the production of milk at a dairy farm during a given observation week. Most people agree that a true value exists, and that, with the good measuring devices that are available, we can observe that value essentially without error. We have more difficulty with the notion of a true value when a person is asked to quantify his/her satisfaction with his/her job on a 10 point scale. The concept is not easy to measure. In this case, nobody can really claim to correctly measure an underlying true value. In our discussion, we assume for simplicity the existence of true values.

Measurement errors complicate estimation from survey data. It is often hard to identify all errors, and there is often insufficient knowledge about their statistical properties to assess their effect on the survey estimates. Each error contributes a component of variance and sometimes a component of bias to the survey estimate. The goal, often hard to achieve, is to estimate the different components. Error modeling is an aid in understanding the effect of measurement errors on survey estimates. One widely cited modeling endeavor is the United States Bureau of the Census measurement error model developed by Hansen, Hurwitz, and Bershad (1961) and used, for example, by Bailar and Dalenius (1969), and Hartley and Rao (1978). We discuss the theory of measurement error models in Chapter 16.

We can classify the causes of *measurement error* into three broad categories as follows:

i. The measurement instrument is inaccurate.
ii. The respondent provides an inaccurate measurement.
iii. The interviewer influences the respondent in such a way that measurement error results.

Let us examine these three categories. First, the accuracy of the measurement depends greatly on the instruments available to measure the variable of interest. Variables such as production of milk in a given week, or yield of wheat in a given year can be measured with great accuracy. Highly accurate measuring instruments (for example, mechanical, electronic) are available for these kinds of variables. Unless there is error in the reporting, the measurement error $y_k - \theta_k$ will be essentially zero. By contrast, measurement errors can be considerable when a human population is surveyed and a questionnaire is used as the measuring device. Some items on the questionnaire may be poorly formulated, they may not be specific enough and lead to misunderstanding on the part of the respondent, and so on. Good questionnaire design is an extremely important part of survey planning. Cannell et al. (1989) identify problems that can arrive with survey questions and propose techniques for pretesting questionnaries.

Second, the respondent may consciously or unconsciously give incorrect answers. This is called *response error*. In the category of unconscious response errors are those due to telescoping, memory error, and conditioning. Take a survey of household expenditures as an example. In some households a meticulous diary is kept of all expenses, so highly accurate responses are obtained. In other households, the expenses become a matter of recall. *Telescoping* is the tendency on the part of a respondent to allocate to the measurement period (say a given month) an amount (say for household electric appliances) that in actual fact belongs to a different period. *Memory error* is the difficulty on the part of respondents to recall *all* their expenditures in the period. An example of *conditioning* arises when respondents, knowing beforehand that they will be called on to report expenditures, will unconsciously change their normal patterns of behavior during the measurement week, for example, for reasons of prestige. Conscious reporting of false values often occurs with personal or sensitive questions. Questions relating to drug use or abortion are typical examples. Although a selected person consents to participate, this person may feel that the anonymity of his/her response is not guaranteed, with incorrect measurement as a result. Thus, the measurement method must be designed to give the respondent confidence that complete anonymity is guaranteed. *Randomized response* is a method (see Section 15.4.4) that tries to obtain both the cooperation of the individual and a truthful answer. In one variation of the technique, the respondent randomly selects and answers one out of a set of questions; only one of this set is the relevant question. The person administering the survey never knows which question the respondent has answered, but the estimation of parameters of interest is still possible.

Lastly, even thoroughly trained interviewers differ in interviewing skill, interviewing experience, the rapport that they establish with the respondent. Such factors introduce variability into the responses. The answer to a given question by a given respondent may vary, depending on who the interviewer is. And even for one and the same interviewer, it is not unrealistic in some cases to think that the answer to a given question by a given respondent may vary, depending on the time of day or other interview conditions. Moreover, experience has shown that responses obtained by the same interviewer from two different respondents tend to be correlated. The reason may be that the interviewer's personal characteristics (age, sex, race, etc.) and interviewing style elicit similar responses from different respondents. Some models appropriate for studying interviewer differences are examined in Chapter 16.

Recent research on measurement errors is reviewed in Groves (1987). Much of the important work in the area is due to researchers in psychology and social psychology.

14.9. Data Processing

For a large household survey, the typical "paper and pencil" data collection and subsequent data processing is as follows. The interviewer meets with an adult household member and records the responses on a questionnaire form conceived for the survey. The completed questionnaires are then forwarded to the coding section of the central statistical office, where they are handled manually. Coding of responses is carried out. That is, verbal information on the questionnaire is transformed into numerical information suitable for estimation, tabulation, and analysis. The information on the amended questionnaires is entered into a mainframe computer for editing, a procedure whose primary aim is to eliminate inconsistencies in the data, to spot suspect data, and to correct erroneous data. The edit program produces a list of errors which is sent to the coding section. The corresponding questionnaires are identified and corrections are made, perhaps after renewed contact with the respondent. If two or three editing cycles are required, this process can take several months.

In recent years this process has become simpler for surveys where computer assisted data entry technology can be used. Coding is sometimes entirely eliminated, and an edit takes place already at the time of the interview. CATI (Computer Assisted Telephone Interviewing) is one of these techniques. While interviewing the respondent over the telephone, the interviewer works in front of a computer monitor on which the interview questions appear in the appropriate order. The interviewer enters the responses, and the computer immediately carries out a series of edit checks. That is, the system screens the responses for inconsistencies with preceding responses. A message on the screen will tell the interviewer if an inconsistency has occured. The interviewer can immediately go back and resolve any problems in the data for the element currently processed. CATI directly produces a data tape or other medium that can be processed, and considerable resources are often saved.

In the CAPI (Computer Assisted Personal Interviewing) technique, the data capture is carried out in the field. The interviewer visits the households, equipped with a lap-top computer. The data entry and an edit is carried out at the time of the interview, as in the CATI method. At the end of the work day, the interviewer transfers the interview data to a computer at the central office, with the aid of a modem and a telephone line. Any remaining coding and editing can take place at the central office, and in a few days, all data from, say, 20,000 interviews will be ready for the estimation stage.

In CATI and CAPI, coding and a preliminary edit take place during data collection. In a more traditional survey, *coding* is the process by which the information on the data collection medium (say, a questionnaire) is translated into numerical information suitable for data processing. The procedure may be as follows. Suppose that for each element the response recorded on the questionnaire is verbal. Furthermore, assume there is a set of "code numbers," each code number denoting a category of the study variable. The coding is carried out with a set of *coding instructions* relating the verbal descriptions with the code numbers. As a result a code number is assigned to each element. A *coding error* occurs if an element is assigned a code number other than the true one. In statistical modeling of the coding operation, one may assume that each coder has a mean error rate, and a variance around that mean (coder variability). Coder performance can be studied in specially arranged experiments. One may, for example, observe the codes assigned by the same coder at two independent trials. With more than one coder, there will ordinarily be coder differences. Each coder has a specific error rate and variability. The errors due to coder differences can be reduced by thorough training of coders, and by giving very precise coding instructions. However, even experienced coders often show considerable variation.

Coding can be entirely manual, but is nowadays more or less automated. The procedures known as *automated coding* are usually a mixture of automatic and manual components. The verbal information for a given element is transferred to, say, a magnetic tape and fed into a computer in which a dictionary is stored. The information is matched against the specifications in the dictionary. When a match occurs, the element is automatically coded. Otherwise, the element is set aside and coded manually. Coding operations, coding error, and automated coding are discussed by Lyberg (1983, 1986), and Lyberg et al. (1978).

Editing is typically carried out on the survey data before the estimation stage. Its traditional role has been to detect and correct suspect individual element values. A preedit, aimed at eliminating gross errors and inconsistencies, can often take place at the time of the coding. In an edit, the responses obtained from a selected element are exposed to a series of *edit checks* that identify logically impossible values or values deemed "almost certainly wrong." Large survey organizations have developed systems that carry out the editing operation by automatic, computer handled programs. If a selected element fails one or more edit checks on one or more questionnaire items, it is put aside for further examination. One may try to recontact the sampled element for clarification. If this is impossible or too costly, one may attempt to correct the suspect values by a mechanical procedure. Such a procedure must be carefully structured, well thought out, and likely to give high-quality replacement values. These artificial substitutes are called *imputed values*. Imputation is also used when values are missing due to nonresponse, see Section 15.7. Aspects of computerized editing are discussed in Granquist (1984), Pullum, Harpham and Ozsever (1986), Hidiroglou and Berthelot (1986), and Giles (1988).

Hidiroglou and Berthelot (1986) distinguish two types of editing: (i) consistency edits and (ii) statistical edits.

Consistency edits are used to ensure that an individual variable value (data field), or a function of data fields, satisfy specific requirements stated in the form of inequalities or logical relationships. For qualitative data, the edit must verify that variable values are well defined, for example, that only the values '0' or '1' are admitted for a dichotomous variable.

Some consistency checks are numerical. Suppose the study variable is multivariate with value $\mathbf{y}_k$ for element k. The components of $\mathbf{y}_k$ (the fields) may be qualitative or quantitative. A numerical consistency edit check for the element can be represented as

$$\mathbf{A}'\mathbf{y}_k \leq \mathbf{c}$$

where $\mathbf{A}$ is a matrix stating the rules that the values in the vector $\mathbf{y}_k$ must obey, and $\mathbf{c}$ is a vector that specifies the constraints. An element that fails the check is set aside for verification.

Other consistency edits are of the "if-then" type. For example, if the questions "What is your age" and "Do you have a driver's licence" are on the questionnaire, an edit check may be: If Age <18, then Licence $=$ No. Any element not verifying the relation is set aside for verification.

Statistical edits aim at spotting outlying values. First, "outlying value" must be defined. If data are available only for the current period, a statistical edit may consist in comparing the ratios of two highly correlated variables, for elements in a subset of elements considered similar. If the ratio is "too high" or "too low" for a certain element, the variable values in question may be deemed "outlying." If we have observations on the same element over time, an outlying value can be defined as one whose trend value, $y_k(t)/y_k(t-1)$ ($=$current period value divided by previous period value), is "too high" or "too low" compared to the collective trend value of a subset of similar elements. Outlier theory is used to fix limits for "too high" or "too low." Such procedures are given in Hidiroglou and Berthelot (1986).

An abnormal value may be due to transcription or incorrect response. Then renewed contact (follow-up) will clear up the case. In other cases, the value is indeed correct, and the apparent abnormality has an unlikely but natural cause such as "sudden growth." For other elements, the suspect values cannot be verified or corrected by follow-ups. In such cases, an imputed value may be substituted. Element values that are imputed or identified as "outlying" should always be "flagged" in the data file, so that different options are kept open for special treatment of these values at the estimation stage.

A large variable value with a large weight can in some cases lead to illogical or ridiculous estimates. A reduced weight is occasionally given to extremely large values. Principles and techniques for modified weighting are given in Hidiroglou and Srinath (1981).

14.10. Nonresponse

Nonresponse is a form of nonobservation present in most surveys. Nonresponse means failure to obtain a measurement on one or more study variables for one or more elements k selected for the survey. The term encompasses a wide variety of reasons for nonobservation: "impossible to contact," "not at home," "unable to answer," "incapacity," "hard core refusal," "inaccessible," "unreturned questionnaire," and others. In the first two cases, contact with the selected element is never established. Nonresponse results in missing data.

An operational definition of nonresponse is as follows. Consider first the case of a single study variable y. If response is obtained from all elements k in a selected sample s, the data on record when the estimation phase is about to begin is $\{y_k: k \in s\}$. If one or more y_k-values in this data set are missing, we have *nonresponse*. Let r be the subset of s for which the value y_k is available. We call r the *response set*.

In a survey with q study variables, $y_1, \ldots, y_j, \ldots, y_q$, let y_{jk} be the value of the variable y_j for the element k. Here complete data requires that y_{jk} is recorded, when the estimation phase begins, for $j = 1, \ldots, q$ and for every k in the sample. If one or more values in this data set are missing, there is nonresponse. For each variable there is now a response set. To have identical response sets would be unusual.

Nonresponse has long been recognized as a major problem in surveys. The uninitiated observer may wonder why nonresponse is such a serious problem. If a survey results in 10% nonresponse, 90% of the observations still remain. Since sample sizes are often large, the number of observations is still considerable, even after a 10% loss. The fact is that there would be minor cause for concern if the only consequence of the nonresponse was an increase in variance. The real reason statisticians fear nonresponse is that nonresponse usually introduces selection bias.

For example, if high-income earners are less inclined to answer, they are likely to be underrepresented in the final data set. This implies that the data set tends to contain too few high values and too many low values on any study variable positively correlated with income, such as savings or consumer durable expenditures. The standard estimators of a population total will then yield underestimates. Standard confidence statements will not meet the desired confidence level, that is, they are invalid.

If standard estimators fail, can modified estimators be created that lead to valid inferences? There exists no perfect method for valid statistical inference when there is nonresponse. The methods in popular use are more or less successful attempts to reduce nonresponse bias. Some of these are presented in Chapter 15.

The problem of nonresponse was recognized already in the 1930s and 1940s, and some early methods for dealing with the problem were suggested. It was realized that the higher the nonresponse rate, the more there is reason to worry about the validity of the survey results. Cochran (1951) states "unfortunately, any sizeable percentage of nonresponse makes the results open to question by anyone who cares to do so."

During the 1950s and 1960s, the general public had an essentially positive attitude to surveys, especially to many of the large-scale surveys conducted by government statistical agencies. Nonresponse rates were low. However, around 1970, the climate changed radically. In the early 1970s, population censuses encountered strong public resistance in several countries, including Britain, Sweden, and the United States. In many important sample surveys, nonresponse rates also reached considerable or even alarmingly high levels. For example, the nonresponse rate for the Swedish Labour Force Survey increased from approximately 1.5% in 1970 to more than 8% in 1975. In certain household surveys, the nonresponse rate was considerably higher.

The trend towards higher nonresponse rates forced major statistical agencies and survey institutes to allocate an increased portion of their resources to studying the causes of nonresponse and to finding ways to deal with the problem. An important research activity developed. In the United States, a Panel on Incomplete Data was established. Its work led to the 1983 publication of *Incomplete Data in Sample Surveys*, an authoritative three volume set showing the state of the knowledge at that time in the area of nonresponse.

A sample survey with nonresponse can be viewed as the selection of a sample s from the population U, followed by a subselection of a response set r from the sample s. Suppose that there exists a probabilistic *response mechanism* that generates r from s. This mechanism is represented by $p(r|s)$, the probability that the response set r is generated, given that the sample s was selected. Except in truly exceptional cases, $p(\cdot|s)$ is an unknown distribution. However, the statistician is perhaps willing to specify certain features of this distribution. Unknown parameters of $p(\cdot|s)$ can then be estimated from the sample. In particular, estimates may be obtained for the unknown response probabilities. In this way, the statistician can proceed with estimation and analysis similarly as if two-phase sampling had taken place. This approach underlies some of the estimation techniques presented in Chapter 15.

The use of (estimated) response probabilities as a tool in treating nonresponse goes back to Politz and Simmons (1949). Their estimation technique was to weight observations by the inverse of the estimated probability that the respondent would be found at home. Likewise, Deming (1953) implicitly uses response probabilities. Later, explicit response probabilities appear in Lindström and Lundström (1974) and in Nargundkar and Joshi (1975).

The approach with response probabilities is now established as a fruitful remedy to a difficult problem. Complete elimination of nonresponse is the ideal (but usually too costly) solution. Instead, a viable solution is to adjust for the nonresponse. The two main techniques are *weighting adjustment* and *imputation*. The objective of a weighting adjustment is to inflate the weight given in the estimator to y-values of the responding elements. This compensates for those values that are lost through nonresponse. Imputation is the procedure whereby artificial variable values are substituted for missing genuine variable values. It is self-evident that the imputed values must be generated by a procedure or a modeling technique that can be trusted to yield high quality substitutes. Several weighting adjustment and imputation methods are considered in Chapter 15.

Remark 14.10.1. At present there is no uniformly accepted terminology for nonsampling errors or nonresponse in particular. Research reports and documentation of surveys should be read with some care. A certain term may be used in the literature with different meanings. Conversely, the same concept may be referred to by different terms. Terminology for nonsampling errors is discussed in U.S. Department of Commerce (1978b), Swensson (1982), Platek and Gray (1983), and Lessler (1982).

Exercises

14.1. Suppose that an *SI* sample s of size $n = fN$ is drawn from the population U. Let $t_{U-s} = \sum_{U-s} y_k$ denote the unknown total of the nonsampled part of the population. Show that

$$E_p[(N - n)\bar{y}_s - t_{U-s}] = 0$$

which implies that the random quantity t_{U-s} is design unbiasedly "estimated" by $\hat{t}_{U-s} = (N - n)\bar{y}_s$. It may seem paradoxical that an estimate is obtained for the total of a subpopulation $U - s$, from which not a single observation is available. Explain how this is possible. How would you estimate t_{U-s} through a simple model-based argument?

14.2. This is a generalization of Exercise 14.1. Suppose that a sample s is drawn from the population U by a probability sampling design $p(\cdot)$ with positive inclusion probabilities π_k; $k \in U$. Let t_{U-s} be defined as in Exercise 14.1. Show that the random quantity t_{U-s} is design unbiasedly "estimated" by $\hat{t}_{U-s} =$

$\sum_s (1 - \pi_k) y_k / \pi_k$ in the sense that

$$E_p[\sum_s (1 - \pi_k) y_k/\pi_k - t_{U-s}] = 0$$

14.3. Verify that the ratio estimator $\hat{t}_{ra}$ in Example 14.5.1 is model unbiased for $t = \sum_U y_k$ under the model shown in (14.5.2) and that its model mean square error is given by (14.5.3).

14.4. Verify that the model parameter σ^2 in the model shown in (14.5.2) is model unbiasedly estimated by

$$\hat{\sigma}^2 = [1/(n-1)] \sum_s (y_k - bx_k)^2/x_k$$

with $b = \bar{y}_s/\bar{x}_s$. Verify that (14.5.4) is also a model unbiased estimator of σ^2.

14.5. Show that under the model $E_\xi(y_k) = \beta x_k$; $V_\xi(y_k) = \sigma^2 x_k$ for $k \in U$, the cut-off estimator $\hat{t}_{\text{cut}} = (\sum_U x_k)\hat{R}_{U_0}$ considered in Section 14.4 is model unbiased for $\sum_U y_k$.

14.6. Under the model $E_\xi(y_k) = \mu_h$; $V_\xi(y_k) = \sigma_h^2$ for elements $k \in U_h$, $h = 1, \ldots, H$, show that the quota sampling estimator $\hat{t}_{qu}$ considered in Example 14.5.2 is model unbiased for $t = \sum_U y_k$ and that the model mean square error is given by

$$\text{MSE}_\xi(\hat{t}_{\text{qu}}) = \sum_{h=1}^{H} N_h^2 \frac{1 - f_h}{n_h} \sigma_h^2$$

14.7. In Section 14.7.2, $\hat{t}_{Fra} = (\sum_U x_k)(\sum_{s_F} y_k/\pi_k)/(\sum_{s_F} x_k/\pi_k)$ was mentioned as a possible estimator of $t = \sum_U y_k$ when the frame has undercoverage. Show that $\hat{t}_{Fra}$ is model unbiased under the model $E_\xi(y_k) = \beta x_k$; $V_\xi(y_k) = \sigma^2 x_k$ for $k \in U$. Find the model mean square error of $\hat{t}_{Fra}$.

14.8. Consider the case (i) in Section 14.7.2, that is, the frame has overcoverage but no other imperfections. Suppose that there is a one-to-one relationship between the units in F and the elements in U_F and that the overcoverage remains undetected, even after inspection of the sample. Let $\hat{t}_{U_F}$ be an unbiased estimator for $t_{U_F} = \sum_{U_F} y_k$. Show that the relative bias of $\hat{t}_{U_F}$ as an estimator for $t_U = \sum_U y_k$,

$$\text{RB}(\hat{t}_{U_F}) = \frac{t_{U_F} - t_U}{t_U}$$

can be written

$$\text{RB}(\hat{t}_{U_F}) = \frac{\bar{y}_{\bar{U}}}{\bar{y}_U} \frac{1 - W_U}{W_U}$$

where $\bar{U} = U_F - U$ and

$$\bar{y}_{\bar{U}} = \frac{\sum_{\bar{U}} y_k}{N_{\bar{U}}}, \qquad \bar{y}_U = \frac{\sum_U y_k}{N} \qquad \text{and} \qquad W_U = N/N_{U_F}.$$

Here N, N_{U_F} and $N_{\bar{U}} = N - N_{U_F}$ are the sizes of U, U_F and $\bar{U} = U_F - U$, respectively. Discuss the factors that affect the relative bias.

14.9. Consider the case (iii) in Section 14.7.2, that is, the frame has undercoverage but no other imperfections.

Let $\hat{t}_{U_{\text{link}}}$ be unbiased for $t_{U_{\text{link}}} = \sum_{U_{\text{link}}} y_k$. Show that the relative bias of $\hat{t}_{U_{\text{link}}}$ as an estimator for t_U,

$$\text{RB}(\hat{t}_{U_{\text{link}}}) = \frac{t_{U_{\text{link}}} - t_U}{t_U}$$

can be written

$$\text{RB}(\hat{t}_{U_{\text{link}}}) = -\frac{rW_{U_{\text{nolink}}}}{1 + (r-1)W_{U_{\text{nolink}}}}$$

where

$$r = \bar{y}_{U_{\text{nolink}}}/\bar{y}_{U_{\text{link}}} \quad \text{and} \quad W_{U_{\text{nolink}}} = N_{U_{\text{nolink}}}/N$$

Discuss the factors that influence the relative bias.

14.10. Consider the case (ii) in Section 14.7.2. The frame has duplicate listings, that is, $L_{\cdot k} > 1$ for some target population elements k, but no other imperfections, that is $L_{i\cdot} = 1$ for every $i \in F$ and $U_F = U$.

A sample s_F^0 is drawn from F with the inclusion probabilities $\pi_i^F > 0$.

(a) Suppose that the survey statistician acts, incorrectly, as if there is a one-to-one relationship between the units in F and the elements in U. Let $z_i = \sum_U L_{ik} y_k$, that is, $z_i = y_k$ for every i linked with k. Show that the expected value of

$$\hat{t}_z = \sum_{i \in s_F^0} z_i/\pi_i^F$$

is given by

$$E(\hat{t}_z) = \textstyle\sum_U L_{\cdot k} y_k$$

and hence that $\hat{t}_z$ is biased.

(b) For every k in the sample, suppose that it is possible to determine $L_{\cdot k}$ and a set of weights ω_{ik} satisfying

$$\sum_{i \in F} \omega_{ik} L_{ik} = 1$$

for example, $\omega_{ik} = 1/L_{\cdot k}$ for every i linked with k. Define

$$z_i^* = \textstyle\sum_U \omega_{ik} L_{ik} y_k$$

Show that an unbiased estimator of $t_U = \sum_U y_k$ is given by

$$\hat{t}_z = \sum_{i \in s_F^0} z_i^*/\pi_i^F$$

CHAPTER 15

Nonresponse

15.1. Introduction

Nonresponse is present in almost all surveys, but the extent and the effect of the nonresponse can vary greatly from one type of survey to another. The nonresponse can also vary considerably over time for a survey that is carried out repeatedly. It is a mark of a well-run survey that effective measures are taken to control the nonresponse, for example, by careful training of interviewers. But even in the best of surveys, nonresponse occurs, and special estimation techniques are required to deal with the problem. Subsampling of nonrespondents, randomized response techniques, methods based on response modeling, and imputation are among the methods examined in this chapter. Nonresponse is the only nonsampling error considered in the chapter. We ignore coverage errors, measurement errors, and data processing errors.

15.2. Characteristics of Nonresponse

15.2.1. Definition of Nonresponse

By nonresponse is meant that the desired data are not obtained for the entire set of elements s, designated for observation. Usually, the set s is a probability sample, but in the following we also admit that $s = U$ with probability one, as in a census. Let us formalize the structure of the data available after nonresponse.

The objective of the survey is to observe the sampled elements with respect to q study variables, $y_1, \ldots, y_j, \ldots, y_q$. These may correspond to q items on a questionnaire. Let y_{jk} be the value of the variable y_j for the element k. Let n_s be the size of s.

By *full response* in the survey we mean that, after data collection and edit, the available data consist, for every $k \in s$, of a complete q-vector of observed values

$$\mathbf{y}_k = (y_{1k}, \ldots, y_{jk}, \ldots, y_{qk})$$

These values form a data matrix of dimension $n_s \times q$, with no value missing.

In all other cases, there is *nonresponse*. That is, after data collection and edit, the $n_s \times q$ data matrix is *incomplete*. One or more of the $n_s \times q$ desired y_{jk} values are missing; there are some "blanks" instead of values in the data matrix.

Full response is seldom realized in a survey. There is a variety of reasons for missing values y_{jk}. In a mail survey, the questionnaire may not be returned, or returned but not completely filled in. In a survey with personal interviews, some individuals refuse to respond to some or all of the questions. Some individuals are not found at home, despite repeated calls. Illness or language problems may make it impossible to carry out the interview. A value supplied on a questionnaire or obtained at an interview may fail an edit check. Such a value may then be regarded as false or strongly suspect and recorded as a blank. Here we consider unacceptable values as part of the nonresponse.

15.2.2. Response Sets

We denote by r_j the jth response set, that is, the subset of the sample s for which acceptable responses to the jth item are recorded. We can thus write

$$r_j = \{k\colon k \in s \text{ and } y_{jk} \text{ is recorded}\}.$$

In the survey, there are q usually nonidentical response sets, $r_1, \ldots, r_q$. In the case of a sample survey, the set s is selected with a known sampling design. The response set r_j is a subselection from s.

Several difficulties are caused by nonresponse.

1. We do not ordinarily know how the response sets r_j are generated. To illustrate, suppose that the response is probabilistic. Then there is a certain probability $p_j(r_j|s)$ that the set r_j is realized, given the selected sample s. However, the response distribution $p_j(\cdot|s)$ is ordinarily unknown. To obtain statistical conclusions about the finite population with the aid of data for the respondents, we must then make assumptions about the distribution $p_j(r_j|s)$, for example, that elements respond independently of each other, or that two or more similar elements have the same probability of

responding. Such assumptions can usually not be verified. Consequently, the validity of our inferences about the finite population depend on the truth of unverifiable assumptions, which is undesirable.
2. The second phase of selection makes the estimation more complex. Both the estimator $\hat{\theta}$ of a parameter θ and the corresponding variance estimator $\hat{V}(\hat{\theta})$ are likely to require heavier calculations than in the case of full response.
3. For data processing, it is a disadvantage if the response sets $r_1, \ldots, r_q$ are different, as they usually are in practice. This may force separate treatment of each study variable in the estimation phase and may cause delay in the publication of the survey results.

15.2.3. Lack of Unbiased Estimators

We adopt a probabilistic outlook on the response mechanism. Then denote by θ_{jks} the probability that the element k responds to item j, given that the probability sample s was selected. Note that θ_{jks} can depend on s. It is a good idea to allow such dependence, because features of the sample, such as the geographic spread, may influence the selection of interviewers, the interviewer workloads, the measurement techniques, and other factors that affect the propensity to respond.

Suppose for a moment that, for every given sample s, the probabilities θ_{jks} are known and strictly positive for all $k \in s$. The theory of two-phase sampling then applies. An unbiased estimator of the jth item total,

$$t_j = \sum_U y_{jk}$$

would be given by the π^*-estimator (see Section 9.3),

$$\hat{t}_{j\pi^*} = \sum_{r_j} y_{jk}/\pi_k \theta_{jks} \tag{15.2.1}$$

It is easy to see that $\hat{t}_{j\pi^*}$ is unbiased for t_j. We have

$$E(\hat{t}_{j\pi^*}|s) = \sum_s y_{jk}/\pi_k = \hat{t}_{j\pi}$$

where $\hat{t}_{j\pi} = \sum_s y_{jk}/\pi_k$ is the π estimator for variable j. (It cannot be used unless there is full response.) If $E_p(\cdot)$ denotes expectation with respect to the sampling design used to draw s,

$$E(\hat{t}_{j\pi^*}) = E_p E(\hat{t}_{j\pi^*}|s) = E_p(\hat{t}_{j\pi}) = t_j$$

That is, $\hat{t}_{j\pi^*}$ would be unbiased. We can regard the quantities $1/\theta_{jks}$ in the estimator (15.2.1) as weights needed to eliminate nonresponse bias. These weights are not available when the θ_{jks} are unknown. The dilemma of nonresponse is in a nutshell that standard techniques for obtaining unbiased estimates do not work.

Estimated response probabilities $\hat{\theta}_{jks}$ can sometimes be obtained, using data for the responding elements, as well as auxiliary data and specific assumptions about the response distribution. Replacing θ_{jks} by $\hat{\theta}_{jks}$ in (15.2.1), we obtain

$$\hat{t}_{j\text{mod}} = \sum_{r_j} y_{jk}/\pi_k \hat{\theta}_{jks}$$

Estimators of this type are frequently used in practice. They are not unbiased, but the bias may be modest. Examples are considered in Section 15.6.

15.3. Measuring Nonresponse

The greater the nonresponse, the more one has reason to worry about its harmful effects on the survey estimates. The bias often increases with the rate of nonresponse. It is hard to get objective measures of the bias, but it is relatively simple to quantify the extent of the nonresponse. Different measures of the nonresponse are usually found in the quality declarations that statistical agencies and survey institutes often publish together with the survey results. The user can take this information into account when judging the credibility of the results.

A number of descriptive measures are used for the response, or its complement, the nonresponse. Before detailing these measures we distinguish two types of missing information for an element k from which response is solicited.

1. The element k is a *unit nonresponse* element if the entire vector of y-values, $\mathbf{y}_k = (y_{1k}, \ldots, y_{jk}, \ldots, y_{qk})$, is missing. An example is when the respondent fails to return the questionnaire, even after one or more reminders, or if he or she refuses to participate in a personal interview.
2. The element k is an *item nonresponse* element if at least one, but not all q, components of the vector $\mathbf{y}_k = (y_{1k}, \ldots, y_{jk}, \ldots, y_{qk})$ are missing. For example, the respondent returns a partially filled in questionnaire, or an interview results in responses to some but not all questions.

Remark 15.3.1. Element nonresponse would be a more appropriate term than unit nonresponse, if we adhere to the terminology in earlier chapters. However, the latter term is so entrenched in the minds of survey statisticians that it can hardly be changed; we also say unit nonresponse about an element for which the entire response vector $\mathbf{y}_k$ is missing.

The response sets $r_1, \ldots, r_q$ defined in Section 15.2 are useful for deriving overall measures of the nonresponse in the survey. Set

$$r_u = r_1 \cup r_2 \cup \cdots \cup r_q = \bigcup_{j=1}^{q} r_j$$

and

$$r_c = r_1 \cap r_2 \cap \cdots \cap r_q = \bigcap_{j=1}^{q} r_j$$

Here, r_u, called the *unit response set*, is the set of elements having responded to at least one item. (It is assumed that at least one r_j is nonempty.) Its complement, $s - r_u$, the *unit nonresponse set*, is composed of the elements having responded to no item whatsoever. They are either "no contact" elements, or contacted elements from which no acceptable responses were obtained.

The set r_c consists of the elements having responded to all items, so for every $k \in r_c$, the vector $\mathbf{y}_k$ is complete, suggesting our index c. Now, $r_u - r_c$, the *item nonresponse set*, is composed of the elements having responded to at least one, but not all q, items.

Looking at a specific item, $r_u - r_j$ defines the *item j nonresponse set*. Illustration by a Venn diagram helps to visualize the various sets.

In the case of $q = 2$ questionnaire items y_1 and y_2, the value y_{1k} is recorded for $k \in r_1$ and the value y_{2k} for $k \in r_2$. This is illustrated by the following diagram. The dark shade area indicates r_c. The set r_u consists of r_c and the light shade area.

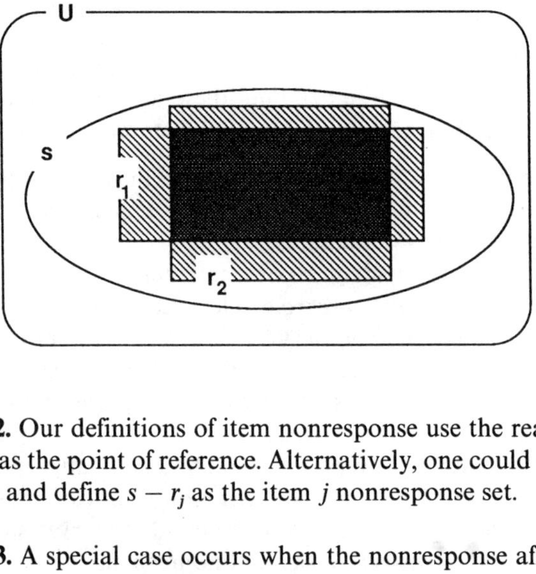

Remark 15.3.2. Our definitions of item nonresponse use the realized unit response set, r_u, as the point of reference. Alternatively, one could start from the total sample s and define $s - r_j$ as the item j nonresponse set.

Remark 15.3.3. A special case occurs when the nonresponse affects all study variables in exactly the same way, so that $r_1 = r_2 = \cdots = r_q = r$. Although rare in practice, this case has pedagogic value. Then, if $r \subset s$, there is unit nonresponse, but no item nonresponse. The sets $r_u - r_j$, $j = 1, \ldots, q$, are empty. For a single-item questionnaire, this implies that any nonresponse that occurs is considered to be unit nonresponse.

It is important in a survey to measure the size of the various sets. Such measures can convey important messages about what has happened in the survey. One type of measure is based on the element counts in the different sets. Denote the respective sizes of the sets s, r_u, r_j, and r_c by n_s, n_{r_u}, n_{r_j}, and n_{r_c}.

i. *Measures of Unit Response and Unit Nonresponse*

A simple measure of the *unit response* is

$$p_{r_u} = n_{r_u}/n_s$$

The *unit nonresponse* is consequently measured by

$$1 - p_{r_u}$$

Here p_{r_u} measures how well the survey has succeeded in obtaining at least partial response from the elements in the selected sample.

Alternative measures are obtained by sample-weighted quantities. The sample-weighted measure of unit response is

$$\tilde{p}_{r_u} = (\textstyle\sum_{r_u} 1/\pi_k)/(\sum_s 1/\pi_k)$$

The corresponding sample-weighted measure of unit nonresponse is

$$1 - \tilde{p}_{r_u}$$

In some cases, $\tilde{p}_{r_u}$ can be interpreted as an estimated average response probability in the population. Some statistical agencies recommend the use of sample-weighted unit nonresponse measures.

Unweighted- and weighted-measures may differ considerably, as seen in the following example.

EXAMPLE 15.3.1. A population of $N = 10{,}000$ elements is stratified into two strata of $N_1 = 1{,}000$ and $N_2 = 9{,}000$ elements. In each stratum, an *SI* sample of 500 elements is drawn. Suppose we have a single item questionnaire, with 250 nonresponding elements in the first stratum and 50 in the second stratum. The unweighted nonresponse rate is then

$$1 - p_{r_u} = (250 + 50)/(500 + 500) = 30\%$$

and the weighted nonresponse rate is computed by noting that $1/\pi_k = 1{,}000/500 = 2$ for all elements k in the first stratum and $1/\pi_k = 9{,}000/500 = 18$ for all elements k in the second stratum, so that

$$1 - \tilde{p}_{r_u} = (250 \times 2 + 50 \times 18)/10{,}000 = 14\%$$

The two rates, 30% and 14%, could give the uninitiated user very different impressions about nonresponse. Note that in this example the strata differ considerably with respect to size, sampling fraction, and response rate.

ii. *Measures of Item Response and Item Nonresponse*

Item response and item nonresponse can be measured analogously. An unweighted measure of the overall *item nonresponse* is

$$1 - p_{r_c}$$

where

$$p_{r_c} = n_{r_c}/n_{r_u}$$

Measures calculated for individual items are also of interest since they may reveal important differences in the sensitivity of the various items. An unweighted measure of the *item j nonresponse* is

$$1 - p_{r_j}$$

where

$$p_{r_j} = n_{r_j}/n_{r_u}$$

Clearly, $1 - p_{r_c}$ is an upper bound for any individual item nonresponse rate, $1 - p_{r_j}$. It is straightforward to define sample-weighted analogs of these item nonresponse measures.

EXAMPLE 15.3.2. In a survey with $q = 3$ questionnaire items, the following counts were recorded:

$$n_s = 1{,}000, \qquad n_{r_u} = 949, \qquad n_{r_c} = 574$$

$$n_{r_1} = 904, \qquad n_{r_2} = 891, \quad \text{and} \quad n_{r_3} = 645$$

Calculation shows the unit nonresponse rate to be

$$1 - p_{r_u} = 0.051$$

and the overall item nonresponse is

$$1 - p_{r_c} = 0.395$$

The individual item nonresponse rates are

$$1 - p_{r_1} = 0.047, \qquad 1 - p_{r_2} = 0.061, \qquad 1 - p_{r_3} = 0.320$$

The individual rates suggest that item 3 is considerably more sensitive than items 1 and 2. On the other hand, the highest individual rate 0.320 is fairly far from the overall rate of 0.395; the difference is attributable to persons who are respondents to item 3 but nonrespondents to at least one of items 1 and 2.

The response and nonresponse rates seen so far are based on counts. When quantitative variables are available, the calculation of value-weighted rates may be an attractive alternative. For example, if x is a positive auxiliary variable that measures size, alternative measures of unit response are

$$\frac{\sum_{r_u} x_k}{\sum_s x_k} \quad \text{and} \quad \frac{\sum_{r_u} x_k/\pi_k}{\sum_s x_k/\pi_k}$$

EXAMPLE 15.3.3. An *SI* sample s with $n = 100$ was drawn from the MU284 population in Appendix B. The variable P75 ($= x$) is known for the elements in the sample. The sample consisted of the element $k = 16$ and 99 other elements. We have $x_{16} = 2340$ and $\sum_s x_k = 3011$. Now suppose that the element $k = 16$ is the only nonrespondent. Then the unit response based on counts is 0.99. However, the value-weighted measures show the response in a strikingly different light, namely, $2340/3011 = 0.78$. This much lower figure reflects the numerical importance of the sole missing element.

Remark 15.3.4. The response and nonresponse rates discussed in this section are useful for describing participation in the survey, but say little about the damage caused by the nonresponse, for example, the selection bias caused by the nonresponse. To eliminate this bias is a difficult methodologic problem; see Section 15.6.

Remark 15.3.5. It is hard to give firm recommendations on what one should consider an acceptable level of nonresponse in a survey. The difficulty is well explained by the Panel on Incomplete Data in Sample Surveys (1983), p. 5:

> Nonresponse rates should be "low," but whether 5%, 10%, 20%, or some other percentage should be an upper bound for acceptable nonresponse rates depends on the survey objectives and is difficult to specify even for a particular survey. Unit and item nonresponse rates must be low for a total survey and for significant domains and subpopulations of interest.
>
> Few surveys can achieve 5% or smaller unit nonresponse rates without either accepting very high costs or defining the survey population especially to achieve that objective, i.e., excluding subsets of the population expected to have high nonresponse rates. Yet even with a low overall nonresponse rate, estimates for some items and domains may have much higher nonresponse rates, and estimates for items characterizing only a small proportion of the population (and other "low" estimates) may possibly be seriously affected when a nonresponse rate is as small as 5%. Thus, the acceptable rates of nonresponse depend on the inferences to be made and the procedures for making those inferences.

15.4. Dealing with Nonresponse

Strategies for dealing with nonresponse can be classified as follows:

a. Before and during data collection, effective measures are taken to reduce the nonresponse to insignificant levels, so that any remaining nonresponse causes little or no harm to the validity of the inferences.
b. Special, perhaps costly techniques for data collection and estimation are used that permit unbiased estimation.
c. Model assumptions about the response mechanism and about relations between variables are used to construct estimators that "adjust" for a nonresponse that cannot be considered harmless.

Approaches under (a) are discussed in Section 15.4.1 and 15.4.2. Two techniques in category (b) are subsampling of nonrespondents and randomized response. These are described in Sections 15.4.3 and 15.4.4. The approaches based on modeling are treated in Sections 15.6 and 15.7.

15.4.1. Planning of the Survey

The ideal survey has no nonresponse. To come close to this ideal requires careful planning and often considerable expense. The nonresponse is affected by a number of the operations that define the survey. Special efforts must be made at the planning stage to foresee how alternative survey operations may influence the response.

The selection, training and supervision of interviewers are factors of great importance. The choice of data collection method (personal interview, telephone interview, mail inquiry, or other) is important, as is the length and the content of the questionnaire or schedule. In a repeated survey, the frequency with which respondents are asked to participate must be considered. Response rates are often adversely affected by a heavy response burden.

The attitude of management toward nonresponse is a factor whose importance should not be underestimated. Dalenius (1976) gives an example where two different survey institutes were asked to carry out the same survey, independently of each other. The nonresponse rates for the two surveys were 40% and 10%. According to Dalenius, the surprisingly large difference reflects the different ambitions in the two institutes in regard to acceptable nonresponse rates.

In this section we focus on a few standard techniques that keep the nonresponse low, namely, callbacks and follow-ups, subsampling of nonrespondents, and randomized response interviews.

15.4.2. Callbacks and Follow-Ups

In surveys with personal interviews, the first contact with a potential respondent may be unsuccessful for a variety of reasons. For example, no one may be at home, the person selected may be sick, or the interview may be broken off at an early stage. If the first attempted contact results in too many unsuccessful interviews, it is common to make one or more callbacks at more convenient times, perhaps using interviewers with special skills and experience. The nonresponse can thereby be considerably reduced if not eliminated.

Callbacks can also give valuable information about the selective effects of nonresponse. In the following examples, the propensity to respond is strikingly related to variables important to the survey.

Example 15.4.1. The following table gives the percentage of response and the associated percentage of households with children under two years of age, by number of call. The data, from Hilgard and Payne (1944) and cited by

P. Rao (1983), are from a survey of 3,265 urban households, through personal interviews.

	Number of the call				
	1	2	>2	Nonresponse	Total
Percentage of response	63.5	22.2	14.3	0	100
Percentage of households with children under 2 years of age	17.2	9.5	6.2	–	13.9

The table shows that adult members of households with young children are more likely to be at home when the interviewer calls. If efforts to elicit a response were stopped after a single call, the response set would tend to over-represent households with small children. If the target population is all households, there will then be a biasing effect on estimates for variables correlated with number of children.

Example 15.4.2. The following data on drug use, from Danermark and Swensson (1987), are from a survey of 528 high school students aged 16. The first call was made at school during a regular classroom session, and the data was collected with anonymous questionnaires. Students that were absent were later given a second or third chance to complete the questionnaire.

	Number of the call			
	1	>1	Nonresponse	Total
Percentage of responses	89.6	8.9	1.5	100
Percentage of students having tried hashish or marihuana at least once	4.9	14.9	unknown	unknown

The percentage of students having tried hashish or marihuana is higher among those absent at the first call, indicating a relation between absenteeism and drug use.

In mail surveys, follow-up letters in combination with telephone interviews and personal interviews are often used as a callback technique. Often two or three reminders are mailed to obtain as many responses as possible from the mail phase. As with personal interviews, the propensity to answer at different calls is often correlated with the study variables.

EXAMPLE 15.4.3. The following data from Finkner (1950) concerning a survey of 3,116 North Carolina Fruit Growers were cited in P. Rao (1983)

	Number of the mailing				
	1	2	3	Nonresponse	Total
Percentage of returns	10	17	14	59	100
Average number of fruit trees per farm	456	386	340	290	329

Late respondents and nonrespondents differ considerably from those responding at the first mailing. There is a strong association between propensity to respond early and size of farm.

In practice, callbacks must for obvious reasons stop after a few attempts. Furthermore, the reduction in the mean square error of the survey estimates will often be small compared to the cost of further callbacks. Deming (1953) developed a model for the study of different callback strategies. The model aids in determining the optimal number of callbacks. Politz and Simmons (1949) proposed a procedure based on weighting the responses with estimated response probabilities to deal with the problem of "not at homes." In some surveys, this is more cost efficient than repeated callbacks. P. Rao (1983) discusses these approaches in detail.

15.4.3. Subsampling of Nonrespondents

Callbacks and follow-ups aim to eliminate or at least greatly reduce nonresponse. In theory, these techniques are commendable, but in practice they are not without problems. A long series of callbacks or follow-ups may prove costly and time-consuming. Nonresponse may still be unacceptably high after the ultimate call or follow-up letter; this is especially common in mail surveys.

An alternative approach is to take a subsample of the nonrespondents, and make every possible effort to obtain responses from *all* elements in the subsample. There are several schemes for subsampling the nonrespondents. This important idea developed by Hansen and Hurwitz (1946) shows that unbiased estimation is possible despite the nonobservation of certain elements in the initial sample. We describe the technique in general terms. That is, an arbitrary design is used to draw the initial sample, and, likewise, the subsampling is by an arbitrary design.

We limit the discussion to the case of a single study variable, denoted y. The objective is to estimate the total $t = \sum_U y_k$. We assume the following situation:

a. In the first phase, a sample s_a of size n_a is drawn according to the design $p_a(\cdot)$ with positive inclusion probabilities π_{ak} and π_{akl}. Let $\Delta_{akl} = \pi_{akl} - \pi_{ak}\pi_{al}$.
b. Despite efforts to obtain responses y_k from all elements in s_a, some nonresponse occurs. Stochastic response is assumed. That is, there exists a response distribution (RD) that governs the dichotomization of the sample s_a into one responding subset, s_{a1}, of size n_{a1}, and one nonresponding subset s_{a2}, of size n_{a2}. This implies that if a given s_a were surveyed repeatedly, the composition of the subsets would vary from one survey to the next.
c. A suitably large subsample s_2 is drawn from s_{a2}, by a design $p(\cdot|s_{a2})$ with positive inclusion probabilities denoted $\pi_{k|s_{a2}}$ and $\pi_{kl|s_{a2}}$. Let $\Delta_{kl|s_{a2}} = \pi_{kl|s_{a2}} - \pi_{k|s_{a2}}\pi_{l|s_{a2}}$. The necessary efforts are made to record a response from every element in s_2.

The requirement of full response in (c) may prove costly, but it is this requirement that makes unbiased estimation possible. Resources must be reserved at the planning stage for the measures that may need to be taken to obtain full response from the subsample.

The procedure resembles two-phase sampling for stratification, with a subdivision of the initial sample s_a into two strata of which only one is subsampled. However, the theory in Section 9.4 can not be applied, because it assumes that the partitioning into strata is always the same, for a given s_a. Here, the partitioning is random.

The set for which y is observed is denoted $s = s_{a1} \cup s_2$. The population total $t = \sum_U y_k$ is estimated by

$$\hat{t} = \sum_s \check{y}_k = \sum_s y_k/\pi_k^* \tag{15.4.1}$$

where $s = s_{a1} \cup s_2$ and

$$\pi_k^* = \begin{cases} \pi_{ak} & \text{if } k \in s_{a1} \\ \pi_{ak}\pi_{k|s_{a2}} & \text{if } k \in s_{a2} \end{cases}$$

Hence the estimator can also be written

$$\hat{t} = \sum_{s_{a1}} \check{y}_{ak} + \sum_{s_2} \check{y}_k \tag{15.4.2}$$

where $\check{y}_{ak} = y_k/\pi_{ak}$. Let $E_{\mathrm{RD}}(\cdot|s_a)$ denote the expectation with respect to the (unknown) response distribution, given s_a. Then,

$$E(\hat{t}) = E_{p_a} E_{\mathrm{RD}} E(\hat{t}|s_a, s_{a2})$$

But

$$E(\hat{t}|s_a, s_{a2}) = \sum_{s_{a1}} \check{y}_{ak} + \sum_{s_{a2}} \check{y}_{ak} = \sum_{s_a} \check{y}_{ak}$$

This expected value does not depend on s_{a2}, so we obtain

$$E(\hat{t}) = E_{p_a} E_{\mathrm{RD}}(\textstyle\sum_{s_a} \check{y}_{ak} | s_a)$$
$$= E_{p_a}(\textstyle\sum_{s_a} \check{y}_{ak}) = \sum_U y_k = t$$

Thus $\hat{t}$ is unbiased, whatever the unknown response distribution. The properties of $\hat{t}$ are highlighted in the following result.

Result 15.4.1. *The estimator shown in* (15.4.1) *is unbiased for the population total* $t = \sum_U y_k$. *The variance is given by*

$$V(\hat{t}) = \textstyle\sum\sum_U \Delta_{akl} \check{y}_{ak} \check{y}_{al} + E_{p_a} E_{\mathrm{RD}}(\sum\sum_{s_{a2}} \Delta_{kl|s_{a2}} \check{\check{y}}_k \check{\check{y}}_l | s_a) \quad (15.4.3)$$

An unbiased variance estimator is given by

$$\hat{V}(\hat{t}) = \sum\sum_s \frac{\Delta_{akl}}{\pi^*_{kl}} \check{y}_{ak} \check{y}_{al} + \sum\sum_{s_2} \frac{\Delta_{kl|s_{a2}}}{\pi_{kl|s_{a2}}} \check{\check{y}}_k \check{\check{y}}_l \quad (15.4.4)$$

where $s = s_{a1} \cup s_2$,

$$\pi^*_{kl} = \begin{cases} \pi_{akl}\pi_{kl|s_{a2}} & \text{if } k, l \in s_{a2} \\ \pi_{akl}\pi_{k|s_{a2}} & \text{if } k \in s_{a2}, l \in s_{a1} \\ \pi_{akl}\pi_{l|s_{a2}} & \text{if } k \in s_{a1}, l \in s_{a2} \\ \pi_{akl} & \text{if } k, l \in s_{a1} \end{cases}$$

and $\pi^*_{kk} = \pi^*_k$.

The derivation of the expressions for the variance and the variance estimator is left as an exercise. Fortunately, it is possible in this case to obtain the unbiased variance estimator shown in (15.4.4) even though the response distribution is unknown. Confidence intervals can thus be obtained without difficulty. However, in the variance expression (15.4.3), the second component is nonexplicit. It cannot be accurately evaluated without knowledge of the response distribution.

Example 15.4.4. Assume that an *SI* sample of n_a elements is drawn in the first phase. Let $w_{a1} = n_{a1}/n_a$ and $w_{a2} = n_{a2}/n_a$ be the responding and nonresponding proportions, respectively. In the subsampling phase, *SI* sampling is also used, with n_2 nonrespondents subsampled from n_{a2}. Let the sampling fractions be $f_a = n_a/N$ in the first phase, and $f_2 = n_2/n_{a2}$ in the subsampling phase. From Result 15.4.1, the estimator is

$$\hat{t} = N(w_{a1}\bar{y}_{s_{a1}} + w_{a2}\bar{y}_{s_2}) \quad (15.4.5)$$

with a variance given by

$$V(\hat{t}) = N^2\left[\frac{1-f_a}{n_a} S^2_{yU} + E_{SI} E_{\mathrm{RD}}\left(w^2_{a2} \frac{1-f_2}{n_2} S^2_{ys_{a2}} | s_a\right)\right] \quad (15.4.6)$$

To derive the variance estimator is left as an exercise.

When subsampling of nonrespondents is planned, decisions are required about the first-phase sample size and the subsampling fraction necessary to attain the desired precision. Suppose that we want to minimize the variance (15.4.6) for a fixed expected sampling cost. This requires explicit expressions for both components of (15.4.6). A solution is to specify one or a few realistic models for the response distribution. Under each model, it may be possible to evaluate explicitly the second variance component and then solve the optimization problem. The results can be used as a guide in the planning process. We give two simple examples.

EXAMPLE 15.4.5. Suppose the following simple response distribution: For every s_a, the response set s_{a1} is generated as the result of n_a independent Bernoulli trials, one for each element k in s_a, with constant probability θ of "success," that is, response. Then, for any initial sample s_a and every k and $l \in s_a$, we have

$$\Pr(k \in s_{a1} | s_a) = \theta_{ks_a} = \theta; \qquad \Pr(k \& l \in s_{a1} | s_a) = \theta_{kl|s_a} = \theta^2$$

Let us assume that the design and the estimator are as in Example 15.4.4, with $f_2 = v$, where v is an a priori fixed constant satisfying $0 < v \leq 1$.

To simplify, assume that $n_2 = vn_{a2}$ is an integer. For a fixed n_{a2}, s_{a2} behaves as an *SI* sample from s_a, so from (15.4.6) we have

$$\begin{aligned} V(\hat{t}) &= N^2 S_{yU}^2 [(1 - f_a)/n_a + (1 - \theta)(1 - v)/vn_a] \\ &= A_0 + A_1/f_a + A_2/vf_a \end{aligned} \tag{15.4.7}$$

where

$$A_0 = -NS_{yU}^2, \qquad A_1 = NS_{yU}^2 \theta, \quad \text{and} \quad A_2 = NS_{yU}^2(1 - \theta)$$

Let the variable cost of the survey be $c_1 n_a + c_2 n_2$, where c_1 and c_2 are the per-element sampling costs in the first and second phases. Here, n_2 is random. The expected variable cost is obtained as

$$\text{EVC} = c_1 n_a + c_2(1 - \theta) n_a v = a_1 f_a + a_2 f_a v$$

where $a_1 = Nc_1$ and $a_2 = N(1 - \theta)c_2$. The problem of minimizing the variance given by (15.4.7) for a fixed expected variable cost now has the same form as the problem solved in Section 12.8. It follows that the optimum subsampling fraction v is determined by

$$v_{\text{opt}} = (a_1 A_2 / a_2 A_1)^{1/2} = (c_1 / c_2 \theta)^{1/2}$$

provided $0 < v_{\text{opt}} \leq 1$. A good guess or an accurate prior knowledge of the unknown θ is required to make use of this result. For example, if $c_1/c_2 = 1/9$ and $\theta = 0.6$, we obtain $v_{\text{opt}} = 0.43$.

EXAMPLE 15.4.6. Suppose the response is deterministic in the following way. The population consists of two groups U_1 and U_2. All elements in U_1 respond with probability 1 if selected, whereas for all elements in U_2 the response probability is 0. Suppose that *SI* sampling is used in both phases. The composition of the groups is fixed once and for all. Suppose that the subsample s_2, drawn from s_{a2}, is of size $n_2 = \nu n_{a2}$, where $0 < \nu \leq 1$ is an a priori fixed constant. Under these assumptions, it follows directly from (9.4.15) that the variance of (15.4.5) takes the form

$$V(\hat{t}) = N^2\left[\frac{1-f_a}{n_a}S_{yU}^2 + \frac{1}{n_a}\left(\frac{1}{\nu}-1\right)W_2 S_{yU_2}^2\right]$$
$$= A_0 + A_1/f_a + A_2/\nu f_a \tag{15.4.8}$$

where $A_0 = -NS_{yU}^2$, $A_1 = N(S_{yU}^2 - W_2 S_{yU_2}^2)$ and $A_2 = NW_2 S_{yU_2}^2$. Here $S_{yU_2}^2$ is the y variance in U_2 and $W_2 = N_2/N$ is the relative size of U_2.

If the variable cost is assumed to be $c_1 n_a + c_2 n_2$, as in Example 15.4.5, the expected variable cost is

$$\text{EVC} = a_1 f_a + a_2 f_a \nu$$

with $a_1 = Nc_1$ and $a_2 = NW_2 c_2$. Again we can use the results in Section 12.8. The optimal values of f_a and ν are derived without difficulty.

15.4.4. Randomized Response

Surveys dealing with sensitive questions, for example, drug use, tax evasion, and AIDS pose particular problems. In such surveys, many respondents either refuse to participate or give false or evasive responses. Methods that protect anonymity are a solution. Such protection is built into the *randomized response* technique, introduced by Warner (1965). The assumption is that randomized response will ensure the cooperation of all selected individuals, as well as truthful responses. The method is designed for surveys where data collection is by personal interview.

Suppose an estimate is required for the proportion of individuals with the sensitive attribute A, for instance, AIDS. Warner's original proposal assures anonymity as follows: Every respondent in a probability sample is asked to select, through the use of a randomized choice device (which may be as simple as an ordinary die), one out of two complementary statements, S and S^*. The statement S is "I have the attribute A," and S^* is "I do not have the attribute A." The respondent is then asked to respond "true" or "false" to the randomly selected statement, without revealing which of the two he or she is answering. A truthful answer is assumed.

The statistician knows the probability P with which the choice device selects the statement S, and it is this knowledge that makes it possible to construct an unbiased estimator of the number or the proportion of individuals with the sensitive attribute A, despite the fact that each respondent's true status with respect to A remains unknown to the statistician. To see how this is achieved, let $y_k = 1$ if the kth individual in the finite population of size N has the attribute A, and $y_k = 0$ otherwise ($k = 1, \ldots, N$). Furthermore, for $k \in s$, define $x_k = 1$ if the kth individual gives the answer "True", and $x_k = 0$ otherwise. Let the index RC indicate "with respect to the randomized choice device." Then, for any element k,

$$\Pr_{\mathrm{RC}}(x_k = 1) = y_k P + (1 - y_k)(1 - P) = 1 - P + (2P - 1)y_k \quad (15.4.9)$$

Assuming $P \neq 1/2$, we define the "pseudo-value"

$$\hat{y}_k = \frac{x_k + P - 1}{2P - 1} \quad (15.4.10)$$

For every k, $\hat{y}_k$ is unbiased for y_k with respect to the choice device, that is, $E_{\mathrm{RC}}(\hat{y}_k) = y_k$, and the variance is

$$V_{\mathrm{RC}}(\hat{y}_k) = \frac{P(1 - P)}{(2P - 1)^2} = V_0 \quad (15.4.11)$$

Suppose that sample selection is by a design $p(s)$ with positive inclusion probabilities π_k and π_{kl}. The ordinary π estimator

$$\hat{t}_\pi = \sum_s y_k/\pi_k$$

is out of question, since direct observation of y_k would cause refusals and false answers. Instead use

$$\hat{t}_{\mathrm{RR}} = \sum_s \hat{y}_k/\pi_k$$

This estimator is easily seen to be unbiased for the population total $t = \sum_U y_k$:

$$\begin{aligned} E(\hat{t}_{\mathrm{RR}}) &= E_p\{E_{\mathrm{RC}}[(\textstyle\sum_s \hat{y}_k/\pi_k)|s]\} \\ &= E_p(\textstyle\sum_s y_k/\pi_k) = \sum_U y_k = t \end{aligned}$$

where the index p indicates "with respect to the sampling design." Moreover, using the customary conditioning argument,

$$V(\hat{t}_{\mathrm{RR}}) = V_p[E_{\mathrm{RC}}(\hat{t}_{\mathrm{RR}}|s)] + E_p[V_{\mathrm{RC}}(\hat{t}_{\mathrm{RR}}|s)]$$

one obtains, since $\mathrm{Cov}_{\mathrm{RC}}(\hat{y}_k, \hat{y}_l|s) = 0$ for all $k \neq l$,

$$V(\hat{t}_{\mathrm{RR}}) = V_p(\hat{t}_\pi) + (\textstyle\sum_U 1/\pi_k)V_0$$

Here,

$$V_p(\hat{t}_\pi) = \sum\sum_U \Delta_{kl}(y_k/\pi_k)(y_l/\pi_l)$$

is the variance of the π estimator, with $\Delta_{kl} = \pi_{kl} - \pi_k\pi_l$, and V_0 is given by equation (15.4.11). The second variance component, $(\sum_U 1/\pi_k)V_0$, can be viewed as the price to pay for randomizing the response. The price may well be worth paying to get an unbiased estimate. The second variance component is large if P lies near 1/2, so P should be chosen well away from 1/2.

An unbiased variance estimator is given by

$$\hat{V}(\hat{t}_{\mathrm{RR}}) = \sum\sum_s \check{\Delta}_{kl} \frac{\hat{y}_k}{\pi_k} \frac{\hat{y}_l}{\pi_l} + N \cdot V_0 \tag{15.4.12}$$

where $\check{\Delta}_{kl} = \Delta_{kl}/\pi_{kl}$. If N is unknown, it should be replaced by the estimate $\hat{N} = \sum_s 1/\pi_k$.

EXAMPLE 15.4.7. A criminologist interested in economic crime wishes to estimate the proportion of owners of single-proprietor businesses that were guilty of tax evasion in the preceding year. He selects an SI sample of $n = 400$ business owners from a population of 2,000. Every person in the sample is asked to randomly pick a card from a deck, in which 90% of the cards read "I cheated on last year's income tax return" and the remaining 10% read "I did not cheat on last year's income tax return." The survey gave 88 "true" responses and 312 "false" responses. Let us compute an approximately 95% confidence interval for the proportion of business owners in the population who were guilty of tax evasion. Truthful responses are assumed. The unknown proportion of tax evaders to estimate is

$$\bar{y}_U = \sum_U y_k/N$$

where $y_k = 1$ if the kth business owner is guilty of tax evation, 0 otherwise. An unbiased estimator is

$$\hat{\bar{y}}_{\mathrm{RR}} = \frac{1}{n}\sum_s \hat{y}_k = \frac{1}{2P-1}\left(\frac{x}{n} + P - 1\right) \tag{15.4.13}$$

where $x = \sum_s x_k$ is the number of "true" answers in the sample. The unbiased variance estimator obtained from (15.4.12) is

$$\hat{V}(\hat{\bar{y}}_{\mathrm{RR}}) = \frac{1}{(2P-1)^2}\left[\frac{1-f}{n-1}\frac{x}{n}\left(1 - \frac{x}{n}\right) + \frac{P(1-P)}{N}\right] \tag{15.4.14}$$

Equations (15.4.13) and (15.4.14) are easily obtained from the general formulas. From the survey results we compute

$$\hat{\bar{y}}_{\mathrm{RR}} = 0.15 \quad \text{and} \quad \hat{V}(\hat{\bar{y}}_{\mathrm{RR}}) = 0.000608$$

which gives the approximate 95% confidence limits 0.15 ± 0.05.

Warner's (1965) seminal paper was followed by a rapid development of randomized response techniques. A number of variations of the basic techniques are now available. A comprehensive review is given in the monograph by Chaudhuri and Mukerjee (1988). Danermark and Swensson (1987) discuss design issues in an experimental survey where randomized response was used. Data collection by randomized response is compared with data collection by anonymous questionnaires.

A further interesting aspect of the randomization device is that it can be used to protect the anonymity of individuals who have provided sensitive information. Prior to release or storage, one can "depersonalize" data on individuals. That is, instead of revealing an individual's true response y_k, one would release or store a value $\hat{y}_k$ coded in the manner of equation (15.4.10). This will permit an analyst, at a later date, to make valid estimates from the stored coded data.

15.5. Perspectives on Nonresponse

In Section 15.4, we considered two special techniques that lead to unbiased estimators, if particular conditions are satisfied. Subsampling of nonrespondents was one of them. An unbiased estimator is obtained with this method provided there is 100% response among elements in the subsample. Randomized response was another technique. The assumption here was that anonymity will guarantee the cooperation of every person selected for the survey. Both subsampling of nonrespondents and randomized response require special arrangements that may be costly and time consuming. Especially in large-scale standardized production of statistics, such arrangements may be impossible. Consequently, the survey statistician must often accept that some nonresponse is inevitable, and, from the data that are actually collected, the best possible estimates must be produced in an efficient, standardized fashion.

If there is full or nearly full response, unbiased or nearly unbiased estimators can easily be constructed, as we know from earlier chapters. Among these estimators, the statistician selects one that has a small variance. However, when the nonresponse is not negligible, there is a shift of emphasis in the search for a good estimator. Small variance remains a desirable goal, but to keep the bias within reasonable limits is an even more important concern. The complete elimination of the bias is often a futile hope, but greatly biased estimators must be avoided as they may lead to grossly invalid confidence statements.

In the following, we construct estimators with the aid of *response models.* Such a model is a set of assumptions about the true unknown *response distribution.* In the planning of a survey, different estimators should be considered and evaluated, using alternative but realistic response models. The goal is to select an estimator that is bias resistant and has low variance. Such estimators ordinarily require strong auxiliary information. The availability of such information should be part of the survey plan.

Assumptions about the response play an important role in justifying the different estimators in common use. Let us examine some of the assumptions that have been invoked.

For a number of years, survey statisticians favored a deterministic model according to which the population is dichotomized into a response stratum and a nonresponse stratum. We used this model in Example 15.4.6. The idea is described by Cochran (1977, p. 359) as follows:

> In the study of nonresponse it is convenient to think of the population as divided into two "strata," the first consisting of all units for which measurements would be obtained if the units happened to fall in the sample, the second of the units for which no measurements would be obtained.

Under this model, elements in the response stratum respond with probability one, all other elements respond with probability zero. Several proposed techniques for nonresponse treatment are based on this model, and it inspired much of the work on nonresponse in central statistical offices in the 1960s and 1970s. Cochran (1977, p. 360) points out the limitations of the model:

> The division into two distinct strata is, of course, an over-simplification. Chance plays a part in determining whether a unit is found and measured in a given number of attempts. In a more complete specification of the problem we would attach to each unit a probability representing the chance that it would be measured by a given field method if it fell in the sample.

Much of the recent work on estimation in the presence of nonresponse has used the idea expressed in this citation, namely, that the response is stochastic, not deterministic. Under such an outlook, a response distribution is assumed to exist. The known sample selection distribution (that is, the sampling design) and the unknown response distribution together form the basis of a randomization theory extended to incorporate nonresponse. Oh and Scheuren (1983) describe this as "quasi-randomization," since the response distribution is an assumption only. The approach has been used to develop general frameworks for estimation with nonresponse and other nonsampling errors, as in Platek and Gray (1983). The techniques given in Section 15.6 are based on the quasi-randomization approach.

The response distribution and the response probabilities are likely to depend on the conditions of the survey. Dalenius (1983) argues:

> ... it appears utterly unrealistic to postulate fixed "response probabilities" which are independent of the varying circumstances under which an effort is made to elicit a response. Whether an individual selected for a survey will respond or not may in many instances be determined by factors external to the individual. In an interview survey, for example, a black individual may respond if the interviewer is also black, but may not respond if the interviewer is white. In other words, for this individual there is no single unique q value, but possibly a wide spectrum of q values.

In this citation, q represents the response probability. Given the particular conditions of the survey, the statistician should try to formulate a good model for the response distribution and use it to build an estimator that adjusts for the nonresponse with the aid of auxiliary variables. Examples of this method can be seen in Section 15.6. However, as Kalton (1983) puts it,

> sampling practitioners do not believe that the nonresponse models on which their adjustments are based hold exactly: they simply hope that they are improvements on the model of data missing at random.

The citation expresses a rather modest hope; in many cases, an adjustment obtained from a realistic response model leads to a substantial reduction of the nonresponse bias. However, the problem of bias is never completely resolved. Kalton and Kasprzyk (1986) warn,

> ...all methods of handling missing survey data must depend upon untestable assumptions. If the assumptions are seriously in error, the analyses may give misleading conclusions. The only secure safeguard against serious nonresponse bias in survey estimates is to keep the amount of missing data small.

The approaches in the following sections can lead to reasonably good estimators, but they are not without bias. The main techniques for dealing with nonresponse are *weighting adjustment* and *imputation*. Weighting adjustment implies that higher than normal weights are applied in the estimation to the y-values of the respondents to compensate for the values that are lost because of nonresponse. Weighting adjustment based on response distribution modeling and estimation of response probabilities is discussed in Section 15.6.

Imputation implies the substitution of "good" artificial values for the missing values. Two principal uses of imputation are the following:

a. Imputation for the item nonresponse only. Imputed values are then provided for missing values corresponding to elements k in the item nonresponse set, $r_u - r_c$. Weighting adjustment is then applied to compensate for the unit nonresponse.
b. Imputation for the item nonresponse as well as for the unit nonresponse. Imputed values are then provided for missing values corresponding to elements k in the set $s - r_c$. No weighting adjustment is applied. Estimates are computed using genuine as well as imputed values.

A complete rectangular data matrix is the result of the imputation, both in case (a) and in case (b). The matrix is of dimension $n_{r_u} \times q$ in case (a) and $n_s \times q$ in case (b). It is a considerable advantage in the estimation phase to be able to work with complete matrices. Imputation is discussed in Section 15.7.

15.6. Estimation in the Presence of Unit Nonresponse

15.6.1. Response Modeling

For simplicity, we concentrate on the case of a single study variable, denoted y. The aim is to estimate the population total $t = \sum_U y_k$. We assume the following survey conditions:

i. A sample s of size n_s is drawn from the finite population $U = \{1, \ldots, k, \ldots, N\}$ according to the design $p(s)$, with positive inclusion probabilities π_k and π_{kl}.

ii. The survey is carefully designed to yield a high response rate, but some nonresponse is nevertheless anticipated. Denote the response set by r and its size by m_r, where $m_r < n_s$.

What estimator of t should be used? The answer will depend on the assumed response model and on the auxiliary information.

It is important now to make the distinction between a *response model* and the actual *response distribution.* A response model is a set of assumptions about the true unknown response distribution. The response model, which will almost certainly differ from the true response distribution, is a tool by which we construct an estimator. Such an estimator will have commendable properties, such as unbiasedness, or approximate unbiasedness, provided the response model coincides with the actual response distribution, but not necessarily otherwise. This is illustrated by the following example.

EXAMPLE 15.6.1. Consider the following naive response model:

$$\Pr(k \in r|s) = \theta_k = \theta; \qquad \Pr(k\&l \in r|s) = \theta_k\theta_l = \theta^2 \tag{15.6.1}$$

for all k and $l \in s$ and every s, and where $\theta > 0$. This can be described as a model for data missing at random throughout the population. It recognizes no differences between elements with regard to their response behavior and is ordinarily much too simple to deal effectively with the response patterns met in practice. Suppose now that the estimator that would have been used in the case of full response is

$$\hat{t} = N\tilde{y}_s = N\frac{\sum_s y_k/\pi_k}{\sum_s 1/\pi_k}$$

If the model (15.6.1) is adopted, despite its simplicity, as a model for the response, then an appropriate modification of $\hat{t}$ is

$$\hat{t}_1 = N\tilde{y}_r = N\frac{\sum_r y_k/\pi_k\theta}{\sum_r 1/\pi_k\theta} = N\frac{\sum_r y_k/\pi_k}{\sum_r 1/\pi_k} \tag{15.6.2}$$

Since the unknown θ conveniently vanishes, this estimator can always be calculated. However, it corresponds to "doing nothing about the nonresponse," in the sense that the response model implies no difference in the weighting of respondent values y_k. The estimator shown in (15.6.2) has the same form as in the case of full response. The only difference is that it is formed over the response set r instead of over the entire sample s. It is of interest to evaluate the bias of $\hat{t}_1$. The bias depends on the true response distribution. Let us consider three cases.

i. The model shown in (15.6.1) is a perfect description of the true response distribution. In this case, $\hat{t}_1$ is essentially unbiased. The negligible bias is not due to the nonresponse per se, but to the fact that the estimator is of the ratio type. This case is highly unlikely. The response behavior ordinarily varies among individuals.

ii. The model shown in (15.6.1) is false, and the true response distribution is such that individual responses are independent but with varying response probabilities, so that

$$\Pr(k \in r|s) = \theta_k; \qquad \Pr(k\&l \in r|s) = \theta_k\theta_l$$

Under this response distribution, the bias of $\hat{t}_1$ is

$$B(\hat{t}_1) = E(\hat{t}_1) - t \doteq N\frac{\sum_U y_k\theta_k}{\sum_U \theta_k} - t$$

$$= (N-1)S_{y\theta U}/\bar{\theta}_U = t(N-1)R_{y\theta U}cv_{yU}cv_{\theta U}/N \qquad (15.6.3)$$

which means that the relative bias of $\hat{t}_1$ is given by

$$RB(\hat{t}_1) = B(\hat{t}_1)/t \doteq R_{y\theta U}cv_{yU}cv_{\theta U} \qquad (15.6.4)$$

where $R_{y\theta U}$ is the finite population correlation between y and θ, and cv_{yU}, $cv_{\theta U}$ are the coefficients of variation of y and θ. The expression (15.6.3) shows that the correlation $R_{y\theta U}$ is a key factor in the bias. The higher the correlation between the study variable y and the response probability θ, the higher the relative bias.

iii. The true response behavior is deterministic, as in Example 15.4.6, with a response stratum U_1 of size N_1 and a nonresponse stratum U_2 of size N_2. Elements in U_1 respond with probability 1, elements in U_2 with probability 0. Let $\bar{y}_{U_1}$ and $\bar{y}_{U_2}$ be the means of the two strata. Then it is easily shown that the bias of $\hat{t}_1$ is approximately given by

$$B(\hat{t}_1) \doteq N_2(\bar{y}_{U_1} - \bar{y}_{U_2}) \qquad (15.6.5)$$

That is, if the model with a response stratum and a nonresponse stratum holds, the bias increases with the difference between the strata means, and with the size of the nonresponse stratum. For the bias to be small, we must have that the two strata means are nearly equal, or that the nonresponse stratum is small, or both.

15.6.2. A Useful Response Model

Example 15.6.1 shows that simplistic response modeling can lead to severely biased estimates. We now consider more realistic response models. In a given survey, the elements of the sample s are exposed to a specific set of survey operations. We assume that a unique response distribution (RD) corresponds to these survey operations. This response distribution generates a response set r. The statistician's task is to formulate a model that describes the unknown response distribution as accurately as possible.

A simple but powerful family of response models is obtained from the following assumptions. The realized sample s is partitioned into H_s groups s_h ($h = 1, \ldots, H_s$). Given s, all elements in s_h are assumed to have the same response probability; different groups have different response probabilities. Elements respond independently of each other. We call this a *response homogeneity group* (RHG) *model*; the sets $s_1, \ldots, s_h, \ldots, s_{H_s}$ are the response homogeneity groups. The RHG model has the following mathematical formulation: For every sample s and for $h = 1, \ldots, H_s$,

$$\begin{cases} \Pr(k \in r|s) = \pi_{k|s} = \theta_{hs} > 0 & \text{for all } k \in s_h \\ \Pr(k\&l \in r|s) = \pi_{kl|s} = \Pr(k \in r|s)\Pr(l \in r|s) & \text{for all } k \neq l \in s \end{cases} \tag{15.6.6}$$

The number of groups H_s need not be the same from one sample s to another. Even if the number of groups is the same for two different samples, a given element k may be assigned to different groups in the two cases. If repeated draws of samples are imagined, the groups are always defined in the same way whenever a specific s is realized. That is, when s is given, the composition of the groups is given.

The groups are to be chosen so that (15.6.6) describes as accurately as possible the unknown response mechanism for the sample s. The statistician must use his/her judgment and survey experience in forming groups that correspond closely to the idea of constant response probability within a group. The number of interviewers, their skill and background, characteristics of a respondent, such as age, sex, and residence (urban or rural), are among factors to consider when the groups are set up. A *response propensity scoring* technique has been proposed for forming the groups (see Rosenbaum and Rubin, 1983; Little, 1986).

Under the response model shown in (15.6.6), data are missing at random within sample subgroups, conditionally on s. This is a substantial improvement on the model (15.6.1), which assumed that data are missing at random throughout the population. To illustrate, suppose that response probability is negatively correlated with income. That is, high-income earners have a lower response probability, on the average, than low-income earners. For any study variable that has a strong positive correlation with income, such as savings or a certain type of consumption, the effect is that persons with large values of the study variable are underrepresented among the respondents. This causes negative bias in the estimates, unless appropriate adjustments can be made.

Adjustments based on constant response probability within well-chosen groups can lead to a considerable reduction of the bias, compared to the assumption of a constant response probability throughout the population. It should be kept in mind that a model is never a perfect image of the real world, but with a good response model, we can take a long step in the direction of unbiasedness and valid inferences.

Let r_h be the responding subset of the group s_h. We use the simplified notation n_h and m_h (rather than n_{s_h} and m_{r_h}) for the sizes of s_h and r_h, respectively. The total set of respondents is denoted r, and its size by m_r. We then have

$$r = \bigcup_{h=1}^{H_s} r_h; \qquad m_r = \sum_{h=1}^{H_s} m_h$$

If the model (15.6.6) holds, the response set r is distributed, for a given s, in accordance with the sampling design *STBE*, that is, the stratified Bernoulli sampling design (see Section 3.7.2). With slight modification, we can use results from the two-phase sampling theory in Section 9.8. The form of the estimator of the total $t = \sum_U y_k$ will depend on the available auxiliary information.

It is convenient to condition on the response count vector $\mathbf{m} = (m_1, \ldots, m_h, \ldots, m_{H_s})$. For $h = 1, \ldots, H_s$, we introduce the conditional response probabilities

$$\pi_{k|s,\,\mathbf{m}} = \Pr(k \in r|s, \mathbf{m}) = \pi_{kk|s,\,\mathbf{m}} = f_h = m_h/n_h; \quad \text{all} \quad k \in s_h \tag{15.6.7a}$$

and

$$\pi_{kl|s,\,\mathbf{m}} = \Pr(k\&l \in r|s, \mathbf{m}) = \begin{cases} f_h(m_h - 1)/(n_h - 1); & \text{all } k \neq l \in s_h \\ f_h f_{h'}; & k \in s_h,\, l \in s_{h'};\, h \neq h' \end{cases} \tag{15.6.7b}$$

The form of these probabilities is familiar from *STSI* sampling. They follow from the important fact that, given s and $\mathbf{m}$, r behaves under the RHG model as an *STSI* selection from s. One consequence is that, given s and $\mathbf{m}$, the probability of response of any element in group h is equal to the group response rate f_h, provided the RHG model holds.

The quantities $1/\pi_{k|s,\,\mathbf{m}}$ play the role of nonresponse adjustment weights in the estimators now to be examined. We treat two cases: (1) estimators that use weighting only, and (2) estimators that use weighting as well as auxiliary variables. A detailed description of these estimators is given in Särndal and Swensson (1987).

Remark 15.6.1. Some authors, for example, Oh and Scheuren (1983), discuss a model called Uniform Response Mechanism within Subpopulations. It bears some resemblance to the RHG model and postulates a set of fixed population groups with uniform response probability in each group. By contrast, in the RHG model shown in (15.6.6), the groups are suitably defined for the realized sample and the particular conditions of the survey.

Remark 15.6.2. The model (15.6.6) is widely used in practice for modeling nonresponse. No practitioner really believes that all elements in a group have exactly the same probability to respond, but the point is that the assumption of constant probability within well-constructed groups removes most of the nonresponse bias and is a vast improvement over more naive models.

Remark 15.6.3. Rubin (1976, 1983, 1987) makes a distinction between ignorable and nonignorable response mechanisms (see also Little and Rubin, 1987, chapter 4). Informally speaking, if the response mechanism is assumed to be unrelated to the values of the study variables, then the response mechanism can be ignored for the Bayesian inferences studied by Rubin, and the observed values can be treated as a random subsample of the values that would have been observed in the hypothetical case of complete response. Both the model "data missing at random throughout the population" and the model "data missing at random within groups" are of the ignorable type.

15.6.3. Estimators That Use Weighting Only

Under the RHG model (15.6.6), a suitable weight for the kth element is

$$1/\pi_k^* = (1/\pi_k)(1/\pi_{k|s,\,\mathbf{m}})$$

where $1/\pi_k$ is the sampling weight and $1/\pi_{k|s,\,\mathbf{m}}$ is the nonresponse adjustment weight defined by (15.6.7a). The nonresponse adjustment weight of an element is the inverse of the response rate in the group to which the element belongs, which is intuitively sound. The weighting estimator of $t = \sum_U y_k$ is therefore

$$\hat{t}_{c\pi^*} = \sum_r \check{y}_k/\pi_{k|s,\,\mathbf{m}} = \sum_{h=1}^{H_s} f_h^{-1} \sum_{r_h} \check{y}_k \tag{15.6.8}$$

where $\check{y}_k = y_k/\pi_k$.

Let us find the expected value and the variance of the estimator (15.6.8). The expected value is in this case to be interpreted as an average over a long series of experiments, where each experiment consists of the following:

1. The selection of a sample s by the given sampling design, followed by the RHG grouping that corresponds to s.
2. Given s, the realization of a response set, $r = \bigcup_{h=1}^{H_s} r_h$, under the conditions of the RHG model shown in (15.6.6).

Suppose the RHG model (15.6.6) coincides with the true response distribution (RD). Starting with the conditional expectation of (15.6.8), given s, we obtain

$$E_{\mathrm{RD}}(\hat{t}_{c\pi^*}|s) = E_{\mathbf{m}}E_{\mathrm{RD}}\left(\sum_{h=1}^{H_s} f_h^{-1} \sum_{r_h} \check{y}_k|s,\,\mathbf{m}\right)$$

$$= E_{\mathbf{m}}\left(\sum_{h=1}^{H_s} \sum_{s_h} \check{y}_k|s\right) = \sum_s \check{y}_k = \hat{t}_\pi$$

This demonstrates an important property of the nonresponse adjusted estimator $\hat{t}_{c\pi^*}$, namely, that, for any given sample s, $\hat{t}_{c\pi^*}$ is on the average equal to the estimate that would have been calculated if every sample element had responded, that is, $\hat{t}_\pi = \sum_s \check{y}_k$.

The reader will notice that a condition of usually minor practical importance is implicit in this result, namely, that the probability under the RHG model of the event

$$\bar{A}_1 = \{m_h = 0 \text{ for some } h = 1, \ldots, H_s\}$$

is negligible. The condition is needed because the estimator (15.6.8) cannot be calculated if $\bar{A}_1$ occurs. A special definition must be given to $\hat{t}_{c\pi^*}$ to cover this event. But if $\bar{A}_1$ has negligible probability, as when the sample size is large, it is of little consequence how this definition is made. Let $E_p(\cdot)$ denote expectation with respect to the sampling design. The unconditional expectation of $\hat{t}_{c\pi^*}$ is now obtained as

$$E(\hat{t}_{c\pi^*}) = E_p[E_{\text{RD}}(\hat{t}_{c\pi^*}|s)] = E_p(\hat{t}_\pi) = t$$

That is, $\hat{t}_{c\pi^*}$ is an unbiased estimator, provided that the RHG model holds and $\bar{A}_1$ has negligible probability. This and other properties of $\hat{t}_{c\pi^*}$ are spelled out in the following result, where the event $\bar{A}_2$ is defined by

$$\bar{A}_2 = \{m_h \leq 1 \text{ for some } h = 1, \ldots, H\}$$

Result 15.6.1. *If the RHG model shown in* (15.6.6) *holds and if the event* $\bar{A}_1$ *has negligible probability, then the direct weighting estimator*

$$\hat{t}_{c\pi^*} = \textstyle\sum_r \check{y}_k/\pi_{k|s,\mathbf{m}} = \displaystyle\sum_{h=1}^{H_s} f_h^{-1} \textstyle\sum_{r_h} \check{y}_k \tag{15.6.8}$$

is unbiased for $t = \sum_U y_k$ *with the variance*

$$V(\hat{t}_{c\pi^*}) = \textstyle\sum\sum_U \Delta_{kl}\check{y}_k\check{y}_l + E_p E_{\mathbf{m}}\left(\displaystyle\sum_{h=1}^{H_s} n_h^2 \frac{1-f_h}{m_h} S^2_{\check{y}s_h}\Big|s\right) \tag{15.6.9}$$

where $S^2_{\check{y}s_h}$ *is the variance in* s_h *of* $\check{y}_k = y_k/\pi_k$, $E_p(\cdot)$ *is with respect to the design, and* $E_{\mathbf{m}}(\cdot|s)$ *is with respect to the distribution of* **m**, *given s. If the model* (15.6.6) *holds and if the event* $\bar{A}_2$ *has negligible probability, then an unbiased variance estimator is given by*

$$\hat{V}(\hat{t}_{c\pi^*}) = \textstyle\sum\sum_r \dfrac{\check{\Delta}_{kl}}{\pi_{kl|s,\mathbf{m}}}\check{y}_k\check{y}_l + \displaystyle\sum_{h=1}^{H_s} n_h^2 \frac{1-f_h}{m_h} S^2_{\check{y}r_h} \tag{15.6.10}$$

where $\pi_{kl|s,\mathbf{m}}$ *is given by* (15.6.7a) *and* (15.6.7b), *and* $S^2_{\check{y}r_h}$ *is the variance in* r_h *of* $\check{y}_k = y_k/\pi_k$.

The result is given in Särndal and Swensson (1987). Singh and Singh (1979) considered the special case of a single group. It is left as an exercise to verify the expressions given in (15.6.9) and (15.6.10) with the aid of conditioning on s and **m**, similarly as in Section 9.8, where s_a and **n** correspond to s and **m**. Note that (15.6.10) corresponds to (9.8.5).

The first component in (15.6.9) is identical to the variance of the π estimator for the complete sample. The second component represents the increase in variance caused by nonresponse, so this component is zero for full response, that is, when $f_h = m_h/n_h = 1$ for all h. Each term of the variance estimator (15.6.10) is unbiased for its counterpart in (15.6.9).

The expression given in (15.6.10) is general and compact. In practice, the variance estimator must be worked out for each specific sampling design. The results are often less elegant, as illustrated by the following example.

EXAMPLE 15.6.2. Suppose *SI* sampling is used to draw the sample s, which is of size n. From (15.6.8) we then obtain

$$\hat{t}_{c\pi*} = \frac{N}{n}\sum_{h=1}^{H_s} n_h \bar{y}_{r_h} = N\hat{\bar{y}}_U \tag{15.6.11}$$

This is sometimes called the *weighting class estimator*. It is widely used in practice. Equations (15.6.9) and (15.6.10) then take the forms

$$\begin{aligned} V(\hat{t}_{c\pi*}) &= N^2\frac{1-n/N}{n}S_{yU}^2 + \frac{N^2}{n^2}E_pE_{\mathbf{m}}\left(\sum_{h=1}^{H_s} n_h^2\frac{1-f_h}{m_h}S_{ys_h}^2 | s\right) \\ &= V_1 + V_2 \end{aligned} \tag{15.6.12}$$

and

$$\hat{V}(\hat{t}_{c\pi*}) = \hat{V}_1 + \hat{V}_2 \tag{15.6.13}$$

with

$$\hat{V}_1 = N^2\frac{1-n/N}{n}\left[\sum_{h=1}^{H_s}\frac{n_h}{n}(1-\delta_h)S_{yr_h}^2 + \frac{n}{n-1}\sum_{h=1}^{H_s}\frac{n_h}{n}(\bar{y}_{r_h}-\hat{\bar{y}}_U)^2\right] \tag{15.6.14}$$

and

$$\hat{V}_2 = N^2\sum_{h=1}^{H_s}\left(\frac{n_h}{n}\right)^2\frac{1-f_h}{m_h}S_{yr_h}^2 \tag{15.6.15}$$

where

$$\delta_h = \frac{1-n_h/n}{m_h}\,\frac{n}{n-1}$$

The first term of (15.6.12) is the familiar variance of $\hat{t}_\pi = N\bar{y}_s$, for an *SI* sample of size n. Note again the nonexplicit form of the second component. The variance estimator is, not unexpectedly, of the same form as for the two-phase sampling design in Example 9.4.2.

Remark 15.6.4. Most surveys have several questionnaire items (that is, several study variables), with nonidentical response sets. We can then apply Result 15.6.1 to the variables one at a time. A careful modeling of the response distribution may require a different set of RHGs for each study variable. This may be time-consuming. In the interest of efficiency and timely data handling, many statisticians prefer to compromise and apply the same adjustment weighting for all variables, after having first created a complete rectangular data set through imputation for the item nonresponse. These matters are discussed in Section 15.7.

15.6.4. Estimators That Use Weighting as Well as Auxiliary Variables

If auxiliary information is available for prediction of y-values, we can often improve substantially over the estimator given by (15.6.8). The improvement is twofold. The major benefit is improved resistance against selection bias when the assumed response model is erroneous. Variance reduction is a second benefit. These ideas are explored in Bethlehem (1988) and Särndal and Swensson (1987).

Result 9.7.1 presented a general regression estimator for two-phase sampling. It can be adapted for use in the presence of nonresponse. Under the RHG model given by (15.6.6), a regression estimator in the image of (9.7.20) is

$$\hat{t}_{cr} = \sum_U \hat{y}_{1k} + \sum_{h=1}^{H_s} \left(\sum_{s_h} \frac{\hat{y}_k - \hat{y}_{1k}}{\pi_k} + f_h^{-1} \sum_{r_h} \frac{y_k - \hat{y}_k}{\pi_k} \right) \tag{15.6.16}$$

This estimator uses two types of regression predictions:

$$\hat{y}_k = \mathbf{x}_k' \hat{\mathbf{B}}_r = \mathbf{x}_k' \left(\sum_{h=1}^{H_s} f_h^{-1} \sum_{r_h} \frac{\mathbf{x}_k \mathbf{x}_k'}{\sigma_k^2 \pi_k} \right)^{-1} \sum_{h=1}^{H_s} f_h^{-1} \sum_{r_h} \frac{\mathbf{x}_k y_k}{\sigma_k^2 \pi_k} \tag{15.6.17}$$

and

$$\hat{y}_{1k} = \mathbf{x}_{1k}' \hat{\mathbf{B}}_{1k} = \mathbf{x}_{1k}' \left(\sum_{h=1}^{H_s} f_h^{-1} \sum_{r_h} \frac{\mathbf{x}_{1k} \mathbf{x}_{1k}'}{\sigma_{1k}^2 \pi_k} \right)^{-1} \sum_{h=1}^{H_s} f_h^{-1} \sum_{r_h} \frac{\mathbf{x}_{1k} y_k}{\sigma_{1k}^2 \pi_k} \tag{15.6.18}$$

Here, $\mathbf{x}_k$ is an auxiliary vector for which there exists information at the sample level. To calculate the estimator (15.6.16), we must know the sums $\sum_{s_h} \mathbf{x}_k$ for all groups h, and have access to the individual values $\mathbf{x}_k$ for elements $k \in r$. Further, $\mathbf{x}_{1k}$ is an auxiliary vector available at the population level. The estimator requires that we know $\sum_U \mathbf{x}_{1k}$ as well as $\sum_{s_h} \mathbf{x}_{1k}$ for all h, and individual values $\mathbf{x}_{1k}$ must be known for elements $k \in r$.

Noteworthy special cases of the estimator (15.6.16) are:

a. The only auxiliary information is at the sample level. The estimator shown in equation (15.6.16) is then

$$\hat{t}_{cr} = \sum_{h=1}^{H_s} \left(\sum_{s_h} \frac{\hat{y}_k}{\pi_k} + f_h^{-1} \sum_{r_h} \frac{y_k - \hat{y}_k}{\pi_k} \right) \tag{15.6.19a}$$

This is an important case in practice. To gather auxiliary information for the elements in the selected sample is quite feasible in many surveys, without excessive effort and cost. The payoff is reduced nonresponse bias as well as reduced variance.

b. Auxiliary information is available at the population level, but there is no further information available at the sample level only. The estimator (15.6.16) is then

$$\hat{t}_{cr} = \sum_U \hat{y}_{1k} + \sum_{h=1}^{H_s} f_h^{-1} \sum_{r_h} \frac{y_k - \hat{y}_{1k}}{\pi_k} \tag{15.6.19b}$$

For the variance and the variance estimator corresponding to (15.6.16) and its special cases, we need the following residuals,

$$\check{E}_k = E_k/\pi_k = (y_k - \mathbf{x}_k'\mathbf{B}_s)/\pi_k \tag{15.6.20}$$

for $k \in s$, with

$$\mathbf{B}_s = (\textstyle\sum_s \mathbf{x}_k\mathbf{x}_k'/\sigma_k^2\pi_k)^{-1} \textstyle\sum_s \mathbf{x}_k y_k/\sigma_k^2\pi_k \tag{15.6.21}$$

and

$$\check{E}_{1k} = E_{1k}/\pi_k = (y_k - \mathbf{x}_{1k}'\mathbf{B}_1)/\pi_k \tag{15.6.22}$$

for $k \in U$, with

$$\mathbf{B}_1 = (\textstyle\sum_U \mathbf{x}_{1k}\mathbf{x}_{1k}'/\sigma_{1k}^2)^{-1} \textstyle\sum_U \mathbf{x}_{1k} y_k/\sigma_{1k}^2 \tag{15.6.23}$$

Moreover, let

$$\check{e}_{kr} = e_{kr}/\pi_k = (y_k - \hat{y}_k)/\pi_k \tag{15.6.24}$$

and

$$\check{e}_{1kr} = e_{1kr}/\pi_k = (y_k - \hat{y}_{1k})/\pi_k \tag{15.6.25}$$

We now have the following result for the regression estimator (15.6.16).

Result 15.6.2. *If the RHG model* (15.6.6) *holds, the regression estimator*

$$\hat{t}_{cr} = \textstyle\sum_U \hat{y}_{1k} + \displaystyle\sum_{h=1}^{H_s} \left(\textstyle\sum_{s_h} \dfrac{\hat{y}_k - \hat{y}_{1k}}{\pi_k} + f_h^{-1} \textstyle\sum_{r_h} \dfrac{y_k - \hat{y}_k}{\pi_k} \right) \tag{15.6.16}$$

is approximately unbiased for the population total $t = \sum_U y_k$. *The values* $\hat{y}_k$ *and* $\hat{y}_{1k}$ *are defined by equations* (15.6.17) *and* (15.6.18). *The approximate variance is*

$$AV(\hat{t}_{cr}) = \textstyle\sum\sum_U \Delta_{kl} \check{E}_{1k} \check{E}_{1l} + E_p E_{\mathbf{m}} \left\{ \displaystyle\sum_{h=1}^{H_s} n_h^2 \frac{1 - f_h}{m_h} S_{\check{E}s_h}^2 | s \right\} \tag{15.6.26}$$

where $S_{\check{E}s_h}^2$ *is the variance in the set* s_h *of* $\check{E}_k$ *given by* (15.6.20) *and where* $\check{E}_{1k}$ *is given by* (15.6.22). *A variance estimator is*

$$\hat{V}(\hat{t}_{cr}) = \textstyle\sum\sum_r \dfrac{\check{\Delta}_{kl}}{\pi_{kl|s,\mathbf{m}}} \check{e}_{1kr} \check{e}_{1lr} + \displaystyle\sum_{h=1}^{H_s} n_h^2 \frac{1 - f_h}{m_h} S_{\check{e}r_h}^2 \tag{15.6.27}$$

where $S_{\check{e}r_h}^2$ *is the variance in the set* r_h *of* $\check{e}_{kr}$ *given by* (15.6.24), *and* $\check{e}_{1kr}$ *and* $\pi_{kl|s,\mathbf{m}}$ *are given by* (15.6.25), (15.6.7*a*), *and* (15.6.7*b*).

A variance estimator with g-weighted residuals is an alternative to (15.6.27). The following examples illustrate the use of Result 15.6.2.

EXAMPLE 15.6.3. An *SI* sample s, of size n, is drawn. Suppose that the positive values x_k are known for the sample elements only, and that the scatter of the points (x_k, y_k) is well described by the model such that $E_\xi(y_k) = \beta x_k$ and $V_\xi(y_k) = \sigma^2 x_k$. As a response model we adopt the RHG model shown in

(15.6.6). We can apply the special case (15.6.19a) and obtain the important *ratio estimator with weighting class adjustment*,

$$\hat{t}_{cr} = \frac{N}{n}(\textstyle\sum_s x_k)\hat{B}_r = \frac{N}{n}(\textstyle\sum_s x_k)\frac{\sum_{h=1}^{H_s} n_h \bar{y}_{r_h}}{\sum_{h=1}^{H_s} n_h \bar{x}_{r_h}} \tag{15.6.28}$$

Its approximate variance is given by

$$\begin{aligned} AV(\hat{t}_{cr}) &= N^2 \frac{1-(n/N)}{n} S_{yU}^2 + \frac{N^2}{n^2} E_p E_{\mathbf{m}} \left\{ \sum_{h=1}^{H_s} n_h^2 \frac{1-f_h}{m_h} S_{Es_h}^2 \middle| s \right\} \\ &= V_1 + AV_2 \end{aligned} \tag{15.6.29}$$

where $S_{Es_h}^2$ is the variance in s_h of the residuals $E_k = y_k - x_k B_s$, with $B_s = \sum_s y_k / \sum_s x_k$. The variance estimator is $\hat{V}(\hat{t}_{cr}) = \hat{V}_1 + \hat{V}_2$, where $\hat{V}_1$ is given by (15.6.14) and $\hat{V}_2$ by the second component of (15.6.27) and the appropriate regression residuals are $e_{kr} = y_k - \hat{B}_r x_k$. The ratio feature of (15.6.28) creates a protection against bias, as illustrated in Example 15.6.6 below. The variance due to nonresponse is also reduced. That is, the component AV_2 in (15.6.29) is often considerably smaller than its counterpart V_2 in (15.6.12).

EXAMPLE 15.6.4. An *SI* sample of size n is drawn, then poststratified. To explain the y variable, consider the group mean model stating that, for $h = 1, \ldots, H$,

$$E_\xi(y_k) = \beta_h; \qquad V_\xi(y_k) = \sigma_h^2$$

for all $k \in s_h$. The population group counts are assumed to be known. Here the vector $\mathbf{x}_{1k}$ is composed of $H - 1$ entries "0" and a single entry "1" to indicate the group to which the element k belongs, and the total $\sum_U \mathbf{x}_{1k} = (N_1, \ldots, N_H)'$ is known. Moreover, the same groups are used for the RHG model. The estimator appropriate for this case is the one given in equation (15.6.19b), and it is left as an exercise to show that

$$\hat{t}_{cr} = \sum_{h=1}^{H} N_h \bar{y}_{r_h}$$

This is sometimes called *the poststratified estimator*. More accurately, it is an estimator that uses a poststratification of the sample s followed by a nonresponse adjustment weighting based on groups that coincide with the poststrata. That is, the poststrata and the RHGs are identical. The case where the poststrata are crossed with the RHGs leads to a different estimator, as the next example shows.

EXAMPLE 15.6.5. Suppose that it is possible to classify the elements in the sample s into groups, $s_1, \ldots, s_g, \ldots, s_{G_s}$, such that the y_k-values of elements in the same group have a modest variation around the group mean. For individuals, this may be accomplished by a set of age/sex groups. Group counts are assumed to be available at the sample level only, not at the population level. The one-way ANOVA model stating that, for elements k in group g,

$$E_\xi(y_k) = \beta_g; \qquad V_\xi(y_k) = \sigma^2$$

is then a fitting description of the y variable. We refer to the groups s_g as strata and denote by $n_{g\cdot}$ the size of s_g. For the response, we assume an RHG model with H_s categories based on a different criterion, for example, income classes. That is, the response is believed to be related to income rather than to age and sex. The cross classification of the strata and the RHGs yields $G_s H_s$ cells. Denote by s_{gh} the subset of s that falls in the cell gh, and let r_{gh} be the responding part of s_{gh}. Let n_{gh} and m_{gh} be the respective sizes of s_{gh} and r_{gh}. The response rate in response group h is $f_h = m_{\cdot h}/n_{\cdot h}$ where

$$n_{\cdot h} = \sum_{g=1}^{G_s} n_{gh}$$

and

$$m_{\cdot h} = \sum_{g=1}^{G_s} m_{gh}$$

The regression estimator (15.6.19a) is appropriate in this case. If sample selection is by the *SI* design with the sampling fraction $f = n/N$, the estimator takes the form

$$\hat{t}_{cr} = \frac{N}{n} \sum_{g=1}^{G_s} n_{g\cdot} \hat{B}_{gr} \tag{15.6.30}$$

where $n_{g\cdot} = \sum_{h=1}^{H_s} n_{gh}$ and

$$\hat{B}_{gr} = \left(\sum_{h=1}^{H_s} f_h^{-1} \sum\nolimits_{r_{gh}} y_k \right) \Bigg/ \left(\sum_{h=1}^{H_s} f_h^{-1} m_{gh} \right) \tag{15.6.31}$$

Note that (15.6.30) has a simple interpretation; it is the sum of the stratum total estimates, $\hat{N}_g \hat{B}_{gr}$, where $\hat{B}_{gr}$ is the nonresponse adjusted estimate of the gth stratum mean, and $\hat{N}_g = N n_{g\cdot}/n$ is the estimated gth stratum count. Since $\hat{B}_{gr}$ combines data from H_s cells, small or zero cell counts cause no computational problem. In most applications, a limited number of RHGs would be used, say eight or less. The estimator of the variance is $\hat{V}(\hat{t}_{cr}) = \hat{V}_1 + \hat{V}_2$, where $\hat{V}_1$ is given by (15.6.14), and $\hat{V}_2$ is given by the second term of (15.6.27), where the appropriate residuals are $e_{kr} = y_k - \hat{B}_{gr}$ for elements k in stratum g ($g = 1, \ldots, G_s$).

Results 15.6.1 and 15.6.2 rest on the assumption that the RHG model shown by (15.6.6) truly represents the unknown response distribution. In practice, the survey statistician can never be certain that his choice of response model is correct. As a result, the estimators based on response modeling always have some bias. However, with well-chosen RHGs, the bias is limited. If the underlying RHG model does not hold, (15.6.8) and (15.6.16) are biased. Analytic expressions for the bias and other important properties are hard to obtain. Monte Carlo simulations can be used to assess the performance of different estimators under varying assumptions about the true response. The following example cites results from a simulation reported in Särndal and Swensson (1987).

EXAMPLE 15.6.6. A population of $N = 1{,}227$ Swedish households was used in an experiment to study the effect of response-model breakdown on three estimators of total disposable household income, namely, (15.6.11), (15.6.28), and (15.6.30). Here, they are denoted $\hat{t}_A$, $\hat{t}_B$, and $\hat{t}_C$, respectively. The auxiliary variable x was the square root of taxable household income. The population correlation coefficient between y and x was 0.83. Four population response groups, U_h, $h = 1, 2, 3, 4$, were formed. All elements in the same group were assigned the same response probability value, denoted θ_h. A true response distribution was fixed for the experiment, namely, that the element k responded with probability θ_h if k was in group h, and that elements responded independently. The population correlation coefficient between y and the response probability was 0.44. Although not strong, this correlation would cause considerable selection bias if the estimates were not adjusted. The population groups are described in the following table:

Population group	N_h	θ_h	$\bar{y}_{U_h}$	$\bar{x}_{U_h}$
U_1	373	0.45	4.18	1.68
U_2	303	0.65	5.65	2.25
U_3	280	0.75	6.73	2.54
U_4	271	0.90	8.31	2.79

The following simulation was carried out. An initial SI sample s of size $n = 400$ was drawn, and a response set r was generated using independent Bernoulli trials with the preassigned response probabilities. That is, the element k responds with the known probability θ_h if k is in group h. Three different response models were considered, called TRUE, FALSE 2, and FALSE 1. The TRUE model coincides with the true response distribution with four groups, as described above. The FALSE 2 model is an RHG model with two groups. Sample elements belonging to U_1 or U_2 form the first group; sample elements in U_3 or U_4 form the second group. In other words, FALSE 2 specifies two groups, whereas the true response distribution has four groups. FALSE 1 is the naïve response model which assumes that all elements respond with the same probability. Each response model leads to a different estimate, since the number of groups differs. For each of the three response models, the point estimates $\hat{t}_A$, $\hat{t}_B$, and $\hat{t}_C$ were calculated. Four strata were used for $\hat{t}_C$. These strata were created by ordering the 400 sample elements from smallest to largest according to the size of x. The first stratum consisted of the 100 smallest elements, the second stratum of the next 100 elements, and so on. Nominal 95% confidence intervals were calculated using the variance estimators described in Examples 15.6.2, 15.6.3, and 15.6.5. The whole procedure was repeated to obtain 1,000 response sets. The associated estimates and confidence intervals were computed, and summary measures for the 1,000 samples were then computed as described in Section 7.9.1. Selected simulation results are given in the following tables, which show that the two regression estimators $\hat{t}_B$ and $\hat{t}_C$ are much less biased than the simple weighting adjustment estimator $\hat{t}_A$.

Estimated Relative Biases (%) $100B(\hat{t})/t$.

Response model	$\hat{t}_A$	$\hat{t}_B$	$\hat{t}_C$
True	0	0	0
False 2	1.5	0.3	0.5
False 1	6.5	1.7	1.6

Estimated Coverage Rates (%) for Nominal 95% Confidence Intervals.

Response model	$\hat{t}_A$	$\hat{t}_B$	$\hat{t}_C$
True	95.2	95.5	95.6
False 2	93.3	95.6	95.6
False 1	46.3	92.6	93.9

Conforming to theory, all three estimators are essentially unbiased for the TRUE response model, and the empirical coverage rates are, as expected, close to the nominal 95%. The main interest here lies in observing the consequences of breakdown of the assumed response model. For the limited model breakdown represented by FALSE 2, the simple weighting class estimator $\hat{t}_A$ already displays a considerable bias, whereas the regression estimators $\hat{t}_B$ and $\hat{t}_C$ strongly resist the tendency toward bias. Under the strong model breakdown of FALSE 1, the weighting class estimator is completely off the mark; as a consequence of the large bias, the coverage rate drops to an unacceptable 46%. The regression estimators $\hat{t}_B$ and $\hat{t}_C$ continue to perform remarkably well. Their coverage rates are only a few percentage points below the nominal 95% rate.

In summary, we conclude the following:

Conclusion 1. To keep the nonresponse bias at low levels, it is essential to identify a response model that describes as accurately as possible the unknown response distribution corresponding to the survey conditions.

Conclusion 2. Since it is virtually certain that the modeler can not exactly identify the true response distribution, information on pertinent auxiliary variables must be collected. Such variables, if they are strong predictors of y, serve to build regression estimators, which are bias resistant.

Remark 15.6.5. Some authors model and estimate individual response probabilities θ_k (as opposed to probabilities that are equal within a group). Cassel, Särndal and Wretman (1983) model θ_k as a parametric function of an auxiliary value $\mathbf{x}_k$, assumed to be known for elements in the selected sample s. Since $\mathbf{x}_k$ is a predictor of y_k, the resulting estimated response probability depends indirectly on y_k. That is, the response probability reflects the value of the study variable. Based on the same type of auxiliary information, Giommi (1988) proposed a nonparametric approach involving density estimation to estimate individual response probabilities as a function of the $\mathbf{x}_k$. Greenlees, Reece, and Zieschang (1982) estimate a model in which θ_k may depend on y_k as well as on $\mathbf{x}_k$.

In this chapter, we have followed the extended probability sampling approach identified in Section 14.6, where it was also pointed out that a fully model-based approach, with superpopulation modeling, is an alternative that some statisticians may prefer. Such an approach was used by Little (1983b) and Pfeffermann (1988). Via the superpopulation, predictions are produced for the nonobserved values. The superpopulation approach is extensively used in the book by Little and Rubin (1987).

15.7. Imputation

We consider now a survey with q study variables $y_1, \ldots, y_j, \ldots, y_q$ corresponding to q questionnaire items. As usual, s denotes a probability sample. Let r_j be the response set for the variable y_j. The unit response set (that is, the set of elements that responded to one or more items) is $r_u = r_1 \cup r_2 \cup \cdots \cup r_q$, and the set of elements that provided a complete response vector $\mathbf{y}_k = (y_{1k}, \ldots, y_{qk})$ is $r_c = r_1 \cap r_2 \cap \cdots \cap r_q$. We assume that both the item nonresponse set $r_u - r_c$ and the unit nonresponse set $s - r_u$ are nonempty.

When there is both unit nonresponse and item nonresponse, the estimation of population parameters of interest is more complex. How should the observed y-data and other information be used? Two options are as follows:

i. A response set approach. The data associated with the jth response set are used to create estimates for the jth variable; $j = 1, \ldots, q$.
ii. A clean data matrix approach. A completely filled data matrix is created and used to calculate estimates for $j = 1, \ldots, q$.

In a response set approach to estimation of the totals $t_j = \sum_U y_{jk}$, $j = 1, \ldots, q$, we can apply, item by item, the adjustment weight approach of Section 15.6. That is, when t_j is estimated, a set of RHGs is defined. The value $\check{y}_{jk}$ will then receive the adjustment weight n_h/m_{jh} if k is in group h, where m_{jh}/n_h is response rate in group h to item j.

The estimation procedure is further complicated if it is deemed necessary to set up the RHGs differently from one y variable to another. This case arises, for instance, if the analyst considers that a model in terms of constant response probability within groups will require age groups for the variable y_1, occupational groups for y_2, and so on.

Another problem with the item-by-item response set approach is that it may lead to impermissible estimates. For example, suppose that $q = 2$ and that we make estimates of $\sum_U y_{1k}$ and $\sum_U y_{1k}^2$ based on r_1, estimates of $\sum_U y_{2k}$ and $\sum_U y_{2k}^2$ based on r_2, and an estimate of the product total $\sum_U y_{1k}y_{2k}$ based on $r_1 \cap r_2$. If these five estimates are used as input for estimating the finite population coefficient of correlation between y_1 and y_2, a value may result that falls outside the permissible range $[-1, 1]$.

The response set approach is thus not without difficulties. To simplify the data handling, survey statisticians usually prefer to work with *a complete rectangular data matrix*. There are several ways to create such a matrix, which is often also called a *clean data matrix*.

One naïve way to obtain such a matrix is to use the observed y data for elements $k \in r_c$ only. By treating r_c as a reduced unit response set, one would then apply the usual techniques for unit nonresponse. In this method one disregards observed y data for elements in $r_u - r_c$. If this set contains few elements, little information is lost. But in other cases, the method could prove wasteful. Instead, imputation is ordinarily used to arrive at a complete matrix.

Imputation implies that an imputed value $\tilde{y}_{jk}$ is produced for a missing value y_{jk}. That is, $\tilde{y}_{jk}$ fills the blank for this missing value in the data matrix. The imputed value is a prediction of the unknown y_{jk}. Auxiliary information is often used to generate the value $\tilde{y}_{jk}$. Different methods are considered later in this section.

We assume from now on that values are imputed for the item nonresponse. That is, an imputation $\tilde{y}_{jk}$ is produced for every missing value y_{jk} such that $k \in r_u - r_c$ and $j = 1, \ldots, q$. This leads to a complete data matrix of dimension $n_{r_u} \times q$. We assume that the unit nonresponse is adjusted for by one of the techniques in Section 15.6.

Remark 15.7.1. A more extensive use of imputation is to impute for the unit nonresponse as well as for the item nonresponse. That is, an imputed value $\tilde{y}_{jk}$ is produced for any missing value y_{jk} such that $k \in s - r_c$ and $j = 1, \ldots, q$. The result is a clean data matrix of dimension $n_s \times q$.

A clean data matrix is practical to work with and avoids the problems with estimation from response sets of varying sizes. The study variables can be treated in a uniform manner in the production of estimates. Most computer packages are constructed for rectangular data files. The wide availability of statistical packages thus entices the statistician to use imputation to create a clean data matrix. Today a common procedure in surveys is to process such a data set as if all values were actual observations. Most computer packages for the analysis of survey data operate according to this principle.

It is always more or less "statistically incorrect" to treat imputed values the same as actual observations for purposes of estimation or other kinds of analysis. Much depends on the quality of the imputed values. The error arising from applying "standard methods," including the estimation methods presented earlier in this book, would be unimportant if the $\tilde{y}_{jk}$ are close to the missing y_{jk}. However, the imputations may be poor substitutes. When they are sample-based predictions, they have their own statistical properties, such as an expected value and a variance. Bias and additional variance usually results from imputation. Lanke (1983) points to some of the difficulties that arise when imputed values are used for statistical analysis.

Remark 15.7.2. All imputed values $\tilde{y}_{jk}$ in a data file should be "flagged," to distinguish them from the observed values y_{jk}. In some countries, legislation rules out the presence of imputed values in data files on individuals, even if such values are flagged to indicate their artificial origin.

A number of imputation methods are in use. These range from simple ad hoc procedures to sophisticated regression-based methods. Some of these methods were proposed in times when nonresponse was less prevalent and less of a problem than today. For example, Pritzker, Ogus and Hansen (1965) describe the policy at the United States Bureau of the Census as follows:

> Basically, our philosophy in connection with the problem of ... imputation is that we should get information by direct measurement on a very high proportion of the aggregates to be tabulated, with sufficient control on quality that almost any *reasonable* rules for ... imputation will yield substantially the same results. ... With respect to imputation in censuses and sample surveys we have adopted a standard that says we have a low level of imputation, of the order of 1 or 2 percent, as a goal.

At that time, no great need for close scrutiny of the proposed imputation methods was felt. With the higher nonresponse rates observed in the last few decades, theoretical as well as applied studies have been undertaken, showing that most of the proposed imputation methods have shortcomings.

A number of imputation techniques have been developed, as discussed by Sande (1982, 1983), Bailar and Bailar (1983), Ford (1983), Kalton and Kasprzyk (1986), Little and Rubin (1987), ch. 4, Little (1988).

Deductive imputation refers to those instances, rare in practice, where a missing value can be filled with a perfect prediction $\tilde{y}_{jk} = y_{jk}$, attained by a logical conclusion. The deduction may be based on responses given to other items on the questionnaire.

Most of the currently used imputation techniques involve the substitution of an "imperfect" predicted value. Some of the main techniques of this kind are mean imputation, hot-deck imputation, cold-deck imputation, regression imputation, and multiple imputation. Let us examine the main features of these techniques.

Overall Mean Imputation

This is a simple method that, for item j, assigns the same value, namely, the respondent item mean, $\bar{y}_{r_j}$, to every missing value y_{jk} in the set $r_u - r_j$. The method may produce a reasonable point estimate of the population total $t_j = \sum_U y_{jk}$, but is less appealing if we wish to compute a confidence interval using a standard variance estimator. As is intuitively clear, to replace all missing values for a given item by the respondent mean for that item will give a set of values with less variability than in a sample of equal size consisting entirely of actually observed values. Unless the nonresponse is negligible, or unless a modified variance estimator is used, the method may easily lead to severely understated variance estimates and to invalid confidence intervals.

Class Mean Imputation

This method works by partitioning the unit response set r_u into imputation classes such that elements in the same class are considered similar. Auxiliary variables are used for classification. For a given item j, and for all elements k in a given imputation class, missing values are replaced by the respondent mean in that class. There will be some distortion of the "natural" distribution of the values, but the distortion is less severe than with overall mean imputation.

Hot-Deck and Cold-Deck Imputation

Improvement on the mean imputation methods is sought by creating a more authentic variability in the imputed values. In *hot-deck* imputation procedures, missing responses are replaced by values selected from respondents in the current survey. *Cold-deck* procedures, on the other hand, use imputations based on other sources than the current survey, for example, earlier surveys or historical data.

A number of hot-deck procedures have been proposed, including random overall imputation, random imputation within classes, sequential hot-deck imputation, hierachical hot-deck imputation, and distance function matching.

Random Overall Imputation

This method works as follows: For item j, a missing value is replaced by an actually observed y_{jk}-value, taken from a respondent, a donor, randomly drawn from the item j response set, r_j. Although the method gives a data set for item j with a close to natural variation, it does not follow that standard techniques can be used straightforwardly, for example, to calculate variance estimates and confidence intervals.

Random Imputation within Classes

This is an alternative to the preceding technique, in which suitable classes are formed, similarly as with class mean imputation. For an element in a given class, an imputed value is obtained from a randomly chosen donor in the same class.

Sequential Hot-Deck Imputation

For example, the United States Current Population Survey uses this method. It has the advantage that a single pass through the data file is sufficient to complete the imputation procedure. When a missing value is spotted on a certain item, a donor is identified by backtracking through the data file to the nearest element that shows a response value for that item and is in the same imputation class as the recipient. The procedure starts with a cold-deck value in each imputation class. One draw-back of this nonrandom procedure is that it often leads to multiple uses of donors. An improvement, called *hierarchical hot-deck imputation*, is used in the March Income Supplement of the United States Current Population Survey. In this random imputation procedure, a very large set of imputation classes is used, so often no donor can be found for an element requiring imputation. To remedy this, imputation classes are collapsed in a hierarchic fashion until a donor is found.

Distance Function Matching

This is another hot-deck procedure. For item j, a missing value y_{jk} is replaced by the value of a respondent classified to be the "nearest," as measured by a distance function defined in terms of known auxiliary variable values.

Regression Imputation

Unlike the hot-deck techniques, *regression imputation* uses estimated relationships between variables. A simple application of this idea due to Buck (1960) uses respondent data to fit a regression of a variable for which one or more imputations are needed on other available variables, assumed to have high predictive value. The predictors can be either study variables (other items on the questionnaire) or auxiliary variables. The fitted regression equation is used to produce imputations. For example, for $j = 5$ items with the y variables $y_1, \ldots, y_5$, let y_{jk} be the value of y_j for the kth element. For a certain element $k \in r_u - r_c$, suppose the values y_{1k} and y_{2k} are missing, so that the recorded information for that element reads (—, —, y_{3k}, y_{4k}, y_{5k}). Imputations for the two blanks are obtained as follows: Let $\tilde{y}_1 = \tilde{f}_1(y_3, y_4, y_5)$ be the regression equation of y_1 on y_3, y_4 and y_5, fitted using data for elements $k \in r_c$. The corresponding estimated regression of y_2 on y_3, y_4 and y_5 is denoted $\tilde{y}_2 = \tilde{f}_2(y_3, y_4, y_5)$. These two equations, and the three recorded values for the element k yield the imputations $\tilde{y}_{1k} = \tilde{f}_1(y_{3k}, y_{4k}, y_{5k})$ and $\tilde{y}_{2k} = \tilde{f}_2(y_{3k}, y_{4k}, y_{5k})$. Imputations are calculated analogously for the blanks corresponding to other elements $k \in r_u$.

Sometimes a randomly produced residual is added to further reflect the uncertainty in the imputed value. The use of multivariate regression has also been advocated for imputation when an element has two or more blanks.

Multiple Imputation

The imputation methods mentioned produce a single imputed value for each missing value y_{jk}. In general, this distorts more or less the natural distribution of values for that item. Rubin (1987) advocates the use of *multiple imputations*, that is, "for each missing datum we impute several values, say *m*.... These *m* values are ordered in the sense that the first set of values imputed for the missing values are used to form the first completed data set, and so on. ... Each completed data set is analyzed using standard complete-data procedures..." From the *m* estimates, a single combined estimate is computed, together with a pooled variance estimate to be used for interval estimation. This variance estimate contains a component that reflects the variability between the completed data sets; it can be seen as an expression of the statistician's uncertainty about which value to impute. A disadvantage with multiple imputation is that it requires more work for data handling and computation of estimates.

Imputation is popular in practice, as evidenced by the many methods that have been proposed. Nevertheless, the literature on the theoretical properties of the various methods and on empirical comparisons is sparse. Ford (1983) states in an overview of hot-deck procedures:

> Because each of these empirical studies is confined to the investigation of one particular survey, broad generalizations of their results are difficult to make. These studies do, however, support the theoretical conclusion that the standard errors of hot-deck estimators are underestimated by computations which assume that the data set was originally complete. They also indicate that there *may* not be much improvement in the mean square error of a hot-deck estimator in comparison to an estimator which omits the missing values unless there is auxiliary information which is highly correlated with the survey data. They do *not* consistently show that one missing data procedure is better than another.

Exercises

15.1. A PO sample with expected size 20 was drawn from the MU281 population of municipalities. The inclusion probabilities were proportional to the value of the variable P75 ($= x$). The objective is to estimate the total of the study variables P85 ($= y_1$) and RMT85 ($= y_2$). The recorded data are found in the following table, where — indicates a value missing due to nonresponse:

k	P75	P85	RMT85	k	P75	P85	RMT85
4	15	19	159	122	11	11	67
8	54	66	517	187	14	—	86
15	43	48	—	188	73	—	—
34	14	13	111	189	28	26	—
37	92	—	—	211	118	118	965
47	119	—	—	214	28	27	—
56	108	107	807	245	28	26	184
70	26	27	183	262	8	8	58
81	16	16	103	265	8	—	61
104	14	14	113	270	72	74	592
117	102	—	—	281	35	39	295

The population total of the variable P75 is 6,818. (a) Compute the unit response measures p_{r_u}, $\tilde{p}_{r_u}$, $(\sum_{r_u} x_k)/(\sum_s x_k)$, $(\sum_{r_u} x_k/\pi_k)/(\sum_s x_k/\pi_k)$, as well as the corresponding unit nonresponse measures. (b) Compute the item response measures p_{r_1}, p_{r_2}, p_{r_c} and the corresponding item nonresponse measures.

15.2. An *SI* sample of 100 municipalities was drawn from the MU200 population to estimate the total of the variable REV84 ($= y$). Mail questionnaires were sent out. After a follow-up reminder, responses were obtained from 43 municipalities. From the 57 nonresponding municipalities, an *SI* subsample s_2 of 28 municipalities was drawn and responses were obtained from all 28. The following survey results were calculated, where s_{a1} is the set of municipalities that responded to the mail inquiry:

$$\bar{y}_{s_{a1}} = 1{,}807, \quad S^2_{ys_{a1}} = 2{,}322{,}828, \quad \bar{y}_{s_2} = 1{,}581, \quad S^2_{ys_2} = 764{,}895$$

Compute an approximately 95% confidence interval for the total of REV84.

15.3. A study designed to estimate the number of HIV infected persons in a population of 10,000 used a survey design based on *BE* sampling with $\pi = 0.1$ and randomized response interviews. The realized sample contained $n_s = 1{,}024$ persons. The randomized response interviews were carried out as follows. Each respondent was handed a plastic cup containing two ordinary six-sided dice. He or she was asked to shake the cup, roll the dice, add the scores, and respond according to the following schedule.

i. If your total score is 4, 5, 6, 7, 8, 9, or 10, then:

If, to the best of your knowledge, you are HIV infected, give the answer *A*.

If, to the best of your knowledge, you are not HIV infected, give the answer *B*.

ii. If your total score is 2, 3, 11, or 12, then:

If, to the best of your knowledge, you are not HIV infected, give the answer *A*.

If, to the best of your knowledge, you are HIV infected, give the answer *B*.

A practical demonstration preceded the actual interview. All necessary measures were taken to prevent the interviewer from finding out the respondent's true status with respect to HIV infection. A total of 315 answers *A* were given. Assuming truthful answers, compute an approximately 95% confidence interval for the total number of HIV infected persons in the population. Use the estimator

$$\hat{t}_{\text{RR}} = N \frac{\sum_s \hat{y}_k/\pi_k}{\sum_s 1/\pi_k}$$

with $\hat{y}_k$ given by (15.4.10).

15.4. Consider the data in Exercise 15.1. In your opinion, can the naive response model shown by (15.6.1) be regarded as an appropriate description of the response distribution that generates the unit response? Justify your answer.

15.5. In a survey of the 200 municipalities in the MU200 population, an *SI* sample of $n = 100$ was drawn with the objective of estimating the total of the variable ME84 ($= y$). Only 58 of the sampled municipalities responded to the mailed questionnaires. A point estimate was calculated from the formula $N\bar{y}_r$, with the result 148,334. However, it was considered out of the question to publish this estimate, since the nonresponse was large, and response propensity was known to be positively correlated with the size of the municipality. A decision was taken to follow up nonrespondents by telephone. Meanwhile, a preliminary estimate based on two RHG groups was calculated as follows: For each of the 100 sampled municipalities, the value of the variable P75 ($= x$) was obtained, and these municipalities were size-ordered according to the value of P75. The 50 smallest municipalities were placed in the first RHG group, and the remaining 50 in the second RHG group. The following results were calculated. (The notation is as in Example 15.6.2.)

Group h	n_h	m_h	$\bar{y}_{r_h}$	$S^2_{yr_h}$
1	50	25	412.04	10,281
2	50	33	991.39	145,820

Calculate a point estimate based on the estimator shown in equation (15.6.11), as well as the corresponding *cve*.

15.6. (Continuation of Exercise 15.5). Use the information that ME84 is approximately proportional to P75 to compute an alternative preliminary estimate based on the estimator given in (15.6.28), as well as the corresponding *cve*. The following additional information is provided:

Group h	$\bar{x}_{r_h}$	$S^2_{xr_h}$	S_{xyr_h}
1	8.68	3.06	156.35
2	18.24	26.94	1,747.81

$\sum_s x_k = 1{,}312$

15.7. Suppose that nonresponse affects all of q study variables in exactly the same way, so that the response sets satisfy $r_1 = r_2 = \cdots = r_q = r$. That is, there is unit nonresponse, but no item nonresponse. Furthermore, suppose that the response distribution is such that $\Pr(k \in r|s) = \theta_k > 0$ for every s and every $k \in U$. Show that the expected value of the sample-weighted measure of unit response $\tilde{p}_r$ is approximately equal to the average response probability in the population, that is,

$$E\left(\frac{\sum_r 1/\pi_k}{\sum_s 1/\pi_k}\right) \doteq \frac{1}{N}\sum_U \theta_k = \bar{\theta}_U$$

15.8. Define sample-weighted analogs of the item nonresponse measures given in Section 15.3.

15.9. Derive the expressions (15.4.3) and (15.4.4) for the variance and the variance estimator in Result 15.4.1.

15.10. Derive an unbiased variance estimator for the variance (15.4.6) given in Example 15.4.4.

15.11. Determine the optimal values of the sampling fraction f_a and the subsampling fraction v defined in Example 15.4.6.

15.12. For the survey with randomized response in Example 15.4.7, verify that (15.4.13) is an unbiased estimator of $\bar{y}_U$ and that (15.4.14) is an unbiased variance estimator.

15.13. Verify the bias expressions (15.6.3) to (15.6.5).

15.14. Show that, given s and $\mathbf{m} = (m_1, \ldots, m_{H_s})$, the response set r, under the RHG model shown in (15.6.6), behaves as an *STSI* selection from s.

15.15. Show that if the RHG model (15.6.6) specifies only one group, then the direct weighting estimator (15.6.8) can be written

$$\hat{t}_{c\pi^*} = (n/m)\sum_r y_k/\pi_k = (n/m)\sum_r \check{y}_k = n\bar{\check{y}}_r$$

which in the case of *SI* sampling takes the form

$$\hat{t}_{c\pi^*} = (N/m)\sum_r y_k = N\bar{y}_r$$

Also show that, for *SI* sampling and under the conditions of Result 15.6.1, the unbiased variance estimator (15.6.10) can be written

$$\hat{V}_{SI}(\hat{t}_{c\pi*}) = N^2 \frac{1 - m/N}{m} S^2_{yr}$$

15.16. Derive a variance estimator with g-weighted residuals as an alternative to the variance estimator (15.6.27) given in Result 15.6.2.

15.17. Consider the regression model $E_\xi(y_k) = \beta$; $V_\xi(y_k) = \sigma^2$, all $k \in U$, and the RHG model with a single group. Show that the regression estimator (15.6.19b) for this case can be written

$$\hat{t}_{cr} = N\tilde{y}_r = N \frac{\sum_r y_k/\pi_k}{\sum_r 1/\pi_k}$$

which is the estimator shown in equation (15.6.2) and discussed in Example 15.6.1.

15.18. Under the conditions of Example 15.6.4, show that the estimator (15.6.19b) takes the form

$$\hat{t}_{cr} = \sum_{h=1}^{H} N_h \bar{y}_{r_h}$$

15.19. Consider the situation described in Example 15.6.5. Present a detailed derivation of the estimator (15.6.30), the variance of this estimator, and the corresponding variance estimator.

15.20. A survey with a single study variable y uses overall mean imputation in such a way that the data after imputation are

$$y_k^* = \begin{cases} y_k & \text{if } k \in r \\ \tilde{y}_r = (1/m) \sum_r y_k/\pi_k & \text{if } k \in s - r \end{cases}$$

Show that substitution of the y_k^* in the standard π estimator formula leads to

$$\hat{t}_{y*\pi} = \sum_s y_k^*/\pi_k = (n/m) \sum_r \check{y}_k = n\tilde{y}_r$$

which is identical to the direct weighting estimator $\hat{t}_{c\pi*}$ given in Exercise 15.15. However, observe that the standard variance estimator calculated on data after imputation, that is,

$$\hat{V}(\hat{t}_{y*\pi}) = \sum\sum_s \check{\Delta}_{kl} \check{y}_k^* \check{y}_l^*$$

will differ from the variance estimator $\hat{V}(\hat{t}_{c\pi*})$ recommended in Result 15.6.1.

15.21. (Continuation of Exercise 15.20). A survey uses *SI* sampling with $f = n/N$ and overall mean imputation as in Exercise 15.20. Then

$$\hat{t}_{y*\pi} = N\bar{y}_r$$

which is identical to $\hat{t}_{c\pi*}$ given in Exercise 15.15. (a) Show that the standard variance estimator formula leads to

$$\hat{V}_{SI}(\hat{t}_{y*\pi}) = N^2 \frac{1 - n/N}{n} S^2_{y*s} = N^2 \frac{1 - n/N}{n} \frac{m-1}{n-1} S^2_{yr}$$

(b) Show that, for every response set r for which $m < n$,

$$\hat{V}_{SI}(\hat{t}_{c\pi*}) = N^2 \frac{1 - m/N}{m} S^2_{yr} > \hat{V}_{SI}(\hat{t}_{y*\pi})$$

This implies that when the single-group RHG model holds, the standard variance estimator calculated on the data after imputation will underestimate the variance and hence produce invalid confidence intervals.

15.22. A survey with a single study variable y uses an alternative to the overall mean imputation in Exercise 15.20 in such a way that the data after imputation are

$$y_k^* = \begin{cases} y_k & \text{if } k \in r \\ \tilde{y}_r = \dfrac{\sum_r y_k/\pi_k}{\sum_r 1/\pi_k} & \text{if } k \in s - r \end{cases}$$

Show that substitution of the y_k^* into the standard π estimator formula leads to

$$\hat{t}_{y*r} = N \frac{\sum_s y_k^*/\pi_k}{\sum_s 1/\pi_k} = N \frac{\sum_r y_k/\pi_k}{\sum_r 1/\pi_k} = N\tilde{y}_r$$

which is identical to $\hat{t}_{cr}$ given in Exercise 15.17.

15.23. A survey with a single study variable y uses SI sampling of n elements from N and class mean imputation. The sample s is partitioned into H_s imputation classes formed in the manner of RHG groups.

(a) The data after imputation are (with notation as in Example 15.6.2)

$$y_k^* = \begin{cases} y_k & \text{if } k \in r_h \\ \bar{y}_{r_h} & \text{if } k \in s_h - r_h \end{cases}$$

Show that substitution of y_k^* into the ordinary π estimator leads to

$$\hat{t}_{y*\pi} = (N/n) \sum_{h=1}^{H_s} n_h \bar{y}_{r_h} = N\hat{\bar{y}}_U$$

which is identical to the weighting class estimator $\hat{t}_{c\pi*}$ given by equation (15.6.11).

(b) Show that substitution of y_k^* into the standard variance estimator leads to

$$\begin{aligned} \hat{V}_{SI}(\hat{t}_{y*\pi}) &= N^2 \frac{1 - n/N}{n} S^2_{y*s} \\ &= N^2 \frac{1 - n/N}{n(n-1)} \left[\sum_{h=1}^{H_s} (m_h - 1) S^2_{yr_h} + \sum_{h=1}^{H_s} n_h (\bar{y}_{r_h} - \hat{\bar{y}}_U)^2 \right] \end{aligned}$$

which differs from the variance estimator $\hat{V}_{SI}(\hat{t}_{c\pi*})$ given by (15.6.13).

(c) Show that for the recommended variance estimator $\hat{V}_{SI}(\hat{t}_{c\pi*})$ given by (15.6.13) we have

$$\hat{V}_{SI}(\hat{t}_{c\pi*}) = N^2 \frac{1 - n/N}{n} \left\{ \sum_{h=1}^{H_s} \frac{n_h}{n} \left[1 - \delta_h + \frac{n_h - m_h}{m_h(1 - n/N)} \right] S^2_{yr_h} \right.$$
$$\left. + \frac{1}{n-1} \sum_{h=1}^{H_s} n_h(\bar{y}_{r_h} - \hat{\bar{y}}_U)^2 \right\}$$
$$> N^2 \frac{1 - n/N}{n(n-1)} \left\{ \sum_{h=1}^{H_s} \frac{n_h(n_h - 1)}{m_h} S^2_{yr_h} + \sum_{h=1}^{H_s} n_h(\bar{y}_{r_h} - \hat{\bar{y}}_U)^2 \right]$$

Conclude that if $m_h < n_h$ for at least one h, then

$$\hat{V}_{SI}(\hat{t}_{c\pi*}) > \hat{V}_{SI}(\hat{t}_{y*\pi})$$

which implies that, if the RHG model holds, the standard variance estimator calculated on the data after imputation will underestimate the variance and hence produce invalid confidence intervals.

CHAPTER 16

Measurement Errors

16.1. Introduction

In the first three parts of this book we worked under an assumption of no measurement errors. More specifically, the assumption was that each element $k \in U$ is equipped with a definite numerical value y_k, which can be observed exactly, provided the element is selected and subject to observation. Then, the observed values y_k for $k \in s$ were used to calculate a point estimate and an associated variance estimate. The rather idealistic assumption of no measurement errors allowed us to concentrate heavily on pure sampling error.

We have seen that survey operations include the following stages:

1. sample selection,
2. data collection,
3. data processing.

We use the term measurement errors to denote those errors in individual data that occur during the data collection stage. By "error" we mean that the recorded value on a study variable for a sampled element differs from the true value. Special studies from many countries indicate that measurement errors may have a considerable effect on survey estimates. Thus, when considering the accuracy of survey results, measurement errors (as well as other non-sampling errors) should be taken into account, in addition to the sampling error.

Remark 16.1.1. Errors in individual data may also occur in the data processing stage of a survey. These errors are called processing errors and may be generated during such operations as coding, editing, and data capture. In this chapter, we only discuss statistical models for measurement errors. Similar models could be developed for processing errors. (The terminology varies somewhat, and some authors use the term measurement errors in a wider sense that includes both what we call measurement errors and what we call processing errors.)

A complete enumeration (a census) of the population has no sampling error at all, but one should not jump to the conclusion that a complete enumeration leads to error-free estimates. The measurement errors (and other nonsampling errors as well, such as nonresponse errors, coverage errors, and processing errors) may be considerable, and even more severe than in a sample survey.

In this chapter, we study the effect of measurement errors on survey estimates. ("Survey" is used here in a wide sense that includes complete enumeration as a special case.) Because survey conditions are so varying, each survey has its own measurement problems, depending on the interviewers, the questionnaire design, and other specific factors.

Our goal is to provide some basic theory for measurement errors in surveys under a fairly general measurement model, and to give some examples of specific applications. The interested reader may adapt the general theory of Sections 16.3 to 16.6 to suit the specific conditions of a particular survey.

Basic contributions to the methodology of measurement error models in survey sampling were given by Mahalanobis (1946), Hansen, Hurwitz and Bershad (1961), and Hansen, Hurwitz and Pritzker (1964). Comprehensive review papers are those by Bailar and Dalenius (1969) and Lessler (1984).

16.2. On the Nature of Measurement Errors

An assumption in this chapter is that it is meaningful to talk about a true value of the study variable y, for each element k in the population U. This true value is denoted θ_k; $k = 1, \ldots, N$. (However, the concept of a true value is not always generally agreed upon; see the discussion below and in Section 14.8.)

We would like to record the true value θ_k for each element k in the sample s. But the data collection operations may generate errors in the individual data, to the effect that the values y_k $(k \in s)$ that are finally recorded and used in the computation of estimates will often differ from the corresponding true values θ_k $(k \in s)$. The difference $y_k - \theta_k$ is called the individual measurement error for element k.

EXAMPLE 16.2.1 Suppose we want to estimate the true population total $t_\theta = \sum_U \theta_k$ using data from a sample s. In the ideal case with no measurement errors, $y_k = \theta_k$ for all $k \in s$, so the π estimate would be

$$\textstyle\sum_s \theta_k/\pi_k \tag{16.2.1}$$

When the imperfect data y_k are substituted, the π estimate is

$$\sum_s y_k/\pi_k \tag{16.2.2}$$

which differs from (16.2.1) when errors are present. The statistical properties of (16.2.2) can only be studied by making assumptions about the errors. It becomes necessary to formulate stochastic measurement error models.

When the variables to be measured are variables such as age, income, expenditures, possession of physical attributes, number of employees, or crop yield, we believe that the concept of "true values" makes good sense and may be of help in planning the measurement process. However, even in these examples, the concept is not without ambiguity. Various authors express different opinions as to whether true values must be operationally defined, whether they could be thought of as existing apart from the actual measurement method, or whether true values exist at all. For a summary, see Lessler (1984).

For many other variables, the concept of a true value is more difficult to accept. Examples are political attitudes, perceived well-being, motivation, intelligence, and so on. Measurement problems for such variables are by tradition not treated in the sampling literature. Although this is a shortcoming, we restrict our discussion to the "hard data" variables mentioned before.

Individual measurement errors may arise at the data collection stage for a variety of reasons. Errors may be due to the respondent, to the interviewer, or to both. For example, an incorrect answer on the part of the respondent may be due to misunderstanding the question, to not knowing the answer, or to unwillingness to provide the true answer. Also, the interviewer may misunderstand the question, put it badly, influence the respondent to give an erroneous answer, or record the information incorrectly. Such errors can never be completely avoided in practice. It is the purpose of questionnaire design and interviewer training to keep these errors under control, so that they will not seriously damage the survey results. Large surveys often have special quality-control programs for the data collection operations. Three examples of measurement errors follow.

Example 16.2.2. A survey of the total crop yield in a district will often proceed as follows. First, a sample of farms is drawn from the district. Then, from each selected farm, a sample of "plots" of equal size are randomly located in the fields where the crop is grown. Finally, these plots are harvested, and the yield from each plot is measured. These data are used to estimate the total crop yield for the district. A particular type of measurement error resulting in a positive bias has sometimes been observed in this type of survey, namely, that the person harvesting a plot has a tendency to include plants growing on the border or just outside the border of the plot. Clearly, the yield on that plot will be overstated.

EXAMPLE 16.2.3. Another well-known type of measurement error occurs in surveys where people are asked to report their ages. Statistics based on data from such surveys sometimes show that there are too many people aged exactly 30, 35, 40, 45, and so on. This indicates a tendency to report age data that are rounded off, for example, to report "30" when the true age is 29 or 31. Table 16.2.1 (from United Nations, 1982) shows how this phenomenon has manifested itself in United States Census data from 1880 to 1960.

Table 16.2.1. Percent of Population with Ages Ending in Each Digit 0–9 in the United States 1880–1960 According to Census Data.

Last digit of age	Year of census								
	1880	1890	1900	1910	1920	1930	1940	1950	1960
0	16.8	15.1	13.2	13.2	12.4	12.3	11.6	11.2	9.9
1	6.7	7.4	8.3	7.7	8.0	8.0	8.5	8.9	9.9
2	9.4	9.7	9.8	10.2	10.2	10.3	10.4	10.2	9.9
3	8.6	9.1	9.3	9.2	9.4	9.4	9.6	9.7	9.8
4	8.8	9.0	9.5	9.4	9.4	9.6	9.7	9.7	10.1
5	13.4	12.3	11.3	11.5	11.3	11.2	10.7	10.6	10.3
6	9.4	9.6	9.4	9.6	9.7	9.6	9.6	9.8	9.9
7	8.5	8.9	9.3	9.1	9.4	9.3	9.6	9.7	10.1
8	10.2	10.4	10.2	10.7	10.6	10.5	10.3	10.2	9.8
9	8.2	8.5	9.7	9.4	9.6	9.8	10.0	10.1	10.3

The table shows a very clear rounding off tendency during the final decades of the 19th century. The gradual establishment of vital registration and the increasing need in modern society for people to know and use their exact birth date has gradually resulted in a virtual elimination of digit preferences in the reporting of age. In developing countries, however, digit preferences as well as other age reporting problems may still cause inaccuracies in census data.

EXAMPLE 16.2.4. This example is based on data from a content evaluation study of the accuracy of the 1980 Census of Population and Housing in the United States, reported in United States Bureau of the Census (1986). The study is based on reinterviewing. That is, for a sample of households, the same question as in the original census is asked once again, and the two answers given by the same household to the same question are then matched. Different summary measures of *response variability* are then calculated. Here, response variability is the tendency of a household to give different answers to the same question, indicating the presence of measurement errors. Table 16.2.2 gives data from the content evaluation study concerning a question about the number of automobiles kept at home for use by members of the same household.

Table 16.2.2. Number of Households in Different Categories According to Responses to the Question: "How Many Automobiles Are Kept at Home for Use by Members of Your Household?" Unedited Data.

	Content evaluation study				
Original census	None	One	Two	Three or more	Total
None	1,050	230	49	6	1,335
One	119	3,308	618	81	4,126
Two	13	339	1,895	248	2,495
Three or more	2	32	171	435	640
Total	1,184	3,909	2,733	770	8,596

The table shows, for example, that of the 4,126 households in the study that answered "One" in the original census, 3,308 also answered "One" in the content evaluation study, whereas 119 answered "None," 618 answered "Two," and 81 answered "Three or more." Main diagonal entries refer to those giving the same answer on the two occasions.

16.3. The Simple Measurement Model

Our objective now is to formulate a statistical model for measurements made on elements of a sample from a finite population. (For a complete enumeration, the "sample" is the whole population.) The basic idea is that the measurements (and consequently the measurement errors) are modeled as random variables. This sets the stage for using standard statistical tools to evaluate the effect of measurement error on the accuracy of the usual estimators. We are concerned in particular with the π estimator.

Consider, as usual, a finite population, $U = \{1, \ldots, k, \ldots, N\}$. It is assumed that for each element $k \in U$, there exists a true value θ_k and that the objective is to estimate the population total of these true values,

$$t_\theta = \textstyle\sum_U \theta_k$$

A sample s of size n_s is selected from U by a given probability sampling design $p(\cdot)$. The ideal is to obtain the true value θ_k for each element $k \in s$, but what we actually obtain through the measurement procedure are the observed values y_k for $k \in s$. The observed y_k is composed of the true value θ_k and a random measurement error $y_k - \theta_k$. For lack of better values, the y_k must nevertheless be used in the estimation. For a given sample s, the random variables y_k $(k \in s)$ are assumed to have a certain joint probability distribution (conditional on s), called a measurement model and denoted m.

Remark 16.3.1. Strictly speaking, we should denote the model as m_s to show that the model is conditional on the obtained sample s. To simplify the notation, we write simply m.

In this chapter, the survey is viewed as a two-stage process, with each stage contributing randomness.

First stage. The sample selection, which results in a selected sample s. Stochastic structure is given by the sampling design, $p(\cdot)$.

Second stage. The measurement procedure, which generates an observed value y_k for each $k \in s$. Stochastic structure is given by the measurement model m.

When evaluating expectations and variances with respect to the two stages jointly, the conditional argument is useful. As for the expected values,

$$E_{pm}(\cdot) = E_p[E_m(\cdot|s)] \tag{16.3.1}$$

where $E_m(\cdot|s)$ denotes conditional expectation with respect to the measurement model m, for a given sample s; $E_p(\cdot)$ denotes expectation with respect to the sampling design $p(\cdot)$; and $E_{pm}(\cdot)$ denotes expectation with respect to sampling design and measurement model jointly.

Similarly, for the joint variance, called the *pm-variance* or the *total variance*, we have

$$V_{pm}(\cdot) = E_p[V_m(\cdot|s)] + V_p[E_m(\cdot|s)] \tag{16.3.2}$$

where $V_m(\cdot|s)$ denotes conditional variance with respect to the model m, given s; $V_p(\cdot)$ denotes variance with respect to $p(\cdot)$; and $V_{pm}(\cdot)$ denotes variance with respect to $p(\cdot)$ and m jointly.

We specify further the model m. For elements k and l belonging to the same sample s, the first and second-order moments are

$$\mu_k = E_m(y_k|s) \tag{16.3.3}$$

$$\sigma_k^2 = V_m(y_k|s) \tag{16.3.4}$$

and

$$\sigma_{kl} = C_m(y_k, y_l|s) \tag{16.3.5}$$

Obviously, $\sigma_{kk} = \sigma_k^2$ for all k. These moments are ordinarily unknown.

The model states that, for any given sample s, the measurement y_k on element k ($k \in s$) has the mean μ_k and the variance σ_k^2, and between y_k and y_l ($k, l \in s$) there is the covariance σ_{kl}. As the notation suggests, the model moments are assumed not to depend on the sample s. Thus, moments according to equations (16.3.3) to (16.3.5) are defined for all k and l in U, provided that $\pi_k > 0$ for all $k \in U$.

This model, which we call the simple measurement model, will be sufficient for many survey situations, and we use it in most of the chapter. A more complex model with moments dependent on s will be considered in Section 16.11.

It is important to have a clear notion of the frequency interpretation of the simple measurement model m given above. There is a given probability sample s and a given measurement procedure that generates an observed value for each element $k \in s$. Suppose measurements could be independently repeated many times on the same sample s, thus generating a long series of measurements on each element $k \in s$. The observed y-values for a particular element k would not necessarily be the same in all repetitions, but would vary in a random fashion, around a "long run" mean value μ_k and with a "long run" variance σ_k^2. Moreover, if two elements k and l were considered together, their respective measurements would, over the long series, show a "long run" covariance of σ_{kl}.

Remark 16.3.2. In practice, it is usually impossible to obtain independent repeated measurements in the hypothetical manner described in our frequency interpretation above. The respondent (and the interviewer, if there is one) would probably be influenced by previously given answers. Thus, even if repeated measurements were obtained, the conditions would not be identical. But in our model, we still like to think of a realized response for the kth element as one of a hypothetical set of possible responses that could have occurred under the actual survey conditions. For a frequency interpretation of this assumption, we must imagine repeated trials under identical survey conditions.

Remark 16.3.3. The same symbol y_k is used in this chapter to denote both the random variable associated with the kth element and a particular value taken by this random variable. The meaning will be clear from the context.

Remark 16.3.4. In survey practice, there are many different kinds of measurement procedures. The simple measurement model presented above is very general in the sense that the measurement procedure is not specified. Sections 16.4 to 16.6 discuss the effect of measurement errors in this general setting. In Section 16.7 to 16.10 we see how these general results can be applied to particular measurement procedures characterized by different ways of assigning respondents to interviewers.

For an estimator $\hat{t}$ of t_θ we now define the estimation error as the difference $\hat{t} - t_\theta$. This is a random variable whose probability distribution is determined jointly by the sampling design p and the measurement model m. We follow the tradition of using the expected square of this error (here with respect to $p(\cdot)$ and m jointly) as a measure of the accuracy of the estimator $\hat{t}$. Formally, the mean square error of $\hat{t}$ is

$$\text{MSE}_{pm}(\hat{t}) = E_{pm}[(\hat{t} - t_\theta)^2] \tag{16.3.6}$$

For a frequency interpretation of this MSE, we think in terms of a hypothetical long sequence of independent repetitions of the survey, each repetition consisting of two steps: (1) a new sample s is selected; and (2) a new measurement y_k is obtained for each element $k \in s$. Then $\text{MSE}_{pm}(\hat{t})$ represents the long-run average over these repetitions of the squared difference $(\hat{t} - t_\theta)^2$.

16.4. Decomposition of the Mean Square Error

Now let us look in particular at the π estimator and the effect that measurement errors will have on its accuracy. We decompose the mean square error shown in (16.3.6) into several components, assuming that the measurements obey the simple measurement model m of the previous section. One can say that the model, through its MSE decomposition, serves as a tool for our understanding of the accuracy.

The mean square error (16.3.6) of $\hat{t}_\pi$ can be written as the sum of total variance and squared bias

$$\text{MSE}_{pm}(\hat{t}_\pi) = V_{pm}(\hat{t}_\pi) + [B_{pm}(\hat{t}_\pi)]^2$$

where the total variance is

$$V_{pm}(\hat{t}_\pi) = E_{pm}\{[\hat{t}_\pi - E_{pm}(\hat{t}_\pi)]^2\}$$

and the bias is

$$B_{pm}(\hat{t}_\pi) = E_{pm}(\hat{t}_\pi) - t_\theta$$

Here

$$E_{pm}(\hat{t}_\pi) = E_p[E_m(\hat{t}_\pi | s)] = E_p(\textstyle\sum_s \mu_k/\pi_k) = \sum_U \mu_k$$

so that

$$B_{pm}(\hat{t}_\pi) = \textstyle\sum_U (\mu_k - \theta_k) = B$$

This is called the measurement bias; it arises when expected measurement values on elements do not agree with true element values.

The variance term can be decomposed as follows

$$\begin{aligned} V_{pm}(\hat{t}_\pi) &= E_p[V_m(\hat{t}_\pi | s)] + V_p[E_m(\hat{t}_\pi | s)] \\ &= V_1 + V_2 \end{aligned} \tag{16.4.1}$$

The first component V_1, called the measurement variance, can be further decomposed:

$$\begin{aligned} V_1 = E_p[V_m(\hat{t}_\pi | s)] &= E_p(\textstyle\sum\sum_s \sigma_{kl}/\pi_k\pi_l) \\ &= \textstyle\sum\sum_U (\pi_{kl}/\pi_k\pi_l)\sigma_{kl} \\ &= \textstyle\sum_U \sigma_k^2/\pi_k + \sum\sum_{U,\, k \neq l} (\pi_{kl}/\pi_k\pi_l)\sigma_{kl} \\ &= V_{11} + V_{12} \end{aligned}$$

If $A \subseteq U$ is a set of elements, we write

$$\sum\sum_{k \neq l} {}_A \quad \text{for} \quad \sum_{k \in A} \sum_{\substack{l \in A \\ k \neq l}}$$

The component V_{11}, called the simple measurement variance, arises from variability in measurements on individual elements. If all $\sigma_k^2 = 0$, that is, if repeated measurements on the same element always yield the same value, then $V_{11} = 0$.

The component V_{12} is called the correlated measurement variance. It depends on the covariances between measurements on different elements. If $\sigma_{kl} = 0$ for all $k \neq l$, then $V_{12} = 0$. (Note that when all $\sigma_k^2 = 0$, then also $\sigma_{kl} = 0$ for all $k \neq l$, and thus $V_{12} = 0$.)

The remaining component of (16.4.1) is

$$\begin{aligned} V_2 = V_p[E_m(\hat{t}_\pi | s)] &= V_p(\textstyle\sum_s \mu_k/\pi_k) \\ &= \textstyle\sum\sum_U \Delta_{kl} \check{\mu}_k \check{\mu}_l. \end{aligned}$$

because $\sum_s \mu_k/\pi_k = \sum_s \check{\mu}_k$ has the form of a π estimator. This component is called the sampling variance. It equals zero under complete enumeration. When there is no measurement variability (all $\sigma_k^2 = 0$), then $V_1 = 0$, and V_2 is the only contribution to the variance $V_{pm}(\hat{t}_\pi)$. If each element is measured without bias ($\mu_k = \theta_k$ for all k), then

$$V_2 = \textstyle\sum_U \Delta_{kl} \check{\theta}_k \check{\theta}_l$$

which is the sampling variance of a π estimator based on the true values. The following result summarizes our findings.

Result 16.4.1. *Jointly under the sampling design $p(\cdot)$ and the simple measurement model m with the moments as given in* (16.3.3) *to* (16.3.5), *the mean square error of the π estimator $\hat{t}_\pi = \sum_s y_k/\pi_k$ can be decomposed as*

$$\text{MSE}_{pm}(\hat{t}_\pi) = V_{pm}(\hat{t}_\pi) + B^2 \tag{16.4.2}$$

where

$$B = \textstyle\sum_U (\mu_k - \theta_k) \tag{16.4.3}$$

is the measurement bias;

$$V_{pm}(\hat{t}_\pi) = V_1 + V_2 = V_{11} + V_{12} + V_2 \tag{16.4.4}$$

is the total variance;

$$V_1 = \textstyle\sum\sum_U (\pi_{kl}/\pi_k \pi_l) \sigma_{kl}$$

is the measurement variance;

$$V_{11} = \textstyle\sum_U \sigma_k^2/\pi_k \tag{16.4.5}$$

is the simple measurement variance;

$$V_{12} = \sum\sum_{k \neq l} {}_U (\pi_{kl}/\pi_k \pi_l) \sigma_{kl} \tag{16.4.6}$$

is the correlated measurement variance; and

$$V_2 = \sum\sum_U \Delta_{kl} \check{\mu}_k \check{\mu}_l \tag{16.4.7}$$

is the sampling variance.

Contrary to what the generally accepted name suggests, the measurement variance V_1 depends both on the measurement model (through σ_k^2 and σ_{kl}) and on the sampling design (through the inclusion probabilities π_k and π_{kl}). It is therefore of interest to isolate a component of V_1 that is unaffected by sampling, that is, a term that would remain even if sampling were pushed to the ultimate limit of complete enumeration. It is an easy exercise to show that

$$V_1 = V_{1\text{cen}} + V_{1\text{sam}} \tag{16.4.8}$$

with

$$V_{1\text{cen}} = \sum\sum_U \sigma_{kl}$$

and

$$V_{1\text{sam}} = \sum\sum_U \Delta_{kl}\sigma_{kl}/(\pi_k \pi_l)$$

The second term, $V_{1\text{sam}}$, is zero under a complete enumeration design, since in that case $\Delta_{kl} = 0$ for all k, l; the first term, $V_{1\text{cen}}$, is independent of the sampling design.

For a clearer picture of the decomposition shown in equation (16.4.8), it is convenient to work with average model moments

$$\mu = \sum_U \mu_k/N \tag{16.4.9}$$

and

$$\sigma^2 = \sum_U \sigma_k^2/N \tag{16.4.10}$$

Also, define

$$\rho = \sum\sum_{U \atop k \neq l} \sigma_{kl}/[N(N-1)\sigma^2] \tag{16.4.11}$$

which is average covariance, divided by average variance, thus a kind of correlation measure. In particular, $\rho = 0$, if $\sigma_{kl} = 0$ for all $k \neq l$ (uncorrelated measurements for all pairs). We also see that $\rho = 1$, if $\sigma_1^2 = \cdots = \sigma_N^2 = \sigma^2$, and $\sigma_{kl} = \sigma_k \sigma_l$ for all $k \neq l$, that is, equal measurement variance for all elements and perfect correlation between measurements on all pairs of elements. In fact, it can be shown that unity is an upper bound to the value of ρ.

In some surveys, $\rho = 0$ may be approximately valid, for example, in a mail survey where respondents reply completely on their own, without an interviewer.

Now, from (16.4.10) and (16.4.11),

$$\begin{aligned} V_{1\text{cen}} = \sum\sum_U \sigma_{kl} &= N[1 + (N-1)\rho]\sigma^2 \\ &= N^2[1 + (N-1)\rho]\sigma^2/N \end{aligned}$$

The factor N^2 in the last expression indicates, as in many other formulas in this book, that the population total is being estimated. The factor σ^2/N indicates variance per measured element. We summarize our latest discussion in the following result.

Result 16.4.2. *An alternative decomposition of the measurement variance under the sampling design* $p(\cdot)$ *and the simple measurement model m is*

$$V_1 = V_{1\text{cen}} + V_{1\text{sam}}$$

where the complete enumeration component is

$$V_{1\text{cen}} = N[1 + (N - 1)\rho]\sigma^2$$

with σ^2 *and* ρ *given respectively by* (16.4.10) *and* (16.4.11), *and the sampling component is*

$$V_{1\text{sam}} = \textstyle\sum\sum_U \Delta_{kl}\sigma_{kl}/(\pi_k\pi_l)$$

Remark 16.4.1. For all essential purposes, $V_{1\text{cen}}$ is nonnegative. However, it does not follow mathematically from the model formulation in Section 16.3 that $V_{1\text{cen}}$ is necessarily nonnegative. The model does imply that, for every possible s, the $n_s \times n_s$ matrix $\{\sigma_{kl}\}$ is positive semidefinite. But this does not ensure that the $N \times N$ matrix $\{\sigma_{kl}\}$ is positive semidefinite, which is required in order that $V_{1\text{cen}} = \sum\sum_U \sigma_{kl} \geq 0$. For an example where $V_{1\text{cen}} < 0$, see Wretman (1983). Under complete enumeration, that is, when $s = U$ with probability one, $V_{1\text{cen}}$ is, of course, nonnegative, and $\rho \geq -1/(N - 1)$.

EXAMPLE 16.4.1. To illustrate Results 16.4.1 and 16.4.2, let us calculate the variance components arising in *SI* sampling with the sampling fraction $f = n/N$. With μ, σ^2, and ρ given by equations (16.4.9) to (16.4.11), the simple measurement variance is

$$V_{11} = N^2\sigma^2/n$$

and the correlated measurement variance is

$$V_{12} = N^2(n - 1)\rho\sigma^2/n$$

for a total measurement variance of

$$V_1 = V_{11} + V_{12} = N^2[1 + (n - 1)\rho]\sigma^2/n \tag{16.4.12}$$

If σ^2 and ρ remain constant, and $\rho > 0$, the interesting conclusion is that an increase in the sample size reduces the simple measurement variance, but has almost no effect on the correlated measurement variance. This shows that correlated measurements can do great harm to the precision of survey estimates. Even a near-zero value of ρ will have considerable effect on V_1, especially in large samples, since the increase due to correlated measurements is measured by the factor $(n - 1)\rho$.

In the alternative decomposition of V_1, given in Result 16.4.2, we have

$$V_{1\text{cen}} = N^2[1 + (N - 1)\rho]\sigma^2/N$$

and

$$V_{1\text{sam}} = N^2(1 - f)(1 - \rho)\sigma^2/n$$

To compare the two components, we form the ratio

$$\frac{V_{1\text{cen}}}{V_{1\text{sam}}} = \frac{f}{1 - f} \frac{1 + (N - 1)\rho}{1 - \rho}$$

Often f is modest and ρ small, but positive. Thus $V_{1\text{cen}}$ is often considerably greater than $V_{1\text{sam}}$ because of the factor $(N - 1)\rho$.

Finally, the sampling variance component is

$$V_2 = N^2(1 - f)S^2_{\mu U}/n$$

with

$$S^2_{\mu U} = \frac{1}{N - 1} \sum_U (\mu_k - \mu)^2$$

Thus, V_2 decreases when n increases, assuming $S^2_{\mu U}$ is constant.

Remark 16.4.2. Should we conclude in Example 16.4.1 that the effect of the measurement error is necessarily reduced by more extensive sampling? The example showed that V_1 and V_2 will in fact decrease as n increases, but this assumes that μ, σ^2, and ρ remain constant. However, when a larger sample is taken, it should perhaps not be taken for granted that μ, σ^2, and ρ remain unchanged. One may argue that σ^2 may actually increase with the sample size, since diverting resources to handling a larger survey may mean a drop in the quality of the measurements. (Exactly how this occurs would be difficult to express in formulas.) Hence, V_1 may not be reduced after all, but rather be increased.

16.5. The Risk of Underestimating the Total Variance

To estimate the *pm* variance of the π estimator meets with considerable difficulties when all that is available is a single observed value y_k for each sampled element, which is the case in most surveys. To illustrate, consider the simple measurement model from Sections 16.3 and 16.4. The *pm* variance is then

$$V = V_{pm}(\hat{t}_\pi) = V_{11} + V_{12} + V_2$$

where the components V_{11}, V_{12}, and V_2 depend on the unknown model moments μ_k, σ^2_k, and σ_{kl} defined by equations (16.3.3) to (16.3.5). To estimate V, one must somehow arrive at estimates of these model parameters for each sample element. Clearly however, one cannot hope to estimate σ_k^2, for example, without repeated observations of the element k. Variance estimation through repeated measurements is considered in Section 16.6. In Section 16.10, we consider another technique, interpenetrating subsamples, which can sometimes be used to estimate V without resorting to repeated measurements.

First, let us examine what "standard" variance estimation will bring, assuming a single available measurement y_k for each sampled element $k \in s$. Suppose we take the variance estimation formula appropriate for the case of no measurement errors, that is,

$$\hat{V}_{\text{stand}} = \sum\sum_s \check{\Delta}_{kl} \check{y}_k \check{y}_l \tag{16.5.1}$$

As one might suspect, $\hat{V}_{\text{stand}}$ will be a biased estimator of the total variance V when the y_k contain measurement error. It is not hard to show that

$$E_{pm}(\hat{V}_{\text{stand}}) = V - V_{1\text{cen}} = V_{1\text{sam}} + V_2 \tag{16.5.2}$$

which is a result found by Koch, Freeman and Freeman (1975). Hence, the bias of $\hat{V}_{\text{stand}}$ as an estimator of V is

$$E_{pm}(\hat{V}_{\text{stand}}) - V = -V_{1\text{cen}} = -\sum\sum_U \sigma_{kl} = -N^2[1 + (N-1)\rho]\sigma^2/N \tag{16.5.3}$$

where σ^2 and ρ are given, respectively, by (16.4.9) and (16.4.10). Assuming $\sigma^2 > 0$, equation (16.5.3) prompts the following conclusions:

1. The bias shown in (16.5.3) is independent of the sampling design, but may be sensitive to the sampling design in that the σ_{kl} may change with the survey conditions, as noted in Remark 16.4.2.
2. The bias shown in (16.5.3) is negative, if

$$\rho > -1/(N-1)$$

   which is likely to be the case in practice (see Remark 16.4.1). Thus, $\hat{V}_{\text{stand}}$ will ordinarily underestimate V.
3. In particular if $\sigma_{kl} = 0$ whenever $k \neq l$, that is, if measurements on different elements are uncorrelated, then $\hat{V}_{\text{stand}}$ underestimates V, since

$$E_{pm}(\hat{V}_{\text{stand}}) - V = -\sum_U \sigma_k^2 < 0$$

4. In practice, one is likely to have $\sigma_{kl} > 0$ for many element pairs (k, l). (Positive covariance of measurements is likely to arise, for example, if k and l are interviewed by the same interviewer.) A likely effect is that ρ becomes positive. Assuming $\sigma^2 = \sum_U \sigma_k^2/N$ is unchanged, the underestimation of V will be more pronounced than in the case of uncorrelated measurements. Positive measurement covariance tends to make the underestimation worse.

Thus, $\hat{V}_{\text{stand}}$ is not large enough, on the average, to cover the total variance V. But is it sufficient to cover the sampling variance V_2? It follows from (16.5.2) that the bias of $\hat{V}_{\text{stand}}$ as an estimator of V_2 is given by

$$E_{pm}(\hat{V}_{\text{stand}}) - V_2 = V_{1\text{sam}} = \sum\sum_U \Delta_{kl}\sigma_{kl}/(\pi_k\pi_l)$$

In many cases, this quantity is positive and $\hat{V}_{\text{stand}}$ will overestimate V_2, as in the following cases:

1. When $\sigma_{kl} = 0$ for all $k \neq l$ (uncorrelated measurements), for then

$$E_{pm}(\hat{V}_{\text{stand}}) - V_2 = \sum_U (1 - \pi_k)\sigma_k^2/\pi_k > 0$$

2. When *SI* sampling is used, and $\rho < 1$, for then, using Example 16.4.1,

$$[E_{pm}(\hat{V}_{\text{stand}}) - V_2]/V_2 = (1 - \rho)\sigma^2/S_{\mu U}^2 > 0$$

To summarize, the standard variance estimator $\hat{V}_{\text{stand}}$ will often estimate a quantity greater than the sampling variance V_2, but smaller than the total variance V. Methods that address these shortcomings are discussed in Sections 16.6 and 16.10.

16.6. Repeated Measurements as a Tool in Variance Estimation

Special types of survey designs are sometimes used to obtain variance estimates in the presence of measurement errors. These designs usually rely on either repeated measurements, or interpenetrating subsamples, or a combination of both. A summary of such methods can be found in Bailar and Dalenius (1969) and Lessler (1984).

This section gives a simple example of how repeated measurements may help in estimating the total variance given in (16.4.4). (Interpenetrating subsamples are considered in Section 16.10.)

Assume that the y measurement is repeated once for each element in a suitably large subsample drawn from the original sample s. As a result, we have two measurements on the same variable y for each subsample element. Several repeats could, of course, be considered, depending on the resources available. Our intention here is not to provide an optimal variance estimator, but rather to provide suggestions for obtaining an unbiased variance estimator.

Consider the following conditions:

1. An original sample s of size n_s is drawn from the population U with a sampling design $p(\cdot)$ having the inclusion probabilities π_k, π_{kl}.
2. From s, a subsample $r (r \subseteq s)$ is drawn by *SI* sampling, in such a way that the size of r is $n_r = fn_s$, where f is a constant fixed in advance. (Thus, n_r is random if n_s is random.)
3. For each element $k \in s$, we observe the y variable; denote the observed value y_{k1}.
4. For each element $k \in r$, we observe the same y variable a second time, thus obtaining the repeat value, denoted y_{k2}.

We assume that the measurement conditions for the two occasions have been identical, or as close to identical as possible. Let M denote the measurement model that we introduce to describe the joint distribution of the $n_s + n_r$ measurements $\{y_{k1}: k \in s\}$ and $\{y_{k2}: k \in r\}$, when s and r are given. This model assumes that the simple measurement model m of Section 16.3 holds for the n_s y_{k1} measurements, and that the same model (with identical first- and second-order model moments) also holds for the n_r y_{k2} measurements. Furthermore, the model M assumes that the y_{k1} measurements are uncorrelated with the y_{k2} measurements. For given s and r (where $r \subseteq s$), let $E_M(\cdot|s, r)$ denote expected value with respect to the model M. This model is then specified as follows

$$E_M(y_{k1}|s, r) = \mu_k \quad \text{for } k \in s$$

$$E_M(y_{k2}|s, r) = E_M(y_{k1}|s, r) = \mu_k \quad \text{for } k \in r \subseteq s$$

$$E_M[(y_{k1} - \mu_k)^2|s, r] = \sigma_k^2 \quad \text{for } k \in s$$

$$E_M[(y_{k2} - \mu_k)^2|s, r] = E_M\{(y_{k1} - \mu_k)^2|s, r\} = \sigma_k^2 \quad \text{for } k \in r \subseteq s$$

$$E_M[(y_{k1} - \mu_k)(y_{l1} - \mu_l)|s, r] = \sigma_{kl} \quad \text{for } k, l \in s$$

$$E_M[(y_{k2} - \mu_k)(y_{l2} - \mu_l)|s, r] = \sigma_{kl} \quad \text{for } k, l \in r \subseteq s$$

$$E_M[(y_{k1} - \mu_k)(y_{l2} - \mu_l)|s, r] = 0 \quad \text{for } k \in s \text{ and } l \in r \subseteq s$$

The last assumption is an important one; it says that all repeat measurements have to be uncorrelated with all original measurements. For example, too high a value in the original measurement must not be followed by a systematic tendency toward too high a value in the repeat measurement. Unfortunately, recall effects may produce this kind of unwanted positive correlation over time. In practice, one must take care to carry out the repeated measurements so that the model assumptions hold.

Assume that the ordinary π estimator, based on s, is used, namely,

$$\hat{t}_\pi = \textstyle\sum_s y_{k1}/\pi_k \tag{16.6.1}$$

We want to estimate the *pm* variance

$$V = V_{pm}(\hat{t}_\pi) = V_{11} + V_{12} + V_2 \tag{16.6.2}$$

where the components are those of Result 16.4.1.

The objective is to create an unbiased estimator of V and of each of the components V_{11} and V_{12}. These estimators should be based on data from s and r. By unbiased we mean unbiased with respect to the randomized sample selection and the model M jointly. Recall that the randomized sample selection has two steps: (i) selection of s by the given design $p(\cdot)$, and (ii) *SI* selection of the subsample r. Denote this selection plan as P; expected value with respect to P is denoted $E_P(\cdot)$. If $\hat{V}$ is an estimator of the *pm* variance (16.6.2), we thus want

$$E_P[E_M(\hat{V}|r, s)] = V$$

In Section 16.5 we considered the standard variance estimator

$$\hat{V}_{\text{stand}} = \sum\sum_s \check{\Delta}_{kl} \check{y}_{k1} \check{y}_{l1} \tag{16.6.3}$$

and found its expected value to be

$$E_{PM}(\hat{V}_{\text{stand}}) = V - V_{1\text{cen}}$$

with

$$V_{1\text{cen}} = \sum\sum_U \sigma_{kl} = N[1 + (N-1)\rho]\sigma^2$$

Now, one can show that under P and M jointly, $V_{1\text{cen}}$ is estimated unbiasedly by

$$\hat{V}_{1\text{cen}} = \frac{n_s}{2n_r} \sum_r z_k^2/\pi_k + \frac{n_s(n_s - 1)}{2n_r(n_r - 1)} \sum\sum_{\substack{r \\ k \neq l}} z_k z_l/\pi_{kl}$$

where, for $k \in r$, z_k is the difference between the two measurements

$$z_k = y_{k1} - y_{k2}$$

Consequently, an unbiased estimator of V is given by

$$\hat{V} = \hat{V}_{\text{stand}} + \hat{V}_{1\text{cen}}$$

and our primary objective is reached.

Unbiased estimation of the components V_{11} and V_{12} in (16.6.2) is also possible. One can show that unbiased component estimators are given by

$$\hat{V}_{11} = \frac{n_s}{2n_r} \sum_r (z_k/\pi_k)^2$$

and

$$\hat{V}_{12} = \frac{n_s(n_s - 1)}{2n_r(n_r - 1)} \{(\textstyle\sum_r z_k/\pi_k)^2 - \sum_r (z_k/\pi_k)^2\}$$

Example 16.6.1. Under *SI* selection of the sample s, the estimators of V_{11} and V_{12} simplify as follows

$$\hat{V}_{11} = \frac{N^2}{2n_s n_r} \sum_r z_k^2$$

and

$$\hat{V}_{12} = \frac{N^2}{2n_s n_r} \frac{n_s - 1}{n_r - 1} [(\textstyle\sum_r z_k)^2 - \sum_r z_k^2]$$

Remark 16.6.1. The preceding analysis used the repeat measurements y_{k2} as a bias-removing remedy in estimating the total variance, and as a tool in estimating the simple and correlated measurement variance components. The

y_{k2} figured neither in the estimator of the population total shown in (16.6.1), nor in the variance estimator $\hat{V}_{\text{stand}}$ given by (16.6.3). It might be possible to replace both (16.6.1) and (16.6.3) by more efficient (but also more complex) estimators that incorporate the repeat observations y_{k2}.

Remark 16.6.2. In practice, the repeated-measurement technique is often used as a tool both for control of interviewer performance and for evaluation of a survey. By evaluation, we mean a special study with the purpose of assessing the quality of results from a survey that has already been carried out. A review of methods for evaluation studies is found in Statistics Canada (1978).

Remark 16.6.3. Estimation of the bias component B given by (16.4.3) would require other techniques than repeated measurements. To get information on B one would need the true values for a subset of the original sample.

16.7. Measurement Models Taking Interviewer Effects into Account

So far, in this chapter, we have used the term *measurement procedure* in a broad sense. When we introduced the simple measurement model, the measurements obtained for a given sample s were described only as numerical values realized on a set of random variables y_k for $k \in s$, with the moments μ_k, σ_k^2, and σ_{kl}. We did not require that the observations should be obtained through any specified measurement procedure, such as, for example, mail questionnaire, telephone interview, or personal interview.

An advantage with this generality is that it can serve in a wide range of practical situations. So far we have been able to give general results on isolating and estimating various components of the total variance.

The next three sections illustrate how the general results of these earlier sections can be applied. We consider examples where data are collected by interviewers. The interviewer may introduce bias, variance, and correlation into the measurements. Such interviewer effects have been seen in many empirical studies.

In the examples to be discussed, each interviewer is assumed to interview several respondents, which may result in correlated answers from these respondents. Differences between examples are due to different rules by which the interviewers are assigned to groups of respondents. In all these examples, the requirements of the simple measurement model are satisfied, that is, the moments μ_k, σ_k^2, and σ_{kl} do not depend on the sample s. We will see how these moments can be expressed under the special assumptions made in each example. Once this is done, we can apply the results from earlier sections to obtain formulae for the variance components V_{11}, V_{12}, and V_2.

16.8. Deterministic Assignment of Interviewers

We assume in this section that there exist a given set of interviewers and a given partitioning of the population into interviewer assignment groups. More formally, the assignment rule is:

1. There is a fixed set of a interviewers labeled $i = 1, \ldots, a$.
2. Prior to the survey, there is a fixed partitioning of the population into a groups of elements $U_1, \ldots, U_a$, so that each interviewer i is linked to a unique group U_i in a deterministic, nonrandom manner.
3. The preassigned interviewer carries out all interviews with sampled elements from his own group.

Thus, for each population element, it is known beforehand who the interviewer will be if that element is selected for the sample. For instance, the population may be geographically divided into districts with one interviewer permanently stationed in each district who carries out all interviews there.

Let N_i be the size of the group U_i. We have

$$U = \bigcup_{i=1}^{a} U_i; \qquad N = \sum_{i=1}^{a} N_i$$

A probability sample s, selected from U, will be partitioned according to

$$s = \bigcup_{i=1}^{a} s_i$$

where $s_i = s \bigcap U_i$ is the ith interviewer assignment. Depending on the sampling design, the sizes of these assignments n_{s_i}, $i = 1, \ldots, a$, may be fixed or random.

Our measurement model m specifies that the measurements y_k, for $k \in s$, consist of a true value plus an error,

$$y_k = \theta_k + \varepsilon_k$$

We now also assume that the ε_k, for $k \in s$, have the following stochastic structure

$$E_m(\varepsilon_k | s) = b_i \quad \text{for } k \in s_i$$

$$V_m(\varepsilon_k | s) = v_i \quad \text{for } k \in s_i$$

$$C_m(\varepsilon_k, \varepsilon_l | s) = \begin{cases} \rho_i v_i & \text{for } k \in s_i, l \in s_i, k \neq l \\ 0 & \text{for } k \in s_i, l \in s_j, i \neq j \end{cases}$$

where b_i, v_i, and ρ_i are unknown constants, and $|\rho_i| < 1$. The true value θ_k is a constant, so the model moments of y_k, defined by (16.3.3) to (16.3.5), are immediately obtained as

$$\mu_k = \theta_k + b_i \quad \text{for } k \in U_i \tag{16.8.1}$$

$$\sigma_k^2 = v_i \quad \text{for } k \in U_i \tag{16.8.2}$$

$$\sigma_{kl} = \begin{cases} \rho_i v_i & \text{for } k \in U_i, l \in U_i, k \neq l \\ 0 & \text{for } k \in U_i, l \in U_j, i \neq j \end{cases} \tag{16.8.3}$$

Our model has two important features:

1. Measurements y_k made by the same interviewer i are affected by the same constant interviewer effect b_i.
2. Measurements made by the same interviewer are correlated (as soon as $\rho_i \neq 0$), whereas measurements made by different interviewers are uncorrelated.

Let us apply results from Section 16.4 to the above model. We insert the moments (16.8.1) to (16.8.3) into Results 16.4.1 and 16.4.2, and obtain the following expressions for the different variance components of the π estimator. The simple measurement variance is

$$V_{11} = \sum_{i=1}^{a} (\textstyle\sum_{U_i} 1/\pi_k) v_i \tag{16.8.4}$$

and the correlated measurement variance is

$$V_{12} = \sum_{i=1}^{a} \left(\sum\sum_{\substack{U_i \\ k \neq l}} \pi_{kl}/\pi_k \pi_l \right) \rho_i v_i \tag{16.8.5}$$

Since

$$\sigma^2 = \sum_{i=1}^{a} N_i v_i / N \tag{16.8.6}$$

and

$$\rho = \frac{\sum_{i=1}^{a} N_i(N_i - 1)\rho_i v_i}{(N-1) \sum_{i=1}^{a} N_i v_i} \tag{16.8.7}$$

the complete enumeration component of the measurement variance is

$$V_{1\text{cen}} = \sum_{i=1}^{a} N_i [1 + (N_i - 1)\rho_i] v_i \tag{16.8.8}$$

The sampling component of the measurement variance is

$$V_{1\text{sam}} = \sum_{i=1}^{a} F_i (1 - \rho_i) v_i + \sum_{i=1}^{a} G_i \rho_i v_i \tag{16.8.9}$$

with

$$F_i = \textstyle\sum_{U_i} \Delta_{kk}/\pi_k^2 = \sum_{U_i} [(1/\pi_k) - 1] \tag{16.8.10}$$

$$G_i = \textstyle\sum\sum_{U_i} \Delta_{kl}/\pi_k \pi_l = \sum\sum_{U_i} [(\pi_{kl}/\pi_k \pi_l) - 1] \tag{16.8.11}$$

The sampling variance is

$$V_2 = \sum_{i=1}^{a} \sum_{j=1}^{a} \sum_{k \in U_i} \sum_{l \in U_j} \Delta_{kl} \frac{(\theta_k + b_i)(\theta_l + b_j)}{\pi_k \pi_l} \tag{16.8.12}$$

and, finally, the bias is

$$B = \sum_{i=1}^{a} N_i b_i \tag{16.8.13}$$

These results lead to some interesting conclusions presented in the following series of remarks.

Remark 16.8.1. An extreme conclusion is the following. From (16.8.7), we see that $\rho = 0$ if all $N_i = 1$, that is, if there is one interviewer for each population element. Provided that all $\rho_i > 0$ and that σ^2 is unaffected by the number of interviewers, this choice of N_i will also give a minimum value of $V_{1\text{cen}}$. From the point of view of cost, one interviewer for each element is obviously unrealistic. In practice, the effect of increasing the number of interviewers is often to decrease $V_{1\text{cen}}$, as long as σ^2 remains constant. Of course, in the case of a self-administered questionnaire, where each respondent is his/her own interviewer, we have one interviewer per element. In this case, however, the value of σ^2 would probably be much higher than in the case when the interviews are carried out by trained interviewers.

Remark 16.8.2. Let us compare the measurement variances V_1 that arise under two different sampling designs, namely, (1) *SI* sampling of $n = fN$ elements, and (2) *STSI* sampling with proportional allocation and with each group U_i serving as a stratum; thus, $n_i = fN_i$ and $\sum_{i=1}^{a} n_i = n$. Let $V_{1,SI}$ denote the measurement variance under *SI* sampling, and $V_{1,STSI,\text{prop}}$ the measurement variance under *STSI* sampling with proportional allocation as described above. Since the complete enumeration component (16.8.8) of the measurement variance is unaffected by the sampling design, the difference between the measurement variances for the two sampling designs can be expressed as

$$V_{1,SI} - V_{1,STSI,\text{prop}} = V_{1\text{sam},SI} - V_{1\text{sam},STSI,\text{prop}} \tag{16.8.14}$$

From Example 16.4.1 we know that

$$V_{1\text{sam},SI} = N^2(1-f)(1-\rho)\sigma^2/n \tag{16.8.15}$$

where σ^2 and ρ are now given by (16.8.6) and (16.8.7). It can also be shown that

$$V_{1\text{sam},STSI,\text{prop}} = N^2 \frac{1-f}{n} \sum_{i=1}^{a} W_i(1-\rho_i)v_i \tag{16.8.16}$$

with $W_i = N_i/N$. It follows that the difference (16.8.14) is

$$V_{1,SI} - V_{1,STSI,\text{prop}} = N^2 \frac{1-f}{n} \frac{N}{N-1} \sum_{i=1}^{a} W_i(1-W_i)\rho_i v_i > 0 \tag{16.8.17}$$

provided $\rho_i > 0$ for all interviewers, which is likely in practice. Thus, proportional *STSI* sampling does give a smaller measurement variance than *SI* sampling.

Remark 16.8.3. Under the present model, the mean measurement μ_k is composed of a true value θ_k and a bias b_i due to the preassigned interviewer. We now see how the sampling variance V_2 can be affected by this interviewer bias. Let us compare the sampling variances under the same two sampling designs as in the preceding remark, namely, (1) *SI* sampling of $n = fN$ elements, and (2) *STSI* sampling with proportional allocation and with each group U_i serving as a stratum ($n_i = fN_i$ and $\sum_{i=1}^{a} n_i = n$).

Under *SI* sampling, the sampling variance V_2 is

$$V_2 = V_{2,SI} = N^2 \frac{1-f}{n}(S_{\theta U}^2 + S_{bU}^2 + 2S_{\theta bU}) \tag{16.8.18}$$

where

$$S_{\theta U}^2 = \frac{1}{N-1} \sum_U (\theta_k - \bar{\theta}_U)^2$$

$$S_{bU}^2 = \frac{1}{N-1} \sum_{i=1}^{a} N_i (b_i - \bar{b}_U)^2$$

and

$$S_{\theta bU} = \frac{1}{N-1} \sum_{i=1}^{a} N_i (\bar{\theta}_{U_i} - \bar{\theta}_U)(b_i - \bar{b}_U)$$

with

$$\bar{\theta}_{U_i} = \sum_{U_i} \theta_k / N_i$$

$$\bar{\theta}_U = \sum_U \theta_k / N = \sum_{i=1}^{a} N_i \bar{\theta}_{U_i} / N$$

and

$$\bar{b}_U = \sum_{i=1}^{a} N_i b_i / N$$

Under the *STSI* design with proportional allocation, the sampling variance is

$$V_2 = V_{2,STSI,\text{prop}} = N^2 \frac{1-f}{n} \sum_{i=1}^{a} W_i S_{\theta U_i}^2 \tag{16.8.19}$$

where

$$S_{\theta U_i}^2 = \frac{1}{N_i - 1} \sum_{U_i} (\theta_k - \bar{\theta}_{U_i})^2$$

Comparing the two sampling variances given by (16.8.18) and (16.8.19), we see that $V_{2,SI}$ is affected by the interviewer bias, whereas $V_{2,STSI,\text{prop}}$ is not, because the interviewer bias is constant within each stratum.

The sampling variance (16.8.18) contains one component, S_{bU}^2, due to the variation of interviewer bias among interviewers, and another component, $S_{\theta bU}$, due to the covariation between interviewer bias and true value. If all interviewers have the same bias b_i, then $S_{bU}^2 = S_{\theta bU} = 0$. But in practice, interviewers differ. We want the disturbing effect of the interviewers on the sampling variance removed; if this cannot be done by stratified sampling, other options exist. Random assignment of interviewers and interpenetration are two such options that will be discussed in the following sections.

16.9. Random Assignment of Interviewers to Groups

In Section 16.8, interviewers were assigned groups of respondents in accordance with a deterministic principle set in advance. In many applications, however, interviewers are randomly assigned to groups. One reason for random assignment is to avoid interaction between interviewer bias and true value, as in Remark 16.8.3. One application of the random assignment of interviewers to groups is the following. Let the population U be partitioned, as in the previous section, into a set of a priori fixed groups, $U_1, \ldots, U_a$ of size $N_1, \ldots, N_a$, respectively, so that

$$U = \bigcup_{i=1}^{a} U_i \quad \text{and} \quad N = \sum_{i=1}^{a} N_i$$

A probability sample s from U will be partitioned as

$$s = \bigcup_{i=1}^{a} s_i$$

where $s_i = s \cap U_i$, and all interviews within s_i will be carried out by the same interviewer. This interviewer is selected at random from a pool of interviewers. For simplicity, we assume that interviewers are selected from an *infinite* pool of possible interviewers.

Remark 16.9.1. The assumption that interviewers are selected at random is often made in practice, even if no randomization mechanism is used. The infinite pool of possible interviewers is then a hypothetical construct, and random selection of interviewers could perhaps be seen as a way of modeling the recruitment of interviewers for a survey.

Remark 16.9.2. Models, similar to the one given below, could also be made assuming random selection of interviewers from a *finite* pool of interviewers.

We let our measurement error model include the random selection of interviewers as part of the measurement procedure, which is thus composed of the following two steps:

i. A randomly selected interviewer is assigned to each sample group s_i, $i = 1, \ldots, a$.
ii. The randomly chosen interviewer carries out an interview with each element in s_i, $i = 1, \ldots, a$.

Whether true random assignment is practical or not will depend on the circumstances of the survey. To move interviewers around to different geographic locations, if the U_i are regions, can be costly. It would be easier to assign interviewers randomly within a limited geographic area, for instance, a city or a town. In that case, one might consider separate random assignment models within local areas. Nevertheless, for a survey organization with a permanently stationed interviewer staff, the deterministic model of Section 16.8 may be more realistic. In telephone interviewing, random assignment of interviewers would be easier.

Our measurement model m is now specified as follows. The measurement y_k, for $k \in s$, is assumed to consist of a true value θ_k and a measurement error ε_k, that is,

$$y_k = \theta_k + \varepsilon_k$$

where the error is composed as

$$\varepsilon_k = B_i + e_k \quad \text{for } k \in s_i$$

Here, B_i is an interviewer effect associated with the randomly chosen interviewer assigned to s_i (and thus common to all measurements on elements in s_i). The second term, e_k, is an error term, representing all other sources of error in the survey. Both B_i and e_k are random variables, with the following statistical properties.

1. $B_1, \ldots, B_a$ are independent and identically distributed random variables with the same expected value μ_B and the same variance v_B.
2. e_k $(k \in s)$ are independent random variables with zero expectation and the same variance v_e.
3. The random variables B_i $(i = 1, \ldots, a)$ are independent of the random variables e_k $(k \in s)$.

Another way of expressing (1) above is that $B_1, \ldots, B_a$ is a random sample of interviewer effects, drawn from an infinite population of interviewer effects. The interviewer effect B_i is a component of the measurement error that is the same for all measurements made by the same interviewer. The remaining error terms e_k are assumed mutually independent, even within the same interviewer assignment.

A special case of interest is when the interviewer variance v_B is zero. The interviewers then introduce the same interviewer effect, $B_i = \mu_B$ (which may or may not be zero) in every measurement. In practice this occurs with questionnaire items that leave little or no room for interviewer influence.

Readers familiar with analysis of variance modeling will recognize our assumptions (1) to (3) as being those traditionally made in a random effects linear model. Step (i) of the measurement procedure corresponds to the realization of a B_i-value for each s_i, whereas the step (ii) corresponds to the realization of a value of e_k (and hence of the total error $\varepsilon_k = B_i + e_k$) for each $k \in s_i$.

The model moments defined by (16.3.1) to (16.3.3) are, under assumptions (1) to (3), easily obtained as

$$\mu_k = \theta_k + \mu_B \quad \text{for all} \quad k \in U \tag{16.9.1}$$

$$\sigma_k^2 = v_B + v_e \quad \text{for all} \quad k \in U \tag{16.9.2}$$

$$\sigma_{kl} = \begin{cases} v_B & \text{for } k \text{ and } l \in U_i, k \neq l \\ 0 & \text{for } k \in U_i, l \in U_j, i \neq j \end{cases} \tag{16.9.3}$$

Let us again study the π estimator $\hat{t}_\pi$ and the corresponding mean square error. Inserting (16.9.1) to (16.9.3) into Results 16.4.1 and 16.4.2 leads to the following expressions for the various components. The simple measurement variance is

$$V_{11} = (v_B + v_e) \textstyle\sum_U 1/\pi_k \tag{16.9.4}$$

and the correlated measurement variance is

$$V_{12} = v_B \sum_{i=1}^{a} \sum\sum_{k \neq l}{}_{U_i} \pi_{kl}/(\pi_k \pi_l) \tag{16.9.5}$$

The complete enumeration component of the measurement variance is

$$V_{1\text{cen}} = v_e N + v_B \sum_{i=1}^{a} N_i^2 \tag{16.9.6}$$

because

$$\sigma^2 = v_B + v_e$$

and

$$\rho = \frac{v_B}{v_B + v_e} \frac{\sum_{i=1}^{a} N_i^2 - N}{N(N-1)}$$

The sampling component of the measurement variance is

$$V_{1\text{sam}} = v_e \sum_{i=1}^{a} F_i + v_B \sum_{i=1}^{a} G_i \tag{16.9.7}$$

where F_i and G_i are given, respectively, by (16.8.10) and (16.8.11). The sampling variance is

$$V_2 = \sum\sum_U \Delta_{kl} \frac{(\theta_k + \mu_B)(\theta_l + \mu_B)}{\pi_k \pi_l} \tag{16.9.8}$$

and, finally, the bias is

$$B = N\mu_B \tag{16.9.9}$$

Remark 16.9.3. Under *SI* sampling (with $f = n/N$), the measurement variance is

$$V_1 = V_{1,SI} = V_{1\text{cen},SI} + V_{1\text{sam},SI}$$

with $V_{1\text{cen},SI}$ given by (16.9.6) and

$$V_{1\text{sam},SI} = N^2 \frac{1-f}{n}\left(v_e + v_B \frac{N^2 - \sum_{i=1}^{a} N_i^2}{N(N-1)}\right) \tag{16.9.10}$$

The sampling variance is

$$V_2 = V_{2,SI} = N^2 \frac{1-f}{n} S_{\theta U}^2 \tag{16.9.11}$$

where $S_{\theta U}^2$ is the population variance of the true values. As a result of randomly assigning interviewers, the sampling variance is no longer dependent on interviewer variability, as it was with the deterministic assignment in the preceding section (see (16.8.18) in Remark 16.8.3).

Remark 16.9.4. Under *STSI* sampling such that each group U_i is a stratum and with proportional allocation ($f_i = f = n/N$), the measurement variance is

$$V_1 = V_{1,STSI,\text{prop}} = V_{1\text{cen},STSI,\text{prop}} + V_{1\text{sam},STSI,\text{prop}}$$

with $V_{1\text{cen},STSI,\text{prop}}$ given by (16.9.6), and

$$V_{1\text{sam},STSI,\text{prop}} = N^2 \frac{1-f}{n} v_e \tag{16.9.12}$$

The sampling variance is

$$V_2 = V_{2,STSI,\text{prop}} = N^2 \frac{1-f}{n} \sum_{i=1}^{a} W_i S_{\theta U_i}^2 \tag{16.9.13}$$

where $S_{\theta U_i}^2$ is the population variance of the true values in the ith stratum. This is the same sampling variance as with the deterministic assignment of interviewers, see (16.8.19).

Remark 16.9.5. Comparing the measurement variances of the two preceding designs, we obtain from (16.9.10) and (16.9.12) that

$$V_{1,SI} - V_{1,STSI} = \frac{1-f}{n} \frac{N}{N-1}\left(N^2 - \sum_{i=1}^{a} N_i^2\right) v_B \geq 0$$

That is, proportional $STSI$ sampling leads to a smaller measurement variance than SI sampling (provided $v_B > 0$ and $a \geq 2$). Comparing the sampling variances, we obtain from (16.9.11) and (16.9.13) that

$$V_{2,SI} - V_{2,STSI,\text{prop}} = N^2 \frac{1-f}{n}\left(S_{\theta U}^2 - \sum_{i=1}^{a} W_i S_{\theta U_i}^2\right)$$

Thus, the sampling variance is ordinarily smaller for the stratified case.

Remark 16.9.6. Under the present model, all σ_{kl} ($k \neq l$), and thus ρ, are nonnegative by definition. Other things being equal, ρ increases with the interviewer variance v_B. If the interviewer effect is constant ($v_B = 0$), then $\rho = 0$, and the measurement variance is reduced to

$$V_1 = v_e \sum\nolimits_U 1/\pi_k$$

Remark 16.9.7. For a fixed number of interviewers a, the value of ρ, and hence the measurement variance V_1, is minimized if the group sizes N_i are equal (supposing that all other factors are constant and that $v_B > 0$).

Remark 16.9.8. In the case of equal group sizes ($N_i = N/a = N_0$, say, for all i),

$$\rho = \frac{N_0 - 1}{N - 1} \frac{v_B}{v_B + v_e}$$

is minimized if the number of interviewers, $a = N/N_0$, is as large as possible. The extreme case is one interviewer per element, which gives $\rho = 0$.

Remark 16.9.9. Consider again SI sampling, with a sampling fraction $f = n/N$ that is negligible. Assume for simplicity equal group sizes, $N_i = N/a = N_0$ for all i. Since $\sigma^2 = v_e + v_B$, it follows from (16.9.10) and (16.9.11) that

$$V_1 + V_2 \doteq N^2 \left(\frac{v_B}{a} + \frac{v_e}{n} + \frac{S_{\theta U}^2}{n}\right) \tag{16.9.14}$$

Here we see that increased sample size has no effect on the component involving v_B. The only efficient way to reduce this term is by increasing the number of interviewers. Alternatively, (16.9.14) can be written as

$$V_1 + V_2 \doteq N^2[1 + (n_0 - 1)\rho_w]\sigma_{\text{tot}}^2/n \tag{16.9.15}$$

where

$$n_0 = n/a$$

is the mean interviewer assignment, and

$$\sigma_{\text{tot}}^2 = v_e + v_B + S_{\theta U}^2$$

represents total variance per element. (It is essentially the variance of y in a sample of size one.) Finally,

$$\rho_w = v_B/\sigma^2_{\text{tot}}$$

has been called the intrainterviewer assignment correlation coefficient.

Here, (16.9.15) shows again the importance of the number of interviewers. For example, suppose $\rho_w = 0.01$, $n = 1{,}000$, and $a = 100$. The mean interviewer assignment is $n_0 = 10$, and

$$1 + (n_0 - 1)\rho_w = 1.09$$

By contrast, if the 1,000 interviews are carried out by only $a = 10$ interviewers, the mean assignment is $n_0 = 100$, and

$$1 + (n_0 - 1)\rho_w = 1.99$$

The total variance $V_1 + V_2$ is larger in the second case by a factor of roughly $1.99/1.09 \doteq 1.8$.

16.10. Interpenetrating Subsamples

This section discusses a third way of assigning interviewers to groups of elements. The set of interviewers is now fixed, and it is the sample that is divided in a random fashion among the interviewers. Unlike with the previous methods, it is now impossible to identify, prior to sampling, groups of elements that would always be assigned to the same interviewer.

Let s be a probability sample of size n, drawn from the population U according to a given sampling design. After s has been obtained, it is subdivided at random into a groups (subsamples), $s_1, \ldots, s_a$, of equal size $m = n/a$. (We assume, for simplicity, that n is a multiple of a.) The random division into groups means that for a given s, each s_i is an SI sample of size m from s. There is a fixed set of a interviewers, who will conduct all interviews in that group. This is an example of a technique known as interpenetrating subsamples, or simply interpenetration. This procedure is practical when interviewing does not entail great travel or other costs.

Remark 16.10.1. Often, when interpenetrating subsamples are used, both data collection and data processing are performed by a separate staff for each subsample. So, measurement and processing errors in different subsamples will be uncorrelated.

We let our measurement error model include the random division of the sample as part of the measurement procedure, which is thus composed of the following two steps.

i. The sample is divided into equal-size random groups, and an interviewer is assigned to each group.
ii. The interviewer conducts the interviews with all elements in his/her group.

For interpenetrating subsamples, the measurement model m is the following. The measurements y_k, for $k \in s$ and $s = \bigcup_{i=1}^{a} s_i$, are assumed to have the structure

$$y_k = \theta_k + \varepsilon_k$$

where ε_k is a random measurement error satisfying

$$E_m(\varepsilon_k | s; s_1, \ldots, s_a) = b_i \quad \text{for } k \in s_i \tag{16.10.1}$$

$$V_m(\varepsilon_k | s; s_1, \ldots, s_a) = v_i \quad \text{for } k \in s_i \tag{16.10.2}$$

$$C_m(\varepsilon_k, \varepsilon_l | s; s_1, \ldots, s_a) = \begin{cases} \rho_i v_i & \text{for } k \neq l, k \in s_i, l \in s_i \\ 0 & \text{for } k \neq l, k \in s_i, l \in s_j, i \neq j \end{cases} \tag{16.10.3}$$

Here, b_i, v_i, and ρ_i are unknown constants, and $|\rho_i| \leq 1$. To reiterate, if element k is interviewed by interviewer i, the measurement error contains an interviewer effect b_i as a constant component, it has a variance v_i, and a covariance $\rho_i v_i$ with measurement errors on other elements in the same interviewer assignment. Measurements in different assignments are assumed uncorrelated.

The assumptions stated in (16.10.1) to (16.10.3) resemble their counterparts in Section 16.8. What is added now is the randomized subsamples. We use the notation

$$\bar{b} = \frac{1}{a} \sum_{i=1}^{a} b_i$$

$$S_b^2 = \frac{1}{a} \sum_{i=1}^{a} (b_i - \bar{b})^2$$

$$\bar{v} = \frac{1}{a} \sum_{i=1}^{a} v_i$$

and

$$\bar{\rho} = \sum_{i=1}^{a} \rho_i v_i \bigg/ \sum_{i=1}^{a} v_i$$

It follows after some calculation that

$$\mu_k = \theta_k + \bar{b} \quad \text{for } k \in U$$

$$\sigma_k^2 = \bar{v} + S_b^2 \quad \text{for } k \in U$$

$$\sigma_{kl} = [(m-1)\bar{\rho}\bar{v} - S_b^2]/(n-1) \quad \text{for } k \neq l \in U$$

Here, μ_k and σ_k^2 resemble their analogs under the randomized assignment of interviewers to fixed groups (Section 16.9). But note that the covariance σ_{kl} is now constant for all pairs of elements throughout the population.

We again use Results 16.4.1 and 16.4.2 to obtain expressions for the mean square error components of the π estimator. With

$$\sigma^2 = \bar{v} + S_b^2$$

and

$$\rho = \frac{(m-1)\bar{\rho}\bar{v} - S_b^2}{(n-1)(\bar{v} + S_b^2)}$$

we can express the simple measurement variance as

$$V_{11} = \sigma^2 \sum_U 1/\pi_k$$

and the correlated measurement variance as

$$V_{12} = \rho\sigma^2 \sum\sum_{U \atop k \neq l} \pi_{kl}/(\pi_k \pi_l)$$

The sampling variance is

$$V_2 = \sum\sum_U \Delta_{kl} \frac{(\theta_k + \bar{b})(\theta_l + \bar{b})}{\pi_k \pi_l}$$

and the bias is

$$B = N\bar{b}$$

Remark 16.10.2. When the sampling design is *SI* (of $n = fN$ elements), the measurement variance is simplified to

$$V_1 = N^2[1 + (m-1)\bar{\rho}]\bar{v}/n \tag{16.10.4}$$

and the sampling variance to

$$V_2 = N^2(1-f)S_{\theta U}^2/n \tag{16.10.5}$$

Although each individual measurement variance σ_k^2 contains the interviewer variability S_b^2 as one component, the measurement variance V_1 is independent of S_b^2 under the *SI* sampling design. Adding equations (16.10.4) and (16.10.5), we obtain the total variance

$$V_{pm}(\hat{t}_\pi) = V_1 + V_2 = N^2[\bar{v} + (m-1)\bar{\rho}\bar{v} + (1-f)S_{\theta U}^2]/n$$

Remark 16.10.3. The random division of the sample s into groups $s_1, \ldots, s_a$ is similar to the division into dependent random groups discussed in Section 11.3.2. The variance estimation procedure described there can be adapted to interpenetrating subsamples to produce an approximately unbiased variance estimator. We assume that the sampling design is *SI* with sampling fraction $f = n/N$. The π estimator of the population total is then

$$\hat{t}_\pi = N\bar{y}_s = N \sum_{i=1}^{a} \bar{y}_{s_i}/a$$

By analogy with the formula in Example 11.3.1, we consider the following estimator of $V_{pm}(\hat{t}_\pi)$,

$$\hat{V} = \frac{N^2}{a(a-1)} \sum_{i=1}^{a} (\bar{y}_{s_i} - \bar{y}_s)^2 = \frac{N^2}{n} MS_b \tag{16.10.6}$$

where MS_b is the mean square between groups,

$$MS_b = \frac{1}{m(a-1)} \sum_{i=1}^{a} (\bar{y}_{s_i} - \bar{y}_s)^2$$

It is left as an exercise to show that

$$E_{pm}(MS_b) = [1 + (m-1)\bar{\rho}]\bar{v} + \frac{n}{n-1}\frac{a-1}{a} S_{\theta U}^2 + \frac{n}{a-1} S_b^2 \quad (16.10.7)$$

It now follows that $\hat{V}$ given by equation (16.10.6) is approximatively unbiased for $V_{pm}(\hat{t}_\pi)$ given in Remark 16.10.2, provided that $S_b^2 = 0$, that is, provided that the interviewers have identical interviewer effects. A remarkable consequence of interpenetration is that we can obtain an approximately unbiased estimate of the total variance, including both measurement variance and sampling variance, without use of repeated measurements.

Remark 16.10.4. With *SI* sampling and $S_b^2 = 0$, it is also possible to estimate the correlated measurement variance. To this end, consider the mean square within groups,

$$\text{MS}_w = \frac{1}{a(m-1)} \sum_{i=1}^{a} \sum_{s_i} (y_k - \bar{y}_{s_i})^2$$

It can be shown that

$$E_{mp}(\text{MS}_w) = (1 - \bar{\rho})\bar{v} + S_{\theta U}^2 \quad (16.10.8)$$

From (16.10.7) and (16.10.8), it follows that $(\text{MS}_b - \text{MS}_w)/m$ is approximately unbiased for $\bar{\rho}\bar{v}$, provided that $S_b^2 = 0$. The correlated measurement variance can be expressed as

$$V_{12} = N^2(m-1)\bar{\rho}\bar{v}/n$$

when $S_b^2 = 0$. Thus, an approximately unbiased estimator of V_{12}, when $S_b^2 = 0$, is given by

$$\hat{V}_{12} = N^2 \frac{m-1}{m} \frac{\text{MS}_b - \text{MS}_w}{n}$$

This is another advantage of the interpenetration technique.

16.11. A Measurement Model with Sample-Dependent Moments

In Section 16.3, the conditional moments μ_k, σ_k^2, and σ_{kl} were assumed to be independent of the sample s to which the elements k and l belong. However, there are cases when such a simple model would not be appropriate. Therefore, we now consider an extended measurement model, where the conditional distribution of the measurement error $y_k - \theta_k$, given a sample s containing k, is allowed to depend on s. For this extended model, we present a new mean square error decomposition.

The extended model is needed when the measurement conditions change from one sample to another. Factors that can affect the measurement conditions are the size of the sample, its geographic spread, the allocation of interviewers, and influence from the respondents in the sample.

Let s be a given sample containing k. Under the extended measurement model, we use the following notation for the conditional first- and second-order moments of y_k, given s,

$$\mu_{ks} = E_m(y_k|s) \tag{16.11.1}$$

$$\sigma_{ks}^2 = V_m(y_k|s) = E_m[(y_k - \mu_{ks})^2|s] \tag{16.11.2}$$

and

$$\sigma_{kls} = C_m(y_k, y_l|s) = E_m[(y_k - \mu_{ks})(y_l - \mu_{ls})|s] \tag{16.11.3}$$

Suppose that s_1 and s_2 are two samples that both contain the element k. Then, for example, the extended measurement model allows $\mu_{ks_1} \neq \mu_{ks_2}$.

For a frequency interpretetation of the mean stated in (16.11.1), we have to imagine a long series of independently repeated y measurements on the same element k, where the sample s to which k belongs is kept fixed. Then these y-values would vary in a random fashion around a "long run" average μ_{ks}, which is now allowed to be a function of the selected sample s.

The design-weighted average model moments are, for $k = 1, \ldots, N$,

$$\mu_k = \sum_{s \ni k} p(s)\mu_{ks}/\pi_k \tag{16.11.4}$$

$$\sigma_k^2 = \sum_{s \ni k} p(s)\sigma_{ks}^2/\pi_k \tag{16.11.5}$$

and, for $k \neq l = 1, \ldots, N$,

$$\sigma_{kl} = \sum_{s \ni k,l} p(s)\sigma_{kls}/\pi_{kl} \tag{16.11.6}$$

The simple measurement model defined in Section 16.3, where, for any fixed k and l, the moments μ_{ks}, σ_{ks}^2, and σ_{kls} were assumed constant for all s, is a special case of the extended model.

Remark 16.11.1. We can interpret (16.11.4) as

$$\mu_k = E_{pm}(y_k|s \ni k)$$

that is, as the conditional expectation of y_k, given that element k was selected. Note that this expectation is not conditional on a particular sample s containing k, but on the event that the sample belongs to the set of possible samples that contain k. The frequency interpretation is as follows. Consider an imagined sequence of independent replications of the experiment consisting of (i) selecting a sample s, and (ii) obtaining measurements from all elements belonging to s. Now, consider the subsequence of these replications such that the particular element k belongs to s. Then μ_k is the long-run average of the measurements on element k over this subsequence. Similar interpretations can be given to (16.11.5) and (16.11.6), namely,

$$\sigma_k^2 = V_{pm}(y_k|s \ni k)$$
$$\sigma_{kl} = C_{pm}(y_k, y_l|s \ni k, l)$$

Our immediate objective under the extended model is to find a decomposition of the mean square error of $\hat{t}_\pi$, by mimicking the procedure used in Section 16.4. We have

$$\text{MSE}_{pm}(\hat{t}_\pi) = V_{pm}(\hat{t}_\pi) + [B_{pm}(\hat{t}_\pi)]^2$$

Here, the measurement bias is

$$B_{pm}(\hat{t}_\pi) = E_{pm}(\hat{t}_\pi) - t_\theta = \textstyle\sum_U (\mu_k - \theta_k) = B \tag{16.11.7}$$

since, using equation (16.11.4),

$$\begin{aligned} E_{pm}(\hat{t}_\pi) &= E_p\{E_m[(\textstyle\sum_s y_k/\pi_k)|s]\} \\ &= \sum_{s \ni \mathscr{S}} p(s) \textstyle\sum_s \mu_{ks}/\pi_k \\ &= \textstyle\sum_U \sum_{s \ni k} p(s)\mu_{ks}/\pi_k \\ &= \textstyle\sum_U \mu_k \end{aligned}$$

Moreover,

$$\begin{aligned} V_{pm}(\hat{t}_\pi) &= E_p[V_m(\hat{t}_\pi|s)] + V_p[E_m(\hat{t}_\pi|s)] \\ &= V_1 + V_2 \end{aligned} \tag{16.11.8}$$

where V_1 is the measurement variance and V_2 the sampling variance. Since

$$\begin{aligned} V_m(\hat{t}_\pi|s) &= \textstyle\sum\sum_s \sigma_{kls}/(\pi_k\pi_l) \\ &= \textstyle\sum_s \sigma_{ks}^2/\pi_k^2 + \sum\sum_{s\,(k \neq l)} \sigma_{kls}/(\pi_k\pi_l) \end{aligned}$$

we have,

$$\begin{aligned} V_1 = E_p[V_m(\hat{t}_\pi|s)] &= E_p(\textstyle\sum_s \sigma_{ks}^2/\pi_k^2) + E_p\left[\sum\sum_{s\,(k \neq l)} \sigma_{kls}/(\pi_k\pi_l)\right] \\ &= \textstyle\sum_U \sigma_k^2/\pi_k + \sum\sum_U \pi_{kl}\sigma_{kl}/(\pi_k\pi_l) \\ &= V_{11} + V_{12} \end{aligned} \tag{16.11.9}$$

where σ_k^2 and σ_{kl} are now the average second-order moments defined by (16.11.5) and (16.11.6), respectively. As before, we call V_{11} the simple measurement variance and V_{12} the correlated measurement variance. The details of the evaluation in equation (16.11.9) are left as an exercise. Here, V_{11} and V_{12} have the same appearance as in Section 16.4, but note that σ_k^2 and σ_{kl} are now average model moments.

As for the sampling variance V_2, note first that

$$E_m(\hat{t}_\pi|s) = \sum_s \mu_{ks}/\pi_k = \sum_s \mu_k/\pi_k + \sum_s (\mu_{ks} - \mu_k)/\pi_k$$

so that

$$\begin{aligned} V_2 = V_p[E_m(\hat{t}_\pi|s)] &= V_p(\textstyle\sum_s \mu_k/\pi_k) + V_p[\textstyle\sum_s (\mu_{ks} - \mu_k)/\pi_k] \\ &\quad + 2C_p[\textstyle\sum_s \mu_k/\pi_k, \textstyle\sum_s (\mu_{ks} - \mu_k)/\pi_k] \\ &= V_{21} + V_{22} + V_{23} \end{aligned}$$

where

$$V_{21} = V_p(\textstyle\sum_s \mu_k/\pi_k) = \sum\sum_U \Delta_{kl}\check{\mu}_k\check{\mu}_l \tag{16.11.10}$$

$$V_{22} = V_p[\textstyle\sum_s (\mu_{ks} - \mu_k)/\pi_k] \tag{16.11.11}$$

and

$$V_{23} = 2C_p[\textstyle\sum_s \mu_k/\pi_k, \textstyle\sum_s (\mu_{ks} - \mu_k)/\pi_k] \tag{16.11.12}$$

Here, V_{21}, V_{22}, and V_{23} are components of the sampling variance; they are all zero under complete enumeration. Compared to Result 16.4.1, there are now two new components, V_{22} and V_{23}, which are somewhat hard to interpret. V_{22} depends on the extent to which the mean measurement μ_{ks} varies across the different samples that contain k.

Finally, V_{23} is an interaction term. It gives a positive contribution in the case where high values of $\sum_s \mu_k/\pi_k$ tend to accompany high values of $\sum_s (\mu_{ks} - \mu_k)/\pi_k$ and vice versa. For example, suppose the study variable is the value of housing units. In a sample s with many expensive housing units, the interviewers may be influenced by the preponderance of high values in the sample, so that exaggerated values are recorded, on the average, for the housing units in the sample. On the other side, in a sample with many inexpensive housing units, understatements may be obtained on the average. A positive value of V_{23} is thus created. Note that $V_{22} = V_{23} = 0$ when the mean measurements are sample-independent, that is, when, for any k, $\mu_{ks} = \mu_k$ for all s containing k. The main findings of this section are now summarized.

Result 16.11.1. *Jointly under the sampling design $p(\cdot)$ and the extended measurement model m, the mean square error of $\hat{t}_\pi = \sum_s y_k/\pi_k$ can be decomposed as*

$$MSE_{pm}(\hat{t}_\pi) = V_1 + V_2 + B^2 = V_{11} + V_{12} + V_{21} + V_{22} + V_{23} + B^2$$

where V_{11} and V_{12} are the simple and the correlated measurement variances given by (16.11.9); *V_{21}, V_{22}, and V_{23} are sampling variance components given,* *respectively, by* (16.11.10), (16.11.11), *and* (16.11.12); *and B is the measurement bias given by* (16.11.7). *In these expressions,* μ_k, σ_k^2, *and* σ_{kl} *are average model moments defined by* (16.11.4) *to* (16.11.6).

Exercises

16.1. Show that unity is an upper bound to the value of the coefficient ρ, defined by equation (16.4.11).

16.2. Do the details of Example 16.4.1.

16.3. Show that under repeated measurements, as described in Section 16.6, $V_{1\text{cen}}$ is estimated unbiasedly by

$$\hat{V}_{1\text{cen}} = (n_s/2n_r) \sum_r z_k^2/\pi_k + [n_s(n_s - 1)/2n_r(n_r - 1)] \sum\sum_{r\, k \neq l} z_k z_l/\pi_{kl}$$

where z_k is the difference between the two measurements,

$$z_k = y_{k1} - y_{k2}$$

16.4. Show that under repeated measurements, as described in Section 16.6, an unbiased estimator of V is given by

$$\hat{V} = \hat{V}_{\text{stand}} + \hat{V}_{1\text{cen}}$$

16.5. Show that under repeated measurements, as described in Section 16.6, unbiased estimators of V_{11} and V_{12} are given by, respectively,

$$\hat{V}_{11} = (n_s/2n_r) \sum_r (z_k/\pi_k)^2$$

and

$$\hat{V}_{12} = [n_s(n_s - 1)/2n_r(n_r - 1)][(\sum_r z_k/\pi_k)^2 - \sum_r (z_k/\pi_k)^2]$$

16.6. With deterministic assignment of interviewers as described in Section 16.8, verify that the model moments shown in equations (16.8.1) to (16.8.3) follow from the assumptions made about the measurement errors ε_k.

16.7. With deterministic assignment of interviewers as described in Section 16.8, verify that the MSE components V_{11}, V_{12}, $V_{1\text{cen}}$, $V_{1\text{sam}}$, V_2, and B are as given by, respectively, (16.8.4), (16.8.5), (16.8.8), (16.8.9), (16.8.12), and (16.8.13).

16.8. Verify the expressions given by (16.8.16) and (16.8.17) in Remark 16.8.2, namely, that

$$V_{1\text{sam},STSI,\text{prop}} = N^2[(1 - f)/n] \sum_{i=1}^{a} W_i(1 - \rho_i)v_i$$

and

$$V_{1,SI} - V_{1,STSI,\text{prop}} = N^2[(1 - f)/n][N/(N - 1)] \sum_{i=1}^{a} W_i(1 - W_i)\rho_i v_i > 0$$

16.9. Verify the expressions given by (16.8.18) and (16.8.19) in Remark 16.8.3, namely, that

$$V_{2,SI} = N^2[(1 - f)/n](S_{\theta U}^2 + S_{bU}^2 + 2S_{\theta bU})$$

and

$$V_{2,STSI,\text{prop}} = N^2[(1-f)/n] \sum_{i=1}^{a} W_i S_{\theta U_i}^2$$

16.10. With random assignment of interviewers as described in Section 16.9, verify that the model moments shown in equations (16.9.1) to (16.9.3) follow from the assumptions made about the measurement errors ε_k.

16.11. With random assignment of interviewers as described in Section 16.9, verify that the MSE components V_{11}, V_{12}, $V_{1\text{cen}}$, $V_{1\text{sam}}$, V_2, and B are as given by, respectively, (16.9.4) to (16.9.9).

16.12. Verify the expressions given by (16.9.10) and (16.9.11) in Remark 16.9.3, namely, that

$$V_{1\text{sam},SI} = N^2[(1-f)/n]\left[v_e + v_B\left(N^2 - \sum_{i=1}^{a} N_i^2\right)\Big/ N(N-1)\right]$$

and

$$V_{2,SI} = N^2[(1-f)/n] S_{\theta U}^2$$

16.13. Verify the expressions given by (16.9.12) and (16.9.13) in Remark 16.9.4, namely, that

$$V_{1\text{sam},STSI,\text{prop}} = N^2[(1-f)/n] v_e$$

and

$$V_{2,STSI,\text{prop}} = N^2[(1-f)/n] \sum_{i=1}^{a} W_i S_{\theta U_i}^2$$

16.14. Verify the expression in Remark 16.9.5 stating that

$$V_{1,SI} - V_{1,STSI} = [(1-f)/n][N/(N-1)]\left(N^2 - \sum_{i=1}^{a} N_i^2\right) v_B \geq 0$$

16.15. Verify the statement of Remark 16.9.6, that if $v_B = 0$, then $\rho = 0$, and the measurement variance is reduced to

$$V_1 = v_e \sum\nolimits_U 1/\pi_k$$

16.16. Verify the statement of Remark 16.9.7, that under random assignment of interviewers to groups and with a fixed number of interviewers, the measurement variance V_1 is minimized if the group sizes N_i are equal.

16.17. Verify the statement of Remark 16.9.8, that under random assignment of interviewers to groups and with equal group sizes, ρ is minimized if the number of interviewers is as large as possible.

16.18. Verify the expressions (16.9.14) and (16.9.15) in Remark 16.9.9, namely, that under random assignment of interviewers to groups and with SI sampling,

$$\begin{aligned} V_1 + V_2 &\doteq N^2[(v_B/a) + (v_e/n) + (S_{\theta U}^2/n)] \\ &= N^2[1 + (n_0 - 1)\rho_w]\sigma_{\text{tot}}^2/n \end{aligned}$$

16.19. With interpenetrating subsamples as described in Section 16.10, verify that the expressions given for the model moments μ_k, σ_k^2, and σ_{kl} follow from the assumptions made about the measurement errors ε_k.

16.20. With interpenetrating subsamples as described in Section 16.10, verify the expressions given for the MSE components V_{11}, V_{12}, V_2, and B.

16.21. Verify the result of Remark 16.10.2, that

$$V_{pm}(\hat{t}_\pi) = N^2[\bar{v} + (m-1)\bar{\rho}\bar{v} + (1-f)S_{\theta U}^2]/n$$

16.22. Verify the statements made in Remark 16.10.3, that

$$E_{pm}(MS_b) = [1 + (m-1)\bar{\rho}]\bar{v} + [n/(n-1)][(a-1)/a]S_{\theta U}^2 + [n/(a-1)]S_b^2$$

16.23. Verify the statements made in Remark 16.10.4.

16.24. Under the extended measurement model of Section 16.11, let the design-weighted average model moments μ_k, σ_k^2, and σ_{kl} be defined by (16.11.4) to (16.11.6). Verify the statements of Remark 16.11.1, namely, that

$$\mu_k = E_{pm}(y_k | s \ni k)$$

$$\sigma_k^2 = V_{pm}(y_k | s \ni k)$$

$$\sigma_{kl} = C_{pm}(y_k, y_l | s \ni k, l)$$

16.25. Verify the details in the derivation of equation (16.11.9).

CHAPTER 17

Quality Declarations for Survey Data

17.1. Introduction

Survey statistics are released through a variety of media, for example, print, computer tape, diskette, computer terminal, to suit the information needs of different categories of users. Such statistics are frequently used for decision making. The user may incur substantial loss if published statistics lead to wrong decisions. It is only fair that he or she be informed of the essential features, including strengths and weaknesses, of the survey from which the statistics were derived.

The published statistics are usually point estimates of different finite population parameters. We have pointed out earlier that such estimates are the end result of a series of survey operations, starting with a statement of survey objectives and continuing with a translation of concepts into definitions, choice of sampling frame, sampling and measurement design, training and supervision of interviewers and coders, editing, imputation, and estimation with appropriate adjustment for nonresponse. Each of these operations will influence the estimate finally produced and published. The estimate itself—the published number—tells nothing about its quality or about the conditions under which it was obtained. The effects of the different survey operations are hidden in the estimate; good statistics and bad statistics look essentially alike.

The producer of the survey has an obligation to describe essential features of the survey. A user can then evaluate and interpret the survey results from his or her own particular objectives. Users may also want to know details of the statistical methodology, such as the choice of point estimators and variance estimators. The information to the user should include (1) a precise description of the population and the domains of the population about which valid statistical inferences can be made, (2) definitions of the study variables, and information on how these definitions were operationalized, for example, in the the form of items on the questionnaire, and (3) good estimates of the total mean square error and its components, the variance and the squared bias.

In a well-run survey, ample information is usually given about the points (1) and (2). It is more difficult to satisfy point (3), which requires that, if $\hat{\theta}$ is an estimator of the parameter θ, reliable estimates must be given for the variance $V(\hat{\theta})$ and the squared bias, $[B(\hat{\theta})]^2$, and for their total,

$$\text{MSE}(\hat{\theta}) = V(\hat{\theta}) + [B(\hat{\theta})]^2$$

By variance is here meant *total* variance, that is, the variance from all sources: sampling error, response error, error due to nonresponse and other forms of missing data, and error from data processing such as edit and imputation. As noted in earlier chapters, it is often hard to identify and properly measure all the errors. Good estimates of the total variance may be obtainable only at great cost. Moreover, the amount of resources required for data quality measurement in a survey usually competes with other strong claims against a limited budget.

Statistical agencies may not always live up to the high ideal represented by points (1) to (3), and point (3) in particular may cause difficulties. However, leading statistical agencies are seriously committed to providing the best possible indicators of survey quality. At present, this is accomplished by following certain *policy standards.* In the following sections, we discuss policy standards and how they have evolved over time.

17.2. Policies Concerning Information on Data Quality

The major statistical agencies have for some time recognized the importance of adhering to a specific information policy when they describe the quality of statistics released. The United Nations has also been at the forefront in promoting information on survey data quality. In 1948, the United Nations Subcommission on Statistical Sampling of the Statistical Commission issued a document entitled *The Preparation of Sampling Survey Reports,* which was revised in 1949 by a group of prominent statisticians (G. Darmois, W. E. Deming, P. C. Mahalanobis, F. Yates, and with R. A. Fisher as a consultant). A revised set of recommendations was released 15 years later (United Nations, 1964), as the result of the work of a group of sampling experts (T. Dalenius, N. Keyfitz, P. C. Mahalanobis, V. Monakhov and P. Thionet).

In 1974 the United States Bureau of the Census issued a technical paper, "Standards for Discussion and Presentation of Errors in Data." A revised version, Gonzalez et al. (1975), subsequently appeared in *The Journal of the*

American Statistical Association with the purpose

> ... to suggest, on the one hand, the type of information about survey errors that should be included in reports containing survey data and to suggest to readers of such reports, on the other hand, the type of information about survey errors they should look for or expect to receive in evaluating the results of a survey.

These developments spurred efforts on the part of some of the leading national survey organizations to prepare their own guidelines for quality declarations. In 1982, an informal meeting on "Timeliness, Cost, and Quality Attributes of Statistics" was organized by the Conference of European Statisticians (Statistical Commission and Economic Commission for Europe, United Nations Economic and Social Council), with attendance from the national statistical agencies in Canada, France, The Netherlands, Spain, Sweden, Switzerland and the United Kingdom, as well as from the International Labor Organization and the European Economic Community. The meeting based its discussion of "Guidelines for the Presentation of the Quality of Statistics" on a report prepared by Statistics Sweden, "Some Reflections on Different Approaches to the Problems of Presenting the Quality of Statistics," which contains a comparison of the quality guidelines used by Canada, Sweden, and the United States. Two documents from the United States also formed part of the basis for discussion, the paper by Gonzalez et al. (1975) and a report on error profiles from the U.S. Department of Commerce (1978a).

The term *error profile* is used about a systematic and comprehensive account of the operations that led to a given set of survey results. The error profile documents what is known in a given survey about each survey operation as a source of error. The magnitude of the error should be given, whenever possible. When no information is available about the source and impact of an error, this should also be noted. The error profile is meant to enhance the user's appreciation of the limitations of the statistics and to guide the producers of statistics in their efforts to identify those survey operations that need to be redesigned or controlled better in order to improve the quality of the survey estimates.

In U.S. Department of Commerce (1978a), the error profile approach is illustrated in the case of the Current Population Survey, an important survey providing national employment statistics for the United States. The following stages are examined.

1. Sampling design
   a. Frames
   b. Sample selection
   c. Quality control of sampling process
2. Observational design
   a. Data collection procedure
   b. Questionnaire design
   c. Data collection staff

   d. Interviewer training
   e. Quality control of field work
3. Data preparation design
   a. Data input operations
   b. Cleaning, editing and imputation
   c. Quality control of data processing
4. Production of estimates
   a. Weighting procedure
   b. Estimation procedure
   c. Quality control of estimation procedure
5. Analysis and publication

The 1982 meeting of the Conference of European Statisticians agreed that summary quality presentations should at a minimum include information on the following items:

a. Basic information on the data source and on definitions (including classifications).
b. Coverage of the data (for example, adequacy of the frame).
c. Short description of the selection and estimation methods.
d. Response rates (including their definition).
e. Sampling error (when applicable) and indication of how the computed standard errors and related measures should be interpreted.
f. Indications of the size and direction of the likely major errors and of their relative importance and impact on the statistics.
g. Information on significant changes in procedures and on other factors that would affect the comparability of statistics over time.
h. Information (if any) on the comparability with statistics on the same subject compiled from other sources.
i. References to the availability of more detailed technical descriptions (for example, technical reports).

This list places considerable emphasis on aspects of the survey proper, namely, (i) target population and definitions, and (ii) aspects related to frame quality, response rates, and sampling error. Comparability is another key concept, in particular comparability over time and comparability with similar statistics from other sources.

As an outgrowth of the agreement, statistical agencies in several countries issued their own policies on information on data quality. For example, Statistics Sweden (1983) adopted a policy focusing on the user's needs for correct interpretation of the published statistics. The policy is based on the following principle:

> The producer shall inform the users on factors of vital importance for a correct interpretation of released statistics. The information shall be easy to access and understand, and it shall be designed with the need of the users in mind.

17.3. Statistics Canada's Policy on Informing Users of Data Quality and Methodology

In 1986, Statistics Canada issued a policy statement with accompanying guidelines on documentation of data quality and statistical methodology. The adoption of a comprehensive policy such as this has far-reaching consequences for the survey organization. High-quality standards must be set for every survey operation, and efforts must be made to systematically control all operations. We end by citing Statistics Canada (1987). This document is reproduced by the kind permission of Statistics Canada and the Chief Editor of the Journal of Official Statistics. First we cite the *general policy statement.*

Introduction

The measurement of data quality is a complex undertaking. There are several dimensions to the concept of quality, many potential sources of error, and typically no comprehensive measure(s) of data quality. A rigid requirement for comprehensive data quality measurement for all bureau products is not achievable given the present state of knowledge. Nevertheless the bureau has an obligation to make its users aware of at least the major describable or quantifiable elements of quality, and of the methods that underly the data being published. Therefore this policy represents a goal toward which all programs should head, while recognizing that budgetary and other constraints may prevent the full attainment of this goal.

Policy

1. Statistics Canada will make available to users indicators of the quality of data it disseminates and descriptions of the underlying concepts, definitions and methods.
2. Statistical products will be accompanied by or make reference to documentation on quality and methodology.
3. Documentation on quality and methodology will conform to such guidelines as shall from time-to-time be issued.
4. Exemption from the requirements of this policy may be sought in special circumstances using procedures described under Responsibilities below.

Scope

1. This policy applies to all data disseminated by Statistics Canada whether collected, or merely assembled, by Statistics Canada. In the latter case, documentation should also describe the particular role of Statistics Canada in the production of the data.

2. This policy applies to all data disseminated outside Statistics Canada through any medium (CANSIM, print, computer tape, micro-film, diskette, etc.) and to any class of user (federal departments, provinces, general public, etc.). The method of making available to users information about the quality of data and about the existence of documentation on methodology may, of course, vary depending on the nature of the dissemination medium.
3. This policy applies to all data disseminated by Statistics Canada however funded. Sponsors of surveys reviewed under the Federal Government Information Collection Policy will be encouraged to conform to this policy.
4. This policy applies to data produced or disseminated in the course of analysis.

Definitions

Indicators of Data Quality

Indicators of data quality are measures or descriptions which summarize the likely magnitude and important sources of differences between the published data and the quantities that the statistical activity was designed to estimate.

Description of the conceptual framework, definitions, methods, external influences on, and other features of the data, may also be relevant to the user's assessment of the suitability of the data for particular purposes.

Documentation and Methodology

Documentation on methodology is the description of the underlying concepts, definitions, and methods used in the production of the data.

The following are Statistics Canada's (1987) *guidelines on the documentation of data quality and methodology.*

Introduction

1. Statistics Canada, as a professional agency in charge of producing official statistics, has the responsibility to inform users of the concepts and methodology used in collecting and processing its data, the quality of the data it produces, and other features of the data which may affect their use or interpretation.
2. Data users have to be able to verify that the concepts they have in mind are the same as, or sufficiently close to, those employed in collecting the data. To do this, a knowledge of the underlying conceptual framework and definitions used in the data collection is required.
3. Users generally recognize that data are subject to error and therefore need

to know whether the data are sufficiently accurate to be useful to them. To make this assessment, they need to be informed of the likely principal sources of error and, where possible, the size of the error. They also need to know of unusual circumstances which might influence the data.

4. Given that indicators of data quality cannot be expected to be comprehensive, data users also require a knowledge of the data collection and processing methodology in order to verify whether the data adequately approximate what they wish to measure and, whether the estimates they wish to use were produced with tolerances acceptable for their intended purpose.

The Guidelines

5. These guidelines are primarily intended for internal use at Statistics Canada when the documentation and dissemination related to a statistical programme are being planned or reviewed.
6. The level of detail to be provided in documentation on data quality or methodology will depend to a considerable extent on the type of data collection, the medium of dissemination, the range and impact of uses of the data, and the total budget of the collection/production process. Managerial discretion is required in determining the level of detail appropriate for a given data set.

Guidelines on the Description of Data Quality

7. Data quality is generally described in terms of sampling and non-sampling errors.
8. Unexpected events which influence the data should be flagged for users to help them in interpretation of the data.
9. It is not generally possible to provide comprehensive measures of data quality. Rather one should aim to identify what are thought to be the most important sources of error and provide quantitative measures where possible or qualitative descriptions otherwise. The result should be a balanced discussion which addresses itself to specific sources of error or bias and is therefore informative to users.
10. For censuses, surveys or administrative data surveys, the description should cover as many as possible of the elements described in paragraph 14 below.
11. Index numbers of product prices or quantities can be treated similarly, but their conceptual basis presents an additional dimension necessary to describe data quality. Particular attention might be given to any substitutions made in developing the estimates with special reference to product changes and changes in product quality. (For the corresponding methodology description, see paragraph 20 below.)
12. In the case of national accounts and data resulting from analytic activities, both the impact of quality problems in the source data, and the impact of the methods of analysis, integration, benchmarking and adjustments used, have to be taken into account. (For the corresponding methodology description, see paragraph 21 below.)

13. Statistics derived from administrative data or data not collected by Statistics Canada can also be dealt with under the guidelines of paragraph 14, but it is likely that the information available will be less detailed. Nevertheless, important issues such as coverage, response errors and comparability over time should be discussed.

Data Quality Descriptions

14. The following aspects of data quality are regarded as basic and, subject to constraints of cost and feasibility, some indication of their level should be provided or made available, where applicable, for every statistical product:
   a. *Coverage*—the quality of the survey frame or list (for surveys or censuses) or source files (for administrative data) as a proxy for the desired universe should be addressed (including gaps, duplications and definitional problems).
   b. *Sampling error*—if the survey is based on a random sample then estimates of the standard error of tabulated data based on the sample should be provided, together with an explanation of how these standard error figures should be used to interpret the data. The method of presentation may vary from explicit estimates of standard errors to use of generalized tables, graphs or other indicators. If the survey is based on a non-random sample, the implications of this on inferences that might be made from the survey should be stated.
   c. *Response rates*—the percentage of the target sample or population from which responses or usable data were obtained (on a question by question basis if appropriate) should be provided. Any known differences in the characteristics of respondents and nonrespondents should also be described as well as a brief indication of the method of imputation or estimation used to compensate for non-response.
   d. *Comparability over time*—it may be appropriate to discuss comparability with the results of the same activity for a previous reference period, especially if there has been a change in methodology, concepts, or definitions. If such a change would affect comparability from one time period to another, a quantitative estimate of this effect should be made whenever possible.
   e. *Benchmarking and revisions*—the effects of benchmarking or revisions on comparability over time should be described. Guidance on the possible impact of future benchmarking should be given based on past experience.
   f. *Comparability with other data sources*—if similar data from other sources exist, they should be identified. Where appropriate, a reconciliation should be attempted describing how the data sets differ and the reasons for these differences. Comments on quality of the other data should be provided if an evaluation is available.
   g. *Other important aspects*—there may be other aspects of data quality that are of prime importance given the objectives of a specific activity.

These should be included with the basic indicators of data quality. Examples are: unusual collection problems, misunderstandings of the intended concepts by respondents, major strikes, changes in classification or in its application, response based on financial years that do not correspond to a fixed reference period. In larger repeated surveys or activities, the most recent available information on more detailed aspects of data quality may also be provided in separate reports on data reliability. Such quality measures will usually be derived as the results of special evaluations. In different surveys and at different levels of aggregation, different sources of error may predominate. Subject to cost limitations, the most important sources of error should be evaluated periodically, and the results made available to users in the most convenient form.

h. *Total variance (or total standard error) and/or its components by source*—the overall variability of the statistics, including the effect of sampling error, response error, and processing error, should be provided if an appropriate model to aggregate these sources of error can be constructed at a cost which is reasonable relative to total program budget.
i. *Non-response bias*—an assessment of the effect of non-response on the statistics should be provided if possible.
j. *Response bias*—evidence of response bias problems stemming from respondent misunderstanding, questionnaire problems, or other sources, should be provided if available.
k. *Edit and imputation effect*—the effect of editing and imputation on the quality of data should be assessed.
l. *Seasonal adjustment*—measures of the impact and significance of the adjustment should be provided together with an explanation of how these measures should be interpreted. Examples of such measures are the mean absolute percent change of the last year's revisions of the seasonal factor, or the MCD (months for cyclical dominance) statistic.
m. *Any other error sources*—if there are particular sources of error or unforeseen events which are relevant to the series or occasion, these should be described.

Guidelines on the Description of Methodology

15. While all users should be provided with some appreciation of the methodology, some will require greater detail. Therefore, two levels of documentation should ideally be available:
    a. General user reports that are prepared for a wide audience in order to assist them in interpreting the data and in deciding on their appropriateness for a particular purpose;
    b. Technical reports that are definitive and exhaustive and give full and detailed information on methods underlying the data.
16. The amount of detail covered in methodology documentation will vary

with the type of data collection (census, sample survey, administrative data survey, index, national accounts), the medium of dissemination, the range and impact of uses of the data, and the total budget of the program. A reference to available documentation may be sufficient, especially when the dissemination medium is a short response to a special request, or a summary report.

17. For data resulting from Statistics Canada surveys or censuses, the methodology reports should provide at least an outline of the main steps in conducting the survey and should provide more detailed information on those aspects of survey methodology which have a direct impact on the quality and applicability of the data produced from the survey. The following topics should be covered where applicable.
    a. Objectives of the survey;
    b. The target universe and any differences between this and the survey frame actually used;
    c. The questionnaire(s) used and all important concepts and definitions (the discussion of concepts and definitions may well be very lengthy and require a separate document, or they may be included with the survey results; in the former case a reference to the separate document should be made);
    d. The sample design and estimation procedures;
    e. The method used for collecting the data (e.g., interview, telephone, mail, etc.) and details of any follow-up procedures for non-respondents;
    f. Any manual processing (e.g., coding) that takes place prior to data capture;
    g. The method of data capture;
    h. Quality control procedures used in connection with operations (e)–(g) above;
    i. Procedures for editing the data and for handling non-response and invalid data;
    j. Benchmarking and revision procedures used;
    k. Seasonal adjustment methodology used;
    l. The form in which the final data are stored and the tabulation or retrieval system, including confidentiality protection procedures;
    m. A brief summary of the results of any evaluation programs; and
    n. Any other special procedures or steps that might be relevant to users of the survey data.
18. For statistical data that are collected (often cooperatively) by agencies other than Statistics Canada, but which are assembled and published as a Statistics Canada product, the methodology report should cover the same points as in paragraph 17 to the extent possible, making a careful distinction between those activities for which the collecting agency is responsible and those for which Statistics Canada is responsible.
19. For administrative data, the original purpose of collection is not generally the same as that for which they are used by Statistics Canada. Although the information described in paragraphs 17 and 18 is desirable, the following topics should at least be covered:
    a. The data sources;

b. The purposes for which the data were originally collected;
c. The merits and shortcomings of the data for the statistical purpose for which they are being used (e.g., in terms of conceptual and coverage biases);
d. How the data are processed after being received and what, if anything, is done to correct problems in the original data set; and
e. The reliability of the estimates, including caveats where necessary.

It will be noted that items (c) and (e) should be covered in a discussion of quality.

20. For index numbers which are based on data collected through specific surveys or derived from administrative or other sources, the corresponding guidelines in paragraphs 17–19 apply. This applies to data on the main variable (i.e., prices in the case of a price index and quantities in the case of a volume index) and on the index weights.

In addition, particular attention should be paid to specific conceptual and methodological aspects of index making. Their proper description, in many cases, may be more important for users than a strict assessment of the quality of input data. The following elements should be developed:

a. *Precise definition of the underlying economic concepts that the index numbers are intended to measure*—reference should be made to any application or class of applications (e.g., deflation of macro-economic aggregates) for which the index numbers are not suitable.
b. *The methodology adopted*—this should cover topics such as the index formula, weighting system, computation of the index at various aggregation levels, basing, re-basing, linking of indices, treatment of changes in the varieties or qualities of goods available on the market. The adopted methodology should be compared with the underlying index concepts and possible distortions discussed.

21. Methodology reports for data resulting from analytic activities (including the System of National Accounts) should cover the following topics:

a. The conceptual framework for the analysis (e.g., the system of national accounts);
b. The major definitions and concepts used and how they are defined operationally;
c. The data sources used, and the extent to which they measure the target concepts, as well as gaps and deficiencies in these data sources. Non-comparability of data elements available from different sources should be noted. For analytical activities, reference should be made to the quality of the primary data underlying the analysis;
d. The methods used in integrating and analysing the data from feeder sources including, where relevant, the adjustments made to data from different sources, the methods used for price deflation, the methods used for seasonal adjustment, and/or benchmarking, and a description of the revision process; and
e. Any discrepancy arising in the integration or analysis of data from different sources, and the procedures by which these discrepancies were handled (e.g., the statistical discrepancy arising in the estimation of income and expenditure accounts).

Exercise

17.1. From a library or a statistical agency, obtain documentation on a survey of your choice, including survey results (estimates) and a description of the survey operations. As far as possible with the information available, evaluate the survey and discuss to what extent it respects Statistics Canada's guidelines on the documentation of data quality and methodology.

APPENDIX A

Principles of Notation

a. Summation

$\sum_A c_k = \sum_{k \in A} c_k$	Sum of c_k values for all population elements k belonging to the index set A. If $A = \{1, 2, \ldots, N\}$, $\sum_A c_k = \sum_{k=1}^N c_k$; if $A = \{1, 3, 7\}$, $\sum_A c_k = c_1 + c_3 + c_7$
$\sum\sum_A c_{kl} = \sum_{k \in A} \sum_{l \in A} c_{kl}$	Sum of c_{kl} values for all k and l such that $k \in A$ and $l \in A$
$\underset{k \neq l}{\sum\sum_A} c_{kl} = \underset{k \neq l}{\sum_{k \in A} \sum_{l \in A}} c_{kl}$	Sum of c_{kl} values for all k and l such that $k \in A$, $l \in A$, and $k \neq l$

b. Population, Variables, Parameters

$U = \{1, 2, \ldots, k, \ldots, N\}$	Population (= set of N elements)
U_h	Stratum h ($h = 1, \ldots, H$)
U_d	Domain d ($d = 1, \ldots, D$)
U_g	Model group g ($g = 1, \ldots, G$)
U_i	Primary sampling unit i ($i = 1, \ldots, N_{\text{I}}$)
y_k, z_k	Value of study variables y and z for element k
x_{jk}	Value of auxiliary variable x_j for element k

$t_y = t_{yU} = \sum_U y_k$	Population total of y
$\bar{y}_U = t_y/N$	Population mean of y
$S^2_{yU} = \sum_U (y_k - \bar{y}_U)^2/(N-1)$	Population variance of y
$S_{yzU} = \sum_U (y_k - \bar{y}_U)(z_k - \bar{z}_U)/(N-1)$	Population covariance of the variables y and z
$r_{yzU} = S_{yzU}/S_{yU}S_{zU}$	Population correlation coefficient for the variables y and z
$R = t_{yU}/t_{zU} = \bar{y}_U/\bar{z}_U$	Population ratio between totals (means) of the variables y and z
$B = S_{yzU}/S^2_{zU}$	Population simple linear regression coefficient

c. Sample

s	Sample (subset of U)
n_s	Sample size (number of elements in s)
$\bar{y}_s = \sum_s y_k/n_s$	Sample mean of y
$S^2_{ys} = \sum_s (y_k - \bar{y}_s)^2/(n_s - 1)$	Sample variance of y
$p(s)$	Probability of selecting the sample s
$p(\cdot)$	Sampling design
π_k	(First-order) inclusion probability of element k
π_{kl}	(Second-order) inclusion probability of elements k and l
$\check{y}_k = y_k/\pi_k$	π-expanded y value for element k
I_k, I_{kl}	Sample membership indicators
$\Delta_{kl} = \pi_{kl} - \pi_k\pi_l$	Covariance of I_k and I_l
$\check{\Delta}_{kl} = \Delta_{kl}/\pi_{kl}$	π-expanded Δ-value
p_k	Drawing probability of element k in with replacement sampling

d. Sampling Designs

SI	Simple random sampling without replacement

SIR	Simple random sampling with replacement
BE	Bernoulli sampling
PO	Poisson sampling
SY	Systematic sampling
πps	Probability-proportional-to-size sampling without replacement
pps	Probability-proportional-to-size sampling with replacement
ST	Stratified sampling
STSI	Stratified sampling with *SI* sampling in each stratum
SIC	Simple random cluster sampling
SI, *SI*	Two-stage sampling with *SI* sampling in both stages

e. Estimators

$\hat{\theta}$	General estimator of a general parameter θ
$E(\hat{\theta})$	Expected value of an estimator $\hat{\theta}$
$V(\hat{\theta})$	Variance of an estimator $\hat{\theta}$
$\hat{V}(\hat{\theta})$	Estimator of $V(\hat{\theta})$
$\hat{t}_{y\pi}$	π estimator of the population total $t_y = \sum_U y_k$
$\hat{t}_{y,\text{pwr}}$	p expanded with-replacement estimator of t_y
$\hat{t}_{yr}$	General (multiple) regression estimator of t_y
$\hat{t}_{yra}$	Ratio estimator of t_y
$\hat{t}_{y,\text{dif}}$	Difference estimator of t_y
$\hat{N} = \sum_s 1/\pi_k$	π estimator of population size
$\tilde{y}_s = \hat{t}_{y\pi}/\hat{N}$	Weighted sample mean estimator of the population mean $\bar{y}_U = t_y/N$

APPENDIX B

The MU284 Population

For administrative purposes, Sweden is divided into 284 municipalities. Typically, a municipality consists of a town and the surrounding area. The municipalities vary considerably in size and other characteristics. We selected a few variables that describe the municipalities in different ways. Data on these variables are readily available from official statistics. The resulting data set is reproduced below and is used in a number of end-of-chapter exercises illustrating various ideas in the book. The data set also provides opportunity for the reader to carry out his or her own experiments in sampling and estimation. The population consisting of the 284 municipalities is referred to as the MU284 populaion. We use the following abbreviated names for the variables.

LABEL	Identifier running from 1 to 284.
P85	1985 population (in thousands).
P75	1975 population (in thousands).
RMT85	Revenues from the 1985 municipal taxation (in millions of kronor).
CS82	Number of Conservative seats in municipal council.
SS82	Number of Social-Democratic seats in municipal council.
S82	Total number of seats in municipal council.
ME84	Number of municipal employees in 1984.
REV84	Real estate values according to 1984 assessment (in millions of kronor).
REG	Geographic region indicator.
CL	Cluster indicator (a cluster consists of a set of neighboring municipalities).

Occasionally, we consider two smaller populations, namely, the MU200 population consisting of the 200 smallest municipalities according to the value of

P75, and the MU281 population consisting of all but the three largest municipalities according to the value of P75. This implies that the MU281 population consists of all municipalities except Stockholm, Göteborg and Malmö. On several variables, the effect of this trimming is to eliminate three extremely large values, and the distributions of these variables become less skewed. We gratefully acknowledge permission from Statistics Sweden to use these data.

LABEL	P85	P75	RMT85	CS82	SS82	S82	ME84	REV84	REG	CL
1	33	27	288	13	24	49	2,135	2,836	1	1
2	19	15	139	14	12	41	957	2,035	1	1
3	26	20	196	12	14	41	1,530	6,030	1	1
4	19	15	159	12	19	41	1,059	4,704	1	1
5	56	52	536	20	27	61	3,951	5,183	1	1
6	16	15	134	16	12	41	918	2,157	1	2
7	70	62	623	18	27	61	4,367	7,072	1	2
8	66	54	517	15	32	61	4,345	5,246	1	2
9	12	12	96	10	12	31	754	951	1	2
10	60	50	467	14	29	61	3,902	6,067	1	2
11	32	29	277	14	20	45	1,993	3,264	1	3
12	20	14	155	10	21	41	1,312	1,899	1	3
13	53	40	386	24	13	51	2,780	5,931	1	3
14	28	27	241	24	8	45	1,649	3,877	1	3
15	48	43	422	19	18	51	2,983	4,968	1	3
16	653	671	6,263	34	41	101	45,324	59,877	1	4
17	79	78	612	14	31	61	5,331	7,027	1	4
18	59	54	532	23	23	61	3,994	6,529	1	4
19	27	28	250	9	22	41	1,616	2,208	1	4
20	49	55	412	20	27	61	3,240	3,976	1	4
21	38	36	339	21	11	51	2,055	4,438	1	5
22	6	6	55	11	12	31	304	960	1	5
23	42	39	290	12	25	57	2,294	7,990	1	5
24	29	27	249	13	23	49	1,899	2,719	1	5
25	21	19	164	8	25	45	1,217	2,389	1	5
26	14	9	97	12	15	35	679	1,462	2	6
27	9	10	74	3	20	31	490	1,751	2	6
28	20	21	144	5	27	49	1,109	2,259	2	6
29	153	138	1,277	21	36	81	7,910	13,205	2	6
30	33	32	240	10	24	51	1,837	3,281	2	6
31	21	19	163	7	23	49	1,176	7,082	2	7
32	10	10	63	4	18	35	454	952	2	7
33	65	62	488	14	33	61	3,254	6,389	2	7
34	13	14	111	4	19	31	759	1,830	2	7
35	17	18	128	6	22	45	871	1,833	2	7
36	32	33	230	9	30	51	1,788	2,914	2	7
37	89	92	720	15	46	79	5,495	6,772	2	7

LABEL	P85	P75	RMT85	CS82	SS82	S82	ME84	REV84	REG	CL
38	25	22	179	11	22	49	1,286	2,628	2	7
39	6	6	37	8	17	41	257	655	2	8
40	4	4	24	5	11	35	177	637	2	8
41	10	11	65	9	19	49	470	1,387	2	8
42	6	6	37	5	23	41	255	742	2	8
43	13	13	89	6	24	45	591	1,307	2	8
44	24	25	187	8	33	55	1,313	2,165	2	9
45	9	9	60	10	21	49	415	1,136	2	9
46	116	108	939	21	36	79	6,313	10,879	2	9
47	118	119	1,008	22	45	85	7,619	12,112	2	9
48	12	10	81	11	17	45	555	1,370	2	9
49	41	42	310	10	31	57	2,402	3,863	2	9
50	8	8	58	7	16	35	369	831	2	9
51	26	25	169	10	26	51	1,311	2,409	2	9
52	7	6	38	6	11	35	280	767	3	10
53	9	8	68	9	13	41	387	790	3	10
54	28	27	190	10	20	49	1,327	2,546	3	10
55	12	11	78	7	14	41	536	1,254	3	10
56	107	108	807	19	38	81	6,107	9,343	3	10
57	31	33	220	10	27	57	1,597	2,653	3	11
58	30	30	226	10	18	49	1,713	2,834	3	11
59	12	12	69	9	14	45	493	1,134	3	11
60	28	29	179	9	18	49	1,349	2,903	3	11
61	18	18	129	8	17	49	970	1,652	3	11
62	18	19	119	9	19	41	963	1,470	3	11
63	11	12	73	6	18	41	450	1,220	3	12
64	9	9	59	6	24	41	458	790	3	12
65	14	15	92	11	17	49	659	1,626	3	12
66	20	19	127	9	20	49	956	1,983	3	12
67	16	15	104	10	21	49	762	1,768	3	12
68	11	12	75	8	22	49	506	1,169	3	12
69	66	62	505	15	23	61	3,789	6,850	3	12
70	27	26	183	9	18	49	1,366	3,053	3	12
71	8	8	49	6	21	41	361	807	3	13
72	8	8	41	7	15	41	349	765	3	13
73	13	11	76	12	19	49	520	1,457	3	13
74	17	18	113	8	20	49	784	1,733	3	13
75	13	13	90	8	24	49	538	2,042	3	13
76	11	12	79	5	21	41	522	1,060	3	14
77	54	52	408	16	29	61	3,095	5,302	3	14
78	21	22	140	7	24	49	1,110	2,010	3	14
79	28	28	200	10	31	59	1,499	5,717	3	14
80	40	42	284	12	38	75	2,087	3,773	3	14
81	16	16	103	10	19	49	736	1,696	3	14
82	11	11	58	10	12	49	442	1,743	3	14

LABEL	P85	P75	RMT85	CS82	SS82	S82	ME84	REV84	REG	CL
83	56	54	654	13	30	71	5,434	6,050	3	15
84	15	18	118	6	28	49	813	1,309	4	15
85	60	60	431	15	39	75	3,285	4,954	4	15
86	30	30	192	8	27	49	1,575	2,553	4	15
87	32	32	233	9	27	51	1,593	4,006	4	15
88	16	16	102	11	24	49	753	1,515	4	15
89	15	15	97	6	24	41	735	1,321	4	16
90	9	9	56	13	14	45	380	894	4	16
91	12	13	69	9	15	41	487	1,343	4	16
92	12	11	73	5	28	41	492	1,222	4	16
93	14	14	86	6	17	41	574	1,306	4	16
94	7	7	53	9	16	35	339	779	4	17
95	16	16	105	13	22	49	852	1,508	4	17
96	13	11	81	9	21	41	598	1,026	4	17
97	12	11	77	16	10	49	481	2,027	4	17
98	70	67	472	17	35	71	3,613	6,317	4	17
99	20	20	129	11	19	49	868	2,216	4	17
100	31	28	194	17	16	49	1,343	3,073	4	17
101	49	48	299	15	25	61	2,212	4,055	4	17
102	13	12	81	7	18	41	528	1,277	4	18
103	17	15	122	14	17	41	894	1,670	4	18
104	14	14	113	10	24	41	729	1,276	4	18
105	25	21	186	24	14	49	1,142	3,252	4	18
106	14	14	91	8	25	41	664	1,380	4	18
107	21	19	145	12	25	49	1,100	4,117	4	19
108	17	15	133	16	17	45	882	1,610	4	19
109	16	13	114	10	22	45	753	1,413	4	19
110	13	11	78	8	15	41	521	1,232	4	19
111	15	14	78	10	17	49	543	1,622	4	19
112	13	12	72	8	12	41	532	1,276	4	20
113	11	10	70	11	13	41	502	1,201	4	20
114	229	247	3,471	20	32	61	24,694	17,949	4	20
115	81	75	641	19	23	65	4,807	6,382	4	20
116	35	38	275	13	30	51	2,112	3,096	4	20
117	105	102	815	20	31	65	6,323	9,371	4	21
118	22	21	149	15	17	41	1,031	2,872	4	21
119	26	26	185	10	23	49	1,495	2,540	4	21
120	24	24	172	13	23	49	1,299	2,355	4	21
121	34	35	240	12	30	51	2,089	3,475	4	21
122	11	11	67	6	20	49	469	1,726	5	22
123	77	74	526	17	32	71	4,245	7,533	5	22
124	21	19	122	10	15	49	917	2,784	5	22
125	36	33	229	10	19	51	1,560	4,014	5	22
126	46	43	298	10	19	51	2,439	10,691	5	22
127	48	37	317	17	15	51	2,115	6,087	5	22

LABEL	P85	P75	RMT85	CS82	SS82	S82	ME84	REV84	REG	CL
128	25	20	179	11	16	41	1,089	2,412	5	23
129	29	27	233	12	16	41	1,377	2,489	5	23
130	10	9	66	14	12	41	485	880	5	23
131	17	14	129	10	16	41	940	3,690	5	23
132	12	10	87	10	10	41	543	1,828	5	23
133	13	10	82	9	15	41	576	2,092	5	24
134	9	9	63	7	19	41	385	1,438	5	24
135	11	10	65	6	17	41	418	1,135	5	24
136	12	11	72	8	12	41	524	1,736	5	24
137	424	446	6,720	21	35	81	47,074	38,945	5	24
138	49	47	381	14	25	61	2,655	4,660	5	25
139	31	28	240	12	19	49	1,677	3,307	5	25
140	15	15	118	7	23	41	781	3,397	5	25
141	46	47	396	12	30	61	2,610	4,623	5	25
142	10	9	66	5	16	39	463	1,190	5	25
143	5	5	35	6	12	41	253	538	5	26
144	7	7	43	6	13	35	344	725	5	26
145	23	22	166	9	26	49	1,122	2,025	5	26
146	31	28	216	12	14	41	1,470	3,065	5	26
147	9	8	53	10	11	41	393	895	5	26
148	12	12	82	9	19	45	539	1,162	5	27
149	12	12	90	6	19	45	534	1,243	5	27
150	11	10	62	8	14	41	414	1,119	5	27
151	12	11	77	6	21	41	570	1,178	5	27
152	32	30	207	10	23	51	1,537	2,955	5	27
153	11	10	68	10	14	41	474	1,224	5	28
154	9	9	52	9	13	41	382	784	5	28
155	36	34	277	10	23	51	1,918	3,376	5	28
156	49	50	396	11	34	61	2,881	4,798	5	28
157	31	28	220	10	18	49	1,579	2,742	5	28
158	100	106	751	20	38	79	5,742	7,710	5	28
159	22	21	146	12	15	49	1,053	2,066	5	28
160	13	13	95	7	20	41	731	1,101	5	28
161	6	5	37	8	9	39	256	658	5	29
162	6	6	35	11	10	41	280	521	5	29
163	7	5	39	9	13	35	329	637	5	29
164	9	6	46	8	12	35	387	813	5	29
165	8	8	59	6	18	35	401	762	5	29
166	6	7	44	8	19	41	311	731	5	30
167	17	17	105	12	12	49	782	1,887	5	30
168	13	12	87	8	17	41	567	1,324	5	30
169	11	11	66	6	18	41	597	1,007	5	30
170	10	10	62	8	16	41	473	971	5	30
171	24	25	175	10	21	49	1,386	2,269	5	31
1.72	35	35	252	10	20	51	1,890	3,335	5	31

LABEL	P85	P75	RMT85	CS82	SS82	S82	ME84	REV84	REG	CL
173	18	17	128	11	19	49	1,058	1,592	5	31
174	46	45	330	12	27	59	2,656	4,041	5	31
175	9	8	57	8	14	33	389	818	5	31
176	13	13	83	5	19	41	623	1,192	5	31
177	32	33	208	11	19	51	1,559	2,681	5	31
178	11	9	74	8	17	39	533	977	6	32
179	9	10	63	6	22	41	411	869	6	32
180	15	16	109	7	25	49	803	2,266	6	32
181	5	5	37	5	21	35	235	513	6	32
182	13	11	105	6	18	31	702	1,366	6	32
183	5	6	38	3	27	39	245	433	6	33
184	12	11	82	6	25	41	646	935	6	33
185	10	11	76	5	27	45	501	1,392	6	33
186	10	10	63	8	13	41	419	1,011	6	33
187	13	14	86	7	15	41	552	1,444	6	33
188	74	73	603	17	28	61	4,270	6,635	6	34
189	26	28	209	9	26	49	1,398	2,291	6	34
190	14	16	100	8	25	41	753	1,509	6	34
191	17	19	121	4	32	49	885	1,868	6	34
192	27	27	191	9	23	49	1,441	2,557	6	34
193	18	20	130	9	22	49	839	1,719	6	34
194	8	9	63	3	23	41	411	950	2	35
195	17	17	128	6	25	45	1,048	1,606	2	35
196	12	12	90	3	26	41	607	1,347	2	35
197	10	11	79	4	22	39	486	1,102	2	35
198	7	7	50	1	18	35	325	688	2	35
199	118	118	1,025	15	32	65	7,700	11,126	2	36
200	18	17	121	5	21	41	938	1,645	2	36
201	12	11	87	7	23	45	552	1,408	2	36
202	35	38	309	10	29	51	2,173	3,635	2	36
203	10	9	75	4	20	35	600	1,064	2	36
204	25	25	189	7	25	49	1,437	2,898	2	36
205	5	5	36	4	24	39	255	692	2	37
206	11	11	82	4	28	39	612	913	2	37
207	13	13	84	4	18	41	564	1,334	2	37
208	8	8	60	6	19	39	390	723	2	37
209	17	19	132	7	31	49	989	1,197	2	37
210	7	7	46	4	22	35	296	605	2	38
211	118	118	965	15	35	65	6,856	10,702	2	38
212	21	20	155	8	21	49	1,150	1,988	2	38
213	14	16	132	6	24	41	769	1,460	2	38
214	27	28	219	6	28	49	1,544	2,441	2	38
215	14	15	112	7	22	41	761	1,238	2	38
216	8	9	53	6	24	49	311	844	6	39
217	12	12	90	6	24	49	543	2,098	6	39

LABEL	P85	P75	RMT85	CS82	SS82	S82	ME84	REV84	REG	CL
218	10	9	66	6	19	41	452	1,113	6	39
219	14	13	100	8	18	49	700	1,841	6	39
220	11	11	77	9	18	49	491	1,416	6	39
221	7	7	52	7	15	31	356	826	6	40
222	8	9	60	5	23	49	422	2,100	6	40
233	13	13	100	5	32	49	665	1,500	6	40
224	20	18	153	4	19	49	1,039	2,223	6	40
225	51	47	418	14	26	61	2,771	5,137	6	40
226	46	46	392	9	36	61	2,711	4,852	6	41
227	11	10	76	7	18	41	503	1,220	6	41
228	17	17	130	6	20	41	922	1,630	6	41
229	25	27	197	5	28	49	1,239	2,461	6	41
230	30	33	262	6	32	51	1,741	2,395	6	41
231	7	6	49	3	23	41	321	694	6	42
232	12	15	105	4	27	41	672	1,008	6	42
233	14	13	87	3	17	41	661	1,447	6	42
234	12	12	70	3	18	41	579	1,150	6	42
235	21	22	147	5	22	49	1,242	2,459	6	42
236	88	85	720	14	44	75	4,758	8,760	6	43
237	41	43	342	8	32	51	2,182	4,146	6	43
238	31	32	233	5	32	51	1,579	3,298	6	43
239	28	28	203	6	25	49	1,524	2,903	6	43
240	38	37	249	5	24	51	2,073	3,882	6	43
241	13	14	92	3	23	41	659	1,989	7	44
242	18	18	134	3	29	49	945	2,616	7	44
243	28	27	231	8	22	49	1,560	2,414	7	44
244	94	93	782	12	46	81	5,779	9,828	7	44
245	26	28	184	5	35	61	1,455	2,067	7	44
246	26	26	182	6	37	61	1,494	6,928	7	44
247	60	60	432	6	33	61	3,070	6,502	7	44
248	7	8	49	3	27	45	427	3,832	7	45
249	9	10	63	5	30	49	502	970	7	45
250	14	13	94	5	23	49	813	2,486	7	45
251	17	18	124	4	29	49	1,061	2,881	7	45
252	10	9	69	7	23	49	593	2,399	7	45
253	9	9	53	7	21	45	484	1,491	7	45
254	13	13	91	5	28	49	736	3,647	7	45
255	56	53	451	12	38	75	3,430	4,677	7	45
256	8	8	53	3	14	31	428	727	8	46
257	3	4	21	5	11	31	173	347	8	46
258	7	7	42	5	14	35	323	903	8	46
259	8	7	50	3	15	41	408	706	8	46
260	5	6	40	3	15	31	287	1,063	8	46
261	4	4	32	2	19	31	236	704	8	47
262	8	8	58	6	20	41	431	2,246	8	47

LABEL	P85	P75	RMT85	CS82	SS82	S82	ME84	REV84	REG	CL
263	4	4	26	2	14	31	202	623	8	47
264	4	4	28	2	19	31	199	422	8	47
265	8	8	61	4	17	35	477	1,321	8	47
266	9	9	60	2	20	35	518	1,313	8	48
267	4	5	35	2	19	31	252	687	8	48
268	84	74	764	10	30	65	5,292	10,827	8	48
269	14	15	105	4	24	45	942	1,911	8	48
270	74	72	592	7	36	65	4,777	7,624	8	48
271	8	8	55	3	25	41	462	690	8	49
272	4	4	28	1	15	31	214	888	8	49
273	7	8	46	4	21	35	414	9,052	8	49
274	5	6	34	1	20	31	280	359	8	49
275	19	18	123	4	27	41	1,175	1,544	8	49
276	6	7	41	3	18	39	302	430	8	50
277	9	10	54	4	19	41	528	551	8	50
278	24	26	207	5	20	41	1,582	3,703	8	50
279	10	9	64	2	24	39	480	689	8	50
280	67	64	562	9	34	61	3,948	6,583	8	50
281	39	35	295	5	32	51	2,227	4 ,033	8	50
282	29	27	226	7	28	49	1,682	2,898	8	50
283	10	9	63	5	19	41	604	594	8	50
284	27	31	233	5	27	45	1,788	2,366	8	50

APPENDIX C

The Clustered MU284 Population

For easy reference, summary statistics are given below for the 50 clusters identified in Appendix B. Here, SIZE is the number of municipalities in the cluster, whereas "T" in front of the name of a variable indicates that the values in the corresponding column are cluster totals for that variable. For example, for the first cluster, labeled 1, the SIZE column shows that there are 5 municipalities in the cluster, whereas the value 153 in the TP85 column indicates a total 1985 population of 153 thousands for the five municipalities that define the cluster.

LABEL	SIZE	TP85	TP75	TRMT85	TCS82	TSS82	TS82	TME84	TREV84
1	5	153	129	1,318	71	96	233	9,632	20,788
2	5	224	193	1,837	73	112	255	14,286	21,493
3	5	181	153	1,481	91	80	233	10,717	19,939
4	5	867	886	8,069	100	144	325	59,505	79,617
5	5	136	127	1,097	65	96	233	7,769	18,496
6	5	229	210	1,832	51	122	247	12,025	21,958
7	8	272	270	2,082	70	213	400	15,083	30,400
8	5	39	40	252	33	94	211	1,750	4,728
9	8	354	346	2,812	99	225	456	20,297	34,765
10	5	163	160	1,181	51	96	247	8,637	14,700
11	6	137	141	942	55	113	290	7,085	12,646
12	8	174	170	1,218	74	163	388	8,946	18,459
13	5	59	58	369	41	99	229	2,552	6,804
14	7	181	183	1,272	70	174	383	9,491	21,301
15	6	209	210	1,730	62	175	344	13,453	20,387
16	5	62	62	381	39	98	209	2,668	6,086

LABEL	SIZE	TP85	TP75	TRMT85	TCS82	TSS82	TS82	TME84	TREV84
17	8	218	208	1,410	107	164	404	10,306	21,001
18	5	83	76	593	63	98	213	3,957	8,855
19	5	82	72	548	56	96	229	3,799	9,994
20	5	369	382	4,529	71	110	259	32,647	29,904
21	5	211	208	1,561	70	124	255	12,237	20,613
22	6	239	217	1,559	70	120	322	11,745	32,835
23	5	93	80	694	57	70	205	4,434	11,299
24	5	469	486	7,002	51	98	245	48,977	45,346
25	5	151	146	1,201	50	113	251	8,186	17,177
26	5	75	70	513	43	76	207	3,582	7,248
27	5	79	75	518	39	96	223	3,594	7,657
28	8	271	271	2,005	89	175	412	14,760	23,801
29	5	36	30	216	42	62	185	1,653	3,391
30	5	57	57	364	42	82	213	2,730	5,920
31	7	177	176	1,233	67	139	333	9,561	15,928
32	5	53	51	388	32	103	195	2,684	5,991
33	5	50	52	345	29	107	207	2,363	5,215
34	6	176	183	1,354	56	156	298	9,586	16,579
35	5	54	56	410	17	114	201	2,877	5,693
36	6	218	218	1,806	48	150	286	13,400	21,776
37	5	54	56	394	25	120	207	2,810	4,859
38	6	201	204	1,629	46	152	280	11,376	18,434
39	5	55	54	386	35	103	237	2,467	7,312
40	5	99	94	783	35	115	239	5,253	11,786
41	5	129	133	1,057	33	134	243	7,116	12,558
42	5	66	68	458	18	107	213	3,475	6,758
43	5	226	225	1,747	38	157	277	12,116	22,989
44	7	265	266	2,037	43	225	403	14,962	32,344
45	8	135	133	994	48	219	410	8,046	22,383
46	5	31	32	206	19	69	169	1,619	3,746
47	5	28	28	205	16	89	169	1,545	5,316
48	5	185	175	1,556	25	129	241	11,781	22,362
49	5	43	44	286	13	108	179	2,545	12,533
50	9	221	218	1,745	45	221	407	13,141	21,847

APPENDIX D

The CO124 Population

Data concerning an alternative population, the CO124 population, consisting of 124 countries are given in the following table. The columns in the table are as follows:

LABEL	Identifier running from 1 to 124.
COUNTRY	The abbreviations, with a few exceptions, are those of the International Olympic Committee.
P83	1983 population (in millions).
IMP	1983 import (in millions of U.S. dollars).
EXP	1983 export (in millions of U.S. dollars).
GNP	1982 gross national product (in tens of millions of U.S. dollars).
MEX	1981 military expenditures (in millions of U.S. dollars).
P80	1980 population (in millions).
CONT	1 = Africa, 2 = North and Central America, 3 = South America, 4 = Asia (non-Soviet part), 5 = Europe (non-Soviet part), 6 = Oceania, 7 = Union of Soviet Socialist Republics.

The data are reproduced here with the kind permission of Professor Leif Johansson, University of Lund, Sweden. They were extracted from a larger computer file compiled by Professor Johansson to be used with a computer package, Herakles, for computer-assisted teaching in the social sciences.

LABEL	COUNTRY	P83	IMP	EXP	GNP	MEX	P80	CONT
1	ALG	20.5	10,395	11,163	4,535	675	18.7	1
2	BEN	3.7	131	63	122	23	3.4	1
3	BUR	6.6	288	57	135	41	6.1	1
4	BRI	4.4	194	76	122	23	4.1	1
5	CMR	9.1	1,217	940	715	82	8.5	1
6	CAF	2.4	127	109	76	12	2.3	1
7	CHA	4.7	109	58	36	62	4.5	1
8	CGO	1.6	807	977	234	68	1.5	1
9	EGY	45.9	10,274	3,215	3,425	1,650	42.3	1
10	ETH	33.6	875	403	465	485	31.1	1
11	GAB	1.1	724	2,161	330	72	1	1
12	GHA	12.2	705	873	416	110	11.5	1
13	GUI	5.1	351	428	172	44	4.8	1
14	CIV	9.3	1,808	2,067	822	111	8.2	1
15	KEN	18.7	1,274	876	698	183	16.7	1
16	LBR	2	422	464	99	16	1.8	1
17	LBA	3.3	8,382	15,576	2,712	3,670	3	1
18	MAD	9.7	540	316	299	71	8.7	1
19	MAW	6.4	312	230	136	22	6	1
20	MLI	7.5	344	167	123	46	7	1
21	MTN	1.8	227	305	81	82	1	1
22	MRI	.9	438	373	123	2	.9	1
23	MAR	21.1	3,599	2,062	1,758	1,005	20.1	1
24	MOZ	13.3	635	132	465	111	12.1	1
25	NIG	5.7	442	333	178	16	5.3	1
26	NGR	89.0	13,440	11,317	7,722	2,037	80.6	1
27	RWA	5.7	279	79	144	18	5.2	1
28	SLE	3.4	171	119	123	19	3.3	1
29	SOM	5.2	330	199	212	150	4.6	1
30	SAF	31.8	14,528	9,671	8,234	2,254	28.6	1
31	SUD	20.3	1,354	624	935	470	18.7	1
32	TOG	2.7	391	177	95	22	2.5	1
33	TUN	6.8	3,117	1,872	922	214	6.4	1
34	UGA	14.6	293	345	325	852	13.2	1
35	TAN	20.3	4,552	1,900	535	285	18.6	1
36	ZAI	31.1	480	569	558	164	26.4	1
37	ZAM	6.2	690	831	386	290	5.8	1
38	ZIM	7.7	1,430	1,115	640	440	7.1	1
39	CAN	24.9	61,325	73,797	27,909	4,227	24	2
40	CRC	2.4	993	867	266	19	2.2	2
41	CUB	10	6,293	5,536	1,740	1,065	9.7	2
42	DOM	5.9	1,282	811	767	92	5.4	2
43	ESA	5.2	891	735	356	86	4.7	2
44	GUA	7.9	1,126	1,184	870	95	6.9	2
45	HAI	5.3	461	154	158	22	5	2

LABEL	COUNTRY	P83	IMP	EXP	GNP	MEX	P80	CONT
46	HON	4	767	692	262	38	3.7	2
47	JAM	2.2	1,531	745	300	29	2.2	2
48	MEX	75.1	8,201	21,012	19,635	782	69.3	2
49	NCA	3	799	411	262	34	2.7	2
50	PAN	2	1,412	304	415	22	2	2
51	TRI	1.1	2,505	2,379	772	12	1.1	2
52	USA	233.9	269,878	200,538	305,690	134,390	227.7	2
53	ARG	29.6	4,486	7,836	8,949	2,241	28.2	3
54	BOL	6	532	789	539	84	5.6	3
55	BRA	129.6	14,494	26,906	27,474	1,234	121.3	3
56	CHI	11.6	2,754	3,836	2,516	225	11.1	3
57	COL	27.5	4,968	3,018	3,819	229	27.1	3
58	ECU	9.2	1,465	2,203	1,288	92	8.1	3
59	GUY	.9	283	256	47	27	.9	3
60	PAR	3.4	506	284	573	40	3.2	3
61	PER	18.7	2,688	3,015	2,204	480	17.3	3
62	URU	2.9	788	1,045	1,001	150	2.9	3
63	VEN	15.1	6,667	15,002	6,907	527	15	3
64	AFG	14.5	695	708	315	85	14.5	4
65	BRN	.4	3,342	3,200	375	115	.3	4
66	BAN	94.6	1,716	690	1,288	140	88.7	4
67	BIR	35.3	268	378	650	225	33.6	4
68	CHN	1,024	21,324	22,151	30,250	37,200	1,002.8	4
69	CYP	.6	1,219	494	239	19	.6	4
70	PRK	19.1	1,880	1,520	1,750	3,424	17.9	4
71	YMD	2.1	1,527	779	93	115	2	4
72	IND	730	13,562	8,304	18,413	3,991	663.6	4
73	INA	159.4	16,352	21,146	8,945	1,426	146.4	4
74	IRN	42	11,539	19,438	6,970	5,092	38.3	4
75	IRQ	14.6	20,500	10,230	2,710	3,759	13.2	4
76	ISR	4	8,587	5,112	2,121	2,750	3.9	4
77	JPN	117.1	126,395	146,676	119,000	9,461	116.8	4
78	JOR	3.2	3,030	579	419	420	2.9	4
79	KUW	1.6	6,980	11,140	3,044	2,031	1.4	4
80	LAO	4.2	125	33	32	21	3.9	4
81	MAL	15.2	13,987	13,917	2,681	1,639	13.9	4
82	MGL	1.8	655	436	165	238	1.6	4
83	NEP	15.7	342	80	247	28	14	4
84	OMA	1.1	2,492	4,058	687	1,444	1	4
85	PAK	92.9	5,341	3,149	3,302	1,307	82.6	4
86	PHI	51.9	7,980	5,005	4,168	688	48.1	4
87	QAT	.2	1,456	3,384	595	893	.2	4
88	KOR	40.5	26,192	24,445	7,509	3,519	38.1	4
89	KSA	10.4	39,206	46,941	15,638	22,458	9.2	4
90	SIN	2.5	28,712	24,108	1,478	556	2.4	4

LABEL	COUNTRY	P83	IMP	EXP	GNP	MEX	P80	CONT
91	SRI	15.4	1,786	1,123	491	35	14.7	4
92	SYR	10.4	4,542	1,900	1,589	2,166	8.7	4
93	THA	49.4	9,159	6,368	3,853	1,036	46.5	4
94	TUR	47.2	9,348	5,694	6,315	3,442	44.4	4
95	UAE	1.1	9,414	17,257	2,917	1,423	1	4
96	YAR	6.2	1,521	39	371	320	5.8	4
97	ALB	2.8	250	200	266	127	2.6	5
98	AUT	7.5	19,368	15,428	7,447	847	7.5	5
99	BEL	9.8	55,269	51,929	10,388	3,690	9.8	5
100	BUL	8.9	12,164	12,130	4,603	964	8.9	5
101	TCH	15.4	16,325	16,522	8,514	2,900	15.3	5
102	DEN	5.1	16,946	16,221	6,314	1,546	5.1	5
103	FIN	4.8	12,854	12,530	5,235	632	4.8	5
104	FRA	55	103,734	93,310	62,731	23,633	53.7	5
105	GDR	16.7	21,254	23,793	11,727	4,394	16.7	5
106	FRG	61	152,899	169,425	75,709	25,509	61.6	5
107	GRE	9.8	9,632	4,459	4,089	2,184	9.6	5
108	HUN	10.6	8,503	8,696	5,542	810	10.7	5
109	IRL	3.5	9,182	8,612	1,757	246	3.4	5
110	ITA	56.8	80,367	72,681	38,223	8,184	56.4	5
111	HOL	14.3	60,743	64,816	15,428	4,931	14.1	5
112	NOR	4.1	13,890	18,920	5,884	1,484	4.1	5
113	POL	36.5	10,179	11,478	14,561	2,467	35.6	5
114	POR	10.1	8,134	4,566	2,472	779	9.9	5
115	ROM	22.7	9,836	11,714	7,109	1,285	22.2	5
116	ESP	38.2	28,812	23,544	20,424	3,682	37.4	5
117	SWE	8.3	25,046	26,313	11,541	3,175	8.3	5
118	SUI	6.4	29,475	25,865	10,856	2,000	6.4	5
119	GBR	55.7	105,477	94,562	53,673	19,901	55.9	5
120	YUG	22.8	11,104	9,038	7,053	2,936	22.3	5
121	AUS	15.3	19,393	20,594	16,904	3,508	14.7	6
122	FIJ	.6	484	240	128	4	.6	6
123	NZL	3.3	5,283	5,272	2,539	393	3.1	6
124	URS	277.4	80,410	91,336	156,300	118,800	265.5	7

References

Altham, P. M. E. (1976). Discrete variable analysis for individuals grouped into families. *Biometrika* **63**, 263–269.

Atmer, J., Thulin, G., and Bäcklund, S. (1975). Samordning av urval med JALES-metoden. *Statistisk Tidskrift* **13**, 443–450.

Bailar, B. A., and Bailar, J. C. (1983). Comparison of the biases of the hot-deck imputation procedure with an "equal-weights" imputation procedure. In: W. G. Madow and I. Olkin (eds.), *Incomplete Data in Sample Surveys*, Vol. 3. New York: Academic Press, pp. 299–311.

Bailar, B. A., and Dalenius, T. (1969). Estimating the response variance components of the US Bureau of the Census' Survey Model. *Sankhya B* **31**, 341–360.

Bankier, M. D. (1988). Power allocations: determining sample sizes for subnational areas. *The American Statistician* **42**, 174–177.

Basu, D. (1958). On sampling with and without replacement. *Sankhya* **20**, 287–294.

Battese, G. E., and Fuller, W. A. (1984). An error components model for prediction of county crop areas using survey and satellite data. Survey Section, Statistical Laboratory, Iowa State University, Ames.

Bean, J. A. (1975). Distribution and properties of variance estimators for complex multistage probability samples. *Vital and Health Statistics*, Series 2, No. 65. Washington, DC: National Center for Health Statistics, Public Health Service.

Bellhouse, D. R. (1977). Optimal designs for sampling in two dimensions. *Biometrika* **64**, 605–611.

Bellhouse, D. R. (1980). Computation of variance-covariance estimates for general multistage sampling designs. *COMPSTAT 1980; Proceedings in Computational Statistics.* Vienna: Physica-Verlag, pp. 57–63.

Bellhouse, D. R. (1981). Spatial sampling in the presence of a trend. *Journal of Statistical Planning and Inference* **5**, 365–375.

Bellhouse, D. R. (1984). A review of optimal designs in survey sampling. *Canadian Journal of Statistics* **12**, 53–65.

Bellhouse, D. R. (1985). Computing methods for variance estimation in complex surveys. *Journal of Offical Statistics* **1**, 323–329.

Bellhouse, D. R. (1988). Systematic sampling. In: P. R. Krishnaiah and C. R. Rao (eds.), *Handbook of Statistics*, Vol. 6. Amsterdam: North-Holland, pp. 125–145.

Bethel, J. (1989a). Sample allocation in multivariate surveys. *Survey Methodology* **15**, 47–57.
Bethel, J. (1989b). Minimum variance estimation in stratified sampling. *Journal of the American Statistical Association* **84**, 260–265.
Bethlehem, J. G. (1988). Reduction of nonresponse bias through regression estimation. *Journal of Official Statistics* **4**, 251–260.
Bethlehem, J. G., and Keller W. J. (1987). Linear weighting of sample survey data. *Journal of Official Statistics* **3**, 141–153.
Bethlehem, J. G., and Kersten, H. M. P. (1985). On the treatment of nonresponse in sample surveys. *Journal of Official Statistics* **1**, 287–300.
Bickel, P. J., and Freedman, D. A (1984). Asymptotic normality and the bootstrap in stratified sampling. *Annals of Statistics* **12**, 470–482.
Binder, D. A. (1983). On the variances of asymptotically normal estimators from complex surveys. *International Statistical Review* **51**, 279–292.
Binder, D. A., and Hidiroglou, M. A. (1988). Sampling in time. In: P. R. Krishnaiah and C. R. Rao (eds.), *Handbook of Statistics*, Vol. 6. Amsterdam: North-Holland, pp. 187–211.
Bondesson, L., and Holm, S. (1985). Bootstrap-estimation of the mean-square error of the ratio estimate for sampling without replacement. In: J. Lanke and G. Lindgren (eds.), *Contributions to Probability and Statistics in Honour of Gunnar Blom*. Department of Statistics, University of Lund, pp. 85–96.
Brewer, K. R. W. (1963a). A model of systematic sampling with unequal probabilities. *Australian Journal of Statistics* **5**, 5–13.
Brewer, K. R. W. (1963b). Ratio estimation and finite population: some results deductible from the assumption of an underlying stochastic process. *Australian Journal of Statistics* **5**, 93–105.
Brewer, K. R. W. (1979). A class of robust sampling designs for large scale surveys. *Journal of the American Statistical Association* **74**, 911–915.
Brewer, K. R. W. (1981). The analytical use of unequal probability samples: A case study. *Bulletin of the International Statistical Institute* **49**, 685–698.
Brewer, K. R. W., and Hanif, M. (1983). *Sampling with Unequal Probabilities*. New York: Springer-Verlag.
Brewer, K. R. W., and Mellor, R. W. (1973). The effect of sample structure on analytical surveys. *Australian Journal of Statistics* **15**, 145–152.
Buck, S. F. (1960). A method of estimation of missing values in multivariate data suitable for use with an electronic computer. *Journal of the Royal Statistical Society B* **22**, 302–306.
Cannell, C., Fowler, F. J., Kalton, G., Oksenberg, L., and Bischoping, K. (1989). New quantitative techniques for presenting survey questions. *Bulletin of the International Statistical Institute* **53**, 2, 481–495.
Cassel, C. M., Särndal, C. E., and Wretman, J. H. (1976). Some results on generalized difference estimation and generalized regression estimation for finite populations. *Biometrika* **63**, 615–620.
Cassel, C. M., Särndal, C. E., and Wretman, J. H. (1977). *Foundations of Inference in Survey Sampling*. New York: Wiley.
Cassel, C. M., Särndal, C. E., and Wretman, J. H. (1979). Prediction theory for finite populations when model-based and design-based principles are combined. *Scandinavian Journal of Statistics* **6**, 97–106.
Cassel, C. M., Särndal, C. E, and Wretman, J. H. (1983). Some uses of statistical models in connection with the nonresponse problem. In: W. G. Madow and I. Olkin (eds.), *Incomplete Data in Sample Surveys*, Vol. 3. New York: Academic Press, pp. 143–160.
Chakrabarti, M. C. (1963). On the use of incidence matrices in sampling from a finite population. *Journal of the Indian Statistical Association* **1**, 78–85.
Chaudhuri, A. (1981). Non-negative unbiased variance estimators. In: D. Krewski,

R. Platek, and J. N. K. Rao (eds.), *Current Topics in Survey Sampling*. New York: Academic Press, pp. 317–328.

Chaudhuri, A., and Mukerjee, R. (1988). *Randomized Response: Theory and Techniques*. New York: Marcel Dekker.

Chaudhuri, A., and Vos, J. W. E. (1988). *Unified Theory and Strategies of Survey Sampling*. Amsterdam: North-Holland.

Chu, A., Mohadjer, L., Morganstein, D., and Rhoades, M. (1985). NASSREG (National Accident Sampling System Regression). Rockville, MD.: Westat, Internal Report.

Cochran, W. G. (1951). Modern methods in the sampling of human populations. *American Journal of Public Health* **41**, 647–653.

Cochran, W. G. (1977). *Sampling Techniques*, 3rd ed. New York: Wiley.

Cohen, J. E. (1976). The distribution of the chi-squared statistic under clustered sampling from contingency tables. *Journal of the American Statistical Association* **71**, 665–670.

Cohen, S. B., Burt, V. L., and Jones, G. K. (1986). Efficiencies in variance estimation for complex survey data. *The American Statistician* **40**, 157–164.

Cohen, S. B., Xanthopoulos, J. A., and Jones, G. K. (1988). An evaluation of statistical software procedures appropriate for the regression analysis of complex survey data. *Journal of Official Statistics* **4**, 17–34.

Cox, L. H., and Boruch, R. F. (1988). Record linkage, privacy and statistical policy. *Journal of Official Statistics* **4**, 3–16.

Dalén, J. (1986). Sampling from finite populations: actual coverage probabilities for confidence intervals on the population mean. *Journal of Official Statistics* **2**, 13–24.

Dalenius, T. (1953). Något om metoder för objektiva skördeberäkningar. (About methods for objective crop estimation.) *Kungliga Lantbruksakademiens Tidskrift* **92**, 99–118.

Dalenius, T. (1957). *Sampling in Sweden*. Stockholm: Almquist and Wiksell.

Dalenius, T. (1974). Ends and means of total survey design. *Forskningsprojektet Fel i Undersökningar*. Stockholm: University of Stockholm.

Dalenius, T. (1976). Bortfallsproblemet vid statistiska undersökningar. *Marknadsvetande* **4**, 3–24.

Dalenius, T. (1983). Some reflections on the problem of missing data. In: W. G. Madow and I. Olkin (eds.), *Incomplete Data in Sample Surveys*, Vol. 3. New York: Academic Press, pp. 411–413.

Dalenius, T. (1986). *Elements of Survey Sampling*. Stockholm: Sarec, Statistics Sweden.

Dalenius, T., and Hodges, J. L. (1959). Minimum variance stratification. *Journal of the American Statistical Association* **54**, 88–101.

Danermark, B., and Swensson, B. (1987). Measuring drug use among Swedish adolescents: randomized response versus anonymous questionnaires. *Journal of Official Statistics* **3**, 439–448.

Danielsson, S. (1975). Optimal allokering vid vissa klasser av urvalsförfaranden. Ph.D. Thesis, Department of Mathematics, University of Linköping, Sweden.

Deming, W. E. (1943). *Statistical Adjustment of Data*. New York: Wiley.

Deming, W. E. (1950). *Some Theory of Sampling*. New York: Wiley.

Deming, W. E. (1953). On a probability mechanism to attain an economic balance between the resultant error of non-response and the bias of non-response. *Journal of the American Statistical Association* **48**, 743–772.

Deming, W. E. (1956). On simplifications of sampling design through replication with equal probabilities and without stages. *Journal of the American Statistical Association* **51**, 24–53.

Deming, W. E., and Stephan, F. F. (1940) On a least squares adjustment of a sampled frequency table when the expected marginal totals are known. *Annals of Mathematical Statistics* **11**, 427–444.

Deng, L. Y., and Wu, C. F. J. (1987). Estimation of variance of the regression estimator. *Journal of the American Statistical Association* **82**, 568–576.

Deville, J. C., Grosbras, J. M., and Roth, N. (1988). Efficient sampling algorithms and balanced samples. *COMPSTAT 1988: Proceedings in Computational Statistics.* Heidelberg: Physica-Verlag, pp. 255–256.

Deville, J. C., and Särndal, C. E. (1992). Calibration estimators in survey sampling. *Journal of the American Statistical Association* **87**, 376–382.

Dippo, C. S., and Wolter, K. M. (1984). A comparison of variance estimators using the Taylor series approximation. *Proceedings of the Section on Survey Research Methods.* American Statistical Association, pp. 113–121.

Doss, D. C., Hartley, H. O., and Somayajulu, G. R. (1979). An exact small sample theory for post-stratification. *Journal of Statistical Planning and Inference* **3**, 235–248.

Drew, J. D., Singh, M. P., and Choudhry, G. H. (1982). Evaluation of small area techniques for the Canadian Labour Force Survey. *Survey Methodology* **8**, 17–47.

DuMouchel, W. H., and Duncan, G. J. (1983). Using sample survey weights in multiple regression analysis of stratified samples. *Journal of the American Statistical Association* **78**, 535–543.

Duncan, G. J., and Kalton, G. (1988). Issues of design and analysis of surveys across time. *International Statistical Review* **55**, 97–117.

Durbin, J. (1953). Some results in sampling theory when the units are selected with unequal probabilities. *Journal of the Royal Statistical Society B* **15**, 262–269.

Durbin, J. (1958). Sampling theory for estimates based on fewer individuals than the number selected. *Bulletin of the International Statistical Institute* **36**, 113–119.

Durbin, J. (1959). A note on the application of Quenouille's method of bias reduction to the estimation of ratios. *Biometrika* **46**, 477–480.

Efron, B. (1979). Bootstrap methods: another look at the jackknife. *Annals of Statistics* **7**, 1–26.

Efron, B. (1981). Nonparametric standard errors and confidence intervals. *Canadian Journal of Statistics* **9**, 139–172.

Efron, B. (1982). The jackknife, the bootstrap and other resampling plans. Philadelphia: SIAM monograph no. 38.

Elvers, E., Särndal, C. E., Wretman, J. H., and Örnberg, G. (1985). Regression analysis and ratio analysis for domains: a randomization theory approach. *Canadian Journal of Statistics* **13**, 185–199.

Ericson, W. A. (1965). Optimum stratified sampling using prior information. *Journal of the American Statistical Association* **60**, 750–771.

Fan, C. T., Muller, M. E., and Rezucha, I. (1962). Development of sampling plans by using sequential (item by item) techniques and digital computers. *Journal of the American Statistical Association* **57**, 387–402.

Fay, R. E. (1985) . A jackknifed chi-square test for complex samples. *Journal of the American Statistical Association* **80**, 148–157.

Fay, R. E., and Herriot, R. A. (1979). Estimates of income for small places. An application of James-Stein procedures to census data. *Journal of American Statistical Association* **74**, 269–277.

Fellegi, I. P. (1980). Approximate tests of independence and goodness of fit based on stratified multistage samples. *Journal of the American Statistical Association* **75**, 261–268.

Fienberg, S. E., and Tanur, J. M. (1987). Experimental and sampling structures: parallels diverging and meeting. *International Statistical Review* **55**, 75–96.

Finkner, A. L. (1950). Methods of sampling for estimating commercial peach production in North Carolina. North Carolina Agricultural Experiment Station Technical Bulletin 91.

Folsom, R. E. (1974). National Assessment approach to sampling error estimation. Sampling error monograph prepared for National Assessment of Educational Progress (First Draft). Research Triangle Park, NC: Research Triangle Institute.

Ford, B. M. (1983). An overview of hot-deck procedures. In: W. G. Madow, I. Olkin, and D. B. Rubin (eds.), *Incomplete Data in Sample Surveys*, Vol. 2. New York: Academic Press, pp. 185–207.

Francis, I. (1981). *Statistical Software: A Comparative Review*. Amsterdam: North-Holland.

Francisco, C. A., and Fuller, W. A. (1986). Estimation of the distribution function with a complex survey. Technical Report, Iowa State University, Ames, Iowa.

Frankel, M. R. (1971). *Inference from Survey Samples: An Empirical Investigation*. Ann Arbor, MI: Institute for Social Research.

Fuller, W. A. (1975). Regression analysis for sample survey. *Sankhya C* **37**, 117–132.

Fuller, W. A. (1987). Estimators of the factor model for survey data. In: I. B. MacNeill and G. J. Umphrey (eds.), *Applied Probability, Statistics and Sampling Theory*. Boston: D. Reidel Publishing Company, pp. 265–284.

Giles, P. (1988). A model for generalized edit and imputation of survey data. *Canadian Journal of Statistics* **16** (Supl) 57–73.

Giommi, A. (1987) Nonparametric methods for estimating individual response probabilities. *Survey Methodology* **13**, 127–134.

Glasser, G. J. (1962). On the complete coverage of large units in a statistical study. *Review of the International Statistical Institute* **30**, 28–32.

Godambe, V. P., and Joshi, V. M. (1965). Admissibility and Bayes estimation in sampling finite populations, 1. *Annals of Mathematical Statistics* **36**, 1707–1722.

Godambe, V. P., and Thompson, M. E. (1986). Parameters of superpopulation and survey population: their relationships and estimation. *International Statistical Review* **54**, 127–138.

Godfrey, J., Roshwalb, A., and Wright, R. L. (1984). Model-based stratification in inventory cost estimation. *Journal of Business and Economic Statistics* **2**, 1–9.

Gonzalez, M. E. (1973). Use and evaluation of synthetic estimates. *Proceedings of the Social Statistics Section*, American Statistical Association, pp. 33–36.

Gonzalez, M. E., Ogus, J. L., Shapiro, G., and Tepping, B. T. (1975). Standards for discussion and presentation of errors in survey and census data. *Journal of the American Statistical Association* **73**, 7–15.

Granquist, L. (1984). On the role of editing. *Statistisk Tidskrift* **22**, 105–118.

Gray, G. B. (1975). Components of variance model in multistage stratified samples. *Survey Methodology* **1**, 27–43.

Gray, G. B., and Platek, R. (1976). Analysis of design effects and variance components in multi-stage sample surveys. *Survey Methodology* **2**, 1–30.

Graybill, F. A. (1983). *Matrices with Applications in Statistics*, 2nd ed. Belmont, CA: Wadsworth.

Greenlees, J. S., Reece, W. S., and Zieschang, K. D. (1982). Imputation of missing values when the probability of response depends on the variable being imputed. *Journal of the American Statistical Association* **77**, 251–261.

Grizzle, J. E., Starmer, C. F., and Koch, G. G. (1969). Analysis of categorical data by linear models. *Biometrics* **25**, 489–504.

Gross, S. T. (1980). Median estimation in sample surveys. *Proceedings of the Section on Survey Research*, American Statistical Association, pp. 181–184.

Groves, R. M. (1987). Research on survey data quality. *Public Opinion Quarterly* **51**, S156–S172.

Hájek, J. (1960). Limiting distributions in simple random sampling from a finite population. *Publ. Math. Inst. Hung. Acad. Sci.* **5**, 361–374.

Hájek, J. (1964). Asymptotic theory of rejective sampling with varying probabilities from a finite population. *Annals of Mathematical Statistics* **35**, 1491–1523.

Hájek, J. (1971). Comment on a paper by D. Basu. In: Godambe, V. P., and Sprott, D. A. (eds.) *Foundations of Statistical Inference*. Toronto: Holt, Rinehart and Winston, p. 236.

Hansen, M. H., and Hurwitz, W. N. (1943). On the theory of sampling from finite populations. *Annals of Mathematical Statistics* **14**, 333–362.

Hansen, M. H., and Hurwitz, W. N. (1946). The problem of non-response in sample surveys. *Journal of the American Statistical Association* **41**, 517–529.

Hansen, M. H., Dalenius, T., and Tepping, B. J. (1985). The development of sample surveys of finite populations. In: A. C. Atkinson and S. E. Fienberg, (eds.), *A Celebration of Statistics: The ISI Centenary Volume*. New York: Springer-Verlag.

Hansen, M. H., Hurwitz, W. N., and Bershad, M. A. (1961). Measurement errors in censuses and surveys. *Bulletin of the International Statistical Institute* **38**, 359–374.

Hansen, M. H., Hurwitz, W. N., and Madow, W. G. (1953). *Sample Survey Methods and Theory*, Vol. I and II. New York: Wiley.

Hansen, M. H., Hurwitz, W. N., and Pritzker, L. (1964). The estimation and interpretation of gross differences and the simple response variance. In: C. R. Rao (ed.), *Contributions to Statistics* (presented to P. C. Mahalanobis on the occasion of his 70th birthday). Calcutta: Statistical Publishing Society.

Hansen, M. H., Hurwitz, W. N., Marks, E. S., and Mauldin, W. P. (1951). Response errors in surveys. *Journal of the American Statistical Association* **46**, 147–190.

Hansen, M. H, Madow, W. G., and Tepping, B. J. (1983). An evaluation of model-dependent and probability-sampling inferences in sample surveys. *Journal of the American Statistical Association* **78**, 776–793.

Hartley, H. O., and Rao, J. N. K. (1978). Estimation of nonsampling variance components in sample surveys. In: N. K. Namboodiri (ed.), *Survey Sampling and Measurement*. New York: Academic Press, 35–43.

Hartley, H. O., and Ross, A. (1954). Unbiased ratio estimators. *Nature* **174**, 270–271.

Hartley, H. O., and Sielken, R. L. (1975). A super population viewpoint for finite population sampling. *Biometrics* **31**, 411–422.

Hedayat, A. (1979). Sampling designs with reduced support size. In: J. S. Rustagi (ed.), *Optimizing Methods in Statistics: Proceedings of an International Conference*. New York: Academic Press, pp. 273–288.

Hidiroglou, M. A. (1974). Estimation of regression parameters for finite populations. Ph.D. thesis, Iowa State University, Ames, Iowa.

Hidiroglou, M. A. (1986a). The construction of a self-representing stratum of large units in survey design. *The American Statistician* **40**, 27–31.

Hidiroglou, M. A. (1986b). Estimation of regression parameters for finite populations: A Monte-Carlo study. *Journal of Official Statistics* **2**, 3–11.

Hidiroglou, M. A., and Berthelot, J. M. (1986). Statistical editing and imputation for periodic business surveys. *Survey Methodology* **12**, 73–83.

Hidiroglou, M. A., and Rao, J. N. K. (1987). Chi-squared tests with categorical data from complex surveys, I and II. *Journal of Official Statistics* **3**, 117–132, 133–140.

Hidiroglou, M. H., and Srinath, K. P. (1981). Some estimators of the population totals from a simple random sample containing large units. *Journal of the American Statistical Association* **76**, 690–695.

Hidiroglou, M. A., Fuller, W. A., and Hickman, R. D. (1980). *SUPERCARP-Sixth Edition*. Statistical Laboratory, Survey Section, Iowa State University, Ames, Iowa.

Hilgard, E. R., and Payne, S. L. (1944). Those not at home: riddle for pollsters. *Public Opinion Quarterly* **8**, 254–261.

Holt, D., and Smith, T. M. F. (1976). The design of surveys for planning purposes. *The Australian Journal of Statistics* **18**, 37–44.

Holt, D., and Smith, T. M. F. (1979). Post-stratification. *Journal of the Royal Statistical Society A* **142**, 33–46.

Holt, D., Scott, A. J., and Ewings, P. O. (1980). Chi-squared tests with survey data. *Journal of the Royal Statistical Society A* **143**, 302–330.

Holt, M. M. (1982). SURREGR: standard errors of regression coefficients from sample survey data. Research Triangle Park, NC: Research Triangle Institute.

Horvitz, D. G., and Thompson, D. J. (1952). A generalization of sampling without replacement from a finite universe. *Journal of the American Statistical Association* **47**, 663–685.

Huddleston, H. F., Claypool, P. L., and Hocking, R. R. (1970). Optimal sample allocation to strata using convex programming. *Applied Statistics* **19**, 273–278.

Hughes, E., and Rao, J. N. K. (1979). Some problems of optimal allocation in sample surveys involving inequality constraints. *Communications in Statistics A* **8**, 1551–1574.

Isaki, C. T., and Fuller, W. A. (1982). Survey design under the regression superpopulation model. *Journal of the American Statistical Association* **77**, 89–96.

Jagers, P. (1986). Post-stratification against bias in sampling. *International Statistical Review* **54**, 159–167.

Jagers, P., Odén, A., and Trulsson, L. (1985). Post-stratification and ratio estimation. *International Statistical Review* **53**, 221–238.

Jessen, R. J. (1978). *Statistical Survey Techniques*. New York: Wiley.

Jönrup, H. (1974). Estimation of variance in multistage sampling. *Statistisk Tidskrift* **12**, 431–436.

Jönrup, H., and Rennermalm, B. (1976). Regression analysis in samples from finite populations. *Scandinavian Journal of Statistics* **3**, 33–37.

Kalton, G. (1983). Models in the practice of survey sampling. *International Statistical Review* **51**, 175–188.

Kalton, G., and Kasprzyk, D. (1986). The treatment of missing survey data. *Survey Methodology* **12**, 1–16.

Kendall, M., and Stuart, A. (1976). *The Advanced Theory of Statistics*, Vol. 3. 3rd ed. London: Griffin.

Kiaer, A. (1897). The representative method of statistical surveys (1976 English translation of the original Norwegian). Oslo: Central Bureau of Statistics of Norway.

Kish, L. (1965). *Survey Sampling*. New York: Wiley.

Kish, L., and Anderson, D. W. (1978). Multivariate and multipurpose stratification. *Journal of the American Statistical Association* **73**, 24–34.

Kish, L., and Frankel, M. R. (1970). Balanced repeated replication for standard errors. *Journal of the American Statistical Association* **65**, 1071–1094.

Kish, L., and Frankel, M. R. (1974). Inference from complex samples (with discussion). *Journal of the Royal Statistical Society B* **36**, 1–37.

Kish, L., Groves, R. M., and Krotki, K. P. (1976). Sampling errors for fertility surveys. Occasional Papers No. 17, World Fertility Survey, London.

Koch, G. G., Freeman, D. H., Jr., and Freeman, J. L. (1975). Strategies in the multivariate analysis of data from complex surveys. *International Statistical Review* **43**, 59–78.

Konijn, H. S. (1962). Regression analysis for sample surveys. *Journal of the American Statistical Association* **57**, 590–606.

Kott, P. S. (1990). Estimating the conditional variance of a design consistent regression estimator. *Journal of Statistical Planning and Inference* **24**, 287–296.

Kovar, J., Rao, J. N. K., and Wu, C. F. J. (1988). Bootstrap and other methods to measure errors in survey estimates. *Canadian Journal of Statistics* **16** (Supl), 25–45.

Krewski, D., and Rao, J. N. K. (1981). Inference from stratified samples: properties of the linearization, jackknife and balanced repeated replication methods. *Annals of Statistics* **9**, 1010–1019.

Kuk, A. Y. C. (1988). Estimation of distribution functions and medians under sampling with unequal probabilities. *Biometrika* **75**, 97–103.

Kumar, S., and Rao, J. N. K. (1984). Logistic regression analysis of Labour Force Survey data. *Survey Methodology* **10**, 62–81.
Lahiri, D. B. (1951). A method of sample selection providing unbiased ratio estimates. *Bulletin of the International Statistical Institute* **33**, 133–140.
Lanke, J. (1975). *Some Contributions to the Theory of Survey Sampling.* Ph.D. thesis, Department of Mathematical Statistics, University of Lund, Sweden.
Lanke, J. (1983). Hot deck imputation techniques that permit standard methods for assessing precision of estimates. *Statistical Review* **21**, no. 5, (Essays in Honour of Tore E. Dalenius), 105–110.
Lavallée, P., and Hidiroglou M. A. (1987). On the stratification of skewed populations. *Survey Methodology* **14**, 33–43.
Leblond, Y. (1990). Contribution à la théorie d'estimation des sous-populations. Ph.D. thesis, Department of Mathematics and Statistics, University of Montreal.
Lessler, J. T. (1982). Frame errors. In: J. T. Lessler, R. E. Folsom, and W. D. Kalsbeek, *A Taxonomy of Error Sources and Error Measures for Surveys, Final Report.* Research Triangle Park, NC: Research Triangle Institute.
Lessler, J. T. (1984). Measurement errors in surveys. In: C. F. Turner and E. Martin (eds.), *Surveying Subjective Phenomena*, Vol.2. New York: Russell Sage Foundation, 405–440.
Levy, P. S. (1977). Optimum allocation in stratified random network sampling for estimating the prevalence of attributes in rare populations. *Journal of the American Statistical Association* **72**, 758–763.
Lindström, H., and Lundström, S. (1974). A method to discuss the magnitude of the nonresponse error. *Statistisk Tidskrift* **12**, 505–520.
Little, R. J. A. (1983a). Estimating a finite population mean from unequal probability samples. *Journal of the American Statistical Association* **78**, 596–604.
Little, R. J. A. (1983b). Superpopulation models for nonresponse, the ignorable case. In: W. G. Madow, I. Olkin, and D. B. Rubin (eds.), *Incomplete Data in Sample Surveys*, Vol. 2. New York: Academic Press, pp. 341–382.
Little, R. J. A. (1986). Survey nonresponse adjustments. *International Statistical Review* **54**, 139–157.
Little, R. J. A. (1988). Missing-data adjustments in large surveys. *Journal of Business and Economic Statistics* **6**, 287–301.
Little, R. J. A., and Rubin, D. B. (1987). *Statistical Analysis with Missing Data.* New York: Wiley.
Liu, T. P., and Thompson, M. E. (1983). Properties of estimators of quadratic finite population functions: the batch approach. *Annals of Statistics* **11**, 275–285.
Lyberg, I., Haglund, T., Lyberg, L., and Wretman, J. (1978). The allocation of resources to production and control steps in the statistical process. *Statistisk Tidskrift* **16**, 253–276.
Lyberg, L. (1983). The development of procedures for industry and occupation coding at Statistics Sweden. *Statistical Review* **21**, no. 5 (Essays in Honour of Tore E. Dalenius), 139–156.
Lyberg, L. (1986). Quality control of coding operations at Statistics Sweden. Stockholm: Statistics Sweden, promemorior från P/STM, no. 22.
Mahalanobis, P. C. (1939). A sample survey of the acreage under jute in Bengal. *Sankhya* **4**, 511–531.
Mahalanobis, P. C. (1944). On large-scale sample surveys. *Philosophical Transactions of the Royal Society of London B* **231**, 329–451.
Mahalanobis, P. C. (1946). Recent experiments in statistical sampling in the Indian Statistical Institute. *Journal of the Royal Statistical Society* **109**, 325–370.
Mahalanobis, P. C. (1952). Some aspects of the design of sample surveys. *Sankhya* **12**, 1–7.
McCarthy, P. J. (1966). Replication: an approach to the analysis of data from complex

surveys. *Vital and Health Statistics*, Series 2, No. 14, Washington, DC: National Center for Health Statistics, Public Health Service.
McCarthy, P. J. (1969). Pseudo-replication: half-samples. *Review of the International Statistical Institute* **37**, 239–264.
McCarthy, P. J., and Snowden, C. B. (1985). The bootstrap and finite population sampling. *Vital and Health Statistics*, Series 2, No. 95. DHHS Pub. No. (PHS) 85–1369. Washington, DC: Public Health Service, US Government Printing Office.
McLeod, A. I., and Bellhouse, D. R. (1983). A convenient algorithm for drawing a simple random sample. *Applied Statistics* **32**, 182–184.
Mickey, M. R. (1959). Some finite population unbiased ratio and regression estimators. *Journal of the American Statistical Association* **54**, 594–612.
Midzuno, H. (1952). On the sampling system with probability proportional to sum of sizes. *Annals of the Institute of Statistical Mathematics* **3**, 99–107.
Montanari, G. E. (1987). Post-sampling efficient prediction in large-scale surveys. *International Statistical Review* **55**, 191–202.
Mulry, M. H., and Wolter, K. M. (1981). The effect of Fisher's Z-transformation on confidence intervals for the correlation coefficient. *Proceedings of the Section on Survey Research Methods*, American Statistical Association, pp. 113–121.
Murthy, M. N. (1983). A framework for studying incomplete data with a reference to the experience in some countries of Asia and the Pacific. In: W. G. Madow and I. Olkin (eds.), *Incomplete Data in Sample Surveys*, Vol. 3. New York: Academic Press, pp. 7–24.
Nargundkar, M. S., and Joshi, G. B. (1975). Nonresponse in sample surveys. 40th Session of the International Statistical Institute, Warsaw, Contributed papers, pp. 626–628.
Nathan, G. (1969). Tests of independence in contingency tables from stratified samples. In: N. L. Johnson and H. Smith (eds.), *New Developments in Survey Sampling*. New York: Wiley, pp. 578–600.
Nathan, G. (1975). Tests of independence in contingency tables from stratified proportional samples. *Sankhya C* **37**, 77–87; corrigendum: (1978), *Sankhya C*, **40**, 190.
Nathan, G. (1981). Notes on inference based on data from complex sample designs. *Survey Methodology* **7**, 110–129.
Nathan, G., and Holt, D. (1980). The effect of survey design on regression analysis. *Journal of the Royal Statistical Society B* **42**, 377–386.
National Center for Health Statistics (1968). *Synthetic State Estimates of Disability*. PHS Publication No. 1759. Washington, DC: Public Health Service, US Government Printing Office.
Nelder, J., and Wedderburn, R. (1972). Generalized linear models. *Journal of the Royal Statistical Society A* **135**, 370–384.
Neyman, J. (1934). On the two different aspects of the representative method: The method of stratified sampling and the method of purposive selection. *Journal of the Royal Statistical Society* **97**, 558–625.
Neyman, J. (1938). Contribution to the theory of sampling human populations. *Journal of the American Statistical Association* **33**, 101–116.
Nordberg, L. (1989). Generalized linear modeling of sample survey data. *Journal of Official Statistics* **5**, 223–239.
Norlén, U., and Waller, T. (1979). Estimation in a complex survey—experiences from a survey of buildings with regard to energy usage. *Statistisk Tidskrift* **17**, 109–124.
Nygård, F., and Sandström, A. (1985). Estimation of the Gini and the entropy inequality parameters in finite populations. *Journal of Official Statistics* **1**, 399–412.
Oh, H. L. and Scheuren, F. J. (1983). Weighting adjustment for unit nonresponse. In: W. G. Madow, I. Olkin, and D. B. Rubin (eds.), *Incomplete Data in Sample Surveys*, Vol. 2. New York: Academic Press, pp. 143–184.
Panel on Incomplete Data in Sample Surveys (1983). Report. In: W. G. Madow,

H. Nisselson, and I. Olkin (eds.), *Incomplete Data in Sample Surveys*, Vol. 1. New York: Academic Press, pp. 1–103.
Pasqual, J. N. (1961). Unbiased ratio estimators in stratified sampling. *Journal of the American Statistical Association* **56**, 70–87.
Pathak, P. K. (1961). On simple random sampling with replacement. *Sankhya A* **24**, 287–302.
Patterson, H. D. (1950). Sampling on successive occasions with partial replacement of units. *Journal of the Royal Statistical Society B* **12**, 241–255.
Pfeffermann, D. (1988). The effect of sampling design and response mechanism on multivariate regression-based predictors. *Journal of the American Statistical Association* **83**, 824–833.
Pfeffermann, D., and Holmes, D. J. (1985). Robustness considerations in the choice of a method of inference for regression analysis of survey data. *Journal of the Royal Statistical Society A* **148**, 268–278.
Pfeffermann, D., and Nathan, G. (1977). Regression analysis of data from complex samples. *Bulletin of the International Statistical Institute* **49**, 699–718.
Pfeffermann, D., and Nathan, G. (1981). Regression analysis of data from a cluster sample. *Journal of the American Statistical Association* **76**, 681–689.
Pfeffermann, D., and Smith, T. M. F. (1985). Regression models for grouped populations in cross-section surveys. *International Statistical Review* **53**, 37–39.
Plackett, R. L., and Burman, J. P. (1946). The design of optimum multifactorial experiments. *Biometrika* **33**, 305–325.
Platek, R., and Gray, G. B. (1983). Imputation methodology. In: W. G. Madow, I. Olkin, and D. B. Rubin (eds.), *Incomplete Data in Sample Surveys*, Vol. 2. New York: Academic Press, pp. 255–293.
Platek, R., Singh, M. P., Rao, J. N. K., and Särndal, C. E., eds. (1987). *Small Area Statistics: An International Symposium.* New York: Wiley.
Politz, A., and Simmons, W. (1949). An attempt to get not-at-homes into the sample without callbacks. *Journal of the American Statistical Association* **44**, 9–31.
Porter, R. M. (1973). On the use of survey sample weights in the linear model. *Annals of Economic and Social Measurement* **2**, 141–158.
Pritzker, L., Ogus, J., and Hansen, M. H. (1965). Computer editing methods: some applications and results. *Bulletin of the International Statistical Institute* **41**, 442–466.
Pullum, T. W., Harpham, T., and Ozsever, N. (1986). The machine editing of large-sample surveys: the experience of the World Fertility Survey. *International Statistical Review* **54**, 311–326.
Purcell, N. J., and Kish, L. (1979). Estimation for small domains. *Biometrics* **35**, 365–384.
Quenouille, M. H. (1949). Problems in plane sampling. *Annals of Mathematical Statistics* **20**, 355–375.
Quenouille, M. H. (1956). Notes on bias in estimation. *Biometrika* **43**, 353–360.
Raj, D. (1964). On forming strata of equal aggregate size. *Journal of the American Statistical Association* **59**, 481–486.
Raj, D., and Khamis, S. H. (1958). Some remarks on sampling with replacement. *Annals of Mathematical Statistics* **29**, 550–557.
Rao, J. N. K. (1965). On two simple schemes of unequal probability sampling without replacement. *Journal of the Indian Statistical Association* **3**, 173–180.
Rao, J. N. K. (1973). On double sampling for stratification and analytic surveys. *Biometrika* **60**, 125–133.
Rao, J. N. K. (1975). Unbiased variance estimation for multistage designs. *Sankhya C* **37**, 133–139.
Rao, J. N. K. (1979a). On deriving mean square errors and their non-negative unbiased estimators. *Journal of the Indian Statistical Association* **17**, 125–136.

Rao, J. N. K. (1979b). Optimization in the design of sample surveys. In: J. S. Rustagi (ed.), *Optimizing Methods in Statistics: Proceedings of an International Conference*. New York: Academic Press, pp. 419–434.

Rao, J. N. K. (1982). Some aspects of variance estimation in sample surveys. *Utilitas Mathematica* **21B**, 205–225.

Rao, J. N. K. (1985). Conditional inference in survey sampling. *Survey Methodology* **11**, 15–31.

Rao, J. N. K., and Bellhouse, D. R. (1990). History and development of survey based estimation and analysis. *Survey Methodology* **16**, 3–29.

Rao, J. N. K., and Hidiroglou, M. A. (1981). Chisquare tests for the analysis of categorical data from the Canada Health Survey. *Bulletin of the International Statistical Institute* **49**, 699–718.

Rao, J. N. K., and Scott, A. J. (1981). The analysis of categorical data from complex surveys: chi-squared tests for goodness of fit and independence in two-way tables. *Journal of the American Statistical Association* **76**, 221–230.

Rao, J. N. K., and Scott, A. J. (1984). On chi-squared tests for multiway contingency tables with cell proportions estimated from survey data. *Annals of Statistics* **12**, 46–60.

Rao, J. N. K., and Scott, A. J. (1987). On simple adjustments to chi-square tests with sample survey data. *Annals of Statistics* **15**, 385–397.

Rao, J. N. K., and Wu, C. F. J. (1984). Bootstrap inference for sample surveys. *Proceedings of the Section on Survey Research Methods*, American Statistical Association, pp. 106–112.

Rao, J. N. K., and Wu, C. F. J. (1987). Methods for standard errors and confidence intervals from sample survey data: some recent work. *Bulletin of the International Statistical Institute* **52**, 5–21.

Rao, J. N. K., Hartley, H. O., and Cochran, W. G. (1962). On a simple procedure of unequal probability sampling without replacement. *Journal of the Royal Statistical Society B* **24**, 482–491.

Rao, J. N. K., Kovar, J. G., and Mantel, H. J. (1990). On estimating distribution functions and quantiles from survey data using auxiliary information. *Biometrika* **77**, 365–375.

Rao, P. S. R. S. (1983). Callbacks, follow-ups, and repeated telephone calls. In: W. G. Madow, I. Olkin, and D. B. Rubin (eds.), *Incomplete Data in Sample Surveys*, Vol. 2. New York: Academic Press, pp. 33–44.

Rao, P. S. R. S. (1988). Ratio and regression estimators. In: P. R. Krishnaiah and C. R. Rao (eds.). *Handbook of Statistics*, Vol. 6. Amsterdam: North-Holland, pp. 449–468.

Robinson, P. M., and Särndal, C. E. (1983). Asymptotic properties of the generalized regression estimator in probability sampling. *Sankhya B* **45**, 240–248.

Rosén, B. (1972). Asymptotic theory for successive sampling with varying probabilities without replacement, I and II. *Annals of Mathematical Statistics* **43**, 373–397, 748–776.

Rosén, B. (1987). Some optimality problems when estimating household data on the basis of a "primary" stratified sample of individuals. R&D Report U/STM 36, Statistics Sweden.

Rosenbaum, P. R., and Rubin, D. B. (1983). The central role of the propensity score in observational studies for causal effects. *Biometrika* **70**, 41–55.

Rossi, P. H., Wright, J. D., and Anderson, A. B., eds. (1983). *Handbook of Survey Research* (Quantitative Studies in Social Relations). San Diego, CA: Academic Press.

Royall, R. M. (1970). On finite population sampling theory under certain linear regression models. *Biometrika* **57**, 377–387.

Royall, R. M., and Cumberland, W. G. (1981a). An empirical study of the ratio estima-

tor and estimators of its variance. *Journal of the American Statistical Association* **76**, 66–77.
Royall, R. M., and Cumberland, W. G. (1981b). The finite-population linear regression estimator and estimators of its variance—an empirical study. *Journal of the American Statistical Association* **76**, 924–930.
Royall, R. M., and Eberhardt, K. R. (1975). Variance estimates for the ratio estimator. *Sankhya C* **37**, 43–52.
Royall, R. M., and Herson, J. (1973a). Robust estimation in finite populations, I. *Journal of the American Statistical Association* **68**, 880–889.
Royall, R. M., and Herson, J. (1973b). Robust estimation in finite populations, II: stratification on a size variable. *Journal of the American Statistical Association* **68**, 890–893.
Rubin, D. B. (1976). Inference and missing data. *Biometrika* **63**, 581–592.
Rubin, D. B. (1978). Multiple imputations in sample surveys—a phenomenological Bayesian approach to nonresponse. In: *Imputation and Editing of Faulty or Missing Survey Data.* Washington, DC: US Department of Commerce, Social Security Administration. Also in (1978) *Proceedings of the Section on Survey Research Methods*, American Statistical Association, pp. 20–34.
Rubin, D. B. (1983). Conceptual issues in the presence of nonresponse. In: W. G. Madow, I. Olkin, and D. B. Rubin (eds.), *Incomplete Data in Sample Surveys*, Vol. 2. New York: Academic Press, pp. 123–142.
Rubin, D. B. (1987). *Multiple Imputation for Nonresponse in Surveys.* New York: Wiley.
Rust, K. (1985). Variance estimation for complex estimators in sample surveys. *Journal of Official Statistics* **1**, 381–397.
Rylett, D. T., and Bellhouse, D. R. (1988). TREES: a computer program for complex surveys. *Proceedings of the Section on Survey Research Methods*, American Statistical Association, pp. 694–697.
Sande, I. G. (1982). Imputation in surveys: coping with reality. *The American Statistician* **36**, 145–152.
Sande, I. G. (1983). Hot-deck imputation procedures. In: W. G. Madow and I. Olkin (eds.), *Incomplete Data in Sample Surveys*, Vol. 3. New York: Academic Press, pp. 339–349.
Sandström, A., Wretman, J. H., and Waldén, B. (1988). Variance estimators of the Gini coefficient—probability sampling. *Journal of Business and Economic Statistics* **6**, 113–119.
Särndal, C. E. (1978). Design-based and model-based inference in survey sampling. *Scandinavian Journal of Statistics* **5**, 27–52.
Särndal, C. E. (1980). On π inverse weighting versus best linear unbiased weighting in probability sampling. *Biometrika* **67**, 639–650.
Särndal, C. E. (1981). Frameworks for inference in survey sampling with applications to small area estimation and adjustment for nonresponse. *Bulletin of the International Statistical Institute* **49**, 494–513.
Särndal, C. E. (1982). Implications of survey design for generalized regression estimation of linear functions. *Journal of Statistical Planning and Inference* **7**, 155–170.
Särndal, C. E. (1984). Design-consistent versus model-dependent estimators for small domains. *Journal of the American Statistical Association* **79**, 624–631.
Särndal, C. E., and Hidiroglou, M. A. (1989). Small domain estimation: a conditional analysis. *Journal of the American Statistical Association* **84**, 266–275.
Särndal, C. E., and Swensson, B. (1987). A general view of estimation for two phases of selection with applications to two-phase sampling and nonresponse. *International Statistical Review* **55**, 279–294.
Särndal, C. E., and Wright, R. L. (1984). Cosmetic form of estimators in survey sampling. *Scandinavian Journal of Statistics* **11**, 146–156.
Särndal, C. E., Swensson, B., and Wretman, J. H. (1989). The weighted residual tech-

nique for estimating the variance of the general regression estimator of the finite population total. *Biometrika* **76**, 527–537.

Satterthwaite, F. E. (1946). An approximate distribution of estimates of variance components. *Biometrics* **2**, 110–114.

Schaible, W. L. (1979). A composite estimator for small area statistics. In: C. Steinberg (ed.), *Synthetic Estimates for Small Areas*, National Institute on Drug Abuse, Research Monograph 24. Washington, DC: US Government Printing Office, pp. 36–83.

Scott, A. J., and Rao, J. N. K. (1981). Chi-squared tests for contingency tables with proportions estimated from survey data. In: D. Krewski, R. Platek, and J. N. K. Rao (eds.), *Current Topics in Survey Sampling*. New York: Academic Press, pp. 247–265.

Sen, A. R. (1953). On the estimate of the variance in sampling with varying probabilites. *Journal of the Indian Society of Agricultural Statistics* **5**, 119–127.

Shah, B. V., Holt, M. M., and Folsom, R. E. (1977). Inference about regression models from sample survey data. *Bulletin of the International Statistical Institute* **47**, 43–57.

Singh, S., and Singh, R. (1979). On random non-response in unequal probability sampling. *Sankhya C* **41**, 127–137.

Sirken, M. G. (1972). Stratified sample surveys with multiplicity. *Journal of the American Statistical Association* **67**, 224–227.

Skinner, C. J., Holmes, D. J., and Smith, T. M. F. (1986). The effect of sample design on principal component analysis. *Journal of the American Statistical Association* **81**, 789–798.

Skinner, C. J., Holt, D., and Smith, T. M. F. (eds.) (1989). *Analysis of Complex Surveys*. Chichester: Wiley.

Solomon, H., and Stephens, M. A. (1977). Distribution of a sum of weighted chi-square variables. *Journal of the American Statistical Association* **72**, 881–885.

Srinath, K. P., and Hidiroglou, M. A. (1980). Estimation of variance in multi-stage sampling. *Metrika* **27**, 121–125.

Statistics Canada (1978). A compendium of methods of error evaluation in censuses and surveys. Ottawa: Statistics Canada.

Statistics Canada (1987). Statistics Canada's policy on informing users of data quality and methodology. *Journal of Official Statistics* **3**, 83–92.

Statistics Sweden (1983). Policy för användarorienterad kvalitetsredovisning av statistik. Meddelanden i samordningsfrågor 1983: 1.

Sukhatme, P. V., and Sukhatme, B. V. (1970). *Sampling Theory of Surveys with Applications*. London: Asia Publishing House.

Sunter, A. B. (1977a). Response burden, sample rotation, and classification renewal in economic surveys. *International Statistical Review* **45**, 209–222.

Sunter, A. B. (1977b). List sequential sampling with equal or unequal probabilities without replacement. *Applied Statistics* **26**, 261–268.

Sunter, A. B. (1986). Solutions to the problem of unequal probability sampling without replacement. *International Statistical Review* **54**, 33–50.

Swensson, B. (1982). A survey of nonresponse terms. *Statistical Journal of the United Nations. ECE* **1**, 241–251.

Ten Cate, A. (1986). Regression analysis using survey data with endogenous design. *Survey Methodology* **12**, 121–138.

Tepping, B. J. (1968). Variance estimation in complex surveys. *Proceedings of the Social Statistics Section*, American Statistical Association, pp. 11–18.

Thomas, D. R., and Rao, J. N. K. (1987). Small-sample comparisons of level and power for simple goodness-of-fit statistics under cluster sampling. *Journal of the American Statistical Association* **82**, 630–636.

Thomsen, I. (1978). Design and estimation problems when estimating a regression coefficient from survey data. *Metrika* **25**, 27–35.

Tin, M. (1965). Comparison of some ratio estimators. *Journal of the American Statistical Association* **60**, 294–307.

Tschuprow, A. (1923). On the mathematical expectation of the moments of frequency distributions in the case of correlated observations. *Metron* **2**, 646–680.

Tukey, J. W. (1958). Bias and confidence in not-quite large samples (abstract). *Annals of Mathematical Statistics* **29**, 614.

United Nations (1964). Recommendations for the preparation of sample survey reports (Provisional Issue). Statistical Papers, Series C, No. 1, Rev. 2. New York: United Nations.

United Nations (1982). Non-sampling errors in household surveys: sources, assessment and control. Preliminary version. National Household Survey Capability Programme, DP/UN/INT-81-041/2. New York: United Nations.

U.S. Department of Commerce (1978a). An error profile: employment as measured by the current population survey. Statistical Policy Working Paper No. 3. Washington, DC: U.S. Department of Commerce. U.S. Government Printing Office.

U.S. Department of Commerce (1978b). Glossary of nonsampling error terms: an illustration of a semantic problem in statistics (authors' preface). Statistical Policy Working Paper No. 4. Washington, DC: U.S. Government Printing Office.

U.S. Bureau of the Census (1986). Content reinterview study: accuracy of data for selected population and housing characteristics as measured by reinterview. 1980 Census of Population and Housing, No. PHC 80-E2. Washington, DC: U.S. Department of Commerce, U.S. Government Printing Office.

Van Eck, N. (1979). Osiris IV User's Manual: Fifth edition. Ann Arbor, MI: Survey Research Center, Institute for Social Research, University of Michigan.

Warner, S. L. (1965). Randomized response: a survey technique for eliminating evasive answer bias. *Journal of the America Statistical Association* **57**, 622–627.

Wolter, K. M. (1985). *Introduction to Variance Estimation.* New York: Springer-Verlag.

Woodruff, R. S. (1952). Confidence intervals for medians and other position measures. *Journal of the American Statistical Association* **47**, 635–646.

Woodruff, R. S. (1971). A simple method for approximating the variance of a complicated estimate. *Journal of the American Statistical Association* **66**, 411–414.

Wretman, J. H. (1983). On variance estimation in survey sampling with measurement errors. *Statistical Review* **21**, no.5 (Essays in Honour of Tore E. Dalenius), 117–124.

Wright, R. L. (1983). Finite population sampling with multivariate auxiliary information. *Journal of the American Statistical Association* **78**, 879–884.

Wright, T., and Tsao, H. J. (1983). A frame on frames: an annotated bibliography. In: T. Wright (ed.), *Statistical Methods and the Improvement of Data Quality*. New York: Academic Press, pp. 25–72.

Wu, C. F. J., and Deng, L. Y. (1983). Estimation of variance of the ratio estimator: an empirical study. In: G. E. P. Box et al. (ed.), *Scientific Inference, Data Analysis and Robustness*. New York: Academic Press, pp. 245–277.

Yates, F., and Grundy, P. M. (1953). Selection without replacement from within strata with probability proportional to size. *Journal of the Royal Statistical Society B* **15**, 235–261.

Answers to Selected Exercises

3.1. (a) 103.9; (b) 101.0

3.2. 5,640; 41,780; 1,421,100; 79,060,680; 21%; 21%

3.3. $(100/124) \cdot (42 \pm 1.96 \cdot 42^{1/2}) = 33.9 \pm 10.2$

3.4. (a) 27; (b) 89; (c) 27)

3.5. $13{,}480 \pm 1{,}004$

3.6. 7.38 ± 0.76

3.7. (a) 108 ± 32; (b) 0.54 ± 0.16

3.8. 21.2; 35.1; 28%

3.10. (a) 13,037; 5,453; 3,352. (b) 9,516. (c) 0.029; −0.022; −0.029; −0.033

3.11. 4,292.2 and 489,456, respectively; 16%

3.12. $V_{BE}(\hat{t}_\pi)/V_{PO}(\hat{t}_\pi) \doteq 8.8 \cdot 10^{11}/9.2 \cdot 10^{10} \doteq 9.6$

3.13. (a) 945,598; (b) $6.53 \cdot 10^{10}$ and 27%; (d) $10.2 \cdot 10^{12}$

3.14. (a) 1,427; (b) 186 and 0.1%

3.15. (a) 6, 24, 8, 2; 5, 21, 10, 4; and 5, 22, 9, 4. (b) 44,093; 45,229; 44,935; and 91,059. (c) 6,487; 46,541; and 3%

3.16. (a) 89,333; (b) 118

4.1. 6,125; 107,254 and 5%

4.2. RMT85, 0.0515, 1.12; CS82, 0.4258, 3.71

4.3. The totals of the three slices of the systematic sample are 650, 899, and 728, which gives (a) 7,590; (b) 378,490 and 8%

4.4. $0.784 \cdot 10^9$

4.5. (a) 160,256 + 356,629 = 516,885; (b) 62,666,646 + 552,114,991 = 614,781,637 and 5%

4.8. (a) 2,775. (b) 234,934; 47,025; 281,959; 19%. (c) 277,256; 308,063; 19%; 20%

4.11. (a) 8,294. (b) 25,376; 2%

4.12. (a) $19.31 \cdot 10^6$; (b) $9.648 \cdot 10^{12}$ and 16%; (c) $9.675 \cdot 10^{12}$ and $9.797 \cdot 10^{12}$; (d) $8.969 \cdot 10^{12}$, $0.556 \cdot 10^{12}$, and $0.159 \cdot 10^{12}$

5.1. $3{,}393 \pm 1.96\,(117{,}242)^{1/2} = 3{,}393 \pm 671$

5.2. 1.099 and 7%

5.3. 20, 70, and 148

5.4. (a) 7.8; (b) 1.2; (c) 12.3 and 1,975

5.5. (a) 719,810 ± 252,721; (b) 818,416 ± 113,023

5.8. $8.00 \pm 1.96\,(0.0371352)^{1/2} = 8.00 \pm 0.38$

5.9. (a) 6.7%, 13.6%, 21.9%, and 73.6%, respectively; (b) 31.7%

5.10. 291 and 17.6%

5.11. 380 and (127, 1423)

5.12. 15 and (10, 18)

5.13. (a) 6,325; 23,719; 2%. (b) 46.8; 1.3; 2%

6.1. The choice $A = 1.045$ gives $\hat{t}_{ydif} = 4{,}460$; $\hat{V}(\hat{t}_{ydif}) = 4{,}202.5$; and $cve = 1.5\%$

6.2. The choice $A = 0.5$ gives $\hat{t}_{ydif} = 6{,}295$, $\hat{V}(\hat{t}_{ydif}) = 37{,}270$ and $cve = 3\%$

6.3. $7{,}386 \pm 1.96\,(319{,}362)^{1/2} = 7{,}386 \pm 1{,}108$

6.4. 1,058, 0.28, 0.08, 0.003, 0.04, 0.19, 0.46, 0.84, and 1.33, respectively

7.1. $136{,}886 \pm 1.96\,(15{,}196{,}356)^{1/2} \doteq 136{,}886 \pm 7{,}641$

7.2. The three approximate variances are, respectively, $6.780 \cdot 10^7$, $12.301 \cdot 10^6$, and $6.947 \cdot 10^6$

7.3. (a) 88.2%; (b) 6.8%

7.5. (a) 653,271; (b) 1,426,120

7.6. (a) 18,831 ± 1,608; (b) 19,124 ± 2,417

7.8. 52,338 ± 1,539

7.9. $27.91 \cdot 10^8$, $19.20 \cdot 10^8$, and $7.52 \cdot 10.^8$, respectively

7.10. 356,373 ± 51,925

7.11. 52,972 ± 3,829

8.2. $V_{SIC}(\hat{t}_{y\pi}) = 112.8 \cdot 10^6$, $AV_{SIC}(\hat{t}_{yAr}) = 11.5 \cdot 10^6$; $V_{BEC}(\hat{t}_{y\pi}) = 207.4 \cdot 10^6$, $AV_{BEC}(\hat{t}_{yAr}) = 11.3 \cdot 10^6$

8.3. 549,626 and 7%, respectively

8.4. 5.1% and 33%

8.5. $V_{SI,SI}(\hat{t}_{y\pi}) = 80.6 \cdot 10^6 + 126.8 \cdot 10^6 = 207.4 \cdot 10^6$
$AV_{SI,SI}(\hat{t}_{yAr}) = 7.6 \cdot 10^6 + 126.8 \cdot 10^6 = 134.4 \cdot 10^6$
$AV_{SI,SI}(\hat{t}_{yBr}) = 7.6 \cdot 10^6 + 14.2 \cdot 10^6 = 21.8 \cdot 10^6$
$AV_{SI,SI}(\hat{t}_{yCr}) = 80.6 \cdot 10^6 + 14.2 \cdot 10^6 = 94.8 \cdot 10^6$

9.1. (a) 829,058; (c) $5.78 \cdot 10^{10}$

9.2. (a) 6,422; (b) 12,391 and 14,534, respectively; (c) 6,422 ± 322; (d) 40,770

9.3. 173

9.4. (a) $\hat{V}(\hat{t}_{dif2}) = N^2 \left\{ \frac{1 - f_a}{n_a} S_{ys}^2 + \frac{1 - f}{n} S_{Ds}^2 \right\}$; (b) 2,646 ± 167

9.5. 57,403 ± 10,014

9.6. 56,476 ± 7,771

9.8. (c) 2,694 ± 83; (d) 366

10.1.

Major region d	$V(\hat{t}_{d\pi})$	$AV(\tilde{t}_d)$	Interval based on $\hat{t}_{d\pi}$	Interval based on $\tilde{t}_d$
1	9,394	2,965	1,308 ± 209	1,152 ± 108
2	7,761	1,316	1,111 ± 175	1,003 ± 62
3	2,387	649	324 ± 73	454 ± 34

10.2. 5.33 ± 0.78

10.3.

Major region d	$V(\hat{t}_{d\pi})$	$AV(\tilde{t}_d)$	$\tilde{t}_d$	cve
1	8,933	2,726	1,021	5%
2	7,734	1,200	1,068	4%
3	2,366	601	499	5%

10.4.

Major region	$AV(\hat{t}_{dra1})$	$AV(\hat{t}_{dra2})$	$\hat{t}_{dra2}$	$\hat{V}(\hat{t}_{dra2})$
1	$1.98 \cdot 10^8$	$1.72 \cdot 10^8$	384,276	$1.79 \cdot 10^8$
2	$1.81 \cdot 10^8$	$1.36 \cdot 10^8$	278,629	$1.12 \cdot 10^8$
3	$0.94 \cdot 10^8$	$1.03 \cdot 10^8$	219,126	$1.64 \cdot 10^8$

12.2. (a) 20, 23, 20, 49, 21 ($=N_5$), and 17 ($=N_6$); (b) 19, 22, 21, 54 ($=N_4$), 21 ($=N_5$), and 17 ($=N_6$)

12.3. The $CV(\hat{t}_h)$ and $CV(\hat{t}_U)$, in percent, are

h	$a = 0$	$a = 1/3$	$a = 1/2$	Neyman
1	12.7	11.5	10.7	8.8
2	12.2	12.2	12.2	13.5
3	11.7	13.3	14.3	15.8
U	7.4	7.1	7.0	6.9

12.4. $n_I \doteq 8$ and $n_{II} \doteq 10$

12.5. (c) $n_a = 160$ and $n = 24$; (d) 0.44

13.1. To arrive at $\hat{u}_k$, replace every population total (including N) in the expression of u_k by its corresponding π estimator.

13.2. 0.61 and 28%

13.3. 0.62 and 7.4%

13.8. $X^2_{WH} = 25.3$, $X^2_{PH} = 7.2$, and $X^2_{MH} = 24.8$

15.1. (a) 0.818, 0.956, 0.624, 0.818, 0.182, 0.044, 0.376, 0.182; (b) 0.889, 0.833, 0.722 and 0.111, 0.167, 0.278

15.2. $335{,}563 \pm 1.96\,(466{,}087{,}327)^{1/2} \doteq 335{,}563 \pm 42{,}315$

15.3. $2{,}114 \pm 417$

15.5. 140,343 and 5.0%

15.6. 136,798 and 4.2%

Author Index

Subject Index

Springer Series in Statistics

(continued from p. ii)

Yaglom: Correlation Theory of Stationary and Related Random Functions I: Basic Results.

Yaglom: Correlation Theory of Stationary and Related Random Functions II: Supplementary Notes and References.